Y0-CRP-368

THEORY OF SIMPLE LIQUIDS

THEORY OF SIMPLE LIQUIDS

Second Edition

JEAN PIERRE HANSEN
Université Pierre et Marie Curie, Paris

IAN R. McDONALD
University of Cambridge, Cambridge

1986

Harcourt Brace Jovanovich, Publishers

London Orlando New York San Diego Austin
Boston Tokyo Sydney Toronto

ACADEMIC PRESS INC. (LONDON) LTD.
24/28 Oval Road
London NW1

United States Edition published by
ACADEMIC PRESS, INC.
Orlando, Florida 32887

Hansen, Jean Pierre
Theory of simple liquids.—2nd ed.
1. Liquids, Kinetic theory of
2. Statistical mechanics
I. Title II. McDonald, Ian R.
530.4′2 Qc175.3

ISBN 0-12-323851-X

PRINTED IN GREAT BRITAIN BY
J. W. ARROWSMITH LTD., BRISTOL BS3 2NT

Preface to the Second Edition

The first edition of this book was written in the wake of an unprecedented advance in our understanding of the microscopic structure and dynamics of simple liquids. The rapid progress made in a number of different experimental and theoretical areas had led to a rather clear and complete picture of the properties of simple atomic liquids. In the ten years that have passed since then, interest in the liquid state has remained very active, and the methods described in our book have been successfully generalized and applied to a variety of more complicated systems. Important developments have therefore been seen in the theory of ionic, molecular and polar liquids, of liquid metals, and of the liquid surface, while the quantitative reliability of theories of atomic fluids has also improved.

In an attempt to give a balanced account both of the basic theory and of the advances of the past decade, this new edition has been rearranged and considerably expanded relative to the earlier one. Every chapter has been completely rewritten, and three new chapters have been added, devoted to ionic, metallic and molecular liquids, together with substantial new sections on the theory of inhomogeneous fluids. The material contained in Chapter 10 of the first edition, which dealt with phase transitions, has been omitted, since it proved impossible to do justice to such a large field in the limited space available. Although many excellent review articles and monographs on specialized topics have appeared in recent years, a comprehensive and up-to-date treatment of the theory of "simple" liquids appears to be lacking, and we hope that the new edition of our book will fill this gap. The choice of material again reflects our own tastes, but we have aimed at presenting the main ideas of modern liquid-state theory in a way that is both pedagogical and, so far as possible, self-contained. The book should be accessible to graduate students and research workers, both experimentalists and theorists, who have a good background in elementary statistical mechanics. We are well aware, however, that certain sections, notably in Chapters 4, 6, 9 and 12, require more concentration from the reader than others. Although the book is not intended to be exhaustive, we give many references to material that is not covered in depth in the text. Even at this level, it is impossible to include all the relevant work. Omissions may reflect either our ignorance or a lack of good judgment, but we consider that our goal will have been achieved if the book serves as an introduction and guide to a continuously growing field.

While preparing the new edition, we have benefited from the advice, criticism and help of many colleagues. We give our sincere thanks to all. There are, alas, too many names to list individually, but we wish to acknowledge our particular debt to Marc Baus, David Chandler, Giovanni Ciccotti, Bob Evans, Paul Madden and Dominic Tildesley, who have read large parts of the manuscript and made suggestions for its improvement; to Susan O'Gorman, for her help with Chapter 4; and to Eduardo Waisman, who wrote the first (and almost final) version of Appendix B. We are also grateful to those colleagues who have supplied references, preprints, and material for figures and tables, and to authors and publishers for permission to reproduce diagrams from published papers. The last stages of the work were carried out at the Institut Laue-Langevin in Grenoble, and we thank Philippe Nozières for the invitations that made our visits possible. Finally, we are greatly indebted to Martine Hansen, Christiane Lanceron, Rehda Mazighi and Susan O'Gorman for their help and patience in the preparation of the manuscript and figures.

May 1986

J. P. HANSEN
I. R. MCDONALD

Preface to the First Edition

The past ten years or so have seen a remarkable growth in our understanding of the statistical mechanics of simple liquids. Many of these advances have not yet been treated fully in any book and the present work is aimed at filling this gap at a level similar to that of Egelstaff's "The Liquid State", though with a greater emphasis on theoretical developments. We discuss both static and dynamic properties, but no attempt is made at completeness and the choice of topics naturally reflects our own interests. The emphasis throughout is placed on theories which have been brought to a stage at which numerical comparison with experiment can be made. We have attempted to make the book as self-contained as possible, assuming only a knowledge of statistical mechanics at a final-year undergraduate level. We have also included a large number of references to work which lack of space has prevented us from discussing in detail. Our hope is that the book will prove useful to all those interested in the physics and chemistry of liquids.

Our thanks go to many friends for their help and encouragement. We wish, in particular, to express our gratitude to Loup Verlet for allowing us to make unlimited use of his unpublished lecture notes on the theory of liquids. He, together with Dominique Levesque, Konrad Singer and George Stell, have read several parts of the manuscript and made suggestions for its improvement. We are also greatly indebted to Jean-Jacques Weis for his help with the sections on molecular liquids. The work was completed during a summer spent as visitors to the Chemistry Division of the National Research Council of Canada; it is a pleasure to have this opportunity to thank Mike Klein for his hospitality at that time and for making the visit possible. Thanks go finally to Susan O'Gorman for her help with mathematical problems and for checking the references; to John Copley, Jan Sengers and Sidney Yip, for sending us useful material; to Martine Hansen for help in preparing the manuscript; and to Mrs K.L. Hales for so patiently typing the many drafts.

A number of figures and tables have been reproduced, with permission, from The Physical Review, Journal of Chemical Physics, Molecular Physics and Physica; detailed acknowledgments are made at appropriate points in the text.

June 1976

J. P. HANSEN
I. R. McDONALD

Contents

CHAPTER 1

Introduction

1.1 THE LIQUID STATE

A convenient starting point in any discussion of the properties of liquids is the relationship between pressure P, number density ρ and temperature T in the various phases, summarized in an equation of state $f(P, \rho, T)=0$. The phase diagram of a typical monatomic substance such as argon is sketched in Figure 1.1; in (a) we show the projection in the ρ–T plane and in (b) the projection in the P–T plane. The region of existence of the liquid phase is bounded above by the critical point (subscript c) and below by the triple point (subscript t) and occupies only a small part of the entire P–ρ–T space; the ratio T_c/T_t generally lies between two and five. In many respects, however, the properties of the dense supercritical fluid are not very different from those of the true liquid, and much of the theory that we outline in later chapters applies equally well to the two cases. Above the critical point there exists only a single fluid phase and there is a continuous path from liquid to fluid to vapour; this is not true of the transition from solid to liquid because the solid–fluid coexistence line, or melting curve, does not terminate at a critical point.

We shall be concerned in this book almost exclusively with classical liquids. For atomic systems, a simple test of the classical hypothesis is obtained by calculating the de Broglie thermal wavelength Λ, defined as

$$\Lambda=\left(\frac{2\pi\beta\hbar^2}{m}\right)^{1/2} \tag{1.1.1}$$

where m is the mass of an atom and $\beta=1/k_BT$; to justify a classical treatment of static properties, it is necessary that $\Lambda/a \ll 1$, where $a \simeq \rho^{-1/3}$ is the mean nearest-neighbour separation. In the case of molecules, we require in addition that $\Theta_{rot} \ll T$, where $\Theta_{rot}=\hbar^2/2Ik_B$ is a characteristic rotational temperature (I is the molecular moment of inertia). Some typical results are shown in Table 1.1, from which we see that quantum effects

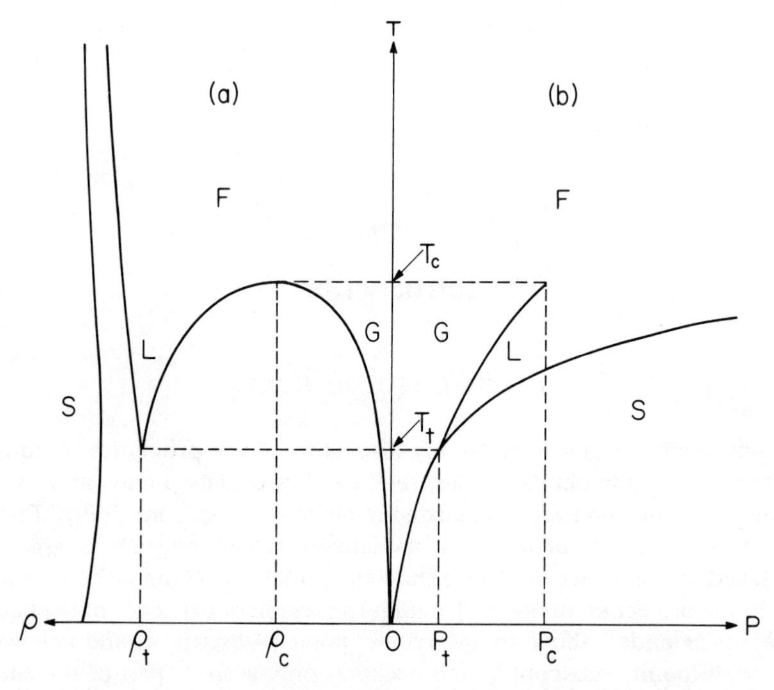

FIG. 1.1. Phase diagram of a typical monatomic substance, showing the boundaries between solid (S), liquid (L) and vapour (G) or fluid (F) phases; (a) is the projection in the $\rho-T$ plane and (b) is that in the $P-T$ plane. The subscript t stands for triple point and c for critical point.

TABLE 1.1. *Test of the classical hypothesis*

Liquid	T_t (K)	Λ (Å)	Λ/a	Θ_{rot}/T_t
H_2	14.05	3.3	0.97	6.1
Ne	24.5	0.78	0.26	
CH_4	90.7	0.46	0.12	0.083
N_2	63.3	0.42	0.11	0.046
Li	454	0.31	0.11	
A	84	0.30	0.083	
HCl	159	0.23	0.063	0.094
Na	371	0.19	0.054	
Kr	117	0.18	0.046	
CCl_4	250	0.09	0.017	0.0009

Λ is the de Broglie thermal wavelength at the triple-point temperature and $a=(V/N)^{1/3}$.

should be small for all the systems listed, with the exceptions of hydrogen and neon.

Use of the classical approximation leads to an important simplification, namely that the contributions to thermodynamic properties that arise from thermal motion can be separated from those due to interactions between particles. The separation of potential and kinetic terms suggests a simple means of characterizing the liquid state. Let V_N be the total potential energy of a system of interacting particles and let K_N be the total kinetic energy. Then we find that in the liquid state, $K_N/|V_N| \sim 1$, whereas $K_N/|V_N| \gg 1$ corresponds to the dilute gas and $K_N/|V_N| \ll 1$ to the low-temperature solid. Alternatively, if we characterize a given system by a length σ and an energy ε, corresponding roughly to the range and strength of the intermolecular forces, we find that in the liquid region of the phase diagram the reduced density $\rho^* = N\sigma^3/V$ and reduced temperature $T^* = k_B T/\varepsilon$ are both of order unity.

Liquids and dense fluids are distinguished from dilute gases by the importance of collisional processes and short-range correlations and from solids by the lack of long-range order; their structure is dominated by the "excluded-volume" effect associated with the packing together of particles with hard cores. A major obstacle to the development of an accurate theory of liquids is the fact that there is no idealized model comparable with the perfect gas or the harmonic solid, both of which can be treated exactly. It is therefore tempting to treat the liquid as an intermediate state between gas and solid. This approach has been widely adopted in the past, but from a theoretical point of view it is not very satisfactory, mainly because it does not take proper account of geometrical factors. Lattice theories, for example, tend to overemphasize the solid-like character of liquids and for that reason have largely fallen out of favour. Methods that rely on expansions in powers of the density are in some respects more useful because they allow the systematic calculation of corrections to ideal-gas behaviour. Nonetheless, they remain essentially theories of the imperfect gas, and cannot be expected to work well under triple-point conditions. We have therefore chosen to place particular emphasis on methods that treat the problem of the liquid state without leaning unduly on concepts taken over from the theory of dilute gases or of solids.

Selected properties of a simple monatomic liquid (argon), a simple molecular liquid (nitrogen) and a simple liquid metal (sodium) are listed in Table 1.2. Not unexpectedly, the properties of the liquid metal are in certain respects very different from those of the other systems, notably in the values of the thermal conductivity, the isothermal compressibility and surface tension, the heat of vaporization, and the ratio of critical to triple-point temperatures. Note, however, that the product of compressibility and

TABLE 1.2. *Selected properties of typical simple liquids*

Property	A	Na	N_2
T_t (K)	84	371	63
T_b (K) ($P = 1$ atm)	87	1155	77
T_c (K)	151	2600	126
T_c/T_t	1.8	7.0	2.0
ρ_t (10^{-3} mol cm^{-3})	35	40	31
C_P/C_V	2.2	1.1	1.6
L_{vap} (kJ mol^{-1})	6.5	99	5.6
χ_T (10^{-12} cm^2 dyn^{-1})	200	19	180
c (m s^{-1})	863	2250	995
γ (dyn cm^{-1})	13	191	12
$\gamma\chi_T$ (Å)	0.26	0.36	0.22
D (10^{-5} cm^2 s^{-1})	1.6	4.3	1.0
η (mg cm^{-1} s^{-1})	2.8	7.0	3.8
λ (mW cm^{-1} K^{-1})	1.3	8800	1.6
$k_B T/2\pi D\eta$ (Å)	4.1	2.7	3.6

χ_T = isothermal compressibility, c = speed of sound, γ = surface tension, D = self-diffusion coefficient, η = shear viscosity and λ = thermal conductivity, all at $T = T_t$; L_{vap} = heat of vaporization at $T = T_b$.

surface tension is roughly the same for the three liquids and, indeed, for many other liquids as well (Egelstaff and Widom, 1970). The quantity $k_B T/2\pi D\eta$ in the table provides a Stokes-law estimate of the molecular diameter.

1.2 INTERMOLECULAR FORCES

The most important feature of the pair potential in simple liquids is the harsh repulsion that appears at short range and has its physical origin in the overlap of the outer electron shells. The effect of these strongly repulsive forces is to create the short-range order that is characteristic of the liquid state, the range of the repulsion being roughly equal to the average nearest-neighbour distance. The attractive forces, which act at long range, vary much more smoothly with the distance between particles and play only a minor role in determining the structure of the liquid. They provide, instead, an essentially uniform attractive background, and give rise to the cohesive energy that is required to stabilize the liquid. This separation of the effects of repulsive and attractive forces is a very old-established concept. It lies at the heart of the ideas of van der Waals, which in turn form the basis of the very successful perturbation theories of the liquid state that we discuss in detail in Chapter 6.

The simplest possible model of a fluid is a system of hard spheres, for which the pair potential $v(r)$ is given by

$$\begin{aligned} v(r) &= \infty, \qquad r < d \\ &= 0, \qquad r > d \end{aligned} \tag{1.2.1}$$

where d is the hard-sphere diameter. This simple model is ideally suited for the study of phenomena in which the hard core of the potential is the dominant factor. "Experimental" information on the hard-sphere model can be obtained by computer simulation, as we shall discuss in Chapter 3. Such calculations have shown very clearly that the structure of a hard-sphere fluid does not differ in any significant way from that corresponding to more complicated interatomic potentials, at least near crystallization; we shall return to this question in Chapter 5. The hard-sphere fluid undergoes a freezing transition at $\rho^* \simeq 0.945$, but the absence of attractive forces means that there is only a single fluid phase (Alder and Wainwright, 1957). A simple model that can describe a true liquid phase is obtained by supplementing the hard-sphere potential by a square-well attraction such that

$$\begin{aligned} v(r) &= \infty, \qquad r < d \\ &= -\varepsilon, \qquad d < r < \gamma d \\ &= 0, \qquad r > \gamma d \end{aligned} \tag{1.2.2}$$

where, typically, $\gamma \simeq 1.5$. The properties of the "square-well" fluid have also been studied extensively by computer simulation.

A more realistic potential for neutral atoms can be constructed theoretically by a detailed quantum-mechanical calculation. At large separations, the dominant contribution to the potential comes from the multipole dispersion interactions between the instantaneous electric moments in one atom and those induced in the other. These moments arise, even though the atoms are electrically neutral, because of spontaneous fluctuations in the electronic charge distributions. All terms in the multipole series represent attractive contributions to the potential. The leading term, varying as r^{-6}, describes the dipole-induced dipole interaction. Higher-order terms represent dipole-quadrupole (r^{-8}), quadrupole-quadrupole (r^{-10}) interactions, and so on, but these are generally small in comparison with the term in r^{-6}.

A rigorous calculation of the short-range interaction presents more difficulty, but over relatively small ranges of r it can be adequately represented by an exponential function of the form $\exp(-r/r_0)$, where r_0 is a range parameter. This approximation must be supplemented by requiring that $v(r) \to \infty$ for r less than some arbitrarily chosen small value. In practice, largely for reasons of mathematical convenience, it is more usual to represent the short-range repulsion by an inverse power law, i.e. r^{-n}, with n lying typically in the range 9 to 15. The behaviour of $v(r)$ in the two limiting cases $r \to \infty$ and $r \to 0$ may therefore be incorporated in a simple potential function of the form

$$v(r) = 4\varepsilon \left[\left(\frac{\sigma}{r} \right)^{12} - \left(\frac{\sigma}{r} \right)^{6} \right] \tag{1.2.3}$$

which is the famous 12-6 potential of Lennard-Jones. Equation (1.2.3) involves two parameters: the collision diameter σ, which is the separation of the particles where $v(r) = 0$; and ε, the depth of the potential well at the minimum in $v(r)$. The Lennard-Jones potential provides a fair description of the interaction between pairs of rare-gas atoms and also of quasispherical molecules such as CH_4. Computer "experiments" (Hansen and Verlet, 1969) show that the triple point of the Lennard-Jones fluid is at $\rho^* \simeq 0.85$, $T^* \simeq 0.68$.

Experimental information on the pair interaction can be extracted from a study of any process that involves collisions between molecules. The most direct method involves measurement of atom–atom scattering cross-sections as a function of the incident energy and scattering angle; inversion of the data allows, in principle, a determination of the pair potential at all separations. A simpler procedure is to assume a specified form for the potential and determine the parameters by measurements in the gas phase either of virial coefficients (see Chapter 4) or, given the theoretical cross-sections, transport properties such as shear viscosity. In this way, the parameters ε and σ in the Lennard-Jones potential have been determined for a large number of gases.

The theoretical and experimental methods we have mentioned all relate to the properties of an isolated pair of molecules. The use of the resulting pair potentials for calculations in the liquid state involves the neglect of many-body forces, an approximation that cannot be wholly justified. In the rare-gas liquids, the three-body, triple–dipole dispersion potential (Axilrod and Teller, 1943) is the most important many-body interaction; the net effect of the triple-dipole forces is predominantly repulsive, amounting in the case of liquid argon to a few per cent of the total potential energy due to pair interactions. Moreover, careful measurements, particularly those of second virial coefficients at low temperatures, have shown that the true pair potential for rare-gas atoms is not of the Lennard-Jones form but has a deeper bowl and weaker tail. Apparently the success of the Lennard-Jones potential in accounting for many of the macroscopic properties of argon-like liquids is the consequence of a fortuitous cancellation of errors. A number of much more accurate potentials have been developed, not only for argon but also for krypton and xenon, but their use in the calculation of condensed-phase properties requires the explicit incorporation of three-body interactions (Maitland *et al.*, 1981).

Although the true pair potential for rare-gas atoms is not the same as the effective pair potential used in liquid-state work, the difference is a relatively minor, quantitative one. The situation in the case of liquid metals is very different, because the form of the effective ion–ion interaction is strongly influenced by effects arising from the presence of a gas of conduction electrons that does not exist before the liquid is formed. The calculation of the ion–ion interaction is a complicated problem, as we shall see in Chapter 11. The ion–electron interaction is first described in terms of a "pseudopotential" that incorporates both the coulombic attraction and the repulsion due to the hard core of the ions. The way in which the pseudopotential is modified by interaction between the conduction electrons is then taken into account, leading finally to a potential that represents the interaction between screened and electrically neutral "pseudoatoms". Irrespective of the detailed assumptions made, the final result has the same qualitative form: a soft repulsion, a deep attractive well, and a long-range oscillatory tail. The potential, and in particular the depth of the well is strongly density dependent, but only weakly dependent on temperature. In Figure 1.2 we plot an effective potential for liquid potassium, together with the Lennard-Jones potential for argon; as the figure shows, the two are very different both in the core region and at long range.

For molten salts and other ionic liquids in which there is no shielding of the electrostatic forces similar to that found in liquid metals, the coulombic interaction provides the dominant contribution to the potential. There must, in addition, be a short-range repulsive force between ions of opposite charge; without such an interaction, the system would collapse. The detailed way in which the repulsive forces are treated is relatively unimportant, and a simple model that we discuss in Chapter 10 is adequate to describe the thermodynamic properties of monovalent molten salts such as the alkali metal halides and nitrates. Polarization of the ions by the internal electric field plays an important part in determining the dynamical properties of many ionic solids and liquids, but such effects are essentially many-body in character and cannot be adequately represented by an additional term in the pair potential.

The description of the interaction between two molecules poses greater problems than for spherical particles because the pair potential is a function both of the separation of the molecules and of their mutual orientation. The model potentials that we shall discuss in this book divide into two classes. The first consists of highly idealized models of polar liquids in which a point dipole-dipole interaction is superimposed on a spherically

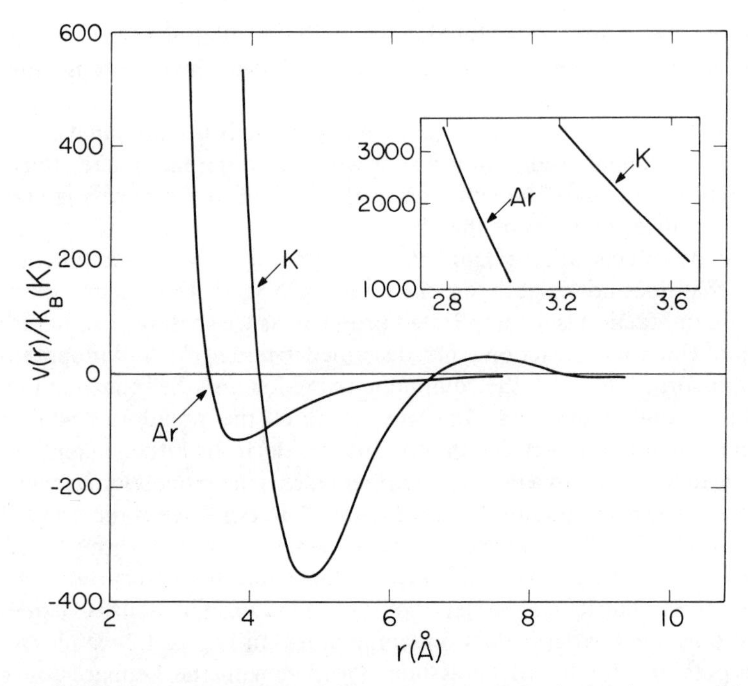

FIG. 1.2. The Lennard-Jones potential for argon and an effective ion–ion potential for liquid potassium (Dagens *et al.*, 1975). The softer core of the potential in potassium is apparent from the inset plot.

symmetric potential. In this case, the pair potentials for molecules labelled 1 and 2 have the general form:

$$v(1,2) = v_0(R) - \boldsymbol{\mu}_1 \cdot \boldsymbol{T}(\mathbf{R}) \cdot \boldsymbol{\mu}_2 \tag{1.2.4}$$

where $\mathbf{R}$ is the vector separation of the molecular centres, $v_0(R)$ is the spherically symmetric term, $\boldsymbol{\mu}_i$ is the dipole moment vector of molecule i, and $\boldsymbol{T}(\mathbf{R})$ is the dipole–dipole interaction tensor, given by

$$\boldsymbol{T}(\mathbf{R}) = \frac{3\mathbf{RR}}{R^5} - \frac{\boldsymbol{I}}{R^3} \tag{1.2.5}$$

where $\boldsymbol{I}$ is the unit tensor. Two examples of (1.2.4) that are of particular interest are those of dipolar hard spheres, where $v_0(R)$ is the hard-sphere interaction (1.2.1), and the Stockmayer potential, where $v_0(R)$ is the Lennard-Jones function (1.2.3). Both these models, together with generalizations that include, for example, dipole–quadrupole and quadrupole–quadrupole terms, have received much attention from theoreticians. Their main limitation when applied to real models is that they ignore the angle dependence of the short-range, shape-determining forces. An easy way to take account of such effects is through the use of model potentials of the second main type with which we shall be concerned. These are models in which the molecule is represented by a set of discrete interaction sites that are commonly, but not invariably, located at the sites of the atomic nuclei. The total potential energy of two interaction-site molecules is obtained as the sum of spherically symmetric interaction-site potentials; the latter are typically of the same functional form as the potentials used for atomic systems, such as (1.2.1) or (1.2.3). Let $\mathbf{r}_{1\alpha}$ be the coordinates of site α in molecule 1 and let $\mathbf{r}_{2\beta}$ be the coordinates of site β in molecule 2, where 1 and 2 are different. Then the total intermolecular potential energy $v(1,2)$ is given by

$$v(1,2) = \sum_{\alpha}\sum_{\beta} v_{\alpha\beta}(|\mathbf{r}_{1\alpha} - \mathbf{r}_{2\beta}|) \tag{1.2.6}$$

where $v_{\alpha\beta}(r)$ is a site–site potential and the sums on α and β run over all interaction sites in the respective molecules.

The use of the interaction-site approach to model a homonuclear diatomic molecule such as N_2 is illustrated in Figure 1.3. The molecule is represented by a pair of interaction sites; these are coincident with the atomic nuclei and are separated by a distance l equal to the chemical bondlength (the

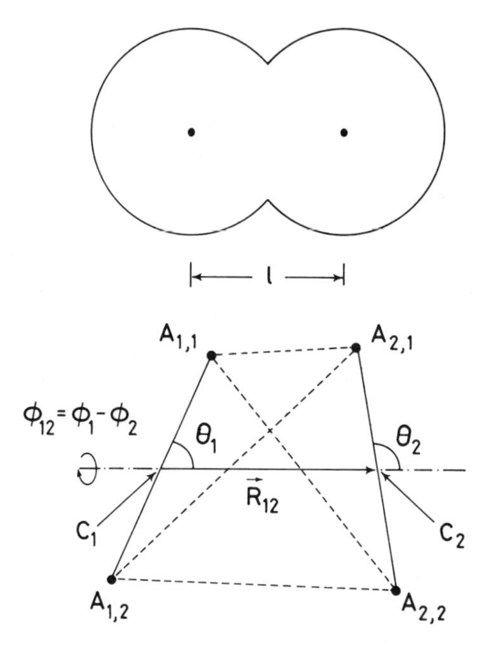

FIG. 1.3. A homonuclear diatomic molecule and its representation by an interaction-site model; $A_{i,\alpha}$ is atom α on molecule i, with coordinates $\mathbf{r}_{i\alpha}$ (see text), and C_1, C_2 are the molecular centres of mass, with vector separation $\mathbf{R}_{12}$. The dashed lines represent the four interaction-site distances.

effects of molecular vibration are ignored). There are four distinct site–site interactions, symbolized by broken lines; because of the symmetry of the molecule, the site–site potential $v_{\alpha\beta}(r)$ is the same for all α, β. Use of an interaction-site model means in this case that the dependence of $v(1,2)$ on the polar coordinates R_{12}, θ_1, θ_2 and ϕ_{12} (defined in the figure) is replaced by a dependence on four independent intermolecular atom–atom separations.

Interaction-site models have been used in calculations of condensed-phase properties of a wide range of molecules; electrostatic interactions are easily allowed for by inclusion of coulombic terms in the site–site potentials. The parameterization of an interaction-site model can be a formidable problem, requiring a process of successive refinements in which computer simulation often plays a major part (Tildesley, 1984). Associated liquids such as water (Rahman and Stillinger, 1971) and liquids composed of flexible molecules, including, for example, *n*-butane (Ryckaert and Bellemans, 1975) have also been successfully described in terms of interaction-site models. These, however, are systems that cannot be regarded as "simple", and we shall have little further to say about them.

1.3 EXPERIMENTAL METHODS

The experimental methods available for studying the properties of simple liquids can be placed in one of two broad categories, depending on whether they are concerned with measurements on a macroscopic or microscopic scale. In general, the calculated microscopic properties are more sensitive to the approximations used in a theory and to the assumptions made about the intermolecular potentials, but the macroscopic properties can usually be determined experimentally with considerably greater accuracy. The two types of measurement are therefore complementary, each providing information that is useful in the development of a statistical-mechanical theory of the liquid state.

The classic macroscopic measurements are those of thermodynamic properties, particularly of the equation of state. Integration of accurate P–ρ–T data yields information on other thermodynamic quantities, such as internal energy and heat capacity, which can be supplemented by calorimetric measurements. For most liquids, the pressure is known as a function of temperature and density only in the vicinity of the liquid–vapour equilibrium line, but for certain systems of particular theoretical interest measurements have been made at much higher pressures; highly specialized techniques are required because of the low compressibility of the liquids near their triple points. The measurement of thermodynamic properties of mixtures of simple liquids also poses a severe experimental problem because the quantities of major interest, i.e. the excess properties, are typically less than one per cent of those of the pure fluids.

The second main class of macroscopic measurements are those relating to transport coefficients. A variety of experimental methods are used. The shear viscosity, for example, can be determined from the observed damping of torsional oscillations or from capillary flow experiments, while the thermal conductivity can be obtained from a steady-state measurement of the transfer of heat between a central filament and surrounding cylinder or between parallel plates. In neither case is it possible to achieve a very high accuracy. The bulk viscosity can, in principle, be obtained from experiments on the attenuation of sound, but the interpretation of the measurements requires an accurate knowledge of certain thermodynamic quantities (specific heat, density and velocity of sound) and also of other transport coefficients (thermal conductivity and shear viscosity), and large uncertainties in the final results seem inevitable. A direct method of determining the coefficient of self-diffusion involves the use of radioactive tracers, which places it in the category of microscopic measurements; in favourable cases, the diffusion coefficient can also be obtained from nuclear magnetic resonance experiments. Nuclear magnetic resonance and other spectroscopic techniques (infrared and Raman) are also useful in the study of reorientational motion in molecular liquids.

Much the most important class of microscopic measurements, at least from the theoretical point of view, are the radiation-scattering experiments. Three particularly valuable techniques can be distinguished: X-ray diffraction, the inelastic scattering of thermal neutrons, and laser-light scattering. The special feature of X-ray scattering experiments is the fact that the energy of the incident radiation is much greater than the thermal energies of the molecules and the scattering process is effectively elastic. This makes it impossible to study time-dependent phenomena, but valuable information can be gained about the static structure of the fluid. By contrast, in the case of thermal neutrons, the energy of the incident particle is comparable with $k_B T$ and the scattering cross-section can therefore be measured as a function of energy transfer as well as momentum transfer. By this means, it is possible to extract information on wavenumber and frequency-dependent fluctuations in liquids at wavelengths comparable with the spacing between particles. This is an extremely powerful method of studying microscopic time-dependent processes in liquids, but the technical difficulties are formidable and the interpretation of many of the experiments reported in the past must be viewed with caution. The relation of neutron-scattering measurements to quantities of statistical-mechanical significance is discussed in Chapter 7. Light-scattering experiments yield similar results to neutron-scattering measurements, but the accessible range of momentum transfer limits the method to the study of fluctuations of wavelengths of order 10^{-5} cm, corresponding to the hydrodynamic regime. Such experiments are particularly useful in the study of critical phenomena.

An important technique of quasi-experimental character is that of *computer simulation.* Two distinct methods are available and have been widely used: the Monte Carlo method of Metropolis *et al.* (1953) and the method of molecular dynamics pioneered by Alder and Wainwright (1959). This type of calculation provides what may be regarded as essentially exact results for a given intermolecular force law, thereby eliminating the ambiguity that invariably arises in the interpretation of experimental data on real liquids. Their usefulness rests ultimately on the fact that a model containing a relatively small number of particles, typically several hundred, is in most cases sufficiently large to simulate the behaviour of a macroscopic system.

In the method of molecular dynamics, the classical equations of motion of a system of interacting particles are solved and equilibrium properties are determined from time averages taken over a sufficiently long time interval (typically $\sim 10^{-10}$–10^{-11} s in real time). The Monte Carlo method involves the generation of a series of configurations of the particles of a model in a way that ensures that the configurations are distributed in phase space according to some prescribed probability density. The mean value of any configurational property determined from a sufficiently large number ($\sim 10^6$) of configurations provides an estimate of the ensemble-averaged value of that quantity, the character of the ensemble average being dependent on the particular sampling procedure that is used. The major advantage of molecular dynamics is that it allows the study of time-dependent processes. For the calculation of thermodynamic properties, however, the Monte Carlo method is often more suitable. A detailed discussion of both methods is given in Chapter 3.

CHAPTER 2

Statistical Mechanics and Molecular Distribution Functions

This chapter is devoted to a brief summary of the principles of classical statistical mechanics, to a discussion of the link between statistical mechanics and thermodynamics, and to the definitions of a variety of equilibrium and time-dependent distribution functions. It also establishes much of the notation that is used in later parts of the book.

2.1 THE LIOUVILLE EQUATION AND THE BBGKY HIERARCHY

Consider an isolated, macroscopic system consisting of N identical particles, each of which has three, translational degrees of freedom. The dynamical state of the system at a given time is completely specified by the $3N$ coordinates $\mathbf{r}^N \equiv \{\mathbf{r}_1, \ldots, \mathbf{r}_N\}$ and $3N$ momenta $\mathbf{p}^N \equiv \{\mathbf{p}_1, \ldots, \mathbf{p}_N\}$ of the particles. The values of these variables define a *phase point* in a $6N$-dimensional *phase space*. Let $\mathcal{H}_N(\mathbf{r}^N, \mathbf{p}^N)$ be the hamiltonian of the system, which we write in the form

$$\mathcal{H}_N(\mathbf{r}^N, \mathbf{p}^N) = \frac{1}{2m} \sum_{i=1}^{N} |\mathbf{p}_i|^2 + V_N(\mathbf{r}^N) \tag{2.1.1}$$

where m is the particle mass and $V_N(\mathbf{r}^N)$ is the total potential energy. Then the motion of the phase point along its *phase trajectory* is determined by Hamilton's equations:

$$\dot{\mathbf{r}}_i = \frac{\partial \mathcal{H}_N}{\partial \mathbf{p}_i} \tag{2.1.2}$$

$$\dot{\mathbf{p}}_i = -\frac{\partial \mathcal{H}_N}{\partial \mathbf{r}_i} \tag{2.1.3}$$

where $i = 1, \ldots, N$. These equations are to be solved subject to $6N$ initial conditions on the coordinates and momenta.

The aim of equilibrium statistical mechanics is to calculate observable properties of the system either as averages over a phase-space trajectory (the method of Boltzmann) or as averages over an ensemble of systems, each of which is a replica of the system of interest (the method of Gibbs). The main features of the two methods are discussed in later sections of this chapter. For the present, it is sufficient to recall that in Gibbs' formulation of statistical mechanics the distribution of phase points of the ensemble is described by a phase-space *probability density* $f^{(N)}(\mathbf{r}^N, \mathbf{p}^N; t)$; the quantity $f^{(N)}\,\mathrm{d}\mathbf{r}^N\,\mathrm{d}\mathbf{p}^N$ is the probability that at time t the physical system is in a microscopic state represented by a phase point lying in the infinitesimal, $6N$-dimensional phase-space element $\mathrm{d}\mathbf{r}^N\,\mathrm{d}\mathbf{p}^N$. Given a complete knowledge of the probability density, it would be possible to calculate the average value of any function of the coordinates and momenta.

The time evolution of the phase-space probability density is governed by the Liouville equation. The latter is the $6N$-dimensional analogue of the equation of continuity of an incompressible fluid; it describes the fact that phase points of the ensemble are neither created nor destroyed as time evolves. The Liouville equation can be written in compact form as

$$\frac{\partial f^{(N)}}{\partial t} = \{\mathscr{H}_N, f^{(N)}\} \tag{2.1.4}$$

where $\{A, B\}$ denotes the Poisson bracket:

$$\{A, B\} \equiv \sum_{i=1}^{N} \left(\frac{\partial A}{\partial \mathbf{r}_i} \cdot \frac{\partial B}{\partial \mathbf{p}_i} - \frac{\partial A}{\partial \mathbf{p}_i} \cdot \frac{\partial B}{\partial \mathbf{r}_i} \right) \tag{2.1.5}$$

Another convenient form is obtained by introducing the Liouville operator $\mathscr{L}$, defined as

$$\mathscr{L} = i\{\mathscr{H}_N, \quad \} \tag{2.1.6}$$

Equation (2.1.4) then becomes

$$\frac{\partial f^{(N)}}{\partial t} = -i\mathscr{L} f^{(N)} \tag{2.1.7}$$

the formal solution to which is

$$f^{(N)}(\mathbf{r}^N, \mathbf{p}^N; t) = \exp(-i\mathscr{L}t) f^{(N)}(\mathbf{r}^N, \mathbf{p}^N; 0) \tag{2.1.8}$$

The time dependence of an arbitrary dynamical variable, A say, can be represented in a manner similar to (2.1.7). Any such variable is a function of the phase-space coordinates $\mathbf{r}^N, \mathbf{p}^N$ and changes in A are associated solely with changes in the independent variables. Thus

$$\frac{\mathrm{d}A}{\mathrm{d}t} = \sum_{i=1}^{N} \left(\frac{\partial A}{\partial \mathbf{r}_i} \cdot \frac{\mathrm{d}\mathbf{r}_i}{\mathrm{d}t} + \frac{\partial A}{\partial \mathbf{p}_i} \cdot \frac{\mathrm{d}\mathbf{p}_i}{\mathrm{d}t} \right) \tag{2.1.9}$$

If we substitute from Hamilton's equations (2.1.2) and (2.1.3) and use the definition (2.1.6), Eqn (2.1.9) becomes

$$\frac{\mathrm{d}A}{\mathrm{d}t} = i\mathscr{L}A \tag{2.1.10}$$

This has the formal solution

$$A(t) = \exp(i\mathscr{L}t)A(0) \tag{2.1.11}$$

The description of the system that is provided by the full phase-space probability density is for many purposes unnecessarily detailed. If we are interested only in the behaviour of a subset of particles of size n, say, the unwanted information can be eliminated by integrating $f^{(N)}$ over the coordinates and momenta of the remaining $(N-n)$ particles. We therefore define a *reduced* phase-space distribution function $f^{(n)}(\mathbf{r}^n, \mathbf{p}^n; t)$ for $n < N$ by

$$f^{(n)}(\mathbf{r}^n, \mathbf{p}^n; t) = \frac{N!}{(N-n)!} \iint f^{(N)}(\mathbf{r}^N, \mathbf{p}^N; t)\, \mathrm{d}\mathbf{r}^{(N-n)}\, \mathrm{d}\mathbf{p}^{(N-n)}, \quad n < N \tag{2.1.12}$$

where we use the notation $\mathbf{r}^n \equiv \{\mathbf{r}_1, \ldots, \mathbf{r}_n\}$, $\mathbf{r}^{(N-n)} \equiv \{\mathbf{r}_{n+1}, \ldots, \mathbf{r}_N\}$, etc. The meaning of $f^{(n)}(\mathbf{r}^n, \mathbf{p}^n; t)$ is that $f^{(n)}\, \mathrm{d}\mathbf{r}^n\, \mathrm{d}\mathbf{p}^n$ is $N!/(N-n)!$ times the probability of finding *any* subset of n particles in the reduced phase-space element $\mathrm{d}\mathbf{r}^n\, \mathrm{d}\mathbf{p}^n$ at time t, irrespective of the coordinates and momenta of the remaining particles; the factor $N!/(N-n)!$ is the number of ways of choosing n particles from N.

The equations of motion of the reduced distribution functions are much more complicated in form than the Liouville equation. Let us suppose that the total force acting on particle i is the sum of an external force $\mathbf{X}_i$ and of pair forces $\mathbf{F}_{ij}$ due to other particles j, with $\mathbf{F}_{ii} = 0$. Then the Liouville equation can be written as

$$\frac{\partial f^{(N)}}{\partial t} + \frac{1}{m} \sum_{i=1}^{N} \mathbf{p}_i \cdot \frac{\partial f^{(N)}}{\partial \mathbf{r}_i} + \sum_{i=1}^{N} \mathbf{X}_i \cdot \frac{\partial f^{(N)}}{\partial \mathbf{p}_i} = -\sum_{i}^{N} \sum_{j} \mathbf{F}_{ij} \cdot \frac{\partial f^{(N)}}{\partial \mathbf{p}_i} \tag{2.1.13}$$

We now multiply through by $N!/(N-n)!$ and integrate over $3(N-n)$ coordinates and momenta. By employing the definition (2.1.12), and exploiting the symmetry of $f^{(N)}$ with respect to interchange of particle labels and the fact that $f^{(N)}$ vanishes as $\mathbf{p}_i \to \pm\infty$ or when $\mathbf{r}_i$ lies outside the volume occupied by the system, we find that

$$\begin{aligned}
\frac{\partial f^{(n)}}{\partial t} &+ \frac{1}{m}\sum_{i=1}^{n} \mathbf{p}_i \cdot \frac{\partial f^{(n)}}{\partial \mathbf{r}_i} + \sum_{i=1}^{n} \mathbf{X}_i \cdot \frac{\partial f^{(n)}}{\partial \mathbf{p}_i} \\
&= -\frac{N!}{(N-n)!}\sum_i^{N}\sum_j \iint \mathbf{F}_{ij} \cdot \frac{\partial f^{(N)}}{\partial \mathbf{p}_i} \, d\mathbf{r}^{(N-n)} \, d\mathbf{p}^{(N-n)} \\
&= -\sum_i^{n}\sum_j \mathbf{F}_{ij} \cdot \frac{\partial f^{(n)}}{\partial \mathbf{p}_i} - \frac{N!}{(N-n)!}\sum_{i=1}^{n}\sum_{j=n+1}^{N} \iint \mathbf{F}_{ij} \cdot \frac{\partial f^{(N)}}{\partial \mathbf{p}_i} \, d\mathbf{r}^{(N-n)} \, d\mathbf{p}^{(N-n)} \\
&= -\sum_i^{n}\sum_j \mathbf{F}_{ij} \cdot \frac{\partial f^{(n)}}{\partial \mathbf{p}_i} - \sum_{i=1}^{n} \iint \mathbf{F}_{i,n+1} \cdot \frac{\partial f^{(n+1)}}{\partial \mathbf{p}_i} \, d\mathbf{r}_{n+1} \, d\mathbf{p}_{n+1}
\end{aligned} \tag{2.1.14}$$

Thus the behaviour of $f^{(n)}$ is linked to that of $f^{(n+1)}$ by the formula

$$\begin{aligned}
\left\{\frac{\partial}{\partial t} + \sum_{i=1}^{n}\left[\frac{1}{m}\mathbf{p}_i \cdot \frac{\partial}{\partial \mathbf{r}_i} + \left(\mathbf{X}_i + \sum_{j=1}^{n} \mathbf{F}_{ij}\right) \cdot \frac{\partial}{\partial \mathbf{p}_i}\right]\right\} f^{(n)}(\mathbf{r}^n, \mathbf{p}^n; t) \\
= -\sum_{i=1}^{n} \iint \mathbf{F}_{i,n+1} \cdot \frac{\partial}{\partial \mathbf{p}_i} f^{(n+1)}(\mathbf{r}^{n+1}, \mathbf{p}^{n+1}; t) \, d\mathbf{r}_{n+1} \, d\mathbf{p}_{n+1}
\end{aligned} \tag{2.1.15}$$

The set of equations for $n = 1, \ldots, N-1$ was first derived by Yvon (1935) and is known as the BBGKY hierarchy (Kirkwood, 1935; Bogolyubov, 1946; Born and Green, 1949).

Equation (2.1.15) is not immediately useful, because it expresses one unknown function, $f^{(n)}$, in terms of another, $f^{(n+1)}$: at some stage an approximation must be made that closes the system of equations. The most important case in practice is that obtained by setting $n = 1$, i.e.

$$\begin{aligned}
\left(\frac{\partial}{\partial t} + \frac{1}{m}\mathbf{p}_1 \cdot \frac{\partial}{\partial \mathbf{r}_1} + \mathbf{X}_1 \cdot \frac{\partial}{\partial \mathbf{p}_1}\right) f^{(1)}(\mathbf{r}_1, \mathbf{p}_1; t) \\
= -\iint \mathbf{F}_{12} \cdot \frac{\partial}{\partial \mathbf{p}_1} f^{(2)}(\mathbf{r}_1, \mathbf{p}_1, \mathbf{r}_2, \mathbf{p}_2; t) \, d\mathbf{r}_2 \, d\mathbf{p}_2
\end{aligned} \tag{2.1.16}$$

The quantity $f^{(1)} \, d\mathbf{r}_1 \, d\mathbf{p}_1$ is N times the probability of finding a particle of the system in the six-dimensional phase element $d\mathbf{r}_1 \, d\mathbf{p}_1$ at time t; $f^{(2)} \, d\mathbf{r}_1 \, d\mathbf{r}_2 \, d\mathbf{p}_1 \, d\mathbf{p}_2$ is $N(N-1)$ times the probability of finding a particle in the phase element $d\mathbf{r}_1 \, d\mathbf{p}_1$ and, simultaneously, another particle in the phase element $d\mathbf{r}_2 \, d\mathbf{p}_2$.

Much effort has been devoted to finding approximate solutions to the BBGKY hierarchy on the basis of expressions that relate $f^{(2)}$ to $f^{(1)}$. The resulting *kinetic equations* are rarely appropriate for the study of liquids, since they mostly treat the pair correlations in a very crude way. The simplest approximation is to ignore the pair correlations altogether by writing

$$f^{(2)}(\mathbf{r}, \mathbf{p}, \mathbf{r}', \mathbf{p}'; t) = f^{(1)}(\mathbf{r}, \mathbf{p}; t) f^{(1)}(\mathbf{r}', \mathbf{p}'; t) \tag{2.1.17}$$

This leads to the Vlasov equation (Vlasov, 1961):

$$\left(\frac{\partial}{\partial t}+\frac{1}{m}\mathbf{p}\cdot\frac{\partial}{\partial \mathbf{r}}+[\mathbf{X}(\mathbf{r},t)+\bar{\mathbf{F}}(\mathbf{r},t)]\cdot\frac{\partial}{\partial \mathbf{p}}\right)f^{(1)}(\mathbf{r},\mathbf{p};t)=0 \qquad (2.1.18)$$

where the quantity

$$\bar{\mathbf{F}}(\mathbf{r},t)=\iint \mathbf{F}(\mathbf{r},\mathbf{r}')f^{(1)}(\mathbf{r}',\mathbf{p}';t)\,\mathrm{d}\mathbf{r}'\,\mathrm{d}\mathbf{p}' \qquad (2.1.19)$$

is the average force exerted by other particles in the system on a particle that at time t is at a point $\mathbf{r}$. Though inadequate for liquids, the Vlasov equation is widely used in plasma physics, since in that case the long range of the Coulomb potential provides a justification for a mean-field treatment of the interactions. A number of essentially intuitive improvements on the Vlasov decoupling scheme have been proposed (Singwi *et al.*, 1968), but the kinetic equations to which they give rise remain applicable only to low-density systems.

The most famous of all kinetic equations is that obtained by Boltzmann more than a century ago. Boltzmann's derivation (Résibois and DeLeener, 1977) was based on two assumptions that, in general, are justified only at low densities, namely that collisions between particles are strictly binary in character and that successive collisions are uncorrelated. The exact kinetic equation for $f^{(1)}(\mathbf{r},\mathbf{p}_1;t)$, say, may be written in schematic form as

$$\left(\frac{\partial}{\partial t}+\frac{1}{m}\mathbf{p}_1\cdot\frac{\partial}{\partial \mathbf{r}}+\mathbf{X}\cdot\frac{\partial}{\partial \mathbf{p}_1}\right)f^{(1)}(\mathbf{r},\mathbf{p}_1;t)=\left(\frac{\partial f^{(1)}}{\partial t}\right)_{\text{collision}} \qquad (2.1.20)$$

where the right-hand side represents the time variation of $f^{(1)}(\mathbf{r},\mathbf{p}_1;t)$ due to collisions between particles. The collision term is given rigorously by the right-hand side Eqn (2.1.16), but with Boltzmann's approximations it becomes

$$\left(\frac{\partial f^{(1)}}{\partial t}\right)_{\text{collision}}=\frac{1}{m}\iint \sigma(\Omega,|\mathbf{p}_1-\mathbf{p}_2|)\,|\mathbf{p}_1-\mathbf{p}_2|\,[f^{(1)}(\mathbf{r},\mathbf{p}_1';t)f^{(1)}(\mathbf{r},\mathbf{p}_2';t)$$
$$-f^{(1)}(\mathbf{r},\mathbf{p}_1;t)f^{(1)}(\mathbf{r},\mathbf{p}_2;t)]\,\mathrm{d}\Omega\,\mathrm{d}\mathbf{p}_2 \qquad (2.1.21)$$

where the primes are used to denote post-collisional momenta and $\sigma(\Omega,|\mathbf{p}_1-\mathbf{p}_2|)$ is the cross-section for scattering into a solid angle $\mathrm{d}\Omega$. As Boltzmann showed, this form of the collision term is able to account for the observed fact that many-particle systems evolve irreversibly towards an equilibrium state. The irreversibility is described by Boltzmann's H-theorem, which is the microscopic equivalent of the Second Law of thermodynamics.

Solution of the Boltzmann equation leads to explicit expressions for the hydrodynamic transport coefficients in terms of certain "collision integrals" (Maitland *et al.*, 1981). The scattering cross-section and hence the collision integrals themselves can be evaluated numerically for a given choice of interparticle potential, though for hard spheres they have a simple analytic form. The results, however, are applicable only to dilute gases. The Boltzmann equation for hard spheres was later modified semiempirically by Enskog (Résibois and DeLeener, 1977) in a manner that extends its range of validity to considerably higher densities. The Enskog theory retains the two main assumptions inolved in the derivation of the Boltzmann equation, but it also corrects for the finite size of the colliding particles in two important ways. First, allowance is made for the fact that at high densities the modification of the collision rate by excluded-volume effects can no longer be ignored. Because the same effects are responsible for the increase in pressure over its ideal-gas value, the enhancement of the collision rate relative to its low-density limit can be calculated if the hard-sphere equation of state is known, as we shall see in Section 3.2. Secondly, *collisional transfer* of energy and momentum is incorporated into the theory. This is achieved by rewriting Eqn (2.1.21) in a form in which the distribution functions for the two colliding particles are evaluated at points in space that are separated by a distance equal to the hard-sphere diameter. The modification is a crucial one, because transport in fluids at high densities is dominated by collisional transfer.

The phase-space probability density of a system in thermodynamic equilibrium is not an explicit function of time. We shall use the symbol $f_0^{(N)}(\mathbf{r}^N, \mathbf{p}^N)$ to denote the equilibrium probability density; it is clear from Eqn (2.1.4) that a sufficient condition for a probability density to be descriptive of a system in equilibrium is that it should be a function solely of the hamiltonian of the system. Integration of $f_0^{(N)}(\mathbf{r}^N, \mathbf{p}^N)$ over a subset of coordinates and momenta in the manner of (2.1.12) yields a set of equilibrium reduced phase-space distribution functions $f_0^{(n)}(\mathbf{r}^n, \mathbf{p}^n)$. The case $n=1$ corresponds to the equilibrium single-particle distribution function; if there is no external field, the distribution is independent of $\mathbf{r}$ and has the familiar Maxwell–Boltzmann form, i.e.

$$f_{\mathrm{MB}}(\mathbf{p}) \equiv f_0^{(1)}(\mathbf{r}, \mathbf{p}) = \frac{\rho}{(2\pi m k_{\mathrm{B}} T)^{3/2}} \exp\left(-\beta\,|\mathbf{p}|^2/2m\right) \tag{2.1.22}$$

with the normalization

$$\iint f_{\mathrm{MB}}(\mathbf{p})\, d\mathbf{r}\, d\mathbf{p} = N \tag{2.1.23}$$

If velocity $\mathbf{u}$ is chosen as the independent variable, Eqn (2.1.22) can be rewritten as

$$f_{\mathrm{MB}}(\mathbf{u}) = \rho \phi_0(\mathbf{u}) \tag{2.1.24}$$

with

$$\phi_0(\mathbf{u}) = \left(\frac{m}{2\pi k_B T}\right)^{3/2} \exp\left(-\tfrac{1}{2}m\beta\,|\mathbf{u}|^2\right) \tag{2.1.25}$$

When the single-particle distribution function is dependent on **r** and t, but the local density $\rho(\mathbf{r}, t)$, local velocity $\mathbf{u}(\mathbf{r}, t)$ and local temperature $T(\mathbf{r}, t)$ vary slowly both in space and time, the distribution of velocities is given by a generalization of Eqn (2.1.24):

$$f_{\mathrm{l.e.}}(\mathbf{u}, \mathbf{r}; t) = \rho(\mathbf{r}, t)\left(\frac{m}{2\pi k_B T(\mathbf{r}, t)}\right)^{3/2} \exp\left(\frac{-m}{2k_B T(\mathbf{r}, t)}|\mathbf{u}-\mathbf{u}(\mathbf{r}, t)|^2\right) \tag{2.1.26}$$

The function $f_{\mathrm{l.e.}}(\mathbf{u}, \mathbf{r}; t)$ is called the "local-equilibrium" Maxwell-Boltzmann distribution (Résibois and DeLeener, 1977).

2.2 TIME AVERAGES AND ENSEMBLE AVERAGES

Thermodynamic properties of a system, with some important exceptions, are expressible as averages of certain functions of the coordinates and momenta of the constituent particles. In a state of thermodynamic equilibrium, the averages must be independent of time. For simplicity, we again suppose that the system consists of N identical, spherical particles. We also assume that the system is isolated from its surroundings, in which case the hamiltonian $\mathcal{H}_N$ (Eqn (2.1.1)) is a constant of the motion. Given the initial coordinates and momenta of the particles, the positions at any later (or earlier) time can in principle be obtained as the solution to Newton's equations of motion, i.e. to a set of $3N$ coupled, second-order, differential equations:

$$m\ddot{\mathbf{r}}_i = -\nabla_i V_N(\mathbf{r}^N) \tag{2.2.1}$$

If $\mathcal{F}(\mathbf{r}^N, \mathbf{p}^N)$ is a function of the $6N$ coordinates and momenta, and if F is the associated thermodynamic property, then

$$F = \langle \mathcal{F}(\mathbf{r}^N, \mathbf{p}^N) \rangle \tag{2.2.2}$$

where the angular brackets denote a statistical average.

Conceptually, the simplest way to view Eqn (2.2.2) is as a time average over the dynamical history of the system. The meaning of the statistical average is then expressed as the integral

$$\langle \mathcal{F} \rangle_t = \lim_{\tau\to\infty} \frac{1}{\tau}\int_0^\tau \mathcal{F}[\mathbf{r}^N(t), \mathbf{p}^N(t)]\,\mathrm{d}t \tag{2.2.3}$$

This interpretation is of great importance for much of what follows in later chapters, since the statistical averages obtained by molecular dynamics simulation are of precisely this type (see Chapter 3).

A simple application of (2.2.3) is the calculation of the temperature of an isolated system as the time average of the kinetic energy. If

$$\mathcal{T}(t) = \frac{1}{3mNk_B} \sum_{i=1}^{N} |\mathbf{p}_i(t)|^2 \tag{2.2.4}$$

then

$$T = \langle \mathcal{T} \rangle_t = \lim_{\tau \to \infty} \frac{1}{\tau} \int_0^{\tau} \mathcal{T}(t)\, dt \tag{2.2.5}$$

As a more interesting example, we can use the definition (2.2.3) and the result (2.2.5) to show that the equation of state of the system is related to the time average of the Clausius virial function. The latter is defined as

$$\mathcal{V}(\mathbf{r}^N) = \sum_{i=1}^{N} \mathbf{r}_i \cdot \mathbf{F}_i \tag{2.2.6}$$

where $\mathbf{F}_i$ is the total force acting on particle *i*. From the previous formulae, together with an integration by parts, it follows that

$$\begin{aligned} \langle \mathcal{V} \rangle_t &= \lim_{\tau \to \infty} \frac{1}{\tau} \int_0^{\tau} dt \sum_{i=1}^{N} \mathbf{r}_i(t) \cdot \mathbf{F}_i(t) \\ &= \lim_{\tau \to \infty} \frac{1}{\tau} \int_0^{\tau} dt \sum_{i=1}^{N} \mathbf{r}_i(t) \cdot m\ddot{\mathbf{r}}_i(t) \qquad (2.2.7) \\ &= -\lim_{\tau \to \infty} \frac{1}{\tau} \int_0^{\tau} dt \sum_{i=1}^{N} m\,|\dot{\mathbf{r}}_i(t)|^2 \\ &= -3Nk_B T \qquad (2.2.8) \end{aligned}$$

The total virial function can be separated into two parts: one, $\mathcal{V}_{int}$, arises from the forces between particles; the other, $\mathcal{V}_{ext}$, comes from the external forces. If the particles are confined to a cubic box of length $L = V^{1/3}$, the external forces are related in a simple way to the pressure exerted by the walls of area $S = L^2$, and their contribution to the virial function is $\mathcal{V}_{ext} = -3PSL = -3PV$. Equation (2.2.8) may therefore be rearranged to give

$$\begin{aligned} PV &= Nk_B T + \tfrac{1}{3}\langle \mathcal{V}_{int} \rangle_t \\ &= Nk_B T - \tfrac{1}{3}\left\langle \sum_{i=1}^{N} \mathbf{r}_i(t) \cdot \nabla_i V_N[\mathbf{r}^N(t)] \right\rangle_t \qquad (2.2.9) \end{aligned}$$

or

$$\frac{\beta P}{\rho} = 1 - \frac{\beta}{3N} \left\langle \sum_{i=1}^{N} \mathbf{r}_i(t) \cdot \nabla_i V_N[\mathbf{r}^N(t)] \right\rangle_t \tag{2.2.10}$$

This important result is often known as the *virial equation.*

The alternative to the time-averaging procedure described by Eqn (2.2.3) is to average over a suitably constructed ensemble. An ensemble is an arbitrarily large collection of imaginary systems, all of which are replicas of the system of interest insofar as they are characterized by the same macroscopic parameters. The systems of the ensemble differ from each other in the assignment of the coordinates and momenta of the particles, and the ensemble is represented by a cloud of phase points distributed in space according to the probability density $f^{(N)}(\mathbf{r}^N, \mathbf{p}^N; t)$ introduced in the previous section. The equilibrium ensemble average of a phase function $\mathscr{F}(\mathbf{r}^N, \mathbf{p}^N)$ is given by

$$\langle \mathscr{F} \rangle_e = \iint \mathscr{F}(\mathbf{r}^N, \mathbf{p}^N) f_0^{(N)}(\mathbf{r}^N, \mathbf{p}^N)\, d\mathbf{r}^N\, d\mathbf{p}^N \tag{2.2.11}$$

where $f_0^{(N)}$ is normalized such that

$$\iint f_0^{(N)}(\mathbf{r}^N, \mathbf{p}^N)\, d\mathbf{r}^N\, d\mathbf{p}^N = 1 \tag{2.2.12}$$

For example, the thermodynamic internal energy U is the ensemble average of the hamiltonian:

$$U = \iint \mathscr{H}_N(\mathbf{r}^N, \mathbf{p}^N) f_0^{(N)}(\mathbf{r}^N, \mathbf{p}^N)\, d\mathbf{r}^N\, d\mathbf{p}^N \tag{2.2.13}$$

The various versions of the Monte Carlo method that we discuss in Chapter 3 provide efficient computational schemes for the determination of ensemble averages.

The explicit form of the equilibrium probability density depends on the macroscopic parameters chosen to characterize the ensemble. A particularly simple case is one where the systems of the ensemble are assumed to have the same number of particles, same volume and same total energy, E_0 say. An ensemble constructed in this way is called a *microcanonical* ensemble, and clearly is representative of a real system that can exchange neither heat nor matter with its surroundings. The equilibrium probability density is

$$f_0^{(N)}(\mathbf{r}^N, \mathbf{p}^N) = C\delta[\mathscr{H}_N(\mathbf{r}^N, \mathbf{p}^N) - E_0] \tag{2.2.14}$$

where $\delta(\ )$ is the Dirac delta-function and C a normalization constant. The systems of a microcanonical ensemble are therefore uniformly distributed on the hypersurface in phase space corresponding to a total energy E_0, while C^{-1} is the total "volume" of that hypersurface; from Eqn (2.2.13) we see that the thermodynamic internal energy is equal to the parameter E_0. The constraint of constant total energy is reminiscent of the condition of conservation of total energy under which time averages are taken. Indeed, time averages and microcanonical ensemble averages are identical if the system is *ergodic*, which means that after a suitable lapse of time the phase trajectory of the system will have passed an equal number of times through every phase-space element lying on the hypersurface defined by (2.2.14).

In the next two sections, we discuss other choices of the ensemble probability density that, in practice, are often more useful than the microcanonical one.

2.3 CANONICAL AND ISOTHERMAL–ISOBARIC ENSEMBLES

A *canonical* ensemble is a collection of systems characterized by the same values of N, V and T; it is therefore sometimes called an NVT-ensemble. In order that the temperature can be assigned a fixed value, the systems of the ensemble are imagined to have been brought into thermal equilibrium with each other by immersing them in a heat bath of temperature T. The canonical equilibrium probability density for a system of N identical, spherical particles is

$$f_0^{(N)}(\mathbf{r}^N, \mathbf{p}^N) = \frac{1}{N!} h^{-3N} \frac{\exp[-\beta \mathcal{H}_N(\mathbf{r}^N, \mathbf{p}^N)]}{Q_N(V, T)} \tag{2.3.1}$$

where h is Planck's constant, the factor $N!$ takes care of the indistinguishability of the particles, and the normalizing factor $Q_N(V, T)$ is the *canonical partition function*, defined as

$$Q_N(V, T) = \frac{h^{-3N}}{N!} \iint \exp[-\beta \mathcal{H}_N(\mathbf{r}^N, \mathbf{p}^N)] \, d\mathbf{r}^N \, d\mathbf{p}^N \tag{2.3.2}$$

Inclusion of the factor h^{-3N} in Eqns (2.3.1) and (2.3.2) ensures that both $f_0^{(N)}(\mathbf{r}^N, \mathbf{p}^N) \, d\mathbf{r}^N \, d\mathbf{p}^N$ and $Q_N(V, T)$ are dimensionless and go over correctly to the corresponding quantities of quantum statistics.

The link between statistical mechanics and thermodynamics is established via the relation

$$A = -k_B T \log Q_N(V, T) \tag{2.3.3}$$

where A is the Helmholtz free energy. The latter is the *thermodynamic potential* for a system of fixed N, V and T; at equilibrium in a system of constant N, V and T, A is a minimum. If A is a known function of the independent variables V and T, all other thermodynamic state functions can be obtained by differentiation. If we start from the defining relation

$$A = U - TS \tag{2.3.4}$$

where U is the internal energy and S is the entropy, it is an elementary task to show that

$$P = -\left(\frac{\partial A}{\partial V}\right)_T \tag{2.3.5}$$

$$S = -\left(\frac{\partial A}{\partial T}\right)_V \tag{2.3.6}$$

$$U = \left(\frac{\partial (A/T)}{\partial (1/T)}\right)_V \tag{2.3.7}$$

Corresponding to each such thermodynamic relation, there exists an equivalent relation in terms of the partition function. For example, it follows from Eqns (2.2.13) and (2.3.1) that

$$\begin{aligned} U &= \langle \mathcal{H}_N \rangle \\ &= \frac{1}{N!\,h^{3N} Q_N(V,T)} \iint \mathcal{H}_N(\mathbf{r}^N, \mathbf{p}^N) \exp[-\beta \mathcal{H}_N(\mathbf{r}^N, \mathbf{p}^N)]\, d\mathbf{r}^N\, d\mathbf{p}^N \\ &= -\left(\frac{\partial}{\partial \beta} \log Q_N(V,T)\right)_V \end{aligned} \tag{2.3.8}$$

Taken together with the fundamental relation (2.3.3), this result is equivalent to the thermodynamic formula (2.3.7). Similarly, Eqn (2.3.5) may be rewritten as

$$P = k_B T \left(\frac{\partial \log Q_N(V,T)}{\partial V}\right)_T \tag{2.3.9}$$

and shown to be equivalent to the virial equation (2.2.10), with the time average replaced by a canonical ensemble average (Hill, 1956). Note that both the free energy and entropy are expressible only in terms of the partition function $Q_N(V, T)$ and cannot be written as phase-space averages of any function of the microscopic variables. Finally, the specific heat at constant volume, defined thermodynamically as

$$C_V = \left(\frac{\partial U}{\partial T}\right)_V \tag{2.3.10}$$

is related to the mean-square deviation of the total energy from its average value:

$$C_V = \frac{1}{k_B T^2}[\langle \mathcal{H}_N^2 \rangle - \langle \mathcal{H}_N \rangle^2] \tag{2.3.11}$$

If the hamiltonian is separated into kinetic and potential terms, as in Eqn (2.1.1), the integrations over momenta in Eqn (2.3.2) can be carried out explicitly, yielding a factor $(2\pi m k_B T)^{1/2}$ for each degree of freedom. The partition function may then be rewritten as

$$Q_N(V, T) = \frac{\Lambda^{-3N}}{N!} Z_N(V, T) \tag{2.3.12}$$

where Λ is the de Broglie thermal wavelength defined by Eqn (1.1.1) and

$$Z_N(V, T) = \int \exp[-\beta V_N(\mathbf{r}^N)] \, d\mathbf{r}^N \tag{2.3.13}$$

is the *configuration integral.* In the case of a perfect gas, $V_N(\mathbf{r}^N) = 0$ and $Z_N(V, T) = V^N$. The partition function of a perfect gas is therefore given by

$$Q_N^{\text{id}}(V, T) = \frac{\Lambda^{-3N}}{N!} V^N \tag{2.3.14}$$

where "id" stands for "ideal". If Stirling's approximation is used for $\log N!$, the corresponding expression for the free energy per particle is

$$\frac{\beta A^{\text{id}}}{N} = \log \rho + 3 \log \Lambda - 1 \tag{2.3.15}$$

When inserted in Eqn (2.3.5), this result leads immediately to the ideal-gas equation of state, $\beta P/\rho = 1$.

The partition function of a system of interacting particles is conveniently written as

$$Q_N(V, T) = Q_N^{\text{id}} \frac{Z_N(V, T)}{V^N} \tag{2.3.16}$$

On taking the logarithm of both sides of Eqn (2.3.16), the free energy separates naturally into "ideal" and "excess" parts:

$$A = A^{\text{id}} + A^{\text{ex}} \tag{2.3.17}$$

where A^{id} is given by (2.3.15) and the excess part is

$$A^{\text{ex}} = -k_B T \log \frac{Z_N(V, T)}{V^N} \tag{2.3.18}$$

The excess part contains all the contributions to A that arise from the interactions between particles. A similar division into ideal and excess parts can be made for any thermodynamic function obtained by differentiation of A with respect to either V or T. For example, the internal energy calculated from Eqns (2.3.7) and (2.3.17) is

$$U = U^{\text{id}} + U^{\text{ex}} \tag{2.3.19}$$

where $U^{\mathrm{id}} = \frac{3}{2}Nk_{\mathrm{B}}T$ and

$$U^{\mathrm{ex}} = \frac{1}{Z_N(V, T)} \int V_N(\mathbf{r}^N) \exp\left[-\beta V_N(\mathbf{r}^N)\right] \mathrm{d}\mathbf{r}^N$$
$$= \langle V_N \rangle \tag{2.3.20}$$

In the *isothermal-isobaric* (or *NPT*-) ensemble, pressure rather than volume is a fixed parameter. The thermodynamic potential is now the Gibbs free energy G, defined as

$$G = A + PV \tag{2.3.21}$$

Other state functions are obtained by differentiation of G with respect to the independent variables P and T. Thus

$$V = \left(\frac{\partial G}{\partial P}\right)_T \tag{2.3.22}$$

$$S = -\left(\frac{\partial G}{\partial T}\right)_P \tag{2.3.23}$$

The link with statistical mechanics is now made through the relation

$$G = -k_{\mathrm{B}}T \log \Delta_N(P, T) \tag{2.3.24}$$

where the isothermal-isobaric partition function $\Delta_N(P, T)$ is the Laplace transform with respect to volume of the canonical partition function $Q_N(V, T)$:

$$\Delta_N(P, T) = \iiint \exp\left[-\beta PV - \beta \mathscr{H}_N(\mathbf{r}^N, \mathbf{p}^N)\right] \mathrm{d}\mathbf{r}^N\, \mathrm{d}\mathbf{p}^N\, \mathrm{d}V$$
$$= \int_0^\infty \exp(-\beta PV) Q_N(V, T)\, \mathrm{d}V \tag{2.3.25}$$

Thermodynamic properties in the *NPT*-ensemble are therefore given by derivatives of $\Delta_N(P, T)$. For example, the average volume of a system in the ensemble is

$$\langle V \rangle = -k_{\mathrm{B}}T\left(\frac{\partial \log \Delta_N(P, T)}{\partial P}\right)_T \tag{2.3.26}$$

or, from the definitions (2.3.2), (2.3.13) and (2.3.25):

$$\langle V \rangle = \frac{\displaystyle\int_0^\infty V \exp(-\beta PV) Z_N(V, T)\, \mathrm{d}V}{\displaystyle\int_0^\infty \exp(-\beta PV) Z_N(V, T)\, \mathrm{d}V} \tag{2.3.27}$$

If the system is an ideal gas, $Z_N(V, T) = V^N$ and Eqn (2.3.27) reduces, in the limit of large N, to the expected result, i.e. $\langle V \rangle = N/\beta P$. Similarly, the enthalpy $H = U + PV$ is given by

$$
\begin{aligned}
H &= G - T\left(\frac{\partial G}{\partial T}\right)_P \\
&= -\left(\frac{\partial \log \Delta_N(P, T)}{\partial \beta}\right)_P \\
&= \frac{1}{\Delta_N(P, T)} \int_0^\infty \mathrm{d}V \exp(-\beta PV) \iint [\mathscr{H}_N(\mathbf{r}^N, \mathbf{p}^N) + PV] \\
&\quad \times \exp[-\beta \mathscr{H}_N(\mathbf{r}^N, \mathbf{p}^N)] \, \mathrm{d}\mathbf{r}^N \, \mathrm{d}\mathbf{p}^N
\end{aligned} \tag{2.3.28}
$$

and the specific heat at constant pressure by

$$
\begin{aligned}
C_P &= \left(\frac{\partial H}{\partial T}\right)_P \\
&= \frac{1}{k_B T^2} [\langle(\mathscr{H}_N + PV)^2\rangle - \langle(\mathscr{H}_N + PV)\rangle^2]
\end{aligned} \tag{2.3.29}
$$

The two specific heats C_P and C_V are related by

$$C_P / C_V = \chi_T / \chi_S \tag{2.3.30}$$

where

$$\chi_T = -\frac{1}{V}\left(\frac{\partial V}{\partial P}\right)_T \tag{2.3.31}$$

$$\chi_S = -\frac{1}{V}\left(\frac{\partial V}{\partial P}\right)_S \tag{2.3.32}$$

are, respectively, the isothermal and adiabatic compressibilities.

The *NPT*-ensemble is much less widely used than the canonical ensemble, but it finds application in the study of phase transitions and of the mixing properties of liquids.

2.4 CHEMICAL POTENTIAL AND THE GRAND CANONICAL ENSEMBLE

The discussion until now has been restricted to one-component systems containing a fixed number of particles. In the more general case, the system of interest contains several species of particles and the particle numbers can vary through interchange with the surroundings. A system of the latter type is called an *open* system, the thermodynamic description of which involves the chemical potentials μ_ν, where ν labels a species. The chemical potentials are the thermodynamic variables conjugate to the numbers of particles N_ν of each species and are expressible as derivatives of any of the functions U, A or G:

$$\mu_\nu = \left(\frac{\partial U}{\partial N_\nu}\right)_{V,S} = \left(\frac{\partial A}{\partial N_\nu}\right)_{V,T} = \left(\frac{\partial G}{\partial N_\nu}\right)_{P,T} \tag{2.4.1}$$

It follows from Eqns (2.3.15) and (2.4.1) that the chemical potentials in a mixture of ideal gases are given by

$$\begin{aligned} \beta\mu_\nu^{\text{id}} &= \left(\frac{\partial \beta A^{\text{id}}}{\partial N_\nu}\right)_{V,T} \\ &= \log \rho_\nu + 3 \log \Lambda_\nu \end{aligned} \tag{2.4.2}$$

where $\rho_\nu = N_\nu / V$ is the number density of species ν.

The thermodynamic potential appropriate to a system characterized by fixed values of V, T and the set $\{\mu_\nu\}$ is the *grand potential* Ω, defined as

$$\begin{aligned} \Omega &= A - \sum_\nu N_\nu \mu_\nu \\ &= G - PV - \sum_\nu N_\nu \mu_\nu \end{aligned} \tag{2.4.3}$$

The change in Ω associated with any infinitesimal change in thermodynamic state is

$$\mathrm{d}\Omega = -P\,\mathrm{d}V - S\,\mathrm{d}T - \sum_\nu N_\nu\,\mathrm{d}\mu_\nu \tag{2.4.4}$$

The thermodynamic properties P, S and N_ν are therefore given as derivatives of Ω by

$$P = -\left(\frac{\partial \Omega}{\partial V}\right)_{T,\{\mu_\nu\}} \tag{2.4.5}$$

$$S = -\left(\frac{\partial \Omega}{\partial T}\right)_{V,\{\mu_\nu\}} \tag{2.4.6}$$

$$N_\nu = -\left(\frac{\partial \Omega}{\partial \mu_\nu}\right)_{V,T} \tag{2.4.7}$$

From the definition of the chemical potentials, it follows that the total Gibbs free energy of the system is

$$G = \sum_{\nu} N_{\nu}\mu_{\nu} \tag{2.4.8}$$

The grand potential (2.4.3) can therefore be identified as

$$\Omega = -PV \tag{2.4.9}$$

For the sake of simplicity, we now restrict the discussion to systems containing only one species of particle. An open, one-component system is characterized by fixed values of V, T and chemical potential μ. The ensemble of systems having the same values of the three fixed parameters is called a *grand canonical* ensemble: the constancy of T and μ is ensured by supposing that the systems of the ensemble are allowed to come to equilibrium with a reservoir with which they can exchange both heat and matter. The phase space of the grand canonical ensemble is the union of the phase spaces corresponding to all values of the variable N, and the ensemble probability density gives the probability both that the system contains N particles and that the coordinates and momenta of the particles are, respectively, $\mathbf{r}^N$ and $\mathbf{p}^N$. At equilibrium, the probability density is

$$f_0(\mathbf{r}^N, \mathbf{p}^N; N) = \frac{1}{N!} h^{-3N} \frac{\exp(N\beta\mu) \exp[-\beta\mathcal{H}_N(\mathbf{r}^N, \mathbf{p}^N)]}{\Xi(\mu, V, T)} \tag{2.4.10}$$

where

$$\begin{aligned} \Xi &= \sum_{N=0}^{\infty} \frac{h^{-3N}}{N!} \exp(N\beta\mu) \iint \exp[-\beta\mathcal{H}_N(\mathbf{r}^N, \mathbf{p}^N)\,\mathrm{d}\mathbf{r}^N\,\mathrm{d}\mathbf{p}^N \\ &= \sum_{N=0}^{\infty} \exp(N\beta\mu) Q_N(V, T) \\ &= \sum_{N=0}^{\infty} \frac{z^N}{N!} Z_N(V, T) \end{aligned} \tag{2.4.11}$$

is the *grand partition function* and

$$z = \Lambda^{-3} \exp(\beta\mu) \tag{2.4.12}$$

is the *activity*. The link between statistical mechanics and thermodynamics is made through the relation

$$\Xi = \exp[-\beta\Omega(\mu, V, T)] = \exp(\beta PV) \tag{2.4.13}$$

The probability $P(N)$ that the system contains precisely N particles, irrespective of their coordinates and momenta, is obtained by integrating

$f_0(\mathbf{r}^N, \mathbf{p}^N; N)$ over $\mathbf{r}^N$ and $\mathbf{p}^N$:

$$\begin{aligned} P(N) &= \iint f_0(\mathbf{r}^N, \mathbf{p}^N; N)\, \mathrm{d}\mathbf{r}^N\, \mathrm{d}\mathbf{p}^N \\ &= \frac{1}{\Xi} \exp(N\beta\mu) Q_N(V, T) \\ &= \frac{1}{\Xi}\left(\frac{z^N}{N!}\right) Z_N(V, T) \end{aligned} \tag{2.4.14}$$

The average number of particles in the system is

$$\langle N \rangle = \sum_{N=0}^{\infty} NP(N) = \frac{1}{\Xi} \frac{\partial \Xi}{\partial \log z} = \frac{\partial \log \Xi}{\partial \log z} \tag{2.4.15}$$

a result that is equivalent to the thermodynamic relation (2.4.7). A measure of the fluctuation in particle number about its average value is provided by the mean-square deviation $(\langle N^2 \rangle - \langle N \rangle^2)$. An expression for the latter is obtained by differentiation of Eqn (2.4.15) with respect to $\beta\mu$:

$$\frac{\partial \langle N \rangle}{\partial \beta\mu} = \frac{\partial^2 \log \Xi}{\partial (\beta\mu)^2} = \langle N^2 \rangle - \langle N \rangle^2 \tag{2.4.16}$$

Thus

$$\frac{\langle N^2 \rangle - \langle N \rangle^2}{\langle N \rangle} = \frac{1}{\langle N \rangle} \frac{\partial \langle N \rangle}{\partial \beta\mu} \tag{2.4.17}$$

The right-hand side of Eqn (2.4.17) is clearly an intensive quantity, and the same must also be true of the left-hand side. Hence the relative mean-square deviation $(\langle N^2 \rangle - \langle N \rangle^2)^{1/2}/\langle N \rangle$ tends to zero as $\langle N \rangle \to \infty$. In the thermodynamic limit, i.e. $\langle N \rangle \to \infty$, $V \to \infty$ with $\rho = \langle N \rangle / V$ held constant, the total number of particles in the system can be identified with the grand canonical ensemble average. In the same limit, thermodynamic properties calculated in different ensembles become identical.

The intensive ratio (2.4.17) is related to the isothermal compressibility. We see from Eqns (2.4.4) and (2.4.9) that

$$\mathrm{d}\Omega = -P\,\mathrm{d}V - V\,\mathrm{d}P = -P\,\mathrm{d}V - S\,\mathrm{d}T - N\,\mathrm{d}\mu \tag{2.4.18}$$

For an infinitesimal, isothermal change, it follows that

$$V\,\mathrm{d}P = N\,\mathrm{d}\mu \tag{2.4.19}$$

If the change also takes place at constant volume, both $\mathrm{d}P$ and $\mathrm{d}\mu$ are proportional to dN:

$$dP = \left(\frac{\partial P}{\partial N}\right)_{V,T} dN \tag{2.4.20}$$

$$d\mu = \left(\frac{\partial \mu}{\partial N}\right)_{V,T} dN \tag{2.4.21}$$

It is then straightforward to show that

$$\begin{aligned} N\left(\frac{\partial \mu}{\partial N}\right)_{V,T} &= V\left(\frac{\partial P}{\partial N}\right)_{V,T} \\ &= \left(\frac{\partial P}{\partial \rho}\right)_{N,T} \\ &= \frac{1}{\rho \chi_T} \end{aligned} \tag{2.4.22}$$

In the thermodynamic limit, N may be replaced by $\langle N \rangle$, and substitution of (2.4.22) in Eqn (2.4.17) gives

$$\frac{\langle N^2 \rangle - \langle N \rangle^2}{\langle N \rangle} = \rho k_B T \chi_T \tag{2.4.23}$$

Since the numerator on the left-hand side of this equation can be rewritten as $\langle (N - \langle N \rangle)^2 \rangle$, it is evident that the compressibility is always non-negative.

In addition to its meaning as a fixed parameter of the grand canonical ensemble, the chemical potential may also be expressed as a canonical ensemble average (Widom, 1963). From Eqns (2.3.18) and (2.4.1) we find that

$$\begin{aligned} \mu^{ex} &= \left(\frac{\partial A^{ex}}{\partial N}\right)_{V,T} \\ &= A^{ex}(N+1, V, T) - A^{ex}(N, V, T) \\ &= k_B T \log\left(\frac{V Z_N(V, T)}{Z_{N+1}(V, T)}\right) \end{aligned} \tag{2.4.24}$$

or

$$\frac{V Z_N(V, T)}{Z_{N+1}(V, T)} = \exp(\beta \mu^{ex}) \tag{2.4.25}$$

The ratio $Z_{N+1}(V, T)/Z_N(V, T)$ is given by

$$\frac{Z_{N+1}(V, T)}{Z_N(V, T)} = \frac{\int \exp[-\beta V_{N+1}(\mathbf{r}^{N+1})] \, d\mathbf{r}^{N+1}}{\int \exp[-\beta V_N(\mathbf{r}^N)] \, d\mathbf{r}^N} \tag{2.4.26}$$

If the potential energy $V_{N+1}(\mathbf{r}^{N+1})$ is written as

$$V_{N+1}(\mathbf{r}^{N+1}) = V_N(\mathbf{r}^N) + \phi \tag{2.4.27}$$

where ϕ is the energy of interaction of the $(N+1)$th particle with all others, Eqn (2.4.26) becomes

$$\frac{Z_{N+1}(V, T)}{Z_N(V, T)} = \frac{\int \exp(-\beta\phi) \exp[-\beta V_N(\mathbf{r}^N)] \, d\mathbf{r}^{N+1}}{\int \exp[-\beta V_N(\mathbf{r}^N)] \, d\mathbf{r}^N} \tag{2.4.28}$$

In a translationally invariant system, the point $\mathbf{r}_{N+1}$ may be taken as the origin for the remaining N position vectors. Integration over $\mathbf{r}_{N+1}$ then yields a factor V and Eqn (2.4.28) reduces to

$$\frac{Z_{N+1}(V, T)}{Z_N(V, T)} = \frac{V \int \exp(-\beta\phi) \exp[-\beta V_N(\mathbf{r}^N)] \, d\mathbf{r}^N}{\int \exp[-\beta V_N(\mathbf{r}^N)] \, d\mathbf{r}^N}$$

$$= V\langle \exp(-\beta\phi) \rangle \tag{2.4.29}$$

where the angular brackets denote a canonical ensemble average for the system containing N particles. Substitution of (2.4.29) in Eqn (2.4.25) gives the required result:

$$\mu^{\text{ex}} = -k_B T \log \langle \exp(-\beta\phi) \rangle \tag{2.4.30}$$

The excess chemical potential is therefore proportional to the mean Boltzmann factor of a test particle introduced randomly into the system (Widom, 1963; Shing and Gubbins, 1982).

Note, finally, that the definitions (2.3.1) and (2.4.10) and the result in Eqn (2.4.14) together show that the canonical and grand canonical equilibrium probability densities are related by

$$f_0(\mathbf{r}^N, \mathbf{p}^N; N) = P(N) f_0^{(N)}(\mathbf{r}^N, \mathbf{p}^N) \tag{2.4.31}$$

A grand canonical ensemble average is therefore equal to a weighted sum of canonical ensemble averages, the weighting factor being the probability $P(N)$ that the system contains exactly N particles. Equation (2.4.15) represents a trivial application of this general result.

2.5 EQUILIBRIUM PARTICLE DENSITIES AND DISTRIBUTION FUNCTIONS

We saw in Section 2.3 that a factorization of the equilibrium phase-space probability density into kinetic and potential terms leads to a separation of thermodynamic properties into ideal and excess parts. A similar factorization can be made of the reduced phase-space distribution functions defined by Eqn (2.1.12). For a system at equilibrium, integration of a reduced distribution function over the remaining momenta yields an equilibrium *particle density* $\rho^{(n)}(\mathbf{r}^n)$; $\rho^{(n)}(\mathbf{r}^n)\,\mathrm{d}\mathbf{r}^n$ is $N!/(N-n)!$ times the probability of finding n particles of the system with coordinates in the element $\mathrm{d}\mathbf{r}^n$ of coordinate space, irrespective of the positions of the remaining particles and irrespective of all momenta. The particle densities and the closely related equilibrium *particle distribution functions* provide a complete but compact description of the structure of a fluid. In addition, knowledge of the low-order particle distribution functions is often sufficient to calculate the equation of state and other thermodynamic properties of the system.

In the canonical ensemble, the n-particle density is defined as

$$\rho_N^{(n)}(\mathbf{r}^n) = \frac{N!}{(N-n)!} \frac{\iint \exp[-\beta\mathcal{H}_N(\mathbf{r}^N, \mathbf{p}^N)]\,\mathrm{d}\mathbf{r}^{(N-n)}\,\mathrm{d}\mathbf{p}^N}{Q_N(V,T)}$$

$$= \frac{N!}{(N-n)!} \frac{\int \exp[-\beta V_N(\mathbf{r}^N)]\,\mathrm{d}\mathbf{r}^{(N-n)}}{Z_N(V,T)} \tag{2.5.1}$$

and is normalized such that

$$\int \rho_N^{(n)}(\mathbf{r}^n)\,\mathrm{d}\mathbf{r}^n = \frac{N!}{(N-n)!} \tag{2.5.2}$$

In particular:

$$\int \rho_N^{(1)}(\mathbf{r})\,\mathrm{d}\mathbf{r} = N \tag{2.5.3}$$

For a homogeneous system, it follows from (2.5.3) that

$$\rho_N^{(1)}(\mathbf{r}) = N/V = \rho \tag{2.5.4}$$

In the special case of an ideal gas, $V_N(\mathbf{r}^N) = 0$ and $Z_N(V,T) = V^N$. Thus

$$\rho_N^{(n)}(\mathbf{r}^n) = \rho^n \frac{N!}{N^n(N-n)!}$$

$$= \rho^n\left(1 + \mathcal{O}\left(\frac{n}{N}\right)\right) \tag{2.5.5}$$

For example, the pair density of an ideal gas is

$$\rho_N^{(2)}(\mathbf{r}_1, \mathbf{r}_2) = \rho^2\left(1 - \frac{1}{N}\right) \tag{2.5.6}$$

The appearance of the term of order N^{-1} in Eqn (2.5.6) reflects the fact that, in a system containing a fixed number of particles, the probability of finding a particle at $\mathbf{r}_2$, given that another particle is at $\mathbf{r}_1$, is proportional to $(N-1)/V$ and not to N/V.

The n-particle distribution function $g_N^{(n)}(\mathbf{r}^n)$ is defined in terms of the corresponding particle density by

$$g_N^{(n)}(\mathbf{r}^n) = \rho_N^{(n)}(\mathbf{r}_1, \ldots, \mathbf{r}_n) \Big/ \prod_{i=1}^{n} \rho_N^{(1)}(\mathbf{r}_i) \tag{2.5.7}$$

or, for a homogeneous system, by

$$\rho^n g_N^{(n)}(\mathbf{r}^n) = \rho_N^{(n)}(\mathbf{r}^n) \tag{2.5.8}$$

The particle distribution functions measure the extent to which the structure of the fluid deviates from complete randomness. If the system is isotropic as well as homogeneous, the pair distribution function $g_N^{(2)}(\mathbf{r}_1, \mathbf{r}_2)$ is a function only of the separation $r_{12} = |\mathbf{r}_2 - \mathbf{r}_1|$; it is then usually called the *radial distribution function* and written simply as $g(r)$. When the separation r_{12} is much larger than the range of the interparticle potential, $g_N^{(2)}(\mathbf{r}_1, \mathbf{r}_2)$ approaches its ideal-gas limit. From Eqn (2.5.6), this limit can be identified as

$$g_N^{(2)}(\mathbf{r}_1, \mathbf{r}_2) \underset{r_{12}\to\infty}{\sim} 1 - \frac{1}{N} \tag{2.5.9}$$

The radial distribution function plays a central role in the physics of monatomic fluids. There are two main reasons for this. First, the radial distribution function is directly measurable by radiation-scattering experiments. The results of such an experiment on liquid argon are pictured in Figure 2.1; $g(r)$ shows a pattern of peaks and troughs that is characteristic of all monatomic liquids. Secondly, provided that the atoms interact through pairwise-additive forces, thermodynamic properties of the fluid can be written as integrals over $g(r)$. Consider a system for which the total potential energy is given as a sum of pair terms:

$$V_N(\mathbf{r}^N) = \tfrac{1}{2} \sum_{i \neq j}^{N}\sum v(r_{ij}) \tag{2.5.10}$$

It follows from Eqn (2.3.20) that the excess internal energy is

$$U^{\text{ex}} = \frac{1}{Z_N(V, T)} \int \exp\left[-\beta V_N(\mathbf{r}^N)\right] \left(\tfrac{1}{2} \sum_{i \neq j}^{N}\sum v(r_{ij}) \right) \mathrm{d}\mathbf{r}^N \tag{2.5.11}$$

The double sum on i, j gives rise to $\frac{1}{2}N(N-1)$ terms, each of which yields the same result after integration. The integral may therefore be rewritten

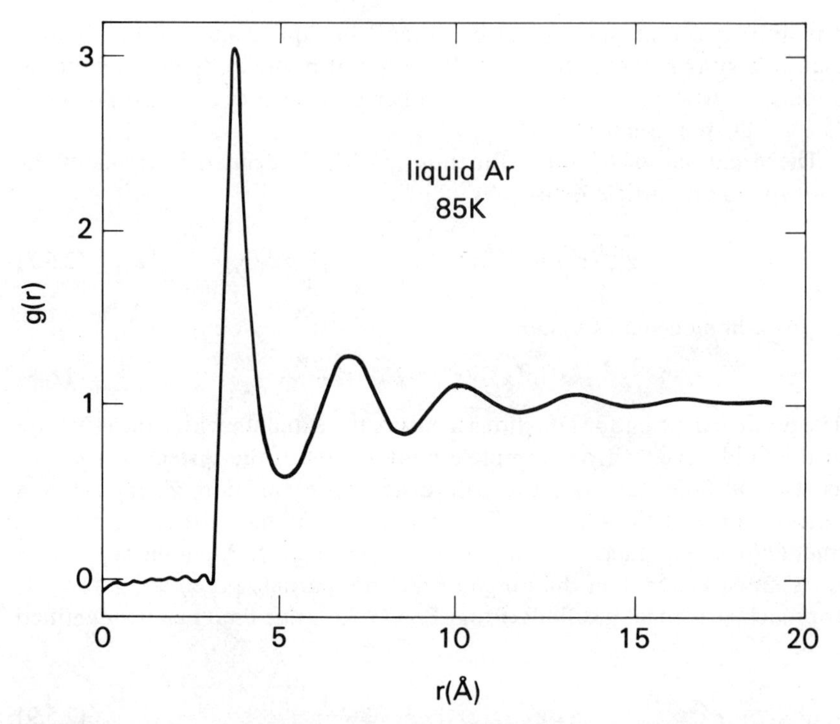

FIG. 2.1. Results of neutron-scattering experiments for the radial distribution function of liquid argon near the triple point. The ripples at small r are artefacts of the data analysis. After Yarnell *et al.* (1973).

as $\frac{1}{2}N(N-1)$ times the sum of any one term. Thus

$$\begin{aligned} U^{\text{ex}} &= \tfrac{1}{2}N(N-1)\iint v(r_{12})\left(\frac{1}{Z_N(V,T)}\int\cdots\int \exp\left[-\beta V_N(\mathbf{r}^N)\right] d\mathbf{r}_3\cdots d\mathbf{r}_N\right) d\mathbf{r}_1\, d\mathbf{r}_2 \\ &= \frac{N^2}{2V^2}\iint v(r_{12})g_N^{(2)}(\mathbf{r}_1,\mathbf{r}_2)\, d\mathbf{r}_1\, d\mathbf{r}_2 \\ &= \frac{N^2}{2V^2}\iint v(r_{12})g(r_{12})\, d\mathbf{r}_1\, d\mathbf{r}_{12} \\ &= \frac{N^2}{2V}\int v(r)g(r)\, d\mathbf{r} \end{aligned} \tag{2.5.12}$$

where we have assumed that the fluid is homogeneous and exploited the definitions (2.5.1) and (2.5.8). Thus, the excess internal energy per particle is given by

$$\frac{U^{\mathrm{ex}}}{N}=2\pi\rho\int_0^\infty v(r)g(r)r^2\,\mathrm{d}r \tag{2.5.13}$$

This result is called the *energy equation.* The same expression can also be obtained in a more intuitive way. The mean number of particles at a distance between r and $r+\mathrm{d}r$ from a reference particle is $n(r)\,\mathrm{d}r=4\pi r^2\rho g(r)\,\mathrm{d}r$; the total potential energy of interaction with the reference particle is $v(r)n(r)\,\mathrm{d}r$, and the excess internal energy per particle is obtained by integrating between $r=0$ and $r=\infty$ and dividing the result by two to avoid counting every interaction twice.

The equation of state (2.2.9) can also be re-expressed in terms of $g(r)$. By following arguments similar to those used to obtain Eqn (2.5.13), we find successively that

$$\begin{aligned}
\frac{\beta P}{\rho}&=1-\frac{\beta}{3NZ_N(V,T)}\int\exp\left[-\beta V_N(\mathbf{r}^N)\right]\sum_{i=1}^N\mathbf{r}_i\cdot\nabla_i\left(\sum_{j\neq i}^N v(r_{ij})\right)\mathrm{d}\mathbf{r}^N\\
&=1-\frac{\beta}{3N}\left(\tfrac{1}{2}\sum_{i\neq j}^N\sum\frac{1}{Z_N(V,T)}\int\exp\left[-\beta V_N(\mathbf{r}^N)\right]\mathbf{r}_{ij}\cdot\nabla_{ij}v(r_{ij})\right)\mathrm{d}\mathbf{r}^N\\
&=1-\tfrac{1}{6}\beta(N-1)\iint r_{12}v'(r_{12})\\
&\quad\times\left(\frac{1}{Z_N(V,T)}\int\cdots\int\exp\left[-\beta V_N(\mathbf{r}^N)\right]\mathrm{d}\mathbf{r}_3\ldots\mathrm{d}\mathbf{r}_N\right)\mathrm{d}\mathbf{r}_1\,\mathrm{d}\mathbf{r}_2\\
&=1-\tfrac{1}{6}\beta\rho\int rv'(r)g(r)\,\mathrm{d}\mathbf{r}\\
&=1-\tfrac{2}{3}\pi\beta\rho\int_0^\infty v'(r)g(r)r^3\,\mathrm{d}r
\end{aligned}\tag{2.5.14}$$

where $v'(r)\equiv\mathrm{d}v/\mathrm{d}r$. Equation (2.5.14) is called either the *pressure equation* or, in common with (2.2.9), the *virial equation.*

Equations (2.5.13) and (2.5.14) are simpler in structure than (2.3.20) and (2.2.9), since they involve only one-dimensional integrals. The simplification is only apparent, however, because the difficulty has shifted to that of determining the radial distribution function from the pair potential via Eqns (2.5.1) and (2.5.7). The problem is even more complicated if there are many-body forces acting between particles or if the pair potential is not spherically symmetric. The presence of three-body forces, for example, leads to the appearance of integrals over the triplet distribution function $g_N^{(3)}(\mathbf{r}_1,\mathbf{r}_2,\mathbf{r}_3)$. We shall not pursue this matter further, since it involves no new point of principle, but the generalization to systems of non-spherical particles will be examined in detail in Chapter 12.

Because the virial equation (2.5.14) involves the derivative of the pair potential, it cannot be used directly in the calculation of the equation of state of hard spheres or for other discontinuous potentials. The problem can be overcome by rewriting the equation in terms of the function $y(r)$ defined as

$$y(r) = \exp[\beta v(r)] g(r) \tag{2.5.15}$$

We shall prove in Chapter 5 that $y(r)$ is a continuous function of r even when both $v(r)$ and $g(r)$ have discontinuities; $y(r)$ has been called both the "background" correlation function and the "indirect" correlation function, but neither name has gained widespread acceptance. On introducing the definition (2.5.15) into Eqn (2.5.14), we find that

$$\begin{aligned} \frac{\beta P}{\rho} &= 1 - \tfrac{2}{3}\pi\beta\rho \int_0^\infty v'(r) y(r) \exp[-\beta v(r)] r^3 \, \mathrm{d}r \\ &= 1 + \tfrac{2}{3}\pi\rho \int_0^\infty r^3 y(r) \frac{\mathrm{d}}{\mathrm{d}r} \exp[-\beta v(r)] \, \mathrm{d}r \end{aligned} \tag{2.5.16}$$

If $v(r)$ is the hard-sphere potential, the Boltzmann factor $\exp[-\beta v(r)]$ is the Heaviside step-function, the derivative of which is the Dirac delta-function. Thus

$$\begin{aligned} \frac{\beta P}{\rho} &= 1 + \tfrac{2}{3}\pi\rho \int_0^\infty r^3 y(r) \delta(r - d) \, \mathrm{d}r \\ &= 1 + \tfrac{2}{3}\pi\rho \lim_{r \to d^+} r^3 y(r) \\ &= 1 + \tfrac{2}{3}\pi\rho d^3 g(d) \end{aligned} \tag{2.5.17}$$

where d is the hard-sphere diameter. The pressure of the hard-sphere fluid is therefore determined by the value of the radial distribution function at contact.

The fact that the pair distribution function $g_N^{(2)}(\mathbf{r}_1, \mathbf{r}_2)$ behaves asymptotically as $(1 - 1/N)$ rather than tending to unity is often irrelevant, since the term of order N^{-1} vanishes in the thermodynamic limit. On the other hand, if a term of order N^{-1} is integrated over the entire volume of the system, a finite result is obtained. The difficulties that this situation sometimes creates can be avoided by working in the grand canonical ensemble. As we shall see in Chapter 5, the grand canonical ensemble also provides a convenient framework for the derivation of density expansions of the equilibrium distribution functions.

In the grand canonical ensemble, the n-particle density is defined as the sum

$$\rho^{(n)}(\mathbf{r}^n) = \sum_{N \geq n}^{\infty} P(N) \rho_N^{(n)}(\mathbf{r}^n)$$

$$= \frac{1}{\Xi} \sum_{N \geq n}^{\infty} \frac{z^N}{(N-n)!} \int \exp\left[-\beta V_N(\mathbf{r}^N)\right] \mathrm{d}\mathbf{r}^{(N-n)} \quad (2.5.18)$$

where $P(N)$ is the probability defined by Eqn (2.4.14). If Eqn (2.5.18) is integrated over the coordinates $\mathbf{r}_1, \ldots, \mathbf{r}_n$, we find that $\rho^{(n)}(\mathbf{r}^n)$ is normalized such that

$$\int \rho^{(n)} \, \mathrm{d}\mathbf{r}^n = \left\langle \frac{N!}{(N-n)!} \right\rangle \quad (2.5.19)$$

In particular:

$$\int \rho^{(1)}(\mathbf{r}) \, \mathrm{d}\mathbf{r} = \langle N \rangle \quad (2.5.20)$$

$$\iint \rho^{(2)}(\mathbf{r}_1, \mathbf{r}_2) \, \mathrm{d}\mathbf{r}_1 \, \mathrm{d}\mathbf{r}_2 = \langle N^2 \rangle - \langle N \rangle \quad (2.5.21)$$

Equation (2.5.20) shows that for a homogeneous system the single-particle density is

$$\rho^{(1)}(\mathbf{r}) = \frac{\langle N \rangle}{V} = \rho \quad (2.5.22)$$

In the case of an ideal gas, the particle densities are

$$\rho^{(n)}(\mathbf{r}^n) = \frac{\sum_{N \geq n}^{\infty} \frac{z^N}{(N-n)!} V^{N-n}}{\sum_{N=0}^{\infty} \frac{z^N}{N!} V^N}$$

$$= z^n \quad (2.5.23)$$

But, for an ideal gas, it follows from Eqns (2.4.2) and (2.4.12) that

$$z = \rho \quad \text{(ideal gas)} \quad (2.5.24)$$

Thus

$$\rho^{(n)}(\mathbf{r}^n) = \rho^n \quad \text{(ideal gas)} \quad (2.5.25)$$

and the terms of order n/N that arise in the canonical ensemble no longer appear.

The relation between the grand canonical n-particle density and the corresponding distribution function is the same as in the canonical ensemble, i.e.

$$g^{(n)}(\mathbf{r}^n) = \rho^{(n)}(\mathbf{r}_1, \ldots, \mathbf{r}_n) \Big/ \prod_{i=1}^{n} \rho^{(1)}(\mathbf{r}_i) \tag{2.5.26}$$

but now $g^{(n)}(\mathbf{r}^n) \to 1$ for all n as the mutual separations of all pairs of particles becomes sufficiently large. In particular, the *pair correlation function*, defined as

$$h(\mathbf{r}_1, \mathbf{r}_2) = g^{(2)}(\mathbf{r}_1, \mathbf{r}_2) - 1 \tag{2.5.27}$$

vanishes in the limit $|\mathbf{r}_1 - \mathbf{r}_2| \to \infty$. If we insert the definition (2.5.18) in Eqn (2.5.26), and set $(N-n) = m$, we find that the n-particle distribution function of a homogeneous system can be expressed as

$$\Xi\left(\frac{\rho}{z}\right)^n g^{(n)}(\mathbf{r}^n) = \exp\left[-\beta V_n(\mathbf{r}^n)\right] + \sum_{m=1}^{\infty} \frac{z^m}{m!} \int \cdots \int \exp\left[-\beta V_{n+m}(\mathbf{r}^{n+m})\right] d\mathbf{r}_{n+1} \ldots d\mathbf{r}_{n+m} \tag{2.5.28}$$

In the low-density limit, $\rho \to 0$, $z \to 0$, $\rho/z \to 1$ and $\Xi \to 1$; the only term on the right-hand side of (2.5.28) that survives is the first one, which corresponds to the case of $n = N$ in Eqn (2.5.18). Thus

$$g^{(n)}(\mathbf{r}^n) \underset{\rho \to 0}{\sim} \exp\left[-\beta V_n(\mathbf{r}^n)\right] \tag{2.5.29}$$

The low-density limit of the pair distribution function is therefore equal to the Boltzmann factor of the pair potential:

$$g(r) \equiv g^{(2)}(\mathbf{r}) \underset{\rho \to 0}{\sim} \exp\left[-\beta v(r)\right] \tag{2.5.30}$$

Note that this result is consistent with our previous assumption that the function $y(r)$ defined by Eqn (2.5.15) is a continuous function of r for all r, irrespective of the form of the pair potential.

The energy and pressure equations (2.5.13) and (2.5.14) were derived in the canonical ensemble, but entirely equivalent results are obtained in the grand canonical ensemble. On the other hand, the *compressibility equation*, which expresses χ_T as an integral over $g(r)$, can be derived only in the grand canonical ensemble, since the compressibility is related to fluctuations in an open system through Eqn (2.4.23). The normalizations (2.5.20) and (2.5.21) show that

$$\iint \left[\rho^{(2)}(\mathbf{r}_1, \mathbf{r}_2) - \rho^{(1)}(\mathbf{r}_1)\rho^{(2)}(\mathbf{r}_2)\right] d\mathbf{r}_1 \, d\mathbf{r}_2 = \langle N^2 \rangle - \langle N \rangle - \langle N \rangle^2 \tag{2.5.31}$$

If the system is homogeneous, it follows immediately that

$$1+\rho\int[g^{(2)}(\mathbf{r})-1]\,d\mathbf{r}=\frac{\langle N^2\rangle-\langle N\rangle^2}{\langle N\rangle}$$
$$=\rho k_B T\chi_T \qquad (2.5.32)$$

Unlike the energy and pressure equations, this result is valid even when the interparticle forces are not pairwise additive.

By working in the canonical ensemble, it is possible to obtain useful expressions for the particle densities in terms of delta-functions of position. We note first that

$$\langle\delta(\mathbf{r}-\mathbf{r}_1)\rangle$$
$$=\frac{1}{Z_N(V,T)}\int\delta(\mathbf{r}-\mathbf{r}_1)\exp[-\beta V_N(\mathbf{r}^N)]\,d\mathbf{r}^N$$
$$=\frac{1}{Z_N(V,T)}\int\cdots\int\exp[-\beta V_N(\mathbf{r},\mathbf{r}_2,\cdots,\mathbf{r}_N)]\,d\mathbf{r}_2\cdots d\mathbf{r}_N \qquad (2.5.33)$$

The statistical average in (2.5.33) is a function of the coordinate **r**, but is independent of the particle label (here taken to be 1). A sum over all particle labels can therefore be written as N times the contribution from any one particle. It follows from the definition (2.5.1) that

$$\rho_N^{(1)}(\mathbf{r})=\left\langle\sum_{i=1}^{N}\delta(\mathbf{r}-\mathbf{r}_i)\right\rangle \qquad (2.5.34)$$

Similarly, the statistical average of a product of two delta-functions is

$$\langle\delta(\mathbf{r}-\mathbf{r}_1)\delta(\mathbf{r}'-\mathbf{r}_2)\rangle$$
$$=\frac{1}{Z_N(V,T)}\int\delta(\mathbf{r}-\mathbf{r}_1)\delta(\mathbf{r}'-\mathbf{r}_2)\exp[-\beta V_N(\mathbf{r}^N)]\,d\mathbf{r}^N$$
$$=\frac{1}{Z_N(V,T)}\int\cdots\int\exp[-\beta V_N(\mathbf{r},\mathbf{r}',\mathbf{r}_3,\cdots,\mathbf{r}_N)]\,d\mathbf{r}_3\cdots d\mathbf{r}_N \qquad (2.5.35)$$

and hence

$$\rho_N^{(2)}(\mathbf{r},\mathbf{r}')=\left\langle\sum_{i\neq j}^{N}\sum\delta(\mathbf{r}-\mathbf{r}_i)\delta(\mathbf{r}'-\mathbf{r}_j)\right\rangle \qquad (2.5.36)$$

Finally, for a system that is homogeneous, use of an elementary property of delta-functions shows that

$$\left\langle \frac{1}{N} \sum\sum_{i \neq j}^{N} \delta(\mathbf{r}+\mathbf{r}_j-\mathbf{r}_i) \right\rangle$$

$$= \left\langle \int \frac{1}{N} \sum\sum_{i \neq j}^{N} \delta(\mathbf{r}'+\mathbf{r}-\mathbf{r}_i)\delta(\mathbf{r}'-\mathbf{r}_j)\, d\mathbf{r}' \right\rangle$$

$$= \frac{1}{N} \int \rho_N^{(2)}(\mathbf{r}'+\mathbf{r}, \mathbf{r}')\, d\mathbf{r}'$$

$$= \rho g_N^{(2)}(\mathbf{r}) \tag{2.5.37}$$

Equations (2.5.34) and (2.5.37) (without the subscripts N) are also valid in the grand canonical ensemble.

2.6 THE YBG HIERARCHY AND THE BORN–GREEN EQUATION

It was shown in Section 2.1 that the non-equilibrium reduced probability densities $f^{(n)}(\mathbf{r}^n, \mathbf{p}^n; t)$ are coupled together by a set of $(N-1)$ equations called the BBGKY hierarchy. A similar hierarchy exists for the equilibrium particle densities; this is generally known as the Yvon–Born–Green or YBG hierarchy.

At equilibrium, the BBGKY hierarchy becomes

$$\sum_{i=1}^{n} \left[\frac{1}{m} \mathbf{p}_i \cdot \frac{\partial}{\partial \mathbf{r}_i} + \left(\mathbf{X}_i + \sum_{j=1}^{n} \mathbf{F}_{ij} \right) \cdot \frac{\partial}{\partial \mathbf{p}_i} \right] f_0^{(n)}(\mathbf{r}^n, \mathbf{p}^n)$$

$$= -\sum_{i=1}^{n} \iint \mathbf{F}_{i,n+1} \cdot \frac{\partial}{\partial \mathbf{p}_i} f_0^{(n+1)}(\mathbf{r}^{n+1}, \mathbf{p}^{n+1})\, d\mathbf{r}_{n+1}\, d\mathbf{p}_{n+1} \tag{2.6.1}$$

where the same restriction to pair forces that was invoked earlier is made and $\mathbf{F}_{ii} = 0$. The probability density $f_0^{(n)}(\mathbf{r}^n, \mathbf{p}^n)$ can be factorized as

$$f_0^{(n)}(\mathbf{r}^n, \mathbf{p}^n) = \mathcal{P}^{(n)}(\mathbf{p}^n)\rho_N^{(n)}(\mathbf{r}^n) \tag{2.6.2}$$

where $\rho_N^{(n)}(\mathbf{r}^n)$ is the equilibrium n-particle density and $\mathcal{P}^{(n)}(\mathbf{p}^n)$ is a product of n independent Maxwell–Boltzmann distributions, i.e.

$$\mathcal{P}^{(n)}(\mathbf{p}^n) = (2\pi m k_B T)^{-3n/2} \exp\left(-\beta \textstyle\sum_{i=1}^{n} |\mathbf{p}_i|^2/2m \right) \tag{2.6.3}$$

with

$$\frac{\partial}{\partial \mathbf{p}_i} \mathcal{P}^{(n)}(\mathbf{p}^n) = -\frac{\beta}{m} \mathbf{p}_i \mathcal{P}^{(n)}(\mathbf{p}^n) \tag{2.6.4}$$

and

$$\int \mathscr{P}^{(n+1)}(\mathbf{p}^{n+1})\,\mathrm{d}\mathbf{p}_{n+1} = \mathscr{P}^{(n)}(\mathbf{p}^n) \tag{2.6.5}$$

On substituting (2.6.2) to (2.6.5) in Eqn (2.6.1) and dividing through by $\mathscr{P}^{(n)}(\mathbf{p}^n)$, we find that

$$\sum_{i=1}^{n} \mathbf{p}_i \cdot \left[\nabla_i - \beta\left(\mathbf{X}_i + \sum_{j=1}^{n} \mathbf{F}_{ij}\right)\right]\rho_N^{(n)}(\mathbf{r}^n)$$
$$= \beta \sum_{i=1}^{n} \mathbf{p}_i \cdot \int \mathbf{F}_{i,n+1}\rho_N^{(n+1)}(\mathbf{r}^{n+1})\,\mathrm{d}\mathbf{r}_{n+1} \tag{2.6.6}$$

Because the particles are identical, the function $\rho_N^{(n)}(\mathbf{r}^n)$ must be symmetric in all $\mathbf{r}_i$ for $1 \leqslant i \leqslant n$. In addition, Eqn (2.6.6) must be independent of the choice of the n momenta $\mathbf{p}^n$. The equality (2.6.6) therefore holds term by term. By picking out the term for $i = 1$, we find that

$$k_\mathrm{B}T\nabla_1\rho_N^{(n)}(\mathbf{r}^n) = \left(\mathbf{X}_1 + \sum_{j=2}^{n} \mathbf{F}_{1j}\right)\rho_N^{(n)}(\mathbf{r}^n)$$
$$+ \int \mathbf{F}_{1,n+1}\rho_N^{(n+1)}(\mathbf{r}^{n+1})\,\mathrm{d}\mathbf{r}_{n+1} \tag{2.6.7}$$

Equation (2.6.7) is the YBG hierachy for a system in an external field. If there is no external field, $\mathbf{X}_1 = 0$, $\rho_N^{(n)}(\mathbf{r}^n) = \rho^n g_N^{(n)}(\mathbf{r}^n)$ and Eqn (2.6.7) becomes

$$-k_\mathrm{B}T\nabla_1 g_N^{(n)}(\mathbf{r}^n) = \sum_{j=2}^{n} \nabla_1 v(r_{1j}) g_N^{(n)}(\mathbf{r}^n)$$
$$+ \rho \int \nabla_1 v(r_{1,n+1}) g_N^{(n+1)}(\mathbf{r}^{n+1})\,\mathrm{d}\mathbf{r}_{n+1} \tag{2.6.8}$$

Equation (2.6.8) is an exact result, but in common with the equations of the BBGKY hierarchy it is useful only if a closure relation between $g_N^{(n)}$ and $g_N^{(n+1)}$ is available. The most important case is that when $n = 2$. This relates the pair and triplet distribution functions in the form

$$-k_\mathrm{B}T\nabla_1 g_N^{(2)}(\mathbf{r}_1, \mathbf{r}_2)$$
$$= \nabla_1 v(\mathbf{r}_1, \mathbf{r}_2) g_N^{(2)}(\mathbf{r}_1, \mathbf{r}_2) + \rho \int \nabla_1 v(\mathbf{r}_1, \mathbf{r}_3) g_N^{(3)}(\mathbf{r}_1, \mathbf{r}_2, \mathbf{r}_3)\,\mathrm{d}\mathbf{r}_3 \tag{2.6.9}$$

In an isotropic fluid

$$\int \nabla_1 v(\mathbf{r}_1, \mathbf{r}_3) g_N^{(2)}(\mathbf{r}_1, \mathbf{r}_3)\,\mathrm{d}\mathbf{r}_3 = 0 \tag{2.6.10}$$

Equation (2.6.9) may therefore be arranged to give

$$-k_B T \nabla_1 \log g_N^{(2)}(\mathbf{r}_1, \mathbf{r}_2)$$
$$= \nabla_1 v(\mathbf{r}_1, \mathbf{r}_2) + \rho \int \nabla_1 v(\mathbf{r}_1, \mathbf{r}_3) \left(\frac{g_N^{(3)}(\mathbf{r}_1, \mathbf{r}_2, \mathbf{r}_3)}{g_N^{(2)}(\mathbf{r}_1, \mathbf{r}_2)} - g_N^{(2)}(\mathbf{r}_1, \mathbf{r}_3) \right) d\mathbf{r}_3 \tag{2.6.11}$$

We now eliminate the triplet distribution function by making the Kirkwood *superposition approximation* (Kirkwood, 1935):

$$g_N^{(3)}(\mathbf{r}_1, \mathbf{r}_2, \mathbf{r}_3) \simeq g_N^{(2)}(\mathbf{r}_1, \mathbf{r}_2) g_N^{(2)}(\mathbf{r}_2, \mathbf{r}_3) g_N^{(2)}(\mathbf{r}_3, \mathbf{r}_1) \tag{2.6.12}$$

When this approximation is used in Eqn (2.6.11), the result is a non-linear, integrodifferential equation for $g(r)$ in terms of $v(r)$:

$$-k_B T \nabla_1 [\log g(\mathbf{r}_1, \mathbf{r}_2) + \beta v(\mathbf{r}_1, \mathbf{r}_2)]$$
$$= \rho \int \nabla_1 v(\mathbf{r}_1, \mathbf{r}_3) g(\mathbf{r}_1, \mathbf{r}_3) [g(\mathbf{r}_2, \mathbf{r}_3) - 1] \, d\mathbf{r}_3 \tag{2.6.13}$$

This is the Born–Green equation (Born and Green, 1949). Given a pair potential $v(r)$, Eqn (2.6.13) can be solved numerically to yield $g(r)$, from which, in turn, all thermodynamic properties of the system may be obtained via Eqns (2.5.13), (2.5.14) and (2.5.32). In practice, the Born–Green equation gives satisfactory results only at low densities (Levesque, 1966). The errors arise entirely from the use of the superposition approximation (Alder 1964a; Rahman, 1964b) and the theory would be much improved if a more accurate but computationally still tractable closure relation could be found (Haymet *et al.*, 1981). The Born–Green equation will be discussed again in Chapter 5, together with other approximate integral equations for $g(r)$.

2.7 FLUCTUATIONS

In earlier sections, we showed that certain thermodynamic properties are expressible in terms of fluctuations in the microscopic variables of a system; examples are the specific heats, Eqns (2.3.11) and (2.3.29), and the compressibility, Eqn (2.4.23). We now re-examine the question of fluctuations from a purely thermodynamic point of view.

Consider a subsystem of macroscopic dimensions that forms part of a much larger thermodynamic system. The subsystem is assumed to be in thermal, mechanical and chemical equilibrium with the remainder of the system which, being much larger, plays the role of a reservoir. The thermodynamic variables of the subsystem fluctuate around the average values characteristic of the total system, and their mean-square deviations can be derived systematically via the thermodynamic theory of fluctuations

(Landau and Lifshitz, 1980). If the total system is isolated from its surroundings, the probability of a fluctuation is

$$w \propto \exp(\Delta S_t / k_B) \tag{2.7.1}$$

where ΔS_t is the entropy change of the total system as the result of the fluctuation. Because S_t is a maximum at equilibrium, the deviations ΔS_t are quadratic functions (with negative coefficients) of the thermodynamic variables, higher-order terms in the expansion of S_t around its maximum value being negligible for large systems. Hence the probability w is a gaussian function of the independent thermodynamic variables. If T, P and μ denote the average temperature, pressure and chemical potential of the reservoir with which the subsystem is in equilibrium, it is easily shown (Landau and Lifshitz, 1980) that $\Delta S_t = R_{min}/T$, where R_{min} is the minimum work required for a reversible transformation corresponding to the fluctuation:

$$R_{min} = \Delta U - T\Delta S + P\Delta V - \mu\Delta N \tag{2.7.2}$$

In (2.7.2), ΔU, ΔS, ΔV and ΔN are the variations, respectively, in the energy, entropy, volume and number of particles of the subsystem. If ΔU is expanded to second order in ΔV, ΔN and ΔS, Eqn (2.7.1) reduces to

$$w \propto \exp[\beta(\Delta P\Delta V - \Delta T\Delta S - \Delta\mu\Delta N)] \tag{2.7.3}$$

The subsystem can be defined either by the fraction of the volume that it occupies in the total system, or by the number of particles it contains. In the second case, $\Delta N = 0$; among the four remaining variables (P, V, T and S), only two are independent. If T and V are chosen as independent thermodynamic variables, and ΔS and ΔP are expressed in terms of ΔT and ΔV, Eqn (2.7.3) becomes

$$w \propto \exp\left[-\frac{\beta C_V}{2T}(\Delta T)^2 + \tfrac{1}{2}\beta\left(\frac{\partial P}{\partial V}\right)_{N,T}(\Delta V)^2\right] \tag{2.7.4}$$

As the earlier discussion should have led us to expect, the probability w is now a gaussian function of the deviations ΔT and ΔV. Equation (2.7.4) shows that the system is stable against temperature and volume fluctuations if $C_V > 0$ and $(\partial P/\partial V)_{N,T} < 0$; these conditions for thermodynamic stability are consistent with the statistical-mechanical results (2.3.11) and (2.4.23). The mean-square fluctuations calculated from Eqn (2.7.4) are

$$\langle \Delta T\, \Delta V \rangle = 0 \tag{2.7.5}$$

$$\langle (\Delta T)^2 \rangle = \frac{k_B T^2}{C_V} \tag{2.7.6}$$

$$\begin{aligned}\langle (\Delta V)^2 \rangle &= -k_B T\left(\frac{\partial V}{\partial P}\right)_{N,T} \\ &= V k_B T \chi_T \end{aligned} \tag{2.7.7}$$

Equation (2.7.5) describes the fact that temperature and volume fluctuations are independent; Eqns (2.7.6) and (2.7.7) relate the mean-square fluctuations in the temperature and volume of the subsystem to the specific heat at constant volume and to the isothermal compressibility. Alternatively, choice of P and S as independent variables transforms Eqn (2.7.4) into

$$w \propto \exp\left[\tfrac{1}{2}\beta\left(\frac{\partial V}{\partial P}\right)_{N,S}(\Delta P)^2 - \frac{1}{2C_P k_B}(\Delta S)^2\right] \tag{2.7.8}$$

and the averages calculated from (2.7.8) are

$$\langle \Delta P\, \Delta S\rangle = 0 \tag{2.7.9}$$

$$\langle (\Delta S)^2\rangle = k_B C_P \tag{2.7.10}$$

$$\langle (\Delta P)^2\rangle = -k_B T\left(\frac{\partial P}{\partial V}\right)_{N,S}$$

$$= \frac{k_B T}{V\chi_S} \tag{2.7.11}$$

Equation (2.7.9) shows that pressure and entropy fluctuations are independent; Eqns (2.7.10) and (2.7.11) relate the pressure and entropy fluctuations to the specific heat at constant pressure and the adiabatic compressibility. Finally, if the subsystem is defined as occupying a fixed fraction of the total volume, the mean-square fluctuation in the number of particles in the subsystem can be calculated, with the help of (2.4.22), to be

$$\langle (\Delta N)^2\rangle = k_B T\left(\frac{\partial N}{\partial \mu}\right)_{V,T}$$

$$= \rho N k_B T \chi_T \tag{2.7.12}$$

Equation (2.7.12) is identical to the statistical mechanical relation (2.4.23), while comparison of (2.7.12) with (2.7.7) shows that volume fluctuations at constant N are equivalent to number fluctuations at constant V.

When the subsystems are of macroscopic size, fluctuations in the thermodynamic variables of neighbouring subsystems will, in general, be uncorrelated. Strong correlations can be expected, however, when the dimensions of the subsystems are comparable with the range of the interparticle forces. In particular, number fluctuations in two infinitesimal volume elements will be highly correlated for small values of the distance between the elements. Such correlations can be expressed in terms of the equilibrium distribution functions introduced in Section 2.5.

CHAPTER 3

Computer "Experiments" on Liquids

In the past two decades, computer "experiments" have come to play a major role in liquid-state physics. Their importance, from the theoretician's point of view, rests on the fact that they provide essentially exact, quasi-experimental data on well-defined models. As there is no uncertainty about the form of the interaction potential, theoretical results can be tested unambiguously in a manner that is generally impossible with data obtained in experiments on real liquids. It is also possible to obtain information on quantities of theoretical importance that are not readily measurable in the laboratory. In this chapter, we give a brief account of how a computer simulation can be carried out, and discuss some of the limitations as well as the advantages of the method.

3.1 METHODS OF COMPUTER SIMULATION

The behaviour of liquids, solids and dense gases can be simulated at the molecular level in one of two ways, by "molecular dynamics" or by a Monte Carlo method. In a conventional molecular dynamics calculation, a system of N particles (atoms, molecules or ions) is placed within a cell of fixed volume, frequently cubic in shape. A set of velocities is also assigned, usually drawn from a Maxwell–Boltzmann distribution appropriate to the temperature of interest and selected in such a way as to make the net linear momentum equal to zero. The subsequent trajectories of the particles are then calculated by integration of the classical equations of motion. The particles are assumed to interact through some prescribed force law and the bulk of the computational labour is concerned with the calculation at every step of the forces acting on each particle. The coordinates and momenta of the particles are usually stored on magnetic tape for later analysis, and thermodynamic and other properties are obtained as time averages over the dynamical history of the system, as outlined in Section 2.2.

Apart from the choice of initial conditions, a molecular dynamics simulation is, in principle, entirely deterministic in nature. By contrast, as the name suggests, a probabilistic element is an essential part of any Monte Carlo computation. In a classical Monte Carlo calculation, a system of N particles interacting through some known potential is again assigned a set of arbitrarily chosen initial coordinates; a sequence of configurations of the particles is then generated by successive random displacements. Not all configurations that occur are accepted, the decision whether to *accept* or *reject* a particular configuration being made in such a way as to ensure that, asymptotically, configuration space is sampled according to the equilibrium probability density appropriate to a chosen ensemble. The ensemble average of a function of the particle coordinates, such as the potential energy, is then obtained as an unweighted average over the resulting set of configurations. The particle momenta do not enter the calculation, there is no time scale involved, and the order in which the configurations occur has no special significance.

In most applications of the Monte Carlo method that have been reported thus far, the goal has been to determine canonical ensemble averages. Results have also been obtained, however, in both the isothermal–isobaric and grand canonical ensembles. In the method of molecular dynamics, apart from certain specialized calculations, the system is free of any external field and the total energy is independent of time. The dynamical states that the method generates therefore represent a sample from a microcanonical ensemble. The solution of the equations of motion is, of course, a particularly efficient way of sampling phase space along a surface of constant total energy. Molecular dynamics has the great advantage that it allows the study of time-dependent phenomena. If static properties alone are required, the Monte Carlo method is often more useful, primarily because temperature and volume (or temperature and pressure in *NPT*-ensemble calculations) are more convenient fixed parameters than volume and total energy.

There are a number of difficulties that are common to both methods. The most obvious of these is the fact that, on a macroscopic scale, the size of sample that can be studied economically is extremely small. Typically, N is of order 10^3 or less. In order to minimize surface effects and thereby to simulate more closely the behaviour of an infinite system, it is customary to use a periodic boundary condition. The way in which the periodic boundary condition works is illustrated for a two-dimensional system in Figure 3.1. The particles of interest lie in the central cell, and this basic unit is surrounded on all sides by periodically repeated images of itself; each image cell contains N particles in the same relative positions as in the central cell. When a particle enters or leaves through one face of a cell, the move is balanced by an image of that particle leaving or entering through

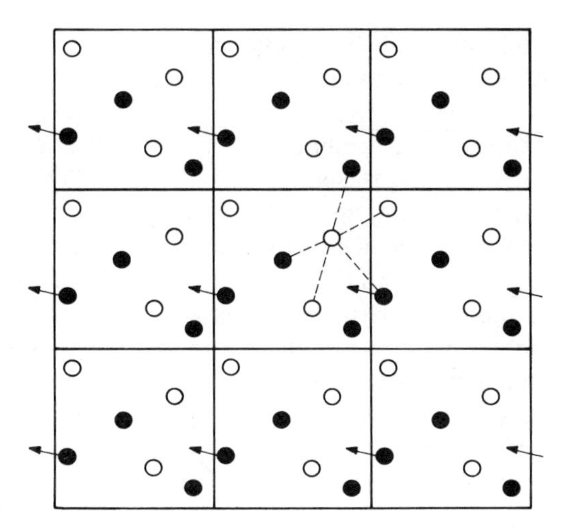

FIG. 3.1. Periodic boundary conditions used in computer "experiments".

the opposite face. It is often advantageous to choose N and the shape of the cells in such a way that the periodic boundary condition generates a perfect lattice appropriate to the physical system under study when the particles in the central cell are arranged in a suitably ordered manner. Argon, for example, crystallizes in a face-centred cubic structure and in that case it is natural to use a cubic cell and take $N = 4n^3$, where n is an integer. This fact accounts for the widespread use of samples containing $N = 32, 108, 256, 500, 864, \ldots$ particles. The alkali metals, on the other hand, crystallize in body-centred cubic form, and it is then more appropriate to choose $N = 2n^3$. For simplicity, we shall always assume the cells to be cubic, although even for fluids there are advantages to the use of cells of other shapes (Adams, 1979).

A key question is whether the properties of a small, periodic sample are truly representative of the macroscopic system that the model is designed to simulate. There is no easy answer to this: it depends very much on the form of the potential and on the properties under investigation. The only safe course in cases of doubt is to make calculations for a range of values of N and look for signs of a systematic trend in the results. Broadly speaking, it appears that bulk properties are only weakly dependent on sample size for $N \gtrsim 100$, and that the errors that do exist, relative to the $N \to \infty$ limit, are no larger than the inevitable statistical uncertainties. Nonetheless, the restriction on sample size does have other, important disadvantages. In particular, it is impossible to study spatial fluctuations having a wavelength larger than L, the length of the cell. For a sample of 864 atoms simulating argon near its triple point, L is approximately 30 Å, corresponding to a minimum accessible wavenumber of approximately 0.2 $Å^{-1}$. Unfortunately, some of the most interesting physics of the liquid state lies in the collective excitations that appear first at about this value of the wavenumber and become more pronounced as the wavelength increases. The suppression of long-wavelength density fluctuations means, in particular, that the study of critical phenomena is largely outside the scope of computer "experiments". The use of a periodic boundary condition also has an effect on time correlations. In a molecular dynamics simulation, a local disturbance can move through the periodic system and reappear at the same place, albeit in attenuated form, after a recurrence time τ_{rec} of order L/c, where c is a speed of propagation that may be roughly equated to the speed of sound. For the example of 864 argon atoms considered earlier, this estimate of the recurrence time gives $\tau_{rec} \simeq 3 \times 10^{-12}$ s. The effects of periodicity will manifest themselves in spurious contributions to time correlations calculated over intervals longer than this.

Another difficulty that is particularly acute for small samples is the so-called quasi-ergodic problem (Wood and Parker, 1957). In the context of computer "experiments", the term refers to the possibility of the system becoming trapped in a small region of phase space. Near the melting temperature, for example, an initial lattice-type arrrangement of particles may persist for a very long time unless the density is some 5% below the freezing density of the model. Whatever the starting conditions, time must be allowed for the system to equilibrate before the "production" stage of the calculation is begun. Throughout the calculation it is important to monitor carefully the bulk properties of the system in order to detect any tendency towards a long-time drift.

The only input information in a computer simulation, apart from the fixed parameters and the chosen initial conditions, are the details of the particle interactions. There is no restriction on the form of the interparticle potential, but in practice it is nearly always assumed to be pairwise additive. For economy in computing time, it is customary to truncate the interaction at a separation $r = r_c \leqslant \frac{1}{2}L$, where the cut-off radius r_c is typically some three molecular diameters. The effect of the neglected interactions on bulk properties of the system can be allowed for by making an appropriate "tail correction". In the case of monatomic fluids, for example, the tail corrections to energy and pressure are usually obtained by evaluating the integrals in Eqns (2.5.13) and (2.5.14) between the limits $r = r_c$ and $r = \infty$ under the assumption that $g(r) = 1$ for $r > r_c$. The tail corrections can be large, particularly that to the pressure. As an example, consider the case of a Lennard-Jones potential truncated at $r_c = 2.5\sigma$. At a reduced density $\rho\sigma^3 = 0.835$ and reduced temperature $k_B T/\varepsilon = 0.72$, i.e. under conditions close to the triple point, it turns out that $U^{ex}/N\varepsilon = -6.12$, to which the tail correction contributes -0.48, and $\beta P/\rho = 0.22$, of which -1.24 comes from the tail. Without the correction, the pressure would be strongly positive.

When a cut-off sphere is used, the interaction of a particle with its neighbours is calculated with a "nearest-neighbour" convention. The principle of the convention is illustrated in Figure 3.1: a particle i lying within the central cell is assumed to interact only with the nearest image of any other particle j (including j itself), the interaction being set equal to zero if the distance from the nearest image is greater than r_c. The significance of the restriction $r_c \leqslant \frac{1}{2}L$ is that there is at most one image of j (including j itself) lying in any sphere of radius $r < \frac{1}{2}L$ centred on i.

Use of a cut-off sphere is inappropriate when the interparticle forces are very long ranged. The problem is most severe for ionic systems, since there is not even a guarantee that the cut-off sphere will be electrically neutral. One way to overcome the difficulty is to calculate the coulombic interaction of a given particle, i say, not only with every other particle in the central cell but also with all images in all other cells. An infinite lattice sum of this type can be evaluated by the method of Ewald (Hansen, 1986). The essence of the Ewald method is to convert the slowly convergent sum in r^{-1} into two series that separately are rapidly convergent. One series is a sum in real space of a short-range potential, which may safely be truncated, while the other is a sum over reciprocal lattice vectors of the periodic array of cells. The simpler alternative of summing over all particles lying in a cube of length L centred on i has been shown to lead to serious errors in both static (Brush *et al.*, 1966; Woodcock and Singer, 1971) and dynamic (Hansen *et al.*, 1979) properties. Related problems arise in the simulation of strongly polar systems, as we shall see in Chapter 12.

Certain properties of a model fluid can be calculated in a simulation much more accurately than others. The mean potential energy, for example, is easily obtained with a statistical uncertainty of less than 1%. Much larger errors are associated with the calculation of thermodynamic quantities, such as the specific heat (Eqn (2.3.11)), that are linked to fluctuations in a microscopic variable. The uncertainties here are typically 10% or more. The pair distribution function can be calculated up to $r = \frac{1}{2}L$, again to an accuracy of about 1%; beyond $r = \frac{1}{2}L$, the distribution is distorted by the periodic boundary condition. The dynamical behaviour of fluids is generally studied by the calculation of the time-correlation functions of the type to be discussed in Chapter 7. This introduces further problems of statistical error: Zwanzig and Ailawadi (1969) have shown that the uncertainty in time correlations between events separated by an interval τ increases as $\tau^{1/2}$. In addition, there are many important dynamical properties that are determined by the correlated motion of large numbers of particles. These collective quantities are subject to much larger errors than is the case for single particle properties, and their calculation inevitably involves the heavy use of a large computer. Finally, standard methods of computer simulation are inappropriate for the study of very slow relaxation processes, since the length of the runs is necessarily limited by economic considerations.

3.2 MOLECULAR DYNAMICS "EXPERIMENTS" ON HARD SPHERES

The molecular dynamics method for the study of condensed phases was devised by Alder and Wainwright (1957, 1959) and applied by them to systems of particles with hard cores: hard discs, hard spheres and square-well molecules. For models such as these, characterized by potentials that have discontinuities and by forces that are purely impulsive, the technical details of the simulations differ in important ways from those appropriate to continuous potentials. The account we give in this section is specific to a one-component system of hard spheres, but it is easily generalized to mixtures and to other discontinuous potentials.

The motion of hard spheres is unusual insofar as the velocities change only at collisions; between collisions, the particles move in straight lines at constant speeds. The time evolution of a many-particle hard-sphere system can therefore be treated as a sequence of strictly binary, elastic collisions. Let $\mathbf{u}_1, \mathbf{u}_2$ be the velocities of two hard spheres before a collision and let $\mathbf{u}_1', \mathbf{u}_2'$ be their velocities afterwards. As the particles are of equal mass, conservation of energy and momentum require that

$$|\mathbf{u}_1'|^2 + |\mathbf{u}_2'|^2 = |\mathbf{u}_1|^2 + |\mathbf{u}_2|^2 \tag{3.2.1}$$

$$\mathbf{u}_1' + \mathbf{u}_2' = \mathbf{u}_1 + \mathbf{u}_2 \tag{3.2.2}$$

Thus

$$\begin{aligned} \Delta\mathbf{u}_1 &\equiv \mathbf{u}_1' - \mathbf{u}_1 \\ &= -(\mathbf{u}_2' - \mathbf{u}_2) \\ &= -\Delta\mathbf{u}_2 \end{aligned} \tag{3.2.3}$$

The projection of the relative velocity onto the vector joining the centres is reversed in an elastic collision, i.e.

$$\begin{aligned} \mathbf{u}_{12}' \cdot \mathbf{r}_{12} &= (\mathbf{u}_2' - \mathbf{u}_1') \cdot (\mathbf{r}_2 - \mathbf{r}_1) \\ &= -(\mathbf{u}_2 - \mathbf{u}_1) \cdot (\mathbf{r}_2 - \mathbf{r}_1) \\ &= -\mathbf{u}_{12} \cdot \mathbf{r}_{12} \end{aligned} \tag{3.2.4}$$

but the orthogonal components are unaltered. Thus the change in velocity at a collision is

$$\begin{aligned}\Delta \mathbf{u}_1 &= -\Delta \mathbf{u}_2 \\ &= \left(\frac{\mathbf{r}_{12} b_{12}}{d^2}\right)_{\text{contact}}\end{aligned} \tag{3.2.5}$$

where $b_{12} = \mathbf{u}_{12} \cdot \mathbf{r}_{12}$, d is the hard-sphere diameter, and the term on the right-hand side must be evaluated at contact. The algorithm for the calculation of the trajectories therefore consists of first advancing the coordinates of all particles until a collision occurs somewhere in the system, and then of calculating the change in velocities of the colliding particles according to Eqn (3.2.5); the procedure is repeated for as many collisions as are necessary to yield results of sufficient statistical reliability for the problem in hand. Since Eqn (3.2.5) is exact, it follows that the trajectories of the particles can be computed with a precision limited only by round-off errors. As the kinetic energy is a conserved quantity, the temperature of the system is rigorously constant.

The natural unit of time for hard spheres is the mean collision time τ_{coll}, i.e. the mean time between collisions suffered by a given particle. The non-ideal contribution to the pressure is proportional to the rate of collision between particles. If, therefore, we take the ratio of $(\beta P/\rho - 1)$ to its low-density limit, found by setting $g(d) = 1$ in Eqn (2.5.17), we obtain a relation between the collision rate $\Gamma = 1/\tau_{\text{coll}}$ and the collision rate Γ_0 in the dilute gas:

$$\begin{aligned}\frac{\Gamma}{\Gamma_0} &= \frac{\beta P/\rho - 1}{\frac{2}{3}\pi \rho d^3} \\ &= g(d)\end{aligned} \tag{3.2.6}$$

The collision rate Γ is the quantity that enters the Enskog kinetic equation (see Section 2.1), and the low-density or Boltzmann collision rate Γ_0 is given (Moore and Pearson, 1981) by

$$\Gamma_0 = 4\rho d^2 \left(\frac{\pi k_B T}{m}\right)^{1/2} \tag{3.2.7}$$

It is instructive to evaluate these expressions for conditions appropriate to real liquids. Let us suppose that an argon atom can be represented as a hard sphere of diameter 3.4 Å. At the triple-point volume ($V = 28.4\ \text{cm}^3\ \text{mol}^{-1}$) and temperature ($T = 84$ K), we find $\Gamma_0 = 2.3 \times 10^{12}\ \text{s}^{-1}$, while computer simulation results for hard spheres show that $g(d) = 4.4$ at the corresponding value of the reduced density ρd^3. We therefore estimate the mean time between "collisions" in liquid argon to be $\tau_{\text{coll}} = 1/g(d)\Gamma_0 \simeq 10^{-13}$ s.

3.3 MOLECULAR DYNAMICS FOR CONTINUOUS POTENTIALS

The extension of the molecular dynamics method to the case where the particles of the fluid interact through continuous potentials was first described by Rahman (1964a, 1966). Rahman's earliest calculations were made for the Lennard-Jones potential and the so-called exp-6 potential (an exponential repulsion and an attraction varying as r^{-6}); in each case, the parameters in the potential were chosen to simulate liquid argon. A more systematic study of the properties of the Lennard-Jones fluid was later reported in a series of papers by Verlet and coworkers (Verlet, 1967, 1968; Levesque and Verlet, 1970; Levesque *et al.*, 1973). Since those pioneering efforts, the method has been applied to a wide range of increasingly complex systems, including liquid metals, molten salts, and molecular liquids of many types.

When the potentials are continuous, the trajectories of the particles, unlike those of hard spheres, can no longer be calculated exactly. We consider first the case of spherically symmetric potentials. The equations of motion are now the $3N$ coupled, second-order, differential equations of Eqn (2.2.1). These equations must be solved numerically by finite difference methods, which leads unavoidably to errors in the particle trajectories. It turns out, however, that use of an elaborate algorithm is rarely necessary. Let the coordinates of particle i at time t be $\mathbf{r}_i(t)$. The coordinates at times $t \pm h$ are given by a Taylor expansion about $\mathbf{r}_i(t)$:

$$\mathbf{r}_i(t \pm h) = \mathbf{r}_i(t) \pm h\dot{\mathbf{r}}_i(t) + \frac{h^2}{2!}\ddot{\mathbf{r}}_i(t) \pm \frac{h^3}{3!}\dddot{\mathbf{r}}_i(t) + \mathcal{O}(h^4) \tag{3.3.1}$$

The *central-difference* prediction for $\mathbf{r}_i(t+h)$ is obtained by adding the two expressions in Eqn (3.3.1) to give

$$\mathbf{r}_i(t+h) \simeq -\mathbf{r}_i(t-h) + 2\mathbf{r}_i(t) + \frac{h^2}{m}\mathbf{F}_i(t) \tag{3.3.2}$$

where $\mathbf{F}_i(t)$ is the total force acting on particle i at time t; in general, $\mathbf{F}_i(t)$ is calculated as a sum of pair forces $\mathbf{F}_{ij}(t)$, with $\mathbf{F}_{ij}(t) = -\mathbf{F}_{ji}(t)$. The error in the predicted coordinates is of order h^4. If we subtract the two expressions in (3.3.1), we obtain an estimate for the velocity of particle i at time t:

$$\dot{\mathbf{r}}_i(t) \simeq \frac{1}{2h}[\mathbf{r}_i(t+h) - \mathbf{r}_i(t-h)] \tag{3.3.3}$$

The error here is of order h^2.

The central-difference algorithm (3.3.2) was first used in molecular dynamics calculations by Verlet (1967) and has been the commonest choice of most later workers. The time step h is taken as constant; this contrasts with the case of hard spheres, for which the analogue of h, namely the interval between successive collisions, is a quantity that varies throughout the calculation. Some care is needed in the choice of h if the errors inherent in the algorithm are to be kept at an acceptable level. As an example, consider a system of Lennard-Jones particles. If we introduce the Lennard-Jones parameter σ as the unit of length, we may rewrite Eqn (3.3.2) in reduced form as

$$\begin{aligned} \mathbf{r}_i^*(t+h) \simeq & -\mathbf{r}_i^*(t-h)+2\mathbf{r}_i^*(t) \\ & - h^2 \frac{48\varepsilon}{m\sigma^2}\left(\frac{\mathbf{r}_{ij}^*}{r_{ij}^*}\right) \sum_{j=1}^{N}{}' (r_{ij}^{*-13} - \tfrac{1}{2} r_{ij}^{*-7}) \end{aligned} \tag{3.3.4}$$

where the prime denotes that the term with $i=j$ is omitted from the sum. The quantity τ_0 defined by

$$\tau_0 = \left(\frac{m\sigma^2}{48\varepsilon}\right)^{1/2} \tag{3.3.5}$$

appears in (3.3.4) as a natural unit of time. On inserting parameters appropriate to argon ($\sigma = 3.405$ Å, $\varepsilon/k_B = 119.8$ K, $m = 6.63 \times 10^{-23}$ g), we find that $\tau_0 = 3.112 \times 10^{-13}$ s. The choice of h is usually made on the basis of conservation of total energy. The energy should be strictly constant, but in practice errors in the integration scheme cause it to fluctuate about a mean value. Fluctuations of order one part in 10^4 are usually considered acceptable. Under liquid-state conditions, this level of energy conservation can be achieved with a reduced time step $h^* \simeq 0.03$, corresponding in the case of argon to $h \simeq 10^{-14}$ s. If we recall the example of the previous section, we see that the criterion based on energy conservation leads to the physically reasonable result that the time step should be roughly an order of magnitude smaller than the mean "collision" time. As the time step is increased, the fluctuations in total energy become larger; eventually, an overall upward drift in energy may develop, though there will still be fluctuations. In most other circumstances, a steady drift in the energy is an almost certain sign of a programming error.

Despite its simplicity, the central-difference algorithm is apparently at least as stable as a variety of higher-order schemes that have been proposed, including a number of so-called predictor-corrector methods (Berendsen, 1986). On the other hand, higher-order algorithms are sometimes useful in obtaining estimates for the particle velocities that improve on the approximation (3.3.3).

A molecular dynamics simulation is initiated by placing the particles at arbitrarily chosen sites $\mathbf{r}_i(0)$. A corresponding set of coordinates at an earlier time $\mathbf{r}_i(-h)$ is then typically obtained by allocating to the particles random velocities drawn from a Maxwell-Boltzmann distribution appropriate to the temperature of interest and chosen such that the net linear momentum is zero. The subsequent calculation is organized as a loop over time. At each step, the time is increased by h, the total force acting on each particle is computed, and the particles are advanced to their new positions. In the early stages of the calculation, it is normal for the temperature to drift away from the value at which it was originally set, and an occasional rescaling of the particle velocities is therefore necessary. Once equilibrium is reached, the system is allowed to evolve undisturbed, with both potential and kinetic energies fluctuating around steady mean values. A run of order 10^3 time steps (excluding the time allowed for equilibration) is generally sufficient to yield the mean potential energy with an uncertainty of about 1% and $\beta P/\rho$ to within ± 0.1. The temperature of the system is calculated from the mean kinetic energy, while the fluctuation in temperature leads to a value for the specific heat via the formula (Lebowitz *et al.*, 1967)

$$\frac{\langle T^2\rangle - \langle T\rangle^2}{\langle T\rangle^2} = \frac{2}{3N}\left(1 - \frac{3Nk_B}{2C_V}\right) \tag{3.3.6}$$

This result is the microcanonical analogue of Eqn (2.3.11) and is subject to similar statistical uncertainties.

The overwhelming bulk of the computing time is spent in the calculation of the pair forces, and it is this that motivates the use of a cut-off sphere with $r_c < \frac{1}{2}L$. However, there is little to be gained from truncating the potential at a separation less than $\frac{1}{2}L$ without at the same time making use of "neighbour tables". Several different schemes have been described (see, for example, Wood, 1968; Hockney and Eastwood, 1981), but the basic idea is very simple: in computing the total force acting on particle i, the loop on j is restricted to those particles that, from calculations at earlier time steps, are known to be relatively close neighbours of i, with due allowance being made for the nearest-neighbour convention. The necessary information is contained in tables that, for each i, list the neighbours j such that $r_{ij} < r_c + \Delta$. The quantity Δ corresponds to a thin shell that covers the cut-off sphere; the tables must be updated at intervals determined by the choice of Δ. If efficient use is made of neighbour tables, the number of pair separations that must be calculated at each time step is reduced from $\frac{1}{2}N(N-1)$ to N times the average number of entries in the neighbour list of a typical particle. This in turn makes the computing time required per step a linear rather than a quadratic function of N.

The methods outlined above are easily extended to molecular systems if the molecules are treated as consisting of independent atoms bound together by continuous intramolecular forces. This is rarely the best way to proceed.

Apart from the problem of treating the vibrational motion classically, the choice of time step is dictated by the timescale of the vibrations rather than the slower and usually more interesting translational and rotational motions. In general, therefore, it is more economic to work with rigid molecules, but the dynamical problem is then more difficult. The conventional approach to the solution of the equations of motion of a rigid body involves a separation of internal and centre-of-mass coordinates. As an illustration, consider a linear molecule with centre-of-mass coordinates $\mathbf{R}_i$ and polar angles θ_i, ϕ_i relative to a laboratory-fixed frame of reference. In terms of these coordinates, the equations of motion of the molecule are

$$\ddot{\mathbf{R}}_i = \frac{1}{M}\mathbf{F}_i \tag{3.3.7}$$

$$\ddot{\theta}_i = \sin\theta_i \cos\theta_i \dot{\phi}_i^2 - \frac{1}{I}\sum_{j\neq i}^{N}\frac{\partial v(i,j)}{\partial\theta_i} \tag{3.3.8}$$

$$\ddot{\phi}_i = -2\cot\theta_i \dot{\theta}_i \dot{\phi}_i - \frac{1}{I\sin^2\theta_i}\sum_{j\neq i}^{N}\frac{\partial v(i,j)}{\partial\phi_i} \tag{3.3.9}$$

where M is the molecular mass and I is the moment of inertia; the angular derivatives of the pair potential energy $v(i,j)$ are related to the torque acting on the molecule (Barojas *et al.*, 1973). The centre-of-mass motion can be handled by the same central-difference algorithm that is used in the atomic case. The difficulty lies in obtaining a numerical solution to the coupled rotational equations (3.3.8) and (3.3.9), because Eqn (3.3.9) has a singularity at $\sin\theta_i = 0$. Since the choice of laboratory-fixed axes is arbitrary, the problem can be overcome by redefining the laboratory frame whenever the singularity is approached; in their simulation of liquid nitrogen, for example, Barojas *et al.* (1973) chose to redefine the polar axes whenever one of the angles θ_i approached within $\pi/10$ of either $\theta_i = 0$ or $\theta_i = \pi$. The alternative is to use an algorithm that is free from singularities at the outset (Cheung and Powles, 1975; Evans and Murad, 1977; Singer *et al.*, 1977; Ryckaert *et al.*, 1977; Ciccotti *et al.*, 1982). The scheme that we describe below is the method of "constraints" due to Ryckaert *et al.* (1977). The method is particularly convenient to use insofar as the equations of motion are solved in cartesian form, no internal coordinates appear, and the computer program is very similar in structure to that required for a monatomic system.

As an illustration of the use of constraint dynamics, consider the example of the non-linear triatomic molecule shown in Figure 3.2, in which each bond is of length l and each atom (labelled 1, 2 and 3) is of mass m. The total force acting on atom 1, say, at time t is the sum of three terms; $\mathbf{f}_1(t)$, the force due to interactions with other molecules; a force of constraint, $\boldsymbol{\gamma}_{12}(t)$, which ensures that the bondvector $\mathbf{r}_{12}(t)$ remains of fixed length l;

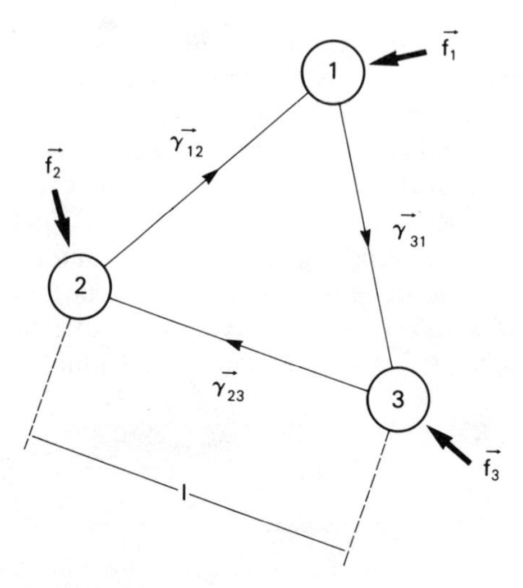

FIG. 3.2. The method of constraints applied to a triatomic molecule: $\mathbf{f}_i$ is the total intermolecular force on atom i and $\boldsymbol{\gamma}_{ij}$ is the force of constraint that maintains rigidity of the bond between i and j.

and a force $\boldsymbol{\gamma}_{13}(t)$ that acts to preserve the bondlength between atoms 1 and 3. Similar considerations apply to atoms 2 and 3. The equations of motion of the atoms may therefore be written as

$$\begin{aligned} m\ddot{\mathbf{r}}_1 &= \mathbf{f}_1(t) + \boldsymbol{\gamma}_{12}(t) + \boldsymbol{\gamma}_{13}(t) \\ m\ddot{\mathbf{r}}_2 &= \mathbf{f}_2(t) + \boldsymbol{\gamma}_{21}(t) + \boldsymbol{\gamma}_{23}(t) \\ m\ddot{\mathbf{r}}_3 &= \mathbf{f}_3(t) + \boldsymbol{\gamma}_{31}(t) + \boldsymbol{\gamma}_{32}(t) \end{aligned} \tag{3.3.10}$$

The forces of constraint are directed along the corresponding bondvectors, i.e.

$$\boldsymbol{\gamma}_{ij}(t) = \lambda_{ij}\mathbf{r}_{ij}(t), \qquad i \neq j \tag{3.3.11}$$

where λ_{ij} is a time-dependent scalar quantity. From the law of action and reaction, it follows that

$$\begin{aligned} \boldsymbol{\gamma}_{ji}(t) &= \lambda_{ji}\mathbf{r}_{ji}(t) \\ &= -\boldsymbol{\gamma}_{ij}(t) \end{aligned} \tag{3.3.12}$$

Thus

$$\lambda_{ij} = \lambda_{ji} \tag{3.3.13}$$

If we now apply the central-difference algorithm (3.3.2) to Eqns (3.3.10), we find that

$$\begin{aligned}
\mathbf{r}_1(t+h) &= \mathbf{r}_1'(t+h) + (h^2/m)[\lambda_{12}\mathbf{r}_{12}(t) + \lambda_{13}\mathbf{r}_{13}(t)] \\
\mathbf{r}_2(t+h) &= \mathbf{r}_2'(t+h) + (h^2/m)[-\lambda_{12}\mathbf{r}_{12}(t) + \lambda_{23}\mathbf{r}_{23}(t)] \\
\mathbf{r}_3(t+h) &= \mathbf{r}_3'(t+h) + (h^2/m)[-\lambda_{13}\mathbf{r}_{13}(t) - \lambda_{23}\mathbf{r}_{23}(t)]
\end{aligned} \tag{3.3.14}$$

where

$$\mathbf{r}_i'(t+h) = -\mathbf{r}_i(t-h) + 2\mathbf{r}_i(t) + (h^2/m)\mathbf{f}_i(t) \tag{3.3.15}$$

are the predicted coordinates of atom i in the absence of any constraint. The set of equations (3.3.14) must be solved subject to the conditions

$$\begin{aligned}
|\mathbf{r}_{12}(t+h)|^2 = |\mathbf{r}_{13}(t+h)|^2 &= |\mathbf{r}_{23}(t+h)|^2 \\
&= l^2
\end{aligned} \tag{3.3.16}$$

This leads to three simultaneous equations for the quantities λ_{ij}, which may be written schematically as

$$\begin{aligned}
\left(\frac{2h^2}{m}\right)
\begin{vmatrix}
2\mathbf{r}_{12}\cdot\mathbf{r}_{12}' & \mathbf{r}_{13}\cdot\mathbf{r}_{12}' & -\mathbf{r}_{23}\cdot\mathbf{r}_{12}' \\
\mathbf{r}_{12}\cdot\mathbf{r}_{13}' & 2\mathbf{r}_{13}\cdot\mathbf{r}_{13}' & \mathbf{r}_{23}\cdot\mathbf{r}_{13}' \\
-\mathbf{r}_{12}\cdot\mathbf{r}_{23}' & \mathbf{r}_{13}\cdot\mathbf{r}_{23}' & 2\mathbf{r}_{23}\cdot\mathbf{r}_{23}'
\end{vmatrix}
\begin{vmatrix} \lambda_{12} \\ \lambda_{13} \\ \lambda_{23} \end{vmatrix} \\
= \begin{vmatrix} l^2 - |\mathbf{r}_{12}'|^2 \\ l^2 - |\mathbf{r}_{13}'|^2 \\ l^2 - |\mathbf{r}_{23}'|^2 \end{vmatrix} + \mathcal{O}(\lambda_{ij}^2)
\end{aligned} \tag{3.3.17}$$

where $\mathbf{r}_{ij} \equiv \mathbf{r}_{ij}(t)$ and $\mathbf{r}_{ij}' \equiv \mathbf{r}_{ij}'(t+h)$. The three equations are easily solved by use of an iterative method; at the first iteration, the terms of order λ_{ij}^2 are omitted, and three to four iterations per molecule are in general sufficient to maintain the bondlengths constant to within one part in 10^4.

The original version of the method of constraints was later reformulated (Ciccotti *et al.*, 1982) in a manner that makes it economical to use even when the number of constraints is very large, and also makes it applicable in certain cases where the treatment given here breaks down. Apart from its simplicity, a great merit of the method is the fact that it can be used for molecules that are flexible as well as for those that are completely rigid. Internal rotations, for example, are straightforwardly allowed for by removal of the appropriate constraints.

3.4 ENSEMBLE AVERAGES AND THE MONTE CARLO METHOD

The canonical ensemble average of any function of the particle coordinates can be written in the general form

$$\langle F\rangle = \frac{\int F(\mathbf{r}^N)\exp[-\beta V_N(\mathbf{r}^N)]\,d\mathbf{r}^N}{\int \exp[-\beta V_N(\mathbf{r}^N)]\,d\mathbf{r}^N} \tag{3.4.1}$$

Equation (2.3.20) for the excess internal energy is an important special case of this relation. Analogous expressions hold in other ensembles, but for the present we confine ourselves to (3.4.1).

Given a fast computer, it is possible to generate a large number of random configurations of particles. We might then hope to obtain an estimate for $\langle F\rangle$ by replacing the integrals in (3.4.1) by sums over the finite number, $\mathcal{N}$ say, of configurations in the sample, i.e.

$$\langle F\rangle \simeq \frac{\sum_m F(m)\exp[-\beta V_N(m)]}{\sum_m \exp[-\beta V_N(m)]} \tag{3.4.2}$$

where the sums run from $m=1$ to $m=\mathcal{N}$ and $V_N(m)$ is the total potential energy of a configuration that we denote symbolically by m. In practice, this crude approach is very inefficient, because a randomly constructed configuration is likely to have a very small Boltzmann factor. If, for example, the centres of 100 hard spheres were placed at random within a given volume, then at liquid-like densities it is overwhelmingly likely that the resulting configuration would contain spheres that overlap. Such configurations make no contribution to the sums in Eqn (3.4.2). An alternative would be to attempt a random, sequential addition of spheres to the system; new spheres would be added randomly to the system, but only at points where they did not overlap with spheres that are already present. A search-and-insert procedure of this type was used in certain early Monte Carlo calculations on hard spheres. As several authors have shown, however, the "jammed" configurations that the method yields are not representative of the hard-sphere fluid at equilibrium, though they do have some relevance to the problem of adsorption (Widom, 1966).

If a Monte Carlo calculation is to be efficient, it is almost invariably necessary to introduce some form of *importance sampling* (Hammersley and Handscomb, 1964). This means that configurations should be sampled in such a way that the regions of configuration space that make the largest contributions to the integrals in (3.4.1) are also the regions that are sampled most frequently. When ensemble averages are calculated, the bias introduced in the sampling is eliminated by attaching an appropriate weight to each configuration. If $W(m)$ is the probability of choosing a configuration m, Eqn (3.4.2) must be replaced by

$$\langle F\rangle \simeq \frac{\sum_m F(m)\exp[-\beta V_N(m)]/W(m)}{\sum_m \exp[-\beta V_N(m)]/W(m)} \tag{3.4.3}$$

A natural choice is to sample on the Boltzmann distribution itself. This is achieved by setting

$$W(m) = \exp[-\beta V_N(m)] \tag{3.4.4}$$

in which case Eqn (3.4.3) reduces to

$$\langle F\rangle \simeq \frac{1}{\mathcal{N}}\sum_{m=1}^{\mathcal{N}} F(m) \tag{3.4.5}$$

The canonical ensemble average is therefore obtained as an unweighted average over configurations in the sample.

The problem of devising a scheme for sampling configuration space according to a specific probability distribution is most easily formulated in terms of the theory of stochastic processes (Wood and Parker, 1957; Wood, 1968). Suppose we have a sequence of random variables. In our case, the random "variable" is the set of all coordinates of the particles and the range of the variable is the configuration space of the system. Instead of speaking of the value of the variable at a particular point in the sequence, we prefer to say that the system occupies a particular state at that point. If the probability of finding the system in a state n at "time" t is determined solely by the state m that it occupies at the previous time $(t-1)$, the sequence is said to form a *Markov chain* (Feller, 1950). The concept of time is introduced merely for descriptive purposes; there is no connection with any molecular timescale.

Let $q_n(t)$ be the probability that the state n occurs at time t, with

$$\sum_n q_n(t) = 1 \tag{3.4.6}$$

for all t. The markovian character of the random process can be expressed by writing

$$q_n(t) = \sum_m P_{nm} q_m(t-1) \tag{3.4.7}$$

where the *transition probability* $P_{nm} \equiv \text{prob}\,(n, t \mid m, t-1)$ is assumed to be independent of t. If we regard the probability distribution $\{q_n(t)\}$ as a column vector $\mathbf{q}(t)$ and the quantities $\{P_{nm}\}$ as a square *transition matrix* $\boldsymbol{P}$, Eqn (3.4.7) can be rewritten in a more compact form as

$$\mathbf{q}(t) = \boldsymbol{P}\mathbf{q}(t-1) \tag{3.4.8}$$

where we adopt the usual convention that P_{nm} is the element in the nth row and mth column of the matrix $\boldsymbol{P}$. Equation (3.4.8) can be generalized immediately to yield the probability distribution at time t in terms of an arbitrary initial distribution $\mathbf{q}(0)$:

$$\mathbf{q}(t) = \boldsymbol{P}^t \mathbf{q}(0) \tag{3.4.9}$$

where $\boldsymbol{P}^t$ is the t-fold product of $\boldsymbol{P}$ with itself. The elements $P_{nm}^{(t)}$ of the matrix $\boldsymbol{P}^t$ are called the "multistep" transition probabilities of order t.

Because of their interpretation as conditional probabilities, the elements of $\boldsymbol{P}$ necessarily satisfy the conditions

$$0 \leqslant P_{nm} \leqslant 1, \qquad \sum_n P_{nm} = 1 \tag{3.4.10}$$

A matrix having such properties is called a *stochastic* matrix. It is easy to see that the matrix $\boldsymbol{P}^t$ is also stochastic, since $P_{nm}^{(t)}$ is merely the probability of state n being reached from state m in t steps rather than in one. If the limits

$$\Pi_n = \lim_{t \to \infty} P_{nm}^{(t)}. \tag{3.4.11}$$

exist and are the same for all m, the Markov chain is said to be *ergodic*; this usage of the term is closely related to the meaning of ergodicity in statistical mechanics. It follows from Eqn (3.4.9) (and the requirement that $\mathbf{q}(0)$ be normalized) that the column vector $\boldsymbol{\Pi} = \{\Pi_n\}$ represents a unique, asymptotic probability distribution, i.e.

$$\boldsymbol{\Pi} = \lim_{t \to \infty} \mathbf{q}(t) \tag{3.4.12}$$

Once the limiting distribution is reached, it will persist, since the transition matrix maps the limiting distribution onto itself:

$$\boldsymbol{P}\boldsymbol{\Pi} = \boldsymbol{\Pi} \tag{3.4.13}$$

This result is called the *steady-state condition.*

Equation (3.4.13) shows that the limiting distribution is a right eigenvector of $\boldsymbol{P}$ corresponding to an eigenvalue of unity. Any stochastic matrix has at least one such eigenvalue, and it is straightforward to show that all the eigenvalues are of modulus one or less (Bartlett, 1966). What is also important to know are the conditions under which the matrix $\boldsymbol{P}$ has a single eigenvalue of unity, since only then will the limiting distribution be unique. If there is more than one such eigenvalue, there exists more than one limiting distribution, and the one that is reached in a given Markov chain will be dependent on the initial distribution. In the latter case, the states of the system fall into different "ergodic classes". From a theorem of Perron and Frobenius (Gantmacher, 1959), a sufficient condition for $\boldsymbol{P}$ to have a non-degenerate eigenvalue of unity is that all elements of $\boldsymbol{P}^t$ must be non-zero for some finite t. In other words, a unique limiting distribution exists if any state of the system can be reached from any other in a finite number of steps.

From the algebraic point of view, the problem of finding an appropriate set of transition probabilities is that of solving the system of linear equations represented by (3.4.13) subject to the conditions (3.4.10). The problem is greatly simplified by looking for a transition matrix to which *microscopic reversibility* applies, i.e.

$$\Pi_m P_{nm} = \Pi_n P_{mn} \tag{3.4.14}$$

In that case, the steady-state condition (3.4.13) is satisfied automatically, irrespective of the form of the limiting distribution $\mathbf{\Pi}$. For our purposes, the relevance of (3.4.14) is that for sampling on the Boltzmann distribution the limits Π_n are known and given by

$$\Pi_n = \frac{\exp[-\beta V_N(n)]}{\sum_m \exp[-\beta V_N(m)]} \tag{3.4.15}$$

Before we can use this result, however, we must briefly consider the way in which a sequence of configurations is generated in practice, though we defer all detailed discussion of the problem to the next section. Suppose that the system is in a state m at a given time. An attempt is made to move to a state n that is *adjacent* to m; the state n is called a *trial* configuration. In most applications of the Monte Carlo method, the trial configuration is generated by displacing one particle of the system within small, prescribed limits, and the term "adjacent to" has the meaning of "not very different from". Let C_{nm} be the conditional probability of choosing n as the trial state. The quantities C_{nm} form the elements of a stochastic matrix $\boldsymbol{C}$, with $C_{nm} = 0$ for all pairs of states that are non-adjacent. We also suppose that $\boldsymbol{C}$ is symmetric. Then $\boldsymbol{P}$ has all the properties that we require if

$$\left.\begin{aligned} P_{nm} &= C_{nm} \quad \text{if } \Pi_n \geqslant \Pi_m,\ n \neq m, \\ &= C_{nm}\Pi_n/\Pi_m \quad \text{if } \Pi_n < \Pi_m \end{aligned}\right\} \tag{3.4.16}$$

$$P_{nn} = 1 - \sum_{m \neq n} P_{nm} \tag{3.4.17}$$

An important point to note is that the limits Π_m, Π_n appear only in the ratio Π_n/Π_m; their absolute values, and in particular the value of the denominator in (3.4.15), are not required.

The transition matrix defined by (3.4.16) and (3.4.17) is the one proposed in the pioneering work of Metropolis *et al.* (1953) and remains much the most commonly used prescription for $\boldsymbol{P}$. Other schemes have been devised, but none has been widely adopted and we shall not consider them further.

3.5. IMPLEMENTATION OF THE MONTE CARLO METHOD

We now consider how the general scheme described in the preceding section can be used in practical calculations. We confine ourselves initially to the case of the canonical ensemble and assume that particles interact via spherically symmetric potentials.

Suppose that at time $t=0$ the system is in a state m. The Monte Carlo process is initiated by selecting a particle i in the central cell of the periodic system and displacing it from its original position $\mathbf{r}_i$ to a trial position $\mathbf{r}_i+\Delta\mathbf{r}_i$. The displacement $\Delta\mathbf{r}_i$ is chosen at random such that $\Delta\mathbf{r}_i=(s_1, s_2, s_3)\xi$, where s_1, s_2 and s_3 are random numbers uniformly distributed in the interval $(-\frac{1}{2}, +\frac{1}{2})$ and ξ is a parameter of the calculation. Any trial configuration, n say, that can be accessed from m in this way is a state adjacent to m in the sense of Section 3.4, i.e. $C_{nm}>0$. The method of constructing the trial configuration ensures that C_{nm} has the same value for all pairs of adjacent states and, in particular, that $C_{nm}=C_{mn}$.

The next step is to determine whether the trial configuration is to be accepted or rejected. Let $\Delta V_N = V_N(n) - V_N(m)$ be the increase in total potential energy resulting from the trial move. According to the Metropolis criterion (3.4.16), a trial configuration is accepted unconditionally if ΔV_N is negative, but it is accepted only with a probability $\exp(-\beta\Delta V_N)$ if ΔV_N is positive. In the latter case, a random number $0 \leqslant s \leqslant 1$ is generated. If $s \leqslant \exp(-\beta\Delta V_N)$, the trial configuration is accepted, otherwise it is rejected; clearly there is a low probability of acceptance of a move that leads to a large increase in potential energy. The procedure takes its simplest form in the case of hard spheres: trial configurations in which two or more spheres overlap are rejected, but all others are accepted. Acceptance of a move means that the system passes from state m at time $t=0$ to state n at $t=1$, while rejection means that the system remains in state m at $t=1$. The entire cycle is then repeated many times. Typically, between 10^5 and 10^6 configurations are sufficient for the determination of thermodynamic properties such as energy and pressure.

The method becomes only slightly more complicated when the interaction between particles is dependent on their mutual orientation. In such cases, particle i is subjected not only to a random displacement but also to a random reorientation. The reorientation can be achieved by first randomly selecting one of the three cartesian axes of a laboratory-fixed coordinate frame and then rotating the particle about that axis through an angle randomly and uniformly distributed in a preset range $(-\phi, +\phi)$.

A number of variations in the sampling scheme are possible. For example, the choice of particle to be displaced may be made either at random, with uniform probability, or in a serial fashion. One can also choose to displace two particles simultaneously or, in the case of molecular systems, to alternate trial displacements with trial rotations. An important question concerns the choice of the parameter ξ (and ϕ, if reorientations are required). If ξ is large, and the system is at a liquid-like density, a very high proportion of trial moves will be rejected. A high rate of acceptance can be ensured by making ξ sufficiently small, but a correspondingly larger number of configurations will need to be generated to ensure an adequate sampling of the relevant region of configuration space. The rule-of-thumb commonly used is to set ξ to a value at which approximately a half of all trial configurations are accepted. In practice, this means that ξ is of order one-tenth of the nearest-neighbour distance. Other technical details of the computations, including the use of neighbour tables, the inclusion of tail corrections, and the handling of long-range interactions, are similar to those described for molecular dynamics calculations in Section 3.3. If full use is made of neighbour tables, the computing time per configuration should be only weakly dependent on sample size.

Only a few minor changes are required in order to adapt the Monte Carlo method to the isothermal–isobaric ensemble (Wood, 1968; McDonald, 1972a). By a generalization of Eqn (2.3.27), the NPT-ensemble average of a function of volume and particle coordinates is given by

$$\langle F\rangle = \frac{\int_0^\infty \mathrm{d}V \exp(-\beta PV) \int_V \mathrm{d}\mathbf{r}^N F(\mathbf{r}^N, V) \exp[-\beta V_N(\mathbf{r}^N)]}{\int_0^\infty \mathrm{d}V \exp(-\beta PV) \int_V \mathrm{d}\mathbf{r}^N \exp[-\beta V_N(\mathbf{r}^N)]} \tag{3.5.1}$$

In a Monte Carlo calculation, the particles are confined within a cubic cell of fluctuating length L. It is convenient to introduce the scaled coordinates $\boldsymbol{\tau}_i$, defined as

$$\boldsymbol{\tau}_i = L^{-1}\mathbf{r}_i \tag{3.5.2}$$

Equation (3.5.1) then becomes

$$\langle F\rangle = \frac{\int_0^\infty \mathrm{d}V \exp(-\beta PV) V^N \int_\Omega \mathrm{d}\boldsymbol{\tau}^N F([L\boldsymbol{\tau}]^N, V) \exp[-\beta V_N([L\boldsymbol{\tau}]^N, L)]}{\int_0^\infty \mathrm{d}V \exp(-\beta PV) V^N \int_\Omega \mathrm{d}\boldsymbol{\tau}^N \exp[-\beta V_N[L\boldsymbol{\tau}]^N, L)]} \tag{3.5.3}$$

where the particle coordinates now range over the unit cube Ω. Equation (3.5.3) represents an average in the $(3N+1)$-dimensional space of the variables $\{V, \boldsymbol{\tau}_1, \ldots, \boldsymbol{\tau}_N\}$ with a probability density proportional to the pseudo-Boltzmann factor

$$W = \exp\left[-\beta PV - \beta V_N([L\boldsymbol{\tau}]^N, L) + N \log V\right] \tag{3.5.4}$$

In the implementation of the Monte Carlo method, a trial configuration n is constructed from the current configuration m by randomly displacing a selected particle and randomly changing the length of the cell within predetermined limits. The trial configuration is accepted unconditionally if W increases and with probability $W(n)/W(m)$ if it decreases. It is not necessary to attempt a volume change at every step, and it may well be computationally more efficient not to do so when the potential is other than a sum of terms that scale in a simple way with volume. The procedure is easily generalized to allow for changes in the shape as well as the volume of the cell; this makes it suitable for the study of structural phase changes in solids (Yashonath and Rao, 1985).

The further extension of the Monte Carlo method to the case of the grand canonical ensemble has been discussed in detail by several authors (Nicholson and Parsonage, 1982); the method that we describe here is due originally to Adams (1974). In the grand canonical ensemble, the average of any function of the particle coordinates is given by

$$\langle F\rangle = \frac{1}{\Xi}\sum_{N=0}^{\infty}\frac{\Lambda^{-3N}}{N!}\exp(N\beta\mu)\int F(\mathbf{r}^N)\exp\left[-\beta V_N(\mathbf{r}^N)\right]\mathrm{d}\mathbf{r}^N \tag{3.5.5}$$

If we introduce the scaled coordinates (3.5.2) and the definitions (2.4.1) and (2.4.2), Eqn (3.5.5) can be rewritten as

$$\langle F\rangle = \frac{1}{\Xi}\sum_{N=0}^{\infty}\exp(N\beta\mu^* - \log N!)\int_{\Omega} F\exp(-\beta V_N)\,\mathrm{d}\boldsymbol{\tau}^N \tag{3.5.6}$$

where μ^* is defined in terms of the excess chemical potential μ^{ex} as

$$\mu^* = \mu^{\text{ex}} + k_{\text{B}}T\log\langle N\rangle \tag{3.5.7}$$

The pseudo-Boltzmann factor analogous to (3.5.4) is now

$$W_N = \exp\left[N\beta\mu^* - \log N! - \beta V_N(\mathbf{r}^N)\right] \tag{3.5.8}$$

The fixed parameters of the calculation are μ^*, V and T, and trial configurations are generated in a two-stage process. The first stage involves the displacement of a selected particle and proceeds in exactly the same way in canonical ensemble calculations, the move being accepted or rejected according to the criteria already discussed. The second stage involves an attempt either to insert a particle at a randomly chosen point in the cell or to remove a particle that is already present (Valleau and Cohen, 1980). The decision whether to attempt an insertion or a removal is itself made randomly with equal probabilities. In either case, the trial configuration is accepted unconditionally if the pseudo-Boltzmann factor increases. If it decreases, the change is accepted with probability equal to either

$$\frac{W_{N+1}}{W_N} = \frac{1}{N+1} \exp[\beta\mu^* - \beta(V_{N+1} - V_N)] \qquad (3.5.9)$$
(insertion)

or

$$\frac{W_{N-1}}{W_N} = N \exp[-\beta\mu^* - \beta(V_{N-1} - V_N)]. \qquad (3.5.10)$$
(removal)

In this approach, the excess chemical potential is not itself an input parameter; at the end of the calculation, when $\langle N \rangle$ is known, μ^{ex} can be obtained from the definition (3.5.7). Such methods work very well at low and intermediate densities, but become increasingly difficult to apply at higher densities because the probability of inserting or removing a particle is then very small. It has, however, been used successfully in the study of a variety of interfacial phenomena (Nicholson and Parsonage, 1982).

One advantage of the grand canonical ensemble calculations is that they provide a direct means of determining the chemical potential and hence the free energy and entropy of the system under study. In other Monte Carlo methods, or in molecular dynamics calculations, a numerical value for the free energy can always be obtained by the integration of thermodynamic relations such as (2.3.5) or (2.3.7) along a path that links the thermodynamic state of interest to one for which the free energy is already known, such as the dilute gas or the low-temperature solid. This requires a very considerable computational effort. A less time-consuming alternative is the *particle-insertion* method, based on Widom's expression (2.4.30) for the chemical potential. Equation (2.4.30) shows that the excess chemical potential can be calculated from the mean Boltzmann factor of a test particle that feels the force field of the "real" particles but does not act upon them. Such a calculation requires only a straightforward addition to a standard (canonical ensemble) Monte Carlo program. At intervals of, say, 100 configurations, the Boltzmann factors of test particles placed on a grid of a few hundred points in the simulation cell are evaluated; the mean Boltzmann factor is then obtained as the average over all grid points and all configurations. The method has been applied with success to both atomic (Adams, 1974, 1975; Powles *et al.*, 1982) and molecular (Romano and Singer, 1979) liquids and has also been implemented in molecular dynamics calculations. Although simple both in conception and in application, the particle-insertion method appears to be at least as useful as any of the more elaborate schemes that have been devised for the calculation of free energies (Bennett, 1976; Frenkel, 1986).

3.6. MOLECULAR DYNAMICS AT CONSTANT PRESSURE AND TEMPERATURE

The molecular dynamics methods discussed so far are limited to the study of systems characterized by fixed values of N, V and E (the total energy). In certain applications, it would be useful to have the temperature and pressure included among the fixed parameters of the calculation. Several schemes have been developed for this purpose, most of which have their inspiration in a paper by Andersen (1980). The work of Andersen, and subsequently that of Nosé (1984a, b) is based on the concept of an "extended" system consisting of the physical system of interest and an external reservoir. The coupling to the reservoir serves to maintain constancy of pressure or temperature (or both) by suitable modification of the equations of motion of the particles in the system of interest.

As an example, we briefly describe the method devised by Andersen (1980) for carrying out molecular dynamics simulations of atomic systems under conditions of constant pressure. The system of interest, consisting of N particles with coordinates $\mathbf{r}_i$, is assumed to be coupled to an external piston. The role of the piston is to contract or expand the system uniformly in response to any imbalance between the instantaneous pressure within the system and an externally set pressure P. The extended system, i.e. the system of interest plus the piston, is described by the fluctuating volume V and scaled coordinates $\boldsymbol{\tau}_i = V^{-1/3}\mathbf{r}_i$ (cf. Eqn (3.5.2)). The lagrangian of the extended system is written as

$$L = \tfrac{1}{2}mV^{2/3}\sum_{i=1}^{N}|\dot{\boldsymbol{\tau}}_i|^2 - V_N + \tfrac{1}{2}W\dot{V}^2 - PV \tag{3.6.1}$$

where V_N is the usual interparticle potential energy, W is an inertial constant, and the terms PV and $\frac{1}{2}W\dot{V}^2$ represent, respectively, the "potential" and "kinetic" energies associated with the variable V. The momenta conjugate to $\boldsymbol{\tau}_i$ and V are

$$\boldsymbol{\pi}_i = \frac{\partial L}{\partial \dot{\boldsymbol{\tau}}_i} = mV^{2/3}\dot{\boldsymbol{\tau}}_i = mV^{1/3}\dot{\mathbf{r}}_i \tag{3.6.2}$$

and

$$p_V = \frac{\partial L}{\partial \dot{V}} = W\dot{V} \tag{3.6.3}$$

and the hamiltonian of the extended system is

$$\mathscr{H} = \frac{1}{2mV^{2/3}} \sum_{i=1}^{N} |\boldsymbol{\pi}_i|^2 + V_N + \frac{p_V^2}{2m} + PV \tag{3.6.4}$$

It then follows straightforwardly from Hamilton's equations (see Eqns (2.1.2) and (2.1.3)), as applied to the conjugate pairs $(\boldsymbol{\tau}_i, \boldsymbol{\pi}_i)$ and (V, p_V), that the time evolution of the variables $\boldsymbol{\tau}_i$ and V is determined by two coupled, second-order, differential equations of the form

$$m\ddot{\boldsymbol{\tau}}_i = V^{-1/3}\mathbf{F}_i - \tfrac{2}{3}mV^{-1}\dot{V}\dot{\boldsymbol{\tau}}_i \tag{3.6.5}$$

$$3WV\ddot{V} = mV^{2/3} \sum_{i=1}^{N} |\dot{\boldsymbol{\tau}}_i|^2 + V^{1/3} \sum_{i=1}^{N} \boldsymbol{\tau}_i \cdot \mathbf{F}_i - 3PV \tag{3.6.6}$$

Andersen (1980) has shown that time averages of static quantities over the resulting phase-space trajectories are equivalent to averages in an isobaric–isoenthalpic (or *NPH*-) ensemble.

The success of the extended-system approach depends on the skill with which the corresponding lagrangian is chosen. The choice is an arbitrary one, in which physical intuition plays a part. As another example, Nosé (1984a) has shown that a lagrangian of the form

$$L = \tfrac{1}{2}ms^2 \sum_{i=1}^{N} |\dot{\mathbf{r}}_i|^2 - V_N + \tfrac{1}{2}Q\dot{s}^2 - (3N+1)k_B T \log s \tag{3.6.7}$$

where T is a preset temperature, provides a scheme for carrying out calculations in the canonical ensemble, while a combination of (3.6.1) and (3.6.7) can be used to generate states of an *NPT*-ensemble. The quantity s in (3.6.7) describes a coupling of the system of interest to an external heat bath and can be interpreted as a velocity-scaling variable; Q is a parameter with a significance similar to that of the "mass" W in the constant-pressure calculations.

An unsatisfactory feature of the methods outlined above is the further arbitrariness involved in the choice of the parameters W and Q. If the values used are too small, the variables V and s are effectively decoupled from the particle coordinates and momenta; if they are too large, phase space is sampled inefficiently. Because the time evolution of the system depends on the choice of W and Q, the meaning of the dynamical correlations is unclear. It has been established empirically, however, that a variety of dynamical properties calculated in constant-pressure and constant-temperature simulations are insensitive to the precise choice of W and Q, and show no significant differences from those obtained by conventional molecular dynamics calculations (Nosé and Klein, 1983a; Nosé, 1984a).

Parrinello and Rahman (1980, 1981) have described an extension of the constant-pressure method that allows for fluctuations in the shape of the molecular dynamics cell, and several authors have developed algorithms suitable for use with molecular systems (Nosé and Klein, 1983b; Ryckaert and Ciccotti, 1983; Ferrario and Ryckaert, 1985). Alternatives to the extended-system methods have also been devised in which the kinetic energy and/or pressure are held strictly constant by building appropriate constraints into the equations of motion of the particles (Hoover *et al.*, 1982; Evans and Morriss, 1984).

CHAPTER 4

Diagrammatic Expansions

The best-known example of the use of diagrammatic methods in statistical mechanics is in the derivation of the density expansion of the equation of state of the classical, imperfect gas (Mayer and Mayer, 1940; Uhlenbeck and Ford, 1962). For liquids, density expansions are generally inadequate. Nonetheless, as we shall see in later chapters, diagrammatic methods feature prominently in the modern theory of dense fluids. The introductory account given here is based largely on the work of Morita and Hiroike (1961), de Dominicis (1962, 1963) and Stell (1964), in which systematic use is made of the technique of functional differentiation. The discussion is self-contained but limited in scope, and no attempt is made at mathematical rigour.

4.1 THE IMPERFECT GAS AND THE SECOND VIRIAL COEFFICIENT

The equation of state of an imperfect gas can be written as a power series in the density in the form

$$\frac{\beta P}{\rho} = 1 + \sum_{i=2}^{\infty} B_i(T)\rho^{i-1} \qquad (4.1.1)$$

This is the *virial series* or *virial expansion*, and the temperature-dependent coefficients $B_i(T)$ are called the *virial coefficients*. Experimental equation-of-state data are often reported in the form of a truncated virial series. The leading term on the right-hand side of Eqn (4.1.1) represents ideal-gas behaviour, valid in the low-density limit. With increasing density, interactions between increasingly large numbers of molecules begin to make significant contributions to the equation of state, and we may therefore expect the leading correction to the ideal-gas law, represented by the second virial coefficient $B_2(T)$, to be determined by the interaction energy of isolated pairs.

As we shall see in Section 4.5, a rigorous derivation of the virial expansion can be given in the framework of the grand canonical ensemble. The derivation is lengthy, and the expression for the second virial coefficient is obtained much more easily by inserting in the virial equation (2.5.14) the low-density limit of $g(r)$ given by Eqn (2.5.30). Then

$$\frac{\beta P}{\rho} \simeq 1 - \frac{2\pi\beta\rho}{3} \int_0^\infty e(r) r^3 v'(r) \, dr \tag{4.1.2}$$

where

$$e(r) = \exp[-\beta v(r)] \tag{4.1.3}$$

is the Boltzmann factor for a pair of particles separated by a distance r. If the potential decays faster than r^{-3} at large r, Eqn (4.1.2) can be integrated by parts to give

$$\frac{\beta P}{\rho} \simeq 1 - 2\pi\rho \int_0^\infty [e(r) - 1] r^2 \, dr \tag{4.1.4}$$

Comparison with Eqn (4.1.1) shows that

$$B_2(T) = -\tfrac{1}{2} \int f(\mathbf{r}) \, d\mathbf{r} \tag{4.1.5}$$

where

$$f(r) = e(r) - 1 \tag{4.1.6}$$

is the *Mayer f-function.* The f-function will appear many times in this and later chapters. It has the important property of decaying rapidly to zero with increasing r for all short-range potentials; this behaviour is evident in Figure 4.1, where the function $e(r)$ for the Lennard-Jones potential is plotted for two values of the reduced temperature $T^* = k_B T/\varepsilon$.

Measurements of the deviation of the equation of state of dilute gases from the ideal-gas law allow the second virial coefficient to be determined experimentally as a function of temperature. Such measurements are an important source of information on the nature of the force law between simple molecules (Dymond and Smith, 1980; Maitland *et al.*, 1981). As an example, consider the case of the Lennard-Jones potential. By introducing the reduced length $r^* = r/\sigma$, the second virial coefficient for this potential may be expressed in the form

$$B_2(T^*) = -2\pi\sigma^3 \int_0^\infty \left\{ \exp\left[-\frac{4}{T^*}(r^{*-12} - r^{*-6}) \right] - 1 \right\} r^{*2} \, dr^* \tag{4.1.7}$$

Equation (4.1.7) shows that the dimensionless quantity $B_2^*(T^*) = B_2(T^*)/\sigma^3$ is a universal function of T^*, independent of the choice of ε and σ; this

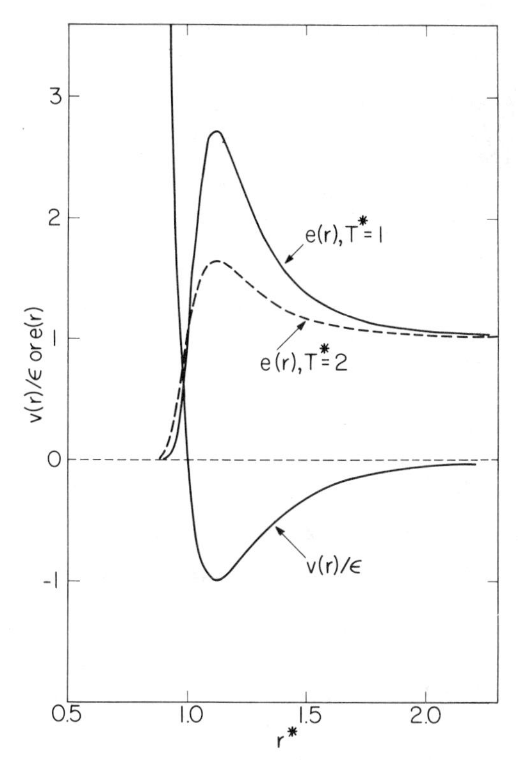

FIG. 4.1. The Lennard-Jones potential in reduced units and the function $e(r) = \exp[-\beta v(r)]$ for two values of $T^* = k_B T/\varepsilon$.

is an example of the law of corresponding states. The integrand in (4.1.7) is negative for $r^* < 1$ and positive for $r^* > 1$. When $T^* \ll 1$, the attractive part of the potential dominates the integrand, with a consequent lowering of the pressure below the ideal-gas result; when $T^* \gg 1$, the repulsive part of the interaction is dominant, the effect of which is to raise the pressure above the ideal value. It follows that $B_2(T^*)$ is negative when the reduced temperature is low and positive when it is high. The temperature at which $B_2(T)$ passes through zero is called the Boyle temperature; for the Lennard-Jones potential, $T_B^* \simeq 3.42$.

In Table 4.1 we list the values of the Lennard-Jones parameters ε and σ that have been obtained by fitting the experimental second virial coefficients of various gases to the universal curve of $B_2^*(T^*)$ versus T^* (Hirschfelder *et al.*, 1954); the results quoted are representative of a large number of similar calculations. If the Lennard-Jones potential correctly described the pair interactions in these systems, the reduced critical constants $T_c^* = k_B T_c/\varepsilon$

TABLE 4.1. *Values of the Lennard-Jones parameters based on second virial coefficient measurements and the corresponding reduced critical constants*

Gas	ε/k_B(K)	σ (Å)	$k_B T_c/\varepsilon$	$\rho_c\sigma^3$	Reference
He	10.2	2.556	0.52	0.174	a
Ne	35.8	2.75	1.24	0.300	b
A	119.8	3.405	1.26	0.316	c
Kr	116.7	3.68	1.26	0.326	d
Xe	225.3	4.07	1.29	0.338	d
H_2	36.7	2.959	0.91	0.240	e
CH_4	148.2	3.817	1.29	0.338	f, g

References: (a) Michels and Wouters (1941); (b) Kihara (1958); (c) Michels *et al.* (1949); (d) Whalley and Schneider (1955); (e) Michels *et al.* (1960); (f) Schamp *et al.* (1958); and (g) Bergeon (1958).

and $\rho_c^* = \rho_c\sigma^3$ would be the same in all cases. The departures from this rule are most pronounced for the lighter gases, particularly for helium, and are linked to deviations from the universal curve at temperatures below T_B^*. The deviations are reduced if account is taken of quantum-mechanical effects (Maitland *et al.*, 1981). Even for the heavier gases, however, careful study reveals that the Lennard-Jones potential is insufficiently flexible in form to provide an accurate fit to experimental data over a wide range of temperature. To overcome this problem, recourse must be made to a more complicated potential energy function, as discussed in Section 1.2.

For long-range potentials, in particular for the Coulomb potential, the second virial coefficient is not defined because the integral in Eqn (4.1.5) diverges. The same is true of the higher-order coefficients, signifying that the equation of state of particles interacting through coulombic forces is a non-analytic function of density (Résibois, 1968).

4.2 FUNCTIONAL DIFFERENTIATION

The grand partition function of an atomic fluid depends on the value of the pair potential $v(r)$ at every point in the range $0 < r < \infty$. It is therefore said to be a *functional* of v, and the relationship between Ξ and $v(r)$ is expressed by writing $\Xi = \Xi[v]$. The pair density $\rho^{(2)}(\mathbf{r})$ is likewise a functional of v, but it is also a function of $\mathbf{r}$; we express the dual dependence on v and $\mathbf{r}$ by writing $\rho^{(2)}(\mathbf{r}) = \rho^{(2)}[v; \mathbf{r}]$. As is clear from these examples, most statistical-mechanical quantities are functionals of some type. This section is devoted to a brief review of the mathematical properties of functionals (Volterra, 1959) and of the rules that govern the related process of *functional differentiation.*

A functional is a natural extension of the familiar mathematical concept of a function. The meaning of a function is that of a mapping from points in n-space to a real or complex number, n being the number of variables on which the function depends. A functional, by contrast, depends on all values of a function $y(x)$, say, in some range $a < x < b$. It can therefore be interpreted as a mapping from ∞-space to a real or complex number, the points in ∞-space being the values of $y(x)$ at the infinite number of points in the relevant range of the variable x. Functions of several variables and functionals are therefore conveniently treated as discrete and continuous versions of the same mathematical concept. By making this identification, it is possible to construct the rules of functional differentiation by analogy with those of elementary calculus. As usual, a sum in the discrete case is replaced by an integral in the limit where the distribution of variables becomes continuous.

If $f(\mathbf{z})$ is a function of the n variables $\mathbf{z} \equiv (z_1, \ldots, z_n)$, then the change in f due to an infinitesimal change in $\mathbf{z}$ is

$$\mathrm{d}f = f(\mathbf{z} + \mathrm{d}\mathbf{z}) - f(\mathbf{z}) = \sum_{i=1}^{n} A_i(\mathbf{z})\, \mathrm{d}\mathbf{z}_i \tag{4.2.1}$$

where

$$A_i(\mathbf{z}) \equiv \frac{\partial f}{\partial z_i} \tag{4.2.2}$$

Similarly, if $\mathscr{F} = \mathscr{F}[u]$, then

$$\begin{aligned} \delta\mathscr{F} &= \mathscr{F}[u + \delta u] - \mathscr{F}[u] \\ &= \int \mathrm{d}x\, A[u; x]\delta u(x) \end{aligned} \tag{4.2.3}$$

and the functional derivative can be identified as

$$A[u; x] \equiv \frac{\delta\mathscr{F}}{\delta u(x)} \tag{4.2.4}$$

The integral in (4.2.3) must be evaluated between limits appropriate to the problem in hand, sometimes covering the full range of the variable x, but for the sake of simplicity we leave the limits of integration unspecified. Equation (4.2.4) shows that the functional derivative determines the change in $\mathscr{F}$ resulting from an infinitesimal change in u at a particular value of x. To obtain the total variation in $\mathscr{F}$ due to a variation in $u(x)$ throughout the range of interest, it is necessary to integrate over x, as in Eqn (4.2.3).

The meaning of functional differentiation is most easily grasped by considering some examples. If f is a linear function of n variables, we know that

$$\left.\begin{aligned} f &= \sum_{i=1}^{n} a_i z_i \\ \mathrm{d}f &= \sum_{i=1}^{n} a_i \, \mathrm{d}z_i \end{aligned}\right\} \tag{4.2.5}$$

Hence

$$\frac{\partial f}{\partial z_i} = a_i, \quad \text{independent of } \mathbf{z} \tag{4.2.6}$$

The analogue of (4.2.5) for a linear functional is

$$\left.\begin{aligned} \mathscr{F} &= \int \mathrm{d}x \, k(x) u(x) \\ \delta\mathscr{F} &= \int \mathrm{d}x \, k(x) \delta u(x) \end{aligned}\right\} \tag{4.2.7}$$

and comparison with (4.2.3) shows that

$$\frac{\delta\mathscr{F}}{\delta u(x)} = k(x), \quad \text{independent of } u \tag{4.2.8}$$

An important special case is that when

$$\begin{aligned} \mathscr{F}[u] &\equiv u(x') \\ &= \int \mathrm{d}x \, \delta(x - x') u(x) \end{aligned} \tag{4.2.9}$$

Then

$$\begin{aligned} \delta\mathscr{F} &= \int \mathrm{d}x \, \delta(x - x') \delta u(x) \\ &= \delta u(x') \end{aligned} \tag{4.2.10}$$

and

$$\frac{\delta\mathscr{F}}{\delta u(x)} = \frac{\delta u(x')}{\delta u(x)} = \delta(x - x') \tag{4.2.11}$$

As a slightly more complicated example, consider the non-linear functional defined as

$$\mathscr{F} = \int \mathrm{d}x \log [1 + u(x)] \tag{4.2.12}$$

for which

$$\begin{aligned} \delta\mathscr{F} &= \int \mathrm{d}x \{\log [1 + u(x) + \delta u(x)] - \log [1 + u(x)]\} \\ &= \int \mathrm{d}x \frac{d \log [1 + u(x)]}{du} \delta u(x) \end{aligned} \tag{4.2.13}$$

and hence

$$\frac{\delta\mathscr{F}}{\delta u(x)} = \frac{1}{1 + u(x)} \tag{4.2.14}$$

The example shows how functional derivatives can be evaluated with the help of rules appropriate to ordinary differentiation.

When u is a function of two variables, the functional derivative is defined through the relation

$$\delta\mathscr{F} = \iint \mathrm{d}x_1 \, \mathrm{d}x_2 \frac{\delta\mathscr{F}}{\delta u(x_1, x_2)} \delta u(x_1, x_2) \tag{4.2.15}$$

In many cases of interest to us, the symmetry of the problem leads to a useful simplification. Consider the functional defined as

$$\mathscr{F} = \iiint \mathrm{d}x_1 \, \mathrm{d}x_2 \, \mathrm{d}x_3 \, k(x_1, x_2, x_3) u(x_1, x_2) u(x_2, x_3) u(x_3, x_1) \tag{4.2.16}$$

where the function $k(x_1, x_2, x_3)$ is symmetric with respect to permutation of the variables x_1, x_2 and x_3. The change in $\mathscr{F}$ due to an infinitesimal variation in the function u is now

$$\begin{aligned} \delta\mathscr{F} = \iiint & \mathrm{d}x_1 \, \mathrm{d}x_2 \, \mathrm{d}x_3 k(x_1, x_2, x_3) u(x_2, x_3) u(x_3, x_1) \delta u(x_1, x_2) \\ & + \text{two other terms} \end{aligned} \tag{4.2.17}$$

The three terms on the right-hand side are all equivalent. Thus

$$\frac{\delta\mathscr{F}}{\delta u(x_1, x_2)} = 3 \int \mathrm{d}x_3 \, k(x_1, x_2, x_3) u(x_2, x_3) u(x_3, x_1) \tag{4.2.18}$$

The rule for differentiation of a functional of a functional is a straightforward extension of the corresponding rule for a function of a function. For example, if

$$\begin{aligned}\mathscr{F}[u] &= \left(\int \mathrm{d}x\, k(x)u(x)\right)^N \\ &= \int \mathrm{d}x_1 \ldots \mathrm{d}x_N\, k(x_1)\ldots k(x_N)u(x_1)\ldots u(x_N) \\ &\equiv (\mathscr{G}[u])^N \end{aligned} \tag{4.2.19}$$

then

$$\begin{aligned}\frac{\delta \mathscr{F}}{\delta u(x)} &= \frac{\mathrm{d}\mathscr{F}}{\mathrm{d}\mathscr{G}} \frac{\delta \mathscr{G}}{\delta u(x)} \\ &= Nk(x)\left(\int \mathrm{d}x\, k(x)u(x)\right)^{N-1}\end{aligned} \tag{4.2.20}$$

Higher-order derivatives are defined in a manner similar to (4.2.3). In particular, the second derivative is defined through the relation

$$\begin{aligned}\delta A[u;x] &= \int \mathrm{d}x' \frac{\delta A[u;x]}{\delta u(x')} \delta u(x') \\ &\equiv \int \mathrm{d}x' \frac{\delta^2 \mathscr{F}}{\delta u(x)\, \delta u(x')} \delta u(x')\end{aligned} \tag{4.2.21}$$

The second derivative of the functional (4.2.19), for example, is

$$\frac{\delta^2 \mathscr{F}}{\delta u(x)\, \delta u(x')} = N(N-1)k(x)k(x')\left(\int \mathrm{d}x\, k(x)u(x)\right)^{N-2} \tag{4.2.22}$$

and is a functional of u and a function of both x and x'. If the derivatives exist, a functional $\mathscr{F}[u]$ can be expanded in a Taylor series about a function u_0:

$$\begin{aligned}\mathscr{F}[u] = \mathscr{F}[u_0] &+ \int \mathrm{d}x \left(\frac{\delta \mathscr{F}}{\delta u(x)}\right)_{u=u_0} [u(x)-u_0(x)] \\ &+ \frac{1}{2!}\iint \mathrm{d}x\, \mathrm{d}x' \left(\frac{\delta^2 \mathscr{F}}{\delta u(x) \delta u(x')}\right)_{u=u_0} [u(x)-u_0(x)][u(x')-u_0(x')] \\ &+ \cdots \end{aligned} \tag{4.2.23}$$

Finally, if $\mathscr{F}$ is not a functional of u, $\delta\mathscr{F}/\delta u(x) = 0$, and the equivalent of the change-of-variable formula is

$$\frac{\delta \mathscr{F}}{\delta u(x)} = \int \mathrm{d}x' \frac{\delta \mathscr{F}}{\delta v(x')} \frac{\delta v(x')}{\delta u(x)} \tag{4.2.24}$$

In statistical-mechanical applications, use of the methods of functional differentiation leads to a number of important relations involving the particle densities. Consider a system of particles interacting through pair forces and subject to an external potential $\phi(\mathbf{r})$. For notational convenience, a position vector $\mathbf{r}_i$ will be represented simply by i. The total potential energy of an N-particle system can then be written as

$$V_N(1, \ldots, N) = \sum_{i=1}^{N} \phi(i) + \sum_{i<j}^{N}\sum v(i, j) \tag{4.2.25}$$

and the grand partition function as

$$\Xi = \sum_{N=0}^{\infty} \frac{1}{N!} \int \cdots \int \prod_{i=1}^{N} z^*(i) \prod_{i<j}^{N} e(i, j) \, \mathrm{d}1 \cdots \mathrm{d}N \tag{4.2.26}$$

where

$$z^*(i) = z \exp[-\beta\phi(i)] \tag{4.2.27}$$

In the absence of an external field, the system is homogeneous; z^* and z are then identical. When $\phi(i)$ is non-zero, the definition of the n-particle density given by Eqn (2.5.18) must be rewritten as

$$\rho^{(n)}(1, \ldots, n) = \frac{1}{\Xi} \sum_{N=n}^{\infty} \frac{1}{(N-n)!} \int \cdots \int \prod_{i=1}^{N} z^*(i) \prod_{i<j}^{N} e(i, j) \, \mathrm{d}(n+1) \cdots \mathrm{d}N \tag{4.2.28}$$

Given the definitions (4.2.26) and (4.2.27), application of the rules described earlier shows that $\rho^{(n)}$ is the nth functional derivative of Ξ with respect to z^*. In particular, from the example (4.2.8), it follows straightforwardly that

$$\begin{aligned} \frac{\delta \Xi}{\delta z^*(1)} &= \sum_{N=0}^{\infty} \frac{1}{N!} \int \cdots \int \prod_{i=2}^{N} z^*(i) \prod_{i<j}^{N} e(i, j) \, \mathrm{d}2 \cdots \mathrm{d}N \\ &= \frac{1}{z^*(1)} \sum_{N=1}^{\infty} \frac{1}{(N-1)!} \int \cdots \int \prod_{i=1}^{N} z^*(i) \prod_{i<j}^{N} e(i, j) \, \mathrm{d}2 \cdots \mathrm{d}N \end{aligned} \tag{4.2.29}$$

or

$$\begin{aligned} \rho^{(1)}(1) &= \frac{z^*(1)}{\Xi} \frac{\delta \Xi}{\delta z^*(1)} \\ &= \frac{\delta \log \Xi}{\delta \log z^*(1)} \end{aligned} \tag{4.2.30}$$

More generally, by differentiating n times, it can be shown that

$$\rho^{(n)}(1, \ldots, n) = \frac{1}{\Xi} z^*(1) \cdots z^*(n) \frac{\delta^n \Xi}{\delta z^*(1) \cdots \delta z^*(n)} \tag{4.2.31}$$

In the absence of an external field, the pair distribution function can be expressed as a functional derivative of the canonical partition function. We first rewrite Eqn (2.3.13) as

$$Z_N(V, T) = \int \cdots \int \prod_{i<j}^{N} e(i, j)\, \mathrm{d}1 \cdots \mathrm{d}N \tag{4.2.32}$$

By a generalization of the example (4.2.18), the functional derivative of Z_N with respect to $e(1, 2)$ is

$$e(1, 2)\frac{\delta Z_N}{\delta e(1, 2)} = \tfrac{1}{2}N(N-1)\int \cdots \int \prod_{i<j}^{N} e(i, j)\, \mathrm{d}3 \cdots \mathrm{d}N \tag{4.2.33}$$

where $\frac{1}{2}N(N-1)$ is the number of equivalent terms that contribute to δZ_N. Comparison of (4.2.33) with the definitions (2.5.1) and (2.5.8) shows that

$$\begin{aligned} \rho^2 g_N^{(2)}(1, 2) &= 2\frac{e(1, 2)}{Z_N}\frac{\delta Z_N}{\delta e(1, 2)} \\ &= 2\frac{\delta \log Z_N}{\delta \log e(1, 2)} \end{aligned} \tag{4.2.34}$$

or, from Eqn (2.3.18):

$$\rho^2 g_N^{(2)}(1, 2) = 2\frac{\delta A^{\mathrm{ex}}}{\delta v(1, 2)} \tag{4.2.35}$$

The meaning of Eqn (4.2.34) is that

$$\begin{aligned} \delta \log Z_N &= \tfrac{1}{2}\rho^2 \iint g_N^{(2)}(1, 2)\delta \log e(1, 2)\, \mathrm{d}1\, \mathrm{d}2 \\ &= \tfrac{1}{2}\rho^2 \iint g_N^{(2)}(\mathbf{r}_{12})\delta \log e(\mathbf{r}_{12})\, \mathrm{d}\mathbf{r}_1\, \mathrm{d}\mathbf{r}_{12} \\ &= \tfrac{1}{2}\rho^2 V \int g_N^{(2)}(\mathbf{r}_{12})\delta \log e(\mathbf{r}_{12})\, \mathrm{d}\mathbf{r}_{12} \end{aligned} \tag{4.2.36}$$

where we have used the fact that the system is translationally invariant in order to integrate over $\mathbf{r}_1$. Equation (4.2.34) may therefore be rewritten as

$$\rho^2 g_N^{(2)}(\mathbf{r}_{12}) = \frac{2}{V}\frac{\delta \log Z_N}{\delta \log e(\mathbf{r}_{12})} \tag{4.2.37}$$

The appearance of the factor V^{-1} on the right-hand side of (4.2.37) and its absence in (4.2.34) is a consequence of the volume integration that is implicit in the use of a functional derivative with respect to $e(\mathbf{r}_{12})$ rather than $e(1, 2)$.

4.3 DIAGRAMS

We saw in Chapter 2 that the partition function, particle densities and particle distribution functions are defined as complicated, many-dimensional integrals over particle coordinates. Such integrals are conveniently represented by *diagrams* or *graphs.*

Consider, for example, the integrals appearing on the right-hand side of Eqn (4.2.28). These are of the general form

$$\int \cdots \int \prod_{i=1}^{N} z^*(i) \prod_{i<j}^{N} e(i,j)\, \mathrm{d}(n+1) \cdots \mathrm{d}N$$

$$= \int \cdots \int \prod_{i=1}^{N} z^*(i) \prod_{i<j}^{N} [1+f(i,j)]\, \mathrm{d}(n+1) \cdots \mathrm{d}N \qquad (4.3.1)$$

If the product in (4.3.1) is expanded, the result is a sum of terms of the type

$$\int \cdots \int \prod_{i=1}^{N} z^*(i) \prod_{\{i,j\}} f(i,j)\, \mathrm{d}(n+1) \cdots \mathrm{d}N \qquad (4.3.2)$$

where $\{i,j\}$ denotes a subset of the $\frac{1}{2}N(N-1)$ pairs (i,j). In most cases, the short-range character of the Mayer f-function makes the integrals in (4.3.2) easier to handle than those in (4.3.1).

To each of the integrals appearing in (4.3.1) and (4.3.2) there corresponds a *labelled diagram.* Such a diagram consists of a number of *circles,* certain pairs of which are linked by *bonds.* Circles represent particle coordinates and carry an appropriate label; for that reason, the diagrams are sometimes called "cluster" diagrams. The circles are of two types: *white circles* (or *root points*), which correspond to coordinates that are not integrated over; and *black circles* (or *field points*), which represent the variables of integration. With a circle labelled i we associate a function of coordinates, $\gamma(i)$ say. The circle is then referred to as a white or black *γ-circle.* Bonds are drawn as lines between circles. With a bond between circles i and j we associate a function $\eta(i,j)$ and refer to it as an *η-bond.* A *simple* diagram is one in which no pair of circles is linked by more than one bond. Thus the integrals in (4.3.2) are represented by simple diagrams consisting of z^*-circles (both white and black) and f-bonds. The *value* of a labelled diagram is the value of the integral that the diagram represents; it is a function of the coordinates attached to the white circle and a functional of the functions associated with the black circles and bonds. For example, three of the integrals in (4.3.2) (for $N=4$ and $n=2$, 2 and 1) are represented in the following way:

$$I_1 = \iint \prod_{i=1}^{4} z^*(i) f(1,3) f(2,4) f(3,4)\, \mathrm{d}3\, \mathrm{d}4$$

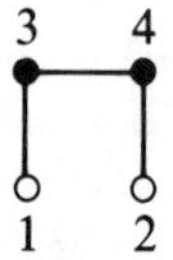

$$= \text{[diagram: black circles 3, 4; white circles 1, 2]} \tag{4.3.3a}$$

$$I_2 = \iint \prod_{i=1}^{4} z^*(i) f(1,3) f(1,4) f(2,3) f(2,4)\, \mathrm{d}3\, \mathrm{d}4$$

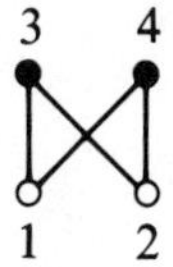

$$= \text{[diagram: black circles 3, 4; white circles 1, 2]} \tag{4.3.3b}$$

$$I_3 = \iiint \prod_{i=1}^{4} z^*(i) f(1,3) f(1,4) f(2,3) f(2,4)\, \mathrm{d}2\, \mathrm{d}3\, \mathrm{d}4$$

$$= \text{[diagram: black circles 3, 2, 4; white circle 1]} \tag{4.3.3c}$$

The black circles in a diagram correspond to the dummy variables of integration. The manner in which they are labelled is therefore irrelevant and the labels may conveniently be omitted altogether. The value of an *unlabelled diagram* involves a combinatorial factor related to the topological structure of the diagram. Consider a labelled diagram containing m black γ-circles. Any of the $m!$ possible permutations of the labels of the black circles leaves the value of the diagram unchanged. However, there is a subgroup of permutations that give rise to diagrams that are *topologically equivalent.* Two labelled diagrams are said to be topologically equivalent if they are characterized by the same set of *connections*, i.e. two circles labelled i and j in one diagram are linked by an η-bond if and only if they are similarly linked in the other. In terms of the integral that the diagrams represent, topological equivalence means that the integrand is unaltered by the relabelling. The subgroup of permutations that leave the connections the same is called the *graph group* of the diagram. In the case when all black circles are associated with the same function, the *symmetry number* of a simple diagram is the order of the graph group. We adopt the convention that when the word "diagram" or the symbol for a diagram appears in an equation, it is the value of that diagram which is referred to. Then, by definition, the value of a simple diagram Γ containing n white circles labelled 1 to n and m unlabelled, black circles is

$$\Gamma = (1/m!)\{\text{the sum of all topologically inequivalent diagrams obtained by attaching labels } n+1, \ldots, n+m \text{ to the black circles}\} \quad (4.3.4)$$

The number of diagrams appearing on the right-hand side of (4.3.4) is equal to $m!/S$, where S is the symmetry number. Each of these diagrams has the same value. The definition (4.3.4) may therefore be reformulated as

$$\Gamma = (1/S)\{\text{any one of the diagrams obtained by attaching labels } n+1, \ldots, n+m \text{ to the black circles}\}$$

$$= (1/S)\{\text{the value of the corresponding integral}\} \quad (4.3.5)$$

In order to illustrate the use of the definitions (4.3.4) and (4.3.5), we return to the examples given in (4.3.3a) to (4.3.3c). It follows from (4.3.4) that

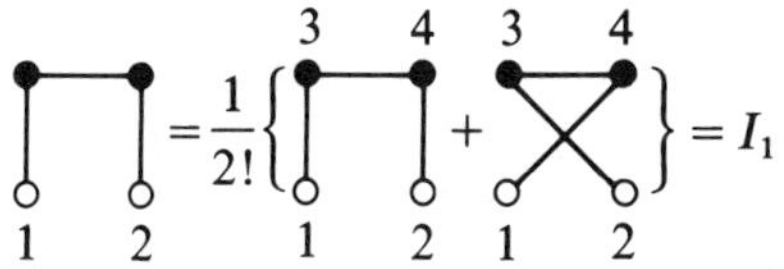

$$[\text{diagram; } 1, 2] = \frac{1}{2!}\{[\text{diagram; } 3, 4, 1, 2] + [\text{diagram; } 3, 4, 1, 2]\} = I_1 \quad (4.3.6a)$$

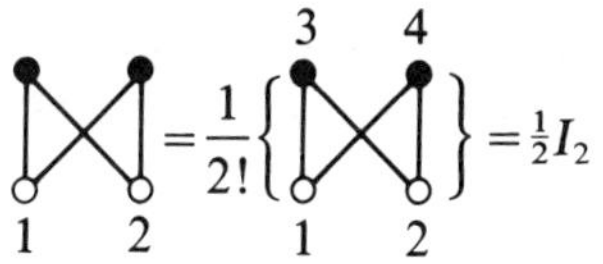

$$[\text{diagram; } 1, 2] = \frac{1}{2!}\{[\text{diagram; } 3, 4, 1, 2]\} = \tfrac{1}{2}I_2 \quad (4.3.6b)$$

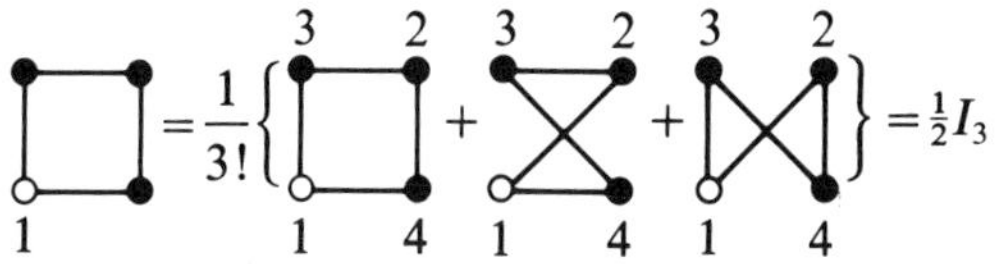

$$[\text{diagram; } 1] = \frac{1}{3!}\{[\text{diagram; } 3, 2, 1, 4] + [\text{diagram; } 3, 2, 1, 4] + [\text{diagram; } 3, 2, 1, 4]\} = \tfrac{1}{2}I_3 \quad (4.3.6c)$$

These results are also consistent with the definition (4.3.5), because the symmetry numbers of the three diagrams are, respectively, 1, 2 and 2. (In both the last two examples, the connections of the first labelled diagram are unaltered by interchange of the labels 3 and 4.)

The difference in the value of labelled and unlabelled diagrams is important because the greater ease with which unlabelled diagrams are manipulated is due precisely to the inclusion of the combinatorial factor S. In all that follows, use of the term "diagram" without qualification should be taken as referring to the unlabelled variety, though the distinction will often be irrelevant. Note, however, that in order to determine the symmetry number of a diagram it is first necessary to label it in an arbitrary way. Two unlabelled diagrams are *topologically distinct* if it is impossible to find a permutation that converts a labelled version of one diagram into a labelled version of the other. Diagrams that are topologically distinct represent different integrals, though they may accidentally have the same value. Statistical-mechanical quantities that are usefully discussed in diagrammatic terms are generally obtained as "the sum of all topologically distinct diagrams" having certain properties. To avoid undue repetition, we shall always replace the cumbersome phrase in quotation marks by the simpler expression "all diagrams".

The definition (4.3.5) can be extended to a wider class of diagrams than those we have discussed, but the definition of symmetry number may need to be modified. For example, if a diagram is *composite* rather than simple, the symmetry number is increased by a factor $n!$ for every pair of circles that are linked by n bonds of the same *species.* On the other hand, if the functions associated with the black circles are not all the same, the symmetry number is reduced.

Two circles are *adjacent* if they are linked directly by a bond. A sequence of adjacent circles is called a *path.* Two paths between a given pair of circles are *independent* if they have no intermediate circle in common. If a diagram is *disconnected,* it consists of two or more *components*; if two circles lie in different components, there is no path between them. A *connected* diagram is either *simply* or *multiply* connected. If there exist (at least) n independent paths between any pair of circles, the diagram is (at least) *n-tuply connected.* In the examples given below, diagram (a) is simply connected, (b) is triply connected, and (c) is a disconnected diagram with two doubly-connected components.

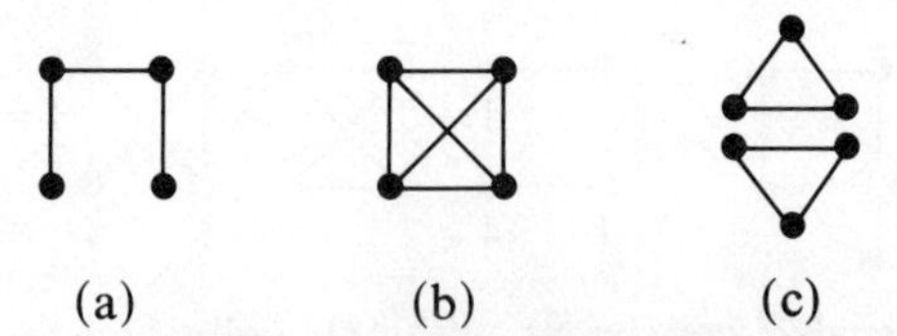

(a) (b) (c)

A bond is said to *intersect* the circles that it links. *Removal* of a circle from a diagram means that the circle and all bonds that intersect it are erased. If removal of a circle from a connected diagram causes the diagram to become disconnected, the circle is called a *connecting* circle. The *multiplicity* of a circle is the number of components into which the diagram separates when the circle is removed; a circle that is not a connecting circle has a multiplicity of one. Removal of an *articulation* circle from a connected diagram causes the diagram to separate into two or more components, of which at least one contains no white circle; an *articulation pair* is a pair of circles whose removal has the same effect. A diagram that is free of articulation circles is said to be *irreducible.* The absence of articulation pairs implies irreducibility, but not vice versa. If a diagram contains at least two white circles, a *nodal* circle is one through which all paths between two particular white circles pass. Clearly there can be no nodal circle associated with two white circles that are connected by a bond. A nodal circle is necessarily also a connecting circle; on its removal, the diagram separates into two or more components, the white circles in question appearing in different components. A nodal circle may also be an articulation circle if its multiplicity is three or more. In the examples shown below, the arrows point (a) to an articulation circle, (b) to an articulation pair, and (c) to a nodal circle.

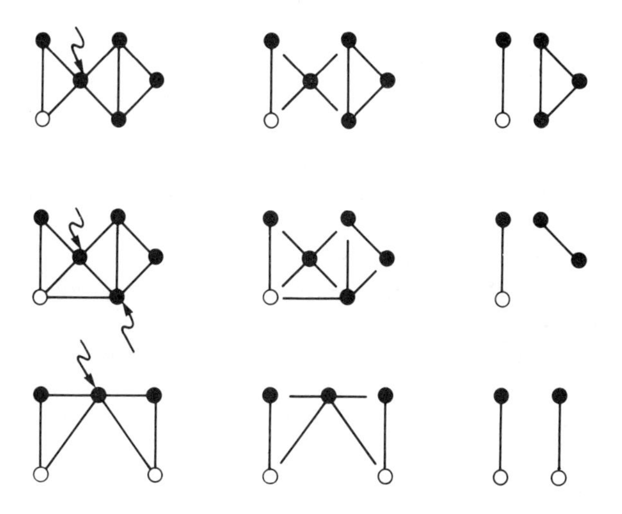

A *subdiagram* of a diagram Γ is any diagram that can be obtained from Γ by removal of circles or erasure of bonds, or by any combination of these operations. A subdiagram is *maximal* with respect to a given property if it is not embedded in any other subdiagram with the same property. Maximal subdiagrams are not necessarily unique, and "maximal" must not be interpreted in the sense of "largest". As a trivial example, the maximal connected subdiagrams of a disconnected diagram are just the components of the diagram. Classes of maximal subdiagrams that are of particular importance are those that are irreducible and those that are free of articulation pairs. The number of maximal, irreducible subdiagrams of a given diagram Γ is related to the multiplicities of the circles of Γ by a simple formula due to de Dominicis (1962).

Finally, there are two special classes of diagrams that are of importance in the theory of fluids. These are the *chain* and *ring* diagrams. In the examples below, (a) is a chain diagram with two *terminal* white circles, and (b) is a ring diagram containing only black circles.

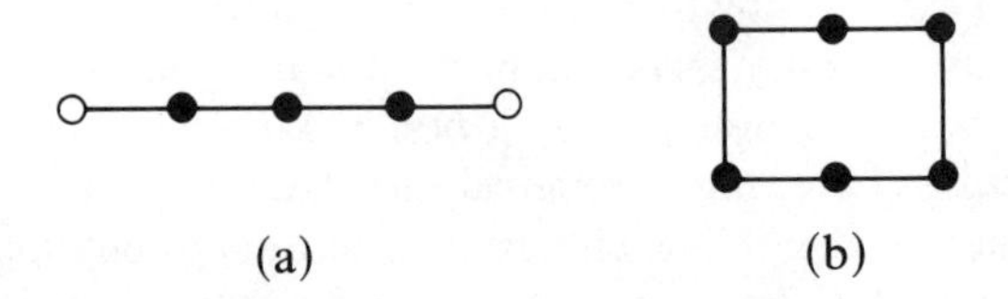

(a) (b)

A *composite chain* is a chain diagram in which circles are linked by multiple bonds. A simple chain diagram has a symmetry number of one; a simple ring diagram in which the circles are all black has a symmetry number equal to $2n$, where n is the number of circles.

4.4 FIVE FUNDAMENTAL LEMMAS

Diagrammatic expressions are manipulated with the aid of certain rules, the most important of which are contained in a series of lemmas derived by Morita and Hiroike (1961) and Stell (1964). In this section, the lemmas are presented without proof and illustrated by simple examples; some details of the proofs are given in Appendix A.

Statement of the first lemma requires the definition of the *star product* of two diagrams. Let Γ_1 be a connected diagram with n_1 white 1-circles and let Γ_2 be a connected and possibly identical diagram with n_2 white 1-circles. The star product $\Gamma_3 = \Gamma_1 * \Gamma_2$ is the diagram obtained by linking together Γ_2 and Γ_2 in such a way that the white circles carrying the same labels coincide. The diagram Γ_3 contains $(n_1 + n_2 - n_3)$ white 1-circles, where n_3 is the number of labels common to both Γ_1 and Γ_2, and Γ_1 and Γ_2 are said to be *connected in parallel* at the n_3 white circles. If Γ_1 and Γ_2 are connected in parallel at white γ-circles, the corresponding circles in Γ_3 are white γ^2-circles. The formation of a star product is illustrated by the following example:

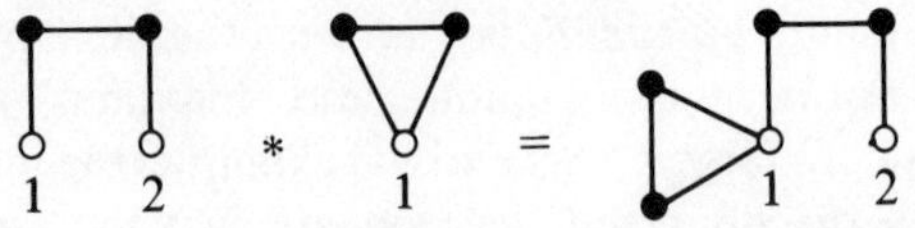

If the white circles of Γ_1 and Γ_2 have no label in common, or if one diagram contains only black circles, the star product is a disconnected diagram having Γ_1 and Γ_2 as its components. *Star-irreducible* diagrams are connnected diagrams that cannot be expressed as the star product of two other diagrams except when one of the two is the diagram consisting of a single white circle. The star product of two star-irreducible diagrams can be uniquely decomposed into the factors that form the product. Star-irreducible diagrams are therefore analogous to prime numbers. The definition of star-irreducibility excludes all diagrams containing white connecting circles or connecting subsets, all diagrams with adjacent white circles and, by convention, the diagram consisting of a single white circle.

Lemma 1. *Let G be a set of topologically distinct, star-irreducible diagrams and let H be the set of all diagrams in G and all possible star products of diagrams in G. Then*

$$\{\textit{all diagrams in } H\} = \exp\{\textit{all diagrams in } G\} - 1$$

Illustration. If G consists of the single diagram $\Gamma =$ ○—●, then

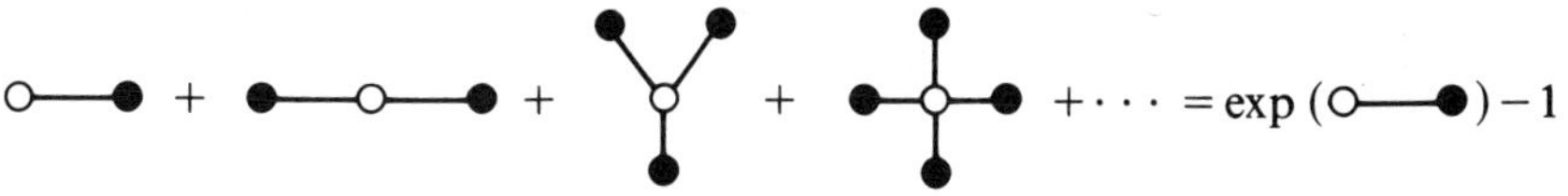

Lemma 1 is often called the "exponentiation theorem". If the diagrams in G consist wholly of black circles and bonds, use of the lemma allows us to express a sum of connected and disconnected diagrams in terms of the subset that are connected.

Lemmas 2 and 3 contain useful rules for the evaluation of certain important functional derivatives.

Lemma 2. *Let* Γ *be a diagram consisting of black* γ*-circles and bonds. Then*

$$\frac{\delta\Gamma}{\delta\gamma(1)} = \{\textit{all diagrams obtained by replacing a black } \gamma\textit{-circle of } \Gamma \textit{ by a white } 1\textit{-circle labelled } 1\}$$

Illustration.

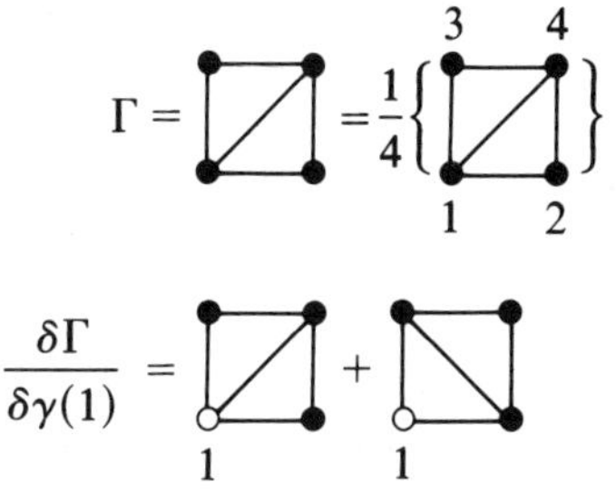

Application of Lemma 2 provides an easy way to derive Eqn (4.2.29). The extension to higher-order derivatives is straightforward.

Lemma 3. *Let* Γ *be a simple diagram consisting of black circles and* η*-bonds. Then*

$$\frac{\delta\Gamma}{\delta\eta(1,2)} = \tfrac{1}{2}\{\textit{all diagrams obtained by erasing an } \eta\textit{-bond of } \Gamma, \textit{ whitening the circles that it intersected, and labelling the whitened circles } 1 \textit{ and } 2\}$$

Illustration.

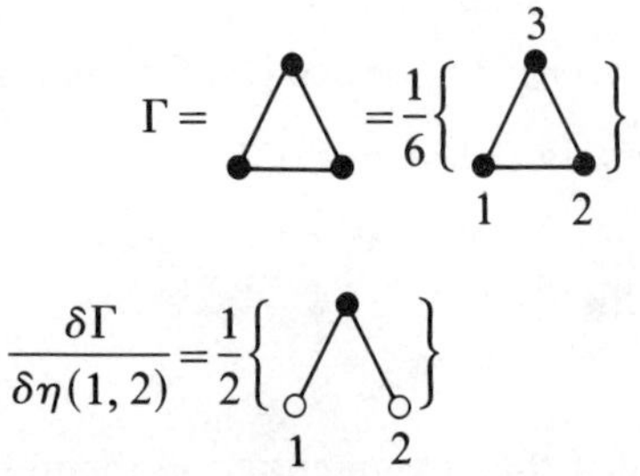

The example illustrated is the diagrammatic representation of Eqn (4.2.18); the numerical factor present in (4.2.18) is taken care of by the different symmetry numbers of the diagram before and after differentiation.

Lemmas 4 and 5 are useful in the process of *topological reduction.*

Lemma 4. *Let* G *be a set of topologically distinct, connected diagrams consisting of a white* γ*-circle labelled* $\mathbf{r}$*, black* γ*-circles and bonds, and let* $\mathscr{G}(\mathbf{r})$ *be the sum of all diagrams in* G*. If* Γ *is a connected diagram, if* H *is the set of all topologically distinct diagrams obtained by decorating all black circles of* Γ *with diagrams in* G*, and if each diagram in* H *is uniquely decomposable, then*

$$\{\textit{all diagrams in } H\} = \{\textit{the diagram obtained from } \Gamma \textit{ by associating the function } \mathscr{G} \textit{ with each of the black circles}\}$$

The process of *decorating* the diagram Γ consists in attaching one of the elements in G such that its white circle is superimposed on a black circle of Γ and then blackened. For the diagrams in H to be *uniquely decomposable* it must be possible, given the structure of Γ, to determine by inspection which diagram in G has been used to decorate each black circle of Γ; this can always be done if Γ is free of black articulation circles.

Illustration. If the set G consists of the two diagrams

and

$(S=1)$ $(S=2)$

and if

$\Gamma =$ $(S=2)$

then the set H consists of the three diagrams

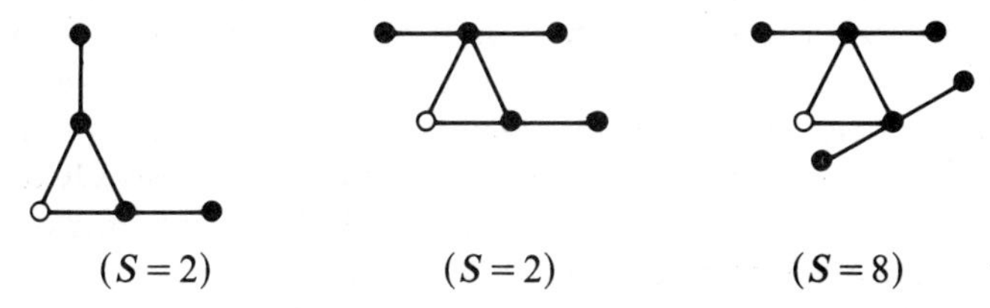

The illustration is a simple one, but it contains the main ingredients of a topological reduction: the sum of a number of diagrams is replaced by a single diagram of simpler structure than those in the original set.

Lemma 5. *Let G be a set of topologically distinct, simple, connected diagrams consisting of two white circles labelled* $\mathbf{r}$ *and* $\mathbf{r}'$, *black circles and bonds, and let $\mathscr{G}(\mathbf{r}, \mathbf{r}')$ be the sum of all diagrams in G. If Γ is a connected diagram containing at least one bond, if H is the set of all topologically distinct diagrams obtained by replacing all bonds in Γ by diagrams in G, and if each diagram in H is uniquely decomposable, then*

{*all diagrams in H*} = {*the diagram obtained from Γ by associating the function $\mathscr{G}$ with each of the bonds*}

Replacement of bonds in Γ involves superimposing the two white circles of the diagram drawn from G onto the circles of Γ that the bond intersects and erasing the bond between them. The circles take the same colour and, if white, the same label as the corresponding circle in Γ.

Illustration. If the set G consists of the two diagrams

and if

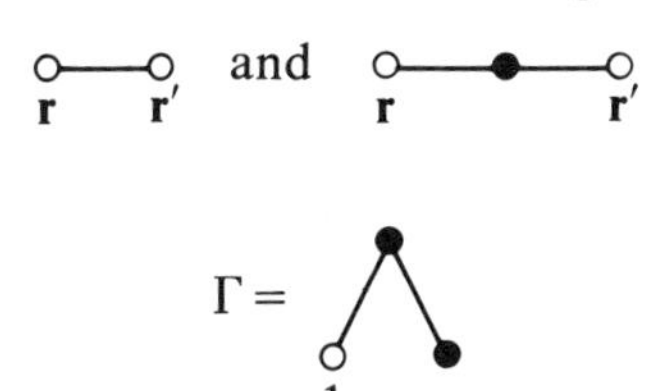

then the set H consists of the three diagrams

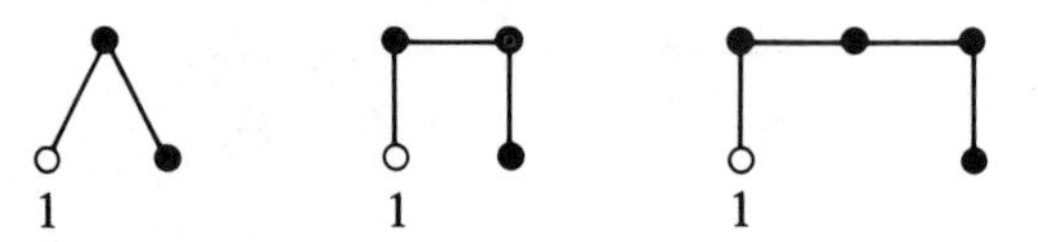

4.5 THE VIRIAL EXPANSION OF THE EQUATION OF STATE

We now give an example of how the definitions and lemmas of the two previous sections can be used to obtain results of physical interest. The example we choose is the classic problem of the density expansion of the equation of state of a one-component, monatomic fluid. The derivation given is for a system of particles interacting through pairwise-additive forces and placed in an external field, but the final expressions take a simpler form if the system is homogeneous.

It follows directly from the definition (4.2.26) of the grand partition function in the presence of an external field that Ξ can be represented diagrammatically as

$$\Xi = 1 + \{\text{all diagrams consisting of black } z^*\text{-circles with an } e\text{-bond linking each pair of circles}\}$$

$$= 1 + \quad + \quad + \quad + \quad + \cdots \qquad (4.5.1)$$

Note that the definition of the value of a diagram takes care of the factors $1/N!$ in Eqn (4.2.26).

The expansion (4.5.1) is not immediately useful. Because $e(i, j) \to 1$ as $|\mathbf{r}_i - \mathbf{r}_j| \to \infty$, the contribution from the Nth term is of order V^N, and problems arise in the thermodynamic limit. It is therefore better to reformulate the series in terms of f-bonds by making the substitution $e(i, j) = f(i, j) + 1$. The expansion then becomes

$$\Xi = 1 + \{\text{all diagrams consisting of black } z^*\text{-circles and at most one } f\text{-bond between each pair of circles}\}$$

$$= 1 + \quad + \quad + \quad + \quad + \quad + \quad + \quad + \cdots \qquad (4.5.2)$$

The disconnected diagrams in (4.5.2) can be eliminated by taking the logarithm of Ξ and applying Lemma 1:

$$\log \Xi = \{\text{all simple, connected diagrams consisting of black } z^*\text{-circles and } f\text{-bonds}\}$$

$$= \bullet + \bullet\!\!-\!\!\bullet + \Lambda + \Delta + \cdots \quad (4.5.3)$$

If the system is homogeneous (no external field), $z^*(\mathbf{r})$ is simply the activity z. In that case, combination of Eqns (2.4.13) and (4.5.3) yields the expansion of the equation of state in powers of z:

$$\beta P = \sum_{i=1}^{\infty} b_i z^i \quad (4.5.4)$$

where

$$b_i = (1/V)\{\text{all simple, connected diagrams consisting of } i \text{ black 1-circles and } f\text{-bonds}\} \quad (4.5.5)$$

The coefficients b_i for $i = 1$ to 4 are

$$b_1 = \frac{1}{V}\{\bullet\} = \frac{1}{V}\int \mathrm{d}\mathbf{r} = 1$$

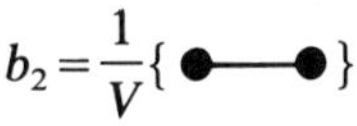

$$b_2 = \frac{1}{V}\{\bullet\!\!-\!\!\bullet\} = \tfrac{1}{2}\int f(\mathbf{r})\,\mathrm{d}\mathbf{r}$$

$$b_3 = \frac{1}{V}\{\Lambda + \Delta\}$$

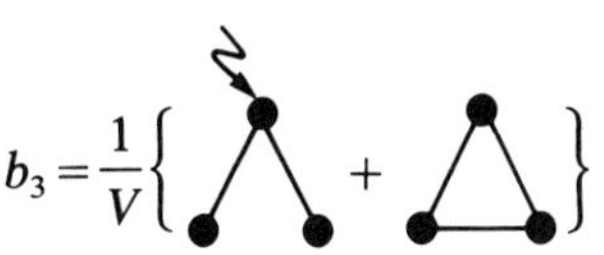

$$= \tfrac{1}{2}\iint f(1,2)f(1,3)\,\mathrm{d}2\,\mathrm{d}3 + \tfrac{1}{6}\iiint f(1,2)f(1,3)f(2,3)\,\mathrm{d}2\,\mathrm{d}3$$

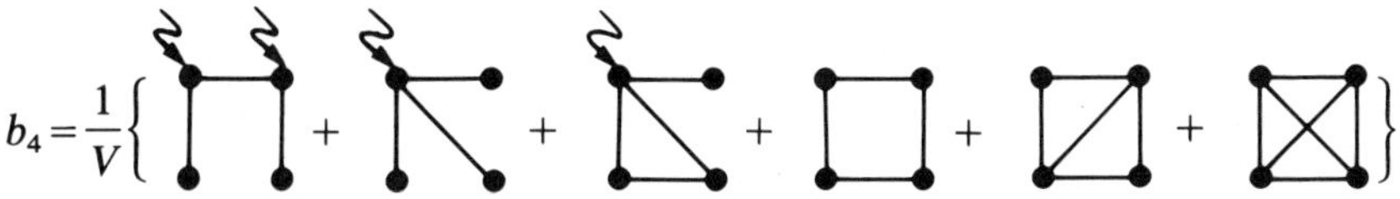

$$b_4 = \frac{1}{V}\{\;[\text{diagram}] + [\text{diagram}] + [\text{diagram}] + \square + \boxslash + \boxtimes\;\}$$

Some of the diagrams in (4.5.5) contain articulation circles. In the case of b_3 and b_4, these are indicated by arrows. Because the system is assumed to be translationally invariant, the articulation circles can be chosen as the origin of coordinates in the corresponding integrals. If this is done, the integrals factorize as products of integrals that have already appeared at lower order in the expansion. For example, the first diagram in the expression for b_3 given above is equal to $2b_2^2$, and the first, second and third diagrams in b_4 are equal, respectively, to $4b_2^3$, $4b_2^3/3$ and $6(b_3-2b_2^2)b_2$.

Elimination of the diagrams that contain articulation circles would clearly lead to a simplification of the problem. The elimination is achieved by switching from an activity to a density expansion and requires, as an intermediate step, the activity expansion of the single-particle density. If the expansion (4.5.3) is inserted in Eqn (4.2.30), it follows from Lemma 2 that

$$\rho^{(1)}(\mathbf{r}) = z^*(\mathbf{r})\frac{\delta \log \Xi}{\delta z^*(\mathbf{r})}$$

$$= z^*(\mathbf{r})[1+\{\text{all simple, connected diagrams consisting of one white 1-circle labelled } \mathbf{r}, \text{ one or more black } z^*\text{-circles, and } f\text{-bonds}\}] \quad (4.5.6)$$

If the system is homogeneous, Eqn (4.5.6) takes the simpler form

$$\rho = \sum_{i=1}^{\infty} i b_i z^i \quad (4.5.7)$$

where the b_i are the coefficients defined by Eqn (4.5.5). The diagrams in (4.5.6) fall into two classes: those in which the white circle is an articulation circle and those in which it is not and are therefore star-irreducible. The first of these classes is just the set of all diagrams that can be expressed as star products of diagrams in the second class. An application of Lemma 1 shows that

$$\log \rho^{(1)}(\mathbf{r}) = \log z^*(\mathbf{r}) + \{\text{all simple, connected diagrams consisting of one white 1-circle labelled } \mathbf{r}, \text{ one or more } z^*\text{-circles, and } f\text{-bonds, such that the white circle is not an articulation circle}\} \quad (4.5.8)$$

The diagrams in (4.5.8) are all star-irreducible, but some contain black articulation circles. To eliminate the latter, we proceed as follows. For each diagram Γ in (4.5.8), we identify a maximal, irreducible subdiagram Γ_m that contains the single white circle.

Illustration.

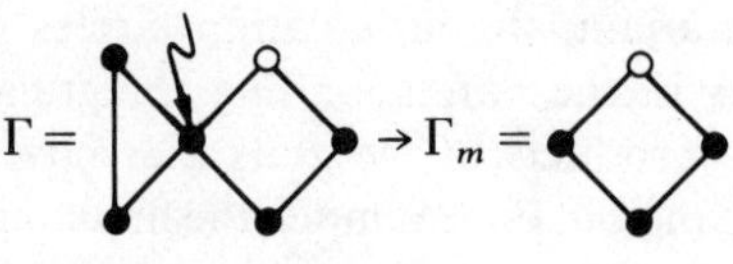

$\Gamma =$ [diagram] $\rightarrow \Gamma_m =$ [diagram]

In the example shown, there is one articulation circle (denoted by the arrow) and there are two maximal, irreducible subdiagrams, only one of which contains the white circle. It is readily proved (McDonald and O'Gorman, 1978) that for each Γ there is a unique choice of Γ_m; if Γ itself is irreducible, Γ and Γ_m are the same. The set $\{\Gamma_m\}$ is a subset of the diagrams in (4.5.8). Given any Γ_m, the diagram from which it derives can be reconstructed by decorating the black circles with diagrams taken from the set (4.5.6). Lemma 4 can therefore be used in a topological reduction whereby the z^*-circles in (4.5.8) are replaced by $\rho^{(1)}$-circles and the black articulation circles disappear. Thus

$$\log[\rho^{(1)}(\mathbf{r})/z^*(\mathbf{r})] = \{\text{all simple, irreducible diagrams consisting of one white 1-circle labelled } \mathbf{r}, \text{ one or more black } \rho^{(1)}\text{-circles, and } f\text{-bonds}\} \tag{4.5.9}$$

In the homogeneous case, Eqn (4.5.9) becomes

$$\log z = \log \rho - \sum_{i=1}^{\infty} \beta_i \rho^i \tag{4.5.10}$$

The coefficients β_i are the irreducible Mayer *cluster integrals* (Uhlenbeck and Ford, 1962); they are represented by diagrams that are irreducible, i.e. free of articulation circles of either colour. The coefficients for $i = 1$ to 3 are

$$\beta_1 = \int f(0,1)\, \mathrm{d}1 \tag{4.5.11}$$

$$\beta_2 = \tfrac{1}{2} \iint f(0,1) f(0,2) f(1,2)\, \mathrm{d}1\, \mathrm{d}2 \tag{4.5.12}$$

$$\beta_3 = \quad + \quad + \quad + \tag{4.5.13}$$

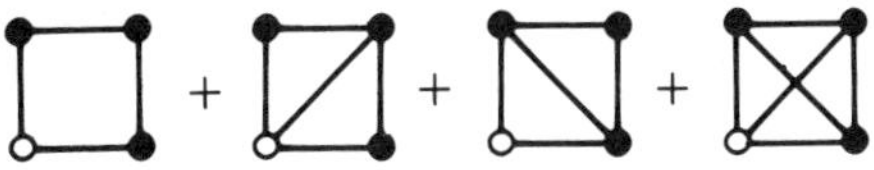

where, in each case, the white circle is labelled 0. Because the system is homogeneous, the coefficients are independent of the label attached to the white circle, and the second and third diagrams in the expression for β_3 are equal in value. In the presence of an external field, the coefficients depend explicitly on the coordinates of the white circle.

Equations (4.5.9) and (4.5.10) express the activity as a function of the density. The final step is to eliminate the activity between Eqns (4.5.3) and (4.5.9), if the system is in an external field, or between Eqns (4.5.4) and (4.5.10) if there is translational invariance. In the homogeneous case, starting from (2.4.13), we find successively that

$$\beta PV = \log \Xi$$

$$\beta V \frac{\mathrm{d}P}{\mathrm{d}z} = \frac{1}{\Xi} \frac{\mathrm{d}\Xi}{\mathrm{d}z} = \frac{1}{\Xi} \sum_{N=0}^{\infty} \frac{N z^{N-1}}{N!} Z_N = \frac{1}{z} \langle N \rangle$$

$$\beta \frac{\mathrm{d}P}{\mathrm{d}z} = \beta \frac{\mathrm{d}P}{\mathrm{d}\rho} \frac{\mathrm{d}\rho}{\mathrm{d}z} = \frac{1}{z} \frac{\langle N \rangle}{V} = \frac{\rho}{z}$$

$$\beta \frac{\mathrm{d}P}{\mathrm{d}\rho} = \frac{\rho}{z} \frac{\mathrm{d}z}{\mathrm{d}\rho}$$

$$\beta P = \int_0^{\rho} \rho' \frac{\mathrm{d}(\log z)}{\mathrm{d}\rho'} \mathrm{d}\rho'$$

and hence, from Eqn (4.5.10):

$$\beta P = \rho - \sum_{i=1}^{\infty} \frac{i}{i+1} \beta_i \rho^{i+1} \tag{4.5.14}$$

If the virial coefficients B_i are defined as

$$\left.\begin{aligned} B_1 &= 1 \\ B_{i+1} &= -\frac{i}{i+1} \beta_i, \qquad i \geq 1 \end{aligned}\right\} \tag{4.5.15}$$

we recover the standard form of the virial expansion given by Eqn (4.1.1). In particular, we find that $B_2 = -\frac{1}{2}\beta_1$, in agreement with (4.1.5). Note that the short-range character of the f-function means that all β_i, and hence all B_i, are independent of volume and functions only of temperature for a given $v(r)$.

The last step in the derivation is only slightly more complicated in the presence of an external field. This less common case is treated in detail by Morita and Hiroike (1961).

4.6 THE EQUATION OF STATE OF THE HARD-SPHERE FLUID

It is clear from the definition of the virial coefficients that the number of diagrams that contribute to the ith coefficient grows rapidly with i, while the associated integrals become increasingly more complicated. For example, the numbers of diagrams entering the expressions for B_3, B_4, B_5, B_6 and B_7 are, respectively, 1, 3, 10, 56 and 468, while the dimensions of the integrals increase each time by three. Not surprisingly, therefore, explicit calculations have been confined to the low-order coefficients.

For hard spheres, the second, third and fourth virial coefficients are known analytically. If d is the hard-sphere diameter, then $f(r) = -1$ for $r < d$ and $f(r) = 0$ for $r > d$. Thus

$$B_2 = -\tfrac{1}{2}\beta_1 = (-\tfrac{1}{2})4\pi \int_0^d (-1)r^2\,\mathrm{d}r = \tfrac{2}{3}\pi d^3 \tag{4.6.1}$$

$$B_3 = -\tfrac{2}{3}\beta_2 = -\tfrac{1}{3}\iint f(r)f(r')f(|\mathbf{r}-\mathbf{r}'|)\,\mathrm{d}\mathbf{r}\,\mathrm{d}\mathbf{r}' \tag{4.6.2}$$

By choosing d as the unit of length, Eqn (4.6.2) may be rewritten as

$$B_3 = -\tfrac{1}{3}d^6 \int f(r)\xi(r)\,\mathrm{d}\mathbf{r} \tag{4.6.3}$$

where now $f(r) = -1$ for $r < 1$ and $f(r) = 0$ for $r > 1$. The integral

$$\xi(r) = \int f(r')f(|\mathbf{r}-\mathbf{r}'|)\,\mathrm{d}\mathbf{r}' \tag{4.6.4}$$

represents the volume of overlap of two spheres of unit radius with centres separated by r. The integral is most easily evaluated by transforming to bipolar coordinates, by which means it is found that

$$\begin{aligned} \xi(r) &= \frac{4\pi}{3}\left(1 - \frac{3r}{4} + \frac{r^3}{16}\right), \qquad r<2 \\ &= 0, \qquad\qquad\qquad\qquad\quad r>2 \end{aligned} \tag{4.6.5}$$

When this result is substituted into Eqn (4.6.3), B_3 becomes

$$\begin{aligned} B_3 &= \frac{4\pi d^6}{3}\int_0^1 f(r)\xi(r)r^2\,\mathrm{d}r \\ &= \frac{5\pi^2 d^6}{18} \end{aligned} \tag{4.6.6}$$

The calculation of the fourth virial coefficient is a mathematical *tour de force* due to Boltzmann (Rowlinson, 1965). The final expression has the form

$$\begin{aligned} B_4/B_2^3 &= -\frac{89}{280} + \frac{219\sqrt{2}}{2240\pi} + \frac{4131}{2240\pi}\arccos(1/\sqrt{3}) \\ &= 0.28695\ldots \end{aligned} \tag{4.6.7}$$

The fifth, sixth and seventh virial coefficients have been evaluated numerically by a Monte Carlo method (Ree and Hoover, 1964, 1967; Kim and Henderson, 1968). The available results for hard spheres are summarized in Table 4.2. Surprisingly, the pressures calculated from the truncated,

TABLE 4.2. *Exact virial coefficients of hard spheres*

$B_2 = b$	$\frac{2}{3}\pi d^3$
B_3/b^2	$\frac{5}{8}$
B_4/b^3	0.28695
B_5/b^4	0.1103 ± 0.0003
B_6/b^5	0.0386 ± 0.0004
B_7/b^6	0.0138 ± 0.0004

seven-term series are in good agreement with the results of computer simulations at all densities up to the fluid–solid transition, as illustrated in Figure 4.2; the phase transition occurs at $\rho \simeq 0.67\rho_0$, where $\rho_0 = \sqrt{2}/d^3$ is the close-packed density. Lebowitz and Penrose (1964) have established a rigorous lower bound on the radius of convergence of the virial expansion.

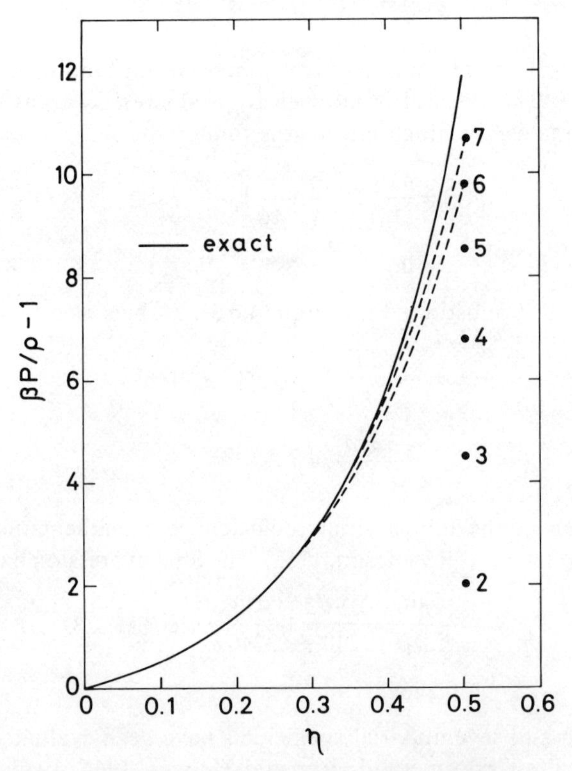

FIG. 4.2. Equation of state of hard spheres calculated from the virial series and from the nearly-exact results of Eqn (4.6.14). The numbered points show the values calculated for $\eta = 0.5$ when different numbers of virial coefficients are included.

For hard spheres, the lower bound is $0.026\rho_0$, but in view of the good results obtained with the seven-term series it seems likely that the true radius of convergence is much larger than this.

If we define the *packing fraction* η as

$$\eta = \tfrac{1}{6}\pi\rho d^3 \tag{4.6.8}$$

the virial expansion for hard spheres can be rewritten as

$$\frac{\beta P}{\rho} = 1 + \sum_{i=1}^{\infty} \mathscr{B}_i \eta^i \tag{4.6.9}$$

with

$$\mathscr{B}_i = \left(\frac{6}{\pi d^3}\right)^i B_{i+1} \tag{4.6.10}$$

The seven-term series is now

$$\frac{\beta P}{\rho} = 1 + 4\eta + 10\eta^2 + 18.365\eta^3 + 28.24\eta^4 + 39.5\eta^5 + 56.5\eta^6 + \cdots \tag{4.6.11}$$

Guided by the form of Eqn (4.6.11), Carnahan and Starling (1969) were able to devise a simple but very accurate hard-sphere equation of state. Noting that $\mathscr{B}_1$ and $\mathscr{B}_2$ are both integers, they chose to replace $\mathscr{B}_3$ by the closest integer, 18, and supposed that $\mathscr{B}_i$ for all i is given by

$$\mathscr{B}_i = a_i i^2 + a_2 i + a_3 \tag{4.6.12}$$

With $\mathscr{B}_1 = 4$, $\mathscr{B}_2 = 10$ and $\mathscr{B}_3 = 18$, the solution to (4.6.12) is $a_1 = 1$, $a_2 = 3$ and $a_3 = 0$. The formula then predicts that $\mathscr{B}_4 = 28$, $\mathscr{B}_5 = 40$ and $\mathscr{B}_6 = 54$, in close agreement with the exact coefficients in Eqn (4.6.11). The expression

$$\frac{\beta P}{\rho} = 1 + \sum_{i=1}^{\infty} (i^2 + 3i)\eta^i \tag{4.6.13}$$

may be written as a linear combination of the first and second derivatives of the geometric series $\sum_{i=1}^{\infty} \eta^i$. It can therefore be summed explicitly to give

$$\frac{\beta P}{\rho} = \frac{1 + \eta + \eta^2 - \eta^3}{(1-\eta)^3} \tag{4.6.14}$$

The Carnahan–Starling equation of state provides an excellent fit to the results of computer simulations over the entire fluid range. Its simple form also makes it convenient for use in thermodynamic calculations. In particular, a closed expression for the Helmholtz free energy is obtained by combining (4.6.14) with the thermodynamic relation (2.3.5):

$$\frac{\beta A^{\text{ex}}}{N} = \int_0^\eta \left(\frac{\beta P}{\rho} - 1\right)\frac{\mathrm{d}\eta'}{\eta'}$$

$$= \frac{\eta(4-3\eta)}{(1-\eta)^2} \tag{4.6.15}$$

The Carnahan–Starling equation of state is widely used in perturbation theories of the type to be discussed in Chapter 6. Another simple but accurate hard-sphere equation of state has been proposed by Shinomoto (1983).

CHAPTER 5

Distribution Function Theories

The greater part of this chapter is devoted to some of the more important theoretical methods used in calculating the pair distribution function of a fluid from a given intermolecular pair potential. Once the pair distribution function is known, thermodynamic properties of the system can be obtained via the energy equation (2.5.13), the virial equation (2.5.14) or the compressibility equation (2.5.32). We begin, however, by describing the way in which the pair distribution function can be measured experimentally.

5.1 THE STATIC STRUCTURE FACTOR

The pair distribution function introduced in Section 2.5 may conveniently be re-expressed in terms of the *local particle density*. The density at a point $\mathbf{r}$ is defined as

$$\rho(\mathbf{r}) = \sum_{i=1}^{N} \delta(\mathbf{r}-\mathbf{r}_i) \tag{5.1.1}$$

From Eqn (2.5.34), we see that the average density at $\mathbf{r}$ is

$$\langle\rho(\mathbf{r})\rangle = \rho^{(1)}(\mathbf{r}) \tag{5.1.2}$$

or, in a homogeneous system (see Eqn (2.5.4)):

$$\langle\rho(\mathbf{r})\rangle = \rho \tag{5.1.3}$$

The correlation between the densities at two points separated by $\mathbf{r}$ in a homogeneous fluid is described by a static density-density autocorrelation function $G(\mathbf{r})$, defined as

$$G(\mathbf{r}) = \frac{1}{N}\int \langle\rho(\mathbf{r}'+\mathbf{r})\rho(\mathbf{r}')\rangle \, d\mathbf{r}'$$

$$=\frac{1}{N}\left\langle\int\sum_{i\neq j}^{N}\sum\delta(\mathbf{r}'+\mathbf{r}-\mathbf{r}_i)\delta(\mathbf{r}'-\mathbf{r}_j)\,d\mathbf{r}'\right\rangle+\delta(\mathbf{r})$$

$$=\frac{1}{N}\left\langle\sum_{i\neq j}^{N}\sum\delta(\mathbf{r}+\mathbf{r}_j-\mathbf{r}_i)\right\rangle+\delta(\mathbf{r})$$

$$=\rho g(\mathbf{r})+\delta(\mathbf{r}) \tag{5.1.4}$$

where we have used the expression (2.5.37) for $g(\mathbf{r})$; $\delta(\mathbf{r})$ represents the contribution from the terms for which $i=j$. The time-dependent generalization of $G(\mathbf{r})$ is called the *van Hove function*; this plays an important role in the description of the dynamical properties of liquids, as we shall see in Chapter 7.

The Fourier components of the density $\rho(\mathbf{r})$ are

$$\rho_{\mathbf{k}}=\int\exp(-i\mathbf{k}\cdot\mathbf{r})\rho(\mathbf{r})\,d\mathbf{r}$$

$$=\sum_{i=1}^{N}\exp(-i\mathbf{k}\cdot\mathbf{r}_i) \tag{5.1.5}$$

with an autocorrelation function defined as

$$S(\mathbf{k})=\frac{1}{N}\langle\rho_{\mathbf{k}}\rho_{-\mathbf{k}}\rangle \tag{5.1.6}$$

The function $S(\mathbf{k})$ is called the *static structure factor.* In the case of a homogeneous fluid, the structure factor can be written, with the help of (2.5.8) and (2.5.36), as the Fourier transform of $g(\mathbf{r})$:

$$S(\mathbf{k})=\frac{1}{N}\left\langle\sum_{i=1}^{N}\sum_{j=1}^{N}\exp(-i\mathbf{k}\cdot\mathbf{r}_i)\exp(i\mathbf{k}\cdot\mathbf{r}_j)\right\rangle$$

$$=1+\frac{1}{N}\left\langle\sum_{i\neq j}^{N}\sum\exp[-i\mathbf{k}\cdot(\mathbf{r}_i-\mathbf{r}_j)]\right\rangle$$

$$=1+\frac{1}{N}\left\langle\sum_{i\neq j}^{N}\sum\iint\exp[-i\mathbf{k}\cdot(\mathbf{r}-\mathbf{r}')]\delta(\mathbf{r}-\mathbf{r}_i)\delta(\mathbf{r}'-\mathbf{r}_j)\,d\mathbf{r}\,d\mathbf{r}'\right\rangle$$

$$=1+\frac{1}{N}\iint\exp[-i\mathbf{k}\cdot(\mathbf{r}-\mathbf{r}')]\rho^{(2)}(\mathbf{r},\mathbf{r}')\,d\mathbf{r}\,d\mathbf{r}'$$

$$=1+\rho\int\exp(-i\mathbf{k}\cdot\mathbf{r})g(\mathbf{r})\,d\mathbf{r} \tag{5.1.7}$$

Conversely, $g(\mathbf{r})$ is the Fourier transform of $S(\mathbf{k})$:

$$\rho g(\mathbf{r})=\frac{1}{(2\pi)^3}\int\exp(i\mathbf{k}\cdot\mathbf{r})[S(\mathbf{k})-1]\,d\mathbf{k} \tag{5.1.8}$$

If the system is also isotropic, $S(\mathbf{k})$ is a function only of the wavenumber $k = |\mathbf{k}|$. Thus

$$S(k) = 1 + 2\pi\rho \int r^2 g(r) \int_{-1}^{1} \exp(-ikr\cos\theta)\, \mathrm{d}(\cos\theta)\, \mathrm{d}r$$

$$= 1 + 4\pi\rho \int r^2 g(r) \frac{\sin kr}{kr}\, \mathrm{d}r \tag{5.1.9}$$

Equation (5.1.7) can also be rewritten in the form

$$S(\mathbf{k}) = 1 + (2\pi)^3 \rho\delta(\mathbf{k}) + \rho\hat{h}(\mathbf{k}) \tag{5.1.10}$$

where

$$\hat{h}(\mathbf{k}) = \int \exp(-i\mathbf{k}\cdot\mathbf{r}) h(\mathbf{r})\, \mathrm{d}\mathbf{r} \tag{5.1.11}$$

and $h(\mathbf{r}) = g(\mathbf{r}) - 1$ is the translationally invariant form of the pair correlation function defined by Eqn (2.5.27). Henceforth, we shall ignore the delta-function term in Eqn (5.1.10), which corresponds experimentally (see below) to the forward scattering of radiation, and the relation between $S(\mathbf{k})$ and $\hat{h}(\mathbf{k})$ will be taken simply as

$$S(\mathbf{k}) = 1 + \rho\hat{h}(\mathbf{k}) \tag{5.1.12}$$

It follows from the compressibility equation (2.5.32) that the limit of $S(\mathbf{k})$ for $k \to 0$ is

$$S(0) = \rho k_\mathrm{B} T \chi_T = \chi_T / \chi_T^0. \tag{5.1.13}$$

where $\chi_T^0 = \beta/\rho$ is the compressibility of the ideal gas.

The structure factor of a fluid can be obtained experimentally from measurements of the cross-section for scattering of neutrons or X-rays as a function of scattering angle. Below we give a simplified treatment of the calculation of the neutron cross-section in terms of $S(\mathbf{k})$.

Consider a neutron that is scattered by the sample through an angle θ. The incoming neutron can be represented as a plane wave, i.e.

$$\psi_1(\mathbf{r}) = \exp(i\mathbf{k}_1\cdot\mathbf{r}) \tag{5.1.14}$$

while at sufficiently large distances from the sample the scattered neutron has the form of a spherical wave:

$$\psi_2(\mathbf{r}) \sim \frac{\exp(ik_2 r)}{r} \tag{5.1.15}$$

Thus, asymptotically, the wave function of the neutron behaves as

$$\psi(\mathbf{r}) \underset{\mathbf{r}\to\infty}{\sim} \exp(i\mathbf{k}_1\cdot\mathbf{r}) + f(\theta)\frac{\exp(ik_2 r)}{r} \tag{5.1.16}$$

and the intensity of the scattered component determines the differential cross-section $d\sigma/d\Omega$ for scattering into a solid angle $d\Omega$ in the direction θ, ϕ:

$$\frac{d\sigma}{d\Omega} = |f(\theta)|^2 \tag{5.1.17}$$

As Eqn (5.1.16) shows, $d\sigma/d\Omega$ has the dimensions of an area.

The geometry of the scattering event is illustrated in Figure 5.1. The momentum transfer from sample to neutron (in units of $\hbar$) is

$$\mathbf{k} = \mathbf{k}_2 - \mathbf{k}_1 \tag{5.1.18}$$

In order to simplify the calculation,we shall assume that the scattering is elastic. Then $|\mathbf{k}_1| = |\mathbf{k}_2|$ and

$$k = 2k_1 \sin \tfrac{1}{2}\theta = \frac{4\pi \sin \frac{1}{2}\theta}{\lambda} \tag{5.1.19}$$

where λ is the neutron wavelength.

The scattering of the neutron occurs as the result of interactions with the atomic nuclei. These interactions are very short ranged, and the total scattering potential $\mathcal{V}(\mathbf{r})$ may therefore be approximated by a sum of delta-function pseudopotentials in the form

$$\mathcal{V}(\mathbf{r}) = \frac{2\pi\hbar^2}{m} \sum_{i=1}^{N} b_i \delta(\mathbf{r} - \mathbf{r}_i) \tag{5.1.20}$$

where b_i is the *scattering length* of the ith nucleus (Squires, 1978). For most nuclei, b_i is positive, but it can also be negative and even complex; it varies both with isotopic species and with the spin state of the nucleus.

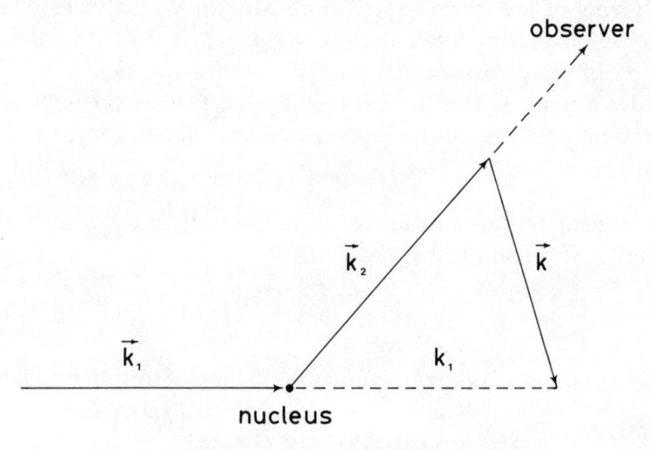

FIG. 5.1. Geometry of a neutron-scattering event.

The wave function $\psi(\mathbf{r})$ must also be a solution of the Schrödinger equation:

$$\left(-\frac{\hbar^2}{2m}\nabla^2+\mathcal{V}(\mathbf{r})\right)\psi(\mathbf{r})=E\psi(\mathbf{r}) \tag{5.1.21}$$

The general solution (Shankar, 1980) that satisfies the boundary condition (5.1.16) in the case when the incoming neutron is described by a plane wave is

$$\psi(\mathbf{r})=\exp(i\mathbf{k}_1\cdot\mathbf{r})-\frac{m}{2\pi\hbar^2}\int\frac{\exp(ik_1|\mathbf{r}-\mathbf{r}'|)}{|\mathbf{r}-\mathbf{r}'|}\mathcal{V}(\mathbf{r}')\psi(\mathbf{r}')\,\mathrm{d}\mathbf{r}' \tag{5.1.22}$$

The second term on the right-hand side of (5.1.22) represents a superposition of spherical waves that originate from every point in the sample.

Equation (5.1.22) is an integral equation for $\psi(\mathbf{r})$. If the interaction $\mathcal{V}(\mathbf{r})$ is assumed to be weak, a solution can be obtained by making the approximation $\psi(\mathbf{r})\simeq\exp(i\mathbf{k}_1\cdot\mathbf{r})$ inside the integral sign. This substitution yields the *first Born approximation* to $\psi(\mathbf{r})$:

$$\psi(\mathbf{r})\simeq\exp(i\mathbf{k}_1\cdot\mathbf{r})-\frac{m}{2\pi\hbar^2}\int\frac{\exp(ik_1|\mathbf{r}-\mathbf{r}'|)}{|\mathbf{r}-\mathbf{r}'|}\mathcal{V}(\mathbf{r}')\exp(i\mathbf{k}_1\cdot\mathbf{r}')\,\mathrm{d}\mathbf{r}' \tag{5.1.23}$$

We can now obtain an expression for $f(\theta)$ by taking the $\mathbf{r}\to\infty$ limit of (5.1.23) and matching the result to the known asymptotic form of $\psi(\mathbf{r})$ given by Eqn (5.1.16). If $|\mathbf{r}|\gg|\mathbf{r}'|$, then

$$\begin{aligned}|\mathbf{r}-\mathbf{r}'|&=(r^2+r'^2-2\mathbf{r}\cdot\mathbf{r}')^{1/2}\\&\simeq r-\hat{\mathbf{r}}\cdot\mathbf{r}'\end{aligned} \tag{5.1.24}$$

where $\hat{\mathbf{r}}$ is a unit vector in the direction of $\mathbf{r}$. Since we have assumed that the scattering is elastic, $k_1\hat{\mathbf{r}}=\mathbf{k}_2$. Thus

$$\psi(\mathbf{r})\underset{\mathbf{r}\to\infty}{\sim}\exp(i\mathbf{k}_1\cdot\mathbf{r})-\frac{\exp(ik_1r)}{r}\frac{m}{2\pi\hbar^2}\int\exp(-i\mathbf{k}_2\cdot\mathbf{r}')\mathcal{V}(\mathbf{r}')\exp(i\mathbf{k}_1\cdot\mathbf{r}')\,\mathrm{d}\mathbf{r}' \tag{5.1.25}$$

By comparison of (5.1.16) and (5.1.25), and remembering that $k_1=k_2$, we find that

$$\begin{aligned}f(\theta)&=-\frac{m}{2\pi\hbar^2}\int\exp(-i\mathbf{k}_2\cdot\mathbf{r})\mathcal{V}(\mathbf{r})\exp(i\mathbf{k}_1\cdot\mathbf{r})\,\mathrm{d}\mathbf{r}\\&=-\frac{m}{2\pi\hbar^2}\int\exp(-i\mathbf{k}\cdot\mathbf{r})\mathcal{V}(\mathbf{r})\,\mathrm{d}\mathbf{r}\end{aligned} \tag{5.1.26}$$

Hence the amplitude of the scattered component is given by the Fourier transform of the scattering potential. The first line of (5.1.26) also shows that $f(\theta)$ can be expressed as a matrix element of the interaction $\mathcal{V}(\mathbf{r})$ between initial and final plane-wave states of the neutron. Use of the first Born approximation is therefore equivalent to calculating the cross-section $\mathrm{d}\sigma/\mathrm{d}\Omega$ by the "golden rule" of quantum-mechanical perturbation theory.

An expression for $\mathrm{d}\sigma/\mathrm{d}\Omega$ can now be derived by substiuting for $\mathcal{V}(\mathbf{r})$ in Eqn (5.1.26), inserting the result in (5.1.17), and taking the thermal average. This yields the expression

$$\begin{aligned}\frac{\mathrm{d}\sigma}{\mathrm{d}\Omega} &= \left\langle \left| \sum_{i=1}^{N} b_i \exp\left(-i\mathbf{k}\cdot\mathbf{r}_i\right) \right|^2 \right\rangle \\ &= \left\langle \sum_{i=1}^{N} \sum_{j=1}^{N} b_i b_j \exp\left[-i\mathbf{k}\cdot(\mathbf{r}_i - \mathbf{r}_j)\right] \right\rangle \end{aligned} \tag{5.1.27}$$

A more useful result is obtained by taking a statistical average of the scattering lengths over both the isotopic species present in the sample and the spin states of the nuclei; this can be done independently of the thermal averaging over the coordinates. We therefore introduce the notation

$$\left.\begin{aligned} \langle b_i^2 \rangle = \langle b^2 \rangle, \qquad & \langle b_i b_j \rangle = \langle b_i \rangle \langle b_j \rangle \\ & \qquad\quad = \langle b \rangle^2 \\ \langle b \rangle^2 = b_{\text{coh}}^2, \qquad & (\langle b^2 \rangle - \langle b \rangle^2) = b_{\text{inc}}^2 \end{aligned}\right\} \tag{5.1.28}$$

and rewrite Eqn (5.1.27) as

$$\begin{aligned}\frac{\mathrm{d}\sigma}{\mathrm{d}\Omega} &= N\langle b^2 \rangle + \langle b \rangle^2 \left\langle \sum_{i \neq j}^{N}\sum \exp\left[-i\mathbf{k}\cdot(\mathbf{r}_i - \mathbf{r}_j)\right] \right\rangle \\ &= N(\langle b^2 \rangle - \langle b \rangle^2) + \langle b \rangle^2 \left\langle \left| \sum_{i=1}^{N} \exp\left(-i\mathbf{k}\cdot\mathbf{r}_i\right) \right|^2 \right\rangle \\ &= N b_{\text{inc}}^2 + N b_{\text{coh}}^2 S(\mathbf{k}) \end{aligned} \tag{5.1.29}$$

The subscripts "coh" and "inc" refer, respectively, to *coherent* and *incoherent* scattering. Information about the structure of the fluid is contained entirely within the coherent contribution to the cross-section; there is no incoherent contribution if the sample consists of a single isotopic species of zero nuclear spin. The amplitude of the wave scattered by a single fixed nucleus is

$$f(\theta) = -b \int \exp\left(-i\mathbf{k}\cdot\mathbf{r}\right)\delta(\mathbf{r})\,\mathrm{d}\mathbf{r} = -b \tag{5.1.30}$$

In the absence of incoherent scattering, the cross-section can be written in the form

$$\frac{\mathrm{d}\sigma}{\mathrm{d}\Omega} = N b^2 S(\mathbf{k}) \tag{5.1.31}$$

where Nb^2 would be the scattering from N independent nuclei and $S(\mathbf{k})$ represents the effect of correlations.

A similar calculation can be made of the cross-section for elastic scattering of X-rays. There is now no separation into coherent and incoherent parts, but the expression for the differential cross-section has the same general form as in Eqn (5.1.31). One important difference is that X-rays are scattered by the atomic electrons and the analogue of b is the atomic *form factor*, $f(\mathbf{k})$. The latter, unlike b, is a function of $\mathbf{k}$, and defined as

$$f(\mathbf{k}) = \left\langle \sum_{n=1}^{Z} \exp\left[-i\mathbf{k}\cdot(\mathbf{r}_i^{(n)} - \mathbf{r}_i)\right] \right\rangle_{\mathrm{Q}} \tag{5.1.32}$$

where the subscript Q denotes a quantum-mechanical expectation value, $\mathbf{r}_i^{(n)}$ represents the coordinates of the nth electron of the ith atom (with nuclear coordinates $\mathbf{r}_i$), and Z is the atomic number.

The pair distribution function can be obtained from a measured structure factor by numerically transforming the experimental data according to Eqn (5.1.8). Difficulties arise in practice because measurements of $S(\mathbf{k})$ necessarily introduce a cut-off at large values of k (Pings, 1968). An example of an experimentally determined structure factor for liquid sodium near its triple point is shown in Figure 5.2 (Greenfield *et al.*, 1971). The dominant feature is a pronounced peak at a value of k approximately equal to $2\pi/\Delta r$, where Δr is the spacing of the peaks in $g(r)$. The experimental structure factor is very well fitted by Monte Carlo results for a potential that varies as r^{-4} (Hansen and Schiff, 1973). Since the r^{-4} potential is only a crude representation of the effective pair potential for liquid sodium, the good agreement shown in Figure 5.2 suggests that the structure factor is insensitive to details of the potential.

The definitions given above are easily extended to systems of more than one component. Consider a system containing N_ν particles of species ν, with $\nu = 1$ to n. If $N = \sum_\nu N_\nu$ is the total number of particles, the number concentration of species ν is $x_\nu = N_\nu/N$. The partial densities $\rho_\nu(\mathbf{r})$ and their Fourier components $\rho_{\mathbf{k}}^{\nu}$ are defined in a manner analogous to (5.1.1) and (5.1.5), except that the sums run only over the particles of species ν. The partial structure factors are defined, by a generalization of (5.1.6), as

$$S_{\nu\mu}(\mathbf{k}) = \frac{1}{N}\langle \rho_{\mathbf{k}}^{\nu} \rho_{-\mathbf{k}}^{\mu} \rangle \tag{5.1.33}$$

and are related to the partial pair distribution functions $g_{\nu\mu}(\mathbf{r})$ by

$$S_{\nu\mu}(\mathbf{k}) = x_\nu \delta_{\nu\mu} + x_\nu x_\mu \rho \int \exp(-i\mathbf{k}\cdot\mathbf{r}) g_{\nu\mu}(\mathbf{r})\, d\mathbf{r} \tag{5.1.34}$$

If the fluid is homogeneous, the function $g_{\nu\mu}(\mathbf{r})$ has a delta-function

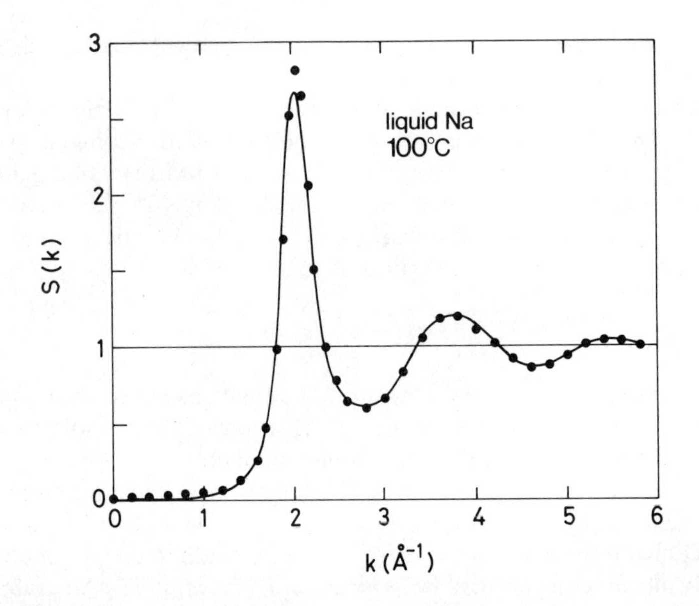

FIG. 5.2. Structure factor of liquid sodium near the triple point. The points are experimental X-ray scattering results (Greenfield *et al.*, 1971) and the curve is obtained from a Monte Carlo calculation for the r^{-4} potential (Hansen and Schiff, 1973).

representation given by

$$x_\nu x_\mu \rho g_{\nu\mu}(\mathbf{r}) = \frac{1}{N}\left\langle \sum_{i=1}^{N_\nu} \sum_{j=1}^{N_\mu}{}' \delta(\mathbf{r}+\mathbf{r}_i-\mathbf{r}_j)\right\rangle \tag{5.1.35}$$

where the prime indicates that the terms for which $i=j$ are omitted when $\nu=\mu$. The corresponding pair correlation functions, defined as

$$h_{\nu\mu}(\mathbf{r}) = g_{\nu\mu}(\mathbf{r}) - 1 \tag{5.1.36}$$

all approach zero as $|\mathbf{r}|\to\infty$; their Fourier transforms are related to the partial structure factors by

$$S_{\nu\mu}(\mathbf{k}) = x_\nu \delta_{\nu\mu} + x_\nu x_\mu \rho \hat{h}_{\nu\mu}(\mathbf{k}) \tag{5.1.37}$$

Definitions of the partial structure factors that differ trivially from (5.1.34) can also be found in the literature.

5.2 THE ORNSTEIN–ZERNIKE DIRECT CORRELATION FUNCTION

We now show that the structure factor (5.1.6) determines the linear response of a fluid to a spatially varying, static, external field (Yvon, 1958). In doing so, we shall find that the link between the field and the response can also be expressed in terms of the *direct correlation function* first introduced by

Ornstein and Zernike (1914) in their investigation of density fluctuations near the critical point.

Consider an open system in which the particles interact via a pair potential $v(|\mathbf{r}_1-\mathbf{r}_2|)$ and are subject to an external field $\phi(\mathbf{r})$. The N-particle potential energy is then given by Eqn (4.2.25), the single-particle density by Eqn (4.2.30), and the pair density, from Eqn (4.2.31), by

$$\rho^{(2)}(1,2)=\frac{1}{\Xi}z^*(1)z^*(2)\frac{\delta^2\Xi}{\delta z^*(1)\delta z^*(2)} \tag{5.2.1}$$

From Eqns (2.5.26) (taken for $n=2$), (2.5.27), (4.2.11), (4.2.30) and (5.2.1), we find that

$$\begin{aligned}\frac{\delta\rho^{(1)}(1)}{\delta\log z^*(2)}&=z^*(2)\frac{\delta}{\delta z^*(2)}z^*(1)\frac{\delta\log\Xi}{\delta z^*(1)}\\&=z^*(2)\frac{\delta\log\Xi}{\delta z^*(1)}\frac{\delta z^*(1)}{\delta z^*(2)}+z^*(1)z^*(2)\frac{\delta^2\log\Xi}{\delta z^*(1)\delta z^*(2)}\\&=\rho^{(1)}(1)\delta(1,2)+\rho^{(2)}(1,2)-\rho^{(1)}(1)\rho^{(1)}(2)\\&=\rho^{(1)}(1)\delta(1,2)+\rho^{(1)}(1)\rho^{(1)}(2)h(1,2)\end{aligned} \tag{5.2.2}$$

It follows from the definition of a functional derivative that the variation in the single-particle density that results from an infinitesimal variation $\delta\phi$ in the external potential is

$$\begin{aligned}\delta\rho^{(1)}(1)&=\int\frac{\delta\rho^{(1)}(1)}{\delta\phi(2)}\delta\phi(2)\,\mathrm{d}2\\&=-\beta\int\frac{\delta\rho^{(1)}(1)}{\delta\log z^*(2)}\delta\phi(2)\,\mathrm{d}2\end{aligned} \tag{5.2.3}$$

We are concerned only with the linear response of the system to a weak external field. We may therefore substitute Eqn (5.2.2) in (5.2.3), but at the same time replace $\rho^{(1)}(1)$ and $h(1,2)$ by their values in the absence of the external field, namely the number density ρ and the translationally-invariant pair correlation function $h(\mathbf{r}_1-\mathbf{r}_2)$. Hence the variation in $\rho^{(1)}(1)$ to first order in $\delta\phi$ is

$$\delta\rho^{(1)}(1)=-\beta\rho\delta\phi(1)-\beta\rho^2\int h(1,2)\delta\phi(2)\,\mathrm{d}2 \tag{5.2.4}$$

This result is called the Yvon equation; it is equivalent to a first-order functional Taylor expansion of $\rho^{(1)}$ in powers of $\delta\phi$.

We now take the Fourier transform of Eqn (5.2.4) and relate the response $\delta\rho^{(1)}(\mathbf{k})$ to the Fourier components of the external potential, defined as

$$\delta\hat{\phi}(\mathbf{k})=\int\exp(-i\mathbf{k}\cdot\mathbf{r})\phi(\mathbf{r})\,\mathrm{d}\mathbf{r} \tag{5.2.5}$$

The result is

$$\begin{aligned}\delta\hat{\rho}^{(1)}(\mathbf{k}) &= -[1+\rho\hat{h}(\mathbf{k})]\beta\rho\delta\hat{\phi}(\mathbf{k}) \\ &= -S(\mathbf{k})\beta\rho\delta\hat{\phi}(\mathbf{k}) \\ &= \chi(\mathbf{k})\delta\hat{\phi}(\mathbf{k}) \end{aligned} \tag{5.2.6}$$

which serves also as the definition of the function $\chi(\mathbf{k})$. The structure factor $S(\mathbf{k})$ appears in (5.2.6) as a generalized response function (Yvon, 1958), analogous to the magnetic susceptibility of spin systems. The linear density response of the system to an external field is therefore determined by the density-density autocorrelation function $G(\mathbf{r})$ in the absence of the field through Eqns (5.1.4), (5.1.7) and (5.2.6). This is an example of a very general relationship in statistical mechanics known as the *fluctuation–dissipation theorem.* We shall have more to say about the fluctuation-dissipation theorem in the discussion of dynamical properties in Chapter 7.

We now reformulate the link between $\delta\rho^{(1)}$ and $\delta\phi$ in terms of the direct correlation function, $c(1,2)$. The latter can be defined through the *Ornstein–Zernike relation* between the functions $c(1,2)$ and $h(1,2)$. For a system in an external field, the Ornstein–Zernike relation has the form

$$h(1,2) = c(1,2) + \int \rho^{(1)}(3)c(1,3)h(3,2)\,\mathrm{d}3 \tag{5.2.7}$$

Equation (5.2.7) can be solved recursively for $h(1,2)$ to give

$$\begin{aligned} h(1,2) = c(1,2) &+ \int \rho^{(1)}(3)c(1,3)c(3,2)\,\mathrm{d}3 \\ &+ \int \rho^{(1)}(3)\rho^{(1)}(4)c(1,3)c(3,4)c(4,2)\,\mathrm{d}3\,\mathrm{d}4 \\ &+ \dots \end{aligned} \tag{5.2.8}$$

In terms of diagrams, this result means that

$h(1,2) = \{$all simple chain diagrams consisting of two white terminal 1-circles labelled 1 and 2, black $\rho^{(1)}$-circles and c-bonds$\}$

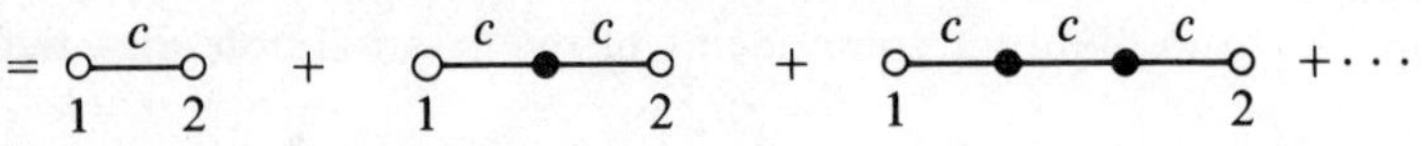

$$= \underset{1\quad 2}{\circ\!\overset{c}{\text{—}}\!\circ} + \underset{1\qquad\quad 2}{\circ\!\overset{c}{\text{—}}\!\bullet\!\overset{c}{\text{—}}\!\circ} + \underset{1\qquad\qquad\quad 2}{\circ\!\overset{c}{\text{—}}\!\bullet\!\overset{c}{\text{—}}\!\bullet\!\overset{c}{\text{—}}\!\circ} + \cdots \tag{5.2.9}$$

Equation (5.2.9) has an obvious physical interpretation. It describes the fact that the *total* correlation between particles 1 and 2, represented by $h(1,2)$, is due in part to the *direct* correlation between 1 and 2 but also to the *indirect* correlation propagated via increasingly large numbers of intermediate particles. It is plausible to suppose that the range of $c(1,2)$ is comparable with that of the pair potential $v(1,2)$, and to ascribe the fact that $h(1,2)$ is generally much longer ranged than $v(1,2)$ to the effects of the indirect correlation. The differences between the two functions for a typical simple liquid are illustrated in Figure 5.3; $c(r)$ is not only shorter ranged than $h(r)$, but is also simpler in structure.

If the system is both homogeneous and isotropic, the Ornstein–Zernike relation becomes

$$h(r)=c(r)+\rho\int c(|\mathbf{r}-\mathbf{r}'|)h(r')\,\mathrm{d}\mathbf{r}' \tag{5.2.10}$$

On taking the Fourier transform of both sides of this equation, we obtain an algebraic relation between $\hat{h}(\mathbf{k})$ and $\hat{c}(\mathbf{k})$:

$$\hat{h}(k)=\frac{\hat{c}(k)}{1-\rho\hat{h}(k)}. \tag{5.2.11}$$

The density response function defined by Eqn (5.2.6) can therefore be written in terms of $\hat{c}(k)$ as

$$\chi(k)=-\beta\rho S(k)=-\beta\rho[1+\rho\hat{h}(k)]=\frac{\beta\rho}{\rho\hat{c}(k)-1} \tag{5.2.12}$$

As an alternative to introducing the direct correlation function via the Ornstein–Zernike relation, we could choose to define $c(1,2)$ as the functional derivative

$$c(1,2)=\frac{\delta\log[\rho^{(1)}(1)/z^*(1)]}{\delta\rho^{(1)}(2)} \tag{5.2.13}$$

This equation can be rearranged with the help of (4.2.11) to give

$$\begin{aligned}\frac{\delta\log z^*(1)}{\delta\rho^{(1)}(2)}&=\frac{1}{\rho^{(1)}(1)}\frac{\delta\rho^{(1)}(1)}{\delta\rho^{(1)}(2)}-c(1,2)\\&=\frac{1}{\rho^{(1)}(1)}\delta(1,2)-c(1,2)\end{aligned} \tag{5.2.14}$$

We now use the change-of-variable formula (4.2.24) to express $\delta(1,2)$ in

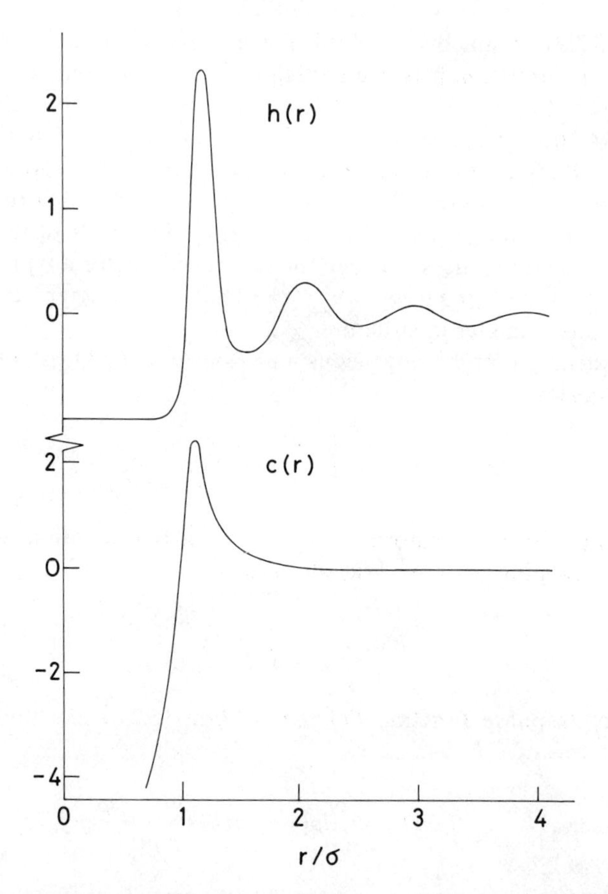

FIG. 5.3. The functions $h(r)$ and $c(r)$ for a typical monatomic liquid near the triple point. After Gray and Gubbins (1984).

the form

$$\delta(1,2) = \frac{\delta \log z^*(1)}{\delta \log z^*(2)}$$

$$= \int \frac{\delta \log z^*(1)}{\delta \rho^{(1)}(3)} \frac{\delta \rho^{(1)}(3)}{\delta \log z^*(2)} \, \mathrm{d}3 \qquad (5.2.15)$$

On inserting the results (5.2.2) and (5.2.14) inside the integral sign in Eqn (5.2.15), we recover the Ornstein–Zernike relation (5.2.7). The definitions of $c(1,2)$ given by (5.2.7) and (5.2.13) are therefore equivalent. The advantage of (5.2.13) is that it leads directly to the density expansion of $c(1,2)$: all that is needed is to substitute for $\log [\rho^{(1)}(1)/z^*(1)]$ from Eqn (4.5.9) and apply Lemma 2 of Chapter 4. The diagrams in (4.5.9) are irreducible; since they contain only one white circle, this is equivalent to saying that they are free of connecting circles. Clearly they remain free of connecting circles when a black circle is whitened as a result of the functional differentiation. It follows that $c(1,2)$ can be expressed diagrammatically as

$c(1,2) = \{$all simple diagrams that consist of two white 1-circles labelled 1 and 2, black $\rho^{(1)}$-circles and f-bonds, and are free of connecting circles$\}$

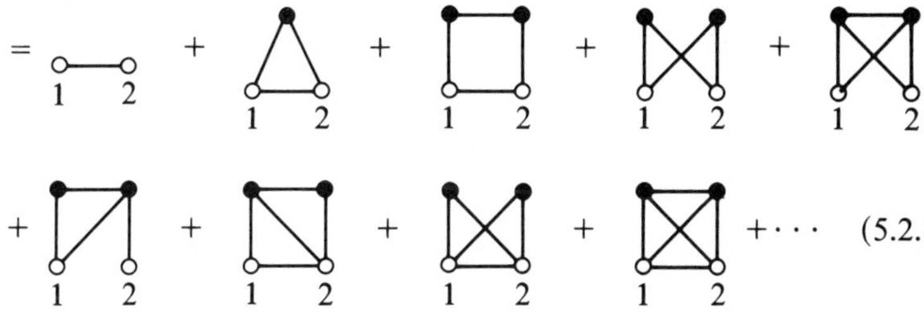

(5.2.16)

This result was first obtained by Rushbrooke and Scoins (1953).

From the expansion (5.2.16) we can deduce that the range of $c(r)$ is roughly the same as the range of the potential, as anticipated in our discussion of the Ornstein–Zernike relation. To lowest order in density, $c(r) \simeq f(r)$, or, at large r, $c(r) \simeq -\beta v(r)$. Since all higher-order diagrams in (5.2.16) are at least doubly connected, their contributions to $c(r)$ decay at least as fast as $[f(r)]^2$, and are therefore negligible in comparison with the leading term in the limit $r \to \infty$. On the other hand, the effects of the indirect correlations are such that $h(r)$ can be significantly different from zero even when $v(r)$ is very small. This contrast in behaviour is particularly evident close to the critical point. As the critical point is approached, the compressibility χ_T becomes very large. It follows from Eqn (5.1.13) that the structure factor $S(k)$ acquires a strong peak at the origin. This, in turn, implies that $h(r)$, the Fourier transform of $S(k)$, becomes very long ranged: in other words, spatial correlations decay very slowly with separation. If, however, we combine Eqns (5.1.12), (5.1.13) and (5.2.11), we find that

$$\rho\hat{c}(0) = 1 - \chi_T^0/\chi_T \tag{5.2.17}$$

Close to the critical point, $\rho\hat{c}(0) \simeq 1$, and therefore $c(r)$ remains short ranged.

The argument concerning the relative ranges of $c(r)$ and $h(r)$ does not apply to ionic fluids. The effect of screening in such systems is to cause $h(r)$ to decay exponentially at large r, whereas $c(r)$ still has the range of the potential and therefore decays as r^{-1}. In this situation, $c(r)$ is of longer range than $h(r)$.

The discussion until now has been limited to one-component systems, but the generalization to mixtures is straightforward. As in Section 5.1, we consider an n-component system in which N_ν particles are of species ν, $N = \sum_\nu N_\nu$ and $x_\nu = N_\nu / N$. The pair structure of the mixture is determined by $\frac{1}{2}n(n+1)$ pair correlation functions $h_{\nu\mu}(r)$ or, alternatively, by $\frac{1}{2}n(n+1)$ direct correlation functions $c_{\nu\mu}(r)$. The two sets of functions are linked by $\frac{1}{2}n(n+1)$ coupled equations, representing a generalization of the Ornstein–Zernike relation (5.2.7). If the system is homogeneous, the multicomponent generalization of (5.2.7) can be written as

$$h_{\nu\mu}(1,2) = c_{\nu\mu}(1,2) + \rho \sum_\lambda x_\lambda \int c_{\nu\lambda}(1,3) h_{\lambda\mu}(3,2) \, \mathrm{d}3 \qquad (5.2.18)$$

where the sum on λ runs over all components in the mixture. The generalization of (5.2.17) in terms of $\hat{c}_{\nu\mu}(k)$ is

$$\sum_\nu \sum_\mu x_\nu x_\mu \rho \hat{c}_{\nu\mu}(0) = 1 - \chi_T^0 / \chi_T \qquad (5.2.19)$$

If the partial structure factors are represented as a matrix $\boldsymbol{S}(k)$, combination of (5.2.18) and (5.2.19), together with a matrix inversion, shows that the corresponding generalization of Eqn (5.1.13) is

$$\rho k_{\mathrm{B}} T \chi_T = \frac{|\boldsymbol{S}(0)|}{\sum_\nu \sum_\mu x_\nu x_\mu \, |\boldsymbol{S}(0)|_{\nu\mu}} \qquad (5.2.20)$$

where $|\boldsymbol{S}(k)|_{\nu\mu}$ denotes the cofactor of $S_{\nu\mu}(k)$ in the determinant $|\boldsymbol{S}(k)|$ (Kirkwood and Buff, 1951).

5.3 DIAGRAMMATIC EXPANSIONS OF THE PAIR FUNCTIONS

In the preceding section, we derived the density expansion of the direct correlation function $c(1,2)$. We wish now to obtain similar expansions for the pair correlation function $h(1,2)$ and the pair distribution function $g(1,2)$. This can be done systematically by following the same steps that lead to the virial expansion of the equation of state (Section 4.5). Taken together, Eqn (5.2.1) and Lemma 2 of Chapter 4 yield the expansion of $\rho^{(2)}(1,2)$ in powers of the activity; this result, combined with the z^*-expansion of $\rho^{(1)}(\mathbf{r})$, Eqn (4.5.8), gives the activity series for $h(1,2)$; finally, an application of Lemma 4 leads to the density expansion of $h(1,2)$ and hence that of $g(1,2)$.

Rather than following the procedure outlined above, we shall use a short-cut to obtain the expansion of $h(1,2)$ directly from the corresponding series for $c(1,2)$, Eqn (5.2.16). To do this, we simply replace the c-bonds in Eqn (5.2.9) by their series expansion. The first term on the right-hand side of (5.2.9) yields the set of diagrams that contribute to $c(1,2)$ and are therefore free of all connecting circles, i.e. they contain neither articulation circles nor nodal circles. Higher-order terms in (5.2.9) contain one or more black nodal circles; these remain nodal circles when the c-bonds are replaced by diagrams drawn from (5.2.16), but no articulation circles appear. The topology of the diagrams that contribute to the density expansion of $h(1,2)$ is therefore similar to that of the diagrams in the series for $c(1,2)$ except that nodal circles are now permitted. Thus

$$h(1,2) = \{\text{all simple diagrams that consist of two white 1-circles labelled 1 and 2, black } \rho^{(1)}\text{-circles and } f\text{-bonds, and are free of articulation circles}\} \tag{5.3.1}$$

This is a result first derived by Mayer and Montroll (1941).

Equation (5.3.1) contains more diagrams that (5.2.16) at each order in density beyond the zeroth-order term. The additional diagrams all contain at least one nodal circle. For example, of the two second-order terms shown below, (a) appears in both expansions but (b) appears only in (5.3.1), because in (b) the black circles are nodal circles.

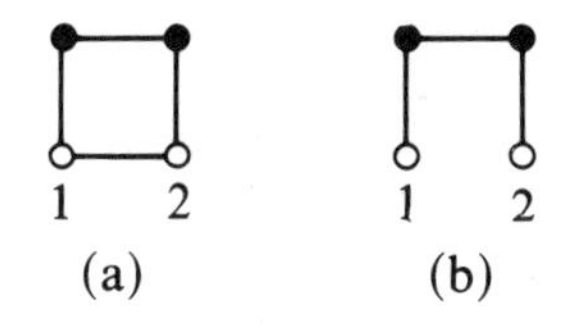

Diagrams (a) and (b) differ only by the presence in (a) of an f-bond between the white circles. If we recall that $e(1,2) = f(1,2) + 1$, we see that the sum of the two diagrams can be represented by a single diagram in which the white circles are linked by an e-bond, but all other bonds are f-bonds; the resulting diagram is free of connecting circles. All the diagrams in (5.3.1) can be paired uniquely in this way, except that the lowest-order diagram

$$\underset{1}{\circ}\!\!-\!\!\!-\!\!\underset{2}{\circ}$$

appears alone. We therefore add to (5.3.1) the diagram consisting of two disconnected white 1-circles

$$\underset{1}{\circ} \quad \underset{2}{\circ} = 1$$

and obtain the density expansion of $g(1,2)$ in the form

$$g(1,2) = h(1,2) + 1$$
$= \{$all simple diagrams that consist of two white 1-circles labelled 1 and 2 and linked by an e-bond, black $\rho^{(1)}$-circles and f-bonds, and are free of connecting circles$\}$ (5.3.2)

If the system is homogeneous, Eqn (5.3.2) can be written as a power series in ρ with coefficients $g_n(1,2)$ such that

$$g(1,2) = \exp[-\beta v(1,2)]\left(1 + \sum_{n=1}^{\infty} \rho^n g_n(1,2)\right) \tag{5.3.3}$$

with

$$g_1(1,2) = \quad = \int f(1,3) f(2,3)\, \mathrm{d}3 \tag{5.3.4}$$

$$g_2(1,2) = \quad + \quad + \quad + \quad + \tag{5.3.5}$$

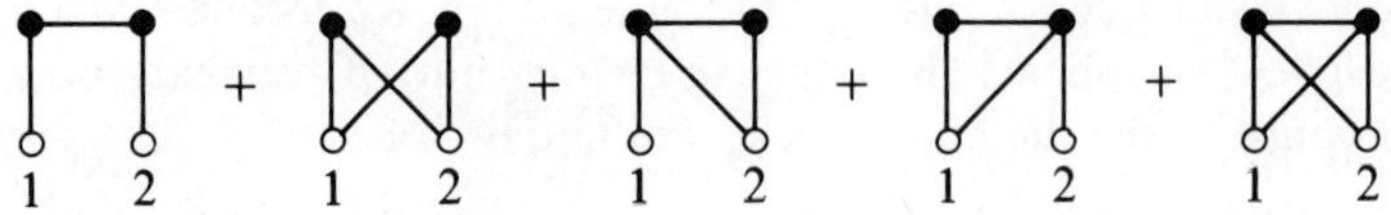

The form of the density series for $g(1,2)$ leads immediately to two important results. First, $g(r)$ behaves as $\exp[-\beta v(r)]$ as $\rho \to 0$, as was proved in a different way in Section 2.5; in the same limit, $h(r)$ and $c(r)$ both behave as $f(r)$. Secondly, the function $y(r)$ defined by Eqn (2.5.15) is a continuous function of r even for hard spheres, for which the discontinuity in $g(r)$ at $r = d$ is wholly contained in the factor $\exp[-\beta v(r)]$. This useful property of $y(r)$ has already been exploited in the derivation of the hard-sphere equation of state (2.5.17); $y(r)$ itself is given diagrammatically by the sum of all diagrams in $g(1,2)$ with the e-bonds removed. The coefficients $g_1(r)$ and $g_2(r)$ in Eqn (5.3.3) have been evaluated analytically for hard spheres (Kirkwood, 1935; Nijboer and van Hove, 1952); $g_1(r)$ is almost identical to the function $\xi(r)$ of Eqn (4.6.4).

The pair distribution function is sometimes written as

$$g(1,2) = \exp[-\beta \psi(1,2)] \tag{5.3.6}$$

where $\psi(1,2)$ is the *potential of mean force.* The form of (5.3.6) is suggested by the known, low-density limit of $g(1,2)$. If we define a function $w(1,2)$ by

$$w(1,2)=\beta[v(1,2)-\psi(1,2)] \quad (5.3.7)$$

then

$$g(1,2)=e(1,2)\exp[w(1,2)] \quad (5.3.8)$$

and therefore

$$w(1,2)=\log y(1,2) \quad (5.3.9)$$

On deleting the e-bonds from the diagrams in (5.3.2), an application of Lemma 1 of Chapter 4 shows that

$w(1,2)=$ {all simple diagrams consisting of two white 1-circles labelled 1 and 2, at least one black $\rho^{(1)}$-circle, and f-bonds, such that the white circles are not an articulation pair and there is no $f(1,2)$ bond} (5.3.10)

We recall from Section 4.3 that the fact that the white circles are not an articulation pair means that there exists at least one path between each pair of black circles that does not pass through either white circle.

From the earlier discussion, we known that $c(1,2)$ is the sum of all diagrams in $h(1,2)$ that are free of nodal circles. Hence we may write

$$h(1,2)=c(1,2)+b(1,2) \quad (5.3.11)$$

where

$b(1,2)=$ {all simple diagrams that consist of two white 1-circles labelled 1 and 2, black $\rho^{(1)}$-circles and f-bonds, and are free of articulation circles but contain at least one nodal circle}

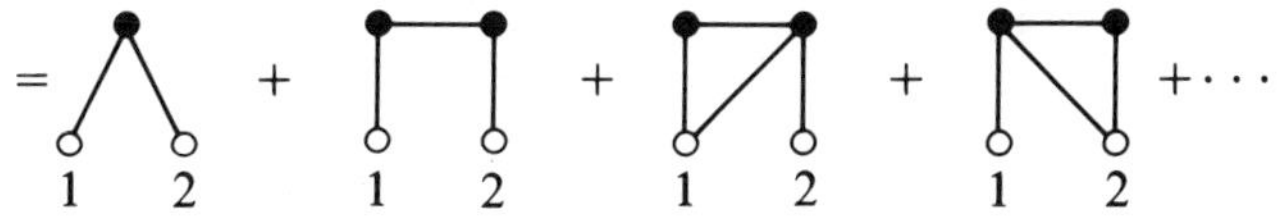

= [diagram] + [diagram] + [diagram] + [diagram] + ⋯

1 2 1 2 1 2 1 2

(5.3.12)

Diagrams that belong to the set (5.3.12) are sometimes called *series* diagrams. All series diagrams are also members of the set (5.3.10). The function $w(1,2)$ can therefore be re-expressed as

$$w(1,2)=b(1,2)+d(1,2) \quad (5.3.13)$$

where $d(1,2)$ is the sum of all diagrams in (5.3.10) that are free of nodal circles. Such diagrams are called the *bridge* or *elementary* diagrams. To second order in density, the only bridge diagram is

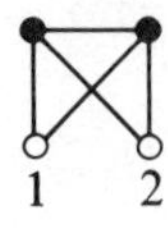

On combining Eqns (5.3.8), (5.3.11) and (5.3.13), we obtain the following, exact relation (van Leeuwen *et al.*, 1959):

$$h(1,2) - c(1,2) - \log [h(1,2)+1] = \beta v(1,2) - d(1,2) \qquad (5.3.14)$$

Because $h(1,2)$ and $c(1,2)$ are linked by the Ornstein–Zernike relation, Eqn (5.3.14) would be transformed into an integral equation for h (or c) if the unknown function $d(1,2)$ were to be replaced by an expression that involved only h (or c). For example, the f-bond expansion of $d(1,2)$ can be rewritten as an h-bond expansion (Stell, 1964) and inserted in (5.3.14). The result, together with the Ornstein–Zernike relation, constitutes an exact integral equation for $h(1,2)$, but the equation is intractable because the h-bond expansion introduces an infinite series of many-dimensional integrals of products of h. If we make the approximation of setting $d(1,2) = 0$ in (5.3.14), we obtain instead an approximate integral equation for $h(1,2)$ called the hypernetted-chain or HNC equation (van Leeuwen *et al.*, 1959; Meeron, 1960; Morita, 1960; Rushbrooke, 1960; Verlet, 1960). We shall rederive the HNC equation in a different way in Section 5.4.

The derivation of the non-linear Debye–Hückel result for the radial distribution function of a system of charged particles provides a simple but useful example of the application of diagrammatic techniques to the calculation of pair functions. We consider a homogeneous, one-component plasma of point charges q, for which the pair potential is

$$v(r) = q^2/r \qquad (5.3.15)$$

Use of (5.3.15) in the expansions of the pair functions leads to divergent integrals, but convergent results can be obtained if entire classes of diagrams are summed. The most strongly divergent integrals in the expansion of $w(1,2)$, Eqn (5.3.10), are those associated with the most weakly connected diagrams, namely the chain diagrams (Résibois, 1968). If the chain diagrams are summed to all orders in ρ, but all other diagrams are ignored, the result is an approximation for $w(1,2)$ of the form

$w(1,2) \simeq$ ○—f—●—f—○ + ○—f—●—f—●—f—○ + ○—f—●—f—●—f—●—f—○ + ...
(end labels: 1, 2 on each diagram)

(5.3.16)

By analogy with Eqns (5.2.7) and (5.2.9), the sum of all chain diagrams is

$$w(1,2)=\rho\int f(1,3)[f(3,2)+w(3,2)]\,\mathrm{d}3 \tag{5.3.17}$$

On taking Fourier transforms, Eqn (5.3.17) becomes

$$\hat{w}(k)=\frac{\rho[\hat{f}(k)]^2}{1-\rho\hat{f}(k)} \tag{5.3.18}$$

with

$$\begin{aligned}\rho\hat{f}(k)&=\rho\int \exp(-i\mathbf{k}\cdot\mathbf{r})[\exp(-\beta q^2/r)-1]\,\mathrm{d}\mathbf{r}\\ &\simeq-\beta\rho q^2\int\frac{\exp(-i\mathbf{k}\cdot\mathbf{r})}{r}\,\mathrm{d}\mathbf{r}\\ &=-\frac{4\pi\beta\rho q^2}{k^2}\\ &=-\frac{k_\mathrm{D}^2}{k^2}\end{aligned} \tag{5.3.19}$$

where

$$k_\mathrm{D}=(4\pi\beta\rho q^2)^{1/2} \tag{5.3.20}$$

is the Debye wavenumber and $\Lambda_\mathrm{D}=1/k_\mathrm{D}$ is the Debye screening length. We now substitute for $\rho\hat{f}(k)$ in Eqn (5.3.18) and find that

$$\rho[\hat{w}(k)-\beta\hat{v}(k)]=\frac{k_\mathrm{D}^2}{k_\mathrm{D}^2+k^2} \tag{5.3.21}$$

or

$$\begin{aligned}w(r)-\beta v(r)&=-\beta\psi(r)\\ &=-\frac{\beta q^2}{r}\exp(-k_\mathrm{D}r)\end{aligned} \tag{5.3.22}$$

It follows from Eqn (5.3.8) that the corresponding approximation for $g(r)$ is

$$g(r)=\exp\left[-\frac{\beta q^2}{r}\exp(-k_\mathrm{D}r)\right] \tag{5.3.23}$$

We see that summing the chain diagrams leads to a potential of mean force or "renormalized" potential equal to $(q^2/r)\exp(-k_\mathrm{D}r)$. The damping of the coulombic term by the factor $\exp(-k_\mathrm{D}r)$ is familar from elementary Debye–Hückel theory and corresponds physically to the effects of screening.

Equation (5.3.23) is better known in its linearized form, valid for $r \gg \Lambda_D$, i.e.

$$g(r) = 1 - \frac{\beta q^2}{r} \exp(-k_D r) \tag{5.3.24}$$

This result could have been obtained more directly either by summing the chain diagrams in (5.3.1) with $f(r)$ replaced by its linear form, $f(r) \simeq -\beta v(r)$, or, equivalently, by setting $c(r) = -\beta v(r)$ in Eqn (5.2.9). A serious weakness of the linearized approximation is the fact that it allows $g(r)$ to become negative at small r; this failing is rectified in the non-linear version (5.3.23).

5.4 FUNCTIONAL EXPANSIONS AND INTEGRAL EQUATIONS

A series of approximate, integral equations for the pair distribution function of a fluid in which the particles interact through pairwise-additive forces can be derived systematically by an elegant method due to Percus (1962, 1964). The basis of the method is the interpretation of the quantity $\rho g(r)$ as the local density at a point $\mathbf{r}$ in the fluid when a particle of the system is known to be located at the origin, $\mathbf{r} = 0$. The particle at the origin, labelled 0, is assumed to be fixed in space, while the other particles move in the force field of particle 0. Then the total potential energy of the system is of the form (4.2.25), with

$$\phi(i) = v(0, i) \tag{5.4.1}$$

Equation (4.2.26) shows that the grand partition function in the presence of the "external field" (5.4.1) is

$$\Xi(\phi) = \sum_{N=0}^{\infty} \frac{z^N}{N!} \int \cdots \int \exp\left(-\beta \sum_{i=1}^{N} \phi(i)\right) \exp(-\beta V_N)\, d1 \cdots dN \tag{5.4.2}$$

where, as usual, V_N is the total potential energy of the $\frac{1}{2}N(N-1)$ pairs. Alternatively, we may treat the particle at the origin as an $(N+1)$th particle. Then

$$\begin{aligned} V_N + \sum_{i=1}^{N} \phi(i) &= \sum_{i<j}^{N}\sum v(i, j) + \sum_{i=1}^{N} \phi(i) \\ &= \sum_{i<j}^{N+1}\sum v(i, j) \\ &= V_{N+1} \end{aligned} \tag{5.4.3}$$

If we denote the partition function in the absence of the external field by $\Xi(0)$, Eqn (5.4.2) can be rewritten as

$$\begin{aligned}
\Xi(\phi) &= \sum_{N=0}^{\infty} \frac{z^N}{N!} \int \cdots \int \exp(-\beta V_{N+1})\, \mathrm{d}1 \cdots \mathrm{d}N \\
&= \frac{\Xi(0)}{z} \sum_{N=0}^{\infty} \frac{1}{\Xi(0)} \frac{z^{N+1}}{N!} \int \cdots \int \exp(-\beta V_{N+1})\, \mathrm{d}1 \cdots \mathrm{d}N \\
&= \frac{\Xi(0)}{z} \sum_{N=1}^{\infty} \frac{1}{\Xi(0)} \frac{z^{N}}{(N-1)!} \int \cdots \int \exp(-\beta V_{N})\, \mathrm{d}1 \cdots \mathrm{d}(N-1)
\end{aligned} \tag{5.4.4}$$

According to Eqn (2.5.18), the sum on N in (5.4.4) is the definition of the single-particle density in a homogeneous system. Thus

$$\Xi(\phi) = \rho \Xi(0)/z \tag{5.4.5}$$

By a similar manipulation, the single-particle density $\rho^{(1)}(1|\phi)$ in the presence of the external field can be related to the pair density $\rho^{(2)}(0, 1|\phi = 0)$ in the absence of the field:

$$\rho^{(1)}(1|\phi) = \rho^{(2)}(0, 1|\phi = 0)/\rho \tag{5.4.6}$$

As the system is spatially uniform in the absence of the field, Eqns (2.5.26) and (5.4.6) together yield the relation

$$\rho^{(1)}(1|\phi) = \rho g(0, 1) \tag{5.4.7}$$

which is the mathematical expression of Percus' idea. The effect of slowly switching on the force field of particle 0 is to change the potential ϕ from zero to $\Delta\phi = v(0, 1)$; the resulting change in the single-particle density is

$$\begin{aligned}
\Delta\rho^{(1)} &= \rho^{(1)}(1|\phi) - \rho^{(1)}(1|0) \\
&= \rho g(0, 1) - \rho \\
&= \rho h(0, 1)
\end{aligned} \tag{5.4.8}$$

If the external field due to particle 0 is regarded as a perturbation, it is natural to consider functional Taylor expansions of various functionals of ϕ with respect either to $\Delta\phi$ or to the response $\Delta\rho^{(1)}$. One obvious choice is to expand $\Delta\rho^{(1)}$ directly in powers of $\Delta\phi$. The first-order result is simply the Yvon equation (5.2.4), with the infinitesimal quantities $\delta\rho^{(1)}$, $\delta\phi$ replaced by $\Delta\rho^{(1)}$, $\Delta\phi$. On combining this result with (5.4.1) and (5.4.8), we find that

$$h(0, 1) = -\beta v(0, 1) + \rho \int h(1, 2)[-\beta v(0, 2)]\, \mathrm{d}2 \tag{5.4.9}$$

Comparison with the Ornstein–Zernike relation (5.2.10) shows that in this approximation

$$c(0,1) = -\beta v(0,1) \tag{5.4.10}$$

If the potential is strongly repulsive at short range, Eqns (5.4.9) and (5.4.10) are very poor approximations, because $\Delta\rho^{(1)}$ is a highly non-linear functional of ϕ. In the case of the Coulomb potential, the approach is more successful; as we have seen in the previous section, Eqn (5.4.10) is equivalent to the linearized Debye–Hückel approximation.

When the pair potential is short ranged, better results are obtained by expanding in powers of the response $\Delta\rho^{(1)}$. In combination with the Ornstein–Zernike relation, each choice of functional to be expanded yields a different, approximate, integral equation for the pair distribution function. For example, we can choose to expand $\rho^{(1)}(1|\phi)/z^*(1)$. Then, to first order, we find by application of (4.2.23) that

$$\begin{aligned}\frac{\rho^{(1)}(1|\phi)}{z^*(1)} &= \frac{\rho^{(1)}(1|0)}{z} \\ &\quad + \int \left(\frac{\delta}{\delta\rho^{(1)}(2|\phi)} \frac{\rho^{(1)}(1|\phi)}{z^*(1)} \right)_{\phi=0} [\rho^{(1)}(2|\phi) - \rho^{(1)}(2|0)] \, \mathrm{d}2 \\ &= \frac{\rho}{z} + \int \left(\frac{\rho^{(1)}(1|\phi)}{z^*(1)} \frac{\delta}{\delta\rho^{(1)}(2|\phi)} \log \frac{\rho^{(1)}(1|\phi)}{z^*(1)} \right)_{\phi=0} \rho h(2,0) \, \mathrm{d}2 \\ &= \frac{\rho}{z} + \frac{\rho^2}{z} \int c(1,2) h(2,0) \, \mathrm{d}2 \end{aligned} \tag{5.4.11}$$

where, in the last line, we have used the definition (5.2.13) of the direct correlation function. On multiplying through by z/ρ, and making use of Eqn (5.4.7) and the relation

$$z^*(1) = z \exp[-\beta\phi(1)] = z \exp[-\beta v(0,1)] \tag{5.4.12}$$

Eqn (5.4.11) becomes

$$g(0,1) = \exp[-\beta v(0,1)] \left(1 + \rho \int c(1,2) h(2,0) \, \mathrm{d}2 \right) \tag{5.4.13}$$

When this result is combined with the Ornstein–Zernike relation, we find that

$$g(0,1) = \exp[-\beta v(0,1)][g(0,1) - c(0,1)] \tag{5.4.14}$$

The Ornstein–Zernike relation can also be used to eliminate the direct correlation function from (5.4.14). The result is a non-linear integral equation for $g(\mathbf{r})$ in terms of $v(\mathbf{r})$, known as the Percus–Yevick or PY equation:

$$\exp[\beta v(\mathbf{r})]g(\mathbf{r}) = 1+\rho\int[g(\mathbf{r}-\mathbf{r}')-1]\{1-\exp[\beta v(\mathbf{r}')]\}g(\mathbf{r}')\,\mathrm{d}\mathbf{r}' \qquad (5.4.15)$$

Equation (5.4.15) was originally derived in a very different way by Percus and Yevick (1958); it can also be obtained by a diagrammatic method (Stell, 1963).

The PY equation is the most successful of the first-order integral equations, at least for short-range potentials; its solution for the special but important case of hard spheres will be discussed in Section 5.5. The PY approximation is equivalent to setting

$$\begin{aligned} c(\mathbf{r}) &= \{1-\exp[\beta v(\mathbf{r})]\}g(\mathbf{r}) \\ &= g(\mathbf{r})-y(\mathbf{r}) \end{aligned} \qquad (5.4.16)$$

so that $c(\mathbf{r})$ is zero wherever the potential vanishes.

Another possible choice of the functional to be expanded is $\log[\rho^{(1)}(1|\phi)/z^*(1)]$. If we proceed along the same lines as before, we now find that

$$\begin{aligned} \log g(0,1) &= -\beta v(0,1)+\rho\int c(1,2)h(2,0)\,\mathrm{d}2 \\ &= -\beta v(0,1)+h(0,1)-c(0,1) \end{aligned} \qquad (5.4.17)$$

which is identical to (5.3.14) if, in the latter, we ignore the contribution of the bridge diagrams. The neglect of the bridge diagrams, as we saw in Section 5.3, is the characteristic approximation of HNC theory, and use of the Ornstein–Zernike relation to eliminate $c(r)$ from (5.4.17) leads to the HNC integral equation:

$$\log g(\mathbf{r})+\beta v(\mathbf{r}) = \rho\int[g(\mathbf{r}-\mathbf{r}')-1][g(\mathbf{r}')-1-\log g(\mathbf{r}')-\beta v(\mathbf{r}')]\,\mathrm{d}\mathbf{r}' \qquad (5.4.18)$$

Equation (5.4.18) is equivalent to the approximation

$$c(\mathbf{r}) = -\beta v(\mathbf{r})+h(\mathbf{r})-\log[h(\mathbf{r})+1] \qquad (5.4.19)$$

For large r, $h(\mathbf{r})\ll 1$; if we expand the logarithmic term in (5.4.19), we recover the Debye–Hückel approximation to $c(\mathbf{r})$, Eqn (5.4.10). As we shall see in Chapter 10, the r^{-1} decay of $c(\mathbf{r})$ at large r is crucial in understanding the properties of ionic fluids. For such systems, we must expect the HNC equation to be superior to the PY approximation, because the PY $c(\mathbf{r})$ has a different asymptotic behaviour. This expectation is borne out by calculations for electrolytes (Rasaiah and Friedman, 1968), molten salts (Hansen and McDonald, 1975) and plasmas (Baus and Hansen, 1980).

If Eqn (5.4.17) is written in the form

$$g(0,1)=\exp[-\beta v(0,1)]\exp[h(0,1)-c(0,1)] \qquad (5.4.20)$$

we see that the PY approximation (5.4.14) is recovered by linearization of the HNC result with respect to $(h-c)$. Not surprisingly, a diagrammatic analysis shows that the PY equation corresponds to summing a smaller class of diagrams than in the HNC approximation (Stell, 1963). To some extent, therefore, the greater success of the PY equation in the case of short-range potentials must be due to a cancellation of errors.

Finally, if the functional $\rho^{(1)}(1|\phi)\nabla_1 \log[1/z^*(1)]$ is expanded to first order in $\log[\rho^{(1)}(1|\phi)/z^*(1)]$, we recover the Born–Green equation of Section 2.6 (Percus, 1964).

We postpone until later sections a discussion of the quantitative reliability of the PY, HNC and Born–Green equations, but it is useful at this stage to consider some of the general features of these approximations. All three equations predict correctly that $g(1,2)$ behaves as $\exp[-\beta v(1,2)]$ in the limit $\rho\to 0$. They also yield the correct expression for the term of order ρ in the density expansion of $g(1,2)$; this can be checked by inserting the expansion (5.3.3) in both sides of Eqns (5.4.15), (5.4.18) and (2.6.13). It follows that they all give the correct second and third virial coefficients in the expansion of the equation of state. Higher-order coefficients are incorrect, because at order ρ^2 and beyond each approximation neglects a certain number (different for each theory) of the diagrams which appear in the exact expansion of $g(1,2)$. Once a solution for the pair distribution function has been obtained, the internal energy, equation of state and compressibility can be calculated, respectively, from Eqns (2.5.13), (2.5.14) and (2.5.32). Internal consistency of the theory may then be tested in two ways. First, the inverse compressibility can be integrated numerically with respect to density to yield the so-called compressibility equation of state, and the calculated pressures compared with those obtained directly from the virial equation. Secondly, the internal energy can be integrated with respect to inverse temperature to give the Helmholtz free energy (see Eqn (2.3.7)); the latter can in turn be differentiated numerically with respect to volume to yield the "energy" equation of state (Chen *et al.*, 1969). The results obtained via the three routes are in general different, sometimes greatly so. This lack of thermodynamic consistency is a common feature of approximate theories. Use of the HNC approximation offers some scope for simplification in numerical work, since it yields an expression for the free energy in terms of $g(1, 2)$; this avoids the need to proceed via integration of the internal energy (Morita, 1960).

5.5 THE PY SOLUTION FOR HARD SPHERES

The PY equation is of special interest in the theory of simple liquids because it is soluble analytically in the important case of the hard-sphere fluid. Written in terms of the function $y(r)$, the PY approximation (5.4.16) is

$$c(r) = y(r)f(r) \tag{5.5.1}$$

For hard spheres of diameter d, Eqn (5.5.1) is equivalent to

$$\left.\begin{aligned} c(r) &= -y(r), \qquad & r<d \\ &= 0, & r>d \end{aligned}\right\} \tag{5.5.2}$$

It follows that $c(r)$ has a discontinuity at $r = d$, since $y(r)$ is continuous everywhere. In addition to (5.5.2), there is a further, exact restriction on the solution, namely that

$$g(r) = 0, \qquad r<d \tag{5.5.3}$$

Given Eqns (5.5.2) and (5.5.3), it is possible to rewrite the Ornstein–Zernike relation as an integral equation for $y(r)$ of the form

$$\begin{aligned} y(r) = 1 + \rho \int_{r'<d} y(r')\,\mathrm{d}\mathbf{r}' \\ - \rho \int_{\substack{r'<d \\ |\mathbf{r}-\mathbf{r}'|>d}} y(r')y(|\mathbf{r}-\mathbf{r}'|)\,\mathrm{d}\mathbf{r}' \end{aligned} \tag{5.5.4}$$

Equation (5.5.4) can be solved by Laplace-transform methods (Thiele, 1963; Wertheim, 1963, 1964). The final result for $c(r)$ is

$$\left.\begin{aligned} c(x) &= -\lambda_1 - 6\eta\lambda_2 x - \tfrac{1}{2}\eta\lambda_1 x^3, \qquad & x<1 \\ &= 0, & x>1 \end{aligned}\right\} \tag{5.5.5}$$

where $x = r/d$, η is the packing fraction, and

$$\lambda_1 = (1+2\eta)^2/(1-\eta)^4 \tag{5.5.6}$$

$$\lambda_2 = -(1+\tfrac{1}{2}\eta)^2/(1-\eta)^4 \tag{5.5.7}$$

In the limit $\eta \to 0$, Eqn (5.5.5) reduces to $c(x) = -1$ for $x<1$ and $c(x) = 0$ for $x>1$, in agreement with the density expansion (5.2.16).

An alternative method of solution has been described by Baxter (1968); details are given in Appendix B. One advantage of Baxter's method is the fact that it is easily generalized to the case when the potential consists of a hard-sphere core plus a "tail".

The compressibility of the hard-sphere fluid is obtained by substitution of (5.5.5) into Eqn (5.2.17), and integration of the compressibility with respect to η yields the compressibility equation of state:

$$\frac{\beta P^{c}}{\rho}=\frac{1+\eta+\eta^{2}}{(1-\eta)^{3}} \tag{5.5.8}$$

Alternatively, substitution of

$$\lim_{r\to d+} g(r)=y(d)=-\lim_{r\to d-} c(r) \tag{5.5.9}$$

in Eqn (2.5.17) leads to the virial equation of state:

$$\frac{\beta P^{v}}{\rho}=\frac{1+2\eta+3\eta^{2}}{(1-\eta)^{2}} \tag{5.5.10}$$

The difference between P^{c} and P^{v} increases with increasing density.

Equations (5.5.8) and (5.5.10) both yield the exact values of the second and third virial coefficients, but give incorrect (and different) values for the higher-order coefficients. The general expressions for the nth coefficient, obtained by expanding the two equations of state in powers of the density, are

$$B_{n}^{c}/b^{n-1}=2[2+3n(n-1)]/4^{n} \tag{5.5.11}$$

$$B_{n}^{v}/b^{n-1}=8[3n-4]/4^{n} \tag{5.5.12}$$

where $b\equiv B_{2}=(2\pi/3)d^{3}$. The numerical values for $4\leqslant n\leqslant 7$ are shown in Table 5.1, together with the exact results of Section 4.6, the predictions of the two corresponding HNC equations of state up to $n=6$ (Rowlinson,

TABLE 5.1. *Virial coefficients of hard spheres: comparison between exact values and predictions of various theories*

	Exact	HNC(c)	HNC(v)	PY(c)	PY(v)	SCA
B_4/b^3	0.28695	0.2092	0.4453	0.2969	0.2500	0.2824
B_5/b^4	0.1103	0.0493	0.1447	0.1211	0.0859	0.1041
B_6/b^5	0.0386	0.0281	0.0382	0.0449	0.0273	0.0341
B_7/b^6	0.0138			0.0156	0.0083	

v = from virial equation; c = from compressibility equation; SCA = self-consistent approximation of Rowlinson (1965).

1964b), and results from a self-consistent approximation that we discuss in Section 5.8 (Rowlinson, 1965).

The PY compressibility and virial equations of state are plotted in Figure 5.4 for comparison with the Carnahan–Starling formula (4.6.14); the latter gives results that are almost indistinguishable from those of molecular dynamics and Monte Carlo calculations. As is the case with the virial coefficients, the two equations of state bracket the "exact" curve, the pressures calculated from the compressibility equation being systematically closer to and above the "exact" results at all densities. It appears that the Carnahan–Starling formula provides an accurate interpolation between the two PY equations of state. Indeed, Eqn (4.6.14) can be recovered by adding together the expressions (5.5.8) and (5.5.10) with weights, respectively, of two-thirds and one-third:

$$\begin{aligned}\frac{\beta P}{\rho} &= \frac{\beta}{3\rho}(2P^{c}+P^{v}) \\ &= \frac{1+\eta+\eta^{2}-\eta^{3}}{(1-\eta)^{3}}\end{aligned} \tag{5.5.13}$$

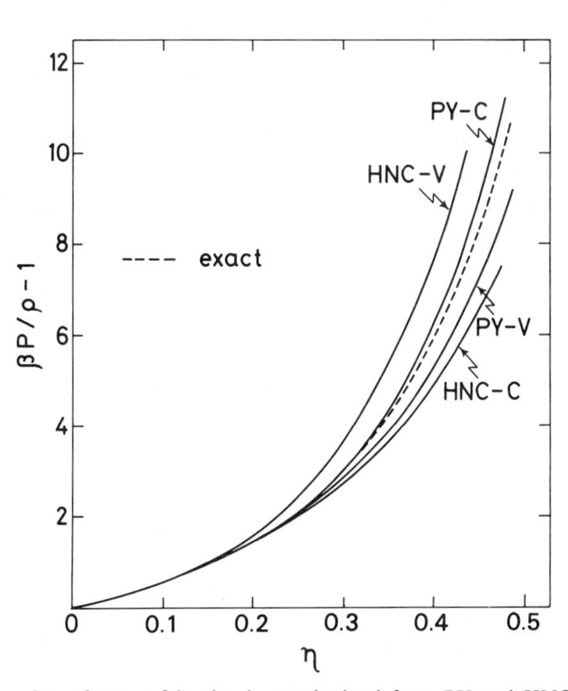

FIG. 5.4. Equation of state of hard spheres obtained from PY and HNC theories via the compressibility (C) and virial-equation (V) routes. The dashed curve is calculated from the Carnahan–Starling equation of state (4.6.14).

Results obtained by numerical solution of the HNC equation are also shown in Figure 5.4. They are clearly inferior to the PY results, while those obtained from the Born–Green equation (not shown) are even worse (Klein, 1963, 1964; Levesque, 1966).

The PY approximation to the pair distribution function is obtained by substitution of (5.5.5) into the Ornstein–Zernike relation. As a consequence of the discontinuity in $c(x)$ at $x=1$, $g(x)$ is only a piecewise-analytic function (Wertheim, 1964); the analytic expressions in the intervals $n \leq x \leq (n+1)$ for $n=1$ to 5 are given in a paper by Smith and Henderson (1970). A comparison of the PY pair distribution function with the results of a Monte Carlo calculation for the hard-sphere fluid at a density ($\eta=0.49$) close to crystallization is shown in Figure 5.5. Although the general agreement is good, the PY solution has two significant defects. First, the value at contact is too low. Secondly, the oscillations are slightly out of phase with the Monte Carlo results; this particular feature of the PY approximation is strongly accentuated when the pair potential has an attractive component (Madden and Rice, 1980). In addition, the amplitude of the oscillations decreases too slowly with increasing distance, with the result that the main peak in the structure factor is too high, reaching a maximum value of 3.05 rather than the value 2.85 obtained by simulation (Hansen and Verlet, 1969). An accurate representation of the pair distribution function of the hard-sphere fluid is an important ingredient of many theories. For that reason, Verlet and Weis (1972b) have devised a simple, semiempirical modification of the PY solution in which the faults of the latter are corrected. The implementation of certain theories also relies on a knowledge of the hard-sphere $y(r)$ inside the hard core ($r<d$). Grundke and Henderson (1972) have therefore developed an approximate form for $y(r)$ in the range $0 \leq r \leq d$ that interpolates smoothly between the Verlet–Weis expression at $r=d$ and exact results for $y(0)$ due to Hoover and Poirier (1962) and Meeron and Siegert (1968); in this region, the PY results for $y(r)$ are seriously in error. A detailed description of the Verlet–Weis and Grundke–Henderson parameterizations is given in Appendix C.

The Thiele–Wertheim solution of the PY equation was later extended by Lebowitz (1964) to the case of binary mixtures of hard spheres with different but additive diameters. The thermodynamic properties obtained from this solution (Lebowitz and Rowlinson, 1964) have been tested against Alder's results for equimolar mixtures of hard spheres that differ in diameter by a factor of three (Alder, 1964b); the agreement is comparable with that obtained in the one-component case. The Carnahan–Starling equation of state has also been generalized to mixtures (Mansoori *et al.*, 1971). Perhaps the most interesting result to emerge from this body of work is the fact that, irrespective of the diameter ratio, there is no phase separation in the fluid

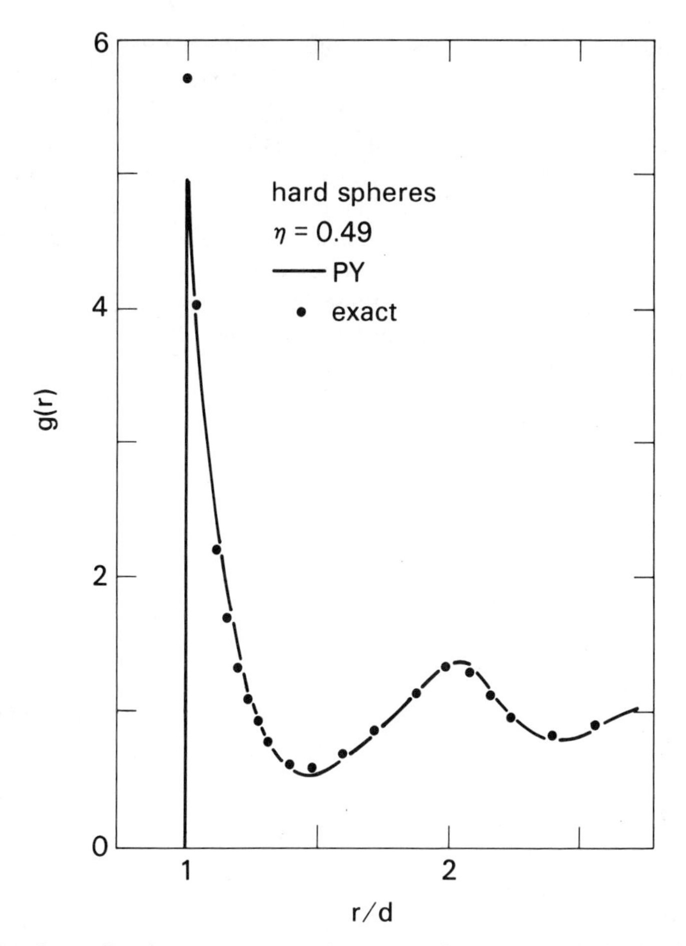

FIG. 5.5. Comparison between "exact" and PY results for the radial distribution function of hard spheres at $\eta = 0.49$.

phase of mixtures of additive hard spheres. The behaviour of non-additive hard spheres, i.e. those for which

$$d_{12} = \tfrac{1}{2}(1+\Delta)(d_{11}+d_{22}) \tag{5.5.14}$$

with $\Delta \neq 0$, is more complicated. Computer simulations have shown that positive non-additivity ($\Delta > 0$) leads to phase separation at a density which is dependent on Δ (Melnyk and Sawford, 1975), while mixtures with negative Δ have been studied as models of systems that have compound-forming tendencies (Adams and McDonald, 1975; Nixon and Silbert, 1984).

It is of interest historically that the result given by the PY compressibility equation of state, Eqn (5.5.8), had been obtained much earlier by "scaled-particle" theory (Reiss *et al.*, 1959). The basis of scaled-particle theory is an expression for the reversible work needed to create a spherical cavity from which the centres of hard spheres are excluded. This expression is then related to the average density of hard-sphere centres that are in contact with the wall of the cavity. The significance of the calculation lies in the fact that a fixed, spherical cavity of radius d affects the surrounding fluid in precisely the same way as an additional hard sphere of diameter d placed at the centre of the cavity. Scaled-particle theory has been extended to fluids of non-spherical, hard particles by Gibbons (1969, 1970), including systems, such as hard spherocylinders, that serve as models for the study of the nematic–isotropic fluid transition of liquid crystals. Gibbons' work has also been the inspiration for the development of a number of semiempirical equations of state for hard, convex bodies of arbitrary shape (Boublik, 1983).

5.6 THE MEAN-SPHERICAL APPROXIMATION

There are a variety of model fluids of interest in the theory of liquids for which the pair potential consists of a hard-sphere interaction plus a "tail". The latter is normally attractive, but not necessarily spherically symmetric. The full potential is therefore given by

$$\begin{aligned} v(1,2) &= \infty, \qquad R_{12} < d \\ &= v_1(1,2), \qquad R_{12} > d \end{aligned} \tag{5.6.1}$$

Such systems, which include both the "square-well" and dipolar hard-sphere fluids of Section 1.2, have been widely studied in the *mean-spherical approximation* or MSA. The name has its origins in the fact that the approximation was first proposed as a generalization of the mean-spherical model of Ising spin systems (Lebowitz and Percus, 1966). The MSA is given in terms of the pair distribution function and direct correlation function by

$$g(1,2) = 0, \qquad R_{12} < d \tag{5.6.2a}$$

$$c(1,2) = -\beta v_1(1,2), \qquad R_{12} > d \tag{5.6.2b}$$

When supplemented by the Ornstein–Zernike relation, these expressions yield an integral equation for $g(1,2)$. The first expression is exact, while the second extends the asymptotic behaviour of $c(1,2)$ to all $R_{12} > d$ and is clearly an approximation. Despite the crude form assumed for $c(1,2)$, the MSA gives good results in many cases. For example, it provides a much better description of the properties of the square-well fluid than is given by either the PY or HNC approximation (Smith *et al.*, 1977). However, the most attractive feature of the MSA is the fact that the integral equation can be solved analytically for a number of potential models of physical interest, including simple models of electrolyte solutions and of polar liquids. These applications of the MSA will be discussed in detail in Chapters 10 and 12.

The PY equation for hard spheres is the special case of the MSA when the tail in the potential is absent. The analytic solution of the MSA for certain pair potentials is therefore closely linked to the Wertheim–Thiele and Baxter methods of solution of the PY hard-sphere problem. The relation between the PY approximation and the MSA is deeper than this, because the two theories have a common diagrammatic structure (Madden, 1981), but the connection between the two can be established in a very simple way. The basic PY approximation (5.4.14) may be re-expressed in the form

$$c(1,2)=f(1,2)+f(1,2)[h(1,2)-c(1,2)] \tag{5.6.3}$$

where $f(1,2)$ is the Mayer function for the potential $v(1,2)$. In the low-density limit, $h(1,2)$ and $c(1,2)$ become the same, and the right-hand side of Eqn (5.6.3) reduces to $f(1,2)$. Equation (5.6.3) can therefore be written as

$$c(1,2)=c_0(1,2)+f(1,2)[h(1,2)-c(1,2)] \tag{5.6.4}$$

where $c_0(1,2)$, the limiting value of $c(1,2)$ at low density, is equal to $f(1,2)$ both in an exact theory and in the PY approximation. If, however, we choose another form for $c_0(1,2)$ in Eqn (5.6.4), we generate a different approximate theory. For a potential of the type defined by Eqn (5.6.1), the exact $c_0(1,2)$ is

$$\begin{aligned} c_0(1,2)&=\exp[-\beta v(1,2)]-1 \\ &=[1+f_d(r)]\exp[-\beta v_1(1,2)]-1 \end{aligned} \tag{5.6.5}$$

where $f_d(r)$ is the Mayer function for hard spheres of diameter d (with $r=R_{12}$). The MSA consists in linearizing (5.6.5) with respect to $v_1(1,2)$ by setting

$$\begin{aligned} c_0(1,2)&\simeq[1+f_d(r)][1-\beta v_1(1,2)]-1 \\ &=f_d(r)-\beta v_1(1,2)[1+f_d(r)] \end{aligned} \tag{5.6.6}$$

while at the same time replacing $f(1,2)$ in (5.6.4) by $f_d(r)$. Taken together, the two approximations lead to

$$f_d(r)[1+h(1,2)]=[c(1,2)+\beta v_1(1,2)][1+f_d(r)] \tag{5.6.7}$$

which is equivalent to the closure relations (5.6.2). This characterization of the MSA shows that it involves approximations that are much more severe than those underlying the PY equation. One would not expect the MSA to be of comparable accuracy to the PY approximation except when $\beta v_1(1,2)$ is very small. As the results for the square-well fluid show, this expectation is not always realized.

The structure of Eqn (5.6.7) suggests a natural way in which the MSA can be extended to a more general class of pair potentials than that defined by (5.6.1) (Chihara, 1973; Madden and Rice, 1980). We suppose that the potential $v(1,2)$ is divided in the form

$$v(1,2) = v_0(1,2) + v_1(1,2) \tag{5.6.8}$$

The conventional MSA is applicable only when $v_0(1,2)$ is the hard-sphere potential. When $v_0(1,2)$ is strongly repulsive but continuous, the natural generalization of the closure relation (5.6.7) is obtained by replacing f_d by f_0, the Mayer function for the potential $v_0(1,2)$. The resulting equation can then be rearranged to give

$$c(1,2) = \{1 - \exp[\beta v_0(1,2)]\}g(1,2) - \beta v_1(1,2) \tag{5.6.9}$$

which becomes the same as the PY approximation (5.4.14) when $v_1(1,2)$ is very weak. Madden and Rice (1980) have applied the "soft-core" MSA to both the Lennard-Jones fluid and a model of liquid potassium (Murphy, 1977). In each case, excellent results are obtained when the potential is divided at its (first) minimum in the manner of Weeks *et al.* (1971) (see Section 6.4).

5.7 NUMERICAL RESULTS

With the exception of those special cases where analytic solutions have been found, the integral equations described in Sections 5.4 and 5.6 must be solved by numerical methods. The solution is usually obtained by an iterative approach, starting with a guess for either of the functions h or c. Perhaps the easiest method is to use the Ornstein–Zernike relation between the Fourier transforms of h and c, Eqn (5.2.11). An initial guess $c^{(0)}(r)$ (or $h^{(0)}(r)$) is made and its Fourier transform $\hat{c}^{(0)}(k)$ is calculated; the latter is then inserted in Eqn (5.2.11) and an inverse transform yields an approximation $h^{(0)}(r)$. The appropriate relation between h and c is then used to obtain an improved guess $c^{(1)}(r)$. For example, in the PY approximation, it follows from Eqn (5.4.16) that

$$c^{(1)}(r) = \{1 - \exp[\beta v(r)]\}[h^{(0)}(r) + 1] \tag{5.7.1}$$

In the case of the HNC approximation, Eqn (5.4.19) must be used. The procedure is then repeated, with $c^{(1)}(r)$ replacing $c^{(0)}(r)$ as the input, and the iteration continues until convergence is achieved. This scheme is an example of what, in numerical analysis, is called a Picard method. In practice, the simple Picard method works only in the weak-coupling limit (low density or high temperature). To ensure convergence, it is generally necessary to mix successive approximations to $c(r)$ before they are used as input at the next level of iteration (Broyles, 1960). It is usually sufficient to mix the output of two previous iterations according to the rule

$$c_{\text{in}}^{(i+1)}(r) = (1-\alpha)c_{\text{out}}^{(i)}(r) + \alpha c_{\text{out}}^{(i-1)}(r) \tag{5.7.2}$$

where the mixing parameter α $(0 < \alpha < 1)$ is adjusted empirically to force the solution to converge. Both the value of α and the number of iterations that are needed tend to increase rapidly as the coupling becomes stronger. As an illustration, for the Lennard-Jones fluid near its triple point, it requires of order 10^3 iterations with $\alpha \simeq 0.95$ to obtain converged solutions to the PY and HNC equations, even when a good guess for $c^{(0)}(r)$ is available.

An iterative method can also be used to extend the pair distribution function obtained from a molecular dynamics or Monte Carlo calculation. Computer "experiments" yield $g(r)$ only for $r < r_c$, where r_c is a cut-off distance equal, at most, to $\frac{1}{2}L$, where L is the length of the central cell (see Section 3.1). Since the number of particles in the central cell is typically of order 10^3 or less, the maximum choice of r_c is only a few interatomic spacings. In a dense fluid, $g(r)$ is far from having reached its asymptotic value at distances of that size, and the problem arises of how to determine the behaviour of $g(r)$ for $r > r_c$. Such information is needed, for example, if accurate values of the structure factor $S(k)$ are to be obtained via its definition as a Fourier transform, Eqn (5.1.9); the problem is analogous to that encountered by experimentalists in passing from $S(k)$ to $g(r)$. The extrapolation scheme proposed by Verlet (1968) is based on the fact that, in most circumstances, the pair correlation function $h(r)$ is small for $r > r_c$, and a first-order, functional Taylor expansion with respect to $h(r)$ is therefore a good approximation in that range of r. Such an expansion corresponds precisely to the PY approximation (see Eqn (5.4.11)), so that, for $r > r_c$, we may use the PY relation between $c(r)$ and $g(r)$, i.e.

$$c(r) = g(r)\{1 - \exp[\beta v(r)]\}, \qquad r > r_c \tag{5.7.3}$$

For shorter distances, $g(r)$ is equal to $g_c(r)$, the function calculated in the computer "experiment":

$$g(r) = g_c(r), \qquad r < r_c \tag{5.7.4}$$

The pair of equations (5.7.3) and (5.7.4), in combination with the Ornstein–Zernike relation, can be solved by iteration, in the manner of an ordinary integral equation, to yield $g(r)$ for $r > r_c$ and $c(r)$ for $r < r_c$. If there is a coulombic term in the potential, Eqn (5.7.3) should be replaced by its HNC counterpart (Hansen and McDonald, 1975, 1981). A more elaborate method of extrapolation of $g(r)$ has been devised by Foiles *et al.* (1984b).

In the numerical solution of an integral equation, the range of r is usually divided into n equally spaced values r_i, where n is typically equal to 1024 (a power of two is required if the "fast Fourier transform" method is to be used). The functions $h(r)$ and $c(r)$ may then be regarded as vectors $\mathbf{h}$ and $\mathbf{c}$, with components $h_i \equiv h(r_i)$ and $c_i \equiv c(r_i)$. Similarly, the Ornstein–Zernike relation becomes a matrix equation between $\mathbf{h}$ and $\mathbf{c}$; this equation, together with an approximate closure relation, yields in turn a set of n non-linear, simultaneous equations for the components of $\mathbf{h}$ (or $\mathbf{c}$). As Gillan (1979) has pointed out, sets of equations of this type are commonly solved by the Newton–Raphson method (Ortega and Rheinboldt, 1970). The Newton–Raphson method is also an iterative scheme, but it converges much more rapidly than the Picard method. Unfortunately, implementation of the Newton–Raphson method requires the inversion of an $n \times n$ matrix at each iteration; when n is of order 10^3, this is a formidable problem. Gillan (1979) has therefore proposed the use of a hybrid approach in which $h(r)$ and $c(r)$ are obtained on a coarse grid of values of r by the Newton–Raphson method, and the calculation is completed by applying the Picard method on a finer grid. The hybrid scheme is more difficult to program than the Picard method, but it reduces the number of iterations that are required by an order of magnitude, as well as being insensitive to the initial approximation. Alternatively, given a sufficiently large computer, it is possible to exploit the numerical techniques that are available for the inversion of large matrices, and do the calculation solely by the Newton–Raphson method (Zerah, 1986).

Solutions to the PY, HNC and Born–Green equations have been obtained for a variety of pair potentials over wide ranges of temperature and density. The most comprehensive set of calculations for the Lennard-Jones potential are those made by Levesque (1966). Comparison of Levesque's results with the results of computer simulations shows that the PY equation is superior to the other two approximations at all thermodynamic states that were studied. At high temperatures ($T^* \gtrsim 2.5$), the PY results from the virial equation of state are in excellent agreement with the simulations at all densities up to $\rho^* \simeq 0.8$; the agreement worsens as the temperature is reduced. The situation is summarized in Table 5.2, where the equation of state and excess internal energy for the PY and HNC approximations are compared with Monte Carlo results along the isotherms $T^* = 2.74$ and $T^* = 1.35$; the latter corresponds to a near-critical temperature. At subcritical temperatures, the agreement with computer "experiments", even for the PY approximation, becomes very poor. The deficiencies of the PY

TABLE 5.2. *Thermodynamic properties of the Lennard-Jones fluid: comparison between Monte Carlo results and the predictions of integral-equation theories*

$T^* = 2.74$

	MC		HNC			PY		
ρ^*	$\dfrac{\beta P}{\rho}$	$\dfrac{U^{ex}}{N\varepsilon}$	$\dfrac{\beta P^c}{\rho}$	$\dfrac{\beta P^v}{\rho}$	$\dfrac{U^{ex}}{N\varepsilon}$	$\dfrac{\beta P^c}{\rho}$	$\dfrac{\beta P^v}{\rho}$	$\dfrac{U^{ex}}{N\varepsilon}$
0.3	1.040	−1.783	1.050	1.083	−1.787	1.056	1.070	−1.791
0.4	1.199	−2.371	1.177	1.279	−2.351	1.186	1.234	−2.368
0.55	1.653	−3.207	1.542	1.901	−3.127	1.614	1.722	−3.238
0.7	2.641	−3.902	2.213	3.160	−3.693	2.437	2.646	−3.931
0.8	3.604	−4.281	2.895	4.528	−3.852	3.345	3.603	
1.0	7.388	−4.180	5.095	9.113	−3.261	6.808	6.670	−4.576

$T^* = 1.35$

	MC		HNC		PY	
ρ^*	$\dfrac{\beta P}{\rho}$	$\dfrac{U^{ex}}{N\varepsilon}$	$\dfrac{\beta P^v}{\rho}$	$\dfrac{U^{ex}}{N\varepsilon}$	$\dfrac{\beta P^v}{\rho}$	$\dfrac{U^{ex}}{N\varepsilon}$
0.3	0.352	−2.090			0.396	−2.18
0.35	0.298	−2.405			0.376	−2.48
0.4	0.272	−2.747	0.388	−2.79	0.386	−2.77
0.45	0.280	−3.030	0.442	−3.08	0.434	−3.07
0.5	0.303	−3.372	0.557	−3.37	0.532	−3.38
0.55	0.415	−3.704	0.755	−3.68	0.692	−3.69
0.65	0.850	−4.343	1.492	−4.26	1.256	−4.32
0.7	1.166	−4.684	2.086	−4.52	1.689	−4.65

MC = Monte Carlo results of Hansen and Verlet (1969); HNC and PY results are from Levesque (1966) and Verlet and Levesque (1967). c = from compressibility equation; v = from virial equation.

equation in the liquid range are also evident in the behaviour of the pair distribution function; the main peak in $g(r)$ is too large in magnitude and occurs at too small a value of r, and the later oscillations, like those in the hard-sphere case, are out of phase with the results of simulations. The "energy" equation of state obtained from the PY solution yields much better results near the triple point than those given by the other two routes to the pressure (Chen *et al.*, 1969), but the improvement is largely fortuitous. The serious thermodynamic inconsistency of the PY approximation at low temperatures shows that it cannot be regarded as a satisfactory theory of the liquid state.

The PY, HNC and Born-Green theories are all exact in the low-density limit. In particular, the pair distribution functions, as noted in Section 5.4, are all exact to order ρ. It is therefore not surprising to find that they all yield increasingly more accurate results as the density is reduced. However, for those model fluids that display a liquid–gas transition, there are some special difficulties associated with the intermediate range of temperature and density in which the critical point is located. It has long been known experimentally that many thermodynamic and transport properties show an anomalous behaviour in the vicinity of the critical point. The anomalies are a direct consequence of the large density fluctuations that are characteristic of the critical region. A good example is provided by the compressibility. At the critical point, χ_T becomes infinite; its behaviour for $T \simeq T_c$ as the critical temperature is approached from above is described (Stanley, 1971) by a simple power law of the form

$$\chi_T \propto (T - T_c)^{-\gamma}, \qquad T > T_c \tag{5.7.5}$$

where γ, a positive number, is called a *critical exponent*; similar power-law behaviour is found for other properties. It is a striking fact that the numerical values of the critical exponents for a given physical property are apparently the same, within experimental uncertainties, for all fluids. This *universality* in the values of the critical exponents extends to all members of what is called the same "universality class". The task of extracting an accurate value for a critical exponent from the numerical solution to an integral equation is computationally a very demanding one, but Green *et al.* (1979, 1980) have been able to show that, in the Born-Green approximation, the exponent γ for the "square-well" fluid has a value equal to 1.24 ± 0.04. This result is in excellent agreement with the accepted experimental value of 1.23 ± 0.02 (Stanley, 1971). The apparent success of the Born–Green equation in this example may, however, be deceptive. Analytic studies (Jones *et al.*, 1981; Fisher and Fishman, 1981) suggest that there is no true criticality in the Born–Green approximation, since the compressibility is always bounded, while the deficiencies of the PY and HNC approximations in the critical region are found to be even more severe.

5.8 EXTENSIONS OF INTEGRAL EQUATIONS

Much ingenuity has been devoted to finding ways of extending the PY and HNC equations so as to achieve better agreement with the results of computer simulations. As we showed in Section 5.4, the PY approximation can be derived from a functional Taylor expansion truncated at first order. The question naturally arises as to whether any significant improvement is obtained when the quadratic term is included. This problem was investigated by Verlet (1964, 1965), who proposed what is called the PY2 approximation. As is to be expected, the integral equation is much more complicated than in first order. A supplementary term must be added to the basic PY relation (5.4.16), and the extra term is a functional, not only of $c(1, 2)$ and $h(1, 2)$, but also of the triplet distribution function $g^{(3)}(1, 2, 3)$. An additional, approximate equation for the triplet function is therefore needed in order to obtain a self-consistent theory. The theory is computationally awkward to handle, but numerical solutions for the Lennard-Jones potential have been obtained (Verlet and Levesque, 1967). At moderate densities and temperatures, the results show a clear improvement over the first-order approximation, but in the liquid range they remain unsatisfactory.

A second-order, HNC2 theory can also be derived (Verlet, 1964, 1965), but the HNC equation is much more easily and successfully extended in the following way. We first rewrite Eqn (5.3.14) in the form

$$c(r) = h(r) - \log\{g(r) \exp[\beta v(r)]\} + d_0(r) \tag{5.8.1}$$

When $d_0(r) = d(r)$, the sum of all bridge diagrams or "bridge function" for the potential $v(r)$, this relation is exact; when $d_0(r) = 0$, it reduces to the HNC approximation. More generally, $d_0(r)$ may be identified with the known bridge function of a suitable reference system, a step that leads to what is called the "reference" HNC or RHNC approximation (Lado, 1973). Because Eqn (5.8.1) can also be written as

$$g(r) = \exp[-\beta\{v(r) - k_B T d_0(r)\} + h(r) - c(r)] \tag{5.8.2}$$

it follows (see Eqn (5.4.17)) that the task of solving the RHNC integral equation is equivalent (Rosenfeld and Ashcroft, 1979) to finding the solution to the HNC equation for an effective potential $v_{\text{eff}}(r)$, defined as

$$v_{\text{eff}}(r) = v(r) - k_B T d_0(r) \tag{5.8.3}$$

The obvious choice of reference system is a fluid of hard spheres, since this is the only potential model for which the bridge function is known with sufficient accuracy. The parameterizations described in Appendix C give $h_0(r)$ and $y_0(r)$ for all r, and the sum of all series diagrams, $b_0(r) = h_0(r) - c_0(r)$, can be calculated from $h_0(r)$ via the Ornstein–Zernike relation. Once $y_0(r)$ and $b_0(r)$ are known, $d_0(r)$ can be obtained from the expression

$$d_0(r) = \log y_0(r) - b_0(r) \tag{5.8.4}$$

Equations (5.8.1) and (5.8.4) together form a one-parameter theory in which the only unknown quantity is the hard-sphere diameter d.

The approximation (5.8.1) was first proposed by Lado (1973) and has been systematically exploited by Rosenfeld and Ashcroft (1979), Lado *et al.* (1983) and Foiles *et al.* (1984a). (The "bridge function" in these papers is usually defined, in the present notation, as $-d(r)$.) Rosenfeld and Ashcroft (1979) point out that, for a variety of model fluids, solution of the HNC equation gives a pair distribution function that, in comparison with the results of computer simulations, is too weakly structured and has its main peak shifted to smaller distances. The consistent pattern of the HNC results suggests that the true bridge function always makes a repulsive contribution to the effective potential (5.8.3). This conclusion is supported by the explicit calculations made for hard spheres, which show not only that $d(r)$ is negative at all densities, but that it is also very short ranged, being essentially zero for $r \gtrsim 1.2d$. Rosenfeld and Ashcroft further argue that the characteristics of the bridge function make it likely that it is insensitive to details of the potential, and that its representation by a hard-sphere function is therefore a good approximation. They have applied the method to a number of different systems, the free parameter d being chosen so as to give thermodynamic consistency to the theory; in later work, a variational criterion has been used (Lado, 1982). The overall agreement with the results of simulations is very good, as exemplified by the results for thermodynamic properties of the Lennard-Jones fluid given in Table 5.3. The errors in the corresponding pair distribution functions are barely discernible, even at conditions ($\rho^* = 0.85$, $T^* = 0.719$) close to the triple point (Lado *et al.*, 1983).

TABLE 5.3. *Thermodynamic properties of the Lennard-Jones fluid: comparison between molecular dynamics results (Verlet, 1967) and predictions of the RHNC approximation. After Lado* et al. *(1983)*

		$\beta P/\rho$		$\beta U^{ex}/N$	
ρ^*	T^*	MD	RHNC	MD	RHNC
0.85	0.719	0.36	0.424	−6.12	−6.116
0.85	2.889	4.36	4.364	−4.25	−4.240
0.75	1.071	0.89	0.852	−5.17	−5.166
0.65	1.036	−0.11	−0.155	−4.52	−4.522
0.65	2.557	2.14	2.136	−3.78	−3.786
0.45	1.552	0.57	0.552	−2.98	−2.982
0.45	2.935	1.38	1.377	−2.60	−2.608
0.40	1.424	0.38	0.382	−2.73	−2.728

Many attempts have been made to combine the PY and HNC approximations in integral-equation schemes that guarantee thermodynamic consistency. In the case of the hard-sphere potential, for example, the PY approximation neglects the direct correlation function for $r > d$, while in HNC theory $c(r)$ has a "tail" for $r > d$, given by

$$c(r) = h(r) - \log [h(r)+1], \qquad r > d \tag{5.8.5}$$

The evidence from computer simulations is that the true strength of the tail is intermediate between these two approximations. It is therefore appropriate to write $c(r)$ in the form suggested by Rowlinson (1965), i.e. as

$$c(r) = f(r)y(r) + t(r) \tag{5.8.6}$$

with

$$t(r) = C(\rho)[y(r) - \log y(r) - 1] \tag{5.8.7}$$

where the first term on the right-hand side of (5.8.6) represents the short-range part of $c(r)$ and the second term represents the tail; the PY and HNC approximations are recovered, respectively, when $C(\rho) = 0$ and $C(\rho) = 1$ for all ρ. The function $C(\rho)$ can now be expanded in powers of ρ and the coefficients chosen such that the same virial coefficients are obtained from the virial and compressibility equations (Rowlinson, 1965). If this procedure were carried through to all orders in ρ, a self-consistent equation of state would be obtained. A similar idea was later applied by Lado (1967) to the calculation of the pair distribution function. Among more recent schemes (Hutchinson and Conkie, 1971; Rogers and Young, 1984; Zerah and Hansen, 1986), perhaps the simplest is that of Rogers and Young. Their method is based on the observation that whereas the HNC approximation is correct at large separations, the PY approximation, since it is much superior for strongly repulsive potentials, is presumably more accurate at short distances. It is therefore plausible to mix the two approximations in such a way that the function $y(r)$ of Eqn (2.5.15) reduces to its PY value for $r \to 0$ and to its HNC value for $r \to \infty$. The parameter that determines the proportions in which the two approximations are mixed at intermediate values of r can then be chosen to force consistency between the compressibility and virial equations of state.

5.9 INTEGRAL EQUATIONS FOR NON-UNIFORM FLUIDS

The applications of distribution-function theories that we have considered so far have been limited to situations in which the fluid is homogeneous. In this section, we show how the basic theory we have already developed can be applied to non-uniform systems, focussing our attention on the calculation of the single-particle density $\rho^{(1)}(\mathbf{r})$. The two examples of special interest are those of a planar liquid–gas interface and of a fluid in contact with a planar wall with which the fluid can interact. The assumption of a planar geometry makes it convenient to work in a system of coordinates in which the z-axis is perpendicular to the interface. The single-particle density $\rho^{(1)}(x, y, z)$ is then a function only of z; $\rho^{(1)}(z)$ is usually called the *density profile.* Even for this limited class of problems, a wide variety of density profiles can be envisaged. Some examples are plotted in Figures 5.6 and 5.7. In Figure 5.6(a) we show the density profile of a typical free-liquid surface. The origin of coordinates, indicated by the dashed line, is at the Gibbs *dividing surface* or, more precisely, the equimolar dividing surface

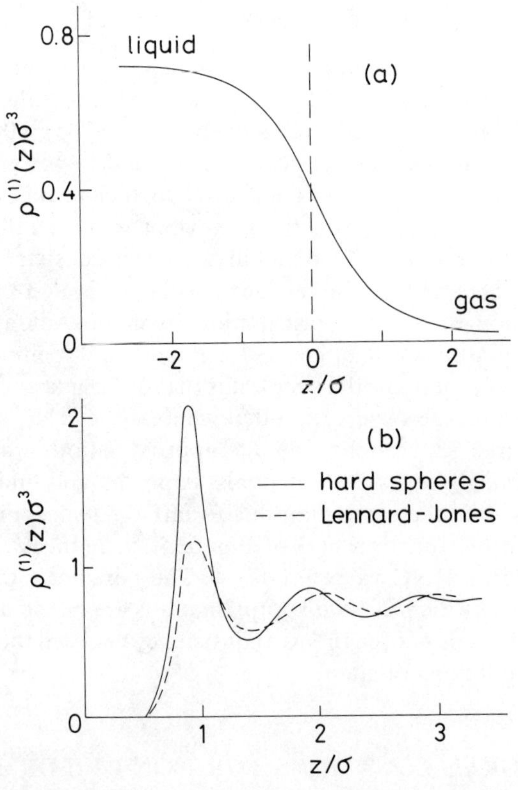

FIG. 5.6. Typical density profiles at (a) the liquid–gas interface of the Lennard-Jones fluid near the triple point and (b) the interfaces between Lennard-Jones or hard-sphere fluids and a 9-3 wall. The quantity σ is the Lennard-Jones length parameter or the hard-sphere diameter, as appropriate; the dashed line in (a) represents the Gibbs dividing surface. After Fischer and Methfessell (1980).

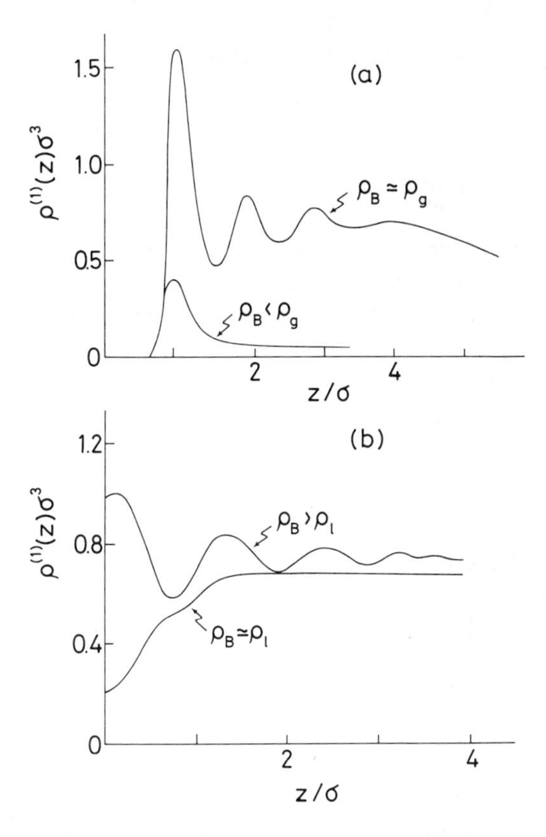

FIG. 5.7. Density profiles of a Lennard-Jones fluid against a wall showing the onset (a) of wetting by liquid at a 9-3 wall and (b) of wetting by gas ("drying") at a hard wall; the quantities ρ_B, ρ_g and ρ_l are, respectively, the densities of the bulk fluid and of the gas (g) and liquid (l) at coexistence. After Meister and Kroll (1985).

(Rowlinson and Widom, 1982). In other words, the plane $z=0$ is chosen such that

$$\int_{-\infty}^{0} [\rho_l - \rho^{(1)}(z)]\, dz = \int_{0}^{\infty} [\rho^{(1)}(z) - \rho_g]\, dz \tag{5.9.1}$$

where ρ_l and ρ_g are the bulk densities, respectively, of the coexisting liquid and gas. The usual convention is followed in which $\rho^{(1)}(z) \to \rho_l$ as $z \to -\infty$ and $\rho^{(1)}(z) \to \rho_g$ as $z \to +\infty$. Near the triple point, the transition from a liquid-like to a gas-like density occurs over an interval of about two molecular diameters, but the interfacial width increases rapidly with temperature. The transition occurs smoothly; although there has been much speculation about the matter, there is no credible evidence from either theory or simulation in favour of the existence of stable density oscillations at the liquid–gas interface.

When a fluid is in contact with a wall, the details of the density profile are sensitive both to the interactions between particles of the fluid and to the form of the fluid–wall potential. In theoretical work, the latter is often represented in a way that treats the wall as a continuum or semicontinuum. A popular choice is to assume that atoms in the wall interact with those in the fluid through a Lennard-Jones 12-6 potential of the type defined by Eqn (1.2.3); when integrated over a continuous distribution of wall atoms, this gives rise to a fluid-wall potential of Lennard-Jones 9-3 form (Steele, 1974), i.e.

$$\phi(z) = \tfrac{2}{3}\pi\rho_W\sigma^3\varepsilon\left[\frac{2}{15}\left(\frac{\sigma}{z}\right)^9 - \left(\frac{\sigma}{z}\right)^3\right] \tag{5.9.2}$$

where ρ_W is the number density of atoms in the wall and $z = 0$ corresponds to the surface of the wall. Thus $\phi(z) \to \infty$ as $z \to 0$ and $\rho^{(1)}(z) \to \rho_B$ as $z \to +\infty$, where ρ_B is the bulk density of the fluid. If ρ_B is sufficiently large, the density profile always has a layered structure that extends several molecular diameters into the bulk fluid and is qualitatively similar to the radial distribution function of a dense uniform liquid. Behaviour of this type is evident in Figure 5.6(b), where the density profiles are shown of both hard-sphere and Lennard-Jones fluids in contact with a 9-3 wall. It is also clear from the figure that the differences between the two interparticle potentials lead to substantial differences in structure in the region close to the wall. Such sensitivity to the potential is in marked contrast to what is observed in the corresponding bulk fluids, as we shall discuss in Chapter 6. In the opposite limit, when the bulk density is sufficiently low, the density profile has only a single peak and resembles the radial distribution function of a uniform gas. An example is shown in Figure 5.7(a). On the other hand, as ρ_B approaches ρ_g from below, ρ_g being again the density of the gas at coexistence with the liquid, we may expect to see changes of the type, also pictured in Figure 5.7(a), in which there is a pile-up of fluid against the wall and the growth of a film of liquid-like density. This behaviour is characteristic of the familiar phenomenon of *wetting* of the solid–gas interface by liquid. For wetting to be observed, it is necessary that the temperature T be such that $T_w < T < T_c$, where T_c is the critical temperature and T_w is the temperature of the *wetting transition.* Complete wetting can be expected to occur as $\rho_B \to \rho_g$, with the film becoming macroscopically thick. Such a transition was first predicted on very general grounds by Cahn (1977). In a similar way, as ρ_B approaches ρ_l from above, a layer of gas can be expected to develop at the interface of a liquid with a purely repulsive wall. The effect is to suppress the oscillations in the density profile, which comes to resemble that of a liquid–gas interface. This phenomenon, the onset of which in a typical case is illustrated in Figure 5.7(b), corresponds to wetting by gas or "drying". Wetting by gas has been observed in computer simulations (Abraham, 1978; Sullivan *et al.*, 1980), but simulations have so far given, at best, only a hint of thick-film formation of the type associated with wetting by liquid (Rowley *et al.*, 1976; Lane *et al.*, 1979; Snook and van Megen, 1982).

In order to provide a framework for the development of approximate theories of the density profile, we first derive some exact integrodifferential equations for $\rho^{(1)}(\mathbf{r})$. We are concerned with the way in which $\rho^{(1)}(\mathbf{r})$ varies with an external potential $\phi(\mathbf{r})$ at fixed values of μ, V and T. In these circumstances, variations in the generalized activity $z^*(\mathbf{r})$ defined by (4.2.27) are equivalent to variations in $\exp[-\beta\phi(\mathbf{r})]$. Equation (5.2.2) may therefore be rewritten as

$$-k_{\mathrm{B}}T\frac{\delta\rho^{(1)}(1)}{\delta\phi(2)}=\rho^{(1)}(1)\delta(1,2)+\rho^{(1)}(1)\rho^{(1)}(2)h(1,2) \tag{5.9.3}$$

The gradient produced in $\rho^{(1)}(\mathbf{r})$ by a gradient in the external potential (cf. Eqn (5.2.3)) is

$$\nabla_1\rho^{(1)}(1)=\int\frac{\delta\rho^{(1)}(1)}{\delta\phi(2)}\nabla_2\phi(2)\,\mathrm{d}2 \tag{5.9.4}$$

If (5.9.3) is substituted in (5.9.4), we find that

$$\begin{aligned}\nabla_1\rho^{(1)}(1)&=-\beta\int\rho^{(1)}(1)\delta(1,2)\nabla_2\phi(2)\,\mathrm{d}2\\&\quad-\beta\int\rho^{(1)}(1)\rho^{(1)}(2)h(1,2)\nabla_2\phi(2)\,\mathrm{d}2\\&=-\beta\rho^{(1)}(1)\nabla_1\phi(1)-\beta\rho^{(1)}(1)\int\rho^{(1)}(2)h(1,2)\nabla_2\phi(2)\,\mathrm{d}2\end{aligned} \tag{5.9.5}$$

or, after division by $-\beta\rho^{(1)}(1)$:

$$-k_{\mathrm{B}}T\nabla_1\log\rho^{(1)}(1)=\nabla_1\phi(1)+\int\rho^{(1)}(2)h(1,2)\nabla_2\phi(2)\,\mathrm{d}2 \tag{5.9.6}$$

Equation (5.9.6) is an exact relation for $\rho^{(1)}(\mathbf{r})$ in which the unknown quantity is the pair correlation function $h(1,2)$. A second equation, which relates $\rho^{(1)}(\mathbf{r})$ to the direct correlation function, can be obtained by rewriting the definition (5.2.13) of $c(1,2)$ to give

$$\delta\log\rho^{(1)}(1)+\beta\delta\phi(1)=\int c(1,2)\delta\rho^{(1)}(2)\,\mathrm{d}2 \tag{5.9.7}$$

or

$$\nabla_1 \log \rho^{(1)}(1) + \beta \nabla_1 \phi(1) = \int c(1,2) \nabla_2 \rho^{(1)}(2) \, d2 \tag{5.9.8}$$

This result can also be written as

$$\nabla_1 \log \rho^{(1)}(1) + \beta \nabla_1 \phi(1) = \nabla_1 c^{(1)}(1) \tag{5.9.9}$$

where we have introduced a function $c^{(1)}(\mathbf{r})$ defined as

$$c^{(1)}(\mathbf{r}) = \log \frac{\rho^{(1)}(\mathbf{r})}{z^*(\mathbf{r})} \tag{5.9.10}$$

When expressed in terms of $c^{(1)}(\mathbf{r})$, Eqn (5.2.13) becomes

$$c(1,2) = \frac{\delta c^{(1)}(1)}{\delta \rho^{(1)}(2)} \tag{5.9.11}$$

The function $c^{(1)}(\mathbf{r})$ is called the single-particle direct correlation function, and Eqn (5.9.11) can be regarded as the first in a hierarchy of direct correlation functions in which an n-particle function $c^{(n)}$ is defined as a functional derivative of $c^{(n-1)}$. A feeling for the physical significance of $c^{(1)}(\mathbf{r})$ can be gained by rewriting the definition (5.9.10) as

$$\rho^{(1)}(\mathbf{r}) = z \exp[-\beta\phi(\mathbf{r}) + c^{(1)}(\mathbf{r})] \tag{5.9.12}$$

For an ideal gas, $c^{(1)}(\mathbf{r})$ is zero, as is obvious from its diagrammatic definition (4.5.9). In that case, Eqn (5.9.12) reduces to the usual formula for the distribution of non-interacting particles in an external field (the barometric law). Viewed in this way, $c^{(1)}(\mathbf{r})$ describes how the density $\rho^{(1)}(\mathbf{r})$ is altered by the interactions, with the function $-k_B T c^{(1)}(\mathbf{r})$ playing the role of an effective, single-particle potential.

Finally, a third exact equation for $\rho^{(1)}(\mathbf{r})$ is provided by the first member of the YBG hierarchy (2.6.7). If the force $\mathbf{X}(\mathbf{r})$ is replaced by $-\nabla\phi(\mathbf{r})$, the equation corresponding to $n = 1$ in (2.6.7) can be written in the form

$$-k_B T \nabla_1 \log \rho^{(1)}(1) = \nabla_1 \phi(1) + \int \rho^{(1)}(2) g(1,2) \nabla_1 v(1,2) \, d2 \tag{5.9.13}$$

This equation is less general than either (5.9.6) or (5.9.8), since its relies for its validity on the assumption of pairwise-additivity of the interparticle forces.

Equations (5.9.6), (5.9.8) and (5.9.13) provide three possible starting points for the calculation of $\rho^{(1)}(\mathbf{r})$. In particular, the YBG equation (5.9.13) can be used to treat the problem of a free-liquid surface by setting the external potential equal to zero. Since we are dealing only with planar interfaces, it is convenient to transform to cylindrical coordinates (Nicholson and Parsonage, 1982). If we make the substitutions $r_{12} = |\mathbf{r}_2 - \mathbf{r}_1|$, $z_{12} = (z_2 - z_1)$ and $g(r_{12}, z_1, z_2) \equiv g(1, 2)$, Eqn (5.9.13) becomes

$$\frac{d \log \rho^{(1)}(z_1)}{dz_1} = 2\pi\beta \int_{-\infty}^{\infty} dz_2 \int_{|z_{12}|}^{\infty} dr_{12} r_{12} \rho^{(1)}(z_2) g(r_{12}, z_1, z_2) \frac{d}{dz_2} v(r_{12})$$

$$= 2\pi\beta \int_{-\infty}^{\infty} dz_2 z_{12} \rho^{(1)}(z_2) \int_{|z_{12}|}^{\infty} dr_{12} g(r_{12}, z_1, z_2) v'(r_{12}) \quad (5.9.14)$$

where $v'(r) \equiv dv(r)/dr$. Equation (5.9.14) can be integrated between $z = -\infty$, where $\rho^{(1)}(z) = \rho_l$, and an arbitrary point in the density profile to give

$$\log \frac{\rho^{(1)}(z)}{\rho_l} = 2\pi\beta \int_{-\infty}^{z} dz_1 \int_{-\infty}^{\infty} dz_2 z_{12} \rho^{(1)}(z_2) \int_{|z_{12}|}^{\infty} dr_{12} g(r_{12}, z_1, z_2) v'(r_{12}) \quad (5.9.15)$$

If an approximation is made for $g(r_{12}, z_1, z_2)$, it is possible to solve Eqn (5.9.15) by iteration to obtain $\rho^{(1)}(z)$ for all z; the first guess usually consists in representing the density profile as a step function. The equation has a trivial solution in which $\rho^{(1)}(z)$ is independent of z, and the iterative procedure must be expected to converge towards this result unless ρ_l is chosen to be the true density of the saturated model liquid at the temperature of interest. When ρ_1 is correctly chosen, the density of the coexisting vapour is given by the limiting value of $\rho^{(1)}(z)$ as $z \to +\infty$. A natural approximation to make is to replace $g(r_{12}, z_1, z_2)$ by the radial distribution function of a uniform fluid at a density $\bar{\rho}$ equal to some average of the local densities at z_1 and z_2. This is a method used by Toxvaerd (1976), whose results for the Lennard-Jones fluid agree with computer simulations (Chapela *et al.*, 1977) about as well as those obtained by Fischer and Methfessel (1980) with the help of a more elaborate approximation to $g(r_{12}, z_1, z_2)$. The comparison with Monte Carlo and molecular dynamics calculations is not straightforward because in the simulated profiles the width of the transition region increases significantly with increase in the size of the system in the xy-plane. The size-dependence of the results is related to the excitation of surface-density fluctuations, or "capillary waves", of progressively longer wavelengths; these lead, in turn, to fluctuations in the position of the dividing surface and hence to a thickening of the interface (Buff *et al.*, 1965). The YBG results fall systematically below the widths obtained for the smallest simulated systems ($5\sigma \times 5\sigma$ in the xy-plane), suggesting that the approximations used for $g(r_{12}, z_1, z_2)$ yield only a "bare" density profile, unthickened by capillary waves.

Once a solution to Eqn (5.9.15) has been obtained, it is possible to calculate the surface tension. When a system includes an interface, the grand potential Ω contains a term proportional to the surface area $\mathscr{A}$, and Eqn (2.4.9) becomes

$$\Omega = -PV + \gamma\mathscr{A} \tag{5.9.16}$$

where γ, the surface tension, is defined thermodynamically as

$$\gamma = \left(\frac{\partial \Omega}{\partial \mathscr{A}}\right)_{\mu, V, T} \tag{5.9.17}$$

The link between (5.9.17) and the structure of the interface can be made through Eqns (2.4.13) and (4.2.38). The latter, suitably generalized, shows that the change in Ω brought about by a change in the pair potential is

$$\delta\Omega = \tfrac{1}{2}\iint \rho^{(2)}(1,2)\delta v(1,2)\,\mathrm{d}1\,\mathrm{d}2 \tag{5.9.18}$$

We now suppose that the change in $v(1,2)$ arises from a scaling of linear dimensions that increases the surface area, but leaves the volume unaltered, as required by the definition (5.9.17). Such an effect can be achieved by a transformation (Evans, 1979) of the form

$$x_i \to (1+\alpha)x_i, \qquad y_i \to y_i, \qquad z_i \to (1+\alpha)^{-1}z_i \tag{5.9.19}$$

In the limit $\alpha \to 0$, the change in $v(1,2)$ is

$$\begin{aligned} \delta v(1,2) &= \alpha\left(x_1\frac{\partial}{\partial x_1} + x_2\frac{\partial}{\partial x_2} - z_1\frac{\partial}{\partial z_1} - z_2\frac{\partial}{\partial z_2}\right)v(1,2) \\ &= \alpha\frac{(x_{12}^2 - z_{12}^2)}{r_{12}}v'(r_{12}) \end{aligned} \tag{5.9.20}$$

If we convert to cylindrical coordinates, $\mathrm{d}\mathbf{r}_1\,\mathrm{d}\mathbf{r}_2$ is replaced by $\mathscr{A}\,\mathrm{d}z_1\,\mathrm{d}\mathbf{r}_{12}$, and Eqn (5.9.18) becomes

$$\delta\Omega = \tfrac{1}{2}\int_{-\infty}^{\infty}\mathrm{d}z_1\alpha\mathscr{A}\int \rho^{(2)}(\mathbf{r}_1,\mathbf{r}_2)\frac{(x_{12}^2 - z_{12}^2)}{r_{12}}v'(r_{12})\,\mathrm{d}\mathbf{r}_{12} \tag{5.9.21}$$

Since the change in area resulting from the scaling of coordinates is $\delta\mathscr{A} = \alpha\mathscr{A}$, an expression for the surface tension is obtained immediately from comparison of Eqn (5.9.21) with the definition (5.9.17):

$$\gamma = \tfrac{1}{2}\int_{-\infty}^{\infty}\mathrm{d}z_1\int \rho^{(1)}(\mathbf{r}_1)\rho^{(1)}(\mathbf{r}_2)g(r_{12}, z_1, z_2)\frac{(x_{12}^2 - z_{12}^2)}{r_{12}}v'(r_{12})\,\mathrm{d}\mathbf{r}_{12} \tag{5.9.22}$$

This result, which can be derived in other ways, is usually called the Kirkwood–Buff formula (Kirkwood and Buff, 1949).

A self-consistent calculation of γ can now be made by substituting in (5.9.22) both the density profile obtained as the solution of Eqn (5.9.15) and the corresponding approximation used for $g(r_{12}, z_1, z_2)$. Results obtained in this way for the Lennard-Jones fluid (Fischer and Methfessell, 1980) are in fair agreement with the values of γ calculated by computer simulation for the same potential. Rowlinson and Widom (1982) have drawn what they consider to be the best smooth curve through the very scattered results of different simulations; the curve suggests that the results of Fischer and Methfessell (1980) are systematically about 10–15% too low. The simulations also show that the proven reliability of the Lennard-Jones potential as an effective pair potential for liquid argon does not extend to interfacial properties. If calculations are made with parameters appropriate to the bulk liquid, the resulting surface tensions at low temperatures are up to 40% larger than the experimental values for argon. By contrast, use of the accurate pair potential of Barker *et al.* (1971), when combined with corrections for three-body interactions, gives results that are in very good agreement with experiment (Lee *et al.*, 1974; Miyazaki *et al.*, 1976).

Equation (5.9.15), generalized to include an external potential, can equally well be used to calculate the density profile of a fluid in contact with a wall (Fischer and Methfessell, 1980). The wall-fluid problem has also been treated by making an approximation to the direct correlation function in Eqn (5.9.8); the latter can then be integrated numerically to yield $\rho^{(1)}(\mathbf{r})$. For example, Nieminen and Ashcroft (1981) have calculated the density profile of a Lennard-Jones gas against a 9-3 wall by combining Eqn (5.9.8) with a closure of the Ornstein–Zernike relation (5.2.7) analogous to the RHNC approximation described in Section 5.8. An alternative and computationally less demanding approach is to treat the wall and the fluid as two species in an infinitely dilute, homogeneous "mixture" in which the particles of the dilute component (the wall) are infinitely large (Perram and White, 1975; Henderson *et al.*, 1976). Consider a mixture of two components, labelled 1 and 2. In the limit $x_2 \to 0$, the coupled Ornstein–Zernike relations (5.2.18) reduce to

$$h_{11}(1,2) = c_{11}(1,2) + \rho \int h_{11}(1,3) c_{11}(2,3)\, \mathrm{d}3 \tag{5.9.23a}$$

$$h_{12}(1,2) = c_{12}(1,2) + \rho \int h_{11}(1,3) c_{12}(2,3)\, \mathrm{d}3 \tag{5.9.23b}$$

$$h_{21}(1,2) = c_{21}(1,2) + \rho \int h_{21}(1,3) c_{11}(2,3)\, \mathrm{d}3 \tag{5.9.23c}$$

$$h_{22}(1,2) = c_{22}(1,2) + \rho \int h_{21}(1,3) c_{12}(2,3)\, \mathrm{d}3 \tag{5.9.23d}$$

Equation (a) is simply the Ornstein–Zernike relation for the pure fluid (species 1); Eqn (d) is of no physical interest; and Eqns (b) and (c) are relations that describe correlations between the particles of the fluid and the wall (species 2) in terms of pair functions of the fluid. Equations (b) and (c) are equivalent, but (c) is the one that is used in numerical work. If the size of the wall "particles" is now allowed to become infinitely large, the wall–fluid interaction becomes a function solely of the distance z from the interface, and we can make the identifications $\rho_B \equiv \rho = \lim_{z\to\infty} \rho^{(1)}(z)$, $\rho h(1,2) \equiv \rho^{(1)}(z) - \rho_B$, and $c(z) \equiv c_{12}(r)$. Then Eqn (5.9.23c) can be rewritten as

$$\rho^{(1)}(z) = c(z) + \rho_B \int d\mathbf{r}' c_B(|\mathbf{r}-\mathbf{r}'|)\rho^{(1)}(z') \tag{5.9.24}$$

where $c_B(r)$ is the direct correlation function of the uniform bulk fluid. To obtain a numerical solution, a closure relation between $c(z)$ and $\rho^{(1)}(z)$ is required. If the analogue of the HNC approximation (5.4.19) is used, the density profile is given by

$$\rho^{(1)}(z) = \rho_B \exp[-\beta\phi(z)] \exp\left(\int d\mathbf{r}' c_B(|\mathbf{r}-\mathbf{r}'|)[\rho^{(1)}(z') - \rho_B]\right) \tag{5.9.25}$$

where it is natural to replace the unknown function $c_B(r)$ by the HNC solution for the bulk fluid. The corresponding PY equation is obtained by linearizing the second exponential in Eqn (5.9.25).

Solutions to the wall-particle Ornstein–Zernike relation have been obtained for a number of different approximate closures, including both HNC and PY, and for a variety of fluid–wall and fluid–fluid potentials (Henderson *et al.*, 1976; Waisman *et al.*, 1976; Snook and Henderson, 1978; Sullivan and Stell, 1978). The method gives very good results for the problem of hard spheres against a hard wall, but it is less satisfactory when the fluid–fluid potential has an attractive part (Smith and Lee, 1979). In particular, none of the approximations considered so far are found to describe either of the wetting processes pictured in Figure 5.7. Complete wetting by liquid at a gas–solid interface has also not been observed in simulations. In the case of a model of argon adsorbed on a carbon dioxide substrate under conditions where wetting might be expected to occur (Ebner and Saam, 1977), the PY approximation gives no evidence of thick-film formation, in agreement with Monte Carlo calculations on the same system (Saam and Ebner, 1978; Lane *et al.*, 1979). The results of the simulations apparently give support to the theory, but the situation is more complicated than this, as we shall see in Section 6.7.

CHAPTER 6

Perturbation Theories

6.1 INTRODUCTION: THE VAN DER WAALS MODEL

The intermolecular pair potential can in many cases be separated in a natural way into a harsh, short-range repulsion and a smoothly varying, long-range attraction. A separation of this type is an explicit ingredient of many empirical representations of the intermolecular forces including, for example, the Lennard-Jones potential. It is now generally accepted that the structure of simple liquids, at least at high density, is largely determined by geometric factors associated with the packing of the molecular hard cores. By contrast, the attractive interactions may, in a first approximation, be regarded as giving rise to a uniform background potential that provides the cohesive energy of the liquid but has little effect on its structure. A further plausible approximation consists in modelling the short-range forces by the infinitely steep repulsion of the hard-sphere potential. In this way, the properties of a given liquid can be related to those of a hard-sphere *reference system*, the attractive part of the potential being treated as a perturbation. The choice of the hard-sphere fluid as a reference system is an obvious one, since its thermodynamic and structural properties are well known.

The idea of representing a liquid as a system of hard spheres moving in a uniform, attractive potential well is an old one, providing as it does the physical basis for the van der Waals equation. At the time of van der Waals, little was known of the true properties of the dense hard-sphere fluid. The approximation that van der Waals made was to take the excluded volume per sphere of diameter d as equal to $\frac{2}{3}\pi d^3$ (or four times the molecular volume), which leads to an equation of state of the form

$$\frac{\beta P_0}{\rho}=\frac{1}{1-4\eta} \tag{6.1.1}$$

where η is the packing fraction defined by Eqn (4.6.8). Equation (6.1.1) gives the second virial coefficient correctly (see (4.6.11)), but it fails badly at high densities. In particular, the pressure diverges as $\eta \to 0.25$, a packing fraction lying well below that of the fluid-solid phase transition of hard spheres ($\eta \simeq 0.49$).

Considerations of thermodynamic consistency (Widom, 1963) show that the equation of state compatible with the hypothesis of a uniform, attractive background is necessarily of the form

$$\frac{\beta P}{\rho} = \frac{\beta P_0}{\rho} - \beta \rho a \tag{6.1.2}$$

where a is a positive constant. The classic van der Waals equation is then recovered by substituting for P_0 from Eqn (6.1.1). Clearly, therefore, a first step towards improving on van der Waals' result is to replace (6.1.1) by a more accurate hard-sphere equation of state, such as that of Carnahan and Starling, Eqn (4.6.14). A calculation along these lines was first carried out by Longuet-Higgins and Widom (1964), who in this way were able to predict successfully the melting properties of rare-gas solids. The work of Longuet-Higgins and Widom provides a simple example of an "augmented" van der Waals theory of the type later elaborated by Alder and collaborators (Alder and Hecht, 1969; Alder *et al.*, 1972).

The sections that follow are devoted to perturbation methods that may be regarded as attempts to improve the theory of van der Waals in a systematic fashion. The methods we describe have as a main ingredient the assumption that the structure of a dense, monatomic fluid resembles that of an assembly of hard spheres. Justification for this intuitively appealing idea is provided by the great success of the perturbation theories to which it gives rise, and which mostly reduce to Eqn (6.1.2) in some well-defined limit, but more direct evidence to support it also exists. For example, Ashcroft and Lekner (1966) have shown that for a variety of liquid metals near their respective triple points, the experimental structure factor can to a good approximation be superimposed on the structure factor of an "equivalent" hard-sphere fluid. Ashcroft and Lekner used the PY solution for the hard-sphere structure factor and chose the packing fraction so as to bring the height of the calculated main peak into agreement with the experimental one. For the metals that were studied, the packing fraction was found to be remarkably constant at $\eta = 0.45 \pm 0.02$; somewhat larger values of η would have been obtained if "exact" hard-sphere results had been used. A similar but more elaborate analysis was later made by Verlet (1968) of "experimental" data obtained by molecular dynamics calculations for the Lennard-Jones potential; the excellence of the fit to hard-sphere results is illustrated in Figure 6.1 for a thermodynamic state that again is close to the triple point.

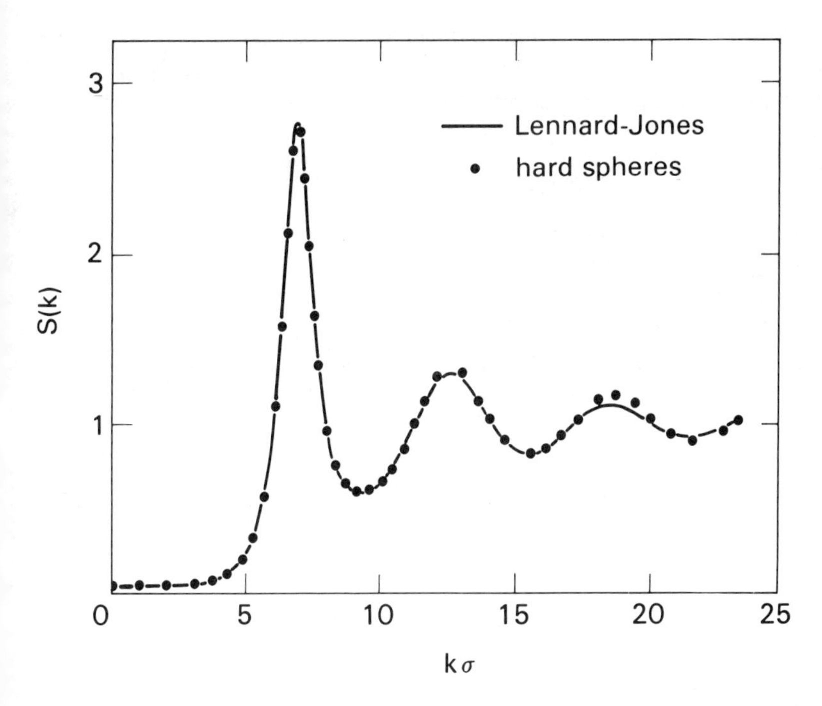

FIG. 6.1. Structure factor of the Lennard-Jones fluid at $\rho^* = 0.844$, $T^* = 0.72$ (curve) and its representation by a hard-sphere model (points). After Verlet (1968).

The fact that the attractive forces play such a minor role in these examples is understandable through the following argument (Verlet, 1968). Equation (5.2.6) shows that the structure factor determines the density response of a system to a weak, external field. If the external potential is identified with the potential due to a test particle placed at the origin, the long-range part of that potential gives rise to a long-wavelength response in the density. In the long-wavelength limit ($k \to 0$), the response is proportional to $S(k=0)$ and hence, through Eqn (5.1.13), to the compressibility. Under triple-point conditions, the compressibility of a liquid is very small (typically $\chi_T/\chi_T^0 \simeq 0.02$), and the effects of long-wavelength perturbations are therefore greatly reduced by what has been called "repulsive-force screening" (Andersen *et al.*, 1976). At lower densities, particularly in the critical region, the compressibility can become very large. The role of the attractive forces is then very important and the simple van der Waals model no longer has a sound physical basis.

We shall assume throughout this chapter that the interactions between particles are pairwise additive, but there is no difficulty in principle in extending the treatment to include three-body and higher-order forces.

Except for the discussion of non-uniform fluids in Section 6.7, we shall also suppose that the system of interest is homogeneous. The basis of all the perturbation theories that we shall discuss is a division of the pair potential of the form

$$v(1,2) = v_0(1,2) + w(1,2) \tag{6.1.3}$$

where $v_0(1,2)$ is the pair potential of the reference system and $w(1,2)$ is the perturbation. The calculation usually then proceeds in two stages. The first step is to compute the effect of the perturbation on the thermodynamic properties and pair distribution function of the reference system. This can be done systematically via an expansion in powers either of inverse temperature (the "λ-expansion") or of a parameter that measures the range of the perturbation (the "γ-expansion"). When hard spheres themselves are the reference system, this completes the calculation. In the more general situation, the properties of the "soft-core" reference system must in turn be related to those of the hard-sphere fluid.

6.2 THE λ-EXPANSION

Consider a pair potential $v_\lambda(1,2)$ of the form

$$v_\lambda(1,2) = v_{\lambda_0}(1,2) + w_\lambda(1,2) \tag{6.2.1}$$

where λ is a parameter that varies between λ_0 and λ_1. When $\lambda = \lambda_0$, $w_\lambda = 0$ and the potential $v_{\lambda_0} \equiv v_0$ reduces to that of a reference system whose properties are known, whereas for $\lambda = \lambda_1$, the potential $v_{\lambda_1} \equiv v$ is the one that characterizes the system of interest. The quantity λ has the meaning of a *coupling parameter*: the effect of varying λ continuously from λ_0 to λ_1 is that of gradually "switching on" the perturbation $w_{\lambda_1}(1,2)$. The commonest example of such a potential is

$$v_\lambda(1,2) = v_0(1,2) + \lambda w(1,2) \tag{6.2.2}$$

with $\lambda_0 = 0$ and $\lambda_1 = 1$; when $\lambda = 1$, the potential is the same as that introduced in Eqn (6.1.3).

Let $V_N(\lambda)$, given by

$$V_N(\lambda) = \sum_{i<j}^{N}\sum v_\lambda(i,j) \tag{6.2.3}$$

be the total potential energy of a system of particles interacting through the potential $v_\lambda(1,2)$. From the definitions of the configuration integral, Eqn (2.3.13), and the excess free energy (here denoted simply by A), Eqn

(2.3.18), it follows immediately that the derivative of $A(\lambda)$ with respect to the coupling parameter is

$$\beta \frac{\partial A(\lambda)}{\partial \lambda} = \frac{1}{Z_N(\lambda)} \int \exp\left[-\beta V_N(\lambda)\right] \beta V_N'(\lambda) \, d\mathbf{r}^N$$

$$= \beta \langle V_N'(\lambda) \rangle_\lambda \tag{6.2.4}$$

where $V_N'(\lambda) \equiv \partial V_N(\lambda)/\partial \lambda$ and $\langle \cdots \rangle_\lambda$ denotes a canonical ensemble average for the system characterized by the potential $v_\lambda(1, 2)$. Integration of (6.2.4) gives

$$\beta A(\lambda_1) = \beta A_0 + \beta \int_{\lambda_0}^{\lambda_1} \langle V_N'(\lambda) \rangle_\lambda \, d\lambda \tag{6.2.5}$$

where $A_0 \equiv A_{\lambda_0}$ is the excess free energy of the reference system. We now make a Taylor's series expansion of the ensemble average $\langle V_N'(\lambda) \rangle_\lambda$ around its value for $\lambda = \lambda_0$:

$$\langle V_N'(\lambda) \rangle_\lambda = \langle V_N'(\lambda) \rangle_0 + (\lambda - \lambda_0) \left(\frac{\partial}{\partial \lambda} \langle V_N'(\lambda) \rangle_\lambda \right)_{\lambda = \lambda_0} + \mathcal{O}(\lambda - \lambda_0)^2 \tag{6.2.6}$$

The calculation of the derivative with respect to λ is straightforward; insertion of the result for $\langle V_N'(\lambda) \rangle_\lambda$ in Eqn (6.2.5) yields an expansion of the free energy in the form

$$\begin{aligned} \beta A(\lambda_1) = \beta A_0 &+ (\lambda_1 - \lambda_0) \beta \langle V_N'(\lambda_0) \rangle_0 \\ &+ \tfrac{1}{2} (\lambda_1 - \lambda_0)^2 \{ \beta \langle V_N''(\lambda_0) \rangle_0 - \beta^2 (\langle [V_N'(\lambda_0)]^2 \rangle_0 \\ &- \langle V_N'(\lambda_0) \rangle_0^2) \} + \mathcal{O}(\lambda_1 - \lambda_0)^3 \end{aligned} \tag{6.2.7}$$

We now restrict ourselves to the important special case when $v_\lambda(1, 2)$ is given by Eqn (6.2.2). If we define the total perturbation energy as

$$W_N = \sum_{i<j}^{N} \sum w(i, j) \tag{6.2.8}$$

and recall that in this case $\lambda_0 = 0$ and $\lambda_1 = 1$, Eqn (6.2.7) simplifies to give

$$\beta A = \beta A_0 + \beta \langle W_N \rangle_0 - \tfrac{1}{2} \beta^2 (\langle W_N^2 \rangle_0 - \langle W_N \rangle_0^2) + \mathcal{O}(\beta^3) \tag{6.2.9}$$

The series (6.2.9) is called the *high-temperature expansion.* The name is not entirely appropriate: although successive terms are multiplied by increasing powers of β, the ensemble averages are also, in general, functions of temperature. When, however, the reference system is a hard-sphere fluid, the averages depend only on density and the λ-expansion reduces to a Taylor's series in T^{-1}. Equation (6.2.9) was first derived by Zwanzig (1954), who showed that the nth term in the series can be written in terms of the mean fluctuations $\langle[(W_N-\langle W_N\rangle_0)]^\nu\rangle_0$, with $\nu\leq n$. In particular:

$$\beta A_3=\frac{\beta^3}{3!}\langle[W_N-\langle W_N\rangle_0]^3\rangle_0 \tag{6.2.10a}$$

$$\beta A_4=-\frac{\beta^4}{4!}(\langle[W_N-\langle W_N\rangle_0]^4\rangle-3\langle[W_N-\langle W_N\rangle_0]^2\rangle_0^2) \tag{6.2.10b}$$

Thus every term in the high-temperature expansion corresponds to a statistical average evaluated in the reference-system ensemble.

The assumption of pairwise additivity of the potential (including the perturbation) means that Eqn (6.2.5) can be rewritten as

$$\frac{\beta A}{N}=\frac{\beta A_0}{N}+\frac{\beta}{2N}\int_0^1 \mathrm{d}\lambda\iint\rho_\lambda^{(2)}(1,2)w(1,2)\,\mathrm{d}1\,\mathrm{d}2 \tag{6.2.11}$$

where $\rho_\lambda^{(2)}(1,2)$ is the pair density for the system with potential $v_\lambda(1,2)$; this result also follows straightforwardly from Eqn (4.2.36). The pair density can then be expanded in powers of λ:

$$\rho_\lambda^{(2)}(1,2)=\rho_0^{(2)}(1,2)+\lambda\left(\frac{\partial\rho_\lambda^{(2)}(1,2)}{\partial\lambda}\right)_{\lambda=0}+\mathcal{O}(\lambda^2) \tag{6.2.12}$$

When this result is inserted in Eqn (6.2.11), the zeroth-order term in λ yields the first-order term in the high-temperature expansion of the free energy:

$$\frac{\beta A_1}{N}=\frac{\beta}{2N}\iint\rho_0^{(2)}(1,2)w(1,2)\,\mathrm{d}1\,\mathrm{d}2$$

$$=\tfrac{1}{2}\beta\rho\int g_0(1,2)w(1,2)\,\mathrm{d}\mathbf{r}_{12} \tag{6.2.13}$$

In this approximation, the structure of the fluid is unaltered by the perturbation. At second order in λ, however, calculation of the free energy involves the derivative $\partial\rho_\lambda^{(2)}/\partial\lambda$. Care is needed in passing to the thermodynamic limit and the differentiation is easier to perform in the grand canonical ensemble (Henderson and Barker, 1971); some details are given in Appendix

D. The final result for A_2 is

$$\begin{aligned}\frac{\beta A_2}{N} = -\frac{1}{2}\beta^2 \Big\{ & \tfrac{1}{2}\rho \int g_0(1,2)[w(1,2)]^2 \,\mathrm{d}2 \\ & + \rho^2 \iint g_0^{(3)}(1,2,3) w(1,2) w(1,3) \,\mathrm{d}2\,\mathrm{d}3 \\ & + \tfrac{1}{4}\rho^3 \iiint [g_0^{(4)}(1,2,3,4) - g_0^{(2)}(1,2) g_0^{(2)}(3,4)] \\ & \times w(1,2) w(3,4) \,\mathrm{d}2\,\mathrm{d}3\,\mathrm{d}4 \\ & - \tfrac{1}{4} S_0(0) \left[\frac{\partial}{\partial \rho} \left(\rho^2 \int g_0(1,2) w(1,2) \,\mathrm{d}2 \right) \right]^2 \Big\} \end{aligned} \tag{6.2.14}$$

where $S_0(k)$ is the structure factor of the reference system. For a finite system, the last term on the right-hand side of (6.2.14) disappears.

We see from Eqn (6.2.14) that a calculation of the second-order term requires a knowledge of the three and four-particle distribution functions of the reference system. The situation is even worse for higher-order terms, since the calculation of A_n requires the distribution functions of all orders up to and including $2n$. By the same rule, calculation of the first-order term requires only the pair distribution function of the reference system. If ε defines the energy scale of the perturbation, truncation at first order is likely to be justified if $\beta\varepsilon \ll 1$. The fact that second and higher-order terms are determined by fluctuations in the total perturbation energy suggests that they should also be small, relative to A_1, whenever the perturbing potential is a very smoothly varying function of particle separation. Schemes that simplify the calculation of A_2 have been devised (Barker and Henderson, 1976), but the high-temperature expansion remains easiest to apply in situations where terms beyond first order are negligible. The question of whether or not a first-order treatment is adequate depends on the thermodynamic state, the form of the potential $v(1,2)$, and the manner in which $v(1,2)$ is divided into a reference-system potential and a perturbation.

If the reference system is the hard-sphere fluid and the perturbation potential $w(1,2)$ is very long ranged, the high-temperature expansion limited to first order reduces to the van der Waals equation (6.1.2). It is necessary only that the range of $w(1,2)$ be large compared with the range of interparticle separations for which $g_0(1,2)$ is significantly different from its asymptotic value. Then, to a good approximation:

$$\begin{aligned}\frac{\beta A_1}{N} &= \tfrac{1}{2}\beta\rho \int g_0(\mathbf{r}) w(\mathbf{r}) \,\mathrm{d}\mathbf{r} \\ &\simeq \tfrac{1}{2}\beta\rho \int w(\mathbf{r}) \,\mathrm{d}\mathbf{r} = -\beta\rho a \end{aligned} \tag{6.2.15}$$

where a is a constant that is positive when the perturbing potential is attractive. On differentiating with respect to density, we recover the equation of state (6.1.2):

$$\frac{\beta P}{\rho} = \rho \frac{\partial}{\partial \rho}\left(\frac{\beta A_0}{N} + \frac{\beta A_1}{N}\right)$$

$$= \frac{\beta P_0}{\rho} - \beta \rho a \tag{6.2.16}$$

Another important feature of the high-temperature expansion is the fact that the first-order approximation yields a rigorous upper bound on the free energy of the system of interest, irrespective of the choice of reference system. This result follows from the so-called Gibbs–Bogoliubov inequalities. These inequalities have been derived many times in the literature and are essentially a consequence of the convexity of the exponential function (Griffiths, 1964). We reproduce here a simple and elegant proof due to Isihara (1968).

Consider two integrable, non-negative but otherwise arbitrary configuration-space functions $F(\mathbf{r}^N)$ and $G(\mathbf{r}^N)$, defined such that

$$\int F(\mathbf{r}^N)\, d\mathbf{r}^N = \int G(\mathbf{r}^N)\, d\mathbf{r}^N \tag{6.2.17}$$

The two functions satisfy the inequality

$$\int F(\mathbf{r}^N) \log F(\mathbf{r}^N)\, d\mathbf{r}^N \geqslant \int F(\mathbf{r}^N) \log G(\mathbf{r}^N)\, d\mathbf{r}^N \tag{6.2.18}$$

This may be checked by writing the difference as

$$\int F(\mathbf{r}^N) \log F(\mathbf{r}^N)\, d\mathbf{r}^N - \int F(\mathbf{r}^N) \log G(\mathbf{r}^N)\, d\mathbf{r}^N$$

$$= \int G(\mathbf{r}^N)\left[\frac{F(\mathbf{r}^N)}{G(\mathbf{r}^N)} \log\left(\frac{F(\mathbf{r}^N)}{G(\mathbf{r}^N)}\right) - \frac{F(\mathbf{r}^N)}{G(\mathbf{r}^N)} + 1\right] d\mathbf{r}^N \tag{6.2.19}$$

The integrand on the right-hand side of (6.2.19) is non-negative, because $x \log x \geqslant (x-1)$ for any x. The inequality (6.2.18) is therefore proved. We now make two particular choices for F and G. First, let

$$\begin{aligned} F(\mathbf{r}^N) &= \exp\{\beta[A_0 - V_N(0)]\} \\ G(\mathbf{r}^N) &= \exp\{\beta[A - V_N(1)]\} \end{aligned} \tag{6.2.20}$$

It follows from (6.2.18) that

$$A \leqslant A_0 + [\langle V_N(1) \rangle_0 - \langle V_N(0) \rangle_0] = A_0 + \langle W_N \rangle_0 \tag{6.2.21}$$

This is precisely the inequality announced earlier. If we interchange the definitions of F and G, i.e. if we set

$$\begin{aligned} F(\mathbf{r}^N) &= \exp\{\beta[A - V_N(1)]\} \\ G(\mathbf{r}^N) &= \exp\{\beta[A_0 - V_N(0)]\} \end{aligned} \tag{6.2.22}$$

then we find from (6.2.18) that

$$A \geqslant A_0 + \langle W_N \rangle_1 \tag{6.2.23}$$

where the average of the perturbation is now taken over the ensemble of the system of interest. The second inequality is less useful than the first, because the properties of the system of interest are in general unknown, but in conjunction with (6.2.21) it has proved valuable in testing the convergence of the high-temperature expansion in specific cases (Verlet and Weis, 1972b). With the assumption of pairwise additivity, (6.2.21) and (6.2.23) can be combined in the form

$$\frac{\beta A_0}{N} + \tfrac{1}{2}\beta\rho \int g(\mathbf{r}) w(\mathbf{r})\, d\mathbf{r} \leqslant \frac{\beta A}{N} \leqslant \frac{\beta A_0}{N} + \tfrac{1}{2}\beta\rho \int g_0(\mathbf{r}) w(\mathbf{r})\, d\mathbf{r} \tag{6.2.24}$$

The inequality (6.2.24) provides the basis for a variational approach to the theory of liquids. The variational procedure consists in choosing a reference-system potential that depends on one or more parameters, and then of minimizing the right-hand side of (6.2.24) with respect to these parameters. If the reference-system potential is chosen in a sufficiently flexible way, the resulting lowest upper bound on the free energy of the system of interest will be close to the exact value. The method has been applied to the case of the Lennard-Jones fluid by Mansoori and Canfield (1969) and by Rasaiah and Stell (1970). A convenient choice of reference system is a fluid of hard spheres, the hard-sphere diameter being treated as the single variational parameter; in the liquid range, the calculated free energies lie about 5% above the values derived from computer simulations.

The rate of convergence of the λ-expansion has been studied for a number of potential models consisting of a hard-sphere potential plus a "tail". The most extensive results are for the "square-well" potential (1.2.2) with $\gamma = 1.5$ (Alder *et al.*, 1972; Barker and Henderson, 1976). In such cases, there are no complications due to the softness of the repulsive part of the potential, since it is natural to take the hard-sphere interaction as the reference-system potential and treat the tail as a perturbation. The high-temperature expansion of the free energy can then be written as

$$\frac{\beta A}{N} = \frac{\beta A_0}{N} + \sum_{n=1}^{\infty} \frac{a_n}{T^{*n}} \tag{6.2.25}$$

where the coefficients a_n depend only on the packing fraction of the hard-sphere reference system. For the square-well fluid, $T^* = k_B T/\varepsilon$, where

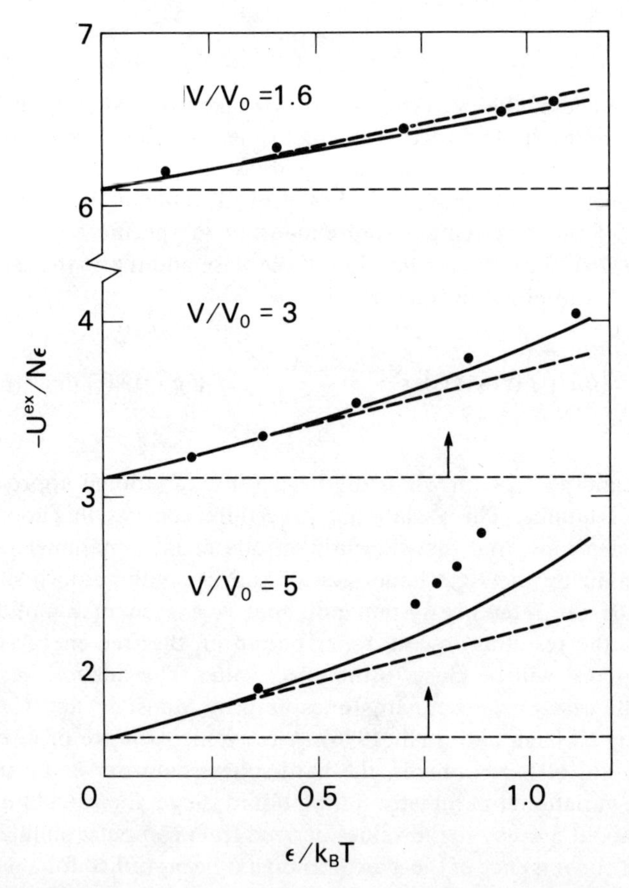

FIG. 6.2. Excess internal energy of the "square-well" fluid calculated from the λ-expansion. The points are molecular-dynamics results, the dashed lines show the second-order results, and the full curves include corrections up to fourth order. The quantity $V_0 = Nd^3/\sqrt{2}$ is the volume at close packing, and the arrows indicate the location of points on the liquid–vapour coexistence curve. After Alder *et al.* (1972).

ε is the depth of the well, while Eqn (6.2.13) becomes

$$a_1 = -2\pi\rho \int_d^{\gamma d} g_0(r) r^2 \, \mathrm{d}r$$

$$= -\frac{\langle M \rangle_0}{N} \tag{6.2.26}$$

where $\langle M \rangle_0$ is the number of pairs in the reference system that, on average, are separated by distances r in the range $d < r < \gamma d$. All other coefficients in (6.2.25) can be expressed in terms of fluctuations in M about its mean value. Alder and coworkers (Alder and Hecht, 1969; Alder *et al.*, 1972) have calculated the "exact" coefficients up to a_4 by molecular dynamics simulation. They find that a_1 is approximately linear in density, in qualitative agreement with the simple, van der Waals picture, but higher-order terms are not negligible. Some results obtained by perturbation theory for the internal energy of the square-well fluid are plotted in Figure 6.2. As the figure shows, the perturbation series converges rapidly only for densities greater than about half that of freezing, i.e. for $V/V_0 \lesssim 3$, where V_0 is the volume of the hard-sphere system at close packing. The rate of convergence is improved when the perturbation is a more smoothly varying function as, for example, in the case of a hard-sphere potential with an r^{-6} tail (Barker and Henderson, 1976).

6.3 TREATMENT OF SOFT CORES

Perturbation theories are useful only if they relate the properties of the system of interest to those of a well-understood reference system. Hard spheres are a natural choice of reference system, for the reasons discussed in Section 6.1. On the other hand, realistic intermolecular potentials do not have an infinitely steep hard core, and there is no natural separation into a hard-sphere part and a weak perturbation. Instead, an arbitrary division of the potential is made, as in Eqn (6.2.1), and the properties of the reference system, with potential $v_0(r)$, are then usually related to those of hard spheres in a step that is independent of the way in which the perturbation is treated. In this section, we discuss how the relation between the reference system and the system of hard spheres can be established, postponing the question of how the potential can best be separated until Section 6.4. We describe in detail only the "blip-function" method of Andersen, Weeks and Chandler (Andersen *et al.*, 1971), but we shall also show how results obtained earlier by Rowlinson (1964a) and by Barker and Henderson (1967) can be recovered from the same analysis. In each case, the free energy of the reference system is equated to that of a hard-sphere fluid at the same density and temperature The hard-sphere diameter is, in general, a functional of $v_0(r)$ and a function of ρ and T, and the various methods of treating the reference system differ from each other primarily in the prescription used to determine d.

Consider a reference-system potential $v_0(r)$ that is harshly repulsive but continuous. The Boltzmann factor $e_0(r) = \exp[-\beta v_0(r)]$ typically has the appearance shown in Figure 6.3, and is not very different from the Boltzmann factor $e_d(r)$ of a hard-sphere potential. Thus, for a properly chosen value of d, the function

$$\Delta e(r) = e_0(r) - e_d(r) \tag{6.3.1}$$

is effectively non-zero over only a small range of r, which we shall denote by ξd. The behaviour of $\Delta e(r)$ as a function of r is sketched in Figure 6.3; the origin of the name "blip function" given to it by Andersen *et al.* (1971) is obvious from the figure.

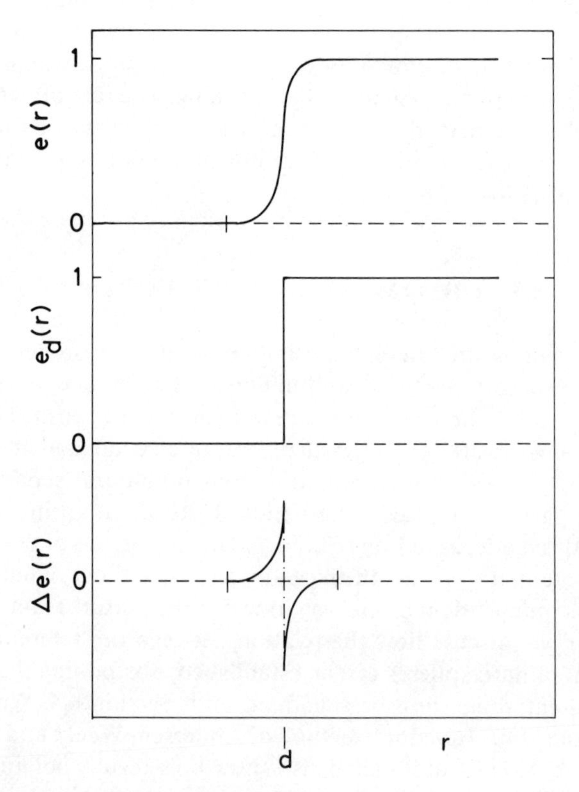

FIG. 6.3. The "blip function". The quantities $e(r)$ and $e_d(r)$ are the Boltzmann factors, respectively, for soft-core and hard-sphere potentials.

If, as we suppose, ξ is small, it is natural to seek an expansion of the properties of the reference system about those of hard spheres in powers of ξ. Such a series can be obtained by making a functional Taylor expansion of $\mathscr{A} = -\beta A^{\text{ex}}/V$ in powers of $\Delta e(r)$ of the form

$$\mathscr{A}_0 = \mathscr{A}_d + \int \frac{\delta \mathscr{A}}{\delta e(\mathbf{r})}\bigg|_{e=e_d} \Delta e(\mathbf{r})\, d\mathbf{r} + \frac{1}{2}\iint \frac{\delta^2 \mathscr{A}}{\delta e(\mathbf{r})\delta e(\mathbf{r}')}\bigg|_{e=e_d} \Delta e(\mathbf{r})\Delta e(\mathbf{r}')\, d\mathbf{r}\, d\mathbf{r}' + \cdots \tag{6.3.2}$$

where $\mathscr{A}_d$ represents the free energy of the hard-sphere fluid. We know from Eqns (2.3.18), (2.5.15) and (4.2.35) that

$$\frac{\delta \mathscr{A}}{\delta e(\mathbf{r})} = \tfrac{1}{2}\rho^2 y(\mathbf{r}) \tag{6.3.3}$$

Equation (6.3.2) can therefore be rewritten as

$$\mathscr{A} = \mathscr{A}_d + \tfrac{1}{2}\rho^2 \int y_d(r)\Delta e(r)\, d\mathbf{r} + \cdots \tag{6.3.4}$$

The expression for the second-order term involves the three and four-particle distribution functions (Andersen *et al.*, 1971). In practice, fortunately, terms beyond first order are never needed for steep potentials.

Since the range of $\Delta e(r)$ is ξd, the first-order term in the expansion (6.3.2) is of order ξ. A natural choice of d is one that causes the first-order term to vanish; d is then determined by the implicit relation

$$\int y_d(r)\Delta e(r)\, d\mathbf{r} = 0 \tag{6.3.5}$$

With this choice of d, the second-order term in (6.3.2), which would normally be of order ξ^2, becomes of order ξ^4 (Andersen *et al.*, 1971). Thus the free energy of the reference system is given in terms of the free energy of hard spheres by

$$\mathscr{A}_0 = \mathscr{A}_d + \mathcal{O}(\xi^4) \tag{6.3.6}$$

where d is defined by Eqn (6.3.5).

Equation (6.3.5) provides only one of many possible prescriptions for calculating the diameter of the "equivalent" hard spheres. Because $\Delta e(r)$ is non-zero only in a narrow range of r, the factor $r^2 y_d(r)$ in (6.3.5) may be expanded in a Taylor's series about $r = d$ (Kim, 1969; Verlet and Weis, 1972a, b) in the form

$$r^2 y_d(r) = \sigma_0 + \sigma_1\left(\frac{r}{d} - 1\right) + \sigma_2\left(\frac{r}{d} - 1\right)^2 + \cdots \tag{6.3.7}$$

with

$$\frac{\sigma_m}{d^m} = \left(\frac{d^m}{dr^m} r^2 y_d(r)\right)_{r=d} \tag{6.3.8}$$

Substitution of the expansion (6.3.7) in Eqn (6.3.5) gives

$$\sum_{m=0}^{\infty} \frac{\sigma_m}{m!} I_m = 0 \tag{6.3.9}$$

where

$$\begin{aligned} I_m &= \int_0^\infty \left(\frac{r}{d} - 1\right)^m \Delta e(r)\, \mathrm{d}(r/d) \\ &= -\frac{1}{m+1} \int_0^\infty \left(\frac{r}{d} - 1\right)^{m+1} \frac{\mathrm{d}}{\mathrm{d}r} \exp\left[-\beta v_0(r)\right] \mathrm{d}r \end{aligned} \tag{6.3.10}$$

If $v_0(r)$ varies very rapidly with r, the derivative in (6.3.10) is approximately a delta-function at $r = d$, and the series (6.3.9) is rapidly convergent. If only the first term is retained, then $I_0 = 0$, and a straightforward integration shows that

$$d = \int_0^\infty \{1 - \exp\left[-\beta v_0(r)\right]\}\, \mathrm{d}r \tag{6.3.11}$$

This expression is identical to one derived earlier by Barker and Henderson (1967). In the case when $v_0(r)$ is an inverse-power potential of the form

$$v_0(r) = \varepsilon(\sigma/r)^n \tag{6.3.12}$$

the integral in (6.3.11) can be evaluated explicitly to give

$$\begin{aligned} d &= \sigma(\varepsilon/k_B T)^{1/n} \Gamma\left(\frac{n-1}{n}\right) \\ &= \sigma(\varepsilon/k_B T)^{1/n}(1 + \gamma/n) + \mathcal{O}(1/n^2) \end{aligned} \tag{6.3.13}$$

where $\gamma = 0.5772\ldots$ is Euler's constant. On discarding terms of order $1/n^2$, we recover an expression given by Rowlinson (1964a). Rowlinson's original calculation was based on an expansion of the free energy in powers of the inverse-steepness parameter $\lambda = 1/n$ about $\lambda = 0$ (hard spheres); the work of Barker and Henderson (1967) may be regarded as a generalization of Rowlinson's method to a repulsive potential of arbitrary form.

The main difference between Eqns (6.3.5) and (6.3.11) is that the former yields a hard-sphere diameter that is a function of both density and temperature, whereas the Barker–Henderson diameter is dependent only on temperature. The greater flexibility provided by the use of (6.3.5) ensures that the predictions of the Andersen–Weeks–Chandler approach are, in general, superior to those of the Barker–Henderson theory (Andersen *et al.*, 1971). The measure of agreement with the results of computer simulation is illustrated for the potential (6.3.12) with $n = 12$ (the r^{-12} fluid) in Table 6.1; the differences between the results of the two theories become smaller as the potential $v_0(r)$ becomes steeper (Barker and Henderson, 1976). For inverse-power potentials of the type (6.3.12), the excess thermodynamic functions have simple scaling properties, and quantities such as $\beta A^{\text{ex}}/N$ and $(\beta P/\rho - 1)$ are functions of the single, dimensionless parameter $(\beta\varepsilon)^{3/n}\rho\sigma^3$. This rule is satisfied by both the Rowlinson and Barker–Henderson approximations, even though the hard-sphere diameters obtained from (6.3.11) and (6.3.13) depend only on $\beta\varepsilon$. It can also be shown that every term in the blip-function expansion (6.3.2) satisfies the scaling rule provided the diameter d is determined by Eqn (6.3.5).

TABLE 6.1. *Thermodynamic properties of the r^{-12} fluid: Comparison between Monte Carlo results and the predictions of perturbation theory. After Andersen* et al. (1971) *and Lado* (1984)

$\rho\sigma^3(\varepsilon/4k_BT)^{1/4}$	$\beta A/N$				$\beta P/\rho$			
	MC	BH	AWC(a)	AWC(b)	MC	BH	AWC(a)	AWC(b)
0.1	0.40	0.39	0.40	0.40	1.45	1.44	1.45	1.45
0.2	0.90	0.89	0.91	0.91	2.12	2.13	2.16	2.12
0.3	1.53	1.54	1.54	1.54	3.12	3.25	3.25	3.12
0.4	2.33	2.42	2.32	2.33	4.58	5.16	4.93	4.56
0.5	3.34	3.68	3.31	3.33	6.66	8.57	7.51	6.59
0.6	4.61	5.58	4.53	4.59	9.56	15.12	11.44	9.39
0.7	6.21		6.03	6.15	13.51		17.32	13.14

MC = Monte Carlo results (Hansen, 1970); BH = predictions of theory of Barker and Henderson (1967) and AWC = predictions of theory of Andersen *et al.* (1971); (a) = based on Eqn (6.3.5); (b) = based on Eqn (6.3.17) (Lado, 1984).

The Andersen–Weeks–Chandler theory also yields a very simple expression for the pair distribution function of the reference system. It follows from Eqns (6.3.3) and (6.3.4) that

$$y_0(r) = y_d(r) + \text{higher-order terms} \tag{6.3.14}$$

where the higher-order terms are of order ξ^2 or smaller if d is chosen to satisfy Eqn (6.3.5) (Andersen *et al.*, 1971). Thus

$$\begin{aligned} g_0(r) &= \exp[-\beta v_0(r)] y_0(r) \\ &= \exp[-\beta v_0(r)] y_d(r) + \mathcal{O}(\xi^2) \end{aligned} \tag{6.3.15}$$

The expression for the reference-system pair distribution function can now be used, via Eqn (6.2.13), to compute the correction to the free energy that results from a perturbing potential $w(r)$. Equation (6.3.15) also allows us to rewrite (6.3.5) in terms of the $k \to 0$ limits of the reference-system and hard-sphere structure factors as

$$S_0(k=0) = S_d(k=0) \tag{6.3.16}$$

Choice of the hard-sphere diameter defined by (6.3.5) therefore has the effect of setting the compressibility of the reference system equal to that of the underlying hard-sphere fluid. The approximation (6.3.15) for $g_0(r)$ is expected to be less accurate than that for the free energy, Eqn (6.3.6), because the neglected terms are now of order ξ^2 rather than ξ^4. This is borne out by calculations made for the r^{-12} fluid; the approximate $g_0(r)$, as given by (6.3.15), is in only moderate agreement with the results of computer simulations (Hansen and Weis, 1972), whereas the agreement obtained for the free energy is very good, as already shown in Table 6.1. We shall see in the next section that the situation improves markedly when a much steeper reference potential is involved.

Although the blip-function method works very well for thermodynamic properties, it is clear from Table 6.1 that there is scope for improvement at high densities. In particular, there is a lack of thermodynamic consistency: results for the pressure calculated from the virial equation differ significantly from those obtained by numerical differentiation of the free energy. The results obtained from the free energy are the more reliable, but they are also more troublesome to compute. Equivalence of the two routes to the equation of state is guaranteed, however, if the hard-sphere diameter is calculated, not from Eqn (6.3.5), but from the relation (Lado, 1984)

$$\int \frac{\partial y_d(r)}{\partial d} \Delta e(\mathbf{r}) \, d\mathbf{r} = 0 \tag{6.3.17}$$

Equation (6.3.17) can be derived by requiring that the free energy of the system of interest be a minimum with respect to variations in the hard-sphere function $y_d(r)$. Thermodynamic properties calculated in this way for the r^{-12} fluid are shown in Table 6.1; they are in excellent agreement with the results of computer simulations. Yet another criterion for the choice of d has been proposed by Kollár (1981); this reduces to the Andersen–Weeks–Chandler result (6.3.5) in the low-density limit.

Apart from the question of thermodynamic consistency, problems are encountered in the application of the blip-function method whenever the reference-system potential is insufficiently steep to justify truncation of the series (6.3.4) after the first-order term. This is true, for example, of calculations on simple liquid metals (Jacobs and Andersen, 1975). Consider a function $B(r)$, defined as

$$B(r) = y_d(r)\Delta e(r) \tag{6.3.18}$$

The effect of neglecting terms of order ξ^2 in Eqn (6.3.15) is to set

$$g_0(r) = g_d(r) + B(r) \tag{6.3.19}$$

or, on taking Fourier transforms:

$$S_0(k) = S_d(k) + \hat{B}(k) \tag{6.3.20}$$

where d is determined by the condition $\hat{B}(0) = 0$. Since $B(r)$ is small except near $r = d$, $\hat{B}(k)$ will have its maximum value at $k \simeq \pi/d$ and will be small elsewhere, except possibly at multiples of π/d. If $v_0(r)$ is not very steeply repulsive, this feature in $\hat{B}(k)$ will manifest itself in a spurious peak in $S(k)$ at $k \simeq \pi/d$. Jacobs and Andersen (1975) have developed an extension of the blip-function theory in which Eqn (6.3.20) is replaced by

$$S_0(k) = \frac{S_d(k)}{1 - \rho\hat{B}(k)S_d(k)} \tag{6.3.21}$$

The justification given for this new approximation is largely intuitive, but it is based on a rigorous diagrammatic analysis of the blip-function expansion. The method works because, at high densities, $S_d(k)$ is always small near $k = \pi/d$; the spurious peak is therefore suppressed.

6.4 AN EXAMPLE: THE LENNARD-JONES FLUID

The λ-expansion described in Section 6.2 is suitable for treating perturbations that vary slowly in space, while the blip-function expansion and related methods of Section 6.3 are appropriate for perturbations that are rapidly varying but localized. In this section, we show how the two approaches may be combined in a case where the potential has both a steep but continuous repulsive part, and a weak, longer ranged attraction. The example we choose is that of the Lennard-Jones fluid. This is a system of practical interest insofar as it provides a fair description of the properties of rare-gas liquids, but it is also one for which there are sufficient data available from computer simulations to allow a complete test to be made of different perturbation schemes.

At first sight, it might appear that the complications due to the softness of the core would make it more difficult to obtain satisfactory results by perturbation theory than in cases where the potential consists of a hard-sphere interaction and a "tail". This is not necessarily true, because there is now the extra flexibility provided by the arbitrary separation of the potential into a reference part $v_0(r)$ and a perturbation $w(r)$; a judicious choice of separation can significantly enhance the convergence of the resulting perturbation series. Three different separations have been proposed for the Lennard-Jones potential, all of which are illustrated in Figure 6.4.

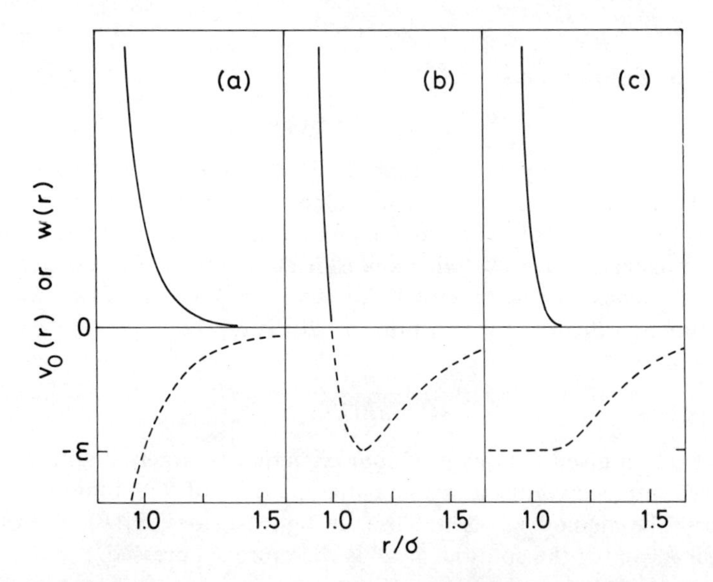

FIG. 6.4. Three separations of the Lennard-Jones potential that have been used in perturbation-theory calculations: (a) McQuarrie and Katz (1966); (b) Barker and Henderson (1967); (c) Weeks *et al.* (1971). In each case, the dashed line represents the perturbation.

In the method of McQuarrie and Katz (1966), the r^{-12} term is chosen as the reference-system potential and the r^{-6} term is treated as a perturbation. In the original calculations, the properties of the reference system were related to those of hard spheres via Rowlinson's inverse-steepness expansion, Eqn (6.3.13), and the effect of the perturbation was calculated to first order in the λ-expansion. Because Rowlinson's method is only moderately successful at high densities, much improved results are obtained when the reference-system properties are calculated by computer simulation. With this refinement of the method, the λ-expansion limited to first order gives an equation of state that at high temperatures ($T^* \gtrsim 3$) is in excellent agreement with Monte Carlo results for the full Lennard-Jones potential, except at very low densities ($\rho^* \lesssim 0.2$) (Hansen, 1970). At lower temperatures, the agreement is much less satisfactory; this is understandable, since the reference-system potential is considerably softer than the full potential in the important region close to the minimum in the latter.

In the separation used by Barker and Henderson (1967), the reference system is defined by the part of the Lennard-Jones potential that is positive ($r < \sigma$), while the perturbation consists of the part that is negative ($r > \sigma$). The reference-system properties are then related to those of hard spheres of diameter d given by Eqn (6.3.11). In contrast to the case of the r^{-12} fluid (see Table 6.1), this treatment of the reference system yields very accurate results. The corrections due to the perturbation are handled in the framework of the λ-expansion; the first-order term is calculated from Eqn (6.2.13), with $g_0(r)$ taken to be the pair distribution function of the hard-sphere fluid. At $T^* = 0.72$ and $\rho^* = 0.85$, which is close to the triple point of the Lennard-Jones fluid, the results are $d/\sigma = 0.9785$, $\beta A_0/N = 3.37$ and $\beta A_1/N = -7.79$. (The last number is the result obtained if the "exact" reference-system radial distribution function is used in Eqn (6.2.13).) The sum of the zeroth and first-order terms is therefore equal to -4.42, whereas the result obtained for the total excess free energy from Monte Carlo calculations is $\beta A/N = -4.87$. The sum of all higher-order terms in the λ-expansion is therefore far from negligible; detailed calculations show that the second-order term accounts for most of the remainder (Levesque and Verlet, 1969). The origin of the large second-order term lies in the way in which the potential is separated. The effect of dividing $v(r)$ at $r = \sigma$ is to include in the perturbation the rapidly varying part of the Lennard-Jones potential between $r = \sigma$ and the minimum at $r = r_m = 2^{1/6}\sigma$. Since the radial distribution function has its maximum value in the same range of r, the fluctuations in the total perturbation energy W_N, and hence the numerical values of A_2, are large.

The work of Barker and Henderson remains a landmark in the development of liquid-state theory, since it demonstrated for the first time that thermodynamic perturbation theory is capable of yielding accurate results even for states close to the triple point of the system of interest. A drawback to their method is the fact that its successful implementation requires a careful evaluation of the second-order term in the λ-expansion. The hard-sphere data needed to calculate A_2 are available in analytical form (Barker and Henderson, 1972), but the theory is inevitably more awkward to handle than is the case when a first-order treatment is adequate. Results for both the first and second-order versions of the theory are given in Table 6.2.

The problem of the second-order term can be overcome by dividing the potential in the manner of Weeks *et al.* (1971), usually called the WCA separation. In this method, the potential is split at $r = r_m$ into its purely repulsive ($r < r_m$) and purely attractive ($r > r_m$) parts; the former defines

TABLE 6.2. *Equation of state of the Lennard-Jones fluid; comparison between the results of computer simulations and the predictions of perturbation theory for the quantity $\beta P/\rho$. After Barker and Henderson* (1976)

T^*	ρ^*	Simul.	BH1	BH2	WCA
2.74	0.65	2.22	2.24	2.22	2.18
	0.75	3.05	3.14	3.10	3.04
	0.85	4.38	4.48	4.44	4.30
	0.95	6.15	6.41	6.40	6.10
1.35	0.10	0.72	0.77	0.74	0.77
	0.20	0.50	0.55	0.52	0.53
	0.30	0.35	0.39	0.36	0.31
	0.40	0.27	0.26	0.26	0.17
	0.50	0.30	0.31	0.27	0.18
	0.55	0.41	0.40	0.35	0.27
	0.65	0.80	0.91	0.74	0.71
	0.75	1.73	1.87	1.64	1.64
	0.85	3.37	3.54	3.36	3.28
	0.95	6.32	6.21	6.32	
1.00	0.65	−0.25	−0.21	−0.36	−0.50
	0.75	0.58	0.71	0.53	0.40
	0.85	2.27	2.48	2.25	2.20
0.72	0.85	0.40	0.70	0.25	0.26

Simul. = simulation results of Verlet and Levesque (1967), Verlet (1967) and Levesque and Verlet (1969); BH = Barker-Henderson theory in first (BH1) and second (BH2) order; WCA = theory of Weeks *et al.* (1971).

the reference system and the latter constitutes the perturbation. Thus

$$\begin{aligned} v_0(r) &= v(r) + \varepsilon, & r < r_m \\ &= 0, & r > r_m \\ w(r) &= -\varepsilon, & r < r_m \\ &= v(r), & r > r_m \end{aligned} \tag{6.4.1}$$

Compared with the Barker–Henderson separation, the perturbation now varies more slowly over the range of r corresponding to the first peak in $g(r)$, and the perturbation series is therefore more rapidly convergent. For example, at $T^* = 0.72$ and $\rho^* = 0.85$, the reference-system free energy is $\beta A_0/N = 4.49$ and first-order correction in the λ-expansion is -9.33; the sum of the two terms is -4.84, which differs by less than 1% from the "exact" result for the full potential (Verlet and Weis, 1972a). Agreement of the same order is found throughout the high-density region, and the perturbation series may therefore confidently be truncated after the first-order term. The difficulties associated with the calculation of the second and higher-order terms are thereby avoided.

In the calculations summarized above, and in most of those based on the WCA separation, the free energy of the reference system is related to that of hard spheres through Eqns (6.3.5) and (6.3.6). At high densities, the order ξ^4 error that this entails is very small. Under the same conditions, use of the approximate relation (6.3.15) to calculate the first-order correction from Eqn (6.2.13) also involves only a very small error; the accuracy of the approximation (6.3.15) is evident from Figure 6.5. Results for the Lennard-Jones fluid are shown in Table 6.2. The general level of agreement with the results of computer simulations is very good, and comparable with that achieved by the Barker–Henderson method taken to second order. The discrepancies are largest at high temperatures ($T^* \gtrsim 3$, not shown in the table), where the McQuarrie–Katz procedure is superior, and at low densities ($\rho^* \lesssim 0.6$). In the latter region, the attractive forces play an important role in determining the structure, and the key assumption of a first-order theory, namely that $g(r) \simeq g_0(r)$, is no longer valid. New methods are then required, as we shall discuss in detail in the next section.

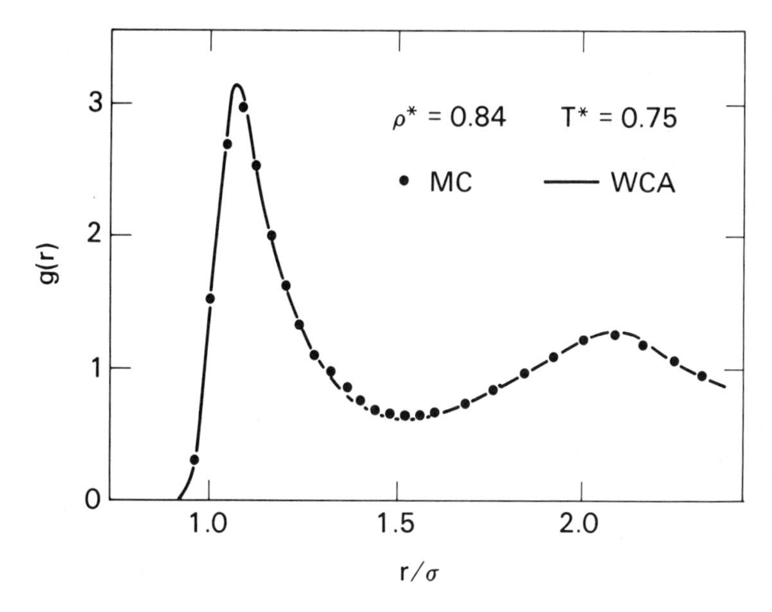

FIG. 6.5. Radial distribution function of the reference system used in the WCA perturbation theory. The points are Monte Carlo results (J. J. Weis, unpublished results) and the curve is calculated from Eqn (6.3.15).

6.5. LONG-RANGE PERTURBATIONS

Situations in which the influence of the attractive forces on the structure cannot be ignored may be treated by methods similar to those used when the perturbation is both weak and very long ranged relative to the reference-system potential. In such cases, the natural expansion parameter is the inverse range rather than the strength of the perturbation; this leads to the so-called γ-expansion (Hemmer, 1964; Lebowitz *et al.*, 1965), the nature of which differs significantly from that of the λ-expansion described in Section 6.2. The early work on the γ-expansion was motivated by the fact that an exact solution can be found for the one-dimensional model of hard rods of length d that attract each other via the Kac potential, i.e.

$$w_\gamma(x) = -a\gamma \exp(-\gamma x), \qquad a\gamma > 0 \tag{6.5.1}$$

where γ is an inverse-range parameter (Kac *et al.*, 1963; Uhlenbeck *et al.*, 1963; Hemmer *et al.*, 1964). The integral of $w_\gamma(x)$ over the entire one-dimensional space is independent of γ and equal to $-a$. In the limit $\gamma \to 0$, taken after the thermodynamic limit, the pressure is given exactly (Kac *et al.*, 1963) by the one-dimensional van der Waals equation

$$\lim_{\gamma \to 0} \frac{\beta P}{\rho} = \frac{1}{1-\rho d} - \beta \rho a \tag{6.5.2}$$

The first term on the right-hand side of (6.5.2) represents the exact equation of state of the hard-rod reference system (Tonks, 1936). The result of Kac *et al.* (1963) was extended to three dimensions by van Kampen (1964) and by Lebowitz and Penrose (1966). It was proved rigorously by these authors that in the limit where the perturbation is both infinitesimally strong and infinitely long ranged, the equation of state is given by the generalized van der Waals equation (6.1.2).

The γ-expansion is obtained by considering perturbations of the general form

$$w_\gamma(r) = -\gamma^3 \phi(\gamma r) \tag{6.5.3}$$

and expanding the properties of the system of interest in powers of γ. If R is the range of the reference-system potential (e.g. the hard-sphere diameter), the dimensionless parameter of the expansion is $\xi = (\gamma R)^3$; ξ is roughly the ratio of the reference-system interaction volume (e.g. the volume of a hard sphere) to the total interaction volume. In most simple liquids, the attractive forces are not truly long ranged in the sense of (6.5.3), but many of the results of the γ-expansion can usefully be carried over to such systems by setting $\gamma = 1$. However, rather than giving the original derivation of the γ-expansion, we shall describe instead the closely related but simpler method of Andersen and Chandler (1972). We assume throughout that the pair potential has the general form given by Eqn (6.1.3).

We first require the diagrammatic expansion of the excess Helmholtz free energy in terms of ρ-circles and f-bonds; this can be derived from the corresponding expansion of the chemical potential, Eqn (4.5.9), taken for the case of zero external field. The thermodynamic relation (2.4.1) shows that

$$\begin{aligned} \beta\mu &= \left(\frac{\partial}{\partial N}\beta A\right)_{V,T} \\ &= \left(\frac{\partial}{\partial \rho}\frac{\beta A}{V}\right)_{V,T} \end{aligned} \tag{6.5.4}$$

It follows, given Eqn (2.4.2) and the definition of z, Eqn (2.4.12), that

$$\log z = \log \rho - \left(\frac{\partial \mathscr{A}}{\partial \rho}\right)_{V,T} \tag{6.5.5}$$

where $\mathscr{A} = -\beta A^{\text{ex}}/V$. The required expansion of $\mathscr{A}$ can then be deduced by comparison of Eqns (4.5.9) and (6.5.5):

$$\mathscr{A} = (1/V)\{\text{all simple, irreducible diagrams consisting of two or more } \rho\text{-circles and } f\text{-bonds}\} \tag{6.5.6}$$

Conversely, if the expansion (6.5.6) is inserted in Eqn (6.5.5), a simple application of Lemma 2 of Chapter 4 leads back to Eqn (4.5.10).

The separation of the pair potential in Eqn (6.1.3) means that the Mayer function $f(1,2)$ can be factorized as

$$f(1,2) = f_0(1,2) + [1 + f_0(1,2)]\{\exp[\Phi(1,2)] - 1\} \tag{6.5.7}$$

where $f_0(1,2)$ is the Mayer function of the reference system and

$$\Phi(1,2) = -\beta w(1,2) \tag{6.5.8}$$

Since the perturbation is weak, the exponential term in (6.5.7) may be expanded to give

$$f(1,2) = f_0(1,2) + [1 + f_0(1,2)] \sum_{n=1}^{\infty} \frac{[\Phi(1,2)]^n}{n!} \tag{6.5.9}$$

The form of Eqn (6.5.9) suggests the introduction of two different types of bond: short-range f_0-bonds, which we shall represent by broken lines, and long-range Φ-bonds, to be denoted by solid lines. The presence of two types of bond transforms the simple diagrams of Eqn (6.5.6) into composite diagrams in which two circles may be connected by at most one f_0-bond and an arbitrary number of Φ-bonds. We recall from the definitions given in Section 4.3 that if two circles in a diagram are connected by n bonds of a given species, the symmetry number of the diagram is increased, and hence its value is decreased, by a factor $n!$ This takes care of the factors $1/n!$ in Eqn (6.5.9). For example, the value of one of the composite diagrams that contribute to $\mathscr{A}$ is

$$= \tfrac{1}{2}\rho^3 \iiint f_0(1,2)\Phi(1,3)[\Phi(2,3)]^2 \,\mathrm{d}1\,\mathrm{d}2\,\mathrm{d}3 \qquad (6.5.10)$$

The complete expansion of $\mathscr{A}$ in terms of composite diagrams is

$\mathscr{A} = (1/V)\{$all irreducible diagrams consisting of two or more black ρ-circles, f_0-bonds and Φ-bonds, such that each pair of circles is connected by at most one f_0-bond and any number of Φ-bonds$\}$ (6.5.11)

The sum of all diagrams in (6.5.11) that consist only of ρ-circles and f_0-bonds yields the free energy $\mathscr{A}_0$ of the reference system. The f_0-bonds in the remaining diagrams can be eliminated in favour of h_0-bonds by a process of topological reduction based on Lemma 5 of Chapter 4. The procedure is similar to that involved in the passage from the z^*-expansion of $\log \rho^{(1)}(\mathbf{r})$, Eqn (4.5.8), to the expansion in terms of $\rho^{(1)}$-circles, Eqn (4.5.9), except that the articulation circle considered there is here replaced by a *reference articulation pair*, i.e. a pair of circles whose removal causes a diagram to separate into two or more components of which at least one contains only black circles and reference-system bonds. The final result of this operation is

$\mathscr{A} = \mathscr{A}_0 + (1/V)\{$all irreducible diagrams consisting of two or more black ρ-circles, h_0-bonds and Φ-bonds, such that each pair of circles is connected by at most one h_0-bond and any number of Φ-bonds, there is at least one Φ-bond, and there is no reference articulation pair$\}$ (6.5.12)

The corresponding expansion of the pair distribution function can be obtained from Eqn (4.2.35). Written in the notation of the present section, the latter becomes

$$\rho^2 g(1,2) = 2V\frac{\delta \mathscr{A}}{\delta \Phi(1,2)} \qquad (6.5.13)$$

and the diagrammatic prescription for $g(1,2)$ follows immediately from application of Lemma 3 of Chapter 4 to the expansion (6.5.12).

There are two diagrams in the expansion (6.5.12) that contain only a single Φ-bond, namely

$$\bullet\!\!-\!\!\!-\!\!\bullet = \tfrac{1}{2}\rho^2 \iint \Phi(1,2)\, \mathrm{d}1\, \mathrm{d}2$$

$$= -\tfrac{1}{2} V\beta\rho^2 \int w(\mathbf{r})\, \mathrm{d}\mathbf{r} \tag{6.5.14}$$

and

$$= \tfrac{1}{2}\rho^2 \iint h_0(1,2)\Phi(1,2)\, \mathrm{d}1\, \mathrm{d}2$$

$$= -\tfrac{1}{2} V\beta\rho^2 \int h_0(\mathbf{r}) w(\mathbf{r})\, \mathrm{d}\mathbf{r} \tag{6.5.15}$$

We write the sum of these two diagrams as

$$Va_{\mathrm{HTA}} = \bullet\!\!-\!\!\!-\!\!\bullet \quad + $$

$$= -\tfrac{1}{2} V\beta\rho^2 \int g_0(\mathbf{r}) w(\mathbf{r})\, \mathrm{d}\mathbf{r} \tag{6.5.16}$$

where HTA stands for "high-temperature approximation". Comparison of Eqns (6.2.13) and (6.5.16) shows that the HTA is equivalent to truncation of the λ-expansion after the first-order term, with

$$a_{\mathrm{HTA}} = -\frac{\beta A_1}{V} \tag{6.5.17}$$

The corresponding approximation to $g(1,2)$ is given by a trivial application of Lemma 3. If $\mathscr{A} = a_{\mathrm{HTA}}$, we find from Eqn (6.5.13) that

$$\rho^2 g(1,2) = 2V \frac{\delta a_{\mathrm{HTA}}}{\delta \Phi(1,2)}$$

$$= \underset{1}{\circ} \quad \underset{2}{\circ} \quad + \quad \underset{1}{\circ}\!- - -\!\underset{2}{\circ}$$

$$= \rho^2 g_0(1,2) \tag{6.5.18}$$

in agreement with the results of Section 6.2.

In order to proceed beyond the HTA, it is necessary to sum a larger class of diagrams in the expansion (6.5.12). An approximation that is similar in spirit to the Debye–Hückel theory of ionic fluids is to write

$$\mathscr{A} = \mathscr{A}_0 + a_{\text{HTA}} + a_{\text{Ring}} \tag{6.5.19}$$

where

$Va_{\text{ring}} =$

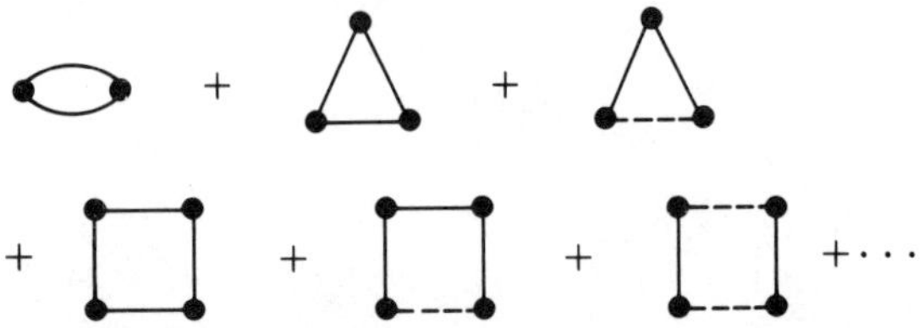

(6.5.20)

is the sum of all the simple ring diagrams in (6.5.12) plus the diagram consisting of two black circles linked by two Φ-bonds (Andersen and Chandler, 1972). Note that the absence of reference articulation pairs means that none of the ring diagrams in (6.5.20) contains two successive h_0-bonds. The approximation to $g(1, 2)$ obtained by applying Lemma 3 is now

$$g(1, 2) = g_0(1, 2) + C(1, 2) \tag{6.5.21}$$

where the function $C(1, 2)$ is given by

$\rho^2 C(1, 2) =$ {all simple chain diagrams consisting of two terminal white ρ-circles labelled 1 and 2, black ρ-circles, Φ-bonds and h_0-bonds, such that there are never two successive h_0-bonds} (6.5.22)

If the reference system is the ideal gas and if $w(r)$ is the Coulomb potential, then $-k_B TC(1, 2)$ is the screened potential of Eqn (5.3.22) and Eqn (6.5.21) reduces to the linearized Debye–Hückel result (5.3.24). For the systems of interest here, $-k_B TC(1, 2)$ is a renormalized potential in which the "bare" perturbation $w(1, 2)$ is screened by the order imposed on the fluid by the short-range interactions between particles.

The function $C(1, 2)$ can be evaluated by Fourier transform techniques similar to those used in the derivation of the Debye–Hückel result in Section 5.3. We first group the chain diagrams in (6.5.22) according to the number of Φ-bonds that they contain. Let $C^{(n)}(1, 2)$ be the sum of all chain diagrams with precisely n Φ-bonds. Then

$$\rho^2 C(1, 2) = \rho^2 \sum_{n=1}^{\infty} C^{(n)}(1, 2) \tag{6.5.23}$$

where

$\rho^2 C^{(1)}(1, 2) =$

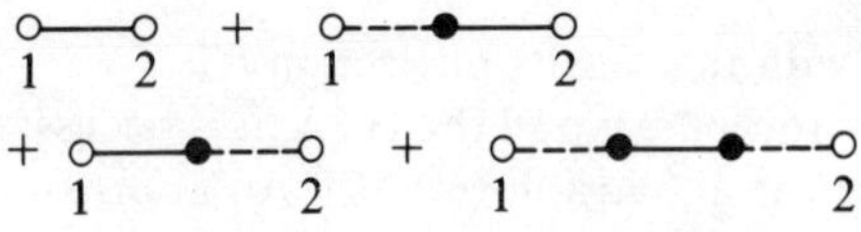

(6.5.24)

$$\rho^2 C^{(2)}(1,2) = \quad (6.5.25)$$

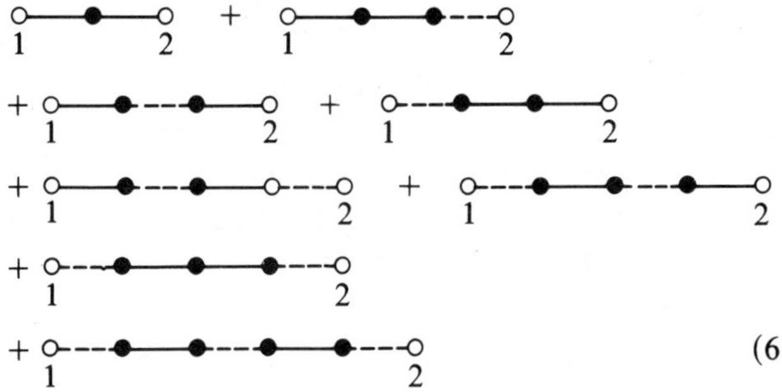

and so on. Each diagram that contributes to $C^{(n)}$ contains at most $(n+1)$ h_0-bonds and $C^{(n)}$ consists of 2^{n+1} topologically distinct diagrams.

The sum of all diagrams in $C^{(n)}(1,2)$ may be represented by a single *generalized chain* in which the circles are replaced by *hypervertices* (Lebowitz et al., 1965). A hypervertex of order n is associated with a function of n coordinates, $F(1,\ldots,n)$, and is pictured as a large circle surrounded by n white or black circles; the latter, correspond, as usual, to the coordinates $\mathbf{r}_1,\ldots\mathbf{r}_n$. For present purposes, we need consider only the hypervertex of order two that is associated with the reference-system function $F_0(1,2)$ defined as

$$F_0(1,2) = \rho\delta(1,2) + \rho^2 h_0(1,2) \quad (6.5.26)$$

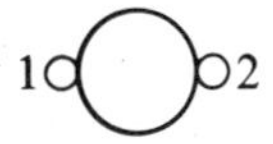

We may then re-express $C^{(n)}(1,2)$ for $n=1$ and $n=2$ as

$$\rho^2 C^{(1)}(1,2) = \iint F_0(1,3)\Phi(3,4)F_0(4,2)\,\mathrm{d}3\,\mathrm{d}4 \quad (6.5.27)$$

$$\rho^2 C^{(2)}(1,2) = \quad (6.5.28)$$

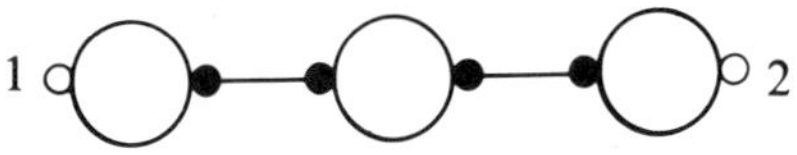

and $C^{(n)}(1,2)$ for any n may be represented as a generalized chain consisting of n Φ-bonds and $(n+1)$ F_0-hypervertices. Each generalized chain corresponds to a simple convolution integral and can therefore be factorized by taking the Fourier transform. Let

$$\hat{\Phi}(\mathbf{k}) = \int \exp(-i\mathbf{k}\cdot\mathbf{r})\Phi(\mathbf{r})\,\mathrm{d}\mathbf{r} = -\beta\hat{w}(\mathbf{k}) \quad (6.5.29)$$

and

$$\hat{F}_0(\mathbf{k}) = \int \exp(-i\mathbf{k}\cdot\mathbf{r})F_0(\mathbf{r})\,d\mathbf{r}$$
$$= \rho S_0(\mathbf{k}) \tag{6.5.30}$$

where $S_0(\mathbf{k})$ is the structure factor of the reference system. Then the Fourier transform of $C^{(n)}(1,2)$ is given by

$$\rho^2\hat{C}^{(n)}(\mathbf{k}) = [F_0(\mathbf{k})\Phi(\mathbf{k})]^n F_0(\mathbf{k}) \tag{6.5.31}$$

and the Fourier transform of the function $C(1,2)$ is obtained as the sum of a geometric series:

$$\rho^2\hat{C}(\mathbf{k}) = \sum_{n=1}^{\infty} \rho^2\hat{C}^{(n)}(\mathbf{k})$$
$$= \frac{[\hat{F}_0(\mathbf{k})^2\hat{\Phi}(\mathbf{k})}{1-\hat{F}_0(\mathbf{k})\hat{\Phi}(\mathbf{k})}$$
$$= -\frac{\beta\rho^2[S_0(\mathbf{k})]^2\hat{w}(\mathbf{k})}{1+\beta\rho S_0(\mathbf{k})\hat{w}(\mathbf{k})} \tag{6.5.32}$$

The derivation of Eqn (6.5.32) tends to obscure the basic simplicity of the theory. If we combine Eqns (5.1.10), (6.5.21) and (6.5.32), we find that the structure factor of the system of interest is related to that of the reference fluid by

$$S(\mathbf{k}) = S_0(\mathbf{k}) - \frac{\beta\rho[S_0(\mathbf{k})]^2\hat{w}(\mathbf{k})}{1+\beta\rho S_0(\mathbf{k})\hat{w}(\mathbf{k})}$$
$$= \frac{S_0(\mathbf{k})}{1+\beta\rho S_0(\mathbf{k})\hat{w}(\mathbf{k})} \tag{6.5.33}$$

On the other hand, we find with the help of Eqn (5.2.11) that the exact relation between the two structure factors is given in terms of the corresponding direct correlation functions by

$$S(\mathbf{k}) = \frac{S_0(\mathbf{k})}{1-\rho[\hat{c}(\mathbf{k})-\hat{c}_0(\mathbf{k})]S_0(\mathbf{k})} \tag{6.5.34}$$

Equation (6.5.33) is therefore equivalent to approximating the true direct correlation function by

$$c(r) \simeq c_0(r) - \beta w(r) \tag{6.5.35}$$

The approximation is asymptotically correct if the perturbation contains the long-range part of the potential. The crude Debye–Hückel approximation corresponds to writing $c(r) \simeq -\beta w(r)$; Eqn (6.5.35) improves on this to the extent that it builds in the exact form of the reference-system direct correlation function.

Equation (6.5.33) has been derived many times in the literature, particularly for the case when the reference system is the ideal gas (Pines and Nozières, 1966; Andersen and Chandler, 1970). For historical reasons, it is known as the "random-phase approximation" or RPA for the fluid structure in the presence of a perturbation. The perturbation must be sufficiently weak, or the density sufficiently low, to ensure that $\beta\rho S_0(\mathbf{k})\hat{w}(\mathbf{k}) > -1$ for all wavenumbers. If this condition is not satisfied, $S(\mathbf{k})$ becomes negative for some range of k, a behaviour that is sometimes called the "RPA catastrophe" (Wheeler and Chandler, 1971).

The RPA approximation for the free energy is obtained by combining Eqns (6.5.16), (6.5.19) and (6.5.20). When functionally differentiated with respect to $\Phi(1,2)$ according to the rule (6.5.13), the total ring-diagram contribution to $\mathscr{A}_{\text{RPA}}$ yields the function $C(1,2)$. It follows that Va_{Ring} can be expressed diagrammatically as

$$Va_{\text{Ring}} = \sum_{n=2}^{\infty} R^{(n)} \tag{6.5.36}$$

where $R^{(n)}$ is a *generalized ring* consisting of F_0-hypervertices and Φ-bonds. A generalized ring can be derived from a generalized chain by inserting a Φ-bond between the white circles and integrating over the coordinates associated with the latter. Thus

$$\begin{aligned} R^{(n)} &= \frac{\rho^2}{2n}\iint C^{(n-1)}(1,2)\Phi(1,2)\,\mathrm{d}1\,\mathrm{d}2 \\ &= \frac{V\rho^2}{2n}\int C^{(n-1)}(\mathbf{r})\Phi(\mathbf{r})\,\mathrm{d}\mathbf{r} \\ &= \frac{V\rho^2}{2n}\left(\frac{1}{2\pi}\right)^3\int \hat{C}^{(n-1)}(\mathbf{k})\Phi(\mathbf{k})\,\mathrm{d}\mathbf{k} \end{aligned} \tag{6.5.37}$$

where the factor $1/2n$ arises from the symmetry number of the generalized ring. If we substitute from Eqn (6.5.31), we find that the ring-diagram contribution to $\mathscr{A}_{\text{RPA}}$ is given by

$$\begin{aligned} a_{\text{Ring}} &= \left(\frac{1}{2\pi}\right)^3\int \sum_{n=2}^{\infty}\frac{1}{2n}[\hat{F}_0(\mathbf{k})\hat{\Phi}(\mathbf{k})]^n\,\mathrm{d}\mathbf{k} \\ &= -\frac{1}{2}\left(\frac{1}{2\pi}\right)^3\int \{\hat{F}_0(\mathbf{k})\hat{\Phi}(\mathbf{k}) + \log[1-\hat{F}_0(\mathbf{k})\hat{\Phi}(\mathbf{k})]\}\,\mathrm{d}\mathbf{k} \end{aligned} \tag{6.5.38}$$

We have seen in Section 5.3 that a defect of the linearized Debye–Hückel approximation is the fact that it yields a pair distribution function that behaves unphysically at small separations. A similar problem arises here. Consider, for simplicity, the case in which the reference system is a fluid of hard spheres of diameter d. In an exact theory, $g(r)$ necessarily vanishes for $r<d$; in the approximation represented by Eqn (6.5.21), there is no guarantee that this will be so, since in general $C(r)$ will be non-zero in that range. There is, however, a flexibility in the choice of $C(r)$ that can usefully be exploited. Although $C(r)$ is a functional of $w(r)$, it is obvious on physical grounds that the true properties of the fluid must be independent of the choice of perturbation for $r<d$. The unphysical behaviour of the RPA can therefore be eliminated by choosing $w(r)$ for $r<d$ in such a way that

$$C(r)=0, \qquad r<d \tag{6.5.39}$$

It follows from Eqn (6.5.13) that this condition implies that

$$\frac{\delta \mathcal{A}_{\mathrm{RPA}}}{\delta \Phi(\mathbf{r})}=\frac{\delta V a_{\mathrm{Ring}}}{\delta \Phi(\mathbf{r})}=\tfrac{1}{2}\rho^2 C(r)=0, \qquad r<d \tag{6.5.40}$$

In other words, the RPA free energy should be stationary with respect to variations in the perturbing potential inside the hard core. Provided $\beta\rho S_0(\mathbf{k})\hat{w}(\mathbf{k})>-1$, a_{Ring} is positive definite (assuming the perturbation to be attractive) and therefore has a non-negative minimum. Consequently, the condition (6.5.40) can be satisfied by expanding $w(r)$ for $r<d$ in a series of suitable basis functions, such as Legendre polynomials, and minimizing a_{Ring} with respect to the coefficients (Andersen *et al.*, 1972). The resulting renormalized potential is then used to calculate both $\mathcal{A}_{\mathrm{RPA}}$ and $g(r)$ for $r>d$.

The RPA together with the condition (6.5.40) is called the "optimized" random-phase approximation or ORPA (Andersen and Chandler, 1972). The ORPA may equally well be viewed as a solution to the Ornstein–Zernike relation that satisfies both the closure approximation (6.5.35) and the restriction that $g(r)=0$ for $r<d$. It is therefore similar in spirit to the MSA of Section 5.6, the difference being that the treatment of the hard-sphere system is exact in the ORPA (Henderson and Smith, 1978; Madden, 1981).

A different method of remedying the unphysical behaviour of the RPA pair distribution function can be developed by extending the analogy with Debye–Hückel theory. If the reference system is the perfect gas, the RPA is

$$g(1,2)=1+\underset{1\qquad 2}{\circ\!\!-\!\!\!-\!\!\!-\!\!\circ} \tag{6.5.41}$$

where the bond is a C-bond and the circles are 1-circles. When $w(r)$ is the Coulomb potential, Eqn (6.5.41) is equivalent to the linearized Debye–Hückel approximation (5.3.24). If we add to the right-hand side of (6.5.41) the sum of all diagrams in the exact expansion of $h(1,2)$ that can be expressed as star products of the diagram $C(1,2)$ with itself, and then apply Lemma 1 of Chapter 4, we obtain an improved approximation in the form

$$g(1,2) = 1 + \underset{1\quad 2}{\circ\!\!-\!\!\!-\!\!\circ} + \underset{1\quad 2}{\circ\!\!=\!\!\circ} + \underset{1\quad 2}{\circ\!\!\equiv\!\!\circ} + \cdots$$

$$= \exp C(1,2) \tag{6.5.42}$$

This result is equivalent to the non-linear equation (5.3.24) (Abe, 1959). In the present case, a generalization of the same approach replaces the RPA of Eqn (6.5.21) by the approximation

$$g(1,2) = g_0(1,2) \exp C(1,2) \tag{6.5.43}$$

Andersen and Chandler (1972) call this the "exponential" or EXP approximation (see also Lado, 1964). Equations (6.5.23) and (6.5.24) show that the renormalized potential behaves at low density as $C(\mathbf{r}) \simeq C^{(1)}(\mathbf{r}) \simeq \Phi(\mathbf{r}) = -\beta w(\mathbf{r})$. In the same limit, $g_0(r) \simeq \exp[-\beta v_0(r)]$. Thus, from Eqn (6.5.43):

$$g(r) \underset{\rho \to 0}{\sim} \exp[-\beta v_0(r)] \exp[-\beta w(r)] = \exp[-\beta v(r)] \tag{6.5.44}$$

The EXP approximation, unlike either the HTA or the ORPA, is therefore exact in the low-density limit. The corresponding expression for the free energy is

$$\mathcal{A}_{\mathrm{EXP}} = \mathcal{A}_0 + a_{\mathrm{HTA}} + a_{\mathrm{Ring}} + B_2 \tag{6.5.45}$$

where

$$B_2 = \tfrac{1}{4}\rho^2 \int h_0(\mathbf{r})[C(\mathbf{r})]^2 \, \mathrm{d}\mathbf{r}$$

$$+ \tfrac{1}{2}\rho^2 \sum_{n=3}^{\infty} \frac{1}{n!} \int g_0(\mathbf{r})[C(\mathbf{r})]^n \, \mathrm{d}\mathbf{r} \tag{6.5.46}$$

Andersen and Chandler (1972) give arguments to show that the contribution from diagrams neglected in the EXP approximation is minimized if the optimized $C(1,2)$ is used in the evaluation of $g(1,2)$ and $\mathcal{A}$; the results are then sometimes referred to as those of the "ORPA+B_2" approximation.

The ORPA and the EXP approximation with optimized $C(1,2)$ both correspond to a truncation of the exact series (6.5.12) (and of the corresponding series for $h(1,2)$) in which the perturbation inside the hard core is chosen so as to increase the rate of convergence. Each is therefore a particular approximation within a general theoretical framework that is called either "optimized cluster theory" or the "optimized cluster expansion". The optimized cluster expansion is not in any obvious way a systematic expansion in powers of a small parameter, but it has the great advantage of yielding successive approximations that are easy to evaluate if the pair distribution function of the reference system is known. By contrast, the γ-expansion provides a natural ordering of the perturbation terms in powers of γ^3, but it leads to more complicated expressions for properties of the system of interest. If the perturbation is of the form of Eqn (6.5.3), the terms of order γ^3 in the expansion of the free energy consist of the diagram (6.5.15) and the sum of all diagrams in Eqn (6.5.20). There is, in addition, a term of zeroth order in γ, given by the diagram (6.5.14):

$$\bullet\!\!-\!\!\!-\!\!\bullet = \tfrac{1}{2}V\beta\rho^2\gamma^3 \int \phi(\gamma^3\mathbf{r})\,\mathrm{d}\mathbf{r}$$

$$= \tfrac{1}{2}V\beta\rho^2 \int \phi(\gamma^3\mathbf{r})\,\mathrm{d}(\gamma^3\mathbf{r})$$

$$= V\beta\rho^2 a \tag{6.5.47}$$

where a is the constant introduced in Eqn (6.2.15). Note that the effect of the volume integration is to reduce the apparent order of the term from γ^3 to γ^0. As a consequence, the free energy does not reduce to that of the reference system in zeroth order. It yields, instead, the van der Waals approximation; the latter is therefore exact in the limit $\gamma \to 0$. Through order γ^3, the free energy (with $\gamma = 1$) is the same as in the RPA. On the other hand, the sum of all terms of order γ^3 in the expansion of $h(1, 2)$ contains diagrams additional to the chain diagrams included in the approximation (6.5.21). To order γ^3 (again with $\gamma = 1$), the result for the pair distribution function is

$$g(1, 2) = g_0(1, 2) + \tfrac{1}{2}C(1, 2)\frac{\partial^2}{\partial\rho^2}\rho^2 g_0(1, 2) \tag{6.5.48}$$

If the density dependence of $g_0(1, 2)$ is ignored, Eqn (6.5.48) becomes

$$g(1, 2) = g_0(1, 2)[1 + C(1, 2)] \tag{6.5.49}$$

which is equivalent to linearizing the right-hand side of Eqn (6.5.43) with respect to $C(r)$ and, for that reason, is called the LIN approximation (Verlet and Weis, 1974).

Results obtained by the optimized cluster approach for a model consisting of a hard-sphere potential plus a Lennard-Jones "tail" are compared with results of Monte Carlo calculations in Figure 6.6 (Stell and Weis, 1980). It is clear from the figure that at the low density to which the results refer the HTA, ORPA and EXP radial distribution functions represent successively improved approximations to the "exact" results. At higher densities, the renormalized potential is very weak, and the HTA, i.e. the distribution function of the hard-sphere reference fluid, is already a good approximation. The figure also shows that the ORPA undercorrects and the EXP approximation overcorrects the hard-sphere results. Similar conclusions have been reached for other potentials (Verlet and Weis, 1974; Henderson and Blum,

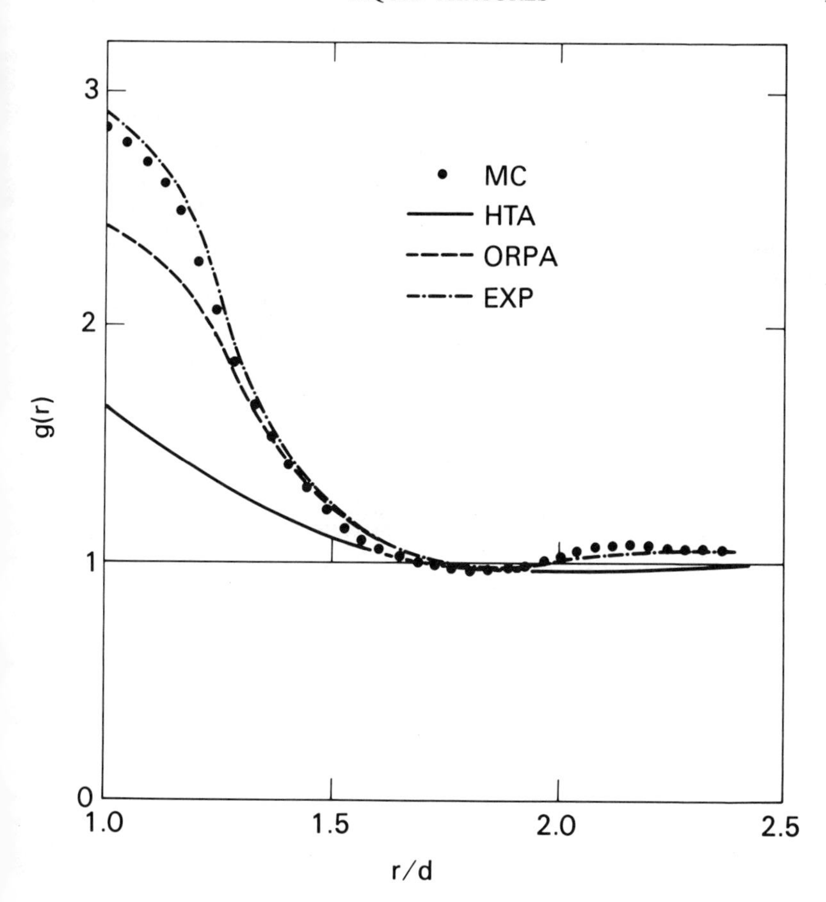

FIG. 6.6. Radial distribution function of a fluid of hard spheres with a Lennard-Jones "tail" at $\rho d^3 = 0.2$, $k_B T/\varepsilon = 1.6$. The curves are calculated from the versions of optimized cluster theory discussed in the text and the points are Monte Carlo results (Stell and Weis, 1980).

1978; Konior and Jedrzejek, 1985), but it does not seem possible to improve the EXP approximation in a systematic way, though *ad hoc* schemes such as (6.5.49) are sometimes successful. Overall, the results obtained by optimized cluster methods are comparable in accuracy with those of conventional perturbation theory taken to second order.

6.6 LIQUID MIXTURES

The perturbation theories described in earlier sections can be generalized without difficulty to multicomponent systems. The methods of Barker and

Henderson (Leonard *et al.*, 1970; Grundke *et al.*, 1973; Gibbons, 1975), of Mansoori and Canfield (Mansoori and Leland, 1970; Mansoori, 1972) and of Weeks, Chandler and Andersen (Lee and Levesque, 1973) have all been extended in this way, with generally satisfactory results. As no new principles are involved, we shall not give any details, and choose instead to give a brief account of those methods, peculiar to mixtures, that are usually grouped together under the heading of *conformal-solution theory* (Longuet-Higgins, 1951).

The goal of a statistical-mechanical theory of mixtures is not so much the calculation of the thermodynamic properties of the mixture itself, but rather of the changes that occur when fluids are mixed. Properties of mixing can be defined in a variety of ways (at constant number density, constant volume, etc.) but in practice, at least for liquids, the most important are those that refer to mixing at the same temperature and pressure, and we shall confine ourselves to these.

As in earlier chapters, we consider a mixture consisting of $N=\sum_\nu N_\nu$ molecules, where ν labels a species. If we denote a function of mixing by a superscript M, then, for any thermodynamic property F, we may write

$$F^{\mathrm{M}}(P, T) = F_{\mathrm{m}}(P, T, \{N_\nu\}) - \sum_\nu x_\nu F_\nu(P, T, N) \tag{6.6.1}$$

where F_{m} is the value for the mixture, F_ν is the value for pure species ν, and $x_\nu = N_\nu/N$ is the fractional concentration of species ν. When a mixture is formed from particles that are labelled but are otherwise identical, the entropy of mixing is given by

$$S^{\mathrm{M}} = -Nk_{\mathrm{B}} \sum_\nu x_\nu \log x_\nu \quad \text{(ideal mixture)} \tag{6.6.2}$$

Such a mixture is said to be *ideal*. In an ideal mixture, the total potential energy depends only on the positions of the particles and not on the assignment of the different species to these positions. The Gibbs free energy of mixing of an ideal mixture is

$$G^{\mathrm{M}} = Nk_{\mathrm{B}}T \sum_\nu x_\nu \log x_\nu \quad \text{(ideal mixture)} \tag{6.6.3}$$

but the volume and enthalpy (heat) of mixing are both zero.

Except for mixtures of isotopes, and then only when quantum effects are negligible, real mixtures are not ideal. It is therefore convenient to introduce a further set of quantities, called the *excess* properties of mixing, which we shall denote by a superscript E. The excess functions are the differences between the actual mixing functions and those of an ideal mixture at the same pressure, temperature and composition. The main quantities of interest are the excess Gibbs free energy G^{E}, the excess enthalpy H^{E}, and the excess volume V^{E}. The simplest model of a real, binary mixture is one in which

G^E is a quadratic function of composition (Rowlinson and Swinton, 1982), i.e.

$$G^E = x_1 x_2 w \tag{6.6.4}$$

where w is a characteristic energy, sometimes called the "interchange" energy. The excess enthalpy and excess volume of a "quadratic" mixture are given by

$$H^E = x_1 x_2 [w - T(\partial w/\partial T)_P] \tag{6.6.5}$$

$$V^E = x_1 x_2 (\partial w/\partial P)_T \tag{6.6.6}$$

Equation (6.6.4) provides a fair fit to the excess free energies of many mixtures. In Hildebrand's theory of "regular solutions", the further approximation is made that w in Eqn (6.6.4) is independent of temperature (Hildebrand *et al.*, 1970). This assumption implies that the excess free energy and excess enthalpy are equal, and that the excess entropy is therefore zero. In practice, such a situation is rarely encountered, even in cases where Eqn (6.6.4) is a good approximation.

Experimentally, most mixtures of simple liquids have positive values of G^E and H^E, but the excess volume may be of either sign. The excess properties are always very small in comparison with the properties of the mixture itself. In the case of the system argon/krypton, for example, $H^E/H_m \simeq 0.006$ and $V^E/V_m \simeq -0.02$ at a temperature corresponding to the triple point of krypton.

The use of conformal solution-theory is restricted to mixtures for which the pair potentials, and those of the pure components, are all of the form

$$v_{\nu\mu}(r) = \varepsilon_{\nu\mu} f(r/\sigma_{\nu\mu}) \tag{6.6.7}$$

where $\varepsilon_{\nu\mu}$ and $\sigma_{\nu\mu}$ are, respectively, a characteristic energy and a characteristic length, and the function f is the same for all potentials: mixtures of hard spheres and mixtures of Lennard-Jones fluids are two important examples. The principle of corresponding states applies rigorously to any group of pure substances whose potentials are conformal in the sense of Eqn (6.6.7). In general, it is assumed that the cross-interaction parameters ($\nu \neq \mu$) are related by some simple prescription to those for like molecules. The most frequently used of these *combining rules* are

$$\sigma_{\nu\mu} = \tfrac{1}{2}(\sigma_{\nu\nu} + \sigma_{\mu\mu}) \quad \text{(Lorentz rule)} \tag{6.6.8}$$

which is exact for hard spheres, and

$$\varepsilon_{\nu\mu} = (\varepsilon_{\nu\nu}\varepsilon_{\mu\mu})^{1/2} \quad \text{(Berthelot rule)} \tag{6.6.9}$$

A mixture of Lennard-Jones fluids to which these rules apply is called a Lorentz-Berthelot mixture. Unfortunately, because the functions of mixing are so small, a few per cent deviation from the Lorentz–Berthelot rules may result in appreciable changes in magnitude and even a change in sign of the calculated excess properties. Comparison with results on real mixtures can therefore be misleading, and the only reliable method of testing theories is by comparison with the results of computer simulations (Alder, 1964b; McDonald, 1972a, b; Singer and Singer, 1972).

The isothermal–isobaric version of the Monte Carlo method (see Section 3.5) is particularly well-suited to the study of mixing at constant pressure. The enthalpy and volume of the mixture at the temperature and pressure of interest are calculated directly, while those of the pure liquids may be obtained either from an empirical equation of state for the particular family of conformal substances, or by separate Monte Carlo computations. It is then a simple task to calculate H^{E} and V^{E}, but the determination of G^{E} is less straightforward (Singer and Singer, 1972). Consider a binary mixture of species 1 and 2 and imagine a "charging" process whereby all characteristic energies $\varepsilon_{\nu\mu}$ are changed to a common value ε_0 and all characteristic lengths $\sigma_{\nu\mu}$ are similarly converted to a common value σ_0. The result of this process is a pure reference liquid (subscript 0). Let the change in excess free energy due to the change in interaction parameters be ΔG. Then

$$G_{\mathrm{m}} = G_0 + Nk_{\mathrm{B}}T(x_1 \log x_1 + x_2 \log x_2) - \Delta G \tag{6.6.10}$$

and

$$G^{\mathrm{E}} = G_0 - x_1 G_1 - x_2 G_2 - \Delta G \tag{6.6.11}$$

The first three terms on the right-hand side of (6.6.11) can all be calculated from a single equation of state. If $\mathbf{X}$ is the array of interaction parameters that characterizes the mixture at an intermediate stage in the charging process, then the quantity ΔG is given by the integral

$$\Delta G = \int_{\mathbf{X}_{\mathrm{m}}}^{\mathbf{X}_0} \frac{\partial G(P, T; \mathbf{X})}{\partial \mathbf{X}} \, \mathrm{d}\mathbf{X} \tag{6.6.12}$$

where $\mathbf{X}_0$ denotes the potential parameters in the reference fluid and $\mathbf{X}_{\mathrm{m}}$ denotes those in the mixture of interest. With the help of Eqns (2.3.25) and (3.5.1), the derivative in (6.6.12) may be expressed as an ensemble average for the system characterized by the array $\mathbf{X}$:

$$\begin{aligned} \frac{\partial G(P, T; \mathbf{X})}{\partial \mathbf{X}} &= -\frac{k_{\mathrm{B}}T}{\Delta_N(P, T)} \frac{\partial \Delta_N(P, T)}{\partial \mathbf{X}} \\ &= \frac{\partial}{\partial \mathbf{X}} \langle V_N(\mathbf{X}) \rangle \end{aligned} \tag{6.6.13}$$

The free energy change ΔG can therefore be calculated by numerical integration of data that are easily obtained from Monte Carlo calculations on a series of mixtures characterized by intermediate values of the interaction parameters.

The simplest form of conformal-solution theory is what is usually called the "one-fluid" approximation. This is a zeroth-order theory in which the properties of the mixture, apart from the ideal terms, are taken to be those of a single, hypothetical, pure fluid. The crudest example is the "random mixing" approximation (Prigogine, 1957); this can be derived from the assumption (Henderson and Leonard, 1971) that all radial distribution functions in the mixture are identical, i.e.

$$g_{\nu\mu}(r) = g_x(r), \quad \text{say} \tag{6.6.14}$$

for all ν, μ. The generalization of the energy equation (2.5.13) to multicomponent systems is

$$\frac{U^{\text{ex}}}{N} = 2\pi\rho \sum_\nu \sum_\mu x_\nu x_\mu \int_0^\infty v_{\nu\mu}(r) g_{\nu\mu}(r) r^2 \, dr \tag{6.6.15}$$

If we make the substitution

$$v_x(r) = \sum_\nu \sum_\mu x_\nu x_\mu v_{\nu\mu}(r) \tag{6.6.16}$$

and use the approximation (6.6.14), Eqn (6.6.15) becomes

$$\frac{U^{\text{ex}}}{N} = 2\pi\rho \int_0^\infty v_x(r) g_x(r) r^2 \, dr \tag{6.6.17}$$

This result has the form of the energy equation for a one-component system with pair potential $v_x(r)$ and radial distribution function $g_x(r)$. Equation (6.6.14) therefore generates a one-fluid theory, but the approximation is a poor one, particularly insofar as it ignores the ordering that occurs when mixtures are formed from particles of different sizes.

An approximation that has a much sounder physical basis than (6.6.14) is to assume (Henderson and Leonard, 1971) that the function $g_{\nu\mu}(r)$ scales with $\sigma_{\nu\mu}$, i.e.

$$g_{\nu\mu}(r/\sigma_{\nu\mu}) = g_x(r/\sigma_x), \quad \text{say} \tag{6.6.18}$$

for all ν, μ. With a change of variable, Eqn (6.6.15) can be rewritten as

$$\begin{aligned} \frac{U^{\text{ex}}}{N} &= 2\pi\rho \sum_\nu \sum_\mu x_\nu x_\mu \varepsilon_{\nu\mu} \sigma_{\nu\mu}^3 \int_0^\infty f(s) g_{\nu\mu}(s) s^2 \, ds \\ &= 2\pi\rho\varepsilon_x\sigma_x^3 \int_0^\infty f(s) g_x(s) s^2 \, ds \end{aligned} \tag{6.6.19}$$

where

$$\varepsilon_x\sigma_x^3 = \sum_\nu \sum_\mu x_\nu x_\mu \varepsilon_{\nu\mu}\sigma_{\nu\mu}^3 \tag{6.6.20}$$

Equation (6.6.19) corresponds to a one-fluid approximation in which $g_x(r)$ is the radial distribution function of a fluid characterized by the interaction parameters ε_x and σ_x. In order to define the hypothetical fluid uniquely, we must supplement Eqn (6.6.20) by a second, independent expression for ε_x or σ_x. Nothing is gained by substitution of (6.6.18) in the multicomponent form of the virial equation (2.5.14), because this leads again to Eqn (6.6.20). One possibility (MacGowan *et al.*, 1985) is to force agreement between the long-wavelength number-density fluctuations in the hypothetical fluid and the mixture by writing

$$\rho \sum_\nu \sum_\mu x_\nu x_\mu \int [g_{\nu\mu}(r) - 1]\,\mathrm{d}\mathbf{r} = \rho \int [g_x(r) - 1]\,\mathrm{d}\mathbf{r} \tag{6.6.21a}$$

or, equivalently:

$$\rho \sum_\nu \sum_\mu x_\nu x_\mu \hat{h}_{\nu\mu}(0) = S_x(0) - 1 \tag{6.6.21b}$$

Substitution of (6.6.18) in Eqn (6.6.21a), combined with a change of variable, gives a second expression in the form

$$\sigma_x^3 = \sum_\nu \sum_\mu x_\nu x_\mu \sigma_{\nu\mu}^3 \tag{6.6.22}$$

Equations (6.6.20) and (6.6.22) together constitute what is called the van der Waals one-fluid or vdW1f approximation (Leland *et al.*, 1968). The name is used because the two equations represent the equivalent in modern terms of the rules used by van der Waals to calculate the constants in his equation of state for mixtures in terms of the corresponding constants for the pure components.

By comparison with the theories described in earlier sections, the vdW1f approximation is a strikingly simple one and very easy to apply. Nevertheless, it is remarkably successful, at least in cases where the interaction parameters for the pure components are not very different. A comparison between Monte Carlo results and theoretical predictions for a number of Lorentz–Berthelot mixtures chosen to simulate real systems is shown in Table 6.3. We see from the table that the results calculated from the vdW1f approximation are comparable in accuracy with those obtained by more complicated routes. Against this, it must be remembered that the vdW1f approximation, though not the one-fluid approach as such, is limited to use with conformal solutions. The approximation also begins to break down as the size differences between components becomes larger (Gubbins *et al.*, 1983). The agreement with experimental data in Table 6.3 is in general

TABLE 6.3. *Excess thermodynamic properties at zero pressure of Lennard-Jones systems chosen to simulate real liquid mixtures*

System	G^E (J mol^{-1})					
	Expt	MC	vdW1f	BH	Var.	WCA
A+Kr(116 K)	+84	+46±7	+46	+33	+47	+36
A+CH_4(91 K)	+74	−14±6	−17	−28	−12	+4
CO+CH_4(91 K)	+115	+77±7	+83	+67	+76	+76
A+N_2(84 K)	+34	+35±5	+39	+29	+42	+40
A+CO(84 K)	+57	+26±5	+29	+21	+28	+34
O_2+N_2(84 K)	+39	+38±5	+43	+30		+47
System	**V^E (cm^3 mol^{-1})**					
	Expt	MC	vdW1f	BH	Var.	WCA
A+Kr(116 K)	−0.52	−0.69	−0.68	−0.73	−0.73	−0.48
A+CH_4(91 K)	+0.17	−0.22	−0.23	−0.36	−0.14	−0.12
CO+CH_4(91 K)	−0.32	−0.76	−0.71	−0.69	−0.75	−0.54
A+N_2(84 K)	−0.18	−0.25	−0.25	−0.30	−0.26	−0.26
A+CO(84 K)	+0.10	−0.17	−0.19	−0.25	−0.17	−0.23
O_2+N_2(84 K)	−0.31	−0.28	−0.28	−0.35		−0.20

MC = Monte Carlo results (McDonald, 1972a); Expt = experimental results from various sources; vdW1f = van der Waals one-fluid approximation; BH = first-order Barker–Henderson perturbation theory (Leonard *et al.*, 1970); Var. = Mansoori–Canfield variational method (Mansoori, 1972); WCA = Weeks–Chandler–Andersen perturbation theory (Lee and Levesque, 1973).

rather poor, but significant improvement is obtained if the cross-interaction parameter $\varepsilon_{\mu\nu}$ is systematically decreased to a value one to two percent smaller than that given by the geometric-mean rule (6.6.9). One feature of note is that all systems listed are predicted to have negative excess volumes. The same appears to be true of all mixtures of Lennard-Jones fluids to which the Lorentz–Berthelot rules apply (McDonald, 1972b; Singer and Singer, 1972), and only by relaxing these rules is it possible to obtain the positive values of V^E that are found experimentally for many real mixtures.

The vdW1f approximation can be derived from a conventional perturbation treatment in which the free energy of the mixture is expanded in powers of $\varepsilon\sigma^3$ and σ^3 about that of a pure reference system with parameters ε_x and σ_x. Use of the rules (6.6.20) and (6.6.22) causes the first-order term in the expansion to vanish and also makes the second-order term small in comparison with the zeroth-order one (Mo *et al.*, 1974). Viewed from this perspective, the success of the vdW1f approximation can in part be ascribed to the fact that the choice of reference system varies with the composition of the mixture under investigation. The original theory of Longuet-Higgins (1951) involved an expansion about the properties of a pure, composition-independent, reference system (subscript 0) in powers of $(\varepsilon_{\nu\mu} - \varepsilon_0)$ and $(\sigma_{\nu\mu} - \sigma_0)$. For a mixture to which the Lorentz rule (6.6.8) applies, the excess free energy is given to first order by

$$G^E(P, T) = U_0^{\text{ex}}(P, T)\Delta\varepsilon/\varepsilon_0 \tag{6.6.23}$$

where U_0^{ex} is the excess internal energy of the reference system and

$$\Delta\varepsilon = \sum_{\nu<\mu}\sum x_\nu x_\mu (2\varepsilon_{\nu\mu} - \varepsilon_{\nu\nu} - \varepsilon_{\mu\mu}) \tag{6.6.24}$$

In the case of a binary mixture, Eqn (6.6.23) has the same form as the free energy of a quadratic mixture, with the interchange energy in (6.6.4) now given explicitly in terms of the interaction parameters in the mixture and the excess internal energy of the reference system. If the Berthelot rule is used, $\Delta\varepsilon$ is necessarily negative and G^{E} is therefore always positive, in fair agreement with the Monte Carlo results discussed earlier. On the other hand, use of Eqn (6.6.6) shows that for $P \simeq 0$:

$$V^{\text{E}}(P, T) \simeq -(VT\alpha_P)\Delta\varepsilon/\varepsilon_0 \tag{6.6.25}$$

where $\alpha_P = (1/V)(\partial V/\partial T)_P$ is the thermal expansion coefficient. Under these conditions, V^{E} should always have the same sign as G^{E} (Rowlinson and Swinton, 1982). As the results of Table 6.3 suggest, this situation is the exception rather than the rule. Nonetheless, the theory has found a number of applications in the study of thermodynamic properties of liquid-metal alloys.

Elaborations of conformal-solution theory have been proposed in which the properties of the system of interest are identified with those of an ideal mixture of two (or more) hypothetical pure fluids (Leland *et al.* 1969), but the results are often worse than those obtained by a one-fluid approach (Gubbins *et al.*, 1983). Higher-order corrections to the vdW1f approximation have also been worked out in certain cases (Mo *et al.*, 1974), but the numerical calculations become more complicated and the simplicity of the theory, which is its main attraction, is therefore lost.

6.7 DENSITY-FUNCTIONAL THEORIES OF INHOMOGENEOUS FLUIDS

As we have emphasized many times in this chapter, the key to the success of perturbation theories of dense fluids is the fact that the attractive forces play only a minor part in determining the structure. This situation no longer applies when the system is inhomogeneous, as is clear from the curves plotted in Figure 5.6(b). Nonetheless, useful theories of non-uniform fluids have been developed in which there is an explicit separation of the effects of short-range, repulsive and long-range, attractive forces of the same type that underlies the perturbation treatment of homogeneous systems. These theories are of a variational character in which the crucial entity is a functional of the single-particle density that has as a lower bound the exact grand potential of the fluid. The application to classical fluids was first described by Ebner *et al.* (1976), but similar methods had been used previously for quantum-mechanical systems by Hohenberg and Kohn (1964) and Mermin (1964). The treatment that we give below draws freely on an article by Evans (1979).

Consider a grand canonical ensemble of systems and let $f(\mathbf{r}^N, \mathbf{p}^N; N)$ be the probability that a system chosen at random contains precisely N particles, with coordinates $\mathbf{r}^N$ and momenta $\mathbf{p}^N$. At equilibrium, $f = f_0$, where f_0 is the probability density defined in Eqn (2.4.10). It is necessary only to assume that f is normalized in the same way as f_0, i.e.

$$\sum_{N=0}^{\infty} \iint f(\mathbf{r}^N, \mathbf{p}^N; N)\, d\mathbf{r}^N\, d\mathbf{p}^N = 1 \tag{6.7.1}$$

but the later argument loses nothing if we also suppose that the particles have a Maxwell–Boltzmann distribution of momenta and that the arbitrariness in the specification of f is confined to its dependence on coordinates (Rowlinson and Widom, 1982). Let $\Omega[f]$ be a functional of f, defined as

$$\Omega[f] = \sum_{N=0}^{\infty} \iint d\mathbf{r}^N\, d\mathbf{p}^N\, (\mathcal{H}_N - N\mu + k_B T \log N!\, h^{3N} + k_B T \log f) f \tag{6.7.2}$$

where $\mathcal{H}_N$ is the hamiltonian. Then

$$\Omega[f_0] = -k_B T \log \Xi = \Omega \tag{6.7.3}$$

where Ξ is the grand partition function (2.4.11), Ω is the grand potential, and

$$\Omega[f] = \Omega[f_0] + k_B T \sum_{N=0}^{\infty} \iint d\mathbf{r}^N\, d\mathbf{p}^N\, (f \log f - f \log f_0) \tag{6.7.4}$$

A slight extension of the method used to derive the inequality (6.2.21) shows that the second term on the right-hand side of (6.7.4) is necessarily positive if $f \neq f_0$. Thus

$$\Omega[f_0] \leqslant \Omega[f] \tag{6.7.5}$$

where the equality holds only if f is the equilibrium probability density.

We now specialize to the case in which the hamiltonian of the system is of the form

$$\mathscr{H}_N(\mathbf{r}^N, \mathbf{p}^N) = K_N(\mathbf{p}^N) + V_N(\mathbf{r}^N) + \int \rho(\mathbf{r})\phi(\mathbf{r})\, \mathrm{d}\mathbf{r} \tag{6.7.6}$$

where K_N is the kinetic energy, V_N is the potential energy due to interactions between particles, $\phi(\mathbf{r})$ is an external potential, and $\rho(\mathbf{r})$ is the local particle density. Let us suppose that $\phi(\mathbf{r})$ changes to $\phi'(\mathbf{r})$, with consequent changes in the hamiltonian, equilibrium probability density and grand potential given, respectively, by $\mathscr{H}_N \to \mathscr{H}'_N$, $f_0 \to f'_0$ and $\Omega \to \Omega'$. Then, since $f'_0 \neq f_0$:

$$\begin{aligned}\Omega' &= \sum_{N=0}^{\infty} \iint \mathrm{d}\mathbf{r}^N\, \mathrm{d}\mathbf{p}^N\, (\mathscr{H}'_N - N\mu + k_\mathrm{B}T \log N!h^{3N} + k_\mathrm{B}T \log f'_0)f'_0 \\ &< \sum_{N=0}^{\infty} \iint \mathrm{d}\mathbf{r}^N\, \mathrm{d}\mathbf{p}^N\, (\mathscr{H}'_N - N\mu + k_\mathrm{B}T \log N!h^{3N} + k_\mathrm{B}T \log f_0)f_0\end{aligned} \tag{6.7.7}$$

and hence

$$\Omega' < \Omega + \int \rho^{(1)}(\mathbf{r})[\phi'(\mathbf{r}) - \phi(\mathbf{r})]\, \mathrm{d}\mathbf{r} \tag{6.7.8}$$

where $\rho^{(1)}(\mathbf{r})$ is the single-particle density for the distribution f_0. If the same argument is carried through with primed and unprimed quantities interchanged, and if $\rho^{(1)}(\mathbf{r})$ remains unaltered, we find that

$$\Omega < \Omega' + \int \rho^{(1)}(\mathbf{r})[\phi(\mathbf{r}) - \phi'(\mathbf{r})]\, \mathrm{d}\mathbf{r} \tag{6.7.9}$$

But Eqns (6.7.8) and (6.7.9) together contain a contradiction, since together they imply that $(\Omega' + \Omega) < (\Omega + \Omega')$. The assumption that $\rho^{(1)}(\mathbf{r})$ remains the same must therefore be false, since no other approximation has been made. The conclusion to be drawn is that for a system with a prescribed interparticle potential there is one and only one external potential that gives rise to a particular single-particle density. From this it follows, since $\phi(\mathbf{r})$ also determines f_0, that f_0 is a unique functional of $\rho^{(1)}(\mathbf{r})$. This result is the classical analogue of the one derived for quantum systems by Mermin (1964). To simplify the argument, we have assumed that the systems with hamiltonians $\mathscr{H}_N$ and $\mathscr{H}'_N$ are at the same chemical potential, but the final conclusion concerning f_0 and $\rho^{(1)}$ remains true even when this restriction is lifted; thus any functional of f_0 can equally well be regarded as a functional of $\rho^{(1)}$.

Consider again an ensemble of systems for which the probability density is $f(\mathbf{r}^N, \mathbf{p}^N; N)$ and let $n(\mathbf{r})$ be the average particle density at a point $\mathbf{r}$; at equilibrium, $f = f_0$ and $n(\mathbf{r}) = \rho^{(1)}(\mathbf{r})$. Bearing in mind the result we have just derived, we define a functional of $n(\mathbf{r})$, or *density functional*, of the form

$$\mathscr{F}[n] = \sum_{N=0}^{\infty} \iint \mathrm{d}\mathbf{r}^N \, \mathrm{d}\mathbf{p}^N (K_N + V_N + k_B T \log N! h^{3N} + k_B T \log f) f \tag{6.7.10}$$

It is easy to show that at equilibrium $\mathscr{F}[n]$ reduces to

$$\mathscr{F}[\rho^{(1)}] = -k_B T \log \Xi + \mu \int \rho^{(1)}(\mathbf{r}) \, \mathrm{d}\mathbf{r} - \int \rho^{(1)}(\mathbf{r}) \phi(\mathbf{r}) \, \mathrm{d}\mathbf{r} \tag{6.7.11}$$

In addition, we define a functional $\Omega_\phi[n]$ as

$$\Omega_\phi[n] = \mathscr{F}[n] - \mu \int n(\mathbf{r}) \, \mathrm{d}\mathbf{r} + \int n(\mathbf{r}) \phi(\mathbf{r}) \, \mathrm{d}\mathbf{r} \tag{6.7.12}$$

which, at equilibrium, becomes

$$\Omega_\phi[\rho^{(1)}] = -k_B T \log \Xi = \Omega \tag{6.7.13}$$

It follows from what has been proved earlier that Ω is also the minimum value of $\Omega_\phi[n]$ with respect to $n(\mathbf{r})$ (Evans, 1979), a fact that we express by writing

$$\left. \frac{\delta \Omega_\phi[n]}{\delta n(\mathbf{r})} \right|_{n=\rho^{(1)}} = 0 \tag{6.7.14}$$

If Eqn (6.7.11) is combined with the necessary generalization of the definition of Ω given in Eqn (2.4.3), i.e.

$$\Omega = A - \mu \int \rho^{(1)}(\mathbf{r}) \, \mathrm{d}\mathbf{r} \tag{6.7.15}$$

we find that

$$A = \mathscr{F}[\rho^{(1)}] + \int \rho^{(1)}(\mathbf{r}) \phi(\mathbf{r}) \, \mathrm{d}\mathbf{r} \tag{6.7.16}$$

The second term on the right-hand side of this equation is the contribution to the Helmholtz free energy that depends explicitly on the external potential, and $\mathscr{F}[\rho^{(1)}]$ is the "intrinsic" free energy. Substitution of (6.7.12) in (6.7.14) shows that

$$\mu = \mu_{\mathrm{int}}[\rho^{(1)}; \mathbf{r}] + \phi(\mathbf{r}) \tag{6.7.17}$$

where

$$\mu_{\text{int}}[\rho^{(1)}; \mathbf{r}] = \left.\frac{\delta \mathscr{F}[n]}{\delta n(\mathbf{r})}\right|_{n=\rho^{(1)}} \tag{6.7.18}$$

is the intrinsic chemical potential. The $\mathbf{r}$-dependence of μ_{int} must be exactly cancelled by that of $\phi(\mathbf{r})$, since the chemical potential itself is a constant.

The functional $\mathscr{F}[n]$ can be divided into ideal and excess parts in the form

$$\mathscr{F}[n] = \mathscr{F}^{\text{id}}[n] + \mathscr{F}^{\text{ex}}[n] \tag{6.7.19}$$

The ideal part is given by

$$\mathscr{F}^{\text{id}}[n] = k_{\text{B}}T \int \mathrm{d}\mathbf{r}\, n(\mathbf{r})[\log \Lambda^3 n(\mathbf{r}) - 1] \tag{6.7.20}$$

which corresponds, when $n(\mathbf{r}) = \rho^{(1)}(\mathbf{r})$, to a straightforward generalization of the expression for the ideal free energy given by Eqn (2.3.15). If Eqns (6.7.17) to (6.7.20) are now combined, we find that the chemical potential can be expressed as

$$\mu = k_{\text{B}}T \log \Lambda^3 \rho^{(1)}(\mathbf{r}) + \left.\frac{\delta \mathscr{F}^{\text{ex}}[n]}{\delta n(\mathbf{r})}\right|_{n=\rho^{(1)}} + \phi(\mathbf{r}) \tag{6.7.21}$$

which is equivalent to (5.9.12) if we make the identification

$$c^{(1)}(\mathbf{r}) = -\beta \left.\frac{\delta \mathscr{F}^{\text{ex}}[n]}{\delta n(\mathbf{r})}\right|_{n=\rho^{(1)}} \tag{6.7.22}$$

This result provides another interpretation of the single-particle direct correlation function $c^{(1)}(\mathbf{r})$: the quantity $-k_{\text{B}}Tc^{(1)}(\mathbf{r})$ is the contribution to the intrinsic chemical potential that arises from interactions between particles. Similarly, by virtue of (5.9.11), the usual (pair) direct correlation function can be viewed as the second functional derivative of $\mathscr{F}^{\text{ex}}$:

$$c(\mathbf{r}, \mathbf{r}') = -\beta \left.\frac{\delta^2 \mathscr{F}^{\text{ex}}[n]}{\delta n(\mathbf{r}) \delta n(\mathbf{r}')}\right|_{n=\rho^{(1)}} \tag{6.7.23}$$

Equations (6.7.12) and (6.7.14) provide the ingredients for a variational calculation of the single-particle density and grand potential of an inhomogeneous fluid. What is required in order to make the theory tractable is a parameterization of $\mathscr{F}[n]$ in terms of $n(\mathbf{r})$; the best estimates of $\rho^{(1)}(\mathbf{r})$ and Ω are then obtained by minimizing the right-hand side of (6.7.12) with respect to variations in $n(\mathbf{r})$. As in any variational calculation, the success achieved depends on the skill with which the trial functional is constructed. One possibility is to expand $\mathscr{F}[n]$ about the free energy of a uniform fluid at a number density that may be a function of $\mathbf{r}$. The result is an expression of the general form

$$\mathscr{F}[n] = \int a(n)\, d\mathbf{r} + \text{other terms} \tag{6.7.24}$$

where $a(n)$ is the free-energy density of a uniform fluid of density $n \equiv n(\mathbf{r})$ and the "other terms" are explicitly non-uniform in character (Tarazona and Evans, 1984). An expression of this type was used in the original density-functional theory of Ebner, Saam and Stroud (Ebner *et al.*, 1976; Ebner and Saam, 1977; Saam and Ebner, 1978) and is a natural choice in cases where the inhomogeneity is weak. Ebner and collaborators approximate the "other terms" in (6.7.24) by an expression that involves the direct correlation function of a uniform fluid of suitably chosen density and also corresponds to a partial summation of the "gradient" expansion used by many earlier workers (Evans, 1979; Rowlinson and Widom, 1982). The difficulty with this approach, as with certain methods of solution of the YBG equation (5.9.13), is that it requires a knowledge of the properties of a uniform fluid over a range of densities that may lie partly in the two-phase region. Partly for this reason, the work has attracted considerable criticism, notably a paper by Ebner and Saam (1977) in which, independently of the work of Cahn (1977) referred to in Section 5.9, a wetting transition of the type shown in Figure 5.7(a) was predicted. The Monte Carlo calculations of Lane *et al.* (1979), also mentioned in Section 5.9, were carried out in an attempt to observe thick-film formation. For the same model system and same state conditions studied by Ebner and Saam, Lane *et al.* found that the adsorption was predominantly monolayer in character and concluded that the thick films obtained by Ebner and Saam were artefacts of the theory. More recent work has cast doubts on this interpretation, as we shall see below.

A different type of trial functional can be constructed by extending to the inhomogeneous case the methods used in perturbation theories of uniform fluids. We divide the functional $\mathscr{F}[n]$ into two parts: the first is a free-energy functional of a reference system characterized by a harshly repulsive pair potential $v_0(r)$, and the second is the contribution from a long-range, attractive perturbation $w(r)$. We suppose that the reference system can be replaced by a fluid of hard spheres of appropriately chosen diameter d, and that the effects of the perturbation can be treated in the spirit of the van der Waals approximation by working at first order in perturbation theory and setting $g_0(1, 2) = 1$. With these simplifications, the functional $\mathscr{F}[n]$ is

$$\mathscr{F}[n] = \mathscr{F}_d[n] + \tfrac{1}{2} \iint n(\mathbf{r}) n(\mathbf{r}') w(|\mathbf{r} - \mathbf{r}'|)\, d\mathbf{r}\, d\mathbf{r}' \tag{6.7.25}$$

where $\mathscr{F}_d[n]$ is the free-energy functional of an inhomogeneous hard-sphere fluid and the term involving $w(r)$ is the non-uniform generalization of (6.2.15).

The functional (6.7.25) is the one introduced implicitly by Sullivan (1979, 1981) and used by him and others (Evans and Tarazona, 1983; Teletzke *et al.*, 1982; Meister and Kroll, 1985) to study a variety of interfacial phenomena. In its general form, the functional is not immediately useful, since the properties of the inhomogeneous hard-sphere fluid, unlike those of the uniform system, are unknown. Some approximation must therefore be made for $\mathscr{F}_d[n]$. One possibility is to write $\mathscr{F}_d[n]$ as the sum of ideal and excess parts, and to replace the latter by an expression similar to the leading term in (6.7.24) but involving a smoothed density $\bar{n}(\mathbf{r})$. Then $\mathscr{F}_d[n]$ becomes

$$\mathscr{F}_d[n] = \mathscr{F}^{\mathrm{id}}[n] + \int n(\mathbf{r}) \frac{a_d^{\mathrm{ex}}[\bar{n}]}{\bar{n}(\mathbf{r})}\, \mathrm{d}\mathbf{r} \tag{6.7.26}$$

where $\mathscr{F}^{\mathrm{id}}[n]$ is given by (6.7.20); $a_d^{\mathrm{ex}}[\bar{n}]$ is the excess free-energy density of a homogeneous hard-sphere fluid at a density $\bar{n} \equiv \bar{n}(\mathbf{r})$, given by

$$\bar{n}(\mathbf{r}) = \int n(\mathbf{r}) \psi(\mathbf{r}+\mathbf{r}')\, \mathrm{d}\mathbf{r}' \tag{6.7.27}$$

with

$$\int \psi(\mathbf{r}+\mathbf{r}')\, \mathrm{d}\mathbf{r}' = 1 \tag{6.7.28}$$

The simplest choice of smoothing function $\psi(\mathbf{r})$ is one that makes $\bar{n} = n$, i.e. $\psi(\mathbf{r}+\mathbf{r}') = \delta(\mathbf{r}-\mathbf{r}')$. This corresponds to the "local-density" approximation of Sullivan (1979, 1981). The local-density approximation is useful for the study of long-range correlations and of problems where density oscillations do not occur, as in the case of the free-liquid surface (Lu *et al.*, 1985). It does not treat the short-range correlations correctly and could not, for example, be expected to reproduce the layered structure of a high-density fluid against a wall (see Figure 5.6(b)). One way to understand the strengths and weaknesses of the local-density approximation is to use Lemma 2 of Chapter 4 to calculate the direct correlation function of the homogeneous fluid via the relation (6.7.23). For the general functional defined by Eqn (6.7.25), the result of the functional differentiation is

$$c(\mathbf{r}, \mathbf{r}') = c_d(\mathbf{r}, \mathbf{r}') - \beta w(|\mathbf{r}-\mathbf{r}'|) \tag{6.7.29}$$

where $c_d(\mathbf{r}, \mathbf{r}')$, the hard-sphere direct correlation function, is the second functional derivative of $-\beta\mathscr{F}_d^{\mathrm{ex}}[n]$. Equation (6.7.29) has the same form as the RPA expression (6.5.35), and is asymptotically correct. If, however, we make the local-density approximation to $\mathcal{F}_d^{\mathrm{ex}}[n]$ and specialize the result to a uniform fluid of density ρ, we find that

$$c(|\mathbf{r}-\mathbf{r}'|) = -\beta a_d''\delta(\mathbf{r}-\mathbf{r}') - \beta w(|\mathbf{r}-\mathbf{r}'|) \tag{6.7.30}$$

where $a_d'' \equiv \mathrm{d}^2 a_d^{\mathrm{ex}}/\mathrm{d}\rho^2$. At short distances, this is an even cruder approximation than the RPA, since it replaces the hard-sphere direct correlation function by a delta-function at the origin (Tarazona and Evans, 1984; Lu *et al.*, 1985).

In order to do justice to the short-range correlations in the fluid, it is necessary to make a better choice of smoothing function in Eqn (6.7.27). Tarazona and Evans (1984) have studied a simple model in which $\bar{n}(\mathbf{r})$ is obtained by smoothing the local density uniformly over a sphere of radius equal to the hard-sphere diameter d (Nordholm *et al.*, 1980). This leads to a hard-sphere direct correlation function that has the correct qualitative features, including a discontinuity at $r = d$. The results obtained for the density profile of hard spheres against a hard wall and for the onset of wetting by gas at the interface of a liquid with a hard wall are in fair agreement with the results of Monte Carlo calculations. Meister and Kroll (1985) have developed a more elaborate scheme in which the grand-potential functional is minimized with respect to both $n(\mathbf{r})$ and $\bar{n}(\mathbf{r})$; their results are similar to those of Tarazona and Evans (1984), but agree better with simulations for the problem of hard spheres against a hard wall (see also Tarazona, 1985).

In the case of wetting by liquid at a gas–solid interface, the local-density approximation and the improvements described above all give results that confirm the general conclusions of Ebner and Saam (1977), while being free of the specific criticisms levelled at the earlier work. The estimates obtained for the wetting temperature do, however, differ widely. It is therefore possible that the Monte Carlo calculations of Lane *et al.* (1979) were carried out for a temperature that lies above the true wetting temperature of the model system that they studied. This would explain the failure to observe the growth of a thick film, which is the signature of complete wetting (Evans and Tarazona, 1983). Care must in any case be taken when interpreting the results of computer "experiments" on fluid–wall interfaces. The simulations are normally carried out for a fluid confined between planar walls separated typically by distances equal to ten or twenty molecular diameters. Under these conditions, effects associated with the phenomenon of "capillary condensation" must be expected to play a crucial role (Lane and Spurling, 1980; Evans and Tarazona, 1984, 1985).

We noted in Section 5.9 that integral-equation theories based on approximate closures of the wall-particle Ornstein–Zernike relation (5.9.24) have so far failed to predict wetting either by liquid or by gas. Certain of these theories can easily be transcribed into density-functional language in a way that makes this failing understandable (R. Evans *et al.*, 1983). For example, the HNC approximation (5.9.25) is equivalent (Grimson and Rickayzen, 1981) to minimizing a grand-potential functional of the form

$$\Omega_{\mathrm{HNC}}[n]=\Omega_{\mathrm{HNC}}(\rho_{\mathrm{B}})+\int \mathrm{d}\mathbf{r}\,[n(\mathbf{r})-\rho_{\mathrm{B}}]\phi(\mathbf{r})\,\mathrm{d}\mathbf{r}$$
$$+k_{\mathrm{B}}T\int \mathrm{d}\mathbf{r}\,n(\mathbf{r})\log\frac{n(\mathbf{r})}{\rho_{\mathrm{B}}}-k_{\mathrm{B}}T\int \mathrm{d}\mathbf{r}\,[n(\mathbf{r})-\rho_{\mathrm{B}}]$$
$$-\tfrac{1}{2}k_{\mathrm{B}}T\iint \mathrm{d}\mathbf{r}\,\mathrm{d}\mathbf{r}'\,c_{\mathrm{B}}(|\mathbf{r}-\mathbf{r}'|)[n(\mathbf{r})-\rho_{\mathrm{B}}][n(\mathbf{r}')-\rho_{\mathrm{B}}] \quad (6.7.31)$$

where $c_{\mathrm{B}}(r)$ is the direct correlation function of a uniform fluid of density ρ_{B}; the equivalence of (5.9.25) and (6.7.31) can be proved by setting $\delta\Omega_{\mathrm{HNC}}[n]/\delta n(\mathbf{r})=0$ and specializing the result to the case of a planar interface. It follows from (6.7.31) that in the HNC approximation the grand potential of a uniform fluid of density ρ is

$$\Omega_{\mathrm{HNC}}(\rho)=\Omega_{\mathrm{HNC}}(\rho_{\mathrm{B}})+Vk_{\mathrm{B}}T\left(\rho\log\frac{\rho}{\rho_{\mathrm{B}}}-\rho+\rho_{\mathrm{B}}\right)$$
$$-\tfrac{1}{2}Vk_{\mathrm{B}}T(\rho-\rho_{\mathrm{B}})^2\int c_{\mathrm{B}}(\mathbf{r})\,\mathrm{d}\mathbf{r} \quad (6.7.32)$$

R. Evans *et al.* (1983) have shown that, irrespective of the value of ρ_{B}, the right-hand side of (6.7.32) has only one minimum as a function of ρ (for $\rho=\rho_{\mathrm{B}}$). Thus the HNC approximation cannot support coexistence and is therefore unable to provide a self-consistent treatment of wetting. The same objection applies to the PY approximation. It does not follow that the HNC and PY approaches are unable to describe partial wetting, but the numerical results are not encouraging. The origin of the problem is the fact that information on pair correlations is fed into these theories solely in the form of the direct correlation function of the bulk fluid.

CHAPTER 7

Time-dependent Correlation Functions and Response Functions

The next three chapters are devoted to a discussion of the microscopic dynamics and transport properties of simple, dense fluids. The present chapter deals with the general formalism of time-correlation functions and with linear response theory; Chapter 8 is concerned with the behaviour of time-dependent fluctuations in the long-wavelength, low-frequency limit, where contact can be made with the macroscopic equations of hydrodynamics; and Chapter 9 describes a number of approximate theories that allow the explicit calculation of time-correlation functions. Other treatments of similar material may be found in the classic review articles of Zwanzig (1965), Kubo (1966), Martin (1968) and Schofield (1975), and in the books by Résibois and DeLeener (1977) and Boon and Yip (1980).

7.1 GENERAL PROPERTIES OF TIME-CORRELATION FUNCTIONS

A dynamical variable, $A(t)$ say, of a system composed of N structureless particles is any function of some or all of the coordinates $\mathbf{r}_i$ and momenta $\mathbf{p}_i$, where $i=1$ to N. In a shorthand notation:

$$A(t) \equiv A[\mathbf{r}^N(t), \mathbf{p}^N(t)] \tag{7.1.1}$$

The time evolution of the variable A is determined by Eqn (2.1.10), where the classical Liouville operator $\mathscr{L}$ is given by Eqn (2.1.6); it is clear from (2.1.6) and (2.1.10) that A has the signature $\varepsilon_A = \pm 1$ under time reversal according to whether it is an even or odd function of the momenta.

In classical statistical mechanics, the equilibrium time-correlation function of two dynamical variables A and B is defined as

$$C_{AB}(t', t'') = \langle A(t')B(t'')\rangle \tag{7.1.2}$$

where the angular brackets denote either an ensemble average over initial conditions (with the convention $t' > t''$):

$$\langle A(t')B(t'')\rangle = \iint A[\mathbf{r}^N(t'), \mathbf{p}^N(t')]B[\mathbf{r}^N(t''), \mathbf{p}^N(t'')] \times f_0^{(N)}[\mathbf{r}^N(t''), \mathbf{p}^N(t'')]\, \mathrm{d}\mathbf{r}^N(t'')\, \mathrm{d}\mathbf{p}^N(t'')$$

$$= \int \{\exp[i\mathscr{L}(t'-t'')]A(t'')\}B(t'')f_0^{(N)}\, \mathrm{d}\Gamma_N \quad (7.1.3)$$

or an average over time:

$$\langle A(t')B(t'')\rangle = \lim_{\tau\to\infty} \frac{1}{\tau}\int_0^\tau A(t'+t)B(t''+t)\, \mathrm{d}t \quad (7.1.4)$$

The averaging in (7.1.3) is carried out over all possible states ($\mathrm{d}\Gamma_N \equiv \mathrm{d}\mathbf{r}^N\, \mathrm{d}\mathbf{p}^N$), i.e. over the total phase of the system at time t''; in the canonical ensemble, the equilibrium probability density $f_0^{(N)}$ is given by Eqn (2.3.1). Equations (7.1.3) and (7.1.4) give the same result in the thermodynamic limit if the system is ergodic.

Since the equilibrium probability density $f_0^{(N)}$ is independent of time, the statistical average in (7.1.3) is independent of the choice of time origin t'', and the correlation function $C_{AB}(t', t'')$ is invariant under time translation. If we put $t'' = s$ and $t' = s + t$, the correlation function is a function only of the time difference t, and is said to be *stationary* with respect to s. It is therefore customary to set $s = 0$ and write

$$C_{AB}(t) = \langle A(t)B(0)\rangle \equiv \langle A(t)B\rangle \quad (7.1.5)$$

where $B \equiv B(0)$. The stationary property of the time-correlation function means that

$$\frac{\mathrm{d}}{\mathrm{d}s}\langle A(t+s)B(s)\rangle = \langle \dot{A}(t+s)B(s)\rangle + \langle A(t+s)\dot{B}(s)\rangle = 0 \quad (7.1.6)$$

and thus

$$\langle \dot{A}(t)B\rangle = -\langle A(t)\dot{B}\rangle \quad (7.1.7)$$

In particular:

$$\langle \dot{A}A\rangle = 0 \quad (7.1.8)$$

Repeated differentiation with respect to s leads to a number of useful relations; these can also be deduced by exploiting the definition (7.1.4).

For example:

$$\begin{aligned}
\frac{d^2}{dt^2}\langle A(t)B\rangle &= \langle \ddot{A}(t)B\rangle \\
&= \lim_{\tau\to\infty}\frac{1}{\tau}\int_0^\tau \ddot{A}(t+t')B(t')\,dt' \\
&= \lim_{\tau\to\infty}\frac{1}{\tau}[\dot{A}(t+t')B(t')]_0^\tau \\
&\quad -\lim_{\tau\to\infty}\frac{1}{\tau}\int_0^\tau \dot{A}(t+t')\dot{B}(t')\,dt' \\
&= -\langle \dot{A}(t)\dot{B}\rangle
\end{aligned} \tag{7.1.9}$$

The definition of the correlation function $C_{AB}(t)$ in Eqn (7.1.3) has the form of an inner product of the two "vectors" $A(t)$ and B in phase space. Given Eqn (2.1.11), relations such as (7.1.7) and (7.1.9) merely express the fact that the Liouville operator $\mathscr{L}$ is hermitian (and hence $i\mathscr{L}$ is antihermitian) with respect to the inner product; the proof that $\mathscr{L}$ is hermitian requires an integration by parts of the derivatives appearing in its Poisson-bracket representation (2.1.6). In the definition given by (7.1.3), the inner product of $A(t)$ and B involves an integration over the initial phase, weighted by the equilibrium probability density $f_0^{(N)}$. The inner product is sometimes defined without the weighting factor, but the Liouville operator retains its hermitian property, since $\mathscr{L}f_0^{(N)}=0$.

The invariance of the correlation function $C_{AB}(t)$ under time translation implies that

$$\begin{aligned}
C_{AB}(t) &= \varepsilon_A\varepsilon_B C_{AB}(-t) = \varepsilon_A\varepsilon_B\langle A(-t)B\rangle \\
&= \varepsilon_A\varepsilon_B\langle AB(t)\rangle = \varepsilon_A\varepsilon_B C_{BA}(t)
\end{aligned} \tag{7.1.10}$$

where ε_A, ε_B are the time-reversal signatures of the variables A, B. If A and B have opposite parities under spatial inversion ($\mathbf{r}_i, \mathbf{p}_i \to -\mathbf{r}_i, -\mathbf{p}_i$), their time-correlation function is zero at all times, provided the hamiltonian, and hence the Liouville operator, have even parity, which is generally the case.

A particularly important class of time-correlation functions are the autocorrelation functions $C_{AA}(t)$, for which A and B are the same variable. Equation (7.1.10) shows that autocorrelation functions are necessarily even functions of time. If A is a complex quantity, the autocorrelation function is conventionally defined as

$$C_{AA}(t) = \langle A(t)A^*\rangle \tag{7.1.11}$$

which ensures that $C_{AA}(t)$ is a real function of t for all times.

It is clear that

$$\lim_{t\to 0} C_{AB}(t) = \langle AB \rangle \tag{7.1.12}$$

where $\langle AB \rangle$ is a static correlation function. It also follows immediately from Schwarz's inequality that

$$|\mathrm{Re}\,\langle A(t)B^*\rangle| \leq [\langle AA^*\rangle\langle BB^*\rangle]^{1/2} \tag{7.1.13}$$

where Re denotes the real part. Hence, for an autocorrelation function:

$$|C_{AA}(t)| \leq \langle AA^* \rangle = C_{AA}(0) \tag{7.1.14}$$

i.e. the magnitude of the function is bounded above by its initial value. Intuitively, this is the result one would expect, since an autocorrelation function reflects how spontaneous (thermal) fluctuations in the system decay in time. In the limit $t \to \infty$, the two dynamical variables become uncorrelated. Thus

$$\lim_{t\to\infty} C_{AB}(t) = \langle A \rangle \langle B \rangle \tag{7.1.15}$$

It is often convenient to define the dynamical variables in such a way as to exclude their average values and to consider only the time correlation of the fluctuating parts, i.e.

$$C_{AB}(t) = \langle [A(t) - \langle A \rangle][B - \langle B \rangle] \rangle \tag{7.1.16}$$

This definition has the advantage that $\lim_{t\to\infty} C_{AB}(t) = 0$, corresponding physically to the complete loss of correlation between the fluctuating parts of the variables.

If $C_{AB}(t)$ is defined as in Eqn (7.1.16), it is also possible to define its Fourier transform $C_{AB}(\omega)$, called either the *power spectrum* or *spectral function*:

$$C_{AB}(\omega) = \frac{1}{2\pi}\int_{-\infty}^{\infty} \exp(i\omega t) C_{AB}(t)\,\mathrm{d}t \tag{7.1.17}$$

and its Laplace transform

$$\tilde{C}_{AB}(z) = \int_0^{\infty} \exp(izt) C_{AB}(t)\,\mathrm{d}t \tag{7.1.18}$$

where z is a complex frequency. Since $C_{AB}(t)$ is bounded, it is clear that $\tilde{C}_{AB}(z)$ is analytic in the upper half of the complex z plane ($\mathrm{Im}\, z > 0$). It also follows at once that $\tilde{C}_{AB}(z)$ is the Hilbert transform of $C_{AB}(\omega)$:

$$\begin{aligned} \tilde{C}_{AB}(z) &= \int_0^{\infty} \mathrm{d}t \exp(izt) \int_{-\infty}^{\infty} \exp(-i\omega t) C_{AB}(\omega)\,\mathrm{d}\omega \\ &= i\int_{-\infty}^{\infty} \frac{C_{AB}(\omega)}{z-\omega}\,\mathrm{d}\omega \end{aligned} \tag{7.1.19}$$

Equation (7.1.10), specialized to the case when $A = B$ (and $\varepsilon_A^2 = 1$), shows that the spectrum of an autocorrelation function is always a real, even function of ω. Furthermore, from the standard relation

$$\lim_{\varepsilon\to 0} \frac{1}{x - i\varepsilon} = \mathcal{P}\left(\frac{1}{x}\right) + i\pi\delta(x) \tag{7.1.20}$$

where $\mathcal{P}$ denotes the principal part, it follows that the spectrum of an autocorrelation function is related to its Laplace transform by

$$C_{AA}(\omega) = \lim_{\varepsilon\to 0} \frac{1}{\pi} \operatorname{Re} \tilde{C}_{AA}(\omega + i\varepsilon) \tag{7.1.21}$$

It can also be shown that the spectrum of an autocorrelation function is non-negative. To prove this important property, we consider an auxiliary variable $A_T(\omega)$, defined as

$$A_T(\omega) = \frac{1}{\sqrt{2T}} \int_{-T}^{T} \exp(i\omega t) A(t)\, dt \tag{7.1.22}$$

Clearly the statistical average $\langle A_T(\omega) A_T^*(\omega)\rangle$ is non-negative:

$$\langle A_T(\omega) A_T^*(\omega)\rangle = \frac{1}{2T} \int_{-T}^{T} dt \int_{-T}^{T} dt' \exp[i\omega(t - t')] \langle A(t) A^*(t')\rangle \geq 0 \tag{7.1.23}$$

Changing from the variables t, t' to t, $\tau = t - t'$ and taking the limit $T \to \infty$, we find that

$$\begin{aligned} \lim_{T\to\infty} \langle A_T(\omega) A_T^*(\omega)\rangle &= \int_{-\infty}^{\infty} \exp(i\omega\tau) C_{AA}(\tau)\, d\tau \\ &= C_{AA}(\omega) \geq 0 \end{aligned} \tag{7.1.24}$$

The experimental significance of time-correlation functions lies in the fact that the spectra measured by various spectroscopic techniques are the power spectra of well-defined dynamical variables. This connection between theory and experiment will be made explicit in Section 7.5 for the important case of inelastic neutron scattering. In addition, as we shall see in Section 7.6 and throughout Chapter 8, the linear transport coefficients of hydrodynamics are related to time integrals of certain autocorrelation functions. Finally, time-correlation functions provide a quantitative description of the microscopic dynamics in liquids. Here, computer simulations play a key role, since they give access to a large variety of correlation functions, many of which are unmeasurable by laboratory experiments. In a molecular dynamics simulation, the value of $C_{AB}(t)$ at time t is calculated by averaging the product $A(t+s)B(s)$ over many choices of the time origin s. In the simpler case of continuous potentials, for which the time step in the integration is constant, the time origins are usually chosen to be uniformly distributed over a finite time interval (see Eqn (7.1.4)). In general, it is sufficient to take every fifth time step, say, as an origin; choosing larger numbers of origins within a given interval τ does not appreciably improve the statistical accuracy. On the other hand, the statistical error decreases as $\tau^{-1/2}$ (Zwanzig and Ailawadi, 1969). Results of simulations will serve frequently as illustrations in later sections.

The properties of time-correlation functions given thus far are completely general. We now restrict the discussion to systems of particles for which the interaction potential is continuous; the hamiltonian is therefore differentiable and the Liouville operator is given by (2.1.6). An autocorrelation function, since it is always even in time, can then be Taylor-expanded in the form

$$\begin{aligned} C_{AA}(t) &= \sum_{n=0}^{\infty} \frac{t^{2n}}{(2n)!} C_{AA}^{(2n)}(0) \\ &= \sum_{n=0}^{\infty} \frac{t^{2n}}{(2n)!} \langle A^{(2n)} A^* \rangle \\ &= \sum_{n=0}^{\infty} \frac{t^{2n}}{(2n)!} (-1)^n \langle A^{(n)} A^{(n)*} \rangle \\ &= \sum_{n=0}^{\infty} \frac{t^{2n}}{(2n)!} (-1)^n \langle |(i\mathscr{L})^n A|^2 \rangle \end{aligned} \tag{7.1.25}$$

where repeated use has been made of Eqn (7.1.9). Differentiation of the inverse Fourier transform of Eqn (7.1.17) $2n$ times with respect to t gives

$$\begin{aligned} \langle \omega^{2n} \rangle_{AA} &\equiv \int_{-\infty}^{\infty} \omega^{2n} C_{AA}(\omega)\, \mathrm{d}\omega \\ &= (-1)^n C_{AA}^{(2n)}(0) \end{aligned} \tag{7.1.26}$$

The frequency moments of the spectral function are therefore directly related to the derivatives of the autocorrelation function taken at $t=0$. The latter quantities are static correlation functions that will generally be expressible as integrals over the particle distribution functions $g^{(n)}(\mathbf{r}^n)$; specific examples will be given in later sections.

On expanding the right-hand side of Eqn (7.1.19) in powers of $1/z$, it becomes clear that the frequency moments defined by (7.1.26) are also the coefficients in the high-frequency expansion of the Laplace transform:

$$\tilde{C}_{AA}(z) = \frac{i}{z} \sum_{n=0}^{\infty} \frac{\langle \omega^{2n} \rangle_{AA}}{z^{2n}} \tag{7.1.27}$$

Expansions of the type displayed in (7.1.25) are not applicable in cases where the interparticle potentials have discontinuities including, in particular, fluids of hard spheres. The impulsive nature of the forces in these systems means that the collisions between particles are of infinitesimally short duration, and the Liouville operator takes on a very different form (Résibois and DeLeener, 1977). As a result, the time correlation functions are non-analytic at $t=0$, and the corresponding power spectra have frequency moments that are infinite; an example is given in the next section.

A close relation exists between the correlation functions of statistical mechanics and those characteristic of gaussian random processes (Résibois and DeLeener, 1977). In particular, the positivity of spectral functions (Eqn (7.1.24)) is known in the theory of stationary random processes as the Wiener–Khinchine theorem. Another important result from the same theory of which we shall make use later is the theorem of Doob, according to which the correlation function of a stochastic variable associated with a gaussian, markovian process decays exponentially with time (Doob, 1942).

7.2 AN ILLUSTRATION: THE VELOCITY AUTOCORRELATION FUNCTION AND SELF-DIFFUSION

The ideas introduced in Section 7.1 can be illustrated by considering one of the simplest but most important examples of a time-correlation function, namely the autocorrelation function of the velocity $\mathbf{u} = \mathbf{p}/m$ of a tagged particle moving through a fluid. For a fluid that is isotropic, the velocity autocorrelation function is defined as

$$Z(t) = \tfrac{1}{3}\langle \mathbf{u}(t) \cdot \mathbf{u} \rangle = \langle u_x(t) u_x \rangle \tag{7.2.1}$$

i.e. $Z(t)$ is a measure of the projection of the particle velocity at time t onto its initial value, averaged over all initial conditions. The value of $Z(t)$ at $t=0$ can be derived immediately from the equipartition theorem:

$$Z(0) = \tfrac{1}{3}\langle u^2 \rangle = \frac{k_B T}{m} \tag{7.2.2}$$

For times long compared with any microscopic relaxation times, the initial and final velocities are expected to be completely decorrelated, so that $Z(t \to \infty) = 0$. Typical computer simulation results on argon-like liquids (Rahman, 1964a; Levesque and Verlet, 1970) show that the velocities are largely decorrelated after about 10^{-12} s, but in general $Z(t)$ also has a slowly decaying part and the detailed behaviour depends on both density and temperature, as is clear from the examples plotted in Figure 7.1. We shall return later to an analysis of the main features of curves such as these, but first we show that there exists a general relationship between the self-diffusion coefficient D and the time integral of $Z(t)$. Consider a set of identical, tagged particles having positions around a point $\mathbf{r}(0)$. If the

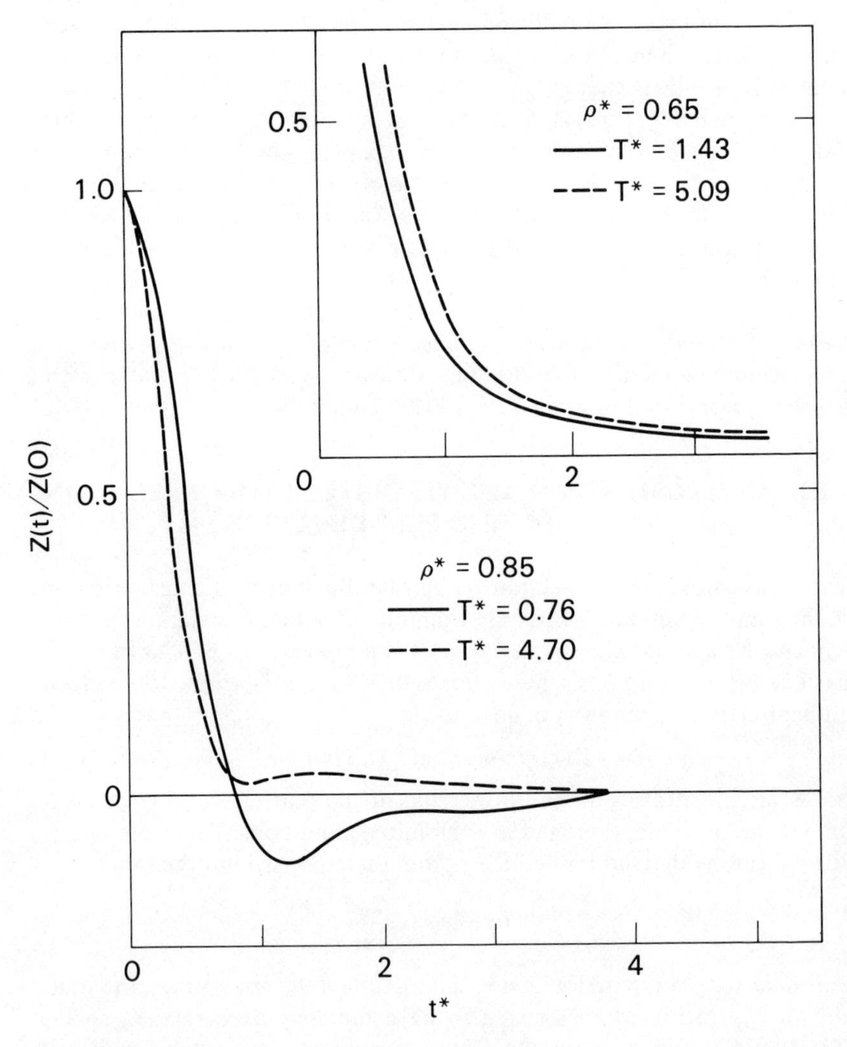

FIG. 7.1. Normalized velocity autocorrelation function of the Lennard-Jones fluid at different reduced temperatures and densities. The unit of time is the quantity τ_0 defined by Eqn. (3.3.5). After Levesque and Verlet (1970).

particles diffuse in time t to positions around $\mathbf{r}(t)$, the self-diffusion coefficient is given by a well-known relation due to Einstein:

$$D = \lim_{t \to \infty} \frac{1}{6t} \langle |\mathbf{r}(t) - \mathbf{r}(0)|^2 \rangle \tag{7.2.3}$$

This result is a direct consequence of Fick's lack of diffusion, as we shall see in Section 8.2. It is also a relation typical of a stochastic "random-walk" process (Résibois and DeLeener, 1977), for which the mean-square displacement of the walker becomes a linear function of time after a sufficiently large number of random displacements have occurred. Similar behaviour is seen in the brownian motion of a particle in a fluid, provided a sufficient time has elapsed for the brownian particle to have suffered many collisions with the particles of the "bath". The nature of the limiting process involved in Eqn (7.2.3) highlights the general importance of taking the thermodynamic limit before the limit $t \to \infty$. For a system of finite volume V, the diffusion coefficient defined by (7.2.3) is strictly zero, since the mean-square displacement must always be less than $V^{2/3}$. In practice, the ratio $\langle |\mathbf{r}(t) - \mathbf{r}(0)|^2 \rangle / 6t$ can be expected to reach a plateau value at times much shorter than those required for the diffusing particles to reach the boundaries of the system; it is the plateau value that provides the definition of D for a finite system.

We now rewrite the Einstein relation (7.2.3) in terms of the velocity autocorrelation function. We note first that

$$\mathbf{r}(t) - \mathbf{r}(0) = \int_0^t \mathbf{u}(t')\, \mathrm{d}t' \tag{7.2.4}$$

When squared and averaged over initial conditions, Eqn (7.2.4) becomes

$$\langle |\mathbf{r}(t) - \mathbf{r}(0)|^2 \rangle = \int_0^t \mathrm{d}t' \int_0^{t'} \mathrm{d}t'' \langle \mathbf{u}(t'') \cdot \mathbf{u}(t') \rangle \tag{7.2.5}$$

The properties of symmetry with respect to time inversion and invariance under time translation mean that Eqn (7.2.5) can be combined with the definition (7.2.1) to give

$$\langle |\mathbf{r}(t) - \mathbf{r}(0)|^2 \rangle = 6 \int_0^t \mathrm{d}t' \int_0^{t'} \mathrm{d}t''\, Z(t' - t'') \tag{7.2.6}$$

If we change from variables t', t'' to t', $s = t' - t''$ and integrate by parts, we find that

$$\begin{aligned} \langle |\mathbf{r}(t) - \mathbf{r}(0)|^2 \rangle &= 6 \int_0^t \mathrm{d}t' \int_0^{t'} \mathrm{d}s\, Z(s) \\ &= 6t \int_0^t (1 - s/t) Z(s)\, \mathrm{d}s \end{aligned} \tag{7.2.7}$$

and substitution in (7.2.3) shows finally that

$$D = \int_0^\infty Z(s)\,\mathrm{d}s \tag{7.2.8}$$

Equation (7.2.8) expresses the self-diffusion coefficient as the time integral of the velocity autocorrelation function. It is an example of an important class of relations, often called "Green–Kubo formulae", whereby a macroscopic, phenomenological transport coefficient is written as the time integral of a microscopic time-correlation function.

If the interparticle potential is continuous, the short-time expansion of $Z(t)$, from (7.1.25), begins as

$$Z(t) = \frac{k_B T}{m}\left(1 - \Omega_0^2 \frac{t^2}{2} + \cdots\right) \tag{7.2.9}$$

The coefficient of $\frac{1}{2}t^2$, which has dimensions of (frequency)^2, is

$$\begin{aligned} \Omega_0^2 &= \tfrac{1}{3}\left(\frac{m}{k_B T}\right)\langle \dot{\mathbf{u}} \cdot \dot{\mathbf{u}} \rangle \\ &= \frac{1}{3mk_B T}\langle |\mathbf{F}|^2 \rangle \end{aligned} \tag{7.2.10}$$

where $\mathbf{F}$ is the total force exerted on the diffusing particle by its neighbours. We consider specifically the case when the tagged particle is identical to all other particles of the fluid; then $\mathbf{F} = -\boldsymbol{\nabla} V_N$, and a straightforward integration by parts shows that

$$\langle |\boldsymbol{\nabla} V_N|^2 \rangle = k_B T \langle \nabla^2 V_N \rangle \tag{7.2.11}$$

Hence

$$\Omega_0^2 = \frac{1}{3m}\langle \nabla^2 V_N \rangle \tag{7.2.12}$$

If V_N is a sum of pair terms, Eqn (7.2.12) can be rewritten as

$$\Omega_0^2 = \frac{\rho}{3m}\int \nabla^2 v(r) g(r)\,\mathrm{d}\mathbf{r} \tag{7.2.13}$$

The quantity Ω_0^2 is therefore expressible in terms of the equilibrium pair distribution function and the interparticle potential; Ω_0 is sometimes called an "Einstein frequency", since it represents the frequency at which the tagged particle would vibrate if it were undergoing small oscillations in the potential well produced by particles of the fluid when maintained at their mean equilibrium positions around the tagged particle. Numerically, Ω_0 is of order $10^{13}\ \mathrm{s}^{-1}$ for an argon-like liquid near its triple point.

Equation (7.2.9) cannot be used for systems of hard spheres; in particular, the quantity Ω_0^2 of Eqn (7.2.13) is not well defined because the hard-sphere potential is not differentiable. In this case, the short-time behaviour of $Z(t)$ is written (Lebowitz *et al.*, 1969) in the form

$$\langle \mathbf{u}(t) \cdot \mathbf{u} \rangle = \langle u^2 \rangle + t \left[\frac{\mathrm{d}}{\mathrm{d}t} \langle \mathbf{u}(t) \cdot \mathbf{u} \rangle \right]_{t=0} + \cdots \tag{7.2.14}$$

where it is now understood that the differentiation with respect to time must be carried out after the ensemble averaging. Equation (7.2.14) becomes

$$Z(t) = \langle u^2 \rangle (1 - \Omega_0 t + \cdots) \tag{7.2.15}$$

where the frequency Ω_0 is now given by

$$\Omega_0 = -\frac{1}{\langle u^2 \rangle} \lim_{\Delta t \to 0} \frac{\langle \Delta \mathbf{u} \cdot \mathbf{u} \rangle}{\Delta t} \tag{7.2.16}$$

Over a sufficiently short time interval, the tagged sphere will suffer at most one collision with another sphere from the bath. It can be shown by elementary kinetic theory (Guggenheim, 1960) that at low density

$$\begin{aligned} \frac{\langle \Delta \mathbf{u} \cdot \mathbf{u} \rangle}{\Delta t} &= \rho \int \mathrm{d}\Omega_\chi \int \mathrm{d}\mathbf{w}\, w \sigma(\Omega_\chi, \mathbf{w}) f_{\mathrm{MB}}(\mathbf{w}) \mathbf{u} \cdot \Delta \mathbf{u} \\ &= 2\pi\rho \frac{d^2}{4} \int_0^\pi \sin\chi (\cos\chi - 1)\, \mathrm{d}\chi \int f_{\mathrm{MB}}(\mathbf{w}) \frac{w^3}{4}\, \mathrm{d}\mathbf{w} \end{aligned} \tag{7.2.17}$$

where $\mathbf{w}$ is the relative velocity of a colliding pair, $f_{\mathrm{MB}}(\mathbf{w})$ is the Maxwell–Boltzmann distribution (2.1.24) (evaluated for a reduced mass equal to $\frac{1}{2}m$), χ is the angle of deflection, and $\sigma(\Omega_\chi, \mathbf{w}) = \frac{1}{4}d^2$ is the differential cross-section for hard spheres of diameter d. Integration of (7.2.17) and substitution in (7.2.16) gives the simple result that

$$\Omega_0 = \tfrac{2}{3}\Gamma_0 = \frac{8\rho d^3}{3} \left(\frac{\pi k_{\mathrm{B}} T}{m} \right)^{1/2} \tag{7.2.18}$$

where Γ_0 is the low-density (Boltzmann) collision rate. At higher densities, it is better to use the Enskog method; this requires the collision rate to be multiplied by the contact value $g(d)$ of the pair distribution function (see Eqn (3.2.6)). In the Enskog approximation, successive binary collisions suffered by the tagged particle are completely uncorrelated. The velocity $\mathbf{u}$ of the tagged sphere may therefore be treated as a gaussian, markovian, random variable, with the consequence, by Doob's theorem, that its time-correlation function is necessarily exponential. By identifying the right-hand side of (7.2.15) as the leading terms in the expansion of an exponential function, we obtain Enskog's approximation to the hard-sphere velocity autocorrelation function:

$$Z_E(t) = \frac{k_B T}{m} \exp(-3|t|/2\tau_E) \tag{7.2.19}$$

where $\tau_E = \Gamma^{-1} = 1/g(d)\Gamma_0$ is the Enskog mean collision time and the absolute value of t appears in the exponential because $Z(t)$ must be an even function of time. The corresponding approximation for the self-diffusion constant of the hard-sphere fluid is obtained by substitution of (7.2.19) in (7.2.8):

$$D_E = \frac{2}{3}\frac{k_B T}{m}\tau_E = \frac{3}{8\rho d^2 g(d)}\left(\frac{k_B T}{\pi m}\right)^{1/2} \tag{7.2.20}$$

This expression becomes exact in the low-density limit; its applicability at higher densities has been tested by comparison with molecular dynamics calculations (Alder *et al.*, 1970). From Figure 7.2, we see that the "exact" D exceeds the Enskog value at intermediate densities, and falls below D_E

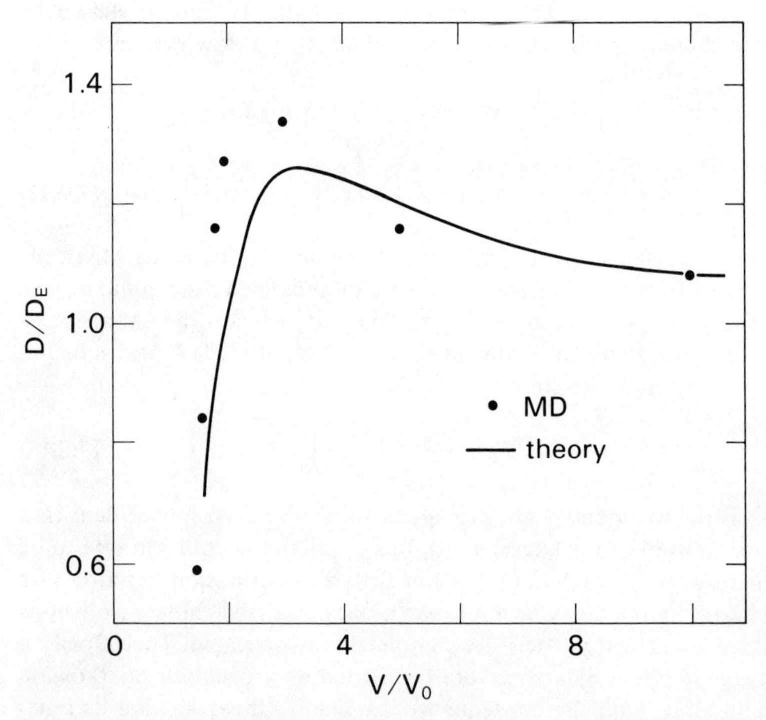

FIG. 7.2. Ratio of the hard-sphere self-diffusion coefficient to its Enskog value. The quantity $V_0 = Nd^3/\sqrt{2}$ is the volume at close packing. The points are molecular dynamics results of Alder *et al.* (1970) and the curve is calculated from a kinetic theory to be discussed in Section 9.9 (Cukier and Mehaffey, 1978).

at densities close to crystallization. The high-density deviations arise from backscattering effects, corresponding to the fact that collisons between near neighbours lead, on average, to the reversal of the velocity of a tagged particle into a comparatively narrow range of angles (Rahman, 1964a); this gives rise to a negative region in $Z(t)$, as shown in Figure 7.1. The increase in D/D_{E} at intermediate densities can be attributed, at least in part, to an enhancement of the velocity correlations due to the excitation of slowly-decaying collective motions in the fluid. The tagged particle induces a

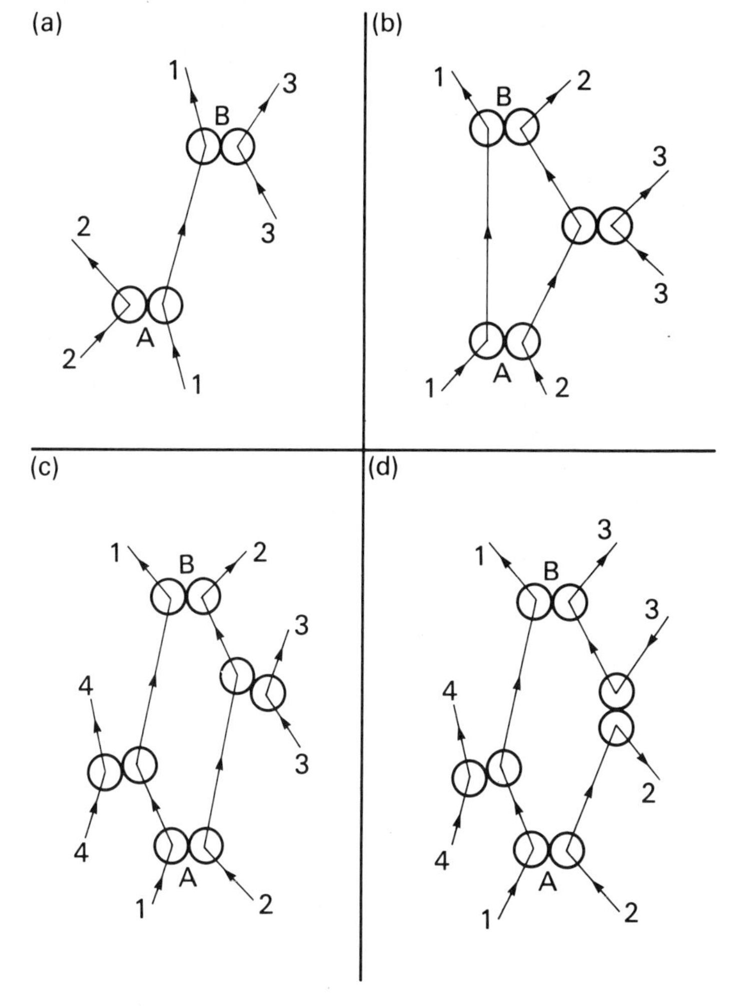

FIG. 7.3. Examples of uncorrelated (a) and correlated (b, c, d) sequences of binary collisions; A and B represent two different space-time points. See text for details.

backflow pattern in the surrounding fluid that reacts back on the particle at a later time, giving rise to persistence (or "memory") effects and an unexpectedly slow ($\sim t^{-3/2}$) decay of $Z(t)$ at long times (Alder and Wainwright, 1970). We shall return to the question of the "long-time tails" of correlation functions in Section 8.7.

A treatment of self-diffusion by kinetic theory that goes beyond the simple Enskog approximation must account for correlated sequences of binary collisions experienced by the tagged particle, among which the so-called ring collisions play a dominant role (Résibois and DeLeener, 1977). In a ring collision, the tagged particle experiences an initial collision with a given bath particle, then propagates through the fluid, suffering collisions with various other bath particles, before recolliding with the same particle that it met initially, or with another bath particle whose motion is correlated in some way with that of the initial collision partner of the tagged particle. Examples of correlated sequences of binary collisions of this general type are illustrated in Figure 7.3; their role in the process of self-diffusion will be considered in more detail in Section 9.6. In each example shown, the tagged particle, labelled 1, collides with particles of the bath, labelled 2, 3 and 4, at space-time points A and B. In (a), the successive collisons are uncorrelated. In (b) and (c), particles 1 and 2 first meet at A, then recollide at B; in (b), the recollision involves one intermediate collision between 2 and 3 (a three-body event), and in (c) it involves intermediate collisions between 1 and 4 and between 2 and 3 (a four-body event). Example (d) is a different type of four-body event in which the initial (at A) and final (at B) collision partners (2 and 3) of the tagged particle are different, but the collisions suffered by 1 at A and B are still correlated. The sequences (b), (c) and (d) are all examples of ring-collision events.

7.3 BROWNIAN MOTION AND THE GENERALIZED LANGEVIN EQUATION

Calculations of the velocity autocorrelation either by the Enskog method or by the other, more sophisticated versions of kinetic theory are limited, strictly speaking, to hard-sphere systems. Attempts have been made to apply similar techniques in calculations for continuous potentials, some of which are briefly discussed in Chapter 9. In this section, however, we describe a different approach that is more phenomenological in character but has found wide application in the treatment of transport processes in liquids. Its basis is the stochastic theory used by Langevin to treat the brownian motion of a large and massive particle in a bath of molecules that are much smaller and lighter than itself. The problem is characterized by two very different timescales, one associated with the slow relaxation of the initial velocity of the brownian particle and another linked to the frequent collisions that the brownian particle suffers with the molecules of the bath. Langevin assumed that the force acting on the brownian particle at any instant consists of two parts: a systematic frictional force proportional to the velocity $\mathbf{u}(t)$, but acting in the opposite sense, and a randomly fluctuating force, $\mathbf{R}(t)$, that arises from collisions with the surrounding molecules. The equation of motion of a brownian particle of mass m is therefore written as

$$m\dot{\mathbf{u}}(t) = -\xi m\mathbf{u}(t) + \mathbf{R}(t) \tag{7.3.1}$$

where ξ is the *friction coefficient.* The random force is assumed to vanish in the mean:

$$\langle \mathbf{R}(t) \rangle = 0 \tag{7.3.2}$$

to be uncorrelated with the velocity:

$$\langle \mathbf{R}(t) \cdot \mathbf{u}(0) \rangle = 0 \tag{7.3.3}$$

and to have a correlation time that is infinitesimally short, i.e.

$$\langle \mathbf{R}(t+s) \cdot \mathbf{R}(s) \rangle = 2\pi R_0 \delta(t) \tag{7.3.4}$$

Equation (7.3.4) implies that the power spectrum of the random force is a constant, R_0 (a "white" spectrum):

$$\frac{1}{2\pi}\int_{-\infty}^{\infty} \langle \mathbf{R}(t) \cdot \mathbf{R}(0) \rangle \exp(i\omega t)\, \mathrm{d}t = R_0 \tag{7.3.5}$$

These are reasonable assumptions when the brownian particle is much larger than its neighbours, because even on a short timescale the motion will be determined by a very large number of essentially uncorrelated collisions. When all particles are of the same size, the assumptions are less well justified, and a generalization is required, of a type to be discussed later.

The two terms on the right-hand side of Eqn (7.3.1) are not independent. To see the connection between them, we first write the solution to (7.3.1) in the form

$$m\mathbf{u}(t) = m\mathbf{u}(0) \exp(-\xi t) + \exp(-\xi t) \int_0^{\infty} \exp(\xi s) \mathbf{R}(s)\, \mathrm{d}s \tag{7.3.6}$$

Squaring and taking the mean, we find, using (7.3.3) and (7.3.4), that

$$\begin{aligned} m^2\langle |\mathbf{u}(t)|^2 \rangle &= m^2\langle |\mathbf{u}(0)|^2 \rangle \exp(-2\xi t) \\ &\quad + \exp(-2\xi t) \int_0^t \mathrm{d}s \int_0^t \mathrm{d}s' \exp[\xi(s+s')] 2\pi R_0 \delta(s-s') \\ &= m^2\langle |\mathbf{u}(0)|^2 \rangle \exp(-2\xi t) + \frac{\pi R_0}{\xi}[1 - \exp(-2\xi t)] \end{aligned} \tag{7.3.7}$$

We now take the limit $t \to \infty$; the brownian particle will then be in thermal equilibrium with the bath, regardless of the intial conditions. Hence

$$\langle |\mathbf{u}(\infty)|^2 \rangle = \frac{3k_B T}{m} \tag{7.3.8}$$

from which it follows that

$$\xi = \frac{\pi \beta R_0}{3m} = \frac{\beta}{3m} \int_0^\infty \langle \mathbf{R}(t) \cdot \mathbf{R}(0) \rangle \, dt \tag{7.3.9}$$

From a physical point of view, it is not unexpected to find a link between the frictional and random forces. If the brownian particle were to be drawn through the medium by an external field, random collisions suffered by the particle would give rise to a systematic retarding force proportional to the particle velocity. This is an illustration of the fluctuation–dissipation theorem to which we have already referred in Section 5.2 and which will be established more generally later in this chapter.

The friction coefficient is also related to the diffusion coefficient. Let us suppose, for simplicity, that the brownian particle is initially ($t = 0$) situated at the origin, $\mathbf{r} = 0$. Our aim is to calculate the mean-square displacement of the particle after a time t. Multiplying through Eqn (7.3.1) by $\mathbf{r}(t)$ and using the results

$$\mathbf{r} \cdot \mathbf{u} = \mathbf{r} \cdot \dot{\mathbf{r}} = \frac{1}{2} \frac{d}{dt} r^2 \tag{7.3.10}$$

$$\mathbf{r} \cdot \dot{\mathbf{u}} = \mathbf{r} \cdot \ddot{\mathbf{r}} = \frac{1}{2} \frac{d^2}{dt^2} r^2 - u^2 \tag{7.3.11}$$

we find that

$$\tfrac{1}{2} m \frac{d^2}{dt^2} |\mathbf{r}(t)|^2 + \tfrac{1}{2} \xi m \frac{d}{dt} |\mathbf{r}(t)|^2 = m |\mathbf{u}(t)|^2 + \mathbf{r}(t) \cdot \mathbf{R}(t) \tag{7.3.12}$$

In the statistical mean, Eqn (7.3.12) becomes

$$\frac{d^2}{dt^2} \langle |\mathbf{r}(t)|^2 \rangle + \xi \frac{d}{dt} \langle |\mathbf{r}(t)|^2 \rangle = \frac{6k_B T}{m} \tag{7.3.13}$$

which must be solved subject to the boundary conditions:

$$\langle |\mathbf{r}(0)|^2 \rangle = 0 \tag{7.3.14a}$$

$$\frac{d}{dt} \langle |\mathbf{r}(0)|^2 \rangle = 2 \langle \mathbf{r}(0) \cdot \mathbf{u}(0) \rangle = 0 \tag{7.3.14b}$$

The solution is

$$\langle |\mathbf{r}(t)|^2 \rangle = \left(\frac{6k_B T}{\xi m}\right) t - \frac{1}{\xi} + \frac{1}{\xi} \exp(-\xi t) \tag{7.3.15}$$

At very short times, such that $\xi t \ll 1$, the solution becomes

$$\langle |\mathbf{r}(t)|^2 \rangle \simeq \left(\frac{3k_B T}{m}\right) t^2 = \langle u^2 \rangle t^2 \tag{7.3.16}$$

which corresponds to free-particle motion. At very large times, such that $\xi t \gg 1$, Eqn (7.3.15) reduces to

$$\langle |\mathbf{r}(t)|^2 \rangle \simeq \left(\frac{6k_B T}{\xi m}\right) t \tag{7.3.17}$$

and comparison with Eqn (7.2.3) leads to Einstein's expression for the diffusion constant in terms of the friction coefficient:

$$D = \frac{k_B T}{\xi m} \tag{7.3.18}$$

An estimate of ξ can be obtained from a hydrodynamic calculation of the frictional force on a sphere of diameter d moving with constant velocity $\mathbf{u}$ in a fluid of shear viscosity η. This leads to a famous result due to Stokes, the precise form of which is determined by the assumptions made about the behaviour, at the surface of the sphere, of the velocity field created in the fluid. If the "stick" boundary condition is used, the fluid velocity at the surface is everywhere taken equal to $\mathbf{u}$; in the "slip" approximation, the normal component of the fluid velocity is set equal to the normal component of $\mathbf{u}$, ensuring that no fluid can enter or leave the sphere, and the tangential force acting on the sphere is assumed to vanish. The stress tensor at the surface can now be obtained by solving the linearized Navier–Stokes equation (see Section 8.3.) subject to one of these boundary conditions, supplemented by the requirement that the fluid velocity must vanish at infinite distance from the sphere. When the stress tensor is known, the total frictional force $\mathbf{F}$ can be calculated by integration over the surface (Landau and Lifshitz, 1963; Zwanzig and Bixon, 1970). The resulting force is found to be of the form $\mathbf{F} = -\xi \mathbf{u}$, with

$$\xi = \frac{3\pi\eta d}{m} \quad \text{(stick)} \tag{7.3.19a}$$

$$\xi = \frac{2\pi\eta d}{m} \quad \text{(slip)} \tag{7.3.19b}$$

Combination of (7.3.19) with (7.3.18) leads to the two familiar forms of Stokes' law:

$$D\eta = \frac{k_B T}{3\pi d} \quad \text{(stick)} \tag{7.3.20a}$$

$$D\eta = \frac{k_B T}{2\pi d} \quad \text{(slip)} \tag{7.3.20b}$$

It is a remarkable feature of Stokes' law that although it is derived from purely macroscopic considerations, and is apparently limited to brownian particles, it also provides a good empirical correlation of experimental data on simple liquids, use of the slip boundary condition generally leading to more reasonable values of the effective diameter *d*.

From Eqn (7.3.1) we can easily deduce the form of the velocity autocorrelation function of the brownian particle. Multiplying through by $\mathbf{u}(0)$ and taking the thermal average, we find that

$$\begin{aligned} Z(t) &= \tfrac{1}{3}\langle \mathbf{u}(t) \cdot \mathbf{u}(0) \rangle \\ &= \left(\frac{k_B T}{m}\right) \exp(-\xi t) \end{aligned} \tag{7.3.21}$$

where $t \geq 0$, and thus

$$\tilde{Z}(z) = \frac{k_B T/m}{-iz + k_B T/mD} \tag{7.3.22}$$

The result (7.3.18) for the diffusion constant is then immediately recovered by inserting (7.3.21) in Eqn (7.2.8). Note that the correlation function (7.3.21) is of the same exponential form as the Enskog result (7.2.19) for the hard-sphere fluid; this is not surprising, since a markovian hypothesis underlies both calculations. In practice, as is evident from Figure 7.1, the velocity autocorrelation function of a typical simple liquid is often very far from exponential. Furthermore, the power spectrum of the function (7.3.21) has an infinite second moment; for continuous potentials, this is in conflict with the result in Eqn (7.2.13). The contradiction seen here is related to the fact that the applicability of Eqn (7.3.21) does not extend to very short times; for t such that $\xi t \ll 1$, the particle experiences very few collisions, and the basic assumptions of the Langevin theory are no longer valid.

When the dimensions of the diffusing particle are similar to those of its neighbours, the weakest part of the theory is the markovian approximation in which the frictional force on the particle at a given time is assumed to be proportional only to its velocity at the same time. The implication is that the motion of the particle adjusts itself instantaneously to changes in the surrounding medium. It would be more realistic to assume that the frictional force acting on a particle reflects the previous history of the system. In other words, we must associate a certain "memory" with the motion of the particle. This can be achieved by introducing a friction coefficient $\xi(t-t')$ that is non-local in time and determines the contribution to the systematic force at time t that comes from the velocity at an earlier time t'. Mathematically, this amounts to writing the frictional force as a convolution in time, giving rise to a non-markovian generalization of the Langevin equation that we write as

$$m\dot{\mathbf{u}}(t) = -m\int_0^t \xi(t-s)\mathbf{u}(s)\,\mathrm{d}s + \mathbf{R}(t) \tag{7.3.23}$$

The properties of $\mathbf{R}(t)$ expressed by Eqns (7.3.2) and (7.3.3) are assumed to be unaltered. If, therefore, we multiply through Eqn (7.3.23) by $\mathbf{u}(0)$ and take the thermal average, we arrive at an equation for the velocity autocorrelation function in the form

$$\dot{Z}(t) = -\int_0^t \xi(t-s)Z(s)\,\mathrm{d}s \tag{7.3.24}$$

The quantity $\xi(t)$ is usually called the *memory function* for the autocorrelation function $Z(t)$. A similar memory-function equation can be written down for the autocorrelation function of an arbitrary dynamical variable, A say. Such a result may be regarded as a generalized Langevin equation in which the random "force" is proportional to the part of $A(t)$ that is uncorrelated with $A(0)$ (cf. Eqn (7.3.3)). All that is lost in extending the use of the generalized Langevin equation to other dynamical variables is an intuitive feeling for the meaning of the "friction" coefficient and the random "force".

If we take the Laplace transform of Eqn (7.3.24), we obtain a simple algebraic relation between $\tilde{Z}(z)$ and $\tilde{\xi}(z)$:

$$\tilde{Z}(z) = \frac{k_\mathrm{B}T/m}{-iz + \tilde{\xi}(z)} \tag{7.3.25}$$

On replacing the frequency-dependent friction coefficient in (7.3.25) by a constant ξ, we recover the exponential form of $Z(t)$ given by Eqn (7.3.21); this amounts to choosing a purely local (markovian) memory function, $\xi(t) = \xi\delta(t)$, which leads back to the original Langevin equation (7.3.1). Equations (7.3.24) and (7.3.25) are exact, however, because they can be regarded as providing a definition of the unknown function $\xi(t)$. What is lacking at this stage is any statistical-mechanical definition of $\mathbf{R}(t)$ or $\xi(t)$, nor is it obvious that $\xi(t)$ is a simpler object to understand than $Z(t)$ itself, in which case Eqn (7.3.24) is of little value. The interpretation of the generalized Langevin equation and the memory-function equation in terms of statistical mechanics, which is the achievement of Zwanzig (1961) and Mori (1965a, b), is discussed in detail in Chapter 9. Here it is sufficient to say that the work of Zwanzig and Mori leads us to expect that $\xi(t)$ will decay much more rapidly than $Z(t)$. The practical significance of this result is that a phenomenological model of a complicated dynamical process can be devised by postulating a rather simple form for the appropriate memory function that satisfies, in particular, the known frequency sum rules on the correlation function. For example, we could suppose that $\xi(t)$ decays exponentially (Berne *et al.*, 1966), i.e.

$$\xi(t) = \xi(0) \exp(-|t|/\tau) \tag{7.3.26}$$

If we differentiate Eqn (7.3.24) with respect to time, set $t = 0$ and use Eqn (7.2.9), we find that

$$\xi(0) = -\frac{\ddot{Z}(0)}{Z(0)} = \Omega_0^2 \tag{7.3.27}$$

Taking the Laplace transform of $\xi(t)$ and substituting the result in (7.3.25), we obtain the expression

$$\tilde{Z}(z) = \frac{k_B T/m}{-iz + \Omega_0^2/(-iz + \tau^{-1})} \tag{7.3.28}$$

From Eqn (7.2.8), the self-diffusion coefficient can be identified as

$$D = \tilde{Z}(0) = \frac{k_B T}{m\Omega_0^2 \tau} \tag{7.3.29}$$

Inverse Laplace transformation of (7.3.28) gives

$$Z(t) = \left(\frac{k_B T/m}{\alpha_+ - \alpha_-}\right)[\alpha_+ \exp(-\alpha_-|t|) - \alpha_- \exp(-\alpha_+|t|)] \tag{7.3.30}$$

where α_+ and α_- are the two poles of $\tilde{Z}(z = i\alpha)$:

$$\alpha_\pm = \frac{1}{2\tau}[1 \mp (1 - 4\Omega_0^2\tau^2)^{1/2}] \tag{7.3.31}$$

If $\tau < 1/2\Omega_0$, the poles are real and positive, and $Z(t)$ decays monotonically with the correct curvature (Ω_0^2) at the origin. On the other hand, if $\tau > 1/2\Omega_0$, which from Eqn (7.3.29) is equivalent to the condition

$$\frac{mD\Omega_0}{k_B T} < 2 \tag{7.3.32}$$

then the poles are a complex-conjugate pair, and the velocity autocorrelation function behaves as

$$Z(t) = \left(\frac{k_B T}{m}\right) \exp(-|t|/2\tau)[\cos \Omega_1 t + (1/2\Omega_1\tau) \sin \Omega_1 t] \tag{7.3.33}$$

where $\Omega_1^2 = \Omega_0^2 - (2\tau)^{-2}$. The function defined by (7.3.33) exhibits a negative region at intermediate times, in qualitative agreement with simulation results on simple liquids near the triple point (see Figure 7.1), where the condition (7.3.32) is indeed well satisfied. The negative region in $Z(t)$ arises from the backscattering effect mentioned earlier.

Although the results are qualitatively satisfactory the argument that leads to Eqn (7.3.30) is inadequate in certain respects. First, it gives no prescription for the relaxation time τ appearing in (7.3.26); τ must be determined empirically from experimental or simulation values of D via Eqn (7.3.29). Secondly, use of the simple memory function (7.3.26) leads to a spectrum $Z(\omega)$ for which the frequency moments beyond the second are all infinite. Both defects can be overcome by postulating a gaussian rather than an exponential memory function, and making use of the fourth frequency moment of $Z(\omega)$, which in turn requires a knowledge of the equilibrium three-particle distribution function (Martin and Yip, 1968). However, none of the phenomenological memory-function calculations that use as their basic ingredients only the short-time behaviour of the correlation function are capable of reproducing the observed slow ($\sim t^{-3/2}$) decay at long time ($t \gg \Omega_0^{-1}$).

7.4 CORRELATIONS IN SPACE AND TIME

A detailed description of the time evolution of spatial correlations in liquids requires the introduction of a time-dependent generalization of the static density–density autocorrelation function defined by Eqn (5.1.4). The relevant dynamical variable is the local density of particles (5.1.1), where account must now be taken of the time-dependence of the coordinates $\mathbf{r}_i$. More generally, we define a local dynamical variable as

$$A(\mathbf{r}, t) = \sum_{i=1}^{N} a_i(t)\delta[\mathbf{r} - \mathbf{r}_i(t)] \tag{7.4.1}$$

where a_i is any physical quantity, such as the mass, velocity, or angular momentum of particle i, and $\mathbf{r}_i(t)$ is the centre-of-mass position of the same particle. The spatial Fourier components of $A(\mathbf{r}, t)$ are

$$A_{\mathbf{k}}(t) = \int A(\mathbf{r}, t) \exp(-i\mathbf{k}\cdot\mathbf{r})\, d\mathbf{r}$$

$$= \sum_{i=1}^{N} a_i(t) \exp[-i\mathbf{k}\cdot\mathbf{r}_i(t)] \tag{7.4.2}$$

A local dynamical variable is said to be *conserved* if it satisfies a continuity equation of the form

$$\dot{A}(\mathbf{r}, t) + \nabla \cdot \mathbf{j}^A(\mathbf{r}, t) = 0 \tag{7.4.3}$$

where $\mathbf{j}^A$ is the "current" associated with the "density" A. Equation (7.4.3) is a local expression of the fact that $\int A(\mathbf{r}, t)\, d\mathbf{r} = \sum_i a_i(t)$ is independent of time; the corresponding equation for the Fourier components of A is

$$\dot{A}_{\mathbf{k}}(t) + i\mathbf{k} \cdot \mathbf{j}_{\mathbf{k}}^A(t) = 0 \tag{7.4.4}$$

An important example of a conserved local variable with which the present section is largely concerned is the microscopic number density $\rho(\mathbf{r}, t)$, for which $a_i = 1$:

$$\rho(\mathbf{r}, t) = \sum_{i=1}^{N} \delta[\mathbf{r} - \mathbf{r}_i(t)] \tag{7.4.5}$$

The associated particle current is

$$\mathbf{j}(\mathbf{r}, t) = \sum_{i=1}^{N} \mathbf{u}_i(t) \delta[\mathbf{r} - \mathbf{r}_i(t)] \tag{7.4.6}$$

with Fourier components

$$\mathbf{j}_{\mathbf{k}}(t) = \sum_{i=1}^{N} \mathbf{u}_i(t) \exp[-i\mathbf{k} \cdot \mathbf{r}_i(t)] \tag{7.4.7}$$

where $\mathbf{u}_i$ is the velocity of particle *i*. The Fourier components can be separated into longitudinal (l) and transverse (t) parts that are parallel and perpendicular, respectively, to the wavevector $\mathbf{k}$; the longitudinal component $\mathbf{j}_{\mathbf{k}l}$ is related to the microscopic density via the continuity equation (7.4.4).

The correlation function of two space-dependent dynamical variables is defined exactly as in Eqns (7.1.3) and (7.1.4), but is now non-local in both space and time:

$$C_{AB}(\mathbf{r}', \mathbf{r}''; t', t'') = \langle A(\mathbf{r}', t') B(\mathbf{r}'', t'') \rangle \tag{7.4.8}$$

The Fourier components of the local variables are complex functions, but the a_i are real physical quantities. Their time-correlation functions are therefore generally defined as

$$\begin{aligned} C_{AB}(\mathbf{k}', \mathbf{k}''; t', t'') &= \langle A_{\mathbf{k}'}(t') B_{\mathbf{k}''}^*(t'') \rangle \\ &= \langle A_{\mathbf{k}'}(t') B_{-\mathbf{k}''}(t'') \rangle \end{aligned} \tag{7.4.9}$$

These correlation functions have all the properties given in Section 7.1, in particular those associated with stationarity. In addition, for homogeneous liquids, translational invariance in space means that the non-local time-correlation function (7.4.8) depends only on the relative coordinates $\mathbf{r} = \mathbf{r}' - \mathbf{r}''$:

$$C_{AB}(\mathbf{r}', \mathbf{r}''; t', t'') = C_{AB}(\mathbf{r}' - \mathbf{r}'', t' - t'') \tag{7.4.10}$$

Translational invariance also implies that correlations between Fourier components $A_{\mathbf{k}'}(t')$ and $B_{\mathbf{k}''}(t'')$ are non-zero only if $\mathbf{k}'=\mathbf{k}''$:

$$C_{AB}(\mathbf{k}',\mathbf{k}'';t',t'')=\langle A_{\mathbf{k}'}(t')B_{-\mathbf{k}''}(t'')\rangle\delta_{\mathbf{k}'\mathbf{k}''}$$
$$=C_{AB}(\mathbf{k}',t'-t'') \tag{7.4.11}$$

Clearly $C_{AB}(\mathbf{k}, t)$ is the spatial Fourier transform of $C_{AB}(\mathbf{r}, t)$:

$$C_{AB}(\mathbf{k},t)=\int C_{AB}(\mathbf{r},t)\exp(-i\mathbf{k}\cdot\mathbf{r})\,\mathrm{d}\mathbf{r} \tag{7.4.12}$$

If the fluid is also isotropic, the correlation functions (7.4.10) and (7.4.11) share with their static counterparts the property that they are functions, respectively, of $r=|\mathbf{r}|$ and $k=|\mathbf{k}|$. The frequency moments of the spectra of the autocorrelation functions $C_{AA}(k, t)$ are defined as in Eqn (7.1.26), but are now dependent on k. The continuity equation (7.4.4) for conserved dynamical variables leads to simple expressions for the second frequency moments, called f-sum rules (Pines and Nozières, 1966). From (7.4.4) and (7.1.26) it follows that

$$\begin{aligned}\langle\omega^2\rangle_{AA}&=-\frac{\mathrm{d}^2}{\mathrm{d}t^2}C_{AA}(k,t)\big|_{t=0}\\&=\langle\dot{A}_{\mathbf{k}}(0)\dot{A}_{-\mathbf{k}}(0)\rangle\\&=\langle-i\mathbf{k}\cdot\mathbf{j}_{\mathbf{k}}^{A}(0)i\mathbf{k}\cdot\mathbf{j}_{-\mathbf{k}}^{A}(0)\rangle\\&=k^2\langle|j_{\mathbf{k}l}^{A}|^2\rangle\end{aligned} \tag{7.4.13}$$

The memory function, M_{AA} say, associated with a space-dependent autocorrelation function C_{AA} must allow for non-local effects in space as well as in time. The generalized Langevin equation satisfied by C_{AA} is therefore written in the general form

$$\dot{C}_{AA}(\mathbf{r},t)+\int_0^t\mathrm{d}t'\int\mathrm{d}\mathbf{r}'\,M_{AA}(\mathbf{r}-\mathbf{r}',t-t')C_{AA}(\mathbf{r}',t')=0 \tag{7.4.14}$$

Given the properties of the convolution product, the spatial Fourier transform of (7.4.14) is

$$\dot{C}_{AA}(\mathbf{k},t)+\int_0^t\mathrm{d}t'\,M_{AA}(\mathbf{k},t-t')C_{AA}(\mathbf{k},t')=0 \tag{7.4.15}$$

We now specialize the discussion to the problem of time-dependent correlations in the microscopic density and particle current and their Fourier components. As a generalization of Eqn (5.1.4), we construct an equilibrium density–density time correlation function $G(\mathbf{r}, \mathbf{r}'; t)$, defined as

$$G(\mathbf{r},\mathbf{r}';t)=\frac{1}{N}\langle\rho(\mathbf{r}'+\mathbf{r},t)\rho(\mathbf{r}',0)\rangle \tag{7.4.16}$$

If we eliminate the dependence on the choice of origin by integrating over $\mathbf{r}'$, we obtain the space and time-dependent distribution function introduced by van Hove (1954):

$$G(\mathbf{r}, t) = \frac{1}{N}\left\langle \sum_{i=1}^{N} \sum_{j=1}^{N} \int \delta[\mathbf{r}' + \mathbf{r} - \mathbf{r}_i(t)]\delta[\mathbf{r} - \mathbf{r}_j(0)]\, d\mathbf{r}' \right\rangle \tag{7.4.17}$$

The integration over $\mathbf{r}'$ is trivially carried out only if the position operators $\mathbf{r}_i(t)$ and $\mathbf{r}_j(0)$ commute. In the quantum-mechanical case, $\mathbf{r}_i(t)$ can be written in the Heisenberg representation as

$$\mathbf{r}_i(t) = \exp(i\mathcal{H}t/\hbar)\mathbf{r}_i \exp(-i\mathcal{H}t/\hbar) \tag{7.4.18}$$

and $\mathbf{r}_i(t)$ does not commute with $\mathbf{r}_j(0)$ when $t \neq 0$. At $t = 0$, however, $\mathbf{r}_i$ and $\mathbf{r}_j$ commute for all i and j, and if the liquid is homogeneous we may integrate over $\mathbf{r}'$ and recover Eqn (5.1.4):

$$\begin{aligned} G(\mathbf{r}, 0) &= \delta(\mathbf{r}) + \frac{1}{N}\left\langle \sum_{i \neq j}^{N}\sum \delta[\mathbf{r} + \mathbf{r}_j(0) - \mathbf{r}_i(0)] \right\rangle \\ &= \delta(\mathbf{r}) + \rho g(\mathbf{r}) \end{aligned} \tag{7.4.19}$$

In the classical case, to which we now restrict ourselves, the integration over $\mathbf{r}'$ in (7.4.17) may be carried out at all times. Thus

$$\begin{aligned} G(\mathbf{r}, t) &= \frac{\langle \rho(\mathbf{r}, t)\rho(\mathbf{0}, 0)\rangle}{\rho} \\ &= \frac{1}{N}\left\langle \sum_{i=1}^{N}\sum_{j=1}^{N} \delta[\mathbf{r} + \mathbf{r}_j(0) - \mathbf{r}_i(t)] \right\rangle \end{aligned} \tag{7.4.20}$$

The function G again separates naturally into two terms, usually called the "self" (s) and "distinct" (d) parts, i.e.

$$G(\mathbf{r}, t) = G_s(\mathbf{r}, t) + G_d(\mathbf{r}, t) \tag{7.4.21}$$

where

$$G_s(\mathbf{r}, t) = \frac{1}{N}\left\langle \sum_{i=1}^{N} \delta[\mathbf{r} + \mathbf{r}_i(0) - \mathbf{r}_i(t)] \right\rangle \tag{7.4.22}$$

$$G_d(\mathbf{r}, t) = \frac{1}{N}\left\langle \sum_{i \neq j}^{N}\sum \delta[\mathbf{r} + \mathbf{r}_j(0) - \mathbf{r}_i(t)] \right\rangle \tag{7.4.23}$$

Hence $G_s(\mathbf{r}, 0) = \delta(\mathbf{r})$ and $G_d(\mathbf{r}, 0) = \rho g(\mathbf{r})$. The physical interpretation of the van Hove correlation function is that $G(\mathbf{r}, t)\, d\mathbf{r}$ is proportional to the probability of finding a particle i in a region $d\mathbf{r}$ around a point $\mathbf{r}$ at time t given that there was a particle j at the origin at time $t = 0$; the division into self and distinct parts corresponds to the possibilities that i and j may be the same particle or different ones. For isotropic fluids, G_s and G_d will both be functions of the scalar quantity r.

The normalization of the van Hove function is determined by the normalization of the delta-function. Integration over $\mathbf{r}$ shows that

$$\int G_s(\mathbf{r}, t)\, d\mathbf{r} = 1 \tag{7.4.24}$$

$$\int G_d(\mathbf{r}, t)\, d\mathbf{r} = N - 1 \tag{7.4.25}$$

These results are valid for all t and represent a formal statement of the fact that the total number of particles is conserved. In the limit of large t, both G_s and G_d become independent of r, while the behaviour at large r is clearly the same as that at large t. It follows from (7.4.24) and (7.4.25) that

$$\lim_{r\to\infty} G_s(\mathbf{r}, t) = \lim_{t\to\infty} G_s(\mathbf{r}, t) = \frac{1}{V} \simeq 0 \tag{7.4.26}$$

$$\lim_{r\to\infty} G_d(\mathbf{r}, t) = \lim_{t\to\infty} G_d(\mathbf{r}, t) \simeq \rho \tag{7.4.27}$$

In fact, as we shall see in Section 8.2, G_s is gaussian at large r and t. The variation of G_s and G_d(in an isotropic liquid) as a function of r for $t \ll \tau$, $t \sim \tau$ and $t \gg \tau$ is illustrated schematically in Figure 7.4; here τ is a characteristic relaxation time, which typically is of order 10^{-12} s.

It is plausible to suppose that G_s is a simpler object to calculate than G_d because G_s is concerned only with single-particle correlations; an example of an approximate but accurate scheme for the calculation of G_s is described in Section 8.2. A number of attempts have therefore been made to relate G_d to G_s under the assumption that the latter is a known function of r and t. Historically, the first of these efforts was the "convolution" approximation of Vineyard (1958). Suppose that at time $t = 0$ a particle i is at the origin. The probability of simultaneously finding a particle j in a region $d\mathbf{r}'$ around $\mathbf{r}'$ is proportional to $\rho g(\mathbf{r}')\, d\mathbf{r}'$. The probability of finding any particle at $\mathbf{r}$ at time t may therefore be written as

$$G(\mathbf{r}, t) = G_s(\mathbf{r}, t) + \rho \int g(\mathbf{r}')\, W(\mathbf{r} - \mathbf{r}', t)\, d\mathbf{r}' \tag{7.4.28}$$

which amounts simply to a definition of the unknown function $W(\mathbf{r}, t)$. Vineyard's approximation consists in replacing W by G_s and writing

$$G_d(\mathbf{r}, t) = \rho \int g(\mathbf{r}') G_s(\mathbf{r} - \mathbf{r}', t)\, d\mathbf{r}' \tag{7.4.29}$$

The weakness in the argument is obvious: it is assumed that the motion of i is uncorrelated with that of j, which cannot be true when $|\mathbf{r} - \mathbf{r}'|$ is small. The convolution approximation has been tested against the "exact" results

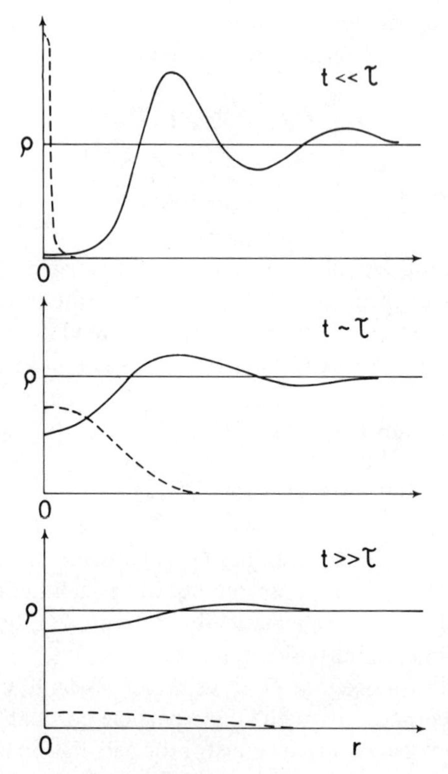

FIG. 7.4. Qualitative behaviour of the van Hove correlation functions. The full curve is $G_d(r, t)$ and the dashed curve is $G_s(r, t)$. After van Hove (1954).

on G_d obtained by Rahman (1964a) in a molecular dynamics simulation of liquid argon, and is found to predict too rapid a decay of G_d with time; the correlations that are ignored in Vineyard's approximation have the effect of inhibiting the rate at which the structure of the liquid breaks up.

Rather than considering the density-density correlation in real space, it is often more convenient to focus attention on the autocorrelation function of the Fourier components $\rho_{\mathbf{k}}$. The latter is usually called the "intermediate scattering function" and given the symbol $F(\mathbf{k}, t)$; the significance of the name will become apparent in the next section. The general properties expressed by (7.4.11) and (7.4.12) enable us to write

$$F(\mathbf{k}, t) = \frac{1}{N}\langle \rho_{\mathbf{k}}(t)\rho_{-\mathbf{k}}\rangle$$

$$= \int G(\mathbf{r}, t) \exp(-i\mathbf{k}\cdot\mathbf{r})\, d\mathbf{r} \qquad (7.4.30)$$

The spectrum of $F(\mathbf{k}, t)$, defined as

$$S(\mathbf{k}, \omega) = \frac{1}{2\pi} \int_{-\infty}^{\infty} F(\mathbf{k}, t) \exp(i\omega t)\, \mathrm{d}t \tag{7.4.31}$$

is called the *dynamic structure factor.* Combination of Eqns (5.1.6) and (7.1.26) leads immediately to an important relation between the static and dynamic structure factors:

$$\begin{aligned} \int_{-\infty}^{\infty} S(\mathbf{k}, \omega)\, \mathrm{d}\omega &= F(\mathbf{k}, 0) \\ &= S(\mathbf{k}) \end{aligned} \tag{7.4.32}$$

The physical meaning of this sum rule will also become clear in the next section. Finally, we define the autocorrelation function of the Fourier components (7.4.7) of the current associated with the microscopic density. Beacuse $\mathbf{j}_\mathbf{k}$ is a vector, the corresponding correlation function is a second rank tensor. However, rotational invariance implies that the longitudinal and transverse projections of the current are uncorrelated in an isotropic fluid. Hence the correlation-function tensor has only two independent components and may conveniently be written in the form

$$\begin{aligned} C_{\alpha\beta}(\mathbf{k}, t) &= \frac{k^2}{N} \langle j_\mathbf{k}^\alpha(t) j_{-\mathbf{k}}^\beta \rangle \\ &= \hat{k}_\alpha \hat{k}_\beta C_l(k, t) + (\delta_{\alpha\beta} - \hat{k}_\alpha \hat{k}_\beta) C_t(k, t) \end{aligned} \tag{7.4.33}$$

where $\alpha, \beta = x, y, z$, $\delta_{\alpha\beta}$ is the Kronecker symbol, and $\hat{k}_\alpha$, $\hat{k}_\beta$ are cartesian components of the unit vector $\hat{\mathbf{k}} = \mathbf{k}/k$. If the z-axis is chosen as the direction of $\mathbf{k}$, the longitudinal and transverse current correlation functions are given by

$$C_l(k, t) = \frac{k^2}{N} \langle j_\mathbf{k}^z(t) j_{-\mathbf{k}}^z \rangle \tag{7.4.34}$$

$$C_t(k, t) = \frac{k^2}{N} \langle j_\mathbf{k}^x(t) j_{-\mathbf{k}}^x \rangle \tag{7.4.35}$$

From the continuity equation (7.4.4) (with $A = \rho$) and the general property (7.1.9) it is clear that the longitudinal-current and density correlation functions are not independent, since

$$\begin{aligned} C_l(k, t) &= \frac{1}{N} \langle \dot{\rho}_\mathbf{k}(t) \dot{\rho}_{-\mathbf{k}} \rangle \\ &= -\frac{\mathrm{d}^2}{\mathrm{d}t^2} F(k, t) \end{aligned} \tag{7.4.36}$$

Written in terms of Laplace transforms, Eqn (7.4.36) becomes

$$\tilde{C}_l(k, z) = z^2 \tilde{F}(k, z) - izS(k) \tag{7.4.37}$$

or, on recalling Eqn (7.1.21) and taking the real part of (7.4.37):

$$C_l(k, \omega) = \omega^2 S(k, \omega) \tag{7.4.38}$$

The function $C_l(k, \omega)$ describes the spectrum of longitudinal-current fluctuations in the liquid. Fluctuations in density are therefore intimately related to fluctuations in longitudinal current, but are independent of the transverse current.

In classical statistical mechanics, positions and velocities at a given instant are statistically independent. From the definitions of the current autocorrelation functions it follows immediately that

$$\begin{aligned} C_l(k, 0) &= C_t(k, 0) \\ &= k^2\left(\frac{k_B T}{m}\right) \\ &= \omega_0^2, \quad \text{say} \end{aligned} \tag{7.4.39}$$

According to the general f-sum rule (7.4.13), which in the present case is a direct consequence of Eqns (7.1.26) and (7.4.36), the second frequency moment of the dynamic structure factor is given by

$$\begin{aligned} \langle \omega^2 \rangle_{\rho\rho} &= \int_{-\infty}^{\infty} \omega^2 S(k, \omega)\, d\omega \\ &= -\ddot{F}(k, 0) \\ &= \omega_0^2 \end{aligned} \tag{7.4.40}$$

The second moment of $S(k, \omega)$ is therefore independent of the interactions between particles; since Eqn (7.4.40) is a consequence of the continuity equation, it is merely an expression of particle conservation. The higher-order moments do depend on the interparticle potential. In the case of the hard-sphere fluid, the third moment is non-zero and the fourth moment diverges (Lebowitz *et al.*, 1969). If the potential is continuous, it follows from the general results embodied in Eqns (7.1.25) and (7.1.26) that the odd frequency moments of $S(k, \omega)$ are all zero, and the fourth moment is equal, by virtue of the relation (7.4.38), to the second moment of $C_l(k, \omega)$. We may therefore base a calculation of the fourth moment of $S(k, \omega)$ on the short-time expansion of $C_l(k, t)$, which we write as

$$C_l(k, t) = \omega_0^2\left(1 - \omega_{1l}^2 \frac{t^2}{2!} + \cdots\right) \tag{7.4.41}$$

where ω_0 is defined by (7.4.40). Equations (7.1.9) and (7.4.39) show that the coefficient of t^2 is given by

$$\begin{aligned}\omega_0^2\omega_{1l}^2 &= -\frac{\mathrm{d}^2}{\mathrm{d}t^2}C_l(k, t)\big|_{t=0}\\ &= \frac{\mathrm{d}^4}{\mathrm{d}t^4}F(k, t)\big|_{t=0}\\ &= \frac{1}{N}\langle \ddot{\rho}_{\mathbf{k}}\ddot{\rho}_{-\mathbf{k}}\rangle\end{aligned} \tag{7.4.42}$$

If we again take the z-axis parallel to $\mathbf{k}$, and make the substitution

$$\dot{u}_{iz} = -\frac{1}{m}\frac{\partial V_N}{\partial z_i} \tag{7.4.43}$$

then (7.4.42) becomes

$$\omega_0^2\omega_{1l}^2 = k^4\langle u_{iz}^4\rangle + k^2\left(\frac{k_\mathrm{B}T}{m}\right)\left\langle \sum_{i=1}^{N}\sum_{j=1}^{N}\frac{\partial V_N}{\partial z_i}\frac{\partial V_N}{\partial z_j}\exp\left[ik(z_i - z_j)\right]\right\rangle \tag{7.4.44}$$

After an integration by parts similar to that used in the derivation of (7.2.11), we arrive at the final result (de Gennes, 1959), i.e.

$$\omega_{1l}^2 = 3\omega_0^2 + (\rho/m)\int g(r)(1-\cos kz)\frac{\partial^2 v(r)}{\partial z^2}\,\mathrm{d}\mathbf{r} \tag{7.4.45}$$

where $v(r)$ is the pair potential. At large k, the kinetic contribution ($\sim k^2$) dominates, corresponding to free-particle behaviour. At small k, the cosine term can be expanded to yield a simpler result:

$$\lim_{k\to 0}\omega_{1l}^2 = 3\omega_0^2 + k^2\left(\frac{2\pi\rho}{15m}\right)\int_0^\infty\left[2r^3\frac{\mathrm{d}v(r)}{\mathrm{d}r} + 3r^4\frac{\mathrm{d}^2 v(r)}{\mathrm{d}r^2}\right]g(r)\,\mathrm{d}r \tag{7.4.46}$$

From Eqn (7.4.38) we see that ω_{1l}^2 is related to the fourth frequency moment of $S(k, \omega)$ by

$$\omega_{1l}^2 = \frac{\langle\omega^4\rangle_{\rho\rho}}{\langle\omega^2\rangle_{\rho\rho}} \tag{7.4.47}$$

A similar calculation can be made of the second moment of the transverse current correlation function. The analogous short-time expansion is

$$C_t(k, t) = \omega_0^2\left(1 - \omega_{1t}^2\frac{t^2}{2!} + \cdots\right) \tag{7.4.48}$$

with

$$\omega_0^2\omega_{1t}^2 = -\frac{\mathrm{d}^2}{\mathrm{d}t^2}C_t(k, t)\big|_{t=0} \tag{7.4.49}$$

By pursuing the methods used in the longitudinal case, we find that

$$\omega_{1t}^2 = \omega_0^2 + (\rho/m) \int g(r)(1 - \cos kz) \frac{\partial^2 v(r)}{\partial x^2} \,\mathrm{d}\mathbf{r} \tag{7.4.50}$$

In the limit of small k, this result reduces to

$$\lim_{k\to 0} \omega_{1t}^2 = \omega_0^2 + k^2 \left(\frac{2\pi\rho}{15m}\right) \int_0^\infty \frac{\mathrm{d}}{\mathrm{d}r}\left[r^4 \frac{\mathrm{d}v(r)}{\mathrm{d}r}\right] g(r)\,\mathrm{d}r \tag{7.4.51}$$

Higher-order moments of C_l and C_t involve correlations between increasingly large numbers of particles and rapidly become very tedious to evaluate (Forster *et al.*, 1968; Bansal and Pathak, 1977).

7.5 INELASTIC NEUTRON SCATTERING

We now show how the Fourier transforms of the van Hove correlation functions $G(\mathbf{r}, t)$ and $G_s(\mathbf{r}, t)$ are related to measurements of the inelastic scattering of slow (thermal) neutrons. To do so, we require a generalization of the calculation of Section 5.1 that allows for the exchange of energy between the neutrons and the target. Neutrons are particularly useful as probes of the microscopic dynamics of liquids because their momentum $\hbar\mathbf{k}$ and energy $E = \hbar\omega$ are related by

$$E = \frac{\hbar^2 k^2}{2m} \tag{7.5.1}$$

where m is the neutron mass. It follows that when E is roughly equal to $k_B T$, and therefore comparable with the thermal energies of particles in the liquid, the wavelength $\lambda = 2\pi/k$ associated with the neutron is approximately 2 Å, which is similar to the distance between neighbouring particles.

In a typical scattering event, a neutron of momentum $\hbar\mathbf{k}_1$ and energy $\hbar\omega_1$ is scattered into a solid angle $\mathrm{d}\Omega$. Let the momentum and energy of the neutron after the event be $\hbar\mathbf{k}_2$ and $\hbar\omega_2$, and let the momentum and energy transfer from neutron to target be $\hbar\mathbf{k}$ and $\hbar\omega$. The dynamical conservation laws require that

$$\hbar\omega = E_2 - E_1 \equiv \hbar\omega_{12} \tag{7.5.2}$$

$$\hbar\mathbf{k} = \hbar\mathbf{k}_1 - \hbar\mathbf{k}_2 \tag{7.5.3}$$

where E_1 and E_2 are the initial and final energies of the target. The probability per unit time for the transition $|1, \mathbf{k}_1\rangle \to |2, \mathbf{k}_2\rangle$, where $|1\rangle$ and $|2\rangle$ denote the initial and final states of the sample, is given by Fermi's "golden rule":

$$W_{12} = \frac{2\pi}{\hbar} |\langle 2, \mathbf{k}_2 | \mathcal{V} | 1, \mathbf{k}_1 \rangle|^2 \delta(\hbar\omega - \hbar\omega_{12}) \tag{7.5.4}$$

where $\mathcal{V}$ represents the perturbation, i.e. the interaction between the neutron and the sample. For simplicity, we have ignored the spin state of the neutron. The partial differential cross-section for scattering into the solid angle $d\Omega$ in a range of energy transfer $d\omega$ is calculated by averaging over all initial states $|1\rangle$ with their statistical weights $p_1 \propto \exp(-\beta E_1)$, summing over all final states $|2\rangle$ allowed by energy conservation, multiplying by the density of final states of the neutron, i.e.

$$d\mathbf{k}_2/(2\pi)^3 = k_2^2\, dk_2\, d\Omega/(2\pi)^3 = (m/\hbar^2)\hbar k_2\, d\omega\, d\Omega/(2\pi)^3,$$

and dividing by the flux $\hbar k_1/m$ of incident neutrons, with the result (Lovesey, 1984a) that

$$\frac{d^2\sigma}{d\Omega\, d\omega} = \frac{k_2}{k_1}\left(\frac{m}{2\pi\hbar^2}\right) \sum_{\{1\}} \sum_{\{2\}} p_1 |\langle 2, \mathbf{k}_2|\mathcal{V}|1, \mathbf{k}_1\rangle|^2 \delta(\omega - \omega_{12}) \tag{7.5.5}$$

The differential cross-section of Eqn (5.1.17) is obtained by integrating over all energy transfers:

$$\frac{d\sigma}{d\Omega} = \int \frac{d^2\sigma}{d\Omega\, d\omega}\, d\omega \tag{7.5.6}$$

The structure and dynamics of the sample, which are the aspects of the problem of interest to us, enter the calculation of the cross-section through the coupling of the neutron to the atomic nuclei. We again assume that $\mathcal{V}$ is given by the sum (5.1.20) of two-body Fermi pseudopotentials between a neutron located at $\mathbf{r}$ and a nucleus at $\mathbf{r}_i$. Since the initial and final states of the neutron are plane-wave states of the form (5.1.14), the matrix element in Eqn (7.5.5) can be rewritten as

$$\langle 2, \mathbf{k}_2|\mathcal{V}|1, \mathbf{k}_1\rangle = \frac{2\pi\hbar^2}{m}\left\langle 2\left| \sum_{i=1}^{N} b_i \exp(i\mathbf{k}\cdot\mathbf{r}_i)\right|1\right\rangle \tag{7.5.7}$$

where $\hbar\mathbf{k}$ is the momentum transfer defined in (7.5.3).

We first consider the case where all nuclei in the sample have the same scattering length b; this is justified only for systems of nuclei of a single isotope and zero spin. By incorporating (7.5.7) into (7.5.5) and exploiting the definition (5.1.5), we obtain an expression for the cross-section in the form

$$\frac{d^2\sigma}{d\Omega\, d\omega} = b^2\left(\frac{k_2}{k_1}\right) \sum_{\{1\}} \sum_{\{2\}} p_1 |\langle 2|\rho_{-\mathbf{k}}|1\rangle|^2 \delta(\omega - \omega_{12}) \tag{7.5.8}$$

or, on introducing the integral representation of the delta-function

$$\frac{d^2\sigma}{d\Omega\, d\omega} = \frac{b^2}{2\pi}\left(\frac{k_2}{k_1}\right) \sum_{\{1\}} \sum_{\{2\}} p_1 \int_{-\infty}^{\infty} |\langle 2|\rho_{-\mathbf{k}}|1\rangle|^2 \times \exp(i\omega t)\exp(-i\omega_{12}t)\, dt \tag{7.5.9}$$

Equation (7.5.9) can be simplified by recognizing that

$$\begin{aligned}
&\exp(-i\omega_{12}t)|\langle 2|\rho_{-\mathbf{k}}|1\rangle|^2 \\
&\qquad = \exp(-iE_2t/\hbar)\exp(iE_1t/\hbar)\langle 1|\rho_{\mathbf{k}}|2\rangle\langle 2|\rho_{-\mathbf{k}}|1\rangle \\
&\qquad = \langle 1|\exp(iE_1t/\hbar)\rho_{\mathbf{k}}\exp(-iE_2t/\hbar)|2\rangle\langle 2|\rho_{-\mathbf{k}}|1\rangle \\
&\qquad = \langle 1|\exp(i\mathcal{H}t/\hbar)\rho_{\mathbf{k}}\exp(-i\mathcal{H}t/\hbar)|2\rangle\langle 2|\rho_{-\mathbf{k}}|1\rangle \\
&\qquad = \langle 1|\rho_{\mathbf{k}}(t)|2\rangle\langle 2|\rho_{-\mathbf{k}}(0)|1\rangle
\end{aligned} \tag{7.5.10}$$

where $\mathcal{H}$ is the hamiltonian of the sample.

It remains only to sum over the intital states of the target, which is equivalent to taking a thermal average, and over the final states, which is done by exploiting the closure property of a complete set of quantum states $|j\rangle$:

$$\sum_j |j\rangle\langle j| = 1 \tag{7.5.11}$$

The final result for the cross-section is

$$\begin{aligned}
\frac{\mathrm{d}^2\sigma}{\mathrm{d}\Omega\,\mathrm{d}\omega} &= b^2\left(\frac{k_2}{k_1}\right)\frac{1}{2\pi}\int_{-\infty}^{\infty}\langle\rho_{\mathbf{k}}(t)\rho_{-\mathbf{k}}\rangle\exp(i\omega t)\,\mathrm{d}t \\
&= b^2\left(\frac{k_2}{k_1}\right)NS(\mathbf{k},\omega)
\end{aligned} \tag{7.5.12}$$

where $S(\mathbf{k}, \omega)$ is the spectrum of the density autocorrelation function (7.4.30), i.e. the dynamic structure factor defined in the preceding section. From Eqns (7.4.30) and (7.4.31) it follows that

$$S(\mathbf{k},\omega) = \frac{1}{2\pi}\int_{-\infty}^{\infty}\mathrm{d}t\exp(i\omega t)\int G(\mathbf{r},t)\exp(-i\mathbf{k}\cdot\mathbf{r})\,\mathrm{d}\mathbf{r} \tag{7.5.13}$$

showing that a measurement of the cross-section (7.5.12) is equivalent, at least in principle, to a determination of the van Hove function $G(\mathbf{r}, t)$. The connection with the elastic cross-section is made via Eqn (7.5.6); comparison of (7.5.12) with (5.1.29), taken for the case $b_{\text{inc}}^2 = 0$, shows that Eqn (7.5.6) provides the physical content of the "elastic" sum rule (7.4.32).

In the classical case, the function $G(\mathbf{r}, t)$ can be separated into the self and distinct parts defined in Eqns (7.4.22) and (7.4.23). By analogy with (7.5.13), it is customary to define a self dynamic structure factor $S_s(\mathbf{k}, \omega)$ as

$$S_s(\mathbf{k},\omega) = \frac{1}{2\pi}\int_{-\infty}^{\infty}\mathrm{d}t\exp(i\omega t)\int G_s(\mathbf{r},t)\exp(-i\mathbf{k}\cdot\mathbf{r})\,\mathrm{d}\mathbf{r} \tag{7.5.14}$$

together with a self intermediate scattering function $F_s(\mathbf{k}, t)$, defined through the transform

$$S_s(\mathbf{k}, \omega) = \frac{1}{2\pi} \int_{-\infty}^{\infty} F_s(\mathbf{k}, t) \exp(i\omega t)\, dt \tag{7.5.15}$$

with $F_s(\mathbf{k}, 0) = 1$. The self functions are important for the discussion of inelastic scattering in situations where the nuclei of the sample belong to different isotopic species or have more than one spin state. In any such case, the scattering lengths must be averaged over both the distribution of isotopes and the internal states of the nuclei, but this can be done independently of the thermal average over nuclear coordinates. In the notation of Eqn (5.1.28), a generalization of the result (5.1.29) allows the inelastic cross-section to be expressed as the sum of coherent and incoherent parts, i.e.

$$\frac{d^2\sigma}{d\Omega\, d\omega} = \left(\frac{d^2\sigma}{d\Omega\, d\omega}\right)^{\text{coh}} + \left(\frac{d^2\sigma}{d\Omega\, d\omega}\right)^{\text{inc}} \tag{7.5.16}$$

with

$$\left(\frac{d^2\sigma}{d\Omega\, d\omega}\right)^{\text{coh}} = \left(\frac{k_2}{k_1}\right) b_{\text{coh}}^2 N S(\mathbf{k}, \omega) \tag{7.5.17a}$$

$$\left(\frac{d^2\sigma}{d\Omega\, d\omega}\right)^{\text{inc}} = \left(\frac{k_2}{k_1}\right) b_{\text{inc}}^2 N S_s(\mathbf{k}, \omega) \tag{7.5.17b}$$

By varying the isotopic composition of the sample, or by using polarized neutrons, it is possible to measure separately the coherent and incoherent cross-sections, and thereby to separate G_s and G_d (Lovesey, 1984a). If b is the same for all nuclei, the scattering is purely coherent. Certain elements, including rubidium, are essentially coherent scatterers, whereas the scattering by protons is almost wholly incoherent.

It is sometimes convenient to define $S(\mathbf{k}, \omega)$ in terms of the fluctuations in density rather than the density itself. We denote the local deviation in density at time t by

$$\delta\rho(\mathbf{r}, t) = \rho(\mathbf{r}, t) - \rho \tag{7.5.18}$$

and its Fourier components by $\delta\rho_{\mathbf{k}}(t)$. Then

$$S(\mathbf{k}, \omega) = (2\pi)^3 \rho\, \delta(\omega)\, \delta(\mathbf{k}) + \frac{1}{2\pi N} \int_{-\infty}^{\infty} \langle \delta\rho_{\mathbf{k}}(t)\, \delta\rho_{-\mathbf{k}} \rangle \exp(i\omega t)\, dt \tag{7.5.19}$$

The first term on the right-hand side of (7.5.19) does not correspond to real scattering, because the energy and momentum transfer are both zero. The observable cross-section is therefore proportional to the spectrum of density fluctuations in the liquid. If we discard the delta-function term in Eqn (7.5.19) we are, in effect, redefining $S(\mathbf{k}, \omega)$ in a way that takes account of the known asymptotic behaviour of $G(\mathbf{r}, t)$ ar large r, i.e.

$$S(\mathbf{k}, \omega)=\frac{1}{2\pi}\int_{-\infty}^{\infty} \mathrm{d}t \exp(i\omega t)\int [G(\mathbf{r}, t)-\rho] \exp(-i\mathbf{k}\cdot\mathbf{r})\, \mathrm{d}\mathbf{r} \tag{7.5.20}$$

The intermediate scattering function is now defined as

$$F(\mathbf{k}, t)=\frac{1}{N}\langle \delta\rho_{\mathbf{k}}(t)\delta\rho_{-\mathbf{k}}\rangle \tag{7.5.21}$$

but the relation between $S(\mathbf{k}, \omega)$ and $F(\mathbf{k}, t)$ remains the same as in Eqn (7.4.31). Because there is no distinction between $\rho_{\mathbf{k}}(t)$ and $\delta\rho_{\mathbf{k}}(t)$ except at $\mathbf{k}=0$, it is convenient for notational reasons to retain the symbol $\rho_{\mathbf{k}}(t)$ for finite-wavelength Fourier components of the fluctuating density.

Returning briefly to the case when quantum effects cannot be ignored, we note that the van Hove function is in general complex, because $\mathbf{r}_i(0)$ and $\mathbf{r}_j(t)$ do not commute. It follows from the definition (7.4.17) that

$$G(-\mathbf{r}, -t)=G^*(\mathbf{r}, t) \tag{7.5.22}$$

In the classical limit, the imaginary part of $G(\mathbf{r}, t)$ vanishes. On the other hand, $S(\mathbf{k}, \omega)$ is a measurable quantity (see Eqn. (7.5.12)), and is therefore necessarily real, a fact already implicit in Eqn (7.5.22). For systems with inversion symmetry, which includes all fluids, $S(\mathbf{k}, \omega)$ is invariant under a change of sign of $\mathbf{k}$. Thus

$$S(-\mathbf{k}, \omega)=S(\mathbf{k}, \omega) \tag{7.5.23}$$

However, the cross-section cannot, strictly speaking, be even with respect to ω. If this were the case, thermal equilibrium between the radiation and the sample would never be reached. The principle of detailed balance requires that the ratio of cross-sections for the scattering processes $|\mathbf{k}_1, 1\rangle \rightarrow |\mathbf{k}_2, 2\rangle$ and $|\mathbf{k}_2, 2\rangle \rightarrow |\mathbf{k}_1, 1\rangle$ be equal to the ratio of the statistical weights of the states $|1\rangle$ and $|2\rangle$, i.e.

$$S(\mathbf{k}, \omega)=\exp(\beta\hbar\omega)S(\mathbf{k}, -\omega) \tag{7.5.24}$$

Experimental scattering data are therefore frequently reported in the form of a "symmetrized" dynamic structure factor $\bar{S}(\mathbf{k}, \omega)$, defined as

$$\bar{S}(\mathbf{k}, \omega)=\exp(-\tfrac{1}{2}\beta\hbar\omega)S(\mathbf{k}, \omega) \tag{7.5.25}$$

This is an even function of frequency for both classical and quantum systems, whereas $S(\mathbf{k}, \omega)$ itself is an even function only in the classical limit ($\hbar \to 0$). In the theoretical discussion that follows, we shall assume that we are dealing with classical liquids, so that $G(\mathbf{r}, t)$ is purely real and $S(\mathbf{k}, \omega)$ is an even function of both $\mathbf{k}$ and ω. In addition, our interest will be focussed on isotropic systems, for which $S(\mathbf{k}, \omega)$ depends only on the magnitude of $\mathbf{k}$ and not on its direction.

The single-particle and collective density fluctuation spectra of a number of simple liquids have been studied by incoherent and coherent inelastic neutron scattering experiments (Copley and Lovesey, 1975). Particularly careful and complete experiments have been carried out on liquid argon (Sköld *et al.*, 1972; van Well *et al.*, 1985), gaseous krypton (Egelstaff *et al.*, 1983), liquid lead (Söderström *et al.*, 1983), liquid rubidium (Copley and Rowe 1974a, b), and on both liquid and dense gaseous neon (Buyers *et al.*, 1975; Bell *et al.*, 1975). Quantum effects are important in the case of liquid neon, and are dominant for normal liquid helium, on which inelastic measurements have been carried out at temperatures above 2 K (Woods *et al.*, 1978). The wavenumbers accessible in inelastic neutron scattering experiments lie typically between 0.1 and 15Å^{-1}, which is the same range of k that is usually studied in molecular dynamics simulations. The most exhaustive computer studies of the dynamical structure factor that have been reported deal with the Lennard-Jones model of argon (Levesque *et al.*, 1973; Schoen and Hoheisel, 1985), a model of liquid rubidium (Rahman, 1974a, b) and the hard-sphere fluid (Alley *et al.*, 1983).

Examples of the dynamic structure factor of simple liquids obtained either by neutron scattering or by computer simulation are plotted in Figures 7.5, 8.5 and 9.4. At reduced wavenumbers $kd \lesssim 1$ (where d is the atomic diameter), $S(k, \omega)$ (or its symmetrized version $\bar{S}(k, \omega)$) has a sharp peak centred at zero frequency and two more or less well-defined side peaks or high-frequency "shoulders", one on each side of the central peak. We shall see in Chapter 8 that the side peaks observed at long wavelengths correspond to propagating sound waves. At shorter wavelengths, the sound waves are strongly damped, and the high-frequency structure disappears when $kd \gtrsim 2$, leaving only a single, lorentzian-like central peak. The width of the central peak first increases with k, but then shows a marked decrease at wavenumbers close to the main peak in the static structure factor. This effect is called "de Gennes narrowing" (de Gennes, 1959); it corresponds to a dramatic slowing down in the decay of the density autocorrelation function $F(k, t)$ and has its origins in the strong spatial correlations that exist at these wavelengths. At still larger wavenumbers, the spectrum broadens again, going over finally to its free-particle limit.

Measurements of $S(k, \omega)$ can also be made by inelastic (Rayleigh) scattering of laser light (Lallemand, 1974; Berne and Pecora, 1976). There are,

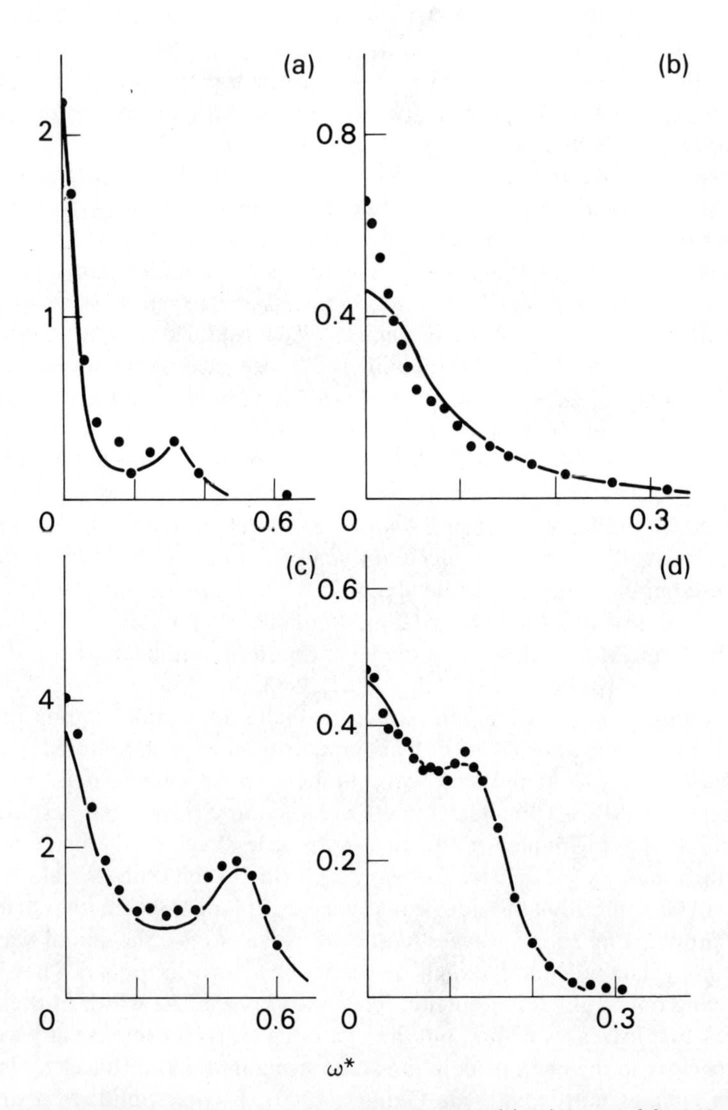

FIG. 7.5. Dynamic structure factor of the hard-sphere fluid for (a) $V/V_0 = 1.6$, $kd = 0.38$; (b) $V/V_0 = 1.6$, $kd = 2.28$; (c) $V/V_0 = 3$, $kd = 0.44$; and (d) $V/V_0 = 10$, $kd = 0.41$ where $V_0 = Nd^3/\sqrt{2}$ is the volume at close packing. The points are molecular dynamics results of Alley *et al.* (1983) and the curves are calculated from the generalized Enskog theory to be discussed in Section 9.9. The unit of frequency is kd/τ, where τ is the mean collision time.

however, some important practical distinctions to be drawn between experiments on the scattering of light and of neutrons. First, light scattering is entirely coherent, so that no information on $G_s(\mathbf{r}, t)$ can be obtained. Secondly, the range of momentum transfer is very different. In slow-neutron scattering experiments, the smallest momentum transfers correspond to wavelengths of the order of the nearest-neighbour separation in the liquid, but in light scattering the wavelengths involved are very much larger, of order 5000 Å. It is therefore possible to calculate the spectral distribution of scattered light from the macroscopic equations of hydrodynamics, as we shall discuss in detail in Chapter 8. Light is scattered by fluctuations of the local dielectric constant $\varepsilon(\mathbf{r}, t)$ of the scattering sample around its macroscopic equilibrium value (Berne and Pecora, 1976). If the system is assumed to be in local equilibrium, the fluctuations in $\varepsilon(\mathbf{r}, t)$ can be expressed in terms of fluctuations in the local density and temperature:

$$\begin{aligned} \delta\varepsilon(\mathbf{r}, t) &= \varepsilon(\mathbf{r}, t) - \varepsilon \\ &= \left(\frac{\partial \varepsilon}{\partial \rho}\right)_T \delta\rho(\mathbf{r}, t) + \left(\frac{\partial \varepsilon}{\partial T}\right)_\rho \delta T(\mathbf{r}, t) \end{aligned} \tag{7.5.26}$$

In most liquids, the variation of the dielectric constant with temperature is negligibly small compared with its density dependence; $\delta\varepsilon$ is therefore directly proportional to the fluctuation in density and the measured spectrum is proportional to $S(k, \omega)$.

In the limit $\mathbf{r}, t \to 0$, the particles in a fluid move as if they were free, with a constant velocity $\mathbf{u}$. These conditions correspond to the limit $k, \omega \to \infty$, where $S(k, \omega)$ behaves in the manner appropriate to an ideal gas. The limiting form of $S(k, \omega)$ is easily derived, since positions of different particles are uncorrelated in an ideal gas ($G_d = \rho$). The calculation of $S(k, \omega)$ is therefore equivalent to a calculation of $G_s(r, t)$, which poses a straightforward problem. The probability that an ideal-gas particle will move a distance r in time t is proportional to the probability of its having a velocity in the range $\mathbf{u}$ to $\mathbf{u} + d\mathbf{u}$, where $\mathbf{u} = \mathbf{r}/t$, as given by the Maxwell–Boltzmann distribution (2.1.25). Thus $G_s(\mathbf{r}, t) \propto \exp(-\frac{1}{2}\beta m r^2/t^2)$ and by exploiting the normalizing condition (7.4.24) we find that

$$G_s(\mathbf{r}, t) = \left(\frac{\beta m}{2\pi t^2}\right)^{3/2} \exp(-\tfrac{1}{2}\beta m r^2/t^2) \tag{7.5.27}$$

The corresponding result for $S(k, \omega)$ is

$$S(k, \omega) = \left(\frac{\beta m}{2\pi k^2}\right)^{1/2} \exp(-\tfrac{1}{2}\beta m \omega^2/k^2) \tag{7.5.28}$$

In practice, Eqn (7.5.28) provides a reasonable fit to data on simple liquids at wavelengths significantly shorter than the spacing between particles, typically for $k \gtrsim 10\ \text{Å}^{-1}$; small deviations from the free-particle limit can be allowed for by calculating the correction to $S(k, \omega)$ due to a single binary collision (Rao, 1974). If quantum effects are too large to be ignored, Eqn (7.5.28) must be replaced by what is called the "impulse" approximation (Sears, 1973). At longer wavelengths, correlations between particles become increasingly important, and the ideal-gas model is no longer valid. Very small values of k correspond to the hydrodynamic régime, where thermodynamic equilibrium is brought about by frequent collisions; this is the opposite extreme to the free-particle limit represented by Eqn (7.5.28).

7.6 LINEAR-RESPONSE THEORY

We turn now to an investigation of the behaviour of a system under the perturbing influence of an external field to which the system is weakly coupled. We shall show that the response of the system can be described entirely in terms of time-correlation functions characteristic of the system at equilibrium, i.e. in the *absence* of the field; the expression already obtained for the inelastic neutron-scattering cross-section in terms of the dynamic structure factor provides an example of this relationship. The derivation of the general result requires only a straightforward calculation of the change produced in a dynamical variable B by an applied space and time-dependent field (or "force") $\mathscr{F}$ conjugate to a variable A (Kubo, 1957, 1966). Both A and B are to be regarded in general as functions of the coordinates and momenta of all particles in the system; the mean value of B in the equilibrium state is assumed to be zero.

The hamiltonian of the system in the presence of the external field is

$$\mathscr{H} = \mathscr{H}_0 + \mathscr{H}'(t) \tag{7.6.1}$$

where $\mathscr{H}_0$ characterizes the unperturbed system and $\mathscr{H}'$ represents the perturbation:

$$\mathscr{H}'(t) = -\int A(\mathbf{r})\mathscr{F}(\mathbf{r}, t)\, d\mathbf{r} \tag{7.6.2}$$

For notational simplicity, we leave undefined the precise form of the coupling between A and $\mathscr{F}$; this is dependent on the particular problem under investigation. The external field can always be considered as a superposition of monochromatic plane waves. Since we are interested in the linear response of the system, it is sufficient to consider a single plane wave:

$$\mathscr{F}(\mathbf{r}, t) = \mathscr{F}_{\mathbf{k}} \exp\left[i(\mathbf{k} \cdot \mathbf{r} - \omega t)\right] \tag{7.6.3}$$

If (7.6.3) is inserted in (7.6.2), we obtain

$$\mathscr{H}'(t) = -A_{\mathbf{k}}^{*} \mathscr{F}_{\mathbf{k}} \exp\left(-i\omega t\right) \tag{7.6.4}$$

As a further simplification, we shall temporarily suppose that the external field is spatially homogeneous and ignore the dependence on $\mathbf{k}$; the latter is trivially reintroduced at a later stage. We also assume that the perturbation vanishes and that the system is in statistical equilibrium in the infinite past ($t = -\infty$). Then $\mathscr{H}'(t)$ may be rewritten as

$$\begin{aligned} \mathscr{H}'(t) &= -A\mathscr{F}(t) \\ &= -A\mathscr{F}_0 \exp\left[-i(\omega + i\varepsilon)t\right] \end{aligned} \tag{7.6.5}$$

where the factor $\exp(\varepsilon t)$ ($\varepsilon > 0$) is included to ensure that $\lim_{t\to-\infty} \mathscr{F}(t) = 0$; the limit $\varepsilon \to 0$ is taken at the end of the calculation. The time evolution of the phase-space probability density $f^{(N)}(t) \equiv f^{(N)}(\mathbf{r}^N, \mathbf{p}^N; t)$ is determined by the Liouville equation (2.1.7). Thus

$$\begin{aligned} \frac{\partial f^{(N)}(t)}{\partial t} &= -i\mathscr{L}f^{(N)}(t) \\ &= \{\mathscr{H}_0 + \mathscr{H}', f^{(N)}(t)\} \\ &= -i\mathscr{L}_0 f^{(N)}(t) - \{A, f^{(N)}(t)\}\mathscr{F}(t) \end{aligned} \tag{7.6.6}$$

where $\mathscr{L}_0$ is the Liouville operator corresponding to the unperturbed hamiltonian. Equation (7.6.6) must be solved subject to the initial condition

$$\begin{aligned} f^{(N)}(-\infty) &= f_0^{(N)} \\ &= C \exp\left(-\beta\mathscr{H}_0\right) \end{aligned} \tag{7.6.7}$$

where the normalization constant C is the same as that in Eqn (2.3.1).

We are interested only in the response to a weak external field. We therefore write the probability density as

$$f^{(N)}(t) = f_0^{(N)} + \Delta f^{(N)}(t) \tag{7.6.8}$$

and linearize Eqn (7.6.6) in the form

$$\frac{\partial}{\partial t} \Delta f^{(N)}(t) = -i\mathscr{L}_0 \Delta f^{(N)}(t) - \{A, f_0^{(N)}\}\mathscr{F}(t) \tag{7.6.9}$$

Equation (7.6.9) can be integrated immediately to give

$$\Delta f^{(N)}(t) = -\int_{-\infty}^{t} \exp\left[-i(t-s)\mathscr{L}_0\right]\{A, f_0^{(N)}\}\mathscr{F}(s)\,\mathrm{d}s \qquad (7.6.10)$$

The mean change in the variable B resulting from the change in the distribution function is therefore given by

$$\begin{aligned}\langle \Delta B(t)\rangle &= \iint \Delta f^{(N)}(t) B(\mathbf{r}^N, \mathbf{p}^N)\,\mathrm{d}\mathbf{r}^N\,\mathrm{d}\mathbf{p}^N \\ &= -\iint \mathrm{d}\mathbf{r}^N\,\mathrm{d}\mathbf{p}^N \int_{-\infty}^{t} \mathrm{d}s \exp\left[-i(t-s)\mathscr{L}_0\right]\{A, f_0^{(N)}\}B\mathscr{F}(s) \\ &= -\iint \mathrm{d}\mathbf{r}^N\,\mathrm{d}\mathbf{p}^N \int_{-\infty}^{t} \mathrm{d}s\{A, f_0^{(N)}\} \exp\left[i(t-s)\mathscr{L}_0\right]B\mathscr{F}(s) \\ &= -\iint \mathrm{d}\mathbf{r}^N\,\mathrm{d}\mathbf{p}^N \int_{-\infty}^{t} \mathrm{d}s\,\{A, f_0^{(N)}\}B(t-s)\mathscr{F}(s) \qquad (7.6.11)\end{aligned}$$

where

$$\begin{aligned}B(t) &\equiv B[\mathbf{r}^N(t), \mathbf{p}^N(t)] \\ &= \exp(i\mathscr{L}_0 t)B(\mathbf{r}^N, \mathbf{p}^N) \qquad (7.6.12)\end{aligned}$$

In going from the second to the third lines of (7.6.11) we have used the hermitian property of the Liouville operator with respect to the inner product of $\{A, f_0^{(N)}\}$ and B. Equation (7.6.11) can be rewritten as

$$\langle \Delta B(t)\rangle = \int_{-\infty}^{t} \Phi_{BA}(t-s)\mathscr{F}(s)\,\mathrm{d}s \qquad (7.6.13)$$

in terms of an *after-effect function* $\Phi_{BA}(t)$, defined as

$$\Phi_{BA}(t) = -\iint \{A, f_0^{(N)}\}B(t)\,\mathrm{d}\mathbf{r}^N\,\mathrm{d}\mathbf{p}^N \qquad (7.6.14)$$

It is sometimes convenient to use as an alternative definition of the after-effect function the expression

$$\theta_{BA}(t) = \Phi_{BA}(t)\theta(t) \qquad (7.6.15)$$

where $\theta(t)$ is the Heaviside step-function. Since $\theta_{BA}(t) = 0$ for $t < 0$, the upper limit of the integral in (7.6.13) can then be extended to $+\infty$.

The physical meaning of (7.6.13) and (7.6.14) is that the response at time t, i.e. the change in the variable B, is a superposition of delayed effects, and the response at time t to a unit delta-function force applied at $t=0$ is simply the after-effect function $\Phi_{BA}(t)$ itself (Kubo, 1957). The definition of $\Phi_{BA}(t)$ can be put in more tractable form by recognizing that

$$\begin{aligned}\{A, f_0^{(N)}\} &= \sum_{i=1}^{N} \left(\frac{\partial A}{\partial \mathbf{r}_i} \cdot \frac{\partial f_0^{(N)}}{\partial \mathbf{p}_i} - \frac{\partial A}{\partial \mathbf{p}_i} \cdot \frac{\partial f_0^{(N)}}{\partial \mathbf{r}_i} \right) \\ &= -\beta \sum_{i=1}^{N} \left(\frac{\partial A}{\partial \mathbf{r}_i} \cdot \frac{\partial \mathcal{H}_0}{\partial \mathbf{p}_i} - \frac{\partial A}{\partial \mathbf{p}_i} \cdot \frac{\partial \mathcal{H}_0}{\partial \mathbf{r}_i} \right) f_0^{(N)} \\ &= -\beta (i\mathcal{L}_0 A) f_0^{(N)} \\ &= -\beta \dot{A} f_0^{(N)} \end{aligned} \tag{7.6.16}$$

The after-effect function is therefore given by

$$\begin{aligned}\Phi_{BA}(t) &= \beta \langle B(t) \dot{A} \rangle \\ &= -\beta \langle \dot{B}(t) A \rangle \end{aligned} \tag{7.6.17}$$

where the statistical averages are taken over the unperturbed system. Finally, a more symmetric expression for $\Phi_{BA}(t)$ can be derived from (7.6.14) by straightforward integration by parts:

$$\Phi_{BA}(t) = -\langle \{B(t), A\} \rangle \tag{7.6.18}$$

In this form, the definition of the after-effect function can be immediately extended to the quantum-mechanical case by replacing the Poisson bracket by the commutator $(-i/\hbar)[B(t), A]$.

The basic result of linear-response theory embodied in Eqns (7.6.13) and (7.6.14) can also be obtained by calculating the changes in the trajectories of the particles in phase space to linear order in the applied force (Hubbard and Beeby, 1969; Schofield, 1975). This method of derivation emphasizes the assumption of mechanical linearity that underlies linear-response theory. Mechanical linearity cannot hold for macroscopic times, however, since it is known that the perturbed and unperturbed phase-space trajectories deviate exponentially from each other on a macroscopic timescale under the influence of even a weak external field (Stoddard and Ford, 1973). On the other hand, the corresponding deviation in the phase-space probability density is expected to behave smoothly as a function of the perturbation. Linearization of the statistically averaged response should therefore be justified, in agreement with experimental observation. The apparent contradiction between mechanical non-linearity and statistical linearity can be resolved by noting that the decay times of the relevant correlations, i.e. the times after which randomization sets in, are generally quite short, and that the linear approximation for the deviations of the trajectories in phase space is valid for time intervals over which the after-effect function $\Phi_{BA}(t)$ differs significantly from zero (van Kampen, 1971; Visscher, 1974).

If the field $\mathscr{F}(t)$ in (7.6.13) has a monochromatic form, the expression for the response becomes

$$\begin{aligned}\langle \Delta B(t) \rangle &= \int_{-\infty}^{t} \Phi_{BA}(t-s)\mathscr{F}_0 \exp\left[-i(\omega+i\varepsilon)s\right] \mathrm{d}s \\ &= \mathscr{F}_0 \exp\left[-i(\omega+i\varepsilon)t\right] \int_{-\infty}^{t} \Phi_{BA}(t-s) \exp\left[-i(\omega+i\varepsilon)(s-t)\right] \mathrm{d}s \\ &= \mathscr{F}_0 \exp\left[-i(\omega+i\varepsilon)t\right] \int_{0}^{\infty} \Phi_{BA}(t) \exp\left[i(\omega+i\varepsilon)t\right] \mathrm{d}t \qquad (7.6.19)\end{aligned}$$

or, taking the limit $\varepsilon \to 0$:

$$\langle \Delta B(t) \rangle = \chi_{BA}(\omega)\mathscr{F}_0 \exp(-i\omega t) \qquad (7.6.20)$$

where $\chi_{BA}(\omega)$ is a complex *dynamic susceptibility* or *response function*:

$$\begin{aligned}\chi_{BA}(\omega) &= \chi'_{BA}(\omega) + i\chi''_{BA}(\omega) \\ &= \lim_{\varepsilon\to 0+} \int_{0}^{\infty} \Phi_{BA}(t) \exp\left[i(\omega+i\varepsilon)t\right] \mathrm{d}t \qquad (7.6.21)\end{aligned}$$

The response function in (7.6.21) can be interpreted as the limit of a Laplace transform $\Phi_{BA}(z)$ defined in the whole upper half of the complex plane ($\mathrm{Im}\, z > 0$):

$$\chi_{BA}(z) = \int_{0}^{\infty} \Phi_{BA}(t) \exp(izt)\, \mathrm{d}t \qquad (7.6.22)$$

Equation (7.6.13) is eaily generalized to the case in which the external field depends both on space and time. If the unperturbed system is spatially uniform, the response can be expressed with the help of an after-effect function $\Phi_{BA}(\mathbf{r}, t)$ as

$$\langle \Delta B(\mathbf{r}, t) \rangle = \int_{-\infty}^{t} \mathrm{d}s \int \Phi_{BA}(\mathbf{r}-\mathbf{r}', t-s)\mathscr{F}(\mathbf{r}', s)\, \mathrm{d}\mathbf{r}' \qquad (7.6.23)$$

or, in terms of spatial Fourier components, as

$$\langle \Delta B_{\mathbf{k}}(t) \rangle = \int_{-\infty}^{t} \Phi_{BA}(\mathbf{k}, t-s)\mathscr{F}_{\mathbf{k}}(s)\, \mathrm{d}s \qquad (7.6.24)$$

where

$$\Phi_{BA}(\mathbf{k}, t) = -\langle \{B_{\mathbf{k}}(t), A_{-\mathbf{k}}\} \rangle \qquad (7.6.25)$$

Equation (7.6.24) shows that a Fourier component of given wavevector can induce in the system a response only of the same wavevector; this is a consequence of the assumed uniformity of the unperturbed system and the property (7.4.11). The corresponding wavenumber and frequency-dependent dynamic susceptibility is the Laplace transform of the **k**-dependent after-effect function (7.6.25). For an isotropic fluid:

$$\chi_{BA}(k, z) = \int_0^\infty \Phi_{BA}(k, t) \exp(izt) \, \mathrm{d}t \tag{7.6.26}$$

The dynamic susceptibility is directly related to the Laplace transform $\tilde{C}_{BA}(k, z)$ of the time-correlation function of the variables A and B. If we substitute in (7.6.26) from (7.6.17) and integrate by parts, we find that

$$\chi_{BA}(k, z) = \beta[C_{BA}(k, t=0) + iz\tilde{C}_{BA}(k, z)] \tag{7.6.27}$$

When $B = A$, it follows from Eqns (7.1.21) and (7.6.27) (with $z = \omega + i\varepsilon$, $\varepsilon \to 0+$) that

$$\begin{aligned} C_{AA}(k, \omega) &= \lim_{\varepsilon \to 0+} \frac{1}{\pi} \operatorname{Re} \tilde{C}_{AA}(k, \omega + i\varepsilon) \\ &= \frac{k_B T}{\pi \omega} \chi''_{AA}(k, \omega) \end{aligned} \tag{7.6.28}$$

Equation (7.6.28) is a particular form of the fluctuation-dissipation theorem; indeed, the name is often applied specifically to this relation between the spectrum of the autocorrelation function of a dynamical variable and the imaginary part of the corresponding dynamic susceptibility. Use of the term "dissipation" is connected with the fact, well known in spectroscopy, that the energy absorbed from the external field and later dissipated as heat is proportional to $\omega\chi''_{AA}(k, \omega)$. The proof is straightforward. The time derivative of the internal energy of a system interacting with an external field $\mathcal{F}(t)$ coupled to the variable A is

$$\begin{aligned} \frac{\mathrm{d}U}{\mathrm{d}t} &= \frac{\mathrm{d}}{\mathrm{d}t} \iint \mathrm{d}\mathbf{r}^N \, \mathrm{d}\mathbf{p}^N f^{(N)}(t)[\mathcal{H}_0 - A\mathcal{F}(t)] \\ &= \iint \mathrm{d}\mathbf{r}^N \, \mathrm{d}\mathbf{p}^N \frac{\partial f^{(N)}}{\partial t} [\mathcal{H}_0 - A\mathcal{F}(t)] \\ &\quad - \frac{\partial \mathcal{F}(t)}{\partial t} \iint \mathrm{d}\mathbf{r}^N \, \mathrm{d}\mathbf{p}^N f^{(N)}(t) A \\ &= \iint \mathrm{d}\mathbf{r}^N \, \mathrm{d}\mathbf{p}^N \{\mathcal{H}_0 - A\mathcal{F}(t), f^{(N)}(t)\}[\mathcal{H}_0 - A\mathcal{F}(t)] - \frac{\partial \mathcal{F}(t)}{\partial t} \langle A(t) \rangle \\ &= -\frac{\partial \mathcal{F}(t)}{\partial t} \langle A(t) \rangle \end{aligned} \tag{7.6.29}$$

where we have omitted the k-dependence of the external field, since this is irrelevant here. We have also used the identity

$$\iint \{A, B\}C \, \mathrm{d}\mathbf{r}^N \, \mathrm{d}\mathbf{p}^N = \iint \{C, A\}B \, \mathrm{d}\mathbf{r}^N \, \mathrm{d}\mathbf{p}^N \tag{7.6.30}$$

which is easily proved by an integration by parts. The energy absorption per period $\tau = 2\pi/\omega$ of the external field (i.e. the absorbed power) is

$$P = \Delta U/\tau = \frac{1}{\tau}\int_0^\tau \frac{\mathrm{d}U}{\mathrm{d}t}\mathrm{d}t = -\frac{1}{\tau}\int_0^\tau \langle A(t)\rangle \frac{\partial \mathscr{F}(t)}{\partial t}\,\mathrm{d}t$$

$$= -\frac{1}{\tau}\int_0^\tau \mathrm{d}t \, \mathrm{Re}\,\chi_{AA}(\omega)\mathscr{F}_0 \exp(-i\omega t)[i\omega\mathscr{F}_0 \exp(i\omega t) - i\omega\mathscr{F}_0 \exp(-i\omega t)]$$

$$= 2\omega\mathscr{F}_0^2\chi''_{AA}(\omega) \tag{7.6.31}$$

where the external field has been made real by adding the complex conjugate to the right-hand side of Eqn (7.6.5). The positivity of P implies that $\omega\chi''_{AA}(\omega) > 0$, in agreement with Eqn (7.6.28) and the inequality (7.1.24).

The properties of the after-effect function Φ_{BA} follow directly from its definition (7.6.17) (or (7.6.25)) and the corresponding properties of time correlation functions. Restricting the discussion to isotropic fluids, we see from Eqns (7.1.10) and (7.6.17) that for $A \neq B$:

$$\Phi_{BA}(k, t) = -\varepsilon_A\varepsilon_B\Phi_{AB}(k, t) \tag{7.6.32}$$

where we have used the fact that $\varepsilon_B = -\varepsilon_{\dot{B}}$. Equation (7.6.32) is an expression of the Onsager *reciprocity relations.* If A and B are real, Φ_{BA} is a real function, and from the definition (7.6.21) we see that on the real axis

$$\chi_{BA}(k, -\omega) = \chi^*_{BA}(k, \omega) = \chi'_{BA}(k, \omega) - i\chi''_{BA}(k, \omega) \tag{7.6.33}$$

Hence the real and imaginary parts of χ_{BA} are, respectively, even and odd functions of frequency.

7.7 PROPERTIES OF THE RESPONSE FUNCTIONS

In this section, we discuss some of the mathematical properties of the dynamic susceptibilities defined in the complex-frequency domain by Eqn (7.6.26). We shall be concerned only with the important special case where the variables A and B are the same. We may therefore temporarily discard the subscripts and consider the behaviour of the susceptibility $\chi(k, z) \equiv \chi_{AA}(k, z)$ as a function of the complex variable $z = \omega + i\varepsilon$, with $\varepsilon > 0$. By restricting ε to positive values, we ensure that $\chi(k, z)$ is analytic in the upper half complex plane of the variable z, but is undefined in the lower half-plane because the integral in Eqn (7.6.26) diverges. Since Eqn (7.6.17) implies that the after-effect function (with $A = B$) is linear in t at short times, it follows that $\chi(k, z)$ behaves asymptotically as z^{-2} at large z (Lighthill, 1958).

We first derive some useful integral representations of $\chi(k, z)$ (Martin, 1968). Let the contour C in the complex plane be $C = C_1 + C_2$, where C_1 is the real axis and C_2 is the infinite semicircle in the upper half-plane. Application of Cauchy's integral formula shows that

$$\chi(k, z) = \frac{1}{2\pi i}\int_C \frac{\chi(k, z')}{z' - z}\,\mathrm{d}z' \tag{7.7.1}$$

where z is any point inside C. On the other hand, because the complex-conjugate variable z^* lies outside C, the function $\chi(k, z')/(z' - z^*)$ is analytic in and on the contour C. It follows from Cauchy's theorem that

$$\int_C \frac{\chi(k, z')}{z' - z^*}\,\mathrm{d}z' = 0 \tag{7.7.2}$$

The contribution to the integrals (7.7.1) and (7.7.2) from the contour C_2 are both zero, because $\chi(k, z)$ vanishes rapidly as $z \to \infty$. By adding quantities that are zero to the right-hand side of Eqn (7.7.1) and discarding the integral around C_2, $\chi(k, z)$ can be expressed as either

$$\chi(k, z) = \frac{1}{2\pi i}\int_{C_1} \chi(k, z')\left(\frac{1}{z' - z} + \frac{1}{z' - z^*}\right)\mathrm{d}z' \tag{7.7.3}$$

or

$$\chi(k, z) = \frac{1}{2\pi i}\int_{C_1} \chi(k, z')\left(\frac{1}{z' - z} - \frac{1}{z' - z^*}\right)\mathrm{d}z' \tag{7.7.4}$$

Two further expressions for $\chi(k, z)$ are obtained by adding the real part of (7.7.3) to i times the imaginary part of (7.7.4) and vice versa:

$$\chi(k, z) = \frac{1}{\pi}\int_{-\infty}^{\infty} \frac{\chi''(k, \omega)}{\omega - z}\,\mathrm{d}\omega \tag{7.7.5}$$

$$\chi(k, z) = \frac{1}{\pi i}\int_{-\infty}^{\infty} \frac{\chi'(k, \omega)}{\omega - z}\,\mathrm{d}\omega \tag{7.7.6}$$

We now let $\varepsilon \to 0$ in Eqn (7.7.5), so that $\chi(k, \omega + i\varepsilon) \to \chi'(k, \omega) + i\chi''(k, \omega)$, and use the identity (7.1.20). Then

$$\chi(k,\omega)=\mathscr{P}\frac{1}{\pi}\int_{-\infty}^{\infty}\frac{\chi''(k,\omega')}{\omega'-\omega}\,d\omega'+i\chi''(k,\omega) \tag{7.7.7}$$

Equation (7.7.7) incorporates the Kramers–Kronig relation for $\chi'(k,\omega)$ in terms of $\chi''(k,\omega)$; the inverse relation, obtained by applying the rule (7.1.20) to Eqn (7.7.6), is

$$\chi''(k,\omega)=-\mathscr{P}\frac{1}{\pi}\int_{-\infty}^{\infty}\frac{\chi'(k,\omega')}{\omega'-\omega}\,d\omega' \tag{7.7.8}$$

The analytic properties of the inverse susceptibility $\chi^{-1}(k,z)$ are also of interest. The fact that $\chi''(k,\omega)$ is an odd function of ω allows us to rewrite Eqn (7.7.5) as

$$\begin{aligned}\chi(k,z)&=\frac{1}{\pi}\int_{-\infty}^{\infty}\frac{1}{2}\left(\frac{\chi''(k,\omega)}{\omega-z}+\frac{\chi''(k,-\omega)}{-\omega-z}\right)d\omega\\&=\frac{1}{\pi}\int_{-\infty}^{\infty}\frac{\omega\chi''(k,\omega)}{\omega^2-z^2}\,d\omega\end{aligned} \tag{7.7.9}$$

If we set $z^2=u+iv$, we find from (7.7.9) that

$$\operatorname{Im}\chi(k,z)=\frac{1}{\pi}\int_{-\infty}^{\infty}\frac{v\omega\chi''(k,\omega)}{(\omega^2-u^2)+v^2}\,d\omega \tag{7.7.10}$$

The positivity of $\omega\chi''(k,\omega)$ implied by (7.6.31) shows that Im (k,z) vanishes only if $v=0$, in which case z is purely imaginary. But from Eqn (7.7.9) it follows that on the upper half imaginary axis $\chi(k,z)$ is not only real but also necessarily non-zero. We therefore conclude that $\chi(k,z)$ has no zeros for Im $z>0$, so that $\chi^{-1}(k,z)$ shares with the susceptibility itself the property of being analytic throughout the upper half-plane (Kadanoff and Martin, 1963). It should also be noted that the spectral representation of $\chi(k,z)$ given by (7.7.5) allows an extension of the complex susceptibility into the lower half-plane. Equation (7.7.5) shows that

$$\chi(k,\omega-i\varepsilon)=\chi^*(k,\omega+i\varepsilon) \tag{7.7.11}$$

Hence the discontinuity in $\chi(k,z)$ across the real axis is related to $\chi''(k,\omega)$ by

$$\chi''(k,\omega)=\lim_{\varepsilon\to 0}\frac{1}{2i}[\chi(k,\omega+i\varepsilon)-\chi(k,\omega-i\varepsilon)] \tag{7.7.12}$$

The response function $\chi_{AA}(k,z)$ has an asymptotic, high-frequency expansion that is linked to the short-time expansion of the corresponding after-effect function. Equation (7.7.9) shows that $\chi_{AA}(k, z)$ may be written as

$$\chi_{AA}(k, z) = -\sum_{n=1}^{\infty} \frac{a_{2n}}{z^{2n}} \tag{7.7.13}$$

A general expression for the coefficients a_{2n} in (7.7.13) can be derived from the fluctuation-dissipation theorem (7.6.28) and the definition (7.1.26)

$$\begin{aligned} a_{2n} &= \int_{-\infty}^{\infty} \frac{d\omega}{\pi} \chi''_{AA}(k, \omega)\omega^{2n-1} \\ &= \beta\langle\omega^{2n}\rangle_{AA} \end{aligned} \tag{7.7.14}$$

Only even terms contribute to the expansion (7.7.13), since $\chi''_{AA}(k, \omega)$ is odd in ω. The expansion is asymptotic in the sense that it is valid only if $|z|$ is large in comparison with all characteristic frequencies of the system.

The zero-frequency limit of $\chi_{AA}(k, z)$, or *static susceptibility* $\chi_{AA}(k)$, is obtained from (7.7.5) or (7.7.9) as

$$\begin{aligned} \chi_{AA}(k) &\equiv \chi_{AA}(k, z=0) \\ &= \int_{-\infty}^{\infty} \frac{d\omega}{\pi} \frac{\chi''(k, \omega)}{\omega} d\omega \\ &= \beta \int_{-\infty}^{\infty} C_{AA}(k, \omega)\, d\omega \\ &= \beta C_{AA}(k, t=0) \end{aligned} \tag{7.7.15}$$

This result has a simple physical significance in the case when A is the local density. Let $\Phi_{\mathbf{k}} \exp(-i\omega t)$ be a Fourier component of a complex external potential that couples to the component $\rho_{\mathbf{k}}$ of the particle density. Use of Eqn (7.6.20) shows that the resulting change in density is

$$\langle\rho_{\mathbf{k}}(t)\rangle = \chi_{\rho\rho}(k, \omega)\Phi_{\mathbf{k}} \exp(-i\omega t) \tag{7.7.16}$$

which is a generalization to non-zero frequencies of the static result (5.2.6). The imaginary part of the density response function $\chi_{\rho\rho}(k, \omega)$ is related to the dynamic structure factor through the fluctuation-dissipation theorem (7.6.28); in this particular case, the latter is often written as

$$\chi''_{\rho\rho}(k, \omega) = -\pi\beta\rho\omega S(k, \omega) \tag{7.7.17}$$

Other forms of the relation between $\chi_{\rho\rho}(k, \omega)$ and $S(k, \omega)$ are to be found in the literature; these differ from (7.7.17) by a sign and/or the absence of the factor ρ. The present form is chosen so as to be consistent with the corresponding static result, Eqn (5.2.12); the change in sign relative to (7.6.28) arises because the density response function is defined in terms of an external potential rather than a force. The static susceptibility is then obtained from (7.7.15) but with a factor $-\rho$ inserted, i.e.

$$\begin{aligned}\chi(k) &\equiv \chi_{\rho\rho}(k, z=0) \\ &= -\beta\rho \int_{-\infty}^{\infty} S(k, \omega)\, \mathrm{d}\omega \\ &= -\beta\rho S(k) \qquad (7.7.18)\end{aligned}$$

in agreement with (5.2.12). Equations (5.1.13) and (5.2.12) show that $\chi(0) = -\rho^2\chi_T$, where χ_T is the isothermal compressibility. The compressibility provides a measure of the macroscopic, static response of a fluid to a change in the applied pressure, and the density response function represents a generalization of the same concept to finite wavelengths and non-zero frequencies.

7.8 APPLICATIONS OF THE LINEAR-RESPONSE FORMALISM

The best known and most important of the applications of linear-response theory is in the derivation of expressions for the transport coefficients of hydrodynamics, through which induced fluxes are related to certain gradients. The simplest example concerns the mobility of a tagged particle under the action of a constant external force $\mathscr{F}$; it is assumed that the force is applied from $t = 0$ onwards, and that it does not affect the other particles of the system. If the force is along the x-direction, the perturbation term in the hamiltonian is $\mathscr{H}' = -\mathscr{F}x_1\theta(t)$, where x_1 is the x-coordinate of the particle under consideration. If the fluid is isotropic, the drift velocity $\mathbf{u}_1$ of the particle will be in the same direction as the applied force. From Eqns (7.6.13) and (7.6.17) it follows that

$$\begin{aligned}\langle u_{1x}(t)\rangle &= \beta \int_{-\infty}^{t} \langle u_{1x}(t')\dot{x}_1\rangle \mathscr{F}\theta(t')\, \mathrm{d}t' \\ &= \beta\mathscr{F} \int_{0}^{t} \langle u_{1x}(t')u_{1x}\rangle\, \mathrm{d}t' \qquad (7.8.1)\end{aligned}$$

This leads to the Einstein relation for the mobility, i.e.

$$\begin{aligned}\mu &= \lim_{t\to\infty} \beta \int_{0}^{t} \langle u_{1x}(t')u_{1x}\rangle\, \mathrm{d}t' \\ &= \frac{D}{k_B T} \qquad (7.8.2)\end{aligned}$$

where D is the self-diffusion constant given by Eqn (7.2.8). Equation (7.8.2) is a manifestation of the fluctuation-dissipation theorem, since D is a quantity that characterizes spontaneous fluctuations in the fluid and μ is a measure of the response of the system to an applied force.

It is instructive to consider an alternative derivation of Eqn (7.8.2). Let $\rho_{\mathbf{k}}^{(s)}$ be a Fourier component of the density of tagged particles, and let $\mathbf{j}_{\mathbf{k}}^{(s)}$ be the associated current; the two quantities are related by the continuity equation (7.4.4). If the particles are subjected to an external force derived from a potential $\phi_{\mathbf{k}} \exp(\varepsilon t)$ $(\varepsilon > 0)$, a concentration gradient is set up and the induced current satisfies Fick's law (see Section 8.2). Thus

$$\langle \mathbf{j}_{\mathbf{k}}^{(s)}(t) \rangle = -i\mu\rho_s \mathbf{k}\phi_{\mathbf{k}} \exp(\varepsilon t) - iD\mathbf{k}\langle \rho_{\mathbf{k}}^{(s)}(t) \rangle \tag{7.8.3}$$

where ρ_s is the number of tagged particles per unit volume; since Eqns (7.4.4) and (7.8.3) are linear, the time dependence of the induced density and current is the same as that of the applied field. If the applied field is turned on sufficiently slowly, i.e. if $\varepsilon \ll Dk^2$, the system has time to readjust and is therefore always in a state of equilibrium. Then $\langle \mathbf{j}_{\mathbf{k}}^{(s)} \rangle = 0$ and Eqn (7.8.3) reduces to

$$\langle \rho_{\mathbf{k}}^{(s)} \rangle = -\frac{\mu\rho_s}{D}\phi_{\mathbf{k}} \tag{7.8.4}$$

The induced density is calculated by averaging over a canonical distribution to give

$$\langle \rho_{\mathbf{k}}^{(s)} \rangle = \frac{\iint \exp[-\beta(\mathcal{H}_0 + \mathcal{H}')]\rho_{\mathbf{k}}^{(s)} \, d\mathbf{r}^N \, d\mathbf{p}^N}{\int \exp[-\beta(\mathcal{H}_0 + \mathcal{H}')] \, d\mathbf{r}^N \, d\mathbf{p}^N} \tag{7.8.5}$$

where the coupling to the external potential is represented by $\mathcal{H}' = \rho_{-\mathbf{k}}^{(s)}\phi_{\mathbf{k}}/V$. Linearizing (7.8.5) with respect to $\phi_{\mathbf{k}}$, and assuming that the tagged particles do not interact, we find that

$$\begin{aligned} \langle \rho_{\mathbf{k}}^{(s)} \rangle &= \langle \rho_{\mathbf{k}}^{(s)} \rangle_0 - \frac{\beta}{V}\langle \rho_{-\mathbf{k}}^{(s)}\rho_{\mathbf{k}}^{(s)} \rangle_0 \phi_{\mathbf{k}} \\ &= -\beta\rho\phi_{\mathbf{k}} \end{aligned} \tag{7.8.6}$$

where the subscript 0 denotes a canonical average over the unperturbed distribution ($\mathcal{H}' = 0$). Combination of (7.8.4) and (7.8.6) leads back to the Einstein result (7.8.2).

The calculation of electrical conductivity provides an example of a different type, in which a collective response of a system to an external field is involved. Suppose that a uniform, time-dependent electric field $\mathbf{E}(t)$ is applied to a system of charged particles. The field gives rise to a charge current, defined as

$$e\mathbf{j}^Z(t) = \sum_{i=1}^{N} z_i e \dot{\mathbf{r}}_i(t) \tag{7.8.7}$$

where $z_i e$ is the charge carried by the ith particle (e is the elementary charge). The interaction with the applied field is described by the hamiltonian

$$\mathcal{H}'(t) = -\sum_{i=1}^{N} z_i e \mathbf{r}_i \cdot \mathbf{E}(t) \tag{7.8.8}$$

If the system is isotropic and the field is applied, say, along the x-axis, then, in the statistical mean, only the x-component of the induced current will survive. The linear response to a real periodic field can therefore be written as

$$e\langle j_x^Z(t)\rangle = \text{Re}\ \sigma(\omega) E_0 \exp(-i\omega t) \tag{7.8.9}$$

where, according to the general formulae (7.6.17) and (7.6.21), the electrical conductivity per unit volume is given by

$$\begin{aligned}\sigma(\omega) &= \frac{\beta e}{V} \int_0^\infty \left\langle j_x^Z(t) \sum_{i=1}^{N} z_i e \dot{x}_i \right\rangle \exp(i\omega t)\, dt \\ &= \frac{\beta e^2}{V} \int_0^\infty \langle j_x^Z(t) j_x^Z\rangle \exp(i\omega t)\, dt\end{aligned} \tag{7.8.10}$$

The usual static electrical conductivity σ is then identified as

$$\sigma = \lim_{\omega \to 0} \sigma(\omega) \tag{7.8.11}$$

The bracketed quantity in the second line of (7.8.10) is the autocorrelation function of the fluctuating charge current in the absence of the electric field. In deriving this result, we have ignored any spatial variation of the electric field, thereby avoiding the difficulties that arise in taking the long-wavelength limit for coulombic systems; we shall return to a discussion of this problem in Chapter 10.

Correlation-function formulae for transport coefficients have been obtained by many authors in a variety of ways. The derivation from linear-response theory is not always as straightforward as it is in the case of electrical conductivity. The difficulty lies in the fact that the dissipative behaviour described by hydrodynamics is generally not induced by external forces, but by gradients of local thermodynamic variables that cannot be represented by a perturbation term in the hamiltonian (Luttinger, 1964; Jackson and Mazur, 1964; Mazur, 1966). The thermal conductivity provides a good example. This is the transport coefficient that relates the induced heat flux to an applied temperature gradient via Fourier's law. A temperature gradient is a manifestation of boundary conditions and obviously cannot be formulated in hamiltonian (or mechanical) terms, since the temperature is a statistical property of the system. A linear-response argument can still be invoked by introducing an inhomogeneous gravitational field that couples to the energy density of the system and sets up a heat flow; Einstein's argument, which relates the diffusion coefficient to the mobility (see Eqns (7.8.3) to (7.8.6)), can then be extended to yield a correlation-function expression for the thermal conductivity (Luttinger, 1964). We postpone a derivation of the microscopic expressions for the thermal conductivity and viscosity coefficients to Chapter 8, where it is shown that both these coefficients are related to the long-wavelength, low-frequency (or "hydrodynamic") limit of certain space and time-dependent correlation functions (Helfand, 1960; Kadanoff and Martin, 1963). The various methods of deriving correlation-function formulae are discussed in a review article by Zwanzig (1965).

The response to a weak, applied field can be measured directly in a molecular dynamics simulation in a way that allows the accurate calculation of transport coefficients with relatively modest computational effort. To understand what is involved, we return to the problem of the electrical conductivity. Clearly we could hope to mimic a real experiment by applying a steady electrical field to the particles of the system and computing the steady-state electric current to which the field gives rise. The same principle has been used with some success in the calculation of the shear viscosity of the Lennard-Jones fluid (Gosling *et al.*, 1973), but its practical value is limited by the fact that a very large field must be applied in order to produce a systematic response that is significantly larger than the statistical noise. Use of a large field leads to a rapid heating-up of the system, non-conservation of energy and other undesirable effects.

The problems associated with the use of large fields can be overcome either by steady-state computer simulations in which constraints are imposed that maintain the system at constant kinetic energy (Hoover, 1983), or by a "subtraction" technique closely related to linear-response theory (Ciccotti and Jacucci, 1975; Ciccotti *et al.*, 1979). In the latter case, the response is computed as the difference along two phase-space trajectories; both start from the same phase point at time $t=0$, but in one case a very small perturbing force is applied. In the example of electrical conduction, the response is the difference in electric current after a time t. For given initial conditions, the difference in mechanical response is

$$\Delta j_x^Z(t) = \exp(i\mathscr{L}t)\, j_x^Z - \exp(i\mathscr{L}_0 t)\, j_x^Z \tag{7.8.12}$$

where $\mathscr{L}$ and $\mathscr{L}_0$ are the Liouville operators that determine, respectively, the perturbed and unperturbed trajectories. The statistical response is obtained by averaging (7.8.12) over initial conditions:

$$\langle \Delta j_x^Z(t) \rangle = \int\int [\exp(i\mathscr{L}t) - \exp(i\mathscr{L}_0 t)] j_x^Z f_0^{(N)} \, \mathbf{dr}^N \, \mathbf{dp}^N$$

$$= \langle j_x^Z(t) \rangle_{\mathscr{L}} - \langle j_x^Z(t) \rangle_{\mathscr{L}_0} \qquad (7.8.13)$$

where the brackets denote averages over the unperturbed equilibrium distribution function and the nature of the mechanical evolution is indicated by the subscripts $\mathscr{L}$ and $\mathscr{L}_0$. Use of this procedure yields the dynamical response we wish to measure with an accuracy that is high if t is not too large. The reason for the high accuracy is the fact that the random fluctuations in the two terms in Eqn (7.8.13) are highly correlated and therefore largely cancel, leaving only the systematic part, i.e. the response to the perturbation. It is therefore possible to use a perturbing force that is very small. In principle, because of the reflection symmetry of the system in the absence of the perturbation, the second term in (7.8.13) should vanish. In practical molecular dynamics simulations, the term may be non-zero, because the average is only taken over a limited sample, typically 50 to 100 trajectories. The form of the statistical response depends on the time-dependence of the applied field. From Eqns (7.8.9) to (7.8.11) we see that if a constant electric field is applied in the x-direction from $t = 0$ onwards, the mean response is proportional to the integral of the current autocorrelation function and the conductivity can be calculated from the plateau value of the mean induced current:

$$\sigma = \frac{e}{EV} \lim_{t \to \infty} \langle \Delta j_x^Z(t) \rangle \qquad (7.8.14)$$

If a delta-function force is applied at $t = 0$, acting in opposite sense on charges of opposite sign, the response is the charge-current autocorrelation function itself. It is, of course, essential that the length of the trajectories should exceed the appropriate relaxation time of the system, in this case the lifetime of spontaneous fluctuations in the electric current. The method is easily generalized to the calculation of wavenumber-dependent currents of number, mass or charge. Results have been obtained in this way for the electrical conductivity of a number of molten alkali halides (Ciccotti *et al.*, 1976) and for the mobility of Ar^+ ions in a non-conducting liquid (Ciccotti and Jacucci, 1975). The subtraction technique has also been used by Gillan and Dixon (1983) in a calculation of the autocorrelation function of the microscopic heat current at zero wavenumber (see Eqn (8.5.26)) and hence of the thermal conductivity. These authors measured the response of the heat current to a weak external field coupled to the energy density and showed that to linear order the effect of the perturbation is to add an extra term to the force acting on each particle; their equations of motion can also be derived from an extension of linear-response theory to non-canonical systems, which do not satisfy the Liouville equation (2.1.7) in its usual form (Evans, 1982a). The use of "non-equilibrium" molecular dynamics methods and their application to the calculation of shear viscosity are discussed in more detail in Section 8.4.

Linear-response theory is usually concerned with the fluxes induced by external perturbations that break the translational and/or rotational invariance of the unperturbed system. On the other hand, changes can also be induced by perturbations that have no effect on the symmetry properties of the system. Such perturbations do not lead to any fluxes, but reflect themselves in differences between the time-correlation functions that characterize spontaneous fluctuations in the perturbed and unperturbed systems. These differences can be evaluated by methods that are closely related both to linear-response theory and to the thermodynamic perturbation theories of Chapter 6. The hamiltonian splits into two parts, as in Eqn (7.6.1), but $\mathscr{H}_0$ now describes a "reference" system and the perturbation $\mathscr{H}'$ is usually independent of time. The Liouville operator is the sum of two terms:

$$\mathscr{L} = \mathscr{L}_0 + \mathscr{L}' = i\{\mathscr{H}_0, \quad \} + i\{\mathscr{H}', \quad \} \tag{7.8.15}$$

and the evolution operator in Eqn (2.1.11) satisfies the integral equation

$$\exp(i\mathscr{L}t) = \exp(i\mathscr{L}_0 t) + \int_0^t ds \exp[i\mathscr{L}(t-s)]i\mathscr{L}' \exp(i\mathscr{L}_0 s) \tag{7.8.16}$$

Equation (7.8.16) is an identity that is easily proved by taking the time derivatives of both sides, since it is obviously true at $t = 0$. A formal, iterative solution of (7.8.16) can be written in powers of $\mathscr{L}'$. If only the linear term is retained, the approximate solution is

$$\exp(i\mathscr{L}t) \simeq \exp(i\mathscr{L}_0 t) + \int_0^t ds \exp[i\mathscr{L}_0(t-s)]i\mathscr{L}' \exp(i\mathscr{L}_0 s) \tag{7.8.17}$$

We wish to relate a time-correlation function $C_{AB}(t)$ of the system described by the full hamiltonian (7.6.1) to the corresponding correlation function, $C_{AB}^{(0)}(t)$ say, of the reference system. The effect of the perturbation $\mathscr{H}'$ consists of two parts. First, there is a change in the equilibrium phase-space probability density and hence also a change in the statistical averages. If we denote a statistical average in the reference system by a subscript 0, it is easy to show that for any property A:

$$\langle A \rangle = \frac{\langle \exp(-\beta\mathscr{H}')A \rangle_0}{\langle \exp(-\beta\mathscr{H}') \rangle_0} \tag{7.8.18}$$

When expanded to terms linear in $\mathcal{H}'$, Eqn (7.8.18) becomes

$$\langle A\rangle \simeq \langle A\rangle_0 - \beta\langle A(\mathcal{H}'-\langle\mathcal{H}'\rangle_0)\rangle_0 \tag{7.8.19}$$

Secondly, the perturbation causes the time evolution of the system to change, as represented by Eqn (7.8.16). Combination of the linearized results (7.8.17) and (7.8.19) shows that the relation between the perturbed and unperturbed correlation functions, valid to first order in $\mathcal{H}'$ (Schofield, 1975), is

$$\langle A(t)B\rangle \simeq \langle A(t)B\rangle_0 - \beta\langle(\mathcal{H}'-\langle\mathcal{H}'\rangle_0)A(t)B\rangle_0 + \int_0^t ds\langle i\mathcal{L}'A(s)B(s-t)\rangle_0 \tag{7.8.20}$$

Note that all variables on the right-hand side of (7.8.20) evolve in time under the influence of the unperturbed evolution operator $\exp(i\mathcal{L}_0 t)$.

An interesting application of perturbation theory is in the investigation of isotope effects on self-diffusion (Ebbsjö *et al.*, 1974; Parrinello *et al.*, 1974). Consider a mixture of two atomic species A and B that differ only in their masses, respectively m_A and m_B. If N_A and N_B are the numbers of atoms of the two species, the hamiltonian of the system is

$$\mathcal{H} = \sum_{i=1}^{N_A} \frac{|\mathbf{p}_i|^2}{2m_A} + \sum_{i=1}^{N_B} \frac{|\mathbf{p}_i|^2}{2m_B} + V_N(\mathbf{r}^N) \tag{7.8.21}$$

An obvious choice of reference system is one made up of N identical atoms, all having some mean mass m and described by the hamiltonian

$$\mathcal{H}_0 = \sum_{i=1}^{N} \frac{|\mathbf{p}_i|^2}{2m} + V_N(\mathbf{r}^N) \tag{7.8.22}$$

where $N = N_A + N_B$. The perturbation is then

$$\mathcal{H}' = \sum_{i=1}^{N} |\mathbf{p}_i|^2\left(\frac{1}{2m_i} - \frac{1}{2m}\right) \tag{7.8.23}$$

where m_i is the mass of the ith atom (m_A or m_B). The Liouville operator splits into two parts:

$$\begin{aligned} i\mathcal{L} &= i\mathcal{L}_0 + i\mathcal{L}' \\ &= \sum_{i=1}^{N}\left(\frac{\mathbf{p}_i}{m}\cdot\frac{\partial}{\partial\mathbf{r}_i} - \frac{\partial V_N}{\partial\mathbf{r}_i}\cdot\frac{\partial}{\partial\mathbf{p}_i}\right) + \sum_{i=1}^{N}\left(\frac{1}{m_i} - \frac{1}{m}\right)\mathbf{p}_i\cdot\frac{\partial}{\partial\mathbf{r}_i} \end{aligned} \tag{7.8.24}$$

Considerable simplification is achieved by introducing the scaled momenta defined as $\boldsymbol{\pi}_i = \mathbf{p}_i/m_i^{1/2}$. The hamiltonian is now independent of the masses, while the Liouville operator can be separated in the form

$$i\mathcal{L} = m^{-1/2}\left(i\mathcal{L}_0 + i \sum_{i=1}^{N} \mu_i \mathcal{L}'_i \right) \tag{7.8.25a}$$

$$i\mathcal{L}_0 = \sum_{i=1}^{N} \left(\boldsymbol{\pi}_i \cdot \frac{\partial}{\partial \mathbf{r}_i} - \frac{\partial V_N}{\partial \mathbf{r}_i} \cdot \frac{\partial}{\partial \boldsymbol{\pi}_i} \right) \tag{7.8.25b}$$

$$i\mathcal{L}'_i = \boldsymbol{\pi}_i \cdot \frac{\partial}{\partial \mathbf{r}_i} - \frac{\partial V_N}{\partial \mathbf{r}_i} \cdot \frac{\partial}{\partial \boldsymbol{\pi}_i} \tag{7.8.25c}$$

where $\mu_i = [(m/m_i)^{1/2} - 1]$ is a natural perturbation parameter; the advantage of the scaling procedure lies in the fact that it automatically takes care of the effect of the perturbation on the thermal average. The next step is to introduce a scaled time $\tau = t/m^{1/2}$, and expand the evolution operator $\exp(i\mathcal{L}t)$ in powers of μ_i. This in turn leads to an expansion of the velocity autocorrelation function in powers of the same quantities; the coefficients in the expansion are correlation functions that can be calculated by molecular dynamics simulation (Ebbsjö *et al.*, 1974; Lantelme *et al.*, 1977). The main conclusion to emerge from this work is that, to lowest order in the perturbation, the difference in the self-diffusion coefficients of the two species is proportional to the difference in the inverse square roots of their masses, but is independent of the composition of the mixture. The numerical data obtained for Lennard-Jones fluids with large mass ratios show that although the velocity autocorrelation functions of the two species may differ considerably, essentially because there is a dilation of the timescale for the heavier atoms compared with that of the lighter species, the two diffusion coefficients are very close in value, even for unphysically large mass ratios. Except for the lightest elements, the mass ratios of isotopes that occur in nature are all close to unity. For most practical purposes, therefore, the isotope effect on self-diffusion is negligible. From Eqns (7.2.12) and (7.3.29) we may conclude that the weakness of the isotope effect implies that the lifetime of the memory function associated with the velocity autocorrelation function should be almost independent of mass.

7.9 MEAN-FIELD THEORIES OF THE DENSITY RESPONSE FUNCTION

We now take up the problem of calculating the density response function on the basis of approximate kinetic equations of the type discussed in Section 2.1 and, in particular, of the Vlasov equation (2.1.18). Since it ignores all correlations between particles, the solution to the Vlasov equation is not immediately applicable to liquids, but its structure is suggestive of generalizations that turn out to be useful even at high densities. A more systematic approach to the calculation of the density response function will be developed in Chapters 8 and 9.

The calculation of a time-dependent density–density correlation function on the basis of a kinetic model requires the solution to an equation for the single-particle phase-space distribution function $f^{(1)}(\mathbf{r}, \mathbf{p}; t)$ defined in Section 2.1. The problem can be posed in the form of a calculation of the response in $f^{(1)}(\mathbf{r}, \mathbf{p}; t)$ to a weak applied field. We begin with a calculation of the density response function of an ideal gas; this serves to illustrate the use of linear-response theory, and also gives a result that is needed in the later discussion.

We suppose that an ideal, monatomic gas is subjected to a weak, external, time-dependent potential $\phi(\mathbf{r}, t)$. The time evolution of the distribution function $f(\mathbf{r}, \mathbf{p}; t)$ (for simplicity we omit the superscript (1)) is determined by the collisionless form of Eqn (2.1.20), i.e.

$$\left(\frac{\partial}{\partial t}+\frac{1}{m}\mathbf{p}\cdot\frac{\partial}{\partial \mathbf{r}}-\frac{\partial \phi(\mathbf{r}, t)}{\partial \mathbf{r}}\cdot\frac{\partial}{\partial \mathbf{p}}\right)f(\mathbf{r}, \mathbf{p}; t)=0 \tag{7.9.1}$$

If we write the function $f(\mathbf{r}, \mathbf{p}; t)$ as

$$f(\mathbf{r}, \mathbf{p}; t)=f_{\mathrm{MB}}(\mathbf{p})+f'(\mathbf{r}, \mathbf{p}; t) \tag{7.9.2}$$

where $f_{\mathrm{MB}}(\mathbf{p})$ is the Maxwell–Boltzmann distribution (2.1.22), then, by hypothesis, the change $f'(\mathbf{r}, \mathbf{p}; t)$ induced by the external field is linear in ϕ. Linearizing Eqn (7.9.1) with respect to small quantities, we find that

$$\left(\frac{\partial}{\partial t}+\frac{1}{m}\mathbf{p}\cdot\frac{\partial}{\partial \mathbf{r}}\right)f'(\mathbf{r}, \mathbf{p}; t)-\frac{\partial \phi(\mathbf{r}, t)}{\partial \mathbf{r}}\cdot\frac{\partial f_{\mathrm{MB}}(\mathbf{p})}{\partial \mathbf{p}}=0 \tag{7.9.3}$$

and a double Fourier transform leads (in an obvious notation) to

$$(\omega+i\varepsilon-\mathbf{p}\cdot\mathbf{k}/m)f'(\mathbf{k}, \mathbf{p}; \omega+i\varepsilon)+\phi(\mathbf{k}, \omega+i\varepsilon)\mathbf{k}\cdot\frac{\partial f_{\mathrm{MB}}}{\partial \mathbf{p}}=0 \tag{7.9.4}$$

The change in particle density at a point $\mathbf{r}$ at time t due to the external field is

$$\langle \delta\rho(\mathbf{r}, t)\rangle=\int f'(\mathbf{r}, \mathbf{p}; t)\,\mathrm{d}\mathbf{p} \tag{7.9.5}$$

or, in terms of Fourier components:

$$\langle \rho_{\mathbf{k}}(\omega)\rangle=\int f'(\mathbf{k}, \mathbf{p}; \omega)\,\mathrm{d}\mathbf{p} \tag{7.9.6}$$

Dividing through Eqn (7.9.4) by $(\omega + i\varepsilon - \mathbf{p}\cdot\mathbf{k}/m)$, integrating over $\mathbf{p}$ and using (7.9.6), we arrive at the result

$$\langle \rho_{\mathbf{k}}(\omega)\rangle = -\phi(\mathbf{k}, \omega + i\varepsilon) \int \frac{\mathbf{k}\cdot(\partial f_{\mathrm{MB}}/\partial \mathbf{p})}{\omega + i\varepsilon - \mathbf{p}\cdot\mathbf{k}/m}\, \mathrm{d}\mathbf{p} \tag{7.9.7}$$

Comparison of (7.9.7) with the definition (7.7.16) shows that the density response function (with the subscripts ρ omitted) is

$$\begin{aligned}\chi_0(k, \omega + i\varepsilon) &= -\int \frac{\mathbf{k}\cdot(\partial f_{\mathrm{MB}}/\partial \mathbf{p})}{\omega + i\varepsilon - \mathbf{p}\cdot\mathbf{k}/m}\, \mathrm{d}\mathbf{p} \\ &= -\rho\beta + (\omega + i\varepsilon)\beta \int \frac{f_{\mathrm{MB}}(\mathbf{p})}{\omega + i\varepsilon - \mathbf{p}\cdot\mathbf{k}/m}\, \mathrm{d}\mathbf{p}\end{aligned} \tag{7.9.8}$$

where we have used Eqn (2.1.22) and an integration by parts. In the limit $\varepsilon \to 0$, the imaginary part of (7.9.8) is

$$\chi_0''(k, \omega) = -\pi\beta\omega \int f_{\mathrm{MB}}(\mathbf{p})\delta(\omega - \mathbf{p}\cdot\mathbf{k}/m)\, \mathrm{d}\mathbf{p} \tag{7.9.9}$$

This follows immediately from the identity (7.1.20). On substituting for $f_{\mathrm{MB}}(\mathbf{p})$ and carrying out the integration, we find that

$$\chi_0''(k, \omega) = -\beta\rho\omega\left(\frac{\pi\beta m}{2k^2}\right)^{1/2} \exp\left(-\tfrac{1}{2}\beta m\omega^2/k^2\right) \tag{7.9.10}$$

Combination of Eqns (7.7.17) and (7.9.10) leads back to the expression for the dynamic structure factor of an ideal gas that we derived earlier in a different way, i.e. Eqn (7.5.28).

The mean-field approximation embodied in the Vlasov equation (2.1.18) provides a particularly easy way to take account of the interactions between particles (Nelkin and Ranganathan, 1967). The effect of the mean-field approximation is to replace the external potential in the problem of the ideal gas by

$$\phi_{\mathrm{tot}}(\mathbf{r}, t) = \phi(\mathbf{r}, t) + \iint v(\mathbf{r} - \mathbf{r}')f(\mathbf{r}', \mathbf{p}'; t)\, \mathrm{d}\mathbf{r}'\, \mathrm{d}\mathbf{p}' \tag{7.9.11}$$

where $\phi_{\mathrm{tot}}(\mathbf{r}, t)$ is a self-consistent potential in the sense that the term involving the pair potential $v(\mathbf{r})$ itself depends upon the perturbed distribution function $f(\mathbf{r}, \mathbf{p}; t)$. The Vlasov equation remains a collisionless approximation; it is of value only for systems such as the low-density Coulomb gas in which short-range correlations are unimportant and the particles can usefully be treated as moving in an average potential provided by their mutual interactions. The linearized Vlasov equation is obtained by adding to the left-hand side of (7.9.3) the term

$$-\frac{\partial f_{\mathrm{MB}}}{\partial \mathbf{p}} \cdot \frac{\partial}{\partial \mathbf{r}} \iint v(\mathbf{r}-\mathbf{r}')f'(\mathbf{r}', \mathbf{p}'; t)\, \mathrm{d}\mathbf{r}'\, \mathrm{d}\mathbf{p}'$$

The calculation then proceeds along the same lines as before, leading to an expression for the dynamic susceptibility of the interacting system in terms of the free-particle response function:

$$\chi(k, \omega) = \frac{\chi_0(k, \omega)}{1-\hat{v}(k)\chi_0(k, \omega)} \tag{7.9.12}$$

where $\hat{v}(k)$ is the Fourier transform of the pair potential. This result is the dynamic generalization of the random-phase approximation discussed in Section 6.5 (Pines and Nozières, 1966).

The mean-field representation given by Eqn (7.9.12) is obviously inappropriate in cases when the interparticle force at short range is strongly repulsive. In particular, quite apart from any physical considerations, the Fourier transform $\hat{v}(k)$ does not exist for a hard-core potential. We could hope to improve upon the Vlasov approximation by replacing the bare potential $v(r)$ by an effective potential $v_{\mathrm{eff}}(r)$ that incorporates some of the effects of short-range correlations. For example, in order to satisfy the elastic sum rule (7.4.32) for $S(k, \omega)$, the static susceptibility $\chi(k)$ must be related to the static structure factor through equation (7.7.18). (The second-moment sum rule is automatically satisfied, irrespective of the choice of $v_{\mathrm{eff}}(r)$). The static susceptibility of the ideal gas is simply $\chi_0(k) = -\beta\rho$. Thus Eqns (7.7.18) and (7.9.12), together with the relation (5.2.12) between $S(k)$ and the direct correlation function, imply (Zwanzig, 1967; Nelkin and Ranganathan, 1967) that

$$v_{\mathrm{eff}}(r) = -k_{\mathrm{B}}Tc(r) \tag{7.9.13}$$

a result that is satisfied only asymptotically by the bare potential $v(r)$. As an alternative to (7.9.12), Singwi *et al.* (1968) have suggested that the Vlasov approximation (2.1.17) be replaced by

$$f^{(2)}(\mathbf{r}, \mathbf{p}, \mathbf{r}', \mathbf{p}'; t) = f^{(1)}(\mathbf{r}, \mathbf{p}; t)f^{(1)}(\mathbf{r}', \mathbf{p}'; t)g(\mathbf{r}-\mathbf{r}') \tag{7.9.14}$$

This is equivalent to choosing $v_{\mathrm{eff}}(r)$ such that

$$\nabla v_{\mathrm{eff}}(r) = g(r)\nabla v(r) \tag{7.9.15}$$

The potential $v_{\mathrm{eff}}(r)$ is now a linear functional of $g(r)$.

Not surprisingly, neither of the equations (7.9.13) and (7.9.15) turn out to be useful approximations for liquids such as argon. For example, use of (7.9.13) leads to a dynamic structure factor having a pronounced and very narrow sidepeak, typical of a propagating sound mode, which in calculations for liquid argon is found to persist up to $k \simeq 1.5\,\text{Å}^{-1}$ (Nelkin and Ranganathan, 1967; Kugler, 1973). Experimentally, no such feature is observed. What is more unexpected than the failure in the case of argon is the fact that the approximations represented by (7.9.13) and (7.9.15) do not work well, except at low densities, even for systems lacking hard cores; this is clearly shown by the results of molecular dynamics calculations on a purely coulombic fluid (Hansen *et al.*, 1975). The reason for their failure is that effective-field approximations account only for static, and not for dynamic correlations betweeen particles. Thus, although they may reproduce reasonably well the dispersion of collective modes, such as the k-dependence of the peak in the spectrum of longitudinal-current fluctuations, they are incapable of accounting correctly for the damping of these modes. The dispersion and damping of collective modes are governed, respectively, by the real and imaginary parts of the poles (corresponding to resonances) of the analytic continuation of the dynamic susceptibility $\chi(k, z)$ into the lower half complex plane (remember that $\chi(k, z)$ is an analytic function of z above the real axis); these poles coincide with the zeros of the inverse susceptibility. In an effective-field approximation based on (7.9.12), the inverse susceptibility is

$$\chi^{-1}(k, z) = \chi_0^{-1}(k, z) - \hat{v}_{\text{eff}}(k) \tag{7.9.16}$$

The distribution of complex roots of the equation $\chi^{-1}(k, z) = 0$ is therefore largely determined by the free-particle dynamics embodied in χ_0^{-1}. The term $v_{\text{eff}}(k)$ in (7.9.16) makes no contribution to the damping of the density fluctuations; damping in the effective field approximation arises solely from the mechanism called Landau damping, well known both from plasma physics (Ichimaru, 1973) and from the theory of Fermi liquids (Pines and Nozières, 1966). At small wavenumbers ($k \lesssim 1\,\text{Å}^{-1}$), and under liquid-state conditions, Landau damping is completely negligible in comparison with collisional effects (Kugler, 1973).

Various generalizations of the effective-field approximation have been proposed in which account is taken of collisional damping. One such approach amounts to choosing for the reference dynamics the exact motion of a single particle (Singwi *et al.*, 1970). This is achieved by replacing $\chi_0(k, z)$ by $\chi_s(k, z)$ in Eqn (7.9.16), where $\chi_s(k, z)$ is the response function of a tagged particle; the fluctuation-dissipation theorem shows that $\chi_s(k, z)$ is related to the self part (7.5.15) of the dynamical structure factor by the analogue of Eqn (7.7.17), i.e.

$$\chi_s''(k, \omega) = -\pi\beta\rho\omega S_s(k, \omega) \tag{7.9.17}$$

Because $\chi_s(k, z=0) = -\beta\rho$, the requirement that the elastic sum rule (7.4.32) be satisfied leads again to Eqn (7.9.13). The mean-field expression (7.9.12) is therefore replaced by

$$\chi(k, z) = \frac{\chi_s(k, z)}{1 + k_B T\hat{c}(k)\chi_s(k, z)} \tag{7.9.18}$$

Given the general result expressed by (7.6.27), it follows that the Laplace transforms of the intermediate scattering function $F(k, t)$ and its self part $F_s(k, t)$ are related in the form

$$\tilde{F}(k, z) = \frac{S(k)\tilde{F}_s(k, z)}{1 - \rho\hat{c}(k)[iz\tilde{F}_s(k, z) + 1]} \tag{7.9.19}$$

Equation (7.9.19) is identical to a result derived by Kerr (1968) from a calculation of the linear response of the density to the time-dependent "external" potential produced by a moving tagged particle. The theory has the virtue of being exact in the limit $k \to \infty$, but its usefulness is restricted by the fact that details of the self motion are assumed to be known. A result similar in structure to (7.9.18) also has been obtained by Hubbard and Beeby (1969).

CHAPTER 8

Hydrodynamics and Transport Coefficients

The material of Chapter 7 dealt largely with the formal definition and general properties of time-correlation functions and with establishing the link between spontaneous time-dependent fluctuations and the response of a fluid to an external probe. The aim of the present chapter is to show how the decay of fluctuations can be described in terms of hydrodynamics and to obtain expressions for the macroscopic transport coefficients (Kadanoff and Martin, 1963; Mountain, 1966). The hydrodynamic approach is valid only on scales of length and time that are much larger than those characteristic of the molecular level. At the end of the chapter, therefore, we show how the gap between the microscopic and macroscopic descriptions can be bridged by an essentially phenomenological extrapolation of the hydrodynamic results to shorter wavelengths and higher frequencies. The same problem is taken up in a more systematic way in Chapter 9.

8.1 THERMAL FLUCTUATIONS AT LONG WAVELENGTHS AND LOW FREQUENCIES

We have seen in Section 5.1 that the microscopic structure of a liquid is revealed experimentally by the scattering of radiation of wavelength comparable with the interparticle spacing. Examination of a typical pair distribution function, such as the one pictured in Figure 2.1, shows that positional correlations decay rapidly beyond a few molecular diameters; from a static point of view, therefore, a fluid behaves, for longer wavelengths, essentially as a continuum. When discussing the dynamics, it is necessary to consider simultaneously the scales of length and time. In keeping with traditional kinetic theory, it is conventional to compare wavelengths with the mean free path l and times with the mean collision time τ. The wavenumber-frequency plane may then be divided into three parts. The region in which $kl \ll 1$, $\omega\tau \ll 1$ corresponds to the *hydrodynamic* regime, where the behaviour of the fluid is described by the phenomenological equations of macroscopic fluid mechanics (Landau and Lifshitz, 1963). The range of intermediate wavenumbers and frequencies ($kl \simeq 1, \omega\tau \simeq 1$) forms the *kinetic* regime, where allowance must be made for the molecular structure of the fluid and where a treatment based on the microscopic equations of motion is required. Finally, the region where $kl \gg 1$, $\omega\tau \gg 1$, represents the *free-particle* regime; here the distances and times involved are so short that the particles move almost independently of each other.

In this chapter, we shall be concerned mostly with the hydrodynamic regime, where the local properties of the fluid vary slowly on microscopic scales of length and time. The set of *hydrodynamic variables* consists of the densities of mass (or particle number), momentum and energy; these are closely related to the conserved microscopic variables introduced in Section 7.4. Like their microscopic counterparts, the hydrodynamic variables satisfy continuity equations of the form of (7.4.3); the latter are merely expressions of the conservation of matter, momentum and energy. In addition, the hydrodynamic variables obey certain *constitutive relations* between the fluxes (or currents) and the gradients of local variables, expressed in terms of phenomenological *transport coefficients.* Fick's law of diffusion and Fourier's law of heat transport are two of the more familiar examples of a constitutive relation.

One of the main tasks of the present chapter is to obtain microscopic expressions for the transport coefficients that are similar in structure to the relation (7.8.10) already derived for the electrical conductivity of an ionic fluid; this will be achieved by calculating the *hydrodynamic limit* of the appropriate time-correlation function. To understand what is involved in such a calculation, it is first necessary to clarify the relation between hydrodynamic and microscopic dynamical variables. As an example, consider the local density. The microscopic particle density $\rho(\mathbf{r}, t)$ is defined by Eqn (7.4.5); its integral over all volume is equal to N, the total number of particles in the system. The hydrodynamic local density $\bar{\rho}(\mathbf{r}, t)$ is obtained by averaging $\rho(\mathbf{r}, t)$ over a volume v that is macroscopically small but still sufficiently large to ensure that the relative fluctuation in the number of particles inside v is negligible. Then

$$\bar{\rho}(\mathbf{r}, t) = \frac{1}{v}\int_v \rho(\mathbf{r}' - \mathbf{r}, t)\,\mathrm{d}\mathbf{r}' \tag{8.1.1}$$

where $\mathbf{r}$ is the centre of the volume v. Strictly speaking, the definition of $\bar{\rho}(\mathbf{r}, t)$ also requires a smoothing ("coarse graining") in time. This can be realized by averaging (8.1.1) over a time interval that is short on a macroscopic scale but long in comparison with the mean collision time. In practice, however, smoothing in time is already achieved by Eqn (8.1.1) if

v is large enough to justify neglect of the fluctuations associated with particles entering or leaving the volume. The Fourier components of the hydrodynamic density are defined as

$$\bar{\rho}_{\mathbf{k}}(t) = \int \bar{\rho}(\mathbf{r}, t) \exp(-i\mathbf{k} \cdot \mathbf{r}) \, d\mathbf{r} \tag{8.1.2}$$

where the wavevector $\mathbf{k}$ is such that $k \lesssim 2\pi/v^{1/3}$. The corresponding density autocorrelation function is then defined as in Eqn (7.4.30), except that the Fourier components of the microscopic density are replaced by $\bar{\rho}_{\mathbf{k}}$. Since we are now working at the macroscopic level, the average to be taken is not an ensemble average, but an average over initial conditions, weighted by the probability density (2.7.1) of thermodynamic fluctuation theory. By forming such an average, we are implicitly invoking the hypothesis of local thermodynamic equilibrium. In other words, we are assuming that although the hydrodynamic densities vary over macroscopic lengths and times, the fluid contained in each of the small subvolumes v is in a state of thermodynamic equilibrium, and hence that the local density, pressure and temperature are properly defined and satisfy the usual relations of equilibrium thermodynamics. This hypothesis is particularly plausible at high densities, since in that case local equilibrium is rapidly brought about by collisions between particles.

Once the calculation that we have described in words has been carried through, the relations of interest are obtained by supposing that in the limit of long wavelengths ($\lambda \gg l$) and long times ($t \gg \tau$) or, equivalently, of small wavenumbers and low frequencies, the correlation functions derived from the hydrodynamic equations are identical to the correlation functions of the corresponding microscopic variables. We thereby assume that spontaneous fluctuations in microscopic quantities decay, in the mean, according to the hydrodynamic laws that govern the behaviour of their macroscopic analogues (Onsager 1931a, b); this is an intuitively appealing hypothesis that can be justified on the basis of the fluctuation-dissipation theorem discussed in Section 7.6 (Kadanoff and Martin, 1963). In the example of the density autocorrelation function, the assumption can be expressed by the statement that

$$\lim_{kl \ll 1, t/\tau \gg 1} \langle \rho_{\mathbf{k}}(t) \rho_{-\mathbf{k}} \rangle = \langle \bar{\rho}_{\mathbf{k}}(t) \bar{\rho}_{-\mathbf{k}} \rangle \tag{8.1.3}$$

with the qualification, explained above, that the meaning of the angular brackets is not the same on the two sides of the equation. As the sections that follow are concerned almost exclusively with the calculation of correlation functions of hydrodynamic variables, no ambiguity is introduced by dropping the bar that distinguishes the latter from the corresponding microscopic quantities.

8.2 SPACE-DEPENDENT SELF MOTION

As an illustration of the general procedure described above, we first consider the relatively simple problem of the diffusion of tagged particles. If the tagged particles are identical to the other particles in the fluid, and if their concentration is sufficiently low that their mutual interactions can be ignored, the problem is equivalent to that of single-particle motion as described by the self part of the van Hove correlation function (see Section 7.4) The macroscopic density $\rho^{(s)}(\mathbf{r}, t)$ and current $\mathbf{j}^{(s)}(\mathbf{r}, t)$ of tagged particles satisfy a continuity equation of the form

$$\dot{\rho}^{(s)}(\mathbf{r}, t) + \nabla \cdot \mathbf{j}^{(s)}(\mathbf{r}, t) = 0 \tag{8.2.1}$$

and the corresponding constitutive relation is provided by Fick's law:

$$\mathbf{j}^{(s)}(\mathbf{r}, t) = -D\nabla\rho^{(s)}(\mathbf{r}, t) \tag{8.2.2}$$

where D is the self-diffusion constant. Combination of (8.2.1) and (8.2.2) yields the *diffusion equation*, i.e.

$$\dot{\rho}^{(s)}(\mathbf{r}, t) = D\nabla^2\rho^{(s)}(\mathbf{r}, t) \tag{8.2.3}$$

or, in reciprocal space:

$$\dot{\rho}_{\mathbf{k}}^{(s)}(t) = -Dk^2\rho_{\mathbf{k}}^{(s)}(t) \tag{8.2.4}$$

Equation (8.2.4) can be integrated immediately to give

$$\rho_{\mathbf{k}}^{(s)}(t) = \rho_{\mathbf{k}}^{(s)} \exp(-Dk^2 t) \tag{8.2.5}$$

where $\rho_{\mathbf{k}}^{(s)}$ is a Fourier component of the tagged-particle density at $t = 0$. If we multiply both sides of Eqn (8.2.5) by $\rho_{-\mathbf{k}}^{(s)}$ and take the thermal average, we find that the normalized correlation function is

$$\begin{aligned} \frac{1}{n}\langle\rho_{\mathbf{k}}^{(s)}(t)\rho_{-\mathbf{k}}^{(s)}\rangle &= \frac{1}{n}\langle\rho_{\mathbf{k}}^{(s)}\rho_{-\mathbf{k}}^{(s)}\rangle \exp(-Dk^2 t) \\ &= \exp(-Dk^2 t) \end{aligned} \tag{8.2.6}$$

where n is the total number of tagged particles and we have used the fact that the coordinates of tagged particles are mutually uncorrelated. It follows from the general hypothesis discussed in the preceding section that the self part of the density autocorrelation function (7.4.30) is given in the hydrodynamic limit by

$$\lim_{kl \ll 1, t/\tau \gg 1} F_s(k, t) = \exp(-Dk^2 t) \tag{8.2.7}$$

The long-wavelength, low-frequency limit of the van Hove self correlation function is the Fourier transform of (8.2.7):

$$G_s(r, t) = \frac{1}{(4\pi Dt)^{3/2}} \exp(-r^2/4Dt) \tag{8.2.8}$$

In the same limit, the self dynamical structure factor is

$$S_s(k, \omega) = \frac{1}{\pi} \frac{Dk^2}{\omega^2 + (Dk^2)^2} \tag{8.2.9}$$

Equation (8.2.9) represents a single lorentzian curve centred at $\omega = 0$ and having a width at half-height equal to $2Dk^2$. A spectrum of this type is typical of a diffusive process described by an equation similar to (8.2.3). Alternatively, the structure of the Laplace transform of (8.2.8), i.e.

$$\tilde{F}_s(k, z) = \frac{1}{-iz + Dk^2} \tag{8.2.10}$$

shows that a diffusive process is characterized by a purely imaginary pole at $z = -iDk^2$. We emphasize again that the simple result expresssed by Eqn (8.2.9) is valid only for $kl \ll 1$, $\omega\tau \ll 1$. In particular, the hydrodynamic result is invalid at high frequencies; this is reflected in the fact that the even frequency moments (beyond zeroth order) of (8.2.9) are all divergent. Note also that the transport coefficient D is related to the behaviour of $S_s(k, \omega)$ in the limit $k, \omega \to 0$. From Eqn (8.2.9) we see that

$$D = \lim_{\omega \to 0} \lim_{k \to 0} \frac{\omega^2}{k^2} \pi S_s(k, \omega) \tag{8.2.11}$$

where it is crucial that the limits are taken in the correct order, i.e. $k \to 0$ before $\omega \to 0$. In principle, Eqn (8.2.11) provides a means of determining D from the results of inelastic neutron scattering experiments.

Equations (7.5.27) and (8.2.8) show that the van Hove self correlation function is a gaussian function of distance both for $t \to 0$ (free-particle behaviour) and $t \to \infty$ (the hydrodynamic limit); it is therefore tempting to suppose that $G_s(r, t)$ is gaussian at all times. To study this point in more detail, we write $G_s(r, t)$ as a generalized gaussian function of r in the form

$$G_s(r, t) = \left(\frac{\alpha(t)}{\pi}\right)^{3/2} \exp[-\alpha(t) r^2] \tag{8.2.12}$$

where $\alpha(t)$ is a function of t but not of r; the hydrodynamic limit corresponds to taking $\alpha(t) = 1/4Dt$ and the ideal gas model to $\alpha(t) = \beta m/2t^2$. We now introduce the even moments of $G_s(r, t)$, defined as

$$\begin{aligned} \langle r^{2n}(t) \rangle &\equiv \langle |\mathbf{r}(t) - \mathbf{r}(0)|^{2n} \rangle \\ &= 4\pi \int_0^\infty r^{2n} G_s(r, t) r^2 \, \mathrm{d}r \end{aligned} \tag{8.2.13}$$

Substitution of (8.2.12) in (8.2.13) shows that the mean-square displacement of tagged particles after a time t is given in terms of the unknown function $\alpha(t)$ by

$$\langle r^2(t)\rangle = \frac{3}{2\alpha(t)} \tag{8.2.14}$$

If we insert (8.2.14) in (8.2.12) and take the Fourier transform, we find that

$$F_s(k, t) = \exp\left[-\tfrac{1}{6}k^2\langle r^2(t)\rangle\right]. \tag{8.2.15}$$

The accuracy of the generalized gaussian approximation has been tested against molecular dynamics data on the Lennard-Jones fluid (Levesque and Verlet, 1970). Some results for a state close to the triple point are shown in Figure 8.1; the quantity plotted there (as a function of k) is the width at half-height of $S_s(k, \omega)$ relative to its value in the hydrodynamic limit (where $\Delta\omega = 2Dk^2$). The approximation is clearly rather good, though at intermediate values of k the predicted width is too large. As use of Eqn (8.2.15) also underestimates the value of $S_s(k, 0)$, the net result is that

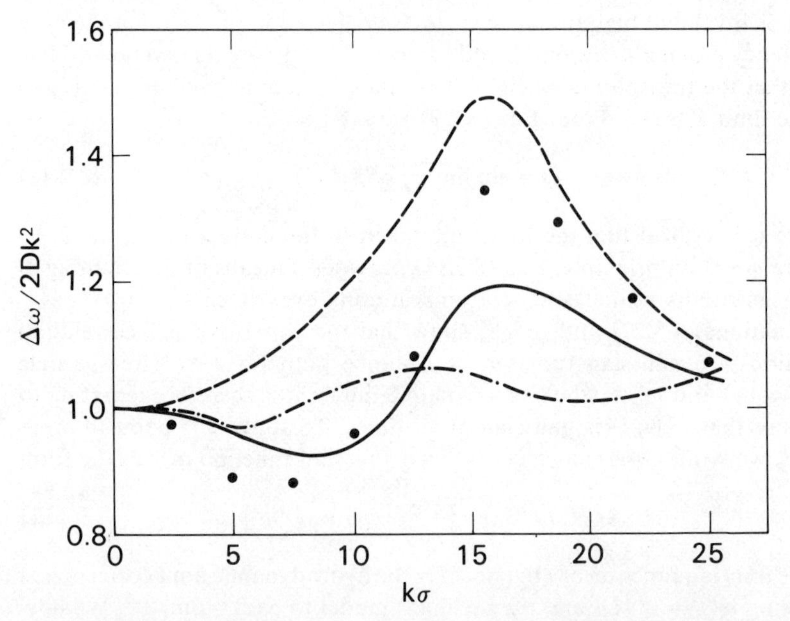

FIG. 8.1. Width at half-height of the self dynamic structure factor relative to its value in the hydrodynamic limit. The points are molecular dynamics results for the Lennard-Jones fluid near the triple point (Levesque and Verlet, 1970). The dashed, dash-dot and solid curves are calculated, respectively, from the gaussian approximation (8.2.15), the single-relaxation-time approximation (9.2.17) and the memory-function analysis of Levesque and Verlet to be discussed in Section 9.2.

$S_s(k, \omega)$ is less sharply defined than it should be. Systematic corrections to the gaussian approximation are formally obtainable from a cumulant expression of $F_s(k, t)$ in powers of k^2 (Nijboer and Rahman, 1966). By taking the Fourier transform Eqn (7.4.22), it can be shown that

$$\begin{aligned} F_s(k, t) &= \langle \exp \{ i\mathbf{k} \cdot [\mathbf{r}(t) - \mathbf{r}(0)] \} \rangle \\ &= \exp(-\tfrac{1}{6}k^2 \langle r^2(t) \rangle) \\ &\quad \times \left[1 + \frac{1}{2} \left(\frac{3\langle r^4(t) \rangle}{5\langle r^2(t) \rangle^2} - 1 \right) (\tfrac{1}{6}k^2 \langle r^2(t) \rangle)^2 + \mathcal{O}(k^6) \right] \end{aligned} \tag{8.2.16}$$

The first correction to the gaussian approximation is positive in argon-like liquids and is typically 10% or less of the leading term; corrections of higher order are even smaller (Rahman, 1964a; Levesque and Verlet, 1970).

Note that the Einstein expression for the long-time limit of the mean-square displacement of a tagged particle is a direct consequence of the hydrodynamic result for $G_s(r, t)$, Eqn (8.2.8). The latter, when substituted into the definition (8.2.13) (taken for $n = 1$) leads immediately to Eqn (7.2.3). The mean-square displacement is also related to the velocity autocorrelation through Eqn (7.2.7), and a close connection therefore exists between the functions $G_s(r, t)$ and $Z(t)$. In fact, in the gaussian approximation represented by Eqn (8.2.15), $G_s(r, t)$ is entirely determined by $Z(t)$, and vice versa. More generally, only the second of these statements is true. If we define the Fourier components of the current associated with a tagged particle i, having velocity $\mathbf{u}_i$, as

$$\mathbf{j}_{\mathbf{k}i}(t) = \mathbf{u}_i(t) \exp[-i\mathbf{k} \cdot \mathbf{r}_i(t)] \tag{8.2.17}$$

and the self-current autocorrelation function as

$$C_s(k, t) = \langle \mathbf{k} \cdot \mathbf{j}_{\mathbf{k}i}(t) \mathbf{k} \cdot \mathbf{j}_{-\mathbf{k}i} \rangle \tag{8.2.18}$$

it is clear that

$$\begin{aligned} Z(t) &= \langle u_{iz}(t) u_{iz} \rangle \\ &= \lim_{k \to 0} \frac{1}{k^2} C_s(k, t) \\ &= -\lim_{k \to 0} \frac{1}{k^2} \frac{\mathrm{d}^2}{\mathrm{d}t^2} F_s(k, t) \end{aligned} \tag{8.2.19}$$

where we have chosen $\mathbf{k}$ along the z-axis and used the single-particle version of Eqn (7.4.36). The relation between the corresponding spectra is

$$\begin{aligned} Z(\omega) &= \frac{\omega^2}{2\pi} \lim_{k \to 0} \frac{1}{k^2} \int_{-\infty}^{\infty} F_s(k, t) \exp(i\omega t) \, \mathrm{d}t \\ &= \omega^2 \lim_{k \to 0} \frac{S_s(k, \omega)}{k^2} \end{aligned} \tag{8.2.20}$$

Equation (8.2.20) can be regarded as a generalization of (8.2.11) to non-zero frequencies in which $Z(\omega)$ appears as a frequency-dependent diffusion coefficient; it also provides a possible route to an experimental determination of the velocity autocorrelation function.

The relationship between $Z(t)$ and $F_s(k, t)$ (or $C_s(k, t)$) is also reflected in the short-time expansions of these functions. By analogy with (7.4.41), the expansion of $C_s(k, t)$ in powers of t may be written as

$$C_s(k, t) = \omega_0^2\left(1 - \omega_{1s}^2\frac{t^2}{2!} + \cdots\right) \tag{8.2.21}$$

From the general result (7.1.25) and the continuity equation (8.2.1) it follows that

$$\begin{aligned}\omega_0^2\omega_{1s}^2 &= -\langle \mathbf{k}\cdot\dot{\mathbf{j}}_{\mathbf{k}i}\mathbf{k}\cdot\dot{\mathbf{j}}_{-\mathbf{k}i}\rangle \\ &= \langle \ddot{\rho}_{\mathbf{k}i}\ddot{\rho}_{-\mathbf{k}i}\rangle \\ &= k^4\langle u_{iz}^4\rangle + k^2\langle \dot{u}_{iz}^2\rangle \\ &= 3\omega_0^4 + (k^2/m^2)\langle F_{iz}^2\rangle \end{aligned} \tag{8.2.22}$$

and hence, from the definition (7.2.10), that

$$\omega_{1s}^2 = 3\omega_0^2 + \Omega_0^2 \tag{8.2.23}$$

The next (order t^4) term in the Taylor expansion of $C_s(k, t)$ involves integrals over the three-particle distribution function (Forster *et al.*, 1968). As we shall see in Section 8.7, short-time expansions such as (8.2.21) are useful in extending the validity of hydrodynamic results to microscopic scales of length and time.

8.3 THE NAVIER–STOKES EQUATION AND HYDRODYNAMIC COLLECTIVE MODES

We turn now to the problem of describing the decay of long-wavelength fluctuations in the collective dynamical variables. The macroscopic local densities associated with the conserved variables are the number density $\rho(\mathbf{r}, t)$, the momentum density $\mathbf{p}(\mathbf{r}, t)$ and the energy density $e(\mathbf{r}, t)$. The local velocity field $\mathbf{u}(\mathbf{r}, t)$ is defined via the relation

$$\mathbf{p}(\mathbf{r}, t) = m\rho(\mathbf{r}, t)\mathbf{u}(\mathbf{r}, t) \tag{8.3.1}$$

where m is the mass of a particle (we assume that the fluid consists of a single chemical species) and $m\rho(\mathbf{r}, t)$ is therefore the mass density. The conservation laws for the local densities are of the form

$$\dot{\rho}(\mathbf{r}, t) + \frac{1}{m}\nabla \cdot \mathbf{p}(\mathbf{r}, t) = 0 \tag{8.3.2}$$

$$\dot{\mathbf{p}}(\mathbf{r}, t) + \nabla \cdot \boldsymbol{\sigma}(\mathbf{r}, t) = 0 \tag{8.3.3}$$

$$\dot{e}(\mathbf{r}, t) + \nabla \cdot \mathbf{J}^e(\mathbf{r}, t) = 0 \tag{8.3.4}$$

where $\boldsymbol{\sigma}$ is the momentum current, or *stress tensor*, and $\mathbf{J}^e$ is the energy current. These equations are supplemented by two constitutive relations in which $\boldsymbol{\sigma}$ and $\mathbf{J}^e$ are expressed in terms of the dissipative processes in the fluid. The stress tensor is given macroscopically (Landau and Lifshitz, 1963) by

$$\sigma^{\alpha\beta}(\mathbf{r}, t) = \delta_{\alpha\beta}P(\mathbf{r}, t) - \eta\left(\frac{\partial u_\alpha(\mathbf{r}, t)}{\partial r_\beta} + \frac{\partial u_\beta(\mathbf{r}, t)}{\partial r_\alpha}\right) + \delta_{\alpha\beta}(\tfrac{2}{3}\eta - \zeta)\nabla \cdot \mathbf{u}(\mathbf{r}, t) \tag{8.3.5}$$

where $P(\mathbf{r}, t)$ is the local pressure, η is the shear viscosity and ζ is the bulk viscosity. We now choose a frame of reference in which the mean velocity of the fluid is zero, and assume that the local deviations of the hydrodynamic variables from their average values are small. The equations above may then be linearized with respect to the deviations. In particular, the momentum density may be expressed as

$$\mathbf{p}(\mathbf{r}, t) = m[\rho + \delta\rho(\mathbf{r}, t)]\mathbf{u}(\mathbf{r}, t) \simeq m\rho\mathbf{u}(\mathbf{r}, t) \tag{8.3.6}$$

Substitution of (8.3.5) and (8.3.6) in Eqn (8.3.3) gives the Navier-Stokes equation in its linearized form:

$$\rho m\dot{\mathbf{u}}(\mathbf{r}, t) + \nabla P(\mathbf{r}, t) - \eta\nabla^2\mathbf{u}(\mathbf{r}, t) - (\tfrac{1}{3}\eta + \zeta)\nabla\nabla \cdot \mathbf{u}(\mathbf{r}, t) = 0 \tag{8.3.7}$$

The second constitutive relation (Landau and Lifshitz, 1963) defines the macroscopic energy current as

$$\mathbf{J}^e(\mathbf{r}, t) = h\mathbf{u}(\mathbf{r}, t) - \lambda\nabla T(\mathbf{r}, t) \tag{8.3.8}$$

where $h = (e + P)$ is the equilibrium enthalpy density, $T(\mathbf{r}, t)$ is the local temperature and λ is the thermal conductivity; terms corresponding to viscous heating have been omitted, since these are quadratic in the local velocity. Substitution of (8.3.8) in Eqn (8.3.4) gives the energy equation, i.e.

$$\frac{\partial}{\partial t}\left[e(\mathbf{r}, t) - \frac{e + P}{\rho}\rho(\mathbf{r}, t)\right] - \lambda\nabla^2 T(\mathbf{r}, t) = 0 \tag{8.3.9}$$

or

$$\dot{Q}(\mathbf{r}, t) - \lambda \nabla^2 T(\mathbf{r}, t) = 0 \tag{8.3.10}$$

where

$$Q(\mathbf{r}, t) = e(\mathbf{r}, t) - \frac{e+P}{\rho}\rho(\mathbf{r}, t) \tag{8.3.11}$$

The quantity $Q(\mathbf{r}, t)$ can be interpreted as a density of heat energy. This identification follows from a simple thermodynamic argument. If the number of particles is held constant, we see that

$$\begin{aligned} T\,\mathrm{d}S &= \mathrm{d}U + P\,\mathrm{d}V \\ &= \mathrm{d}(eV) + P\,\mathrm{d}V \\ &= V\,\mathrm{d}e - eV\frac{\mathrm{d}\rho}{\rho} - PV\frac{\mathrm{d}\rho}{\rho} \end{aligned} \tag{8.3.12}$$

Thus

$$\begin{aligned} \frac{T}{V}\mathrm{d}S &= \mathrm{d}e - \frac{e+P}{\rho}\,\mathrm{d}\rho \\ &= \mathrm{d}Q \end{aligned} \tag{8.3.13}$$

and a change in $Q(\mathbf{r}, t)$ is therefore equal to T multiplied by the change in entropy density.

If we invoke the hypothesis of local thermodynamic equilibrium, the deviations of the local thermodynamic variables from their average values can be expressed in terms of a set of statistically independent quantities. We choose as independent variables the density and temperature (cf. Eqn (2.7.5)), and expand $P(\mathbf{r}, t)$ and $Q(\mathbf{r}, t)$ to first order in the deviations $\delta\rho(r, \mathbf{t}) = \rho(\mathbf{r}, t) - \rho$ and $\delta T(\mathbf{r}, t) = T(\mathbf{r}, t) - T$. Then

$$\delta P(\mathbf{r}, t) = \left(\frac{\partial P}{\partial \rho}\right)_T \delta\rho(\mathbf{r}, t) + \left(\frac{\partial P}{\partial T}\right)_\rho \delta T(\mathbf{r}, t) \tag{8.3.14}$$

$$\begin{aligned} \delta Q(\mathbf{r}, t) &= \frac{T}{V}\left(\frac{\partial S}{\partial \rho}\right)_T \delta\rho(\mathbf{r}, t) + \frac{T}{V}\left(\frac{\partial S}{\partial T}\right)_\rho \delta T(\mathbf{r}, t) \\ &= \rho T\left(\frac{\partial(S/N)}{\partial \rho}\right)_T \delta\rho(\mathbf{r}, t) + \rho c_V \delta T(\mathbf{r}, t) \end{aligned} \tag{8.3.15}$$

where c_V is the specific heat per particle at constant volume. By making use of the thermodynamic relation

$$\left(\frac{\partial(S/N)}{\partial \rho}\right)_T = -\frac{1}{\rho^2}\left(\frac{\partial P}{\partial T}\right)_\rho \tag{8.3.16}$$

and substituting (8.3.14) and (8.3.15) in Eqns (8.3.7) and (8.3.10), we obtain a closed set of linear equations for the variables $\delta\rho(\mathbf{r}, t)$, $\delta T(\mathbf{r}, t)$ and $\mathbf{j}(\mathbf{r}, t) = \rho\mathbf{u}(\mathbf{r}, t)$ of the form

$$\frac{\partial}{\partial t}\delta\rho(\mathbf{r}, t) + \boldsymbol{\nabla}\cdot\mathbf{j}(\mathbf{r}, t) = 0 \tag{8.3.17}$$

$$\left(\frac{\partial}{\partial t} - a\nabla^2\right)\delta T(\mathbf{r}, t) - \frac{T}{\rho^2 c_V}\left(\frac{\partial P}{\partial T}\right)_\rho \frac{\partial}{\partial t}\delta\rho(\mathbf{r}, t) = 0 \tag{8.3.18}$$

$$\left(\frac{\partial}{\partial t} - \frac{\eta}{\rho m}\nabla^2 - \frac{\frac{1}{3}\eta + \zeta}{\rho m}\boldsymbol{\nabla}\boldsymbol{\nabla}\cdot\right)\mathbf{j}(\mathbf{r}, t)$$

$$+\left(\frac{\partial P}{\partial \rho}\right)_T \boldsymbol{\nabla}\delta\rho(\mathbf{r}, t) + \left(\frac{\partial P}{\partial T}\right)_\rho \boldsymbol{\nabla}\delta T(\mathbf{r}, T) = 0 \tag{8.3.19}$$

where

$$a = \frac{\lambda}{\rho c_V} \tag{8.3.20}$$

Equations (8.3.17) to (8.3.19) are readily solved by taking the double transforms with respect to space (Fourier) and time (Laplace) to give

$$-iz\tilde{\rho}_\mathbf{k}(z) + i\mathbf{k}\cdot\tilde{\mathbf{j}}_\mathbf{k}(z) = \rho_\mathbf{k} \tag{8.3.21}$$

$$(-iz + ak^2)\tilde{T}_\mathbf{k}(z) + \frac{iT}{\rho^2 c_V}\left(\frac{\partial P}{\partial T}\right)_\rho \tilde{\mathbf{j}}_\mathbf{k}(z) = T_\mathbf{k} \tag{8.3.22}$$

$$\left(-iz + \frac{\eta}{\rho m}k^2 + \frac{\frac{1}{3}\eta + \zeta}{\rho m}\mathbf{k}\mathbf{k}\cdot\right)\tilde{\mathbf{j}}_\mathbf{k}(z)$$

$$+\frac{i\mathbf{k}}{m}\left(\frac{\partial P}{\partial \rho}\right)_T \tilde{\rho}_\mathbf{k}(z) + \frac{i\mathbf{k}}{m}\left(\frac{\partial P}{\partial T}\right)_\rho \tilde{T}_\mathbf{k}(z) = \mathbf{j}_\mathbf{k} \tag{8.3.23}$$

where, for example:

$$\tilde{\rho}_\mathbf{k}(z) = \int_0^\infty \mathrm{d}t\, \exp(izt) \int \delta\rho(\mathbf{r}, t)\exp(-i\mathbf{k}\cdot\mathbf{r})\,\mathrm{d}\mathbf{r} \tag{8.3.24}$$

and $\rho_\mathbf{k}$, $T_\mathbf{k}$ and $\mathbf{j}_\mathbf{k}$ are the spatial Fourier components at $t = 0$. We now separate the components of the current $\mathbf{j}_\mathbf{k}$ into their longitudinal and tranverse parts. Choosing $\mathbf{k}$ along the z-axis, we rewrite Eqn (8.3.23) as

$$(-iz + bk^2)\tilde{j}^z_\mathbf{k}(z) + \frac{ik}{m}\left(\frac{\partial P}{\partial T}\right)_\rho \tilde{T}_\mathbf{k}(z) + \frac{ik}{m}\left(\frac{\partial P}{\partial \rho}\right)_T \tilde{\rho}_\mathbf{k}(z) = j^z_\mathbf{k} \tag{8.3.25a}$$

$$(-iz + \nu k^2)\tilde{j}^\alpha_\mathbf{k}(z) = j^\alpha_\mathbf{k}, \qquad \alpha = x, y \tag{8.3.25b}$$

where

$$b = \frac{\frac{4}{3}\eta + \zeta}{\rho m} \tag{8.3.26}$$

is the kinematic longitudinal viscosity and

$$\nu = \frac{\eta}{\rho m} \tag{8.3.27}$$

is the kinematic shear viscosity. Equations (8.3.21), (8.3.22) and (8.3.25) are conveniently summarized in matrix form as

$$\begin{pmatrix} -iz & 0 & ik & 0 & 0 \\ 0 & -iz + ak^2 & \dfrac{ikT}{\rho^2 c_V}\left(\dfrac{\partial P}{\partial T}\right)_\rho & 0 & 0 \\ \dfrac{ik}{m}\left(\dfrac{\partial P}{\partial \rho}\right)_T & \dfrac{ik}{m}\left(\dfrac{\partial P}{\partial T}\right)_\rho & -iz + bk^2 & 0 & 0 \\ 0 & 0 & 0 & -iz + \nu k^2 & 0 \\ 0 & 0 & 0 & 0 & -iz + \nu k^2 \end{pmatrix} \begin{pmatrix} \tilde{\rho}_{\mathbf{k}}(z) \\ \tilde{T}_{\mathbf{k}}(z) \\ \tilde{j}^z_{\mathbf{k}}(z) \\ \tilde{j}^x_{\mathbf{k}}(z) \\ \tilde{j}^y_{\mathbf{k}}(z) \end{pmatrix} = \begin{pmatrix} \rho_{\mathbf{k}} \\ T_{\mathbf{k}} \\ j^z_{\mathbf{k}} \\ j^x_{\mathbf{k}} \\ j^y_{\mathbf{k}} \end{pmatrix} \tag{8.3.28}$$

The matrix of coefficient in (8.3.28) is called the *hydrodynamic matrix.* Its block-diagonal structure shows that the transverse-current fluctuations are completely decoupled from the fluctuations in the other (longitudinal) variables. The determinant of the hydrodynamic matrix therefore factorizes into the product of purely longitudinal (l) and purely transverse (t) parts, i.e.

$$D(k, z) = D_l(k, z) D_t(k, z) \tag{8.3.29}$$

with

$$D_l(k, z) = -iz(-iz + ak^2)(-iz + bk^2) + (-iz + ak^2)\frac{k^2}{m}\left(\frac{\partial P}{\partial \rho}\right)_T - iz\frac{k^2 T}{\rho^2 m c_V}\left[\left(\frac{\partial P}{\partial T}\right)_\rho\right]^2 \tag{8.3.30}$$

and

$$D_t(k, z) = (-iz + \nu k^2)^2 \tag{8.3.31}$$

The frequency/wavenumber or *dispersion* relation for the collective modes is given by the poles of the inverse of the hydrodynamic matrix and hence by the complex roots of the equation

$$D(k, z) = 0 \tag{8.3.32}$$

The factorization in (8.3.29) shows that Eqn (8.3.32) has a double root associated with transverse modes, namely

$$z = -i\nu k^2 \tag{8.3.33}$$

while the complex frequencies corresponding to longitudinal modes are obtained as the solution to the cubic equation

$$iz^3 - z^2(a+b)k^2 - iz(abk^2 + c_s^2)k^2 + (a/\gamma)c_s^2 k^4 = 0 \tag{8.3.34}$$

where $\gamma = c_P/c_V$, c_s is the adiabatic speed of sound, given by

$$c_s^2 = \frac{\gamma}{m}\left(\frac{\partial P}{\partial \rho}\right)_T \tag{8.3.35}$$

and use has been made of the thermodynamic relation

$$\left[\left(\frac{\partial P}{\partial T}\right)_\rho\right]^2 = \frac{\rho^2}{T}(c_P - c_V)\left(\frac{\partial P}{\partial \rho}\right)_T \tag{8.3.36}$$

Since the hydrodynamic calculation is valid only in the long-wavelength limit, it is sufficient to calculate the complex frequencies to order k^2. The algebra is simplified by introducing the reduced variable $s = z/c_s k$; it is then straightforward to show (Mountain, 1966) that the approximate solution to Eqn (8.3.34) is of the form

$$z_0 = -iD_T k^2 \tag{8.3.37a}$$

$$z_\pm = \pm c_s k - i\Gamma k^2 \tag{8.3.37b}$$

where

$$D_T = \frac{a}{\gamma} = \frac{\lambda}{\rho c_P} \tag{8.3.38}$$

is the thermal diffusivity and

$$\Gamma = \tfrac{1}{2}[a(\gamma - 1)/\gamma + b] \tag{8.3.39}$$

is the sound-attenuation coefficient. The purely imaginary roots in (8.3.33) and (8.3.37a) represent diffusive processes of the type already discussed in the preceding section, and the pair of complex roots in (8.3.37b) correspond to propagating sound waves, as we shall discuss in Section 8.5.

8.4 TRANSVERSE CURRENT CORRELATIONS

It follows from Eqn (8.3.25b) that the hydrodynamic behaviour of the transverse current fluctuations in the time domain is governed by a first-order differential equation of the form

$$\frac{\partial}{\partial t} j_{\mathbf{k}}^{x}(t) + \nu k^2 j_{\mathbf{k}}^{x}(t) = 0 \tag{8.4.1}$$

This result has exactly the same structure as the diffusion equation (8.2.4); the quantity ν has the same dimensions as the self-diffusion constant but is typically two orders of magnitude larger than D for, say, an argon-like liquid near its triple point. If we multiply through Eqn (8.4.1) by $j_{-\mathbf{k}}^{x}$ and take the thermal average, we find that the transverse-current autocorrelation function satisfies the equation

$$\frac{\partial}{\partial t} C_t(k, t) + \nu k^2 C_t(k, t) = 0 \tag{8.4.2}$$

Equation (8.4.2) is easily solved to give

$$\begin{aligned} \lim_{kl \ll 1, t/\tau \gg 1} C_t(k, t) &= C_t(k, 0) \exp(-\nu k^2 t) \\ &= \omega_0^2 \exp(-\nu k^2 t) \end{aligned} \tag{8.4.3}$$

The exponential decay in (8.4.3) is typical of a diffusive process (see Section 8.2).

The diffusive character of the hydrodynamic "shear" mode is also apparent in the fact that the Laplace transform of $C_t(k, t)$ has a purely imaginary pole corresponding to the root (8.3.33) of $D(k, z)$:

$$\tilde{C}_t(k, z) = \frac{\omega_0^2}{-iz + \nu k^2} \tag{8.4.4}$$

Let $z = \omega + i\varepsilon$ approach the real axis from above ($\varepsilon \to 0+$). Then $\tilde{C}_t(k, \omega)$ at small k is given approximately by

$$\begin{aligned} \tilde{C}_t(k, \omega) &= \frac{\omega_0^2}{-i\omega}\left(1 - \frac{\nu k^2}{i\omega}\right)^{-1} \\ &\simeq \frac{\omega_0^2}{-i\omega}\left(1 + \frac{\nu k^2}{i\omega}\right) \end{aligned} \tag{8.4.5}$$

If we substitute for ω_0^2 from Eqn (7.4.39), and recall the definition (8.3.27), we find that the shear viscosity, which must be real, is related to the long-wavelength, low-frequency behaviour of $\tilde{C}_t(k, \omega)$ by

$$\begin{aligned} \eta &= \beta \rho m^2 \lim_{\omega \to 0} \lim_{k \to 0} \frac{\omega^2}{k^4} \operatorname{Re} \tilde{C}_t(k, \omega) \\ &= \pi \beta \rho m^2 \lim_{\omega \to 0} \lim_{k \to 0} \frac{\omega^2}{k^4} C_t(k, \omega) \end{aligned} \tag{8.4.6}$$

where $C_t(k, \omega)$ is the spectrum of transverse-current fluctuations, i.e. the Fourier transform of $C_t(k, t)$; this result is the analogue of the expression (8.2.11) for the self-diffusion coefficient. From the properties of the Laplace transform and the definition of $C_t(k, t)$, it follows that

$$\int_0^\infty \frac{k^2}{N}\langle \dot{j}_{\mathbf{k}}^x(t)\dot{j}_{-\mathbf{k}}^x\rangle \exp(i\omega t)\,\mathrm{d}t$$

$$= -\int_0^\infty \frac{d^2}{dt^2}C_t(k, t)\exp(i\omega t)\,\mathrm{d}t$$

$$= \omega^2\tilde{C}_t(k, \omega) - i\omega\omega_0^2 \tag{8.4.7}$$

We may therefore rewrite (8.4.6) as

$$\eta = \beta\rho m^2 \lim_{\omega\to 0}\int_0^\infty \lim_{k\to 0}\frac{1}{Nk^2}\langle \dot{j}_{\mathbf{k}}^x(t)\dot{j}_{-\mathbf{k}}^x\rangle \exp(i\omega t)\,\mathrm{d}t \tag{8.4.8}$$

The derivative of the transverse current can be expressed in terms of the stress tensor via the conservation law (8.3.3). Taking the Fourier transform of the latter, and remembering that the z-axis is the direction of $\mathbf{k}$ and that $\mathbf{p}(\mathbf{r}, t) = m\mathbf{j}(\mathbf{r}, t)$, we find that

$$\dot{j}_{\mathbf{k}}^x(t) + \frac{ik}{m}\sigma_{\mathbf{k}}^{xz}(t) = 0 \tag{8.4.9}$$

Combination of Eqns (8.4.8) and (8.4.9) shows that the shear viscosity is given by the time integral of the autocorrelation function of an off-diagonal element of the stress tensor in the limit $k \to 0$:

$$\eta = \frac{\beta}{V}\int_0^\infty \langle \sigma_0^{xz}(t)\sigma_0^{xz}\rangle\,\mathrm{d}t$$

$$= \int_0^\infty \eta(t)\,\mathrm{d}t \tag{8.4.10}$$

where

$$\eta(t) = \frac{\beta}{V}\lim_{k\to 0}\langle \sigma_{\mathbf{k}}^{xz}(t)\sigma_{-\mathbf{k}}^{xz}\rangle \tag{8.4.11}$$

In order to relate the shear viscosity to the intermolecular forces, it is necessary to have a microscopic expression for the stress tensor. It follows from the definition (7.4.7) of the particle current that

$$m\dot{j}_{\mathbf{k}}^{\alpha} = m\sum_{i=1}^{N}(\dot{u}_{i\alpha} - ik_\beta u_{i\alpha}u_{i\beta})\exp(-i\mathbf{k}\cdot\mathbf{r}_i) \tag{8.4.12}$$

where α denotes any of x, y and z, and β is the direction of $\mathbf{k}$; the relation to the stress tensor is then established by use of Eqn (8.4.9) with $\alpha = x$ and $\beta = z$. To introduce the pair potential $v(r)$, we note that $\mathbf{r}_{ji} = -\mathbf{r}_{ij} = \mathbf{r}_i - \mathbf{r}_j$, and rewrite the first term on the right-hand side of Eqn (8.4.12) successively as

$$\begin{aligned} m \sum_{i=1}^{N} \dot{u}_{i\alpha} \exp(-i\mathbf{k}\cdot\mathbf{r}_i) &= \sum_{i\neq j}^{N}\sum \frac{r_{ij\alpha}}{|\mathbf{r}_{ij}|} v'(r_{ij}) \exp(-i\mathbf{k}\cdot\mathbf{r}_i) \\ &= \tfrac{1}{2}\sum_{i\neq j}^{N}\sum \frac{r_{ij\alpha}}{|\mathbf{r}_{ij}|} v'(r_{ij})[\exp(-i\mathbf{k}\cdot\mathbf{r}_i) - \exp(-i\mathbf{k}\cdot\mathbf{r}_j)] \\ &= \tfrac{1}{2} ik_\beta \sum_{i\neq j}^{N}\sum \frac{r_{ij\alpha} r_{ij\beta}}{ik_\beta r_{ij\beta}|\mathbf{r}_{ij}|}[\exp(-i\mathbf{k}\cdot\mathbf{r}_i) \\ &\quad - \exp(-i\mathbf{k}\cdot\mathbf{r}_j)]\, v'(r_{ij}) \end{aligned} \tag{8.4.13}$$

where $v'(r_{ij}) = \mathrm{d}v(r_{ij})/\mathrm{d}r_{ij}$; the second step is taken by writing each term in the double sum as a half the sum of two equal terms. Introducing a quantity $\Phi_{\mathbf{k}}(\mathbf{r})$, defined as

$$\Phi_{\mathbf{k}}(\mathbf{r}) = r\frac{\mathrm{d}v(r)}{\mathrm{d}r}\frac{\exp(i\mathbf{k}\cdot\mathbf{r}) - 1}{i\mathbf{k}\cdot\mathbf{r}} \tag{8.4.14}$$

we finally obtain an expression for $\sigma_{\mathbf{k}}^{\alpha\beta}$ in the form

$$\sigma_{\mathbf{k}}^{\alpha\beta} = \sum_{i=1}^{N}\left(m u_{i\alpha} u_{i\beta} + \tfrac{1}{2}\sum_{j\neq i}^{N} \frac{r_{ij\alpha} r_{ij\beta}}{r_{ij}^2}\Phi_{\mathbf{k}}(\mathbf{r}_{ij})\right)\exp(-i\mathbf{k}\cdot\mathbf{r}_i) \tag{8.4.15}$$

The Green–Kubo relation for the shear viscosity is then obtained by inserting (8.4.15) (taken for $\mathbf{k} = 0$) in Eqn (8.4.11). Note that it follows directly from the virial theorem that

$$\langle \sigma_0^{\alpha\alpha} \rangle = PV \tag{8.4.16}$$

whereas

$$\langle \sigma_0^{\alpha\beta} \rangle = 0, \qquad \alpha \neq \beta \tag{8.4.17}$$

Equation (8.4.11) is not directly applicable to a hard-sphere fluid because the potential $v(r)$ has a discontinuity at $r = d$ (the hard-sphere diameter). However, the microscopic expression for the shear viscosity and, for that matter, the formulae to be derived later for other transport coefficients, can be recast in a form that is similar to the Einstein relation (7.2.3) for the self-diffusion coefficient and is valid even for hard spheres. A Green–Kubo formula for a transport coefficient K, including both Eqns (7.8.10) and (8.4.10), can always be written as

$$K = \frac{\beta}{V}\int_0^\infty \langle \dot{A}(t)\dot{A}\rangle \,\mathrm{d}t \tag{8.4.18}$$

where A is a microscopic dynamical variable. The argument used to derive Eqn (7.2.8) from (7.2.3) shows that Eqn (8.4.18) is equivalent to writing

$$K = \frac{\beta}{V}\lim_{t\to\infty}\frac{1}{2t}\langle[A(t)-A(0)]^2\rangle \tag{8.4.19}$$

which may be regarded as a generalized form of the Einstein relation (7.2.3). In the case of the shear viscosity, we see from Eqn (8.4.8) that the variable $A(t)$ is

$$\begin{aligned} A(t) &= \lim_{k\to 0}\frac{im}{k} j_{\mathbf{k}}^x(t) \\ &= \lim_{k\to 0}\frac{im}{k}\sum_{i=1}^N u_{ix}(t)[1 - ikr_{iz}(t) + \mathcal{O}(k^2)] \\ &= m\sum_{i=1}^N u_{ix}(t) r_{iz}(t) \end{aligned} \tag{8.4.20}$$

where a frame of references has been chosen in which the total momentum of the particles (a conserved quantity) is zero. Hence the generalized Einstein relation for the shear viscosity is

$$\eta = \frac{\beta m^2}{V}\lim_{t\to\infty}\frac{1}{2t}\left\langle\left\{\sum_{i=1}^N [u_{ix}(t)r_{iz}(t) - u_{ix}(0)r_{iz}(0)]\right\}^2\right\rangle \tag{8.4.21}$$

The quantity σ_0^{xz} that appears in the Green–Kubo formula (8.4.10) is the sum of a kinetic part and a potential part. Consequently, there are three distinct contributions to the shear viscosity: a purely kinetic term, corresponding to the transport of transverse momentum via the displacement of particles; a purely potential term, arising from the action of the interparticle forces ("collisional" transport); and a cross term. At liquid densities, the potential term is by far the most important of the three. In the Enskog approximation, the shear viscosity of a hard-sphere fluid is

$$\frac{\eta_{\mathrm{E}}}{\eta_{\mathrm{B}}} = \frac{2\pi\rho d^3}{3}\left(\frac{1}{y} + 0.8 + 0.761y\right) \tag{8.4.22}$$

where $y = \beta p/\rho - 1$ and $\eta_{\mathrm{B}} = (5/16d^2)(mk_{\mathrm{B}}T/\pi)^{1/2}$ is the density-independent Boltzmann result (Résibois and DeLeener, 1977). The three terms between brackets in Eqn (8.4.22) represent, successively, the kinetic, cross and potential contributions; the last of these is clearly dominant close to the fluid-solid transition, where $y \simeq 10$. Figure 8.2 compares the results of molecular dynamics calculations for hard spheres with the predictions of the Enskog theory for each of the three terms in (8.4.22) (Alder *et al.*, 1970). The Enskog approximation is very successful for $V \geqslant 5V_0$, where V_0 is the close-packed volume. At higher densities, there are significant deviations;

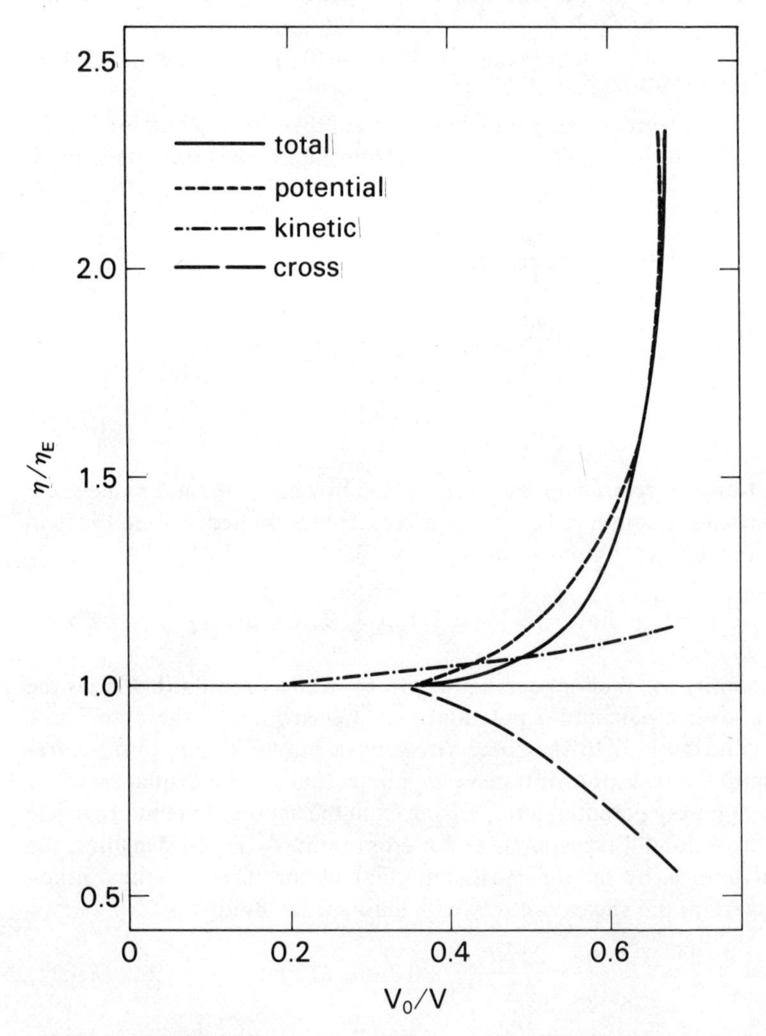

FIG. 8.2. Shear viscosity of the hard-sphere fluid relative to its Enskog value: $V_0 = Nd^3/\sqrt{2}$ is the volume at close packing. The curves represent different contributions to η (see text). After Alder *et al.* (1970).

near solidification, the theory underestimates the shear viscosity by a factor of approximately two. The behaviour of the self-diffusion coefficient at high densities is the reverse of this, as we saw in Section 7.2. The net result is that the product $D\eta$ calculated from the molecular dynamics data is almost constant for $V \leqslant 5V_0$; its value is within 10% of that predicted by Stoke's law (7.3.20) with slip boundary conditions. The enhancement of the shear viscosity at high densities relative to the Enskog result is linked numerically to the presence of a slowly decaying, positive "tail" in the stress-tensor autocorrelation function. Very careful molecular dynamics studies of hard spheres have shown that $\eta(t)$ decays as $t^{-3/2}$ for t in the range from 10 to 30 mean collision times (Erpenbeck and Wood, 1981). This result is in qualitative agreement with calculations based on the hydrodynamic "mode-coupling" theory to be discussed in Section 8.7 (Ernst *et al.*, 1971, 1976a,b). However, the observed coefficient of the $t^{-3/2}$ tail is two orders of magnitude larger than the theoretical value; a similar discrepancy has been found in molecular dynamics calculations for the Lennard-Jones fluid (Evans, 1980; Kirkpatrick, 1984). The shear-stress autocorrelation function for the Lennard-Jones fluid at a state close to the triple point is shown in Figure 8.3. A long-time tail is clearly present (Heyes, 1984; Schoen and Hoheisel, 1985), but at higher temperatures and lower densities the tail, if any, is not perceptible (Holian and Evans, 1983).

Viscous flow in simple, model fluids has also been studied by a variety of "non-equilibrium" molecular dynamics (NEMD) techniques (Evans and Morriss, 1984). In the simplest method, an external shearing force is applied to the particles (Gosling *et al.*, 1973). Since the shear viscosity determines the rate at which, say, x-momentum diffuses in the z-direction, it is natural to apply the shearing force in the x-direction with a magnitude varying with z. To be consistent with the periodic boundary conditions, the force $\mathbf{F} = (F_x, 0, 0)$ is chosen to have the form $F_x(z) = F_0 \sin(2\pi nz/L)$, where n is an integer and L is the length of the simulation cell in the z-direction. Such a force will give rise to a drift velocity in the x-direction that varies in magnitude with z. The Navier–Stokes equation (8.3.7) now takes the form

$$\rho m \frac{\partial u_x}{\partial t} = \eta \frac{\partial^2 u_x}{\partial z^2} + \rho F_x(z) \tag{8.4.23}$$

to which the steady-state solution satisfying the boundary conditions $u_x(0) = u_x(L) = 0$ is

$$u_x(z) = \frac{\rho L^2 F_0}{4\pi^2 n^2 \eta} \sin(2\pi nz/L) \tag{8.4.24}$$

By fitting the observed steady-state drift-velocity profile to the sinusoidal form $u_x(z) = u_0 \sin(2\pi nz/L)$, a k-dependent shear viscosity can be obtained

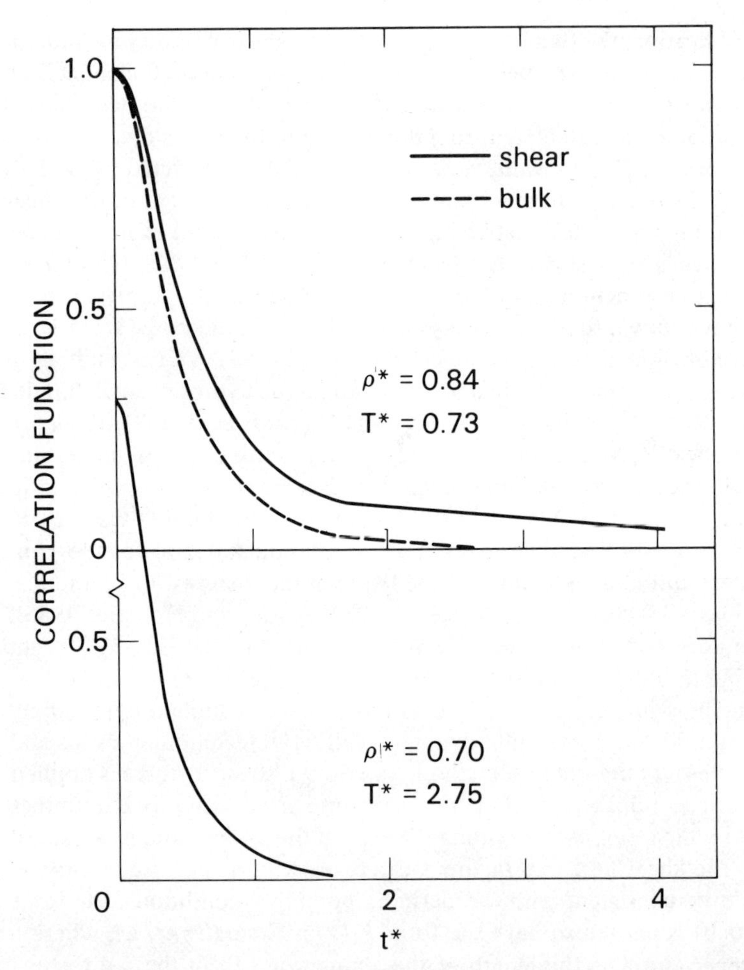

FIG. 8.3. Normalized Green–Kubo integrands for the shear and bulk viscosities of the Lennard-Jones fluid. The unit of time is the quantity τ_0 defined by Eqn (3.3.5). After Holian and Evans (1983) and Heyes (1984).

as

$$\eta(k=2\pi n/L)=\frac{\rho L^2 F_0}{4\pi^2 n^2 u_0} \tag{8.4.25}$$

The main disadvantage of the method is the fact that the macroscopic shear viscosity must be estimated from an extrapolation of $\eta(k)$ to $k=0$. The heat produced by viscous dissipation leads to a steady increase in the temperature of the system, but this problem can be overcome either by including an *ad hoc* mechanism for heat absorption or by using the subtraction technique described in Section 7.8 (Ciccotti *et al.*, 1976).

A second class of NEMD methods involves the simulation of Couette flow. In the calculations of Ashurst and Hoover (1975), for example, shear flow is maintained by external momentum reservoirs composed of layers of particles having a fixed drift velocity and temperature. The presence of the external layers means that the boundary conditions are no longer periodic; partly for this reason, the computed shear viscosities show a significant dependence on system size. Periodicity is preserved if a "homogeneous-shear" algorithm is used (Lees and Edwards, 1972; Hoover and Ashurst, 1975; Singer *et al.*, 1980; Trozzi and Ciccotti, 1984). In particular, it is possible to work with time-varying, non-orthogonal, periodic boundaries of a nature such that the entire periodic array of image cells undergoes a steady, planar, shearing motion. The strain rate

$$\gamma = \frac{\partial u_x}{\partial z} \tag{8.4.26}$$

is maintained by inclining the z-axis of the periodic lattice at an angle θ with respect to the x-axis, where θ is a function of time given by

$$\theta(t) = \tfrac{1}{2}\pi - \tan^{-1} \gamma t \tag{8.4.27}$$

Additionally, if a particle with velocity u_x leaves the basic cell through the upper or lower face in the z-direction, the corresponding image particle enters the cell through the opposite face with a velocity $u'_x = u_x \pm \gamma L$. The shear viscosity is obtained by averaging the microscopic expression for the xz-component of the stress tensor (8.4.15), i.e.

$$\sigma_0^{xz} = \sum_{i=1}^{N} \left(m[u_{ix} - u_x(z_i)]u_{iz} + \tfrac{1}{2} \sum_{j \neq i}^{N} x_{ij} \frac{\partial v(r_{ij})}{\partial z_{ij}} \right) \tag{8.4.28}$$

and identifying the result with the macroscopic expression (8.3.5):

$$\sigma^{xz} = -\eta \frac{\partial u_x}{\partial z} = -\eta\gamma \tag{8.4.29}$$

Note that Eqn (8.4.28) differs from (8.4.15) through the subtraction of the streaming velocity $u_x(z_i)$ from the instantaneous velocity of particle i. The measured shear viscosity is a function of the strain rate γ; to obtain a value for the linear shear viscosity, it is necessary to extrapolate $\eta(\gamma)$ to $\gamma = 0$. Results obtained for Lennard-Jones and other similar potentials show that $\eta(\gamma)$ decreases with increasing γ (the phenomenon of "shear thinning") and is linear in $\gamma^{1/2}$ over a considerable range , i.e.

$$\eta(\gamma) = \eta(0) - \eta_\gamma \gamma^{1/2} \tag{8.4.30}$$

where η_γ is a positive quantity (Evans, 1981). If a time-dependent, periodic strain rate of the form $\gamma = \gamma_0 \cos \omega t$ is used, it is possible to determine the complex, frequency-dependent shear viscosity $\eta(\omega) = \eta'(\omega) + i\eta''(\omega)$ (Evans, 1980, 1981). The imaginary part of $\eta(\omega)$ determines the out-of-phase component of the shear stress; this component is responsible for the observed viscoelastic behaviour of liquids at high frequencies (see Section 8.7). The real part is found to be a non-analytic function of frequency for $\omega \to 0$, behaving as

$$\eta'(\omega) = \eta(0) - \eta_{1/2}\omega^{1/2} + \cdots \tag{8.4.31}$$

The function $\eta(\omega)$ is simply the Laplace transform of the shear-stress autocorrelation function $\eta(t)$ of Eqn (8.4.10); the low-frequency behaviour represented by (8.4.31) is a reflection of the $t^{-3/2}$ decay of the time function.

Instead of inducing planar Couette flow by the use of moving boundaries, a non-equilibrium shear deformation of the system can be achieved by appropriately modifying the equations of motion of the particles (Hoover *et al.*, 1980). If the strain rate $\nabla \mathbf{u}$ is of the simple form (8.4.26), the hamiltonian of the system is

$$\mathcal{H} = \mathcal{H}_0 + \sum_{i=1}^{N} x_i p_{iz} \frac{\partial u_x(z_i)}{\partial z_i} \tag{8.4.32}$$

where $\mathcal{H}_0$ is the usual sum of kinetic and intermolecular potential energy terms. The equations of motion are then (Hoover *et al.*, 1982)

$$\dot{x}_i = p_{ix}/m + \gamma z_i, \qquad \dot{y}_i = p_{iy}/m, \qquad \dot{z}_i = p_{iz}/m \tag{8.4.33a}$$

and

$$\dot{p}_{ix} = F_{ix}, \qquad \dot{p}_{iy} = F_{iy}, \qquad \dot{p}_{iz} = F_{iz} - \gamma p_{ix} \tag{8.4.33b}$$

A non-equilibrium state of constant temperature can be maintained either by scaling the particle velocities or, more elegantly, by applying a constraint of constant total kinetic energy (D. J. Evans *et al.*, 1983). The kinetic-energy constraint gives rise to additional frictional forces in the equations of motion of a type that also appears in the constant-temperature algorithm of Section 3.6. The shear viscosity is finally calculated from Eqns (8.4.28) and (8.4.29) in the manner already described.

The results obtained by different NEMD schemes for the shear viscosity of the Lennard-Jones fluid are in fair agreement with each other when due account is taken of the dependence on system size, shear rate and wavenumber. Reasonable agreement has also been achieved with experimental data for argon along the saturated vapour-pressure line (Hoover and Ashurst, 1975). The NEMD methods are computationally more efficient than equilibrium calculations based on the Green–Kubo formula (8.4.10). It should not be overlooked, however, that a single equilibrium simulation gives results for all transport coefficients, but the study of each different transport process by a NEMD method requires its own algorithm.

8.5 LONGITUDINAL COLLECTIVE MODES

The longitudinal modes are those associated with fluctuations in density, temperature and the projection of the momentum along the direction of **k**. It is clear from the structure of the hydrodynamic matrix (8.3.28) that the variables $\tilde{\rho}_{\mathbf{k}}(z)$, $\tilde{T}_{\mathbf{k}}(z)$ and $\tilde{j}^z_{\mathbf{k}}(z)$ are coupled together. The analysis is therefore more complicated than in the case of the transverse-current fluctuations discussed in the preceding section. There are three longitudinal modes, corresponding to the roots z_0, z_+ and z_- displayed in Eqns (8.3.37). The significance of the different roots is most easily grasped by solving the system of coupled longitudinal equations contained in (8.3.28) to obtain the hydrodynamic limiting form of the dynamic structure factor $S(k, \omega)$ (Mountain, 1966). The solution for $\tilde{\rho}_{\mathbf{k}}(z)$ involves terms proportional, respectively, to the initial values $\rho_{\mathbf{k}}$, $T_{\mathbf{k}}$ and $j^z_{\mathbf{k}}$. We omit the term proportional to $j^z_{\mathbf{k}}$, because **k** can always be chosen to make $\mathbf{u}_{\mathbf{k}}$ (the Fourier transform of the initial velocity field $\mathbf{u}(\mathbf{r}, 0)$) perpendicular to **k**, thereby ensuring that $j^z_{\mathbf{k}} = 0$. We also ignore the term proportional to $T_{\mathbf{k}}$; this contributes nothing to the final expression for $S(k, \omega)$, since the fluctuations in temperature and density are instantaneously uncorrelated, i.e. $\langle T_{\mathbf{k}}\rho_{-\mathbf{k}}\rangle = 0$ (cf. Eqn (2.7.5)). With these simplifications, the solution for $\tilde{\rho}_{\mathbf{k}}(z)$ can be written as

$$\tilde{\rho}_{\mathbf{k}}(z) = \rho_{\mathbf{k}} \frac{(-iz + ak^2)(-iz + bk^2) + (\gamma - 1)c_s^2 k^2/\gamma}{D_l(k, z)} \tag{8.5.1}$$

where all quantities are as defined in Section 8.3. Separation of the right-hand side of Eqn (8.5.1) into partial fractions shows that on the real axis $\tilde{\rho}_{\mathbf{k}}$ is given by

$$\tilde{\rho}_{\mathbf{k}}(\omega) = \rho_{\mathbf{k}} \left[\left(\frac{\gamma - 1}{\gamma} \right) \frac{1}{-i\omega + D_T k^2} + \frac{1}{2\gamma} \left(\frac{1}{-i\omega + \Gamma k^2 - ic_s k} + \frac{1}{-i\omega + \Gamma k^2 + ic_s k} \right) \right] \tag{8.5.2}$$

which, via an inverse transform, yields an expression for $\rho_{\mathbf{k}}(t)$ of the form

$$\rho_{\mathbf{k}}(t) = \rho_{\mathbf{k}} \left[\left(\frac{\gamma - 1}{\gamma} \right) \exp(-D_T k^2 t) + \frac{1}{\gamma} \exp(-\Gamma k^2 t) \cos c_s k t \right] \tag{8.5.3}$$

It is clear from the structure of Eqn (8.5.3) that the purely imaginary root in (8.3.37) corresponds to a fluctuation that decays without propagating, the lifetime of the fluctuation being determined by the thermal diffusivity D_T. By contrast, the complex roots correspond to a fluctuation that propagates through the fluid at the speed of sound, eventually decaying through the combined effects of viscosity and thermal conduction, as represented by the expression (8.3.39) for Γ; note that the thermal damping of the sound mode is small when $\gamma \simeq 1$. On multiplying through Eqn (8.5.3) by $\rho_{-\mathbf{k}}$, dividing by N and taking the thermal average, we obtain an expression for the density autocorrelation function $F(k, t)$; this is easily Fourier transformed to give

$$S(k, \omega) = \frac{1}{2\pi} S(k) \left[\left(\frac{\gamma - 1}{\gamma} \right) \frac{2 D_T k^2}{\omega^2 + (D_T k^2)^2} + \frac{1}{\gamma} \left(\frac{\Gamma k^2}{(\omega + c_s k)^2 + (\Gamma k^2)^2} + \frac{\Gamma k^2}{(\omega - c_s k)^2 + (\Gamma k^2)^2} \right) \right] \quad (8.5.4)$$

The spectrum of density fluctuations therefore consists of three components: the Rayleigh line, centred at $\omega = 0$, and two Brillouin lines at $\omega = \pm c_s k$; a typical spectrum is pictured in Figure 8.4. The two shifted components correspond to the propagating mode, and are analogous to the longitudinal acoustic phonons of a solid, whereas the central line represents the steadily decaying thermal mode. The total integrated intensity of the Rayleigh line is easily calculated to be

$$\mathcal{I}_R = \frac{\gamma - 1}{\gamma} S(k) \quad (8.5.5)$$

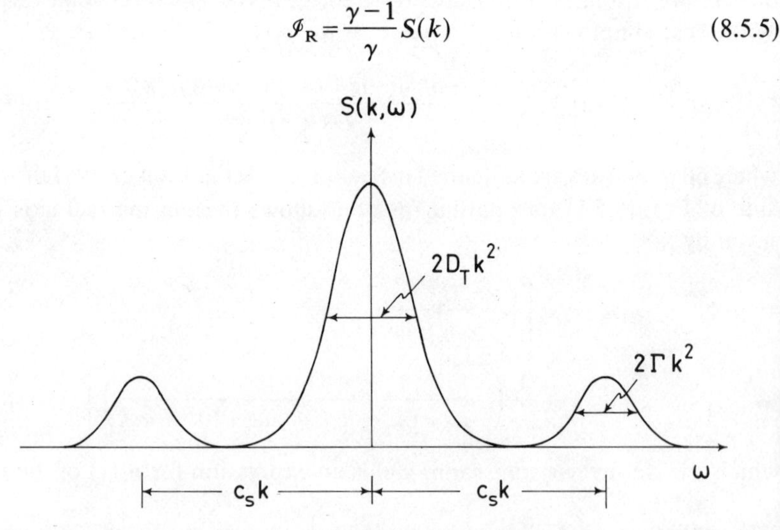

FIG. 8.4. Dynamic structure factor in the hydrodynamic limit; D_T is the thermal diffusivity, Γ is the sound-attenuation coefficient and c_s is the adiabatic speed of sound.

and that of each of the Brillouin lines is

$$\mathcal{I}_{\mathrm{B}} = \frac{1}{2\gamma} S(k) \tag{8.5.6}$$

Thus

$$\mathcal{I}_{\mathrm{R}} + 2\mathcal{I}_{\mathrm{B}} = S(k) \tag{8.5.7}$$

which is a particular case of the sum rule (7.4.32). The ratio

$$\frac{\mathcal{I}_{\mathrm{R}}}{2\mathcal{I}_{\mathrm{B}}} = \gamma - 1 \tag{8.5.8}$$

is called the Landau–Placzek ratio. In passing from Eqn (8.5.1) to (8.5.2) we have, for the sake of simplicity, omitted a non-lorentzian term; in practice, this term makes only a negligibly small, asymmetric correction to the Brillouin lines (Boon and Yip, 1980).

We have chosen to discuss the behaviour of the longitudinal modes in terms of local fluctuations in density and temperature, but it would have been equally appropriate to choose the pressure and entropy as variables, since these are also statistically independent (cf. Eqn (2.7.9)). The calculation is instructive, since it shows that the first term in (8.5.2) can be identified with the decay of entropy fluctuations (McIntyre and Sengers, 1968). It follows that the Brillouin doublet is associated with propagating pressure fluctuations at constant entropy (hence the appearance of the adiabatic speed of sound), while the Rayleigh line corresponds to non-propagating fluctuations in entropy at constant pressure.

The wavelength of visible light is much greater than the nearest-neighbour interparticle spacing in liquids. Light-scattering experiments are therefore ideally suited to measurements of the Rayleigh–Brillouin spectrum at long wavelengths; with the use of lasers and high-resolution Fabry–Perot interferometers, such experiments are capable of yielding accurate values of the thermal diffusivity and of the speed and attenuation coefficient of sound waves at frequencies up to 10 GHz (Lallemand, 1974; Berne and Pecora, 1976). The light scattered by a fluid of optically isotropic molecules is expected to have the same polarization as the incident beam. However, the wings of the Rayleigh line contain a weak depolarized component that arises from the small anisotropies induced by local density fluctuations and by collisions; the spectrum of depolarized light scattered by atomic fluids has been studied both experimentally (Fleury *et al.*, 1971) and by molecular dynamics simulations (Alder *et al.*, 1973).

A question of great interest is whether the propagating density fluctuations characteristic of the hydrodynamic regime can also be supported in simple liquids at wavelengths comparable with the spacing between particles. We shall consider this problem in detail in Chapter 9; here we merely record the fact that well-defined collective excitations of the hydrodynamic type, manifesting themselves in a three-peak structure in $S(k, \omega)$, have been observed in neutron scattering experiments on rubidium (Copley and Rowe, 1974a, b) and lead (Söderström *et al.*, 1980), for $k \leqslant 1$ Å^{-1}, and on neon, for $k \leqslant 0.15$ Å (Bell *et al.*, 1975). Some of the results obtained for liquid rubidium are plotted in Figure 8.5; the good agreement between energy-loss and energy-gain data shows that the measurements are consistent with the principle of detailed balance (see Section 7.5). Neutron experiments at low momentum transfer are particularly difficult, not least because $S(k)$ is very small, i.e. the integrated intensity of scattered radiation is very weak. Brillouin-type sidepeaks have also been observed in molecular dynamics calculations on a variety of systems. A problem with the simulations, already mentioned in Section 3.1, is the fact that use of a periodic boundary condition places a lower limit on the accessible, non-zero values of k. In all cases in which sidepeaks have been seen, the frequency of the mode is close to what is expected on the basis of the macroscopic speed of sound, but in other respects the spectrum is not well described by Eqn (8.5.4). As we shall see in Section 8.7, and again in Chapter 9, a generalization of the hydrodynamic approach is needed; the effect of this is to replace the transport coefficients in (8.5.4) by quantities that are functions of both frequency and wavenumber. The dominant corrections to the dispersion relations (8.3.37) can be calculated by taking account of the coupling between pairs of hydrodynamic modes (Ernst and Dorfman, 1975); these corrections are of order $k^{5/2}$. Kinetic theory predicts that a remnant of a strongly damped sound mode should persist to high wavenumbers with a

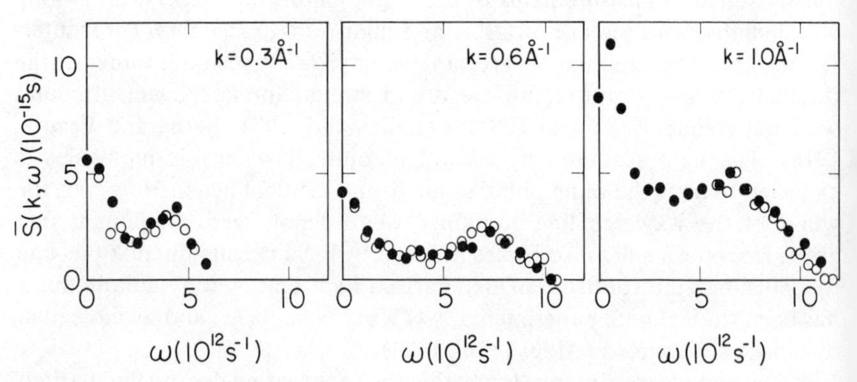

FIG. 8.5. Neutron-scattering results for the symmetrized dynamic structure factor of liquid rubidium. The open circles are energy-loss data and the filled circles represent energy gain. After Copley and Rowe (1974a).

dispersion relation not unlike that of longitudinal phonons in the solid (deSchepper and Cohen, 1980, 1982). The mode is not expected to be visible as a peak in $S(k, \omega)$; some evidence for its existence has been found in the analysis of neutron scattering experiments on liquid argon (deSchepper *et al.*, 1983), but the results are open to other interpretations (Lovesey, 1984b).

For later purposes, we also require an expression for the hydrodynamic limit of the longitudinal-current autocorrelation function $C_l(k, t)$. We proceed, as before, by solving the system of equations (8.3.28) for the variable of interest, which in this case is the longitudinal particle current $\tilde{j}^z_{\mathbf{k}}(z)$. The terms in $\rho_{\mathbf{k}}$ and $T_{\mathbf{k}}$ may be omitted, since they are uncorrelated with $\tilde{j}^z_{-\mathbf{k}}$. For z on the real axis, the result is

$$\tilde{j}^z_{\mathbf{k}}(\omega) = j^z_{\mathbf{k}} \frac{-i\omega(-i\omega + ak^2)}{D_l(k, \omega)} \tag{8.5.9}$$

Thus

$$\tilde{C}_l(k, \omega) = \frac{\omega_0^2}{-i\omega + bk^2 + c_s^2 k^2 \left(\dfrac{1}{-i\omega} + \dfrac{\gamma - 1}{-i\omega + ak^2} \right)} \tag{8.5.10}$$

The same result can be obtained from Eqn (7.4.37) and the hydrodynamic result (8.5.4) for $S(k, \omega)$.

According to Eqn (8.5.10), the spectrum of longitudinal current fluctuations behaves at small k as

$$\begin{aligned} C_l(k, \omega) &= \frac{1}{\pi} \operatorname{Re} \tilde{C}_l(k, \omega) \\ &\simeq \frac{\omega_0^2}{\pi\omega^2} \left[bk^2 + \frac{(\gamma - 1) a c_s^2 k^4}{\omega^2 + (ak^2)^2} \right] \end{aligned} \tag{8.5.11}$$

Hence the longitudinal viscosity $\rho m b = (\frac{4}{3}\eta + \zeta)$ is given by a limiting operation entirely analogous to Eqn (8.4.6) for the shear viscosity, i.e.

$$\tfrac{4}{3}\eta + \zeta = \lim_{\omega \to 0} \lim_{k \to 0} \frac{\omega^2}{k^4} C_l(k, \omega) \tag{8.5.12}$$

If we proceed in the manner that leads to the Green–Kubo formula (8.4.10), we find that the longitudinal viscosity can be expressed in terms of the autocorrelation function of a diagonal element of the stress tensor (8.4.15):

$$\tfrac{4}{3}\eta + \zeta = \lim_{\omega \to 0} \frac{\beta}{V} \int_0^\infty \langle \sigma_0^{zz}(t) \sigma_0^{zz} \rangle \exp(i\omega t) \, dt \tag{8.5.13}$$

In taking the limit $\omega = 0$ in (8.5.13), we find a discontinuity, because the thermal average of σ_0^{zz} is non-zero (see Eqn (8.4.16)). The problem is overcome by subtracting the invariant part, the transport coefficient being linked only to fluctuations in the local variables. Equation (8.5.13) then becomes

$$\tfrac{4}{3}\eta+\zeta=\frac{\beta}{V}\int_0^\infty \langle[\sigma_0^{zz}(t)-PV][\sigma_0^{zz}-PV]\rangle\,\mathrm{d}t \tag{8.5.14}$$

which is the longitudinal analogue of (8.4.11).

To obtain the Green–Kubo relation for the thermal conductivity, we require an expression for the rate of decay of a fluctuation in $Q(\mathbf{r}, t)$, the macroscopic density of heat energy; $Q(\mathbf{r}, t)$ is related to the entropy density by Eqn (8.3.13). We first use Eqn (8.3.15) and a standard thermodynamic chain rule to eliminate the local temperature from the energy conservation equation (8.3.18). The result, after transformation to Fourier–Laplace variables, is

$$(-iz+ak^2)\tilde{Q}_{\mathbf{k}}(z)+\lambda k^2\left(\frac{\partial T}{\partial\rho}\right)_S\tilde{\rho}_{\mathbf{k}}(z)=Q_{\mathbf{k}} \tag{8.5.15}$$

Next, an equation that relates $\tilde{\rho}_{\mathbf{k}}(z)$ to $\tilde{P}_{\mathbf{k}}(z)$ is obtained by taking the divergence of the Navier–Stokes equation (8.3.7), eliminating the particle current with the help of (8.3.17) and transforming again to the variables k and z; the result in this case is

$$izm(-iz+bk^2)\tilde{\rho}_{\mathbf{k}}(z)-k^2\tilde{P}_{\mathbf{k}}(z)=-m(-iz+bk^2)\rho_{\mathbf{k}} \tag{8.5.16}$$

where $\mathbf{k}$ is again chosen perpendicular to the initial particle current. Equation (8.5.16) can now be converted into a relation for $\tilde{Q}_{\mathbf{k}}(z)$ by making the substitutions

$$\tilde{P}_{\mathbf{k}}(z)=\left(\frac{\partial P}{\partial\rho}\right)_S\tilde{\rho}_{\mathbf{k}}(z)+\frac{V}{T}\left(\frac{\partial P}{\partial S}\right)_\rho\tilde{Q}_{\mathbf{k}}(z) \tag{8.5.17}$$

and

$$\rho_{\mathbf{k}}=\left(\frac{\partial\rho}{\partial P}\right)_S P_{\mathbf{k}}+\frac{V}{T}\left(\frac{\partial\rho}{\partial S}\right)_P Q_{\mathbf{k}} \tag{8.5.18}$$

The final step is to eliminate $\tilde{\rho}_{\mathbf{k}}(z)$ between Eqns (8.5.15) and (8.5.16). The resulting expression for $\tilde{Q}_{\mathbf{k}}(z)$ has some similarities with that obtained previously for $\tilde{\rho}_{\mathbf{k}}(z)$ in Eqn (8.5.2). In particular, there are two complex-conjugate poles and a single imaginary pole (Kadanoff and Martin, 1963). At small k, the local pressure and density are uncorrelated (cf. Eqn (2.7.9)); we may therefore simplify the problem by discarding terms proportional to $P_{\mathbf{k}}$. The lowest-order solution for $\tilde{Q}_{\mathbf{k}}(z)$ then reduces to

$$\tilde{Q}_{\mathbf{k}}(z)=\frac{Q_{\mathbf{k}}}{-iz+D_{\mathrm{T}}k^2} \tag{8.5.19}$$

This represents a purely diffusive mode. The form of Eqn (8.5.19) confirms our earlier remark that the Rayleigh peak in $S(k, \omega)$ is associated with the decay of non-propagating entropy fluctuations.

Our main concern is with the behaviour at small k. We therefore replace $Q_{\mathbf{k}}$ by $Q_0 = T\Delta S$ and $\langle Q_{\mathbf{k}} Q_{-\mathbf{k}} \rangle$ by

$$\begin{aligned} \langle Q_0^2 \rangle &= T^2 \langle (\Delta S)^2 \rangle \\ &= T^2 N k_{\mathrm{B}} c_P \end{aligned} \tag{8.5.20}$$

where we have used the result in Eqn (2.7.10). We now proceed as in the earlier cases of the shear and longitudinal viscosities; multiplying (8.5.19) by $Q_{-\mathbf{k}}$ and taking the thermal average, we obtain the relation

$$\lambda = \frac{1}{V k_{\mathrm{B}} T^2} \lim_{\omega \to 0} \lim_{k \to 0} \frac{\omega^2}{k^2} \langle \tilde{Q}_{\mathbf{k}}(\omega) Q_{-\mathbf{k}} \rangle \tag{8.5.21}$$

If we introduce a fluctuating heat current $\mathbf{J}_{\mathbf{k}}^{Q}(t)$ defined, by virtue of (8.3.11), as the Fourier transform of

$$\mathbf{J}^{Q}(\mathbf{r}, t) = \mathbf{J}^{e}(\mathbf{r}, t) - \frac{e + P}{\rho} \mathbf{j}(\mathbf{r}, t) \tag{8.5.22}$$

we see that the energy conservation equation (8.3.4) can be expressed as

$$\dot{Q}_{\mathbf{k}}(t) + i\mathbf{k} \cdot \mathbf{J}_{\mathbf{k}}^{Q}(t) = 0 \tag{8.5.23}$$

Hence, if the z-axis is parallel to $\mathbf{k}$, we can rewrite Eqn (8.5.21) in typical Green–Kubo form:

$$\lambda = \lim_{\omega \to 0} \frac{1}{V k_{\mathrm{B}} T^2} \int_0^\infty \langle J_0^{Qz}(t) J_0^{Qz} \rangle \exp(i\omega t) \, \mathrm{d}t \tag{8.5.24}$$

To make use of Eqn (8.5.24), we require a microscopic expression for the heat current. Taking the Fourier transform of (8.3.4), we find that the component of $\mathbf{J}_{\mathbf{k}}^{e}$ in the direction of $\mathbf{k}$ is given by

$$\begin{aligned} -ikJ_{\mathbf{k}}^{ez} &= \dot{e}_{\mathbf{k}} \\ &= \frac{\mathrm{d}}{\mathrm{d}t} \sum_{i=1}^{N} \left[\tfrac{1}{2} m |\mathbf{u}_i|^2 + \tfrac{1}{2} \sum_{j \neq i}^{N} v(r_{ij}) \right] \exp(-i\mathbf{k} \cdot \mathbf{r}_i) \end{aligned} \tag{8.5.25}$$

where we adopt the convention that the potential energy of interaction of two particles is shared equally between them. Differentiation of the quantity inside square brackets gives rise to a term that can be treated by the method used in calculating the microscopic stress tensor; the final result for $\mathbf{k}=0$ is

$$J_0^{ez} = \sum_{i=1}^{N} u_{iz}\left[\tfrac{1}{2}m|\mathbf{u}_i|^2 + \tfrac{1}{2}\sum_{j\neq i}^{N} v(r_{ij})\right] - \tfrac{1}{2}\sum_{i\neq j}^{N}\sum \mathbf{u}_i \cdot \mathbf{r}_{ij}\frac{\partial v(r_{ij})}{\partial z_{ij}} \qquad (8.5.26)$$

The current J_0^{Qz} is obtained from J_0^{ez} by subtracting the term $(e+P)\sum_i u_{iz}$; with a suitable choice of frame of reference, this term will be zero and J_0^{ez} will also vanish. Thus we can equally well (Green, 1954; Mori, 1958; Kadanoff and Martin, 1963) write the Green–Kubo formula for λ as

$$\lambda = \frac{1}{Vk_B T^2}\int_0^\infty \langle J_0^{ez}(t) J_0^{ez}\rangle \, \mathrm{d}t \qquad (8.5.27)$$

The correlation-function formulae for the various transport coefficients, expressed in the general form of Eqn (8.4.18), are summarized in Table 8.1. These formulae, or the equivalent Einstein expressions (8.4.19), have been used to determine the transport coefficients of various model fluids by molecular dynamics simulations; results obtained for the shear viscosity have already been discussed in Section 8.4. Alder *et al.* (1970) have shown that the thermal conductivity and bulk viscosity (ζ) of the hard-sphere fluid are accurately given by the Enskog theory at all densities. This is in marked contrast to the behaviour of the shear viscosity and self-diffusion coefficient; as we have seen in earlier sections, both η and D differ from their Enskog values by factors of nearly two (but in opposite senses) at densities close to freezing. The good agreement found for λ may be attributed to the absence of a significant "tail" in the corresponding autocorrelation function as confirmed, for example, by the calculations of Levesque *et al.* (1973) for the Lennard-Jones fluid near its triple point. There is also no appreciable tail in the correlation function associated with ζ, as illustrated in Figure 8.3 (Heyes, 1984). However, the bulk viscosity of the Lennard-Jones fluid has also been determined by a non-equilibrium molecular dynamics method involving the application of an oscillatory adiabatic deformation (Hoover *et al.*, 1980); the observed $\omega^{1/2}$ dependence of the low-frequency response is consistent with a $t^{-3/2}$ decay of the corresponding time function.

Tenenbaum *et al.* (1982) have used a molecular dynamics method to simulate stationary non-equilibrium states of a Lennard-Jones fluid in which very large thermal gradients exist. The gradients are maintained by "stochastic" thermal walls corresponding to widely different temperatures; atoms colliding with the walls are reflected with velocities redistributed according to a Maxwell–Boltzmann distribution appropriate to the wall temperature

TABLE 8.1. *Green–Kubo relations for the transport coefficients in the form of Eqn* (8.4.18)

$$K = \int_0^\infty \langle J(t)J(0)\rangle \, dt$$

K	$J(t)$	Name of current	Eqn
D	$u_{ix}(t) = \dfrac{d}{dt} x_i(t)$	Particle velocity	(7.2.8)
$Vk_B T\eta$	$\sigma_0^{xz}(t) = \dfrac{d}{dt} m \sum_{i=1}^{N} u_{ix}(t) z_i(t)$	Off-diagonal component of stress tensor	(8.4.10)
$Vk_B T(\frac{4}{3}\eta + \zeta)$	$\sigma_0^{zz}(t) - PV = \dfrac{d}{dt} m \sum_{i=1}^{N} u_{iz}(t) z_i(t) - PV$	Diagonal component of stress tensor	(8.5.13)
$Vk_B T^2 \lambda$	$J_0^{ez}(t) = \dfrac{d}{dt} \sum_{i=1}^{N} z_i(t) \left\{ \frac{1}{2} m u_i^2(t) + \frac{1}{2} \sum_{j \neq i}^{N} v[r_{ij}(t)] \right\}$	Energy current	(8.5.27)
$\left(\dfrac{\partial^2(\beta G/N)}{\partial c^2} \right)_{P,T} D_{12}$	$j_x^c(t) = \dfrac{d}{dt} \left\{ (1-c) \sum_{i=1}^{N_1} x_{i1}(t) - c \sum_{i=1}^{N_2} x_{i2}(t) \right\}$	Interdiffusion current	(8.6.31)
$Vk_B T\sigma$	$j_x^Z(t) = \dfrac{d}{dt} \sum_{i=1}^{N} q_i x_i(t)$	Electrical current	(7.8.10)

Note: $c = N_1/(N_1 + N_2)$; q_i is the charge carried by particle i.

(Lebowitz and Spohn, 1978). The work shows that Fourier's linear law (8.3.8) remains valid even for temperature gradients as large as 10^9 K cm^{-1}! However, gradients much smaller than this are sufficient to modify the Rayleigh–Brillouin spectrum and cause the Brillouin lines to become asymmetric (Beysens *et al.*, 1980). The asymmetry is a consequence of the long-range spatial correlations that are induced by gradients of hydrodynamic variables. The existence of long-range correlations in systems with broken space-translational symmetry is linked to hydrodynamic mode-coupling effects (Ronis *et al.*, 1979; Tremblay *et al.*, 1981; Kirkpatrick *et al.*, 1982a, b, c). The same effects are responsible for the divergences observed in transport coefficients near the critical point (Fixman, 1962; Kadanoff and Swift, 1968; Kawasaki, 1970) and the long-time "tails" in certain time-correlation functions (see Section 8.7).

8.6 HYDRODYNAMIC FLUCTUATIONS IN BINARY MIXTURES

The problems involved in extending the hydrodynamic description to multi-component systems are technical rather than conceptual ones. We restrict the discussion to binary mixtures; this requires the introduction of only one additional local variable, namely the concentration. We also follow the usual convention (Landau and Lifshitz, 1963; Cohen *et al.*, 1971) whereby the thermodynamic variables are expressed per unit mass (rather than unit volume); the extra local variable is then the mass fraction of one species, which we shall denote by $c(\mathbf{r}, t)$. There are now four linearized longitudinal equations; these link the fluctuations in local mass density, $\delta\rho_{\rm m}(\mathbf{r}, t)$, concentration, $\delta c(\mathbf{r}, t)$, and entropy per unit mass, $\delta s(\mathbf{r}, t)$, to the divergence of the velocity field, $\psi(\mathbf{r}, t) = \nabla \cdot \mathbf{u}(\mathbf{r}, t)$, the fluctuations in local pressure, $\delta P(\mathbf{r}, t)$, and temperature, $\delta T(\mathbf{r}, t)$, the diffusion current $\mathbf{J}^c(\mathbf{r}, t)$ and the heat current $\mathbf{J}^Q(\mathbf{r}, t)$. The continuity equation for $\delta\rho_{\rm m}$ and the longitudinal projection of the Navier-Stokes equation are the same as in the one-component case (see Eqns (8.3.7) and (8.3.17)), i.e.

$$\frac{\partial}{\partial t}\delta\rho_{\rm m}(\mathbf{r}, t) + \rho_{\rm m}\psi(\mathbf{r}, t) = 0 \tag{8.6.1}$$

$$\rho_{\rm m}\frac{\partial}{\partial t}\psi(\mathbf{r}, t) + \nabla^2\delta P(\mathbf{r}, \mathrm{t}) - \rho_{\rm m}b\nabla^2\psi(\mathbf{r}, t) = 0 \tag{8.6.2}$$

where b is the quantity defined by Eqn (8.3.26). The continuity equation for δc is

$$\rho_{\rm m}\frac{\partial}{\partial t}\delta c(\mathbf{r}, t) + \nabla \cdot \mathbf{J}^c(\mathbf{r}, t) = 0 \tag{8.6.3}$$

and the constitutive relations for $\mathbf{J}^c$ and $\mathbf{J}^Q$ are (Landau and Lifshitz, 1963)

$$\mathbf{J}^c(\mathbf{r}, t) = -\alpha \nabla \delta\mu(\mathbf{r}, t) - \beta \nabla \delta T(\mathbf{r}, t) \tag{8.6.4}$$

$$\mathbf{J}^Q(\mathbf{r}, t) - \mu \mathbf{J}^c(\mathbf{r}, t) = -\delta \nabla \delta\mu(\mathbf{r}, t) - \gamma \nabla \delta T(\mathbf{r}, T) \tag{8.6.5}$$

In these equations, $\delta\mu(\mathbf{r}, t)$ is the fluctuation in local chemical potential, defined through the differential of the Gibbs free energy per unit mass as

$$\mathrm{d}g = -s\,\mathrm{d}T + \frac{1}{\rho_{\mathrm{m}}}\mathrm{d}P + \mu\,dc \tag{8.6.6}$$

In other words, μ is the difference in the chemical potential per unit mass of the two components:

$$\mu = \frac{\mu_1}{m_1} - \frac{\mu_2}{m_2} \tag{8.6.7}$$

The quantities α, β, δ and γ are transport coefficients; the coefficients β and δ, which describe, respectively, the Soret effect (a concentration flux induced by a temperature gradient) and the Dufour effect (a heat flux induced by a gradient in chemical potential or concentration), are related by an Onsager reciprocity relation in the form

$$\delta = \beta T \tag{8.6.8}$$

If $\delta\mu$ is eliminated between Eqns (8.6.4) and (8.6.5), the latter can be rewritten as

$$\mathbf{J}^Q(\mathbf{r}, t) = (\mu + \beta T/\alpha)\mathbf{J}^c(\mathbf{r}, t) - \lambda \delta T(\mathbf{r}, t) \tag{8.6.9}$$

where

$$\lambda = \gamma - \frac{\beta^2 T}{\alpha} \tag{8.6.10}$$

is the thermal conductivity of the mixture; this determines the heat current induced by a gradient in temperature in the absence of a concentration flux.

The fluctuation in local chemical potential can be expressed in terms of the independent variables c, P and T in the form

$$\delta\mu(\mathbf{r}, t) = \left(\frac{\partial \mu}{\partial c}\right)_{P,T} \delta c(\mathbf{r}, t) + \left(\frac{\partial \mu}{\partial P}\right)_{T,c} \delta P(\mathbf{r}, t) + \left(\frac{\partial \mu}{\partial T}\right)_{P,c} \delta T(\mathbf{r}, t) \tag{8.6.11}$$

Equation (8.6.11) can be combined with (8.3.13), (8.5.23) and (8.6.5) to give the entropy equation:

$$\rho_{\mathrm{m}} T \frac{\partial}{\partial t} \delta s(\mathbf{r}, t) = \lambda \nabla^2 \delta T(\mathbf{r}, t) - \left[k_T \left(\frac{\partial \mu}{\partial c} \right)_{P,T} - T \left(\frac{\partial \mu}{\partial T} \right)_{P,c} \right] \nabla \cdot \mathbf{J}^c(\mathbf{r}, t) \quad (8.6.12)$$

and the diffusion equation is obtained by combining (8.6.11) with (8.6.3) and (8.6.4):

$$\frac{\partial}{\partial t} \delta c(\mathbf{r}, t) = D[\nabla^2 \delta c(\mathbf{r}, t) + (k_{\mathrm{T}}/T) \nabla^2 \delta T(\mathbf{r}, t) + (k_P/P) \nabla^2 \delta P(\mathbf{r}, t)] \quad (8.6.13)$$

Equations (8.6.12) and (8.6.13) involve three transport coefficients: the thermal conductivity λ, the interdiffusion coefficient D, and the thermal-diffusion ratio k_T; the last of these is related to the coefficients α and β by

$$\frac{\rho_{\mathrm{m}} k_T D}{T} = \alpha \left(\frac{\partial \mu}{\partial T} \right)_{P,c} + \beta \quad (8.6.14)$$

and vanishes as the concentration of one species approaches zero. The quantity k_P in Eqn (8.6.13) is given in terms of purely thermodynamic quantities by the expression

$$k_P = P(\partial \mu / \partial P)_{T,c} / (\partial \mu / \partial c)_{P,T} \quad (8.6.15)$$

and is therefore small whenever the pressure is low. When both k_T and k_P are negligible, Eqn (8.6.13) takes the same form as the self-diffusion equation (8.2.3).

The solution to the set of linearized equations (8.6.1), (8.6.2), (8.6.12) and (8.6.13) is obtained along similar lines to those followed in the one-component case; the main difference is the fact that the longitudinal part of the hydrodynamic matrix is now of dimension 4×4 rather than 3×3. In order to obtain a matrix of static correlation functions that is diagonal, it is necessary to work with a set of statistically independent variables. This can be achieved (Mountain and Deutch, 1969) by replacing the variables δP, δc and δT by δP, δc and

$$\delta \phi = \delta T - \frac{T \alpha_T}{\rho c_P} \delta P \quad (8.6.16)$$

where $\alpha_T = -(\partial \rho / \partial T)_{P,c} / \rho$ is the thermal expansion coefficient and c_P is the specific heat per particle at constant pressure and concentration. An alternative choice consists of the set δP, δc and

$$\delta s' = \delta s - \left(\frac{\partial s}{\partial c} \right)_{P,T} \delta c \quad (8.6.17)$$

The frequencies of the longitudinal modes are the complex roots of the dispersion relation $D_l(k, z) = 0$, where $D_l(k, z)$ is the determinant of the 4×4 matrix. To order k^2, there are two purely imaginary roots; these represent a generalization of (8.3.37a) and correspond (Cohen *et al.*, 1971) to modes in which the processes of thermal diffusion and concentration are coupled together. The two roots are given explicitly by

$$z_0^{\pm} = -\frac{ik^2}{2}\{(D_{\mathrm{T}} + \mathscr{D}) \pm [(D_{\mathrm{T}} + \mathscr{D})^2 - 4D_{\mathrm{T}}D]^{1/2}\} \tag{8.6.18}$$

where D_{T} is the thermal diffusivity defined by Eqn (8.3.38) and

$$\mathscr{D} = D\left[1 + \frac{\rho_{\mathrm{m}} k_T^2}{\rho c_P T}\left(\frac{\partial \mu}{\partial c}\right)_{P,T}\right] \tag{8.6.19}$$

There are, in addition, two complex conjugate roots, analogous to (8.3.37b), that correspond to propagating sound waves, i.e.

$$z_{\pm} = \pm c_{\mathrm{s}} k - i\Gamma k^2 \tag{8.6.20}$$

where c_{s} is the usual adiabatic speed of sound (see Eqn 8.3.35)) and the sound attenuation coefficient Γ is

$$\Gamma = \tfrac{1}{2}[a(\gamma - 1)/\gamma + b + D\rho_{\mathrm{m}}^2 c_{\mathrm{s}}^2 (\partial\mu/\partial c)_{P,T} \mathscr{P}^2] \tag{8.6.21}$$

with

$$\mathscr{P} = \frac{k_P}{P} + \frac{k_T \alpha_T}{\rho c_P} \tag{8.6.22}$$

The extra term in (8.6.21) compared with the one-component result (8.3.39) represents the damping of sound waves by interdiffusion of the two species.

As in the one-component case discussed in Section 8.5, the results of the hydrodynamic analysis can be used to describe the spectrum of light scattered by a mixture of fluids. The intensity of scattered light is proportional to the spectral function of the autocorrelation function of the fluctuating local dielectric constant, i.e.

$$\mathscr{I}(k, \omega) \propto \mathrm{Re}\,\langle \delta\tilde{\varepsilon}_{\mathbf{k}}(\omega)\delta\varepsilon_{-\mathbf{k}}\rangle \tag{8.6.23}$$

For a binary mixture, the fluctuation $\delta\tilde{\varepsilon}_{\mathbf{k}}(\omega)$ is expressible as

$$\delta\tilde{\varepsilon}_{\mathbf{k}}(\omega) = \left(\frac{\partial\varepsilon}{\partial P}\right)_{T,c} \delta\tilde{P}_{\mathbf{k}}(\omega) + \left(\frac{\partial\varepsilon}{\partial c}\right)_{P,T} \delta\tilde{c}_{\mathbf{k}}(\omega) + \left(\frac{\partial\varepsilon}{\partial T}\right)_{P,c} \delta\tilde{T}_{\mathbf{k}}(\omega) \tag{8.6.24}$$

and $\mathscr{I}(k, \omega)$ is the sum of six independent contributions. The spectrum is therefore more complicated than in the one-component case; although it again consists of a Brillouin doublet and a central Rayleigh peak, the central peak is now a superposition of two lorentzian lines, representing the coupled effects of fluctuations in entropy and concentration. In the low-concentration limit, where $k_T \to 0$, the widths of the two lorentzians are determined by D and D_T respectively, so that values for both these transport coefficients can, in principle, be obtained by spectroscopic methods (Bergé *et al.*, 1970; Dubois and Bergé, 1971).

Explicit formulae for the hydrodynamic limiting form of the various time correlation functions are given in the papers of Cohen *et al.* (1971) and Lekkerkerker and Laidlaw (1973). The result for the concentration autocorrelation function is

$$\frac{\langle c_{\mathbf{k}}(t)c_{-\mathbf{k}}\rangle}{\langle|c_{\mathbf{k}}|^2\rangle} = (z_0^- - z_0^+)^{-1}[(z_0^- + iDk^2)\exp(-iz_0^+ t) - (z_0^+ + iDk^2)\exp(-iz_0^- t)] \tag{8.6.25}$$

By taking the Laplace transform of (8.6.25), it is possible to derive a Green–Kubo relation for the interdiffusion coefficient. Direct transformation shows immediately that

$$D = \lim_{\omega\to 0}\lim_{k\to 0}\frac{i\omega}{k^2}\int_0^\infty \frac{\mathrm{d}}{\mathrm{d}t}\left[\frac{\langle c_{\mathbf{k}}(t)c_{-\mathbf{k}}\rangle}{\langle|c_{\mathbf{k}}|^2\rangle}\right]\exp(i\omega t)\,\mathrm{d}t \tag{8.6.26}$$

from which, following an integration by parts and use of the continuity equation (8.6.3), we find that

$$D = \lim_{\omega\to 0}\lim_{k\to 0}\frac{1}{3\rho_\mathrm{m}^2\langle|c_{\mathbf{k}}|^2\rangle}\int_0^\infty \langle \mathbf{J}_{\mathbf{k}}^c(t)\cdot\mathbf{J}_{-\mathbf{k}}^c\rangle\exp(i\omega t)\,\mathrm{d}t \tag{8.6.27}$$

To obtain a microscopic expression for $\mathbf{J}_0^c$, we first relate the fluctuation in local concentration to the fluctuations in local number densities by writing

$$\delta c(\mathbf{r},t) = \frac{m_1 m_2}{\rho_\mathrm{m}^2}[\rho_2\delta\rho_1(\mathbf{r},t) - \rho_1\delta\rho_2(\mathbf{r},t)] \tag{8.6.28}$$

The microscopic particle currents of the two species are defined as in Eqn (7.4.7) and are related to the corresponding particle densities by continuity equations similar to (7.4.4). Given Eqn (8.6.3), it follows that

$$\mathbf{j}_{\mathbf{k}}^c(t) = \frac{m_1 m_2}{\rho_\mathrm{m}}[\rho_2\mathbf{j}_{\mathbf{k}}^1(t) - \rho_1\mathbf{j}_{\mathbf{k}}^2(t)] \tag{8.6.29}$$

Next, thermodynamic fluctuation theory of the type outlined in Section 2.7 can be used to show that the small-k limit of the denominator in (8.6.27) is

$$\lim_{k\to 0}\langle|c_{\mathbf{k}}|^2\rangle = \frac{1}{\rho\rho_\mathrm{m}\left(\dfrac{\partial\beta\mu}{\partial c}\right)_{P,T}} = \left(\frac{m_1 m_2}{\rho_\mathrm{m}\bar{m}}\right)^2\Bigg/\left(\frac{\partial^2(\beta G/N)}{\partial x_1^2}\right)_{P,T} \tag{8.6.30}$$

where $\bar{m} = \rho_m/\rho$ is the mean mass of the particles and x_1 is the mole fraction of species 1. On combining Eqns (8.6.27) to (8.6.30) we arrive at the required result, namely

$$D = \frac{1}{3}\left(\frac{\partial^2(\beta G/N)}{\partial x_1^2}\right)_{P,T} \int_0^\infty \langle \mathbf{j}^c(t) \cdot \mathbf{j}^c \rangle \, dt \tag{8.6.31}$$

where $\mathbf{j}^c$ is the microscopic interdiffusion current:

$$\mathbf{j}^c(t) = x_2 \mathbf{j}^1(t) - x_1 \mathbf{j}^2(t) \tag{8.6.32}$$

with

$$\mathbf{j}^\nu(t) = \sum_{i=1}^{N_\nu} \mathbf{u}_i(t), \qquad \nu = 1, 2. \tag{8.6.33}$$

Many mixtures of simple liquids are nearly ideal in the sense of Section 6.6. In any such case, the dominant contribution to the second derivative of the Gibbs free energy with respect to concentration is the term appropriate to a mixture of ideal gases, i.e.

$$\left(\frac{\partial^2(\beta G^{\text{id}}/N)}{\partial x_1^2}\right)_{P,T} = \frac{1}{x_1 x_2} \tag{8.6.34}$$

Equation (8.6.31) shows that the interdiffusion coefficient is a collective property of the system. If, however, cross-correlations of the type $\langle \mathbf{u}_i(t) \cdot \mathbf{u}_j(0) \rangle$ $(i \neq j)$ are negligible for all t, the autocorrelation function in (8.6.31) can be written in terms of single-particle properties in the form

$$\langle \mathbf{j}^c(t) \cdot \mathbf{j}^c \rangle \simeq 3 N x_1 x_2 [x_2 Z_1(t) + x_1 Z_2(t)] \tag{8.6.35}$$

where $Z_\nu(t)$ is the velocity autocorrelation function of particles of species ν. It then follows immediately that for a nearly ideal mixture

$$\begin{aligned} D &\simeq \left(\frac{\partial^2(\beta G^{\text{id}}/N)}{\partial x_1^2}\right)_{P,T} x_1 x_2 (x_2 D_1 + x_1 D_2) \\ &= x_2 D_1 + x_1 D_2 \end{aligned} \tag{8.6.36}$$

Molecular dynamics calculations have shown that the approximate relations (8.6.35) and (8.6.36) are well satisfied for mixtures of Lennard-Jones fluids representative of the system argon–krypton (Jacucci and McDonald, 1975; Schoen and Hoheisel, 1984a, b); the deviation from (8.6.36) due to cross correlations of velocities is small (about 5%) and positive. For highly dissymmetric mixtures, the deviations are much larger, as illustrated by the correlation functions plotted in Figure 8.6.

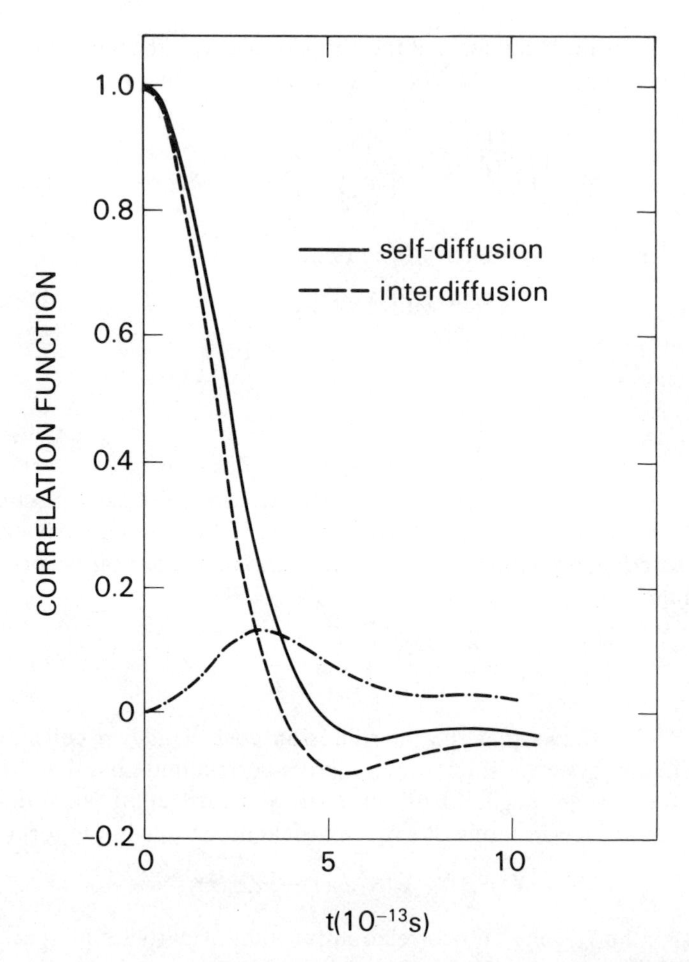

FIG. 8.6. Normalized interdiffusion-current and velocity autocorrelation functions for an equimolar binary mixture of Lennard-Jones particles of equal mass (61.875 a.m.u.) and size ($\sigma_{11} = \sigma_{22}$), but having very different potential-well depths ($\varepsilon_{22}/\varepsilon_{11} = 3$, $\varepsilon_{12} = (\varepsilon_{11}\varepsilon_{22})^{1/2}$). The reduced density is $N\sigma_{11}^3/V = 0.75$ and the reduced temperature is $k_B T/\varepsilon_{11} = 1.70$. The full curve is the mean normalized velocity autocorrelation function and the dash-dot curve shows the difference between the latter and the interdiffusion-current autocorrelation function (the dashed curve). After Schoen and Hoheisel (1984b).

8.7 GENERALIZED HYDRODYNAMICS AND LONG-TIME TAILS

In the earlier sections of this chapter we have shown in some detail how the equations of hydrodynamics can be used to calculate the time-correlation functions of conserved variables in the long-wavelength, low-frequency limit. Two questions then arise. First, what are the scales of length and time over which a continuum description can be maintained? Secondly, how must the hydrodynamic equations be modified such that their predictions can be applied on the atomic scale, where lengths are typically of order a few ångström units and times are of order 10^{-13} s? We have seen in Chapter 7 that the behaviour of the correlation functions at short times is related to frequency sum rules that involve static distribution functions descriptive of the molecular structure of the fluid. It is precisely these sum rules that are violated by hydrodynamic expressions such as (8.4.5) and (8.5.4), because the resulting frequency moments beyond zeroth order all diverge. Similarly, an exponential decay, as in Eqn (8.4.3), is incompatible with certain of the general properties of time-correlation functions discussed in Section 7.1. The failure of the hydrodynamic approach at short times (or high frequencies) is linked to the presence of dissipative terms in the basic hydrodynamic equations; the latter, unlike the microscopic equations of motion, are not invariant under time reversal. In this section, we describe some phenomenological generalizations of the hydrodynamic equations, based on the introduction of frequency and wavenumber-dependent transport coefficients, which have been used in attempts to bridge the gap between the hydrodynamic (small k, ω) and kinetic (large k, ω) regimes (Chung and Yip, 1969; Ailawadi *et al.*, 1971). The use of non-local transport coefficients is closely related to the memory-function formalism that we shall develop in a more systematic fashion in the next chapter. We also discuss the origin of the "long-time tails" in the autocorrelation functions of certain non-conserved variables including both the particle velocity and the components of the stress tensor. As we shall see, the existence of such tails can be explained by simple arguments concerning the decay of fluctuations into pairs of hydrodynamic modes (Ernst *et al.*, 1971, 1976a, b). Again, however, we postpone the development of a more fundamental approach to the problem until Chapter 9.

The ideas of *generalized hydrodynamics* can perhaps most easily be illustrated by considering the example of the transverse current correlations. Equation (8.4.3) shows that in the hydrodynamic limit the correlation function $C_t(k, t)$ decays exponentially with a relaxation time equal to $(\nu k^2)^{-1}$, where ν is the kinematic shear viscosity. The corresponding power spectrum is of lorentzian form:

$$C_t(k, \omega) = \frac{1}{\pi} \operatorname{Re} \tilde{C}_t(k, \omega)$$

$$= \frac{\omega_0^2}{\pi} \frac{\nu k^2}{\omega^2 + (\nu k^2)^2} \qquad (8.7.1)$$

As we have already emphasized, the ω^{-2} behaviour for large ω is in contradiction with the high-frequency sum rules such as (7.4.50). In particular, at large wavenumbers, Eqn (8.7.1) does not yield the correct free-particle limit of $C_t(k, \omega)$; the latter is gaussian in form, similar to the longitudinal free-particle limit in (7.5.28). Moreover, molecular dynamics calculations (Rahman, 1968; Levesque *et al.*, 1973; Alley and Alder, 1983), which are the only source of "experimental" information on transverse-current fluctuations in atomic liquids, show that in an intermediate wavenumber range the funcction $C_t(k, t)$ is not exponential or even monotonically decaying; on the contrary, it displays the type of oscillatory behaviour that is illustrated in Figure 8.7. The corresponding spectrum has

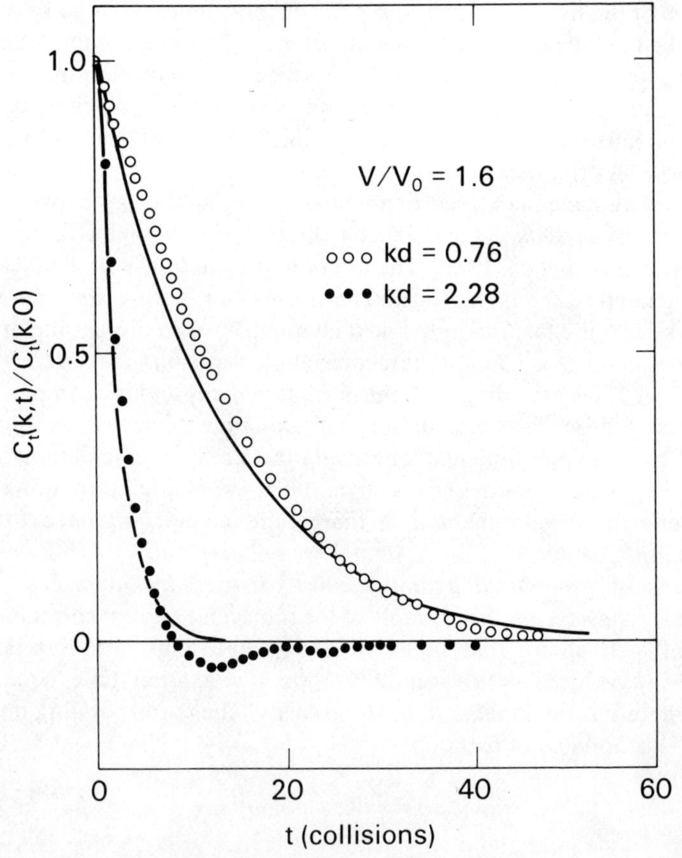

FIG. 8.7. Normalized transverse-current autocorrelation function for the hard-sphere fluid; $V_0 = Nd^3/\sqrt{2}$ is the volume at close packing. The points are molecular dynamics results and the curves are calculated from the hydrodynamic result (8.4.3), but with a wavenumber-dependent shear viscosity. The unit of time is the mean collision time. After Alder and Alley (1984).

a peak at non-zero frequency, suggestive of the existence of a propagating shear wave. What this means physically is that at high frequencies the fluid has insufficient time to flow in response to an applied strain rate, but instead reacts elastically in the way that a solid would. The appearance of shear waves can be accounted for by the incorporation of viscoelastic effects (Frenkel, 1955) into the hydrodynamic description of the transverse currents (see, for example, Schofield, 1966). Suppose that a shearing force is applied to a fluid. The strain at a point (x, y, z) can be expressed in terms of the displacement $\mathbf{r}$ at that point, and the rate of strain is expressible in terms of the velocity $\dot{\mathbf{r}}$. If the flow is purely viscous, the shearing stress is proportional to the rate-of-strain tensor and can be written as

$$\sigma^{xz} = -\eta \frac{\partial}{\partial t}\left(\frac{\partial r_x}{\partial z} + \frac{\partial r_z}{\partial x}\right) \tag{8.7.2}$$

which is the hydrodynamic form (see Eqn (8.3.5)). By contrast, if the force is applied suddenly, the instantaneous displacement is determined by the stress through a typical stress–strain relation, i.e.

$$\sigma^{xz} = -G_\infty\left(\frac{\partial r_x}{\partial z} + \frac{\partial r_z}{\partial x}\right) \tag{8.7.3}$$

where G_∞ is an instantaneous (high-frequency) modulus of rigidity. We can interpolate between these two extremes by making a *viscoelastic* approximation such that

$$\left(\frac{1}{\eta} + \frac{1}{G_\infty}\frac{\partial}{\partial t}\right)\sigma^{xz} = -\left(\frac{\partial \dot{r}_x}{\partial z} + \frac{\partial \dot{r}_z}{\partial x}\right) \tag{8.7.4}$$

By taking the Laplace transform of (8.7.4), it is easy to show that the viscoelastic approximation is equivalent to replacing η in Eqn (8.7.2) by a frequency-dependent shear viscosity given by

$$\eta(\omega) = \frac{G_\infty}{-i\omega + G_\infty/\eta} \tag{8.7.5}$$

The constant $\tau_{\mathrm{M}} = \eta/G_\infty$ is called the Maxwell relaxation time; the limit $\omega\tau_{\mathrm{M}} \ll 1$ corresponds to viscous flow, while the limit $\omega\tau_{\mathrm{M}} \gg 1$ corresponds to elastic waves propagating at a speed $(G_\infty/\rho m)^{1/2}$, as becomes clear when (8.7.5) is substituted in Eqn (8.4.4).

If account is also to be taken of non-local effects in space, the generalized shear viscosity must be a function of wavenumber as well as of frequency. The rigidity modulus is also dependent on k, and is related in a simple way to the second frequency moment ω_{1t}^2, as we shall see below (Zwanzig and Mountain, 1965; Schofield, 1966). These ideas can be formalized by writing down a phenomenological generalization (Ailawadi *et al.*, 1971) of the hydrodynamic equation (8.4.2) such that

$$\frac{\partial}{\partial t} C_t(k, t) + k^2 \int_0^t \nu(k, t-s) C_t(k, s)\, ds = 0 \tag{8.7.6}$$

The kernel $\nu(k, t)$ is a memory function; it describes a response that is non-local in both space and time and its Laplace transform $\tilde{\nu}(k, \omega)$ plays the role of a generalized kinematic viscosity. If we take the Laplace transform of (8.7.6) and compare the result with Eqn (8.4.4), we find that the function $\tilde{\nu}(k, \omega)$ must satisfy the constraint that

$$\lim_{\omega \to 0} \lim_{k \to 0} \tilde{\nu}(k, \omega) = \nu \tag{8.7.7}$$

where ν is the macroscopic kinematic viscosity, given by the Green–Kubo relation (8.4.10). If, on the other hand, we differentiate equation (8.7.6) with respect to t, set $t=0$ and use Eqn (7.4.49), we find that

$$\begin{aligned} \nu(k, t=0) &= \omega_{1t}^2 / k^2 \\ &= \frac{G_\infty(k)}{\rho m} \end{aligned} \tag{8.7.8}$$

which provides a second constraint on the generalized shear viscosity and also acts as a definition of the k-dependent shear modulus $G_\infty(k)$. Equations (8.7.7) and (8.7.8) are useful in the construction of approximate forms of $\nu(k, t)$ that reduce to the hydrodynamic and viscoelastic expressions in the limits, respectively, of low and high frequency (Chung and Yip, 1969; Ailawadi *et al.*, 1971; Levesque *et al.*, 1973).

If molecular dynamics results for $C_t(k, t)$ are available, values of the generalized transport coefficient $\tilde{\nu}(k, \omega)$ can be obtained by numerical inversion of Eqn (8.7.6). Calculations of this type have been carried out for the hard-sphere fluid by Alley and Alder (1983). The infinite-wavelength result $\tilde{\nu}(k=0, \omega)$ is also expressible in terms of the Laplace transform of the stress-tensor autocorrelation function (see Eqns (8.4.10) and (8.4.11)). The same is not true for $k>0$; in other words, $\tilde{\eta}(k>0, \omega) = \rho m \tilde{\nu}(k, \omega)$ is not given by a Green–Kubo relation similar to (8.4.10) in which σ_0^{xz} is replaced by $\sigma_{\mathbf{k}}^{xz}$. This is evident from the manipulations that lead from (8.4.7) to (8.4.11), since ω_0^2 in Eqn (8.4.7) does not vanish for non-zero k. In addition, the generalized shear viscosity $\tilde{\eta}(k, \omega)$ is a non-analytic function of both k and ω. As we have seen in Section 8.4, $\eta(k=0, \omega)$ is believed to behave as $\omega^{1/2}$ at low frequencies, corresponding to a $t^{-3/2}$ decay of the stress-tensor correlation function. On the other hand, the zero-frequency shear viscosity $\eta(k) = \rho m \tilde{\nu}(k, \omega = 0)$ could perhaps be expanded in a Taylor series in k around its macroscopic limit $\eta \equiv \eta(k=0)$, i.e.

$$\eta(k) = \eta + \eta_2 k^2 + \cdots \tag{8.7.9a}$$

where η_2 is called a Burnett coefficient; invariance under space inversion guarantees that only even powers of k can appear. Burnett coefficients are introduced in attempts to extend the range of validity of hydrodynamic equations such as Fick's law or the Navier–Stokes equation through the addition of terms of higher order in the gradients of the hydrodynamic fields. However, the indications from mode-coupling theories of the type to be discussed later in this section are that the Burnett coefficients diverge and that $\eta(k)$ is therefore non-analytic in k (Keyes and Oppenheim, 1973). The results of computer simulations of a "soft-sphere" fluid (Evans, 1982b) are compatible with a small-k behaviour of the form

$$\eta(k) = \eta - \eta_{3/2} k^{3/2} + \cdots \tag{8.7.9b}$$

where $\eta_{3/2}$ is a positive quantity. These calculations, together with results obtained for hard spheres (Alley and Alder, 1983), show that $\eta(k)$ and other generalized transport coefficients decrease smoothly with increasing wavenumber and are an order of magnitude smaller than their macroscopic ($k=0$) values when the wavelength is comparable with the interparticle spacing.

The longitudinal projections of the hydrodynamic equations can be treated in a similar way through the introduction of wavenumber and frequency-dependent quantities that are generalizations of the coefficients a and b defined by Eqns (8.3.20) and (8.3.26). Similarly, the hydrodynamic derivatives, which are related to static correlation functions (Schofield, 1966, 1968), become functions dependent on wavelength. In particular, the macroscopic compressibility must be replaced by its k-dependent generalization, i.e. the structure factor $S(k)$ (see Eqn (5.1.13)), while the temperature derivative of the pressure, which determines the coupling between momentum and energy, now contains a part that is explicitly dependent on frequency and vanishes in the limit $k \to 0$ (Alley and Alder, 1983).

The methods of generalized hydrodynamics have been used in the calculation of the longitudinal current correlation function $C_l(k, \omega)$ and dynamic structure factor $S(k, \omega)$ of simple, monatomic liquids by, amongst others, Ailawadi *et al.* (1971) and Alley *et al.* (1983). In the case of the hard-sphere fluid, Alley *et al.* (1983) have shown that the molecular dynamics results for $S(k, \omega)$ can in large part be satisfactorily reproduced by a scheme in which the various thermodynamic and transport coefficients are assumed to be functions only of wavenumber and not of frequency. This approach breaks down, however, both for wavelengths shorter than the mean free path (corresponding to free-particle behaviour) and at densities close to crystallization; at high densities, viscoelastic effects are important for wavelengths in the range from approximately one to ten particle diameters. Generalized hydrodynamics is equivalent to the damping-function formalism of Kadanoff and Martin (1963). In the Kadanoff–Martin approach, the dependence of the generalized transport coefficients on k and ω is contained in a "damping function" $D(k, \omega)$. In applications (Chung and Yip, 1969), $D(k, \omega)$ is approximated in a manner consistent with the known low-frequency and high-frequency limits; this is very similar in spirit to the type of memory-function analysis that we discuss in Section 9.4.

We turn now to the problem of "long-time tails" to which we have already alluded in earlier sections. Fluctuations in conserved hydrodynamic variables decay infinitely slowly in the long-wavelength limit. The rates of relaxation are determined by the hydrodynamic eigenvalues (8.3.33), (8.3.37a) and (8.3.37b) (multipled by $-i$), all of which vanish with k. No such property holds for the non-conserved currents that enter the Green–Kubo integrands for the transport coefficients; if it did, the transport coefficients would not be well defined. Until the late 1960s, it was generally believed that away from critical points the correlation functions of non-conserved variables decay exponentially at long times. This, for example, is the behaviour predicted by the Boltzmann and Enskog equations. It therefore came as a great surprise when analysis of molecular dynamics results on self diffusion in the hard-disc ($d = 2$) and hard-sphere ($d = 3$) fluids showed that the velocity correlation function apparently decays asymptotically as $t^{-d/2}$, where d denotes the dimensionality of the system (Alder and Wainwright 1967, 1968, 1970). Later simulations of hard-core fluids and other systems have confirmed that there exist long-time tails both in $Z(t)$ (Levesque and Ashurst, 1974; Erpenbeck and Wood, 1982) and in the stress-tensor autocorrelation function (Evans, 1980; Erpenbeck and Wood, 1981). Here we focus attention on the simpler problem of the velocity autocorrelation function.

The presence of a slowly decaying tail in $Z(t)$ suggests that highly collective effects make a significant contribution to the process of self diffusion. The apparent involvement of large numbers of particles makes it natural to discuss the long-time behaviour in hydrodynamic terms, and Alder and Wainwright (1970) were led in this way to a simple but convincing explanation of the molecular dynamics results. Underlying their argument is the idea that the initial motion of a tagged particle creates around that particle a vortex or backflow, which in turn causes a retarded current to develop in the direction of the initial velocity. At low densities, where the initial direction of motion is likely to persist, the effect of this current is to reduce the drag on the particle, thereby "pushing" it on in the initial direction. This results in a long-lasting, positive correlation in the velocity of the particle. At high densities, on the other hand, the initial direction of motion is on average soon reversed. In this case, the retarded current gives rise to an extra drag at later times, and hence to an extended negative region in $Z(t)$; at very large times, an enhancement in the forward motion can again be expected, but the effect is likely to be undetectable. That this physical picture is basically correct has been confirmed in striking fashion by observation of the velocity field that forms around a moving particle in a fluid of hard discs (Alder and Wainwright, 1970). A vortex pattern quickly develops; this, after a few mean collision times, matches closely the pattern obtained by numerical solution of the Navier-Stokes equation. The persistence of the tail in $Z(t)$ is therefore associated with a coupling between the motion of single particle and the hydrodynamic modes of the fluid. The argument can be formalized in such a way as to predict the observed $t^{-d/2}$ decay of the long-time tail; in doing so below, we follow closely the very clear account given by Pomeau and Résibois (1975).

Suppose that at time $t=0$ a particle i has a velocity $u_{ix}(0)$ in the x-direction. After a short time, τ say, the effect of collisions will have been such that the initial momentum of particle i is shared among the ρV_τ particles in a d-dimensional volume V_τ centred on i. Local equilibrium now exists within the volume V_τ, and particle i will be moving with a velocity

$$u_{ix}(\tau) \simeq u_{ix}(0)/\rho V_\tau \tag{8.7.10}$$

(We have assumed, for simplicity, that the neighbours of i were initially at rest.) Further decay in the velocity $u_{ix}(t)$ $(t>\tau)$ will occur as the result of enlargement of the volume V_τ, i.e. from the spread of the velocity field around particle i. At large times, the dominant contribution to the growth of V_τ will come from diffusion of the transverse component of the velocity field, and the radius of V_τ will therefore increase as $(\nu t)^{1/2}$. Thus $V_t \sim (\nu t)^{3/2}$ in the three-dimensional case. It follows immediately that $Z(t) \sim (\nu t)^{-3/2}$. This assumes, however, that particle i remains at the centre of V_t; if the diffusive motion of i is also taken into account, it can be shown that

$$Z(t) \sim [(D+\nu)t]^{-3/2} \tag{8.7.11}$$

The analogous result in two dimensions implies that a self-diffusion coefficient does not exist, because the integral of $Z(t)$ diverges logarithmically; this poses a delicate problem of self-consistency (for a discussion, see Pomeau and Résibois, 1975).

The form of Eqn (8.7.11) has been confirmed by a number of more sophisticated calculations. For example, Dorfman and Cohen (1972, 1975) have used a microscopic approach based on kinetic theory to predict the long-time behaviour of correlation functions in the low-density limit. They have shown that if correlated collision sequences (the "ring" collisions of Section 7.2) are included in the collision operator along with the usual uncorrelated, binary-collision Boltzmann term, the velocity, stress-tensor and energy-current autocorrelation functions of hard-core fluids all decay as $t^{-d/2}$; they also obtain explicit expressions for the coefficients of the long-time tails. A more phenomenological approach, similar in spirit to the hydrodynamic calculations described in Sections 8.3 and 8.4, has been developed by Ernst *et al.*, (1971, 1976*a*, *b*) and Pomeau (1972). The physical content of this work is closely related to the mode-coupling formalism to be discussed in Section 9.5. For that reason, we give a brief derivation of the result obtained for the velocity autocorrelation function (Ernst *et al.*, 1971; see also Erpenbeck and Wood, 1982).

The definition (7.1.3) of a time correlation function involves an equilibrium ensemble average over the initial total phase of the system. This average can be replaced by a constrained ensemble average, characterized by an initial position $\mathbf{r}_0$ and initial velocity $\mathbf{u}_0$ of a tagged particle i, which is then integrated over all $\mathbf{r}_0$ and $\mathbf{u}_0$. The definition of $Z(t)$ is thereby reformulated as

$$
\begin{aligned}
Z(t) &= \langle u_{ix}(t) u_{ix} \rangle \\
&= \int d\mathbf{r}_0 \int d\mathbf{u}_0 u_{0x} \langle u_{ix}(t) \delta(\mathbf{u}_i - \mathbf{u}_0) \delta(\mathbf{r}_i - \mathbf{r}_0) \rangle
\end{aligned} \tag{8.7.12}
$$

The constrained average in (8.7.12) can be written as a non-equilibrium ensemble average (subscript n.e.) defined by

$$
\langle u_{ix}(t) \rangle_{\text{n.e.}} = \frac{\langle u_{ix}(t) \delta(\mathbf{u}_i - \mathbf{u}_0) \delta(\mathbf{r}_i - \mathbf{r}_0) \rangle}{\langle \delta(\mathbf{u}_i - \mathbf{u}_0) \delta(\mathbf{r}_i - \mathbf{r}_0) \rangle} \tag{8.7.13}
$$

In the canonical ensemble, the equilibrium average in the denominator of (8.7.13) is equal to $1/N$ times the single-particle distribution function defined by Eqn (2.1.12) (taken for $n = 1$). We can therefore combine Eqns (8.7.12) and (8.7.13) to give

$$
Z(t) = \frac{1}{V} \int d\mathbf{r}_0 \int d\mathbf{u}_0 \, \phi_0(\mathbf{u}_0) u_{0x} \langle u_{ix}(t) \rangle_{\text{n.e.}} \tag{8.7.14}
$$

where $\phi_0(\mathbf{u}_0)$ is the Maxwell–Boltzmann distribution (2.1.25). If we define a tagged-particle distribution function in the non-equilibrium ensemble as

$$
f^{(s)}(\mathbf{r}, \mathbf{u}; t) = \langle \delta[\mathbf{r}_i(t) - \mathbf{r}] \delta[\mathbf{u}_i(t) - \mathbf{u}] \rangle_{\text{n.e.}} \tag{8.7.15}
$$

we can rewrite the non-equilibrium average in (8.7.14) as

$$
\langle u_{ix}(t) \rangle_{\text{n.e.}} = \int d\mathbf{r} \int d\mathbf{u} \, u_x f^{(s)}(\mathbf{r}, \mathbf{u}; t) \tag{8.7.16}
$$

The development thus far is exact. At this point, Ernst *et al.* (1971) assume that the non-equilibrium distribution function (8.7.15) relaxes towards the corresponding local-equilibrium function on a timescale that is fast in comparison with the rate of decay of $Z(t)$. The long-time behaviour of the non-equilibrium average (8.7.16) is then obtained by replacing $f^{(s)}(\mathbf{r}, \mathbf{u}, t)$ by the tagged-particle analogue of (2.1.26), i.e.

$$f^{(s)}_{\text{l.e.}}(\mathbf{r}, \mathbf{u}; t) = \rho^{(s)}(\mathbf{r}, t)\left(\frac{m}{2\pi k_B T(\mathbf{r}, t)}\right)^{3/2} \times \exp\left(-\frac{m}{2k_B T(\mathbf{r}, t)}|\mathbf{u} - \mathbf{u}(\mathbf{r}, t)|^2\right) \tag{8.7.17}$$

where $\rho^{(s)}(\mathbf{r}, t)$ is the local density of tagged particles. Substitution of (8.7.17) in Eqn (8.7.16) gives an expression for the non-equilibrium average in the form

$$\langle u_{ix}(t)\rangle_{\text{n.e.}} = \int \rho^{(s)}(\mathbf{r}, t) u_x(\mathbf{r}, t)\, d\mathbf{r} \tag{8.7.18}$$

If this result is in turn substituted in Eqn (8.7.14), and the hydrodynamic fields $\mathbf{u}_0$ and $\rho^{(s)}(\mathbf{r})$ are replaced by the sums of their Fourier components, we find that

$$Z(t) = \frac{1}{3V}\int d\mathbf{r}_0 \int d\mathbf{u}_0\, \phi_0(\mathbf{u}_0)\frac{1}{V}\sum_{\mathbf{k}} \rho^{(s)}_{-\mathbf{k}}(t)\mathbf{u}_{\mathbf{k}}(t)\cdot \mathbf{u}_0 \tag{8.7.19}$$

Equation (8.7.19) is said to be of "mode-coupling" form because $Z(t)$ is expressed as a sum of products of two hydrodynamic fields.

The further assumption is now made (Ernst *et al.*, 1971) that at times much longer than the mean collision time, the decay of $Z(t)$ is dominated by the long-wavelength components of the hydrodynamic fields, and that the time evolution of the latter is determined by the linearized hydrodynamic equations. The quantity $\rho^{(s)}_{-\mathbf{k}}(t)$ is then given by Eqn (8.2.5), while the hydrodynamic velocity field is conveniently divided into its longitudinal and transverse parts:

$$\mathbf{u}_{\mathbf{k}}(t) = \mathbf{u}_{\mathbf{k}l}(t) + \mathbf{u}_{\mathbf{k}t}(t) \tag{8.7.20}$$

The term $\mathbf{u}_{\mathbf{k}t}(t)$ satisfies the transverse-current diffusion equation (8.4.1) (with $\mathbf{j}_{\mathbf{k}t} = \rho\mathbf{u}_{\mathbf{k}t}$), the solution to which is

$$\mathbf{u}_{\mathbf{k}t}(t) = \mathbf{u}_{\mathbf{k}t} \exp(-\nu k^2 t) \tag{8.7.21}$$

The longitudinal velocity field can be treated in a similar way, but its contribution to $Z(t)$ turns out to decay exponentially; the physical reason for this is the fact that the momentum of the tagged particle is carried away by the propagating sound waves. The long-time behaviour of $Z(t)$ is therefore entirely determined by the transverse velocity field. Finally, the choice of initial conditions implies that

$$\rho_{-\mathbf{k}}^{(s)} = \exp(i\mathbf{k} \cdot \mathbf{r}_0) \tag{8.7.22a}$$

$$\rho \mathbf{u}_{\mathbf{k}} = \mathbf{u}_0 \exp(-i\mathbf{k} \cdot \mathbf{r}_0) \tag{8.7.22b}$$

An expression for $Z(t)$ can now be obtained by substituting (8.2.5), (8.7.21), (8.7.22a) and the transverse projection of (8.7.22b) into Eqn (8.7.19), and then integrating over $\mathbf{r}_0$ and $\mathbf{u}_0$; the result is

$$Z(t) = \frac{2k_B T}{3\rho m V} \sum_{\mathbf{k}} \exp[-(D+\nu)k^2 t] \tag{8.7.23}$$

or, in the thermodynamic limit:

$$Z(t) = \frac{2k_B T}{3\rho m} \int \frac{d\mathbf{k}}{(2\pi)^3} \exp[-(D+\nu)k^2 t] \tag{8.7.24}$$

The correctness of integrating over all wavevectors must be questioned, since the hydrodynamic equations on which (8.7.24) is based are invalid when k is large. However, we are interested only in the asymptotic form of $Z(t)$, and the main contribution to the integral in (8.7.24) comes from wavenumbers such that $k \simeq ((D+\nu)t)^{-1/2}$; this is in the hydrodynamic range whenever t is much larger than typical microscopic times ($\sim 10^{-13}$ s). Alternatively, a natural upper limit can be introduced into the integration by a more careful choice of the initial spatial distribution of tagged particles (Ernst *et al.*, 1971; Erpenbeck and Wood, 1982); use of such a cut-off has no effect on the predicted long-time behaviour that results from carrying out the integration in (8.7.24), i.e.

$$Z(t) \underset{t \to \infty}{\sim} \frac{2k_B T}{3\rho m} [4\pi(D+\nu)t]^{-3/2} \tag{8.7.25}$$

This is of the same general form as Eqn (8.7.11), but we now have an explicit expression for the coefficient of the long-time tail. The result in (8.7.25) has been confirmed by molecular dynamics calculations for systems of particles interacting through a truncated Lennard-Jones potential (Levesque and Ashurst, 1974) and of hard discs (Erpenbeck and Wood, 1982). The simulations are difficult to carry out with the necessary accuracy. Not only is the tail very weak, but the recurrence-time problem discussed in Section 3.2 means that very large systems (several thousands of particles) must be used to avoid distortion of the asymptotic behaviour (Michels and Trappeniers, 1978). Even worse, the observed tails may be artefacts of numerical error propagation (Fox, 1983)!

Expressions for the coefficients of the $t^{-3/2}$ tails in the stress-tensor and heat-current autocorrelation functions have also been obtained (Pomeau, 1972; Ernst *et al.*, 1976a, b). The derivations are much more complicated than the one sketched above, because the currents involved contain pair-interaction terms. As we mentioned in Section 8.4, the predicted amplitude of the tail in the shear-stress autocorrelation function is much weaker than that observed in computer simulations. The origin of the discrepancy is the fact that the standard mode-coupling approach fails at intermediate times, where structural relaxation plays an important role (Kirkpatrick, 1984). The results of the theory are therefore inapplicable at times that are accessible by present-day computer "experiments".

The mode-coupling calculation of the velocity autocorrelation function is limited to long times, where $Z(t)$ is positive and decays monotonically. At high densities, however, back-scattering effects become important, and $Z(t)$ can take negative values (see Figure 7.1). To describe such behaviour, we must invoke concepts similar to those elaborated earlier in this section. We consider first a simple viscoelastic model based on a generalization of the Stokes relations (7.3.19). What we require is an expression for the frictional force on a sphere of diameter d moving in an incompressible viscous fluid with an arbitrarily varying velocity $\mathbf{u}(t)$. As this is a problem discussed in detail by Landau and Lifshitz (1963), we need quote only the final result; this, for stick boundary conditions is

$$\mathbf{F}(t) = -2\pi\rho m d^3 \left[(3\nu/d^2)\mathbf{u}(t) - \tfrac{1}{3}\dot{\mathbf{u}}(t) - (3/d)(\nu/\pi)^{1/2} \int_{-\infty}^{t} \dot{\mathbf{u}}(s)(t-s)^{-1/2}\,\mathrm{d}s \right] \tag{8.7.26}$$

If we define a friction coefficient in terms of the Fourier components of $\mathbf{F}(t)$ and $\mathbf{u}(t)$ by

$$\mathbf{F}(\omega) = -\tilde{\xi}(\omega)\mathbf{u}(\omega) \tag{8.7.27}$$

we may identify the complex friction coefficient $\tilde{\xi}(\omega)$ as

$$m\tilde{\xi}(\omega) = 3\pi\eta d - \tfrac{1}{12}\pi\rho m d^3 i\omega - \tfrac{3}{2}\pi d^2 i(i\omega\eta\rho m)^{1/2} \tag{8.7.28}$$

which reduces to (7.3.19) (stick approximation) when $\omega = 0$. The analogous result for slip boundary conditions (Zwanzig and Bixon, 1970; Albano *et al.*, 1975) is

$$m\tilde{\xi}(\omega) = 2\pi\eta d - \tfrac{1}{12}\pi\rho m d^3 i\omega - \frac{2\pi d^2 i(i\omega\eta\rho m)^{1/2}}{3 - di(i\omega\rho m/4\eta)^{1/2}} \tag{8.7.29}$$

The corresponding form of the velocity autocorrelation function is then obtained from Eqn (7.3.25).

In practice, neither (8.7.28) or (8.7.29) gives results for $Z(t)$ that are in good agreement with molecular dynamics calculations. Each, however, yields the correct long-time behaviour, since both equations show that $\tilde{\xi}(\omega)$ behaves as $\omega^{1/2}$ as $\omega \to 0$. From the properties of the Laplace transform, it follows that $Z(t)$ behaves as $t^{-3/2}$ at large t, the constant of proportionality being dependent on the assumed boundary conditions (Zwanzig and Bixon, 1970; Widom, 1971); the result in either case differs from (8.7.25) by omission of the quantity D. The predictions for $Z(t)$ at intermediate times can be improved by taking account of the compressibility of the fluid and by replacing the macroscopic viscosity by its frequency-dependent generalization (8.7.5) (Zwanzig and Bixon, 1970). The results are then in fair agreement with molecular dynamics calculations, except at short times (Metiu *et al.*, 1977). Substantially better agreement with simulation data has been achieved by Gaskell and Miller (1978a, b), who use an extension of the mode-coupling calculation of Ernst *et al.* (1971) in which some allowance is made for contributions from wavenumbers outside the hydrodynamic range; their approach is closely related to the mode-coupling formalism to be discussed in Section 9.5.

CHAPTER 9

Microscopic Theories of Time-correlation Functions

We turn now to the problem of devising a general theoretical scheme for the calculation of time-correlation functions at wavelengths and frequencies of a molecular scale. As we shall see, memory functions play a key role in the theoretical development, and we begin by showing how the memory-function approach can be formalized through the use of projection-operator methods (Zwanzig, 1961; Mori, 1965a, b). The calculation of the memory function in a specific problem can be tackled along two different lines. The first represents a systematic extension of the ideas of "generalized hydrodynamics" introduced in Chapter 8; the second is of a more microscopic nature, and includes both the mode-coupling approach discussed in Section 9.5 and the kinetic theories of Sections 9.6 to 9.9.

9.1 THE PROJECTION-OPERATOR FORMALISM

Let A be an arbitrary dynamical variable, dependent in general on all coordinates and momenta of the system. The time variation of $A(t)$ is determined by Eqn (2.1.10), which has a formal solution in terms of the initial value $A \equiv A(0)$:

$$A(t) = \exp(i\mathscr{L}t)A \tag{9.1.1}$$

The definition of A is assumed to be made in such a way that its mean value is zero, but this involves no loss of generality. If the phase function A is represented by a vector in a Hilbert space of dynamical variables, Eqn (9.1.1) has a simple geometric interpretation: the effect of the Liouville operator $\mathscr{L}$ is to "rotate" A through an angle $\mathscr{L}t$. We can also use a vector in such a space to represent a set of dynamical properties of the system, but we restrict ourselves for the present to the single-variable description.

Our aim is to find an exact equation of motion for $A(t)$ that is more readily usable than Eqn (9.1.1). We proceed by considering the time evolution of the projection of $A(t)$ onto A (the *projected* part) and also that of the component of $A(t)$ normal to A (the *orthogonal* part); the latter we shall denote by the symbol $A'(t)$. The projection of an arbitrary variable B onto A can be written formally in terms of a linear projection operator $\mathscr{P}$ as

$$\mathscr{P}B(t) = A(A, A)^{-1}(A, B(t)) \tag{9.1.2}$$

where parentheses $(\cdots, \cdots)$ are used to denote an inner product; in the one-variable case under consideration, the quantities on the right-hand side of Eqn (9.1.2) are all scalars. The operator $\mathscr{Q} = 1 - \mathscr{P}$ projects onto the subspace orthogonal to A. Hence the orthogonal component of $A(t)$ is

$$A'(t) = \mathscr{Q}A(t) \tag{9.1.3}$$

Both $\mathscr{P}$ and $\mathscr{Q}$ satisfy the fundamental properties of projection operators, i.e.

$$\mathscr{P}^2 = \mathscr{P}, \qquad \mathscr{Q}^2 = \mathscr{Q}, \qquad \mathscr{P}\mathscr{Q} = \mathscr{Q}\mathscr{P} = 0 \tag{9.1.4}$$

The geometrical concept of "projection" is to be identified with that of statistical-mechanical "correlation." Specifically, the inner product of the variables A and B is identified as the equilibrium time-correlation function of the two quantities:

$$(A, B(t)) \equiv \langle B(t)A^* \rangle \tag{9.1.5}$$

It follows that the usual requirements of an inner product are satisfied. As we have already pointed out in Section 7.1, the Liouville operator is hermitian in the Hilbert space of dynamical variables. Thus

$$(A, \mathscr{L}B) = (\mathscr{L}A, B) \tag{9.1.6}$$

The interpretation we have given of the inner product means that the projection of $A(t)$ along A is proportional to $Y(t)$, the normalized time autocorrelation function of the variable A, i.e.

$$\mathscr{P}A(t) = Y(t)A \tag{9.1.7}$$

with

$$\begin{aligned} Y(t) &= (A, A(t))(A, A)^{-1} \\ &\equiv \langle A(t)A^* \rangle \langle AA^* \rangle^{-1} \\ &= C_{AA}(t)/C_{AA}(0) \end{aligned} \tag{9.1.8}$$

The definitions (9.1.2) to (9.1.5) also ensure that

$$(A, A'(t)) = 0 \tag{9.1.9}$$

We first derive an equation for the time-evolution of the projected part $Y(t)$. In the notation of this section, the Laplace transform of Eqn (2.1.10) is

$$\begin{aligned} (z + \mathscr{L})\tilde{A}(z) &\equiv (z + \mathscr{L})(\mathscr{P} + \mathscr{Q})\tilde{A}(z) \\ &= iA(0) \end{aligned} \tag{9.1.10}$$

so that

$$\tilde{Y}(z) = i(A, (z+\mathscr{L})^{-1}A)(A, A)^{-1}$$
$$= i(A, \tilde{A}(z))(A, A)^{-1} \tag{9.1.11}$$

We now project Eqn (9.1.10) parallel and perpendicular to A by application, respectively, of the operators $\mathscr{P}$ and $\mathscr{Q}$:

$$z\mathscr{P}\tilde{A}(z) + \mathscr{P}\mathscr{L}\mathscr{P}\tilde{A}(z) + \mathscr{P}\mathscr{L}\mathscr{Q}\tilde{A}(z) = iA \tag{9.1.12}$$

$$z\mathscr{Q}\tilde{A}(z) + \mathscr{Q}\mathscr{L}\mathscr{P}\tilde{A}(z) + \mathscr{Q}\mathscr{L}\mathscr{Q}\tilde{A}(z) = 0 \tag{9.1.13}$$

Equation (9.1.13) can be used to eliminate $\mathscr{Q}\tilde{A}(z)$ from Eqn (9.1.12). With the help of the properties (9.1.4), we find, on taking the inner product of both sides of Eqn (9.1.12) with A, that

$$z(A, \tilde{A}(z)) + (A, \mathscr{L}\mathscr{P}\tilde{A}(z))$$
$$-(A, \mathscr{L}\mathscr{Q}(z+\mathscr{Q}\mathscr{L}\mathscr{Q})^{-1}\mathscr{Q}\mathscr{L}\mathscr{P}\tilde{A}(z)) = i(A, A) \tag{9.1.14}$$

It follows from Eqns (2.1.10), (9.1.2) and (9.1.11) that this result can also be written in the form

$$(-iz - i\Omega)\tilde{Y}(z) + i(K, (z+\mathscr{Q}\mathscr{L}\mathscr{Q})^{-1}K)(A, A)^{-1}\tilde{Y}(z) = 1 \tag{9.1.15}$$

where we have defined a frequency Ω as

$$i\Omega = (A, \dot{A})(A, A)^{-1} = \dot{Y}(0) \tag{9.1.16}$$

and the projection of $\dot{A}$ orthogonal to A as

$$K = \mathscr{Q}\dot{A} \tag{9.1.17}$$

The projection K is often called the "random force". The terminology is conventional: if A is the momentum of particle i, then $\dot{A}$ is the total force acting on i and K is the part of the force that is uncorrelated with the momentum, in which case K is the random force of classic Langevin theory (see Section 7.3). In general, of course, K is not a force in the mechanical sense. Note that in the present, single-variable description, the frequency Ω is identically zero for systems with continuous interactions, since all autocorrelation functions are even functions of time. We retain the term in order to facilitate the later generalization to a multivariable description, where Ω, in general, is a matrix with non-zero off-diagonal elements.

The time evolution of the random force is determined by the equation

$$R(t) = \exp(i\mathscr{Q}\mathscr{L}\mathscr{Q}t)K \tag{9.1.18}$$

or, after a Laplace transform:

$$\tilde{R}(z) = i(z+\mathscr{Q}\mathscr{L}\mathscr{Q})^{-1}K \tag{9.1.19}$$

Because of the special form of its propagator, $R(t)$ remains at all times in the subspace orthogonal to A, i.e.

$$(A, R(t)) = 0 \tag{9.1.20}$$

This is easily proved by expanding the right-hand side of Eqn (9.1.18) in powers of t, since it is clear by inspection that every term in the series is orthogonal to A. Note also that $R(t=0) \equiv K$.

The autocorrelation function of the random force defines the *memory function* for the evolution of the dynamical variable A:

$$M(t) = (R, R(t))(A, A)^{-1} \tag{9.1.21}$$

or

$$\begin{aligned} \tilde{M}(z) &= (R, \tilde{R}(z))(A, A)^{-1} \\ &= i(K, (z + \mathcal{Q}\mathcal{L}\mathcal{Q})^{-1} K)(A, A)^{-1} \end{aligned} \tag{9.1.22}$$

Equation (9.1.15) can be rewritten in terms of the memory function as

$$[-iz - i\Omega + \tilde{M}(z)]\tilde{Y}(z) = 1 \tag{9.1.23}$$

or

$$\dot{Y}(t) - i\Omega Y(t) + \int_0^t M(t-s) Y(s) \, ds = 0 \tag{9.1.24}$$

The time-evolution equation for the orthogonal component A' is obtained along similar lines. From Eqn (9.1.13) we find that for $\tilde{A}'(z) = \mathcal{Q}\tilde{A}(z)$:

$$\begin{aligned} (z + \mathcal{Q}\mathcal{L}\mathcal{Q})\tilde{A}'(z) &= -\mathcal{Q}\mathcal{L}\mathcal{P}\tilde{A}(z) \\ &= -\mathcal{Q}\mathcal{L}\tilde{Y}(z)A \\ &= i\tilde{Y}(z)K \end{aligned} \tag{9.1.25}$$

and hence from Eqn (9.1.19):

$$\begin{aligned} \tilde{A}'(z) &= i\tilde{Y}(z)(z + \mathcal{Q}\mathcal{L}\mathcal{Q})^{-1} K \\ &= \tilde{Y}(z)\tilde{R}(z) \end{aligned} \tag{9.1.26}$$

If we substitute for $\tilde{Y}(z)$ from (9.1.23), Eqn (9.1.26) becomes

$$[-iz - i\Omega + \tilde{M}(z)]\tilde{A}'(z) = \tilde{R}(z) \tag{9.1.27}$$

or

$$\dot{A}'(t) - i\Omega A'(t) + \int_0^t M(t-s) A'(s) \, ds = R(t) \tag{9.1.28}$$

Equations (9.1.24) and (9.1.28) are projections parallel and perpendicular to the variable A of a generalized Langevin equation for A:

$$\dot{A}(t) - i\Omega A(t) + \int_0^t M(t-s)A(s)\,\mathrm{d}s = R(t) \qquad (9.1.29)$$

This is of the same form as the Langevin equation (7.3.23), except that in Eqn (9.1.29) the random force $R(t)$ and memory function $M(t)$ have the explicit statistical-mechanical definitions given by (9.1.18) and (9.1.21).

There is a close connection between the behaviour of the functions $Y(t)$ and $M(t)$ at short times, a fact that we have already exploited in Section 7.3. When differentiated with respect to time, Eqn (9.1.24) becomes

$$\ddot{Y}(t) - i\Omega\dot{Y}(t) + M(0)Y(t) + \int_0^t \dot{M}(t-s)Y(s)\,\mathrm{d}s = 0 \qquad (9.1.30)$$

Since $Y(0) = 1$ and $\dot{Y}(0) = i\Omega$, we see that

$$\begin{aligned} M(0) &= -\ddot{Y}(0) - \Omega^2 \\ &= (\dot{A}, \dot{A})(A, A)^{-1} - \Omega^2 \end{aligned} \qquad (9.1.31)$$

Repeated differentiation leads to relations between the initial time derivatives of $Y(t)$ and $M(t)$ or, equivalently, given Eqn (7.1.26), between the frequency moments of the spectral functions $Y(\omega)$ and $M(\omega)$. These relations are useful in constructing simple forms for $M(t)$ that satisfy the low-order sum rules on $Y(t)$.

Equation (9.1.21) shows that the memory function is the autocorrelation function of the random force; the latter evolves in time under the action of the projected Liouville operator $\mathscr{QLQ}$ (see Eqn (9.1.18)). Retaining the analogy with the generalized Langevin equation for the motion of a brownian particle, it is instructive to establish the link between the time-autocorrelation function of the *random* force (i.e. the memory function), and that of the *total* force $\dot{A}$, which evolves in time according to the full Liouville operator $\mathscr{L}$. If the autocorrelation function of $\dot{A}$ is denoted by $\Phi(t)$, then

$$\Phi(t) = \frac{\langle \dot{A}(t)\dot{A}\rangle}{\langle AA\rangle} = -\ddot{Y}(t) \qquad (9.1.32)$$

and an integration by parts shows that the Laplace transforms $\tilde{\Phi}(z)$ and $\tilde{Y}(z)$ are related in the form

$$\tilde{\Phi}(z) = i\Omega - iz + z^2\tilde{Y}(z) \qquad (9.1.33)$$

Combination of Eqns (9.1.33) and (9.1.23) leads, after some algebra, to the expression

$$\tilde{M}(z) - i\Omega = \frac{\tilde{\Phi}(z) - i\Omega}{1 + [\tilde{\Phi}(z) - i\Omega]/iz} \qquad (9.1.34)$$

The structure of Eqn (9.1.34) emphasizes the fact that the two autocorrelation functions evolve in time in completely different ways, except in the high-frequency (short-time) limit.

There are two important ways in which the projection operator formalism can be extended. First, Eqn (9.1.24) may be regarded as the leading member in a hierarchy of memory-function equations (Mori, 1965b). If we apply the methods already used to the case when R is treated as the dynamical variable, we obtain an equation, similar to (9.1.24), for the time evolution of the projection of $R(t)$ along R. The kernel of the integral equation is now the autocorrelation function of a second-order random force that is orthogonal at all times both to R and to A. As an obvious generalization of this procedure, we can write that

$$\dot{M}_n(t) - i\Omega_n M_n(t) + \int_0^t M_{n+1}(t-s)\Delta_{n+1}^2 M_n(s)\, ds = 0 \qquad (9.1.35)$$

where

$$M_n(t) = (R_n, R_n(t))(R_n, R_n)^{-1} \qquad (9.1.36a)$$

$$R_n(t) = \exp(i\mathcal{Q}_n\mathcal{L}\mathcal{Q}_n t)\mathcal{Q}_n\dot{R}_{n-1} \qquad (9.1.36b)$$

and

$$\Delta_n^2 = (R_n, R_n) \qquad (9.1.37)$$

The operator $\mathcal{P}_n$ projects an ordinary dynamical variable along R_{n-1} according to the definition already given in Eqn (9.1.2). By construction, the operator

$$\mathcal{Q}_n = 1 - \sum_{j=1}^{n} \mathcal{P}_j \qquad (9.1.38)$$

projects onto a subspace that is orthogonal to all R_j for $j<n$. Thus the nth-order random force $R_n(t)$ is uncorrelated at all times with random forces of lower order. Equation (9.1.24) is a special case of (9.1.35), with $\mathcal{P}\equiv\mathcal{P}_1$, $A\equiv R_0$, $R\equiv R_1$, $\Omega\equiv\Omega_0$, $Y\equiv M_0$ and $M\equiv M_1$. Repeated application of the Laplace transform to equations of the hierarchy leads to an expression for $\tilde{Y}(z)$ in the form of a continued fraction:

$$\tilde{Y}(z) = \cfrac{1}{-iz - i\Omega_0 + \cfrac{\Delta_1^2}{-iz - i\Omega_1 + \cfrac{\Delta_2^2}{-iz - i\Omega_2 + \cfrac{\Delta_3^2}{-iz\cdots}}}} \qquad (9.1.39)$$

A second extension of the method, which has proved particularly useful in the study of collective modes in liquids, is one we have already mentioned. This is the generalization to the multivariable case, where the dynamical property of interest is not a single fluctuating property of the system, but a set of independent variables $\{A_1, A_2, \ldots, A_n\}$. We represent this set by a column vector $\mathbf{A}$ and its hermitian conjugate by the row vector $\mathbf{A}^*$. The derivation of the generalized Langevin equation for $\mathbf{A}$ follows the lines already laid down, due account being taken of the fact that the quantities involved are no longer scalars. The result can be written in matrix form as

$$\dot{\mathbf{A}}(t) - i\boldsymbol{\Omega} \cdot \mathbf{A}(t) + \int_0^t \boldsymbol{M}(t-s) \cdot \mathbf{A}(s)\,\mathrm{d}s = \mathbf{R}(t) \tag{9.1.40}$$

The definitions of the random force vector $\mathbf{R}(t)$, the frequency matrix $\boldsymbol{\Omega}$ and the memory-function matrix $\boldsymbol{M}(t)$ are analogous to those of $R(t)$, Ω and $M(t)$ in the single-variable case, the scalars A and A^* being replaced by the vectors $\mathbf{A}$ and $\mathbf{A}^*$. If we multiply Eqn (9.1.40) from the right by $\mathbf{A}^* \cdot (\mathbf{A}, \mathbf{A})^{-1}$ and take the statistical average, we find that

$$\dot{\boldsymbol{Y}}(t) - i\boldsymbol{\Omega} \cdot \boldsymbol{Y}(t) + \int_0^t \boldsymbol{M}(t-s) \cdot \boldsymbol{Y}(s)\,\mathrm{d}s = \boldsymbol{0} \tag{9.1.41}$$

where $\boldsymbol{Y} \equiv [Y_{ij}]$ is the correlation-function matrix. Equation (9.1.41) is the multivariable generalization of (9.1.24); its solution in terms of Laplace transforms is

$$\tilde{\boldsymbol{Y}}(z) = [-iz\boldsymbol{I} - i\boldsymbol{\Omega} + \tilde{\boldsymbol{M}}(z)]^{-1} \tag{9.1.42}$$

where $\boldsymbol{I}$ is the identity matrix. Note that the diagonal elements of $\boldsymbol{Y}$ are autocorrelation functions and the off-diagonal elements are cross correlation functions.

The value of the memory-function formalism is most easily appreciated by considering specific examples of its use. Before doing so, however, it is helpful to look at the problem from a wider point of view. The generalized Langevin equation (9.1.40) may be thought of as arising from a functional expansion of $\mathbf{A}(t)$ and the separation out of the term that is linear in $\mathbf{A}$ and can be represented by a convolution in time. The random force vector may then be regarded as describing the effects of non-linear terms, initial transient processes, and the explicit dependence of $\mathbf{A}(t)$ on variables not included in the set $\{A_i\}$. If $\mathbf{R}(t)$ is ignored, the equation represents the secular motion of the variables $\{A_i\}$; by including $\mathbf{R}(t)$, the fluctuations about the smoothed-out motion are taken into account. It turns out that the extraction of the linear term is equivalent to projecting $\mathbf{A}(t)$ onto the subspace spanned by $\mathbf{A}$ in a Hilbert space of dynamical variables. Such a separation of effects is most useful in cases where the random force fluctuates rapidly and the memory function decays much faster than the correlation function of interest. It is then not unreasonable to represent $\boldsymbol{M}(t)$ in some simple way, in particular by invoking a markovian approximation whereby the non-zero elements are replaced by delta-functions in t. For this representation to be successful, the vector $\mathbf{A}$ should contain as its components, not only the variables of immediate interest, but also those to which they are strongly coupled. If the set of variables is well-chosen, the effect of the operator is to project out all the slowly-varying properties of the system and the markovian assumption can then be used with greater confidence in approximating the memory-function matrix. By extending the dimensionality of $\mathbf{A}$, an increasingly detailed description can be obtained, without departing from the basic markovian hypothesis. In practice, this ideal state of affairs is often difficult to achieve, and some of the elements of $\boldsymbol{M}(t)$ may not be truly short-ranged in time. The calculation of the frequency matrix $\boldsymbol{\Omega}$ is generally a straightforward problem, since it involves only static quantities; the same is true of the static correlation matrix $(\mathbf{A}, \mathbf{A})$.

As an alternative to the multidimensional description it is possible to work with a smaller set of variables and exploit the continued-fraction expansion, truncating the hierarchy at a suitable point in some simple, approximate way. This approach is particularly useful when insufficient is known about the dynamical behaviour of the system to permit an informed choice of a larger set of variables. Its main disadvantage is the fact that the physical significance of the memory functions becomes increasingly more obscure as the expansion is carried to higher orders.

9.2 SELF CORRELATION FUNCTIONS

We consider first the application of projection-operator methods to the calculation of the self intermediate scattering function $F_s(k, t)$. This function is of interest because of its direct link to the velocity autocorrelation function (see Eqn (8.2.19)), and because its spectrum, the self dynamic structure factor $S_s(k, \omega)$, is closely related to the cross-section for incoherent scattering of neutrons.

A straightforward approach to the problem is to choose as the variable A the fluctuating density $\rho_{\mathbf{k}i}$ of a tagged particle i and write a memory function equation for $\tilde{F}_s(k, z)$ in the form

$$\tilde{F}_s(k, z) = \frac{1}{-iz + \tilde{M}_s(k, z)} \tag{9.2.1}$$

We see from Eqns (7.4.40) and (9.1.31) that the effect of setting $M_s(k, t=0) = \omega_0^2 = k^2 k_B T/m$ is to ensure that $S_s(k, \omega)$ has the correct second moment.

We may therefore rewrite $\tilde{M}_s$ in the form $\tilde{M}_s(k, z) = k^2 \tilde{D}(k, z)$ where, by analogy with (8.2.10), $\tilde{D}(k, z)$ plays the role of a generalized self-diffusion coefficient with the property that $\lim_{z\to 0} \lim_{k\to 0} \tilde{D}(k, z) = D$. On extending the continued fraction expansion to second order we find that

$$\tilde{F}_s(k, z) = \cfrac{1}{-iz + \cfrac{\omega_0^2}{-iz + \tilde{N}_s(k, z)}} \tag{9.2.2}$$

The second-order memory function $\tilde{N}_s(k, z)$ can be written as

$$\tilde{N}_s(k, z) = (2\omega_0^2 + \Omega_0^2)\tilde{n}_s(k, z) \tag{9.2.3}$$

where Ω_0^2 is given by (7.2.10) and $n_s(k, t=0) = 1$; the resulting expression for $S_s(k, \omega)$ now also has the correct fourth moment, regardless of the time dependence of $n_s(k, t)$. Equation (9.2.2) can also be derived from the first-order memory function equation for the self-current autocorrelation function $C_s(k, t)$ via the relation (8.2.19).

As an alternative to making a continued fraction expansion of $\tilde{F}_s(k, z)$, we can consider the multivariable description of the problem that results from the choice

$$\mathbf{A} = \begin{bmatrix} \rho_{\mathbf{k}i} \\ \dot{\rho}_{\mathbf{k}i} \\ \sigma_{\mathbf{k}i} \end{bmatrix} \tag{9.2.4}$$

where the variable $\sigma_{\mathbf{k}i}$, given by

$$\sigma_{\mathbf{k}i} = \ddot{\rho}_{\mathbf{k}i} - (\rho_{\mathbf{k}i}, \ddot{\rho}_{\mathbf{k}i})(\rho_{\mathbf{k}i}, \rho_{\mathbf{k}i})^{-1}\rho_{\mathbf{k}i} \tag{9.2.5}$$

is orthogonal to both $\rho_{\mathbf{k}i}$ and $\dot{\rho}_{\mathbf{k}i}$. From results obtained in Sections 7.4 and 8.2, it is straightforward to show that the corresponding static correlation matrix is diagonal and given by

$$(\mathbf{A}, \mathbf{A}) = \begin{bmatrix} 1 & 0 & 0 \\ 0 & \omega_0^2 & 0 \\ 0 & 0 & \omega_0^2(2\omega_0^2 + \Omega_0^2) \end{bmatrix} \tag{9.2.6}$$

whereas the frequency matrix is purely off-diagonal:

$$i\boldsymbol{\Omega} = (\mathbf{A}, \dot{\mathbf{A}}) \cdot (\mathbf{A}, \mathbf{A})^{-1} = \begin{bmatrix} 0 & 1 & 0 \\ -\omega_0^2 & 0 & 1 \\ 0 & -2\omega_0^2 - \Omega_0^2 & 0 \end{bmatrix} \tag{9.2.7}$$

Both $\dot{A}_1$ and $\dot{A}_2$ form part of the space spanned by the vector $\mathbf{A}$. In the case of $\dot{A}_1$, this is easy to see, because $\dot{A}_1 = A_2$. To understand why it is also true for $\dot{A}_2$, it is sufficient to note that the projection $\dot{A}_2$ along A_1 is obviously part of the space of $\mathbf{A}$, whereas the part orthogonal to A_1 is, according to the definition (9.2.5), the same as A_3. It follows that the random-force vector has only one non-zero component and the memory-function matrix has only one non-zero entry:

$$\boldsymbol{M}(t)=\begin{bmatrix} 0 & 0 & 0 \\ 0 & 0 & 0 \\ 0 & 0 & \mathscr{M}(k,t) \end{bmatrix} \tag{9.2.8}$$

On collecting the various results and inserting them in Eqn (9.1.42), we find that the correlation-function matrix has the form

$$\tilde{\boldsymbol{Y}}(k,z)=\begin{bmatrix} -iz & -1 & 0 \\ \omega_0^2 & -iz & -1 \\ 0 & 2\omega_0^2+\Omega_0^2 & -iz+\tilde{\mathscr{M}}(k,z) \end{bmatrix} \tag{9.2.9}$$

Inversion of (9.2.9) shows that $\tilde{F}_s(k,z)$ is given by

$$\tilde{F}_s(k,z)=\tilde{Y}_{11}(k,z)=\cfrac{1}{-iz+\cfrac{\omega_0^2}{-iz+\cfrac{2\omega_0^2+\Omega_0^2}{-iz+\tilde{\mathscr{M}}(k,z)}}} \tag{9.2.10}$$

and comparison with Eqns (9.2.2) and (9.2.3) makes it possible to identify $\mathscr{M}$ as the memory function of N_s. Thus the Laplace transform of $C_s(k,t)$ can be written as

$$\begin{aligned}\tilde{C}_s(k,z)&=\omega_0^2\tilde{Y}_{22}(k,z)\\ &=\frac{\omega_0^2}{-iz+(2\omega_0^2+\Omega_0^2)\tilde{n}_s(k,z)+\omega_0^2/-iz}\end{aligned} \tag{9.2.11}$$

In the limit $k\to 0$, the memory function $n_s(k,t)$ is directly related to the memory function of the velocity autocorrelation function $Z(t)$. From Eqns (7.2.9), (8.2.19) and (9.2.11) we find that

$$\tilde{Z}(z)=\frac{k_B T/m}{-iz+\Omega_0^2\tilde{n}_s(0,z)} \tag{9.2.12}$$

Thus

$$N_s(0,t)=\Omega_0^2 n_s(0,t) \tag{9.2.13}$$

is the memory function of $Z(t)$. Since $N_s(k,t)$ is also the memory function of $M_s(k,t)$, and $M_s(k,t=0)=k^2Z(t=0)$, we see that $k^2Z(t)$ is the memory function of $F_s(k,t)$ in the limit of very small k. In addition, the hydrodynamic limit (8.2.10) imposes the restriction that

$$\Omega_0^2\tilde{n}_s(0,0)=\frac{k_B T}{mD} \tag{9.2.14}$$

A particularly simple approximation is to replace $N_s(k, t)$ by a constant, $1/\tau_s(k)$ say, which is equivalent to assuming an exponential form for $M_s(k, t)$:

$$M_s(k, t) = \omega_0^2 \exp[-t/\tau_s(k)] \tag{9.2.15}$$

with the limitation, required to satisfy (9.2.14), that

$$\tau_s(0) = \beta m D \tag{9.2.16}$$

This approximation suffers from several defects, chief among which is the fact that it leads to an exponential velocity autocorrelation function of the Langevin type, the quantity $1/\tau_s(0)$ appearing as a frequency-independent friction coefficient. Better results are obtained by choosing an exponential form for $N_s(k, t)$, i.e.

$$N_s(k, t) = (2\omega_0^2 + \Omega_0^2) \exp[-t/\tau_s(k)] \tag{9.2.17}$$

with

$$\tau_s(0) = \frac{k_B T}{m D \Omega_0^2} \tag{9.2.18}$$

This second approximation, which is equivalent to neglecting the frequency-dependence of $\tilde{\mathcal{M}}(k, z)$, has the merit of leading to a simple analytic form for $S_s(k, \omega)$ that has the correct zeroth, second and fourth moments:

$$S_s(k, \omega) = \frac{1}{\pi} \frac{\tau_s(k)\omega_0^2(2\omega_0^2 + \Omega_0^2)}{\omega^2 \tau_s^2(k)(\omega^2 - 3\omega_0^2 - \Omega_0^2) + (\omega^2 - \omega_0^2)^2} \tag{9.2.19}$$

The corresponding expression for $\tilde{Z}(z)$ is that given in Eqn (7.3.28) (with $\tau \equiv \tau_s(0)$).

In the absence of any well-based microscopic theory, it is arguably wisest to treat the relaxation time $\tau_s(k)$ as an adjustable parameter. Nonetheless, it is tempting to look for some relatively simple prescription for this quantity. For example, Lovesey (1973) has used an argument based on a scaling of the memory function $M_s(k, t)$ to derive the expression

$$\tau_s^{-1}(k) = \gamma(2\omega_0^2 + \Omega_0^2)^{1/2} \tag{9.2.20}$$

The assumption that γ is independent of k is, in practice, reasonably well borne out. If, in the limit $k \to 0$, we require Eqn (9.2.20) to yield the correct diffusion coefficient, it follows that $\gamma = \beta \Omega_0 m D$; this leads to a value of γ of about 0.9 at the triple point of argon. On the other hand, for large wavenumbers, we can ensure that $S_s(k, 0)$ goes over correctly to the ideal-gas result by choosing $\gamma = 2/\pi^{1/2} \simeq 1.13$.

Although the exponential approximation (9.2.17) has been used with some success in the interpretation of experimental neutron-scattering data on argon (Lovesey, 1973), it is nevertheless clear that the real situation is much less simple, particularly at small wavevectors. For example, analysis of the molecular-dynamics calculations of Levesque and Verlet (1970) (see Figure 7.1) has shown that the details of $Z(t)$ for the Lennard-Jones fluid can be accounted for only by employing a rather complicated expression for the memory function $N_s(0, t)$. It is obvious, of course, that a single-relaxation-time model cannot describe both the short-time behaviour, dominated by binary collisions, and phenomena that are essentially collective in crigin, such as the negative plateau that appears in $Z(t)$ at high density and the positive, hydrodynamic $(\sim t^{-3/2})$ "tail." A similar analysis has been made of $N_s(k, t)$ at non-zero wavenumbers. Some of the results, together with those obtained from the approximation (9.2.17) are shown in Figure 8.1, where $\Delta\omega$, the width at a half-height of $S_s(k, \omega)$ is plotted as a function of k. Comparison with the molecular dynamics data shows that the Levesque–Verlet form for $N_s(k, t)$ gives much the better results; it also reproduces the k-dependence of the width more closely than the corresponding curve based on the gaussian approximation (8.2.15); the latter, like the Levesque–Verlet scheme, requires a complete knowledge of $Z(t)$ for its implementation.

Given an accurate representation of $F_s(k, t)$, it is natural to look for methods of relating the total dynamic structure factor to its self part. We have already given some examples of this approach, the convolution approximation of Vineyard (1958) being the simplest, and we now show how the memory-function formalism can be used to generate a hierarchy of similar relationships (Kim and Nelkin, 1971).

If we regard particle i as tagged, we can define the self part of the fluctuating density as $\rho_{\mathbf{k}}^{(s)}(t) \equiv \rho_{\mathbf{k}i}(t)$ and the distinct part as

$$\rho_{\mathbf{k}}^{(d)}(t) = \sum_{j \neq i}^{N} \exp[-i\mathbf{k} \cdot \mathbf{r}_j(t)]$$
$$= \rho_{\mathbf{k}}(t) - \rho_{\mathbf{k}}^{(s)}(t) \tag{9.2.21}$$

From our previous definitions, we know that the correlation of fluctuations in $\rho_{\mathbf{k}}^{(s)}$ is described by the self correlation function $F_s(k, t)$ and of those in $\rho_{\mathbf{k}}$ by the density-density function $F(k, t)$; we also introduce the cross-correlation function:

$$F_d(k, t) = \langle \rho_{\mathbf{k}}^{(s)}(t) \rho_{-\mathbf{k}}^{(d)} \rangle \tag{9.2.22}$$

which is the spatial Fourier transform of $G_d(\mathbf{r}, t)$. From the definitions (9.2.21) and (9.2.22), it follows immediately that

$$F(k, t) = \frac{1}{N}\langle \rho_{\mathbf{k}}(t)\rho_{-\mathbf{k}}\rangle$$
$$= F_d(k, t) + F_s(k, t) \tag{9.2.23}$$

while the autocorrelation function of the density $\rho_{\mathbf{k}}^{(d)}$, i.e.

$$\langle \rho_{\mathbf{k}}^{(d)}(t)\rho_{-\mathbf{k}}^{(d)}\rangle = NF(k, t) - 2F_d(k, t) - F_s(k, t) \tag{9.2.24}$$

can be approximated, for large N, by $NF(k, t)$. From Eqn (9.2.23) and the known sum rules on $F_s(k, t)$ and $F(k, t)$, we see that the initial value of $F_d(k, t)$ is

$$F_d(k, 0) = S(k) - 1 \tag{9.2.25}$$

Now consider a two-component vector **A**, defined as

$$\mathbf{A} = \begin{bmatrix} \rho_{\mathbf{k}}^{(s)}(t) \\ \rho_{\mathbf{k}}^{(d)}(t) \end{bmatrix} \tag{9.2.26}$$

For this choice of **A**, the frequency matrix (9.2.7) vanishes, and the Laplace transform of the memory-function equation takes the form

$$\begin{bmatrix} -iz + \tilde{M}_{11} & \tilde{M}_{12} \\ \tilde{M}_{21} & -iz + \tilde{M}_{22} \end{bmatrix} \begin{bmatrix} \tilde{F}_s & \tilde{F}_d \\ \tilde{F}_d & N\tilde{F} \end{bmatrix} = \begin{bmatrix} 1 & S(k)-1 \\ S(k)-1 & NS(k) \end{bmatrix} \tag{9.2.27}$$

Since there are only two independent correlation functions, it follows that Eqn (9.2.27) represents only two independent equations of motion, and there must exist some relations between the different elements of the memory-function matrix. When (9.2.27) is written in component form, it is easy to see that

$$\tilde{M}_{12}(k, z) = \frac{1}{N}[\tilde{M}_{22}(k, z) - \tilde{M}_{11}(k, z)] \tag{9.2.28}$$

$$\tilde{M}_{21}(k, z) = N\tilde{M}_{12}(k, z) \tag{9.2.29}$$

The two independent equations of motion can therefore be written as

$$-iz\tilde{F}_s(k, z) - 1 + \tilde{M}_{11}(k, z)\tilde{F}_s(k, z) = 0 \tag{9.2.30}$$

$$-iz\tilde{F}(k, z) - S(k) + \tilde{M}_{22}(k, z)\tilde{F}(k, z) = 0 \tag{9.2.31}$$

when a term is neglected that is of order N times smaller than those retained.

We can now obtain a closed expression for $\tilde{F}(k, z)$ in terms of $\tilde{F}_s(k, z)$ by postulating some relationship between the unknown memory functions M_{11} and M_{22}. The simplest assumption to make is that the two memory functions are identical. It follows immediately that

$$\tilde{F}(k, z) = S(k)\tilde{F}_s(k, z) \tag{9.2.32}$$

which, in different variables, is just Vineyard's approximation (7.4.29). A major objection to this result is the fact that the second moment of $S(k, \omega)$

is given incorrectly; in other words, particle conservation is not properly described. This defect can be remedied by scaling the memory functions in such a way that they have the correct value at $t = 0$, i.e. by writing $M_{22} = S(k)M_{11}$. Elimination of the memory functions between (9.2.30) and (9.2.31) leads in this case to Kerr's expression (7.9.19) (Kerr, 1968). The method can be systematically extended by including, for example, the first time derivatives of $\rho_{\mathbf{k}}^{(s)}$ and $\rho_{\mathbf{k}}^{(d)}$ in the specification of the vector **A** (Kim and Nelkin, 1971), but the approach is unlikely to yield useful results at small k because in the hydrodynamic limit the self and longitudinal currents obey very different equations. At very large k, the relationship between $S_s(k, \omega)$ and $S(k, \omega)$ becomes trivial, because single-particle and collective modes merge and the "distinct" contribution to $S(k, \omega)$ disappears.

9.3 TRANSVERSE COLLECTIVE MODES

As we saw in Section 8.7, the appearance of propagating shear waves in dense fluids can be explained in qualitative terms by a simple viscoelastic treatment based on a generalization of the hydrodynamic approach. In this section, we show how a viscoelastic theory can be developed in a systematic manner through use of the projection-operator formalism.

Taking the viscoelastic relation (8.7.4) as a guide, we choose (Akcasu and Daniels, 1970) as the components of the vector **A** the x-component of the mass current and the xz-component of the stress tensor, assuming as usual that the direction of **k** is the z-axis. Thus

$$\mathbf{A} = \begin{bmatrix} mj_{\mathbf{k}}^{x} \\ \sigma_{\mathbf{k}}^{xz} \end{bmatrix} \tag{9.3.1}$$

and

$$(\mathbf{A}, \mathbf{A}) = Vk_{\mathrm{B}}T \begin{bmatrix} \rho m & 0 \\ 0 & G_{\infty}(k) \end{bmatrix} \tag{9.3.2}$$

where $G_{\infty}(k)$ is the generalized elastic constant defined by Eqn (8.7.8). To calculate the frequency matrix we use the relations

$$(A_1, \dot{A}_1) = (A_2, \dot{A}_2) = 0 \tag{9.3.3}$$

$$(A_2, \dot{A}_1) = (A_1, \dot{A}_2) = -ikVk_{\mathrm{B}}TG_{\infty}(k) \tag{9.3.4}$$

and find that

$$i\boldsymbol{\Omega} = \begin{bmatrix} 0 & -ik \\ \dfrac{-ikG_{\infty}(k)}{\rho m} & 0 \end{bmatrix} \tag{9.3.5}$$

Because $\dot{A}_1$ is proportional to A_2, the projection of $\dot{A}_1$ orthogonal to $\mathbf{A}$ is identically zero. The memory function matrix therefore has only one non-zero element, which we denote by $M_t(k, t)$:

$$\mathbf{M}(k, t) = \begin{bmatrix} 0 & 0 \\ 0 & M_t(k, t) \end{bmatrix} \tag{9.3.6}$$

When these results are substituted in Eqn (9.1.42), we obtain an expression for the Laplace transform of the normalized correlation-function matrix having the form

$$\tilde{\mathbf{Y}}(k, z) = \begin{bmatrix} -iz & ik \\ \dfrac{ikG_\infty(k)}{\rho m} & -iz + \tilde{M}_t(k, z) \end{bmatrix} \tag{9.3.7}$$

In particular, for the Laplace transform of the transverse-current auto-correlation function, we find that

$$\begin{aligned} \tilde{C}_t(k, z) &= \omega_0^2 \tilde{Y}_{11}(k, z) \\ &= \cfrac{\omega_0^2}{-iz + \cfrac{\omega_{1t}^2}{-iz + \tilde{M}_t(k, z)}} \end{aligned} \tag{9.3.8}$$

where ω_{1t}^2 is given by Eqn (7.4.50) and is related to $G_\infty(k)$ through (8.7.8). The description given by (9.3.8) must coincide with the hydrodynamic result (8.4.4) in the limit $k \to 0$. This condition enables us to make the identification

$$\tilde{M}_t(0, 0) = \frac{G_\infty(0)}{\eta} \tag{9.3.9}$$

The function $\tilde{M}_t(k, z)$ is simply the memory function of the generalized kinematic shear viscosity $\tilde{\nu}(k, z)$ introduced in Section 8.7, as can be seen immediately from comparison of Eqn (9.3.8) with the Laplace transform of Eqn (8.7.6):

$$\tilde{C}_t(k, z) = \frac{\omega_0^2}{-iz + k^2 \tilde{\nu}(k, z)} \tag{9.3.10}$$

The markovian approximation here corresponds to ignoring the frequency dependence of $\tilde{M}_t(k, z)$ and replacing it by a constant, $1/\tau_t(k)$ say. This is equivalent to assuming that $\nu(k, t)$ relaxes exponentially with a characteristic decay time $\tau_t(k)$. Hence, from Eqn (8.7.8):

$$\nu(k, t) = \frac{G_\infty(k)}{\rho m} \exp[-t/\tau_t(k)] \tag{9.3.11}$$

Use of (9.3.11) ensures that the spectrum of transverse-current fluctuations has the correct second moment, irrespective of the choice of $\tau_t(k)$, and is given explicitly by

$$C_t(k, \omega) = \frac{1}{\pi} \operatorname{Re} \tilde{C}_t(k, \omega)$$

$$= \frac{1}{\pi} \frac{\omega_0^2 \omega_{1t}^2 \tau_t(k)}{\omega^2 + \tau_t^2(k)(\omega_{1t}^2 - \omega^2)^2} \tag{9.3.12}$$

If, as in Section 8.7, we define a wavenumber-dependent shear viscosity $\eta(k)$ as the zero-frequency limit of $\rho m \tilde{\nu}(k, \omega)$, we find in the approximation represented by Eqn (9.3.11) that

$$\eta(k) = \tau_t(k) G_\infty(k) \tag{9.3.13}$$

so that $\tau_t(k)$ appears as a k-dependent Maxwell relaxation time. In particular

$$\eta \equiv \eta(0) = \tau_t(0) G_\infty(0) \tag{9.3.14}$$

in agreement with (9.3.9).

It is easy to establish the criterion for the existence of propagating transverse modes within the context of the single-relaxation-time approximation represented by Eqn (9.3.12). The condition for $C_t(k, \omega)$ to have a peak at non-zero frequency at a given value of k is

$$\omega_{1t}^2 \tau_t^2(k) > \tfrac{1}{2} \tag{9.3.15}$$

The peak, if it exists, is at a frequency $(\omega_t)_{\max}$ such that

$$(\omega_t)_{\max} = [\omega_{1t}^2 - \tfrac{1}{2}\tau_t^{-2}(k)]^{1/2} \tag{9.3.16}$$

It follows from the inequality (9.3.15) that shear waves will appear for values of k greater than k_c, where k_c is a critical wavevector given by

$$k_c = \tau_t^{-1}(k) \left(\frac{\rho m}{2G_\infty(k)} \right)^{1/2} \tag{9.3.17}$$

We can obtain an estimate for k_c by taking the $k=0$ limit of (9.3.17); this gives

$$k_c = \frac{1}{\eta} [\tfrac{1}{2} \rho m G_\infty(0)]^{1/2} \tag{9.3.18}$$

On inserting the values of η and $G_\infty(0)$ obtained by molecular dynamics calculations for the Lennard-Jones fluid, we find that $k_c \sigma = 0.79$, where σ is the length parameter in Eqn (1.2.3) (Levesque *et al.*, 1973). This is apparently a rather good guide to what occurs in practice, since the same computer experiments show that shear waves first appear somewhere in the range between $k\sigma = 0.7$ and $k\sigma = 1.3$.

At large values of k, the shear waves again disappear, because the particles behave as if they were free and the result for $\tilde{C}_t(k, z)$ must approach that for the perfect gas. This suggests the use of a scheme that interpolates between the known small-k (see Eqn (9.3.14)) and large-k (i.e. free-particle) limits of $\tau_t(k)$; in this way, good results have been obtained for the dispersion of the shear-wave peak as a function of k (Akcasu and Daniels, 1970).

Given its simplicity, the viscoelastic approximation provides a very satisfactory description of the transverse-current fluctuations over a wide range of wavelength. A careful study reveals, however, that there are some systematic discrepancies with the molecular dynamics data, even when the parameter $\tau_t(k)$ is chosen to fit the observed spectrum rather than calculated from some semiempirical presciption. In particular, at long wavelengths, the shear-wave peaks are significantly too flat and broad, as the results shown in Figure 9.1 indicate. Given the choice of vector **A** in (9.3.1), the element $\tilde{Y}_{22}(k, z)$ of the resulting correlation-function matrix is the Laplace

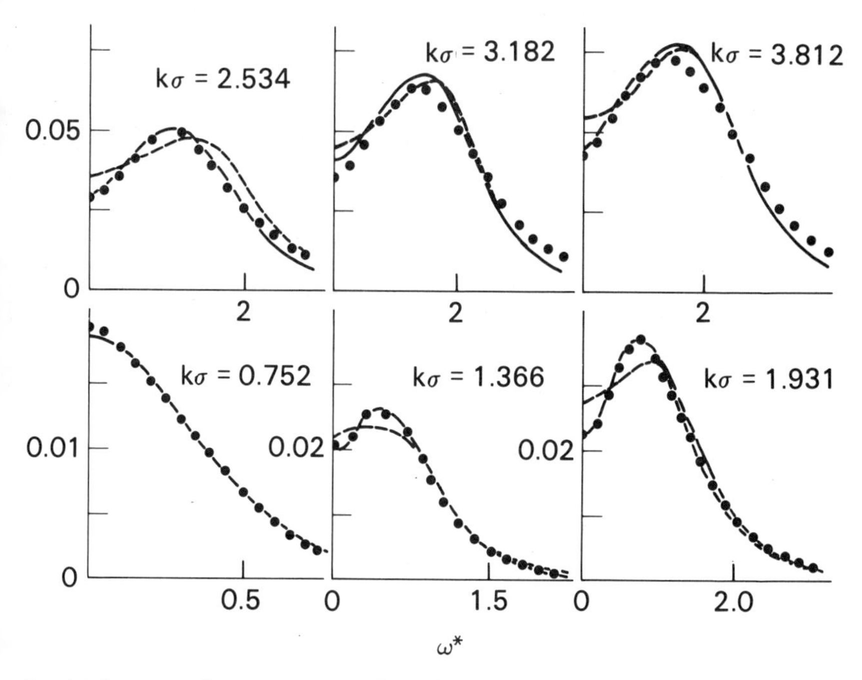

FIG. 9.1. Spectrum of transverse-current fluctuations for the Lennard-Jones fluid near the triple point. The points are molecular-dynamics results and the curves are calculated from Eqns (9.3.12) (dashed lines) and (9.3.21) (full lines). The unit of frequency is τ_0^{-1}, where τ_0 is defined by Eqn (3.3.5). After Levesque *et al.* (1973).

transform of the normalized autocorrelation function of the xz-component of the stress tensor. Thus we can write immediately that

$$\tilde{Y}_{22}(k, z) = \frac{\beta}{VG_\infty(k)} \int_0^\infty \langle \sigma_{\mathbf{k}}^{xz}(t) \sigma_{-\mathbf{k}}^{xz} \rangle \exp(izt)\, dt$$

$$= \frac{1}{-iz + \tilde{M}_t(k, z) + \omega_{1t}^2 / -iz} \tag{9.3.19}$$

If we again replace $\tilde{M}_t(k, z)$ by $1/\tau_t(k)$ and take the limit $k \to 0$, Eqn (9.3.19) can be inverted to give

$$\eta(t) = G_\infty(0)\, Y_{22}(0, t)$$

$$= G_\infty(0) \exp[-G_\infty(0) t / \eta] \tag{9.3.20}$$

which is consistent with (8.4.10), as of course it must be. As we have already pointed out in Section 8.7, the memory function $\nu(k, t)$ and the stress autocorrelation function $G_\infty(k) Y_{22}(k, t)/\rho m$ become identical as k tends to zero; within the single-relaxation-time approximation, the identity is immediately apparent from comparison of Eqns (9.3.11), (9.3.14) and (9.3.20). The identity also provides a clue to the origin of the comparative failure of the approximation (9.3.12). As Figure 8.3 shows, the correlation function $\eta(t)$ has a "tail" that decays very slowly, particularly at high densities, and we may reasonably suspect that the transverse-current fluctuations at small wavevectors can be adequately represented only if a similar long-time contribution is included in the memory function $\nu(k, t)$. Levesque *et al.* (1973) have therefore suggested the use of a two-exponential memory function of the form

$$\nu(k, t) = \frac{G_\infty(k)}{\rho m} \{[1 - \alpha(k)] \exp[-t/\tau_<(k)]$$

$$+ \alpha(k) \exp(-t/\tau_>)\} \tag{9.3.21}$$

where $\tau_>$ is practically independent of k and some seven times larger than $\tau_<(0)$. The mixing parameter $\alpha(k)$ has a value of approximately 0.1 at $k = 0$, and decreases rapidly with increasing k, so that for $k\sigma \gtrsim 3$ the single-relaxation-time approximation is recovered. The consequence of including the tail in the memory function is a marked enhancement of the shear-wave peaks at small k, leading to significantly improved agreement with the computer results, as shown by the curves plotted in Figure 9.1; the price paid is the introduction of a further two parameters. The Fourier transform of $\nu(k, t)$ is proportional to the frequency-dependent shear viscosity (8.7.5), and inclusion of the tail in the memory function is also essential for the correct description of the negative plateau in the velocity autocorrelation function $Z(t)$ when the latter is calculated by the method of Zwanzig and Bixon (1970), as discussed in Section 8.7 (Levesque *et al.*, 1973).

9.4 DENSITY FLUCTUATIONS

The description of the longitudinal-current fluctuations on the basis of the generalized Langevin equation is necessarily a more difficult task than in the case of the transverse modes. This is obvious from the much more complicated structure of the hydrodynamic limiting formula (8.5.10) compared with (8.4.4). The problem of particular interest is to account for the dispersion and eventual disappearance of the collective mode associated with sound-wave propagation.

In discussion of the longitudinal modes, a natural choice for the components of the dynamical vector **A** is the set of conserved variables consisting of $\rho_{\mathbf{k}}$, $\mathbf{j}_{\mathbf{k}}$, and the microscopic energy density $e_{\mathbf{k}}$, defined via Eqn (8.5.25). The variables $\rho_{\mathbf{k}}$ and $e_{\mathbf{k}}$ are both orthogonal to $\mathbf{j}_{\mathbf{k}}$. In place of $e_{\mathbf{k}}$, it is more convenient to choose that part which is also orthogonal to $\rho_{\mathbf{k}}$ and plays the role of a microscopic temperature fluctuation; this we write as $T_{\mathbf{k}}$, by analogy with the macroscopic temperature fluctuations considered in Section 8.3 (Schofield, 1966, 1968). The static correlation matrix is then diagonal. Since our attention is focussed on the longitudinal fluctuations, we include only the projection of the current along **k**, which we label $j_{\mathbf{k}}^z$. The vector **A** specified in this way, namely

$$\mathbf{A} = \begin{bmatrix} \rho_{\mathbf{k}} \\ j_{\mathbf{k}}^z \\ T_{\mathbf{k}} \end{bmatrix} \tag{9.4.1}$$

differs only trivially from that discussed by Schofield (1966, 1975), Murase (1970), and Copley and Lovesey (1975). Larger sets of variables that include both the stress tensor and the heat current have been used by Akcasu and Daniels (1970) and by Tong and Desai (1970). The static correlation matrix associated with the choice (9.4.1) is

$$(\mathbf{A}, \mathbf{A}) = \begin{bmatrix} NS(k) & 0 & 0 \\ 0 & N\left(\dfrac{k_{\mathrm{B}}T}{m}\right) & 0 \\ 0 & 0 & \langle T_{\mathbf{k}} T_{-\mathbf{k}} \rangle \end{bmatrix} \tag{9.4.2}$$

and the corresponding frequency matrix is

$$i\boldsymbol{\Omega} = \begin{bmatrix} 0 & -ik & 0 \\ -\dfrac{ik}{S(k)}\left(\dfrac{k_B T}{m}\right) & 0 & \dfrac{\langle j_{\mathbf{k}}^z T_{-\mathbf{k}}\rangle}{\langle T_{\mathbf{k}} T_{-\mathbf{k}}\rangle} \\ 0 & -\dfrac{\langle T_{\mathbf{k}} j_{-\mathbf{k}}^z\rangle}{N(k_B T/m)} & 0 \end{bmatrix} \tag{9.4.3}$$

It is unnecessary for our purposes to write more explicit expressions for the matrix elements of $(\mathbf{A}, \mathbf{A})$ and $i\boldsymbol{\Omega}$.

Since $\dot{A}_1$ is proportional to A_2, it follows that the component R_1 of the random-force vector is zero, and the memory-function matrix reduces to

$$\boldsymbol{M}(k, t) = \begin{bmatrix} 0 & 0 & 0 \\ 0 & M_{22}(k, t) & M_{23}(k, t) \\ 0 & M_{32}(k, t) & M_{33}(k, t) \end{bmatrix} \tag{9.4.4}$$

The correlation function matrix is therefore given by

$$\tilde{\boldsymbol{Y}}(k, z) = \begin{bmatrix} -iz & ik & 0 \\ \dfrac{ik}{S(k)}\left(\dfrac{k_B T}{m}\right) & -iz + \tilde{M}_{22}(k, z) & -i\Omega_{23} + \tilde{M}_{23}(k, z) \\ 0 & -i\Omega_{32} + \tilde{M}_{32}(k, z) & -iz + \tilde{M}_{33}(k, z) \end{bmatrix} \tag{9.4.5}$$

and the longitudinal-current autocorrelation function by

$$\begin{aligned} \tilde{C}_l(k, z) &= \omega_0^2 \tilde{Y}_{22}(k, z) \\ &= \frac{\omega_0^2}{-iz + \dfrac{\omega_0^2}{-izS(k)} + \tilde{N}_l(k, z)} \end{aligned} \tag{9.4.6}$$

where the memory function $N_l(k, t)$ is defined through its Laplace transform as

$$\tilde{N}_l(k, z) = \tilde{M}_{22}(k, z) - \Theta(k, z)[-iz + \tilde{M}_{33}(k, z)]^{-1} \tag{9.4.7}$$

with

$$\Theta(k, z) = \left(M_{23}(k, z) - \frac{\langle j_{\mathbf{k}}^z T_{-\mathbf{k}}\rangle}{\langle T_{\mathbf{k}} T_{-\mathbf{k}}\rangle}\right)\left(\tilde{M}_{32}(k, z) + \frac{\langle T_{\mathbf{k}} j_{-\mathbf{k}}^z\rangle}{N(k_B T/m)}\right) \tag{9.4.8}$$

The physical significance of the four unknown memory functions in (9.4.4) is easily inferred from their definitions in terms of the components of the random forces $\mathcal{Q}\dot{j}_{\mathbf{k}}^z$ and $\mathcal{Q}\dot{T}_{\mathbf{k}}$. The functions M_{23} and M_{32} describe a coupling between the momentum current and the heat flux, whereas M_{22}

and M_{33} represent, respectively, the relaxation processes associated with viscosity and thermal conduction. By comparison of Eqns (9.4.6) to (9.4.8) with the hydrodynamic result (8.5.10), we can make the following identification in the limit $k \to 0$:

$$\lim_{k\to 0} \tilde{M}_{22}(k, 0) = \frac{k^2(\frac{4}{3}\eta + \zeta)}{\rho m} = k^2 b \tag{9.4.9}$$

$$\lim_{k\to 0} \tilde{M}_{33}(k, 0) = \frac{k^2 \lambda}{\rho c_V} = k^2 a \tag{9.4.10}$$

and

$$\lim_{k\to 0} \frac{|\langle j_{\mathbf{k}}^z T_{-\mathbf{k}}\rangle|^2}{\langle T_{\mathbf{k}} T_{-\mathbf{k}}\rangle} = Nk^2 \left(\frac{k_B T}{m}\right)^2 \frac{\gamma - 1}{S(k)} \tag{9.4.11}$$

Finally, by requiring that

$$N_l(k, t=0) = \omega_{1l}^2 - \frac{\omega_0^2}{S(k)} \tag{9.4.12}$$

with ω_{1l}^2 given by (7.4.45), we guarantee that the first three non-zero moments of $S(k, \omega)$ are correct.

This particular example brings out clearly the advantages of working in terms of a multivariable description such as that provided by (9.4.1). For instance, we can immediately write down an expression for the temperature fluctuations analogous to Eqn (9.4.6) for the current fluctuations. Defining

$$C_T(k, t) = \langle T_{\mathbf{k}}(t) T_{-\mathbf{k}}\rangle \tag{9.4.13}$$

we find from Eqn (9.4.5) that

$$\begin{aligned}\tilde{C}_T(k, z) &= \langle T_{\mathbf{k}} T_{-\mathbf{k}}\rangle \tilde{Y}_{33}(k, z) \\ &= \frac{\langle T_{\mathbf{k}} T_{-\mathbf{k}}\rangle}{-iz + \tilde{M}_{33}(k, z) - \dfrac{\Theta(k, z)}{-iz + (\omega_0^2/-izS(k)) + \tilde{M}_{22}(k, z)}}\end{aligned} \tag{9.4.14}$$

The important point to note is that $\tilde{C}_T(k, z)$ can be expressed in terms of the same memory functions that are used to describe $\tilde{C}_l(k, z)$. Similarly, by solving for $\tilde{Y}_{11}(k, z)$, we obtain an expression for the density autocorrelation function in the form

$$\begin{aligned}\tilde{F}(k, z) &= S(k) \tilde{Y}_{11}(k, z) \\ &= \frac{S(k)}{-iz + \dfrac{1}{S(k)}\left(\dfrac{\omega_0^2}{-iz + \tilde{N}_l(k, z)}\right)}\end{aligned} \tag{9.4.15}$$

This is a less interesting result than that obtained for $C_\mathrm{T}(k, z)$ because $F(k, t)$ and $C_l(k, t)$ are in any case related by Eqn (7.4.36). It nevertheless brings out a second important feature of the multivariable approach. An expression for $\tilde{F}(k, z)$ having the same form as (9.4.15) can be more easily obtained by setting $A = \rho_\mathbf{k}$ and making a continued fraction expansion of $\tilde{F}(k, z)$, truncated at second order. What the more elaborate calculation yields is detailed information on the structure of the memory function $N_l(k, t)$, enabling contact to be made with the hydrodynamic result and allowing approximations to be introduced in a controlled way.

If we write the complex function $\tilde{N}_l(k, z)$ on the real axis ($z = \omega + i\varepsilon$, $\varepsilon \to 0+$) as the sum of its real and imaginary parts, i.e.

$$\tilde{N}_l(k, \omega) = N_l'(k, \omega) + iN_l''(k, \omega) \tag{9.4.16}$$

we find from (9.4.6) that the spectrum of longitudinal-current fluctuations is given by

$$C_l(k, \omega) = \frac{1}{\pi} \frac{\omega^2 \omega_0^2 N_l'(k, \omega)}{[\omega^2 - \omega_0^2/S(k) - \omega N_l''(k, \omega)]^2 + [\omega N_l'(k, \omega)]^2} \tag{9.4.17}$$

If the memory function were small, there would be a resonance at a frequency determined by the static structure of the fluid, i.e. at $\omega^2 = \omega_0^2/S(k)$. The physical role of the memory function is therefore to shift and damp the resonance.

The effective-field theories discussed in Section 7.9 can also be formulated in terms of approximations made for the memory function. For example, if the quantity inside the large round brackets in the denominator of (9.4.15) is replaced by the memory function for the density autocorrelation function of the ideal gas, we recover the mean-field result represented by Eqns (7.9.12) and (7.9.13). This procedure is equivalent to scaling the ideal-gas memory function in such a way as to satisfy the elastic sum rule on $S(k, \omega)$. If, in addition, agreement is forced with the fourth-moment sum rule, the result is the generalized mean-field theory developed by Pathak and Singwi (1970). The point that emerges most clearly is that a large number of apparently independent theories of collective motion in liquids can be treated on a unified basis in the framework of the memory-function formalism.

The task of calculating the function $N_l(k, t)$ remains a formidable one, even with the restrictions we have discussed, and some recourse to modelling is inevitable if tractable expressions for $C_l(k, \omega)$ and $S(k, \omega)$ are to be obtained. The limiting form of $\tilde{N}_l(k, \omega)$ when $k, \omega \to 0$ (hydrodynamic limit) follows from Eqns (9.4.7) to (9.4.11), i.e.

$$\lim_{\omega \to 0} \lim_{k \to 0} \tilde{N}_l(k, \omega) = bk^2 + \left(\frac{\omega_0^2}{S(k)}\right) \frac{\gamma - 1}{-i\omega + ak^2} \tag{9.4.18}$$

where the first term describes viscous relaxation and corresponds to $\tilde{M}_{22}(k, \omega)$ in Eqn (9.4.7), while the second term arises from temperature fluctuations. We now require a generalization of (9.4.18) that is valid for microscopic wavelengths and frequencies. An obvious first approximation is to assume that the coupling between the stress tensor and the heat current makes no appreciable contribution to the density fluctuations. This is true in the hydrodynamic limit, and it is true instantaneously at all values of k because the random forces $\mathscr{Q}\dot{j}_{\mathbf{k}}^{z}$ and $\mathscr{Q}\dot{T}_{\mathbf{k}}$ are instantaneously uncorrelated. Thus

$$M_{23}(k, t=0) = M_{32}(k, t=0) = 0 \tag{9.4.19}$$

If we also assume that the effect of thermal fluctuations is negligible, an approximation that can be justified at large wavenumbers (Ailawadi *et al.*, 1971), we are left only with the problem of representing the generalized longitudinal viscosity $\tilde{M}_{22}(k, \omega)$. In view of the success of the viscoelastic model (9.3.11) in the case of the transverse currents, it is natural to make a similar approximation here by writing

$$N_l(k, t) = \left(\omega_{1l}^2 - \frac{\omega_0^2}{S(k)}\right) \exp\left[-t/\tau_l(k)\right] \tag{9.4.20}$$

which is compatible with the constraint (9.4.12). The corresponding result for $S(k, \omega)$ itself is

$$S(k, \omega) = \frac{1}{\pi} \frac{\tau_l(k)\omega_0^2[\omega_{1l}^2 - \omega_0^2/S(k)]}{[\omega\tau_l(k)(\omega^2 - \omega_{1l}^2)]^2 + [\omega^2 - \omega_0^2/S(k)]^2} \tag{9.4.21}$$

As in the case of transverse currents, a number of proposals have been made for the calculation of the relaxation time $\tau_l(k)$. Akcasu and Daniels (1970), for example, have suggested an expression that interpolates between the known small-k and large-k limits of $\tau_l(k)$. Lovesey (1971) has proposed an even simpler approximation, based on arguments similar to those leading to (9.2.20), namely

$$\tau_l^{-1}(k) = \frac{2}{\pi^{1/2}} \left(\omega_{1l}^2 - \frac{\omega_0^2}{S(k)}\right)^{1/2} \tag{9.4.22}$$

This yields the correct ideal-gas value for $S(k, 0)$ in the limit of very large k. Other ways of estimating $\tau_l(k)$ have been discussed by Chung and Yip (1969) and by Machida and Murase (1973). The usefulness of this approach is illustrated in Figure 9.2, which shows the dispersion of the sound-wave peak in liquid rubidium as obtained experimentally (Copley and Rowe, 1974a, b) and by molecular dynamics simulation (Rahman, 1974a, b), and compares the results with calculations based on the approximation (9.4.21) in conjunction with (9.4.22) (Copley and Lovesey, 1975). The agreement is

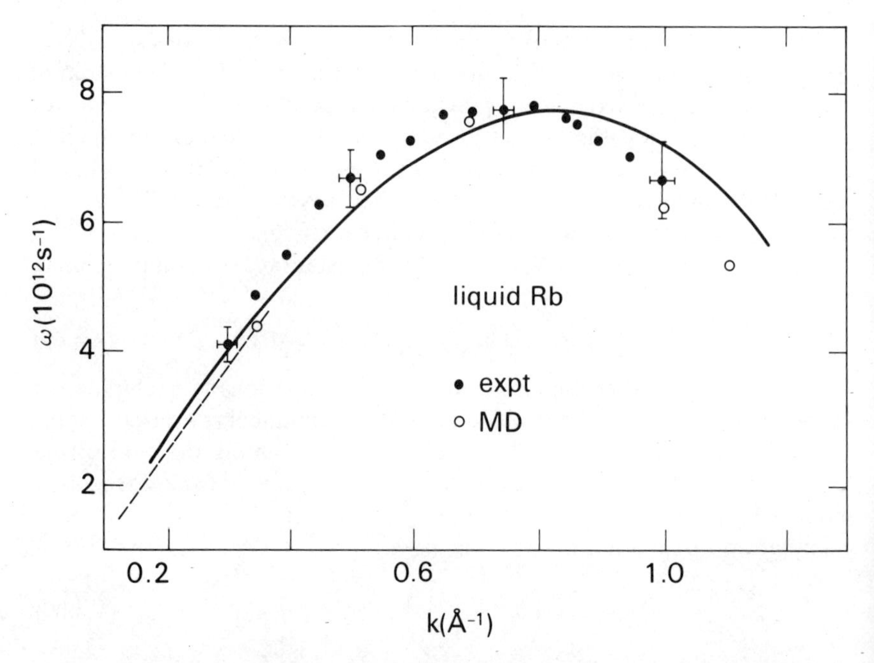

FIG. 9.2. Dispersion of the Brillouin peak in liquid rubidium, obtained from neutron-scattering results (Copley and Rowe, 1974a) and molecular dynamics calculations (Rahman, 1974a, b). The full curve is calculated from the viscoelastic approximation (9.4.21) in conjunction with Eqn (9.4.22); the dashed curve shows the predicted dispersion based on the experimental speed of sound. After Copley and Lovesey (1975).

remarkably good. The detailed shape of $S(k, \omega)$ is less well reproduced, at least at small k; the example plotted in Figure 9.3 shows that the discrepancies occur mostly at low frequencies.

Despite its success in the case of rubidium, and also for argon-like liquids at relatively large wavevectors (Rowe and Sköld, 1972; Schofield, 1975), the type of scheme described above is clearly an oversimplification and in some circumstances can yield qualitatively incorrect results. For example, in the approximation represented by Eqn (9.4.22), viscoelastic theory is unable to account for the Brillouin peak observed at small k in molecular dynamics calculations on the Lennard-Jones fluid near its triple point (Levesque *et al.*, 1973). Given (9.4.22), it is easy to show that viscoelastic theory predicts the existence of such a collective excitation whenever (Lovesey, 1971)

$$\omega_{1l}^2 < \frac{3\omega_0^2}{S(k)} \tag{9.4.23}$$

In the Lennard-Jones system, this inequality is unsatisfied even at $k = 0$,

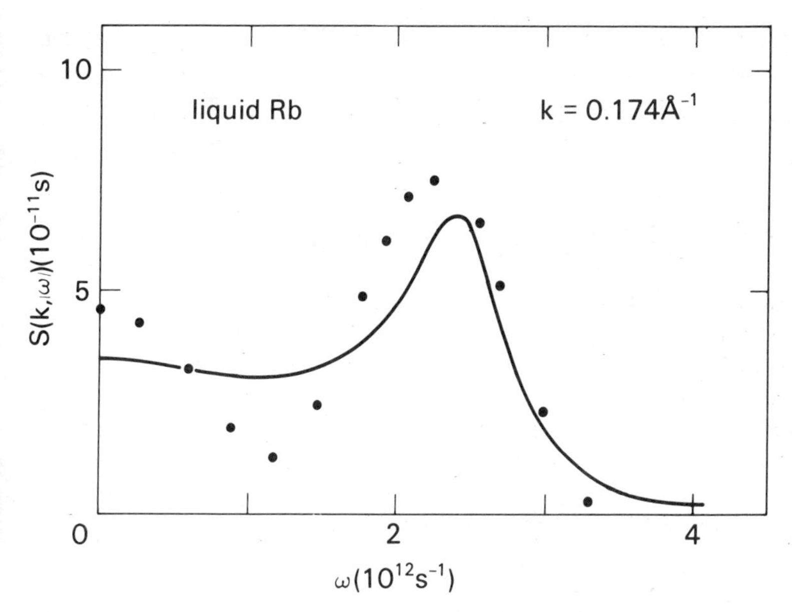

FIG. 9.3. Dynamic structure factor of a model of liquid rubidium. The points are molecular dynamics results (Rahman, 1974a, b) and the curve is calculated from the viscoelastic approximation (9.4.21) in conjunction with Eqn (9.4.22). After Copley and Lovesey (1975).

where it takes the form

$$\chi_{\mathrm{T}}[\tfrac{4}{3}G_\infty(0)+K_\infty(0)]<3 \tag{9.4.24}$$

when ω_{1l}^2 is expressed in terms of the long-wavelength limits of the instantaneous shear modulus of Eqn (8.7.8) and the instantaneous bulk modulus $K_\infty(k)$, defined by

$$\tfrac{4}{3}G_\infty(k)+K_\infty(k)=\frac{\rho m}{k^2}\omega_{1l}^2(k) \tag{9.4.25}$$

In the case of rubidium, under the conditions of the experiment mentioned above, the left-hand side of Eqn (9.4.24) is equal to 2.1. For the Lennard-Jones fluid in the state investigated by Levesque *et al.* (1973), the left-hand side is equal to 4.9, and therefore viscoelastic waves cannot be sustained. It seems plausible to conclude that the persistence of the sound-wave peak in liquid metals to relatively much larger wavenumbers than in rare-gas liquids (Levesque *et al.*, 1973; Bell *et al.*, 1975) is associated with the lower compressibility (greater rigidity) of metals. This in turn can be correlated with the "softer" interatomic potentials in the metals compared with the rare gases (Lewis and Lovesey, 1977).

In order to describe the small-k behaviour of the Lennard-Jones system, it is necesssary to go beyond the viscoelastic approximation (9.4.20) by including the effect of temperature fluctuations. A generalization of the hydrodynamic result (9.4.18) that satisfies the short-time constraint (9.4.12) is obtained by setting

$$\tilde{N}_l(k,\omega)=\left(\omega_{1l}^2-\frac{\omega_0^2\gamma(k)}{S(k)}\right)\tilde{n}_{1l}(k,\omega)+\left(\frac{\omega_0^2}{S(k)}\right)\frac{\gamma(k)-1}{-i\omega+a(k)k^2} \tag{9.4.26}$$

with $n_{1l}(k,t=0)=1$; in this approximation, we neglect the frequency dependence of the generalized thermal diffusivity $a(k)$ (the quantity $a(0)$ is defined by Eqn (8.3.20)). If we set $\gamma(k)$ (a k-dependent ratio of specific heats) equal to one, the term that represents temperature fluctuations disappears and we recover the viscoelastic approximation; the latter, as we have seen, works reasonably well for liquid metals, where $\gamma(0)$ is rather close to one. The first term on the right-hand side of (9.4.26) can be identified as $\tilde{M}_{22}(k,\omega)$; if we assume a simple exponential form for $n_{1l}(k,t)$, i.e.

$$n_{1l}(k,t)=\exp\left[-t/\tau_l(k)\right] \tag{9.4.27}$$

we find that in the hydrodynamic limit $\tilde{M}_{22}(k,0)$ approaches the value

$$\lim_{k\to 0}\frac{\tilde{M}_{22}(k,0)}{k^2}=\frac{1}{\rho m}\left[\tfrac{4}{3}G_\infty(0)+K_\infty(0)-\chi_s^{-1}\right]\tau_l(0) \tag{9.4.28}$$

where $\chi_s=(\beta/\rho\gamma)S(0)$ is the adiabatic compressibility (2.3.32). Comparison of (9.4.28) with (9.4.9) shows that $\tau_l(0)$ is given by

$$\tau_l(0)=\frac{\tfrac{4}{3}\eta+\zeta}{\tfrac{4}{3}G_\infty(0)+K_\infty(0)-\chi_s^{-1}} \tag{9.4.29}$$

Equations (9.4.26) to (9.4.29) make up the set of generalized hydrodynamic equations used by Levesque *et al.* (1973) in analysing their molecular dynamics data on the Lennard-Jones fluid; similar schemes have been proposed by Chung and Yip (1969), Sears (1970), and Copley and Lovesey (1975). Among the satisfying features of this calculation is the fact that at long wavelengths $\tau_l(k)$, as determined by a least-squares fit to the simulation data, tends correctly to its hydrodynamic limiting value, Eqn (9.4.29). Furthermore, $\gamma(k)$ is found to tend rapidly to unity as k increases. The large-k behaviour of $\gamma(k)$ implies that the viscoelastic model is a good approximation at large wavevectors, because the coupling with the thermal modes becomes negligible. On the other hand, at small k, $\gamma(k)$ tends to a value that is larger by a factor of approximately two than the thermodynamic result obtained from the observed fluctuations in temperature (see Eqn (3.3.6)). This fault can be eliminated by inclusion of a slowly relaxing part in the generalized longitudinal viscosity $M_{22}(k, t)$. If a two-exponential form similar to (9.3.21) is used for $n_{1l}(k, t)$, and if the short and long decay times are chosen, for convenience, to be identical to the corresponding transverse relaxation times, an excellent fit to the data is obtained, as Figure 9.4 shows, for which $\gamma(k)$ tends to its thermodynamic value as $k \to 0$. The good agreement obtained with a single exponential is to some extent fortuitous, the omission of the long-time part of the viscous contribution to the memory function being offset by an increase in the size of the thermal contribution.

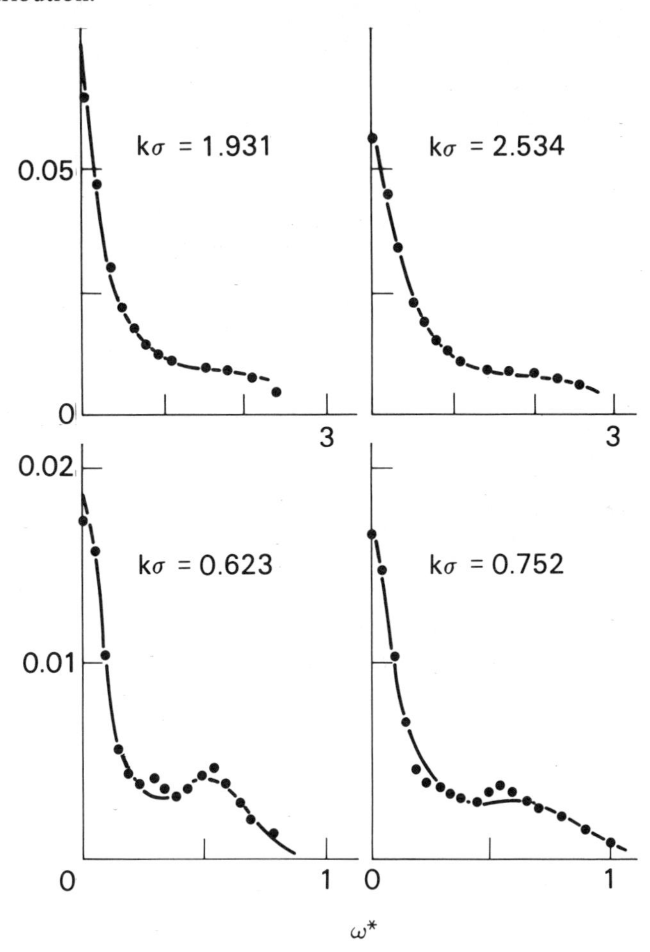

FIG. 9.4. Dynamic structure factor of the Lennard-Jones fluid near the triple point. The points are molecular dynamics results and the curves are calculated from Eqn (9.4.26) with a two-exponential approximation to $n_{1l}(k, t)$. The unit of frequency is τ_0^{-1}, where τ_0 is defined by Eqn (3.3.5). After Levesque *et al.* (1973).

9.5 MODE-COUPLING THEORIES

The applications of the projection-operator formalism studied thus far are largely phenomenological in nature in the sense that a simple functional form has generally been assumed to describe the decay of the various memory functions. Such descriptions can be looked upon as interpolation schemes between the short-time behaviour of correlation functions, which is introduced via frequency sum rules, and the hydrodynamic regime, which governs the choice of the vector **A** of dynamical variables. A more ambitious programme would be to derive expressions for the memory functions from first principles, starting from the formally exact definitions of Section 9.1. A possible route towards such a microscopic theory is provided by the mode-coupling approach; this has already been used in Section 8.7 to investigate the slow decay of correlation functions at long times. In this section, we show how mode-coupling concepts can be applied to the calculation of time-correlation functions and their associated memory functions within the framework of the projection-operator formalism (Götze and Lücke, 1975; Munakata and Igarashi, 1977, 1978; Bosse *et al.*, 1978a). The basic idea behind mode-coupling theory is that the fluctuation (or "excitation") of a given dynamical variable decays predominantly into pairs of hydrodynamic modes associated with conserved single-particle or collective dynamical variables. The possible "decay channels" of a fluctuation are determined by "selection rules" based, for example, on time-reversal symmetry or on physical considerations. In this way, correlation functions can generally be expressed as sums of products of the correlation functions of conserved variables, provided a decoupling (or factorization) approximation is made.

In order to illustrate the method, we first use the projection-operator formalism to rederive the asymptotic form (8.7.24) of the velocity autocorrelation function. Let u_{0x} be the x-component of the velocity of a tagged particle. With the notation of Section 9.1, the velocity autocorrelation function (7.2.1) can be written as

$$Z(t) = (u_{0x}, \exp(i\mathcal{L}t)u_{0x}) \tag{9.5.1}$$

From the considerations of Section 8.7, we expect the tagged-particle velocity to be strongly coupled to the longitudinal and transverse components of the collective particle current (7.4.7). Equation (8.7.18) suggests that we take the tagged-particle density $\rho_{\mathbf{k}'0}$ and the current $\mathbf{j}_{-\mathbf{k}''}$ to be the relevant pair of modes. Translational invariance implies that the only products of Fourier components whose inner product with the tagged-particle velocity are non-zero are those for which $\mathbf{k}' = \mathbf{k}''$. In addition, time-reversal symmetry means that all $\rho_{\mathbf{k}0}\rho_{-\mathbf{k}}$ can be discarded as possible product variables, while the products $\rho_{\mathbf{k}}\mathbf{j}_{-\mathbf{k}0}$ (where $\mathbf{j}_{\mathbf{k}0}$ is the tagged-particle current) are left out for reasons of simplicity and by analogy with Eqn (8.7.18). Hence the first approximation of the mode-coupling treatment corresponds to replacing the full evolution operator $\exp(i\mathcal{L}t)$ by its projection on the subspace of product variables $\rho_{\mathbf{k}0}\mathbf{j}_{-\mathbf{k}}$, i.e.

$$\exp(i\mathcal{L}t) \simeq \mathcal{P}^{(2)} \exp(i\mathcal{L}t)\mathcal{P}^{(2)} \tag{9.5.2}$$

The projector $\mathcal{P}^{(2)}$ is defined, as in Eqn (9.1.2), by its action on a dynamical variable B:

$$\mathcal{P}^{(2)}B = \sum_{\mathbf{k}}\sum_{\alpha} (\rho_{\mathbf{k}0}j^{\alpha}_{-\mathbf{k}}, B)(\rho_{\mathbf{k}0}j^{\alpha}_{-\mathbf{k}}, \rho_{\mathbf{k}0}j^{\alpha}_{-\mathbf{k}})^{-1}\rho_{\mathbf{k}0}j^{\alpha}_{-\mathbf{k}} \tag{9.5.3}$$

where the sum on α runs over all cartesian components. Thus Eqn (9.5.1) for $Z(t)$ becomes

$$\begin{aligned} Z(t) = \sum_{\mathbf{k}}\sum_{\mathbf{k}'}\sum_{\alpha}\sum_{\beta} & (u_{0x}, \rho_{\mathbf{k}0}j^{\alpha}_{-\mathbf{k}})(\rho_{\mathbf{k}0}j^{\alpha}_{-\mathbf{k}}, \rho_{\mathbf{k}0}j^{\alpha}_{-\mathbf{k}})^{-1} \\ & \times (\rho_{\mathbf{k}0}j^{\alpha}_{-\mathbf{k}}, \exp(i\mathcal{L}t)\rho_{\mathbf{k}'0}j^{\beta}_{-\mathbf{k}'})(\rho_{\mathbf{k}'0}j^{\beta}_{-\mathbf{k}'}, \rho_{\mathbf{k}'0}j^{\beta}_{-\mathbf{k}'})^{-1} \\ & \times (\rho_{\mathbf{k}'0}j^{\beta}_{-\mathbf{k}'}, u_{0x}) \end{aligned} \tag{9.5.4}$$

In the sum on the right-hand side of Eqn (9.5.4), the time-correlation functions of the product variables $\rho_{\mathbf{k}0}j^{\alpha}_{-\mathbf{k}}$ are bracketed by two time-independent "vertices" that arise from the projection of u_{0x} onto the product variables. The vertices are easily evaluated. For example

$$\begin{aligned} (u_{0x}, \rho_{\mathbf{k}0}j^{\alpha}_{-\mathbf{k}})(\rho_{\mathbf{k}0}j^{\alpha}_{-\mathbf{k}}, \rho_{\mathbf{k}0}j^{\alpha}_{-\mathbf{k}})^{-1} &= \delta_{\alpha x}\frac{k_{\mathrm{B}}T}{m}\left(\frac{Nk_{\mathrm{B}}T}{m}\right)^{-1} \\ &= \frac{1}{N}\delta_{\alpha x} \end{aligned} \tag{9.5.5}$$

A similar result holds for the second vertex in (9.5.4).

The time-correlation functions appearing on the right-hand side of Eqn (9.5.4) are of an unusual type, since they involve four, rather than two dynamical variables. The second approximation usually made is to assume that the two modes appearing in the product variables propagate independently of each other. This means that the four-variable correlation functions are factorized into products of two-variable correlation functions; in the present case:

$$\begin{aligned} &(\rho_{\mathbf{k}0}j^{\alpha}_{-\mathbf{k}}, \exp(i\mathcal{L}t)\rho_{\mathbf{k}'0}j^{\beta}_{-\mathbf{k}'}) \\ &\quad \simeq (\rho_{\mathbf{k}0}, \exp(i\mathcal{L}t)\rho_{\mathbf{k}'0})(j^{\alpha}_{-\mathbf{k}}, \exp(i\mathcal{L}t)j^{\beta}_{-\mathbf{k}'}) \\ &\quad = \langle\rho_{\mathbf{k}0}(t)\rho_{-\mathbf{k}0}\rangle\langle j^{\beta}_{\mathbf{k}}(t)j^{\alpha}_{-\mathbf{k}}\rangle \end{aligned} \tag{9.5.6}$$

Substitution of Eqns (9.5.5) and (9.5.6) into (9.5.4) leads to the expression

$$Z(t) = \frac{1}{N^2} \sum_{\mathbf{k}} \langle \rho_{\mathbf{k}0}(t) \rho_{-\mathbf{k}0} \rangle \langle j_{\mathbf{k}}^x(t) j_{-\mathbf{k}}^x \rangle \tag{9.5.7}$$

The first factor is the self intermediate scattering function, and the second is the current autocorrelation function; the latter can be decomposed into its longitudinal and transverse parts, as in Eqn (7.4.33). On switching from a sum over wavevectors to an integral, Eqn (9.5.7) becomes

$$\begin{aligned} Z(t) &= \frac{1}{N} \frac{V}{(2\pi)^3} \int d\mathbf{k}\, F_s(k, t) \frac{1}{k^2} \\ &\quad \times \left[\frac{k_x^2}{k^2} C_l(k, t) + \left(1 - \frac{k_x^2}{k^2}\right) C_t(k, t) \right] \\ &= \frac{1}{3\rho(2\pi)^3} \int d\mathbf{k}\, F_s(\mathbf{k}, t) \\ &\quad \times \frac{1}{k^2} [C_l(k, t) + 2C_t(k, t)] \end{aligned} \tag{9.5.8}$$

If the functions F_s, C_l and C_t are replaced by their hydrodynamic limits, Eqn (9.5.8) leads back to the result (8.7.24), which is valid for long times. At short times, however, Eqn (9.5.8) breaks down. In particular, since $F_s(k, t=0) = 1$ and $C_l(k, t=0) = C_t(k, t=0) = \omega_0^2 \propto k^2$, it is clear that $Z(t)$ diverges as $t \to 0$. To overcome this difficulty, a cut-off at large wavenumbers must be introduced in the integration over $\mathbf{k}$ (Munakata and Igarashi, 1977; Dubey *et al.*, 1980). Such a cut-off occurs naturally in the work of Gaskell and Miller (1978a, b), who obtain a result very similar to Eqn (9.5.8) by constructing a microscopic form for the tagged-particle velocity field involving a "form factor" $f(r)$ that vanishes for distances larger than the particle diameter. In this way they arrive at an expression for $Z(t)$ of the form

$$\begin{aligned} Z(t) &= \frac{1}{3\rho(2\pi)^3} \int d\mathbf{k}\, F_s(k, t) \\ &\quad \times \frac{1}{k^2} \hat{f}(k) [C_l(k, t) + 2C_t(k, t)] \end{aligned} \tag{9.5.9}$$

Gaskell and Miller use simple viscoelastic approximations for $C_l(k, t)$ and $C_t(k, t)$ and the gaussian approximation (8.2.15) for $F_s(k, t)$. This leads, together with Eqn (7.2.7), to a system of equations for $Z(t)$ that must be solved self-consistently. In practice, the integral in (9.5.9) turns out to be not very sensitive to the precise form of $Z(t)$ appearing in (7.2.7), and it is therefore sufficient to use the single-relaxation-time approximation

(7.3.30) in the gaussian representation of $F_s(k, t)$. Results for the velocity autocorrelation function of liquid rubidium and the coresponding spectrum are compared with the molecular dynamics data of Rahman (1974b) in Figure 9.5; it is evident from part (b) of the figure that the coupling to the transverse current is dominant at low frequencies, in agreement with the hydrodynamic result discussed in Section 8.7.

Another method whereby the short-time behaviour of the mode-coupling approximation (9.5.8) can be improved is to include the exact frequency moments of $Z(\omega)$ in a systematic way by working in the continued fraction representation. Truncation of (9.1.39) at second order (Bosse *et al.*, 1978d) gives

$$Z(\omega) = \frac{1}{\pi} \operatorname{Re} \tilde{Z}(z=\omega)$$

$$= \frac{k_B T}{\pi m} \frac{\Omega_0^2 M_2'(\omega)}{[\omega^2 - \Omega_0^2 - \omega M_2''(\omega)]^2 + [\omega M_2'(\omega)]^2} \qquad (9.5.10)$$

where $\Omega_0^2 = \Delta_1^2$ is the square of the Einstein frequency (7.2.13) and $M_2'(\omega)$ and $M_2''(\omega)$ are the real and imaginary parts, respectively, of the second-order memory function $\tilde{M}_2(z=\omega)$. From Eqns (9.1.36) to (9.1.38), the exact expression for $M_2(t)$ is

$$M_2(t) = \frac{m}{\Omega_0^2} (\mathcal{Q}_1 \mathcal{L}^2 u_{0x}, \exp(i\mathcal{Q}_2 \mathcal{L} \mathcal{Q}_2 t) \mathcal{Q}_1 \mathcal{L}^2 u_{0x}) \qquad (9.5.11)$$

where $\mathcal{Q}_1 = 1 - \mathcal{P}_1$ projects onto the subspace orthogonal to u_{0x} while $\mathcal{Q}_2 = \mathcal{Q}_1 - \mathcal{P}_2$ projects onto the subspace orthogonal to both u_{0x} and the acceleration $\dot{u}_{0x} = i\mathcal{L} u_{0x}$; note that $\mathcal{Q}_2$ can be replaced by $\mathcal{Q}_1$ in front of $\mathcal{L}^2 u_{0x}$, since the latter is automatically orthogonal to $\mathcal{L} u_{0x}$. If the product variables $\rho_{\mathbf{k}0} \mathbf{j}_{-\mathbf{k}}$ are again chosen as the basis set, use of the approximation (9.5.2) now leads to

$$M_2(t) = \frac{m}{\Omega_0^2} \sum_{\mathbf{k}} \sum_{\mathbf{k}'} \sum_{\alpha} \sum_{\beta} (\mathcal{Q}_1 \mathcal{L}^2 u_{0x}, \rho_{\mathbf{k}0} j_{-\mathbf{k}}^{\alpha})(\rho_{\mathbf{k}0} j_{-\mathbf{k}}^{\alpha}, \rho_{\mathbf{k}0} j_{-\mathbf{k}}^{\alpha})^{-1}$$

$$\times (\rho_{\mathbf{k}0} j_{-\mathbf{k}}^{\alpha}, \exp(i\mathcal{Q}_2 \mathcal{L} \mathcal{Q}_2 t) \rho_{\mathbf{k}'0} j_{-\mathbf{k}'}^{\beta})(\rho_{\mathbf{k}'0} j_{-\mathbf{k}'}^{\beta}, \rho_{\mathbf{k}'0} j_{-\mathbf{k}'}^{\beta})^{-1}$$

$$\times (\rho_{\mathbf{k}'0} j_{-\mathbf{k}'}^{\beta}, \mathcal{Q}_1 \mathcal{L}^2 u_{0x}) \qquad (9.5.12)$$

The assumption of independent propagation of the two modes $\rho_{\mathbf{k}0}$ and $j_{\mathbf{k}}^{\alpha}$ that make up the product variables is now used to factorize both the time correlation functions and the vertices appearing on the right-hand side of Eqn (9.5.12), and the projected Liouville operator $\mathcal{Q}_2 \mathcal{L} \mathcal{Q}_2$ is replaced by the

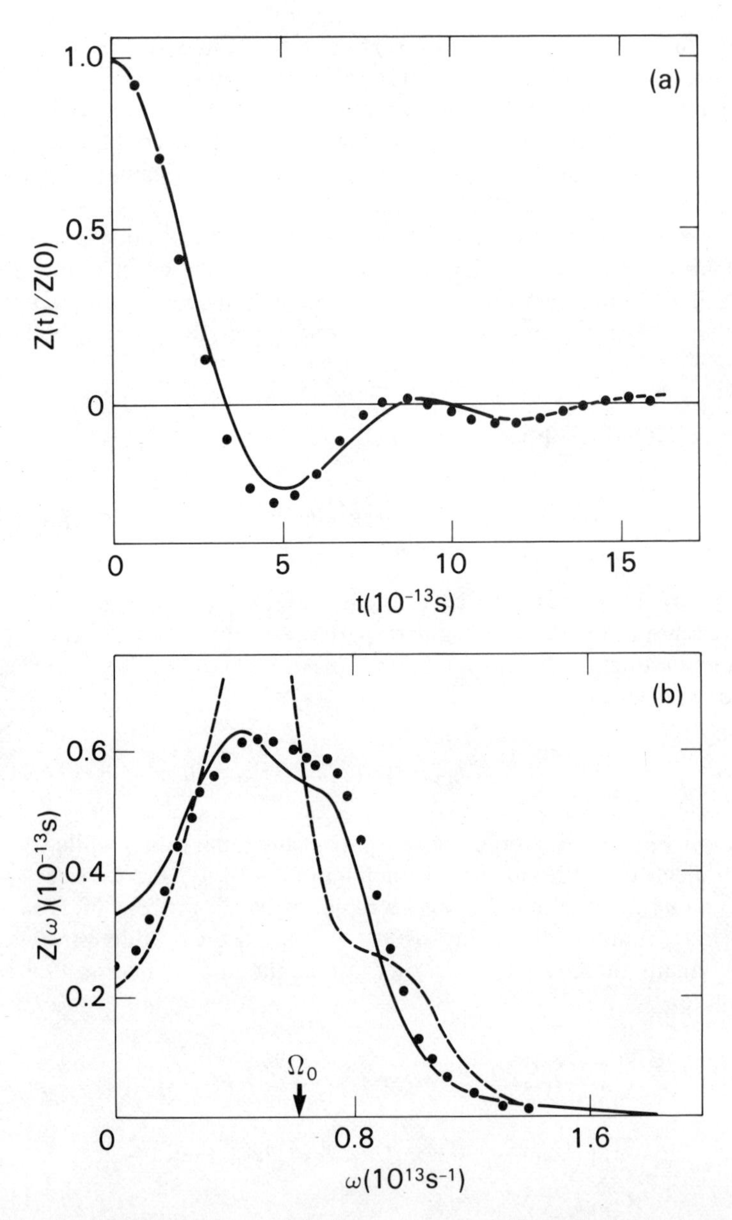

FIG. 9.5. Velocity autocorrelation function (a) and the associated power spectrum (b) of a model of liquid rubidium. The points are molecular dynamics results (Rahman, 1974b), the full curves correspond to the theory of Gaskell and Miller (1978a) (see Eqn (9.5.9)) and the dashed curve in (b) is calculated from the theory of Bosse *et al.* (1978d) (see Eqns (9.5.16)). The low-frequency peak in $Z(\omega)$ arises from the coupling to the transverse current and the shoulder at higher frequencies comes from the coupling to the longitudinal current.

full Liouville operator in the propagator governing the time evolution of the factorized correlation functions. Equation (9.5.12) then becomes

$$M_2(t) = \frac{m}{\Omega_0^2} \sum_{\mathbf{k}} \sum_{\alpha} \sum_{\beta} (\mathcal{Q}_1 \mathcal{L}^2 u_{0x}, \rho_{\mathbf{k}0} j^{\alpha}_{-\mathbf{k}})[(\rho_{\mathbf{k}0}, \rho_{\mathbf{k}0})(j^{\alpha}_{-\mathbf{k}}, j^{\alpha}_{-\mathbf{k}})]^{-1}$$
$$\times (\rho_{\mathbf{k}0}, \rho_{\mathbf{k}0}(t))(j^{\alpha}_{-\mathbf{k}}, j^{\beta}_{-\mathbf{k}}(t))[(\rho_{\mathbf{k}0}, \rho_{\mathbf{k}0})(j^{\beta}_{-\mathbf{k}}, j^{\beta}_{-\mathbf{k}})]^{-1}$$
$$\times (\rho_{\mathbf{k}0} j^{\beta}_{-\mathbf{k}}, \mathcal{Q}_1 \mathcal{L}^2 u_{0x}) \tag{9.5.13}$$

The vertices appearing on each side of the time-dependent correlation functions are easily evaluated (Bosse *et al.*, 1978d), with the result

$$(\mathcal{Q}_1 \mathcal{L}^2 u_{0x}, \rho_{\mathbf{k}0} j^{\beta}_{-\mathbf{k}})[(\rho_{\mathbf{k}0}, \rho_{\mathbf{k}0})(j^{\beta}_{-\mathbf{k}}, j^{\beta}_{-\mathbf{k}})]^{-1}$$
$$= \frac{\Omega_0^2}{m} \mathcal{V}_{\alpha\beta}(k) \left(\frac{N k_{\mathrm{B}} T}{m} \right)^{-1} \tag{9.5.14}$$

where $\mathcal{V}_{\alpha\beta}(k)$ denotes the normalized vertex function

$$\mathcal{V}_{\alpha\beta}(k) = \frac{\rho}{m\Omega_0^2} \int d\mathbf{r} \exp(-i\mathbf{k} \cdot \mathbf{r}) g(r) \nabla_{\alpha} \nabla_{\beta} v(r) \tag{9.5.15}$$

It follows from Eqn (7.2.13) that $\lim_{k\to\infty} \mathcal{V}_{\alpha\beta}(k) = 0$ and $\lim_{k\to 0} \mathcal{V}_{\alpha\beta}(k) = 1$.

The time-correlation functions that appear in Eqn (9.5.13) are the same as in Eqn (9.5.8), i.e. the self intermediate scattering function $F_s(k, t)$ and the longitudinal and transverse components $C_l(k, t)$ and $C_t(k, t)$ of the current correlation function (7.4.33). On gathering together the results, we arrive at the final expression for the second-order memory function (or "relaxation kernel"):

$$M_2(t) = M_{2l}(t) + 2M_{2t}(t) \tag{9.5.16a}$$

$$M_{2l,t}(t) = \frac{m\Omega_0^2}{3\rho k_{\mathrm{B}} T} \frac{1}{(2\pi)^3} \int d\mathbf{k}\, \mathcal{V}^2_{l,t}(k) F_s(k, t) \frac{1}{k^2} C_{l,t}(k, t) \tag{9.5.16b}$$

where $\mathcal{V}_{l,t}$ are the longitudinal and transverse components of the vertex tensor $\boldsymbol{\mathcal{V}}$, defined in a manner analogous to (7.4.33).

The structure of Eqn (9.5.16b) is strikingly similar to the earlier result (9.5.8) for $Z(t)$, except that the formula for $M_2(t)$ contains the vertex factors $\mathcal{V}_{l,t}$. These factors ensure that the integral over wavevectors converges for all t; they therefore play a role similar to the function $\hat{f}(k)$ in the result (9.5.9) of Gaskell and Miller (1978a, b), but have the advantage of being defined unambiguously through Eqn (9.5.15). Bosse *et al.* (1978d) have calculated the memory function $M_2(t)$, and the resulting velocity autocorrelation function and first-order memory function $M_1(t)$, for liquid argon and liquid rubidium near their triple points. They use the gaussian approximation for $F_s(k, t)$, and the results of a self-consistent mode-coupling calculation of $C_l(k, t)$ and $C_t(k, t)$ (Bosse *et al.*, 1978a, b, c). Some of their results are shown in Figure 9.5. Although the agreement with simulation is less satisfactory than in the calculations of Gaskell and Miller (1978a), the theory of Bosse *et al.* is fully self-consistent, since the correlation functions used as input are determined by a mode-coupling approach of the same type. The main conclusions of these calculations are the following. In liquid argon, the "relaxation spectrum" $M_2(\omega)$ has a broad peak around the frequency Ω_0, while in rubidium the peak is significantly sharper and shifted well above Ω_0. The spectrum of the first-order memory function (or "friction spectrum"), $M_1(\omega)$, is highly non-lorentzian, in conflict with the assumption of an exponential form for $M_1(t)$ (Berne *et al.*, 1966). Finally, the power spectrum $Z(\omega)$ has a peak that in liquid rubidium lies close to Ω_0, (see Figure 9.5(b)), whereas in argon it appears at a frequency of about $\frac{1}{2}\Omega_0$ (Bosse *et al.*, 1978d; Gaskell and Miller, 1978b). The derivation of Eqns (9.5.9) and (9.5.16) shows that the difference in behaviour between the two liquids can, as discussed earlier, be ascribed to the lower compressibility of the metal, which gives rise to a higher sound velocity and a weaker damping of the longitudinal collective mode than in argon.

As another application of their method, Bosse *et al.* (1979) have shown that the frequency dependence of the shear viscosity severely limits the range over which the low-frequency square-root cusp in $Z(\omega)$, associated with the $t^{-3/2}$ "tail" in $Z(t)$ can be observed. This makes it unlikely that the tail could be detected experimentally by extrapolation of the incoherent dynamical structure factor $S_s(k, \omega)$ to $k=0$ (see Eqn (8.2.20)) (Carneiro, 1976). A $t^{-3/2}$ tail has, however, been observed by photon correlation spectroscopy for systems of spherical particles (polystyrene latex balls) suspended in a viscous fluid (Bouiller *et al.*, 1978; Paul and Pusey, 1981; Ohbayashi *et al.*, 1983).

9.6 PHASE-SPACE DESCRIPTION OF TIME-DEPENDENT FLUCTUATIONS

The remaining sections of this chapter are concerned with the description of wavenumber and frequency-dependent fluctuations in dense fluids in terms of kinetic theory. As we have emphasized earlier, the classic kinetic equations, such as that of Boltzmann, are useful only for dilute gases, since they either ignore or treat only crudely the memory effects and spatially non-local correlations that play a dominant role under physical conditions appropriate to liquids. A special feature of kinetic theory is the appearance of correlation functions defined in phase space rather than in configuration space alone; retention of the momentum variables is essential if a detailed description of the collision dynamics is to be achieved.

The fundamental dynamical variable in kinetic theory is the *phase-space density* $f(\mathbf{r}, \mathbf{p}; t)$; this is a generalization of the microscopic configuration-space density of Eqn (7.4.5). For a system of N identical, structureless particles, the phase-space density is defined as

$$f(\mathbf{r}, \mathbf{p}; t) = \sum_{i=1}^{N} \delta[\mathbf{r} - \mathbf{r}_i(t)]\delta[\mathbf{p} - \mathbf{p}_i(t)] \tag{9.6.1}$$

where $(\mathbf{r}, \mathbf{p})$ is a field point in the six-dimensional, single-particle phase space. Since two or more field points will appear in the correlation functions built from products of f's, it is sometimes convenient to use a shorthand notation in which the six coordinates of a field point are designated by a single number, i.e. $1 \equiv (\mathbf{r}_1, \mathbf{p}_1)$, $2 \equiv (\mathbf{r}_2, \mathbf{p}_2)$, etc. These labels are not to be confused with the particle labels i, etc. With this convention, the phase-space density is written as $f(1; t) \equiv f(\mathbf{r}_1, \mathbf{p}_1; t)$. The microscopic particle density defined by Eqn (7.4.5) is the momentum integral of $f(\mathbf{r}, \mathbf{p}; t)$;

$$\rho(\mathbf{r}, t) = \int f(\mathbf{r}, \mathbf{p}; t)\, d\mathbf{p} \tag{9.6.2}$$

and the associated current is

$$\mathbf{j}(\mathbf{r}, t) = \frac{1}{m} \int f(\mathbf{r}, \mathbf{p}; t)\mathbf{p}\, d\mathbf{p} \tag{9.6.3}$$

As with any other dynamical variable, the time dependence of the phase-space density is determined by the time-evolution equation (2.1.10), which in this case can be written in two equivalent forms. First, Eqns (2.1.9) and (2.1.10) together show that

$$\frac{\partial f}{\partial t} = i\mathscr{L}_{\mathrm{O}} f + i\mathscr{L}_{\mathrm{I}} f \tag{9.6.4}$$

where $\mathscr{L}_{\mathrm{O}}$ and $\mathscr{L}_{\mathrm{I}}$ are, respectively, the "free-streaming" and "interaction" parts of the Liouville operator:

$$i\mathscr{L}_{\mathrm{O}} = \sum_{i=1}^{N} \frac{\mathbf{p}_i}{m} \cdot \frac{\partial}{\partial \mathbf{r}_i} \tag{9.6.5a}$$

$$i\mathscr{L}_{\mathrm{I}} = -\sum_{i \neq j}^{N}\sum \frac{\partial}{\partial \mathbf{r}_i} v(r_{ij}) \cdot \frac{\partial}{\partial \mathbf{p}_i} \tag{9.6.5b}$$

The operators $\mathscr{L}_{\mathrm{O}}$ and $\mathscr{L}_{\mathrm{I}}$ act on the particle coordinates and momenta. Alternatively, the delta-function properties of $f(\mathbf{r}, \mathbf{p}; t)$ make it possible to represent the action of the Liouville operator in terms of operators L_{O} and

L_{I} that act on the field variables $(\mathbf{r}, \mathbf{p})$. Equation (9.6.4) can thereby be re-expressed as

$$\begin{aligned}\frac{\partial}{\partial t} f(\mathbf{r}_1, \mathbf{p}_1; t) &= iL_{\mathrm{O}}(\mathbf{r}_1, \mathbf{p}_1) f(\mathbf{r}_1, \mathbf{p}_1; t) \\ &+ \iint \mathrm{d}\mathbf{r}_2\, \mathrm{d}\mathbf{p}_2\, iL_{\mathrm{I}}(\mathbf{r}_1, \mathbf{p}_1; \mathbf{r}_2, \mathbf{p}_2) f(\mathbf{r}_1, \mathbf{p}_1; t) f(\mathbf{r}_2, \mathbf{p}_2; t) \qquad (9.6.6)\end{aligned}$$

or, in the shorthand notation introduced above:

$$\frac{\partial}{\partial t} f(1; t) = iL_{\mathrm{O}}(1) f(1; t) + \int \mathrm{d}2\, iL_{\mathrm{I}}(1, 2) f(1; t) f(2; t) \qquad (9.6.7)$$

where

$$iL_{\mathrm{O}}(1) = \frac{-\mathbf{p}_1}{m} \cdot \frac{\partial}{\partial \mathbf{r}_1} \qquad (9.6.8a)$$

$$iL_{\mathrm{I}}(1, 2) = \frac{\partial}{\partial \mathbf{r}_1} v(r_{12}) \cdot \left(\frac{\partial}{\partial \mathbf{p}_1} - \frac{\partial}{\partial \mathbf{p}_2} \right), \qquad 1 \neq 2 \qquad (9.6.8b)$$

The equilibrium average of $f(\mathbf{r}, \mathbf{p}; t)$ is given by Eqn (2.1.22), i.e.

$$\langle f(\mathbf{r}, \mathbf{p}; t) \rangle = \rho \phi_0(\mathbf{p}) \qquad (9.6.9)$$

where $\phi_0(\mathbf{p})$ is the Maxwell–Boltzmann distribution (2.1.25); the fluctuation around the equilibrium value will be denoted by

$$\delta f(\mathbf{r}, \mathbf{p}; t) = f(\mathbf{r}, \mathbf{p}; t) - \rho \phi_0(\mathbf{p}) \qquad (9.6.10)$$

In the study of self motion, the relevant dynamical variable is the tagged-particle phase-space density $f_{\mathrm{s}}(\mathbf{r}, \mathbf{p}; t)$, defined as

$$f_{\mathrm{s}}(\mathbf{r}, \mathbf{p}; t) = \delta[\mathbf{r} - \mathbf{r}_i(t)] \delta[\mathbf{p} - \mathbf{p}_i(t)] \qquad (9.6.11)$$

where i is the singled-out (tagged) particle. In the thermodynamic limit, the equilibrium average of $f_{\mathrm{s}}(\mathbf{r}, \mathbf{p}; t)$ is zero:

$$\langle f_{\mathrm{s}}(\mathbf{r}, \mathbf{p}; t) \rangle = 0 \qquad (9.6.12)$$

The single-particle phase-space distribution function $f^{(1)}(\mathbf{r}, \mathbf{p}; t)$ defined by Eqn (2.1.12) is a non-equilibrium average of the phase-space density $f(\mathbf{r}, \mathbf{p}; t)$ of Eqn (9.6.1). The time evolution of $f^{(1)}(\mathbf{r}, \mathbf{p}; t)$ and, in particular, its relaxation towards equilibrium from a non-equilibrium state, is governed by a kinetic equation. At sufficiently low densities, the time-dependence of $f^{(1)}(\mathbf{r}, \mathbf{p}; t)$ close to equilibrium, and also the equilibrium time-correlation functions, defined below, of the densities f and f_s, can all be obtained from the solution to the Boltzmann equation. This route has been used for the calculation of the dynamic structure factors $S(k, \omega)$ and $S_s(k, \omega)$ of atomic gases (Nelkin and Gathak, 1964; Yip and Nelkin, 1964) in the wavenumber and frequency range accessible by light-scattering experiments (Greytak and Benedek, 1966). However, since the Boltzmann equation is based on the assumption of uncorrelated binary collisions, it is intrinsically markovian in time and local in space. It is therefore inadequate to describe the dynamics of liquids. Much effort has been devoted to the construction of more general kinetic equations that take account both of static correlations and of the correlated collision events responsible for memory and mode-coupling effects. In Enskog's modification of the Boltzmann equation, the effect of static correlations is incorporated, but the theory retains its markovian character; it gives reasonable values for the transport coefficients, even at high densities (see Figures 7.2 and 8.2), though its scope is limited to hard spheres (Résibois and DeLeener, 1977; Dorfman and van Beijeren, 1977). We shall see in Section 9.9 that a generalization of the Enskog equation also yields an accurate description of the dynamical structure factor of the dense hard-sphere fluid (Alley *et al.*, 1983).

A different extension of the Boltzmann equation has been sought in the form of a systematic density expansion, similar to the virial expansion of Chapter 4. This aims at improving the binary collision approximation by considering successively the dynamics of isolated clusters involving an increasing number of particles. The hope is to obtain an expansion of the deviation of a transport coefficient from its Boltzmann value in powers of the density. For example, the shear viscosity would be expanded in the form

$$\eta/\eta_B = 1 + a_1\rho + a_2\rho^2 + \cdots \qquad (9.6.13)$$

where η_B is the Boltzmann value and the coefficients $a_1, a_2, \ldots$ are determined by the dynamics of clusters of $3, 4, \ldots$ particles. In practice, an expansion of this type leads to divergences, starting with a_1 in two and a_2 in three dimensions (Dorfman and van Beijeren, 1977; Dorfman, 1981; Cohen, 1983). The source of these divergences lies in "ring-collision" events of the type pictured in Figure 7.3. Since the clusters are isolated in space, the distance covered by two particles between their first and final encounters can be arbitrarily large; this leads to divergent integrals in the evaluation of the coefficients in Eqn (9.6.13). The divergences are, of course, unphysical. In a real fluid, the flight between two collisions is limited, on average, to a mean-free path; the procedure of considering *isolated* clusters of particles is therefore unrealistic. The recognition of this fact is one of the cornerstones of modern kinetic theory. The formal resummation of the most divergent contributions from ring-collision events of all orders leads to finite but non-analytic contributions to the series (9.6.13), the first of these (in three dimensions) being of the form $\rho^2 \log \rho$ (Kawasaki and Oppenheim, 1965). Ring collisions are also ultimately responsible for the hydrodynamic mode coupling that leads to the slow decay of the time-correlation functions associated with linear transport coefficients and for the divergence of higher-order (Burnett) transport coefficients. The divergence of the Burnett coefficients indicates that the relation between applied gradients and induced hydrodynamic fluxes is non-analytic in character (Pomeau and Résibois, 1975).

We now turn our attention to the equilibrium time-correlation functions of the phase-space densities f and f_s. We restrict the discussion initially to the so-called "two-point" functions; these involve products of two f's taken at different field points and times. At equilibrium, in a homogeneous system, the two-point correlation functions depend only on $\mathbf{r}=\mathbf{r}_1-\mathbf{r}_2$, $t=t_1-t_2$, and the two momenta $\mathbf{p}_1$ and $\mathbf{p}_2$. Thus

$$C(\mathbf{r}_1-\mathbf{r}_2, t_1-t_2; \mathbf{p}_1\mathbf{p}_2) = \langle \delta f(\mathbf{r}_1, \mathbf{p}_1; t)\delta f(\mathbf{r}_2, \mathbf{p}_2; t)\rangle \quad (9.6.14)$$

$$C_s(\mathbf{r}_1-\mathbf{r}_2, t_1-t_2; \mathbf{p}_1\mathbf{p}_2) = \langle f_s(\mathbf{r}_1, \mathbf{p}_1; t) f_s(\mathbf{r}_2, \mathbf{p}_2; t)\rangle \quad (9.6.15)$$

It is often more convenient to work with the dimensionless Fourier–Laplace transforms of these functions, defined in the case of $C(\mathbf{r}, t; \mathbf{p}_1\mathbf{p}_2)$ by

$$\tilde{C}(\mathbf{k}, z; \mathbf{p}_1\mathbf{p}_2) = \frac{1}{\rho}\int d\mathbf{r} \exp(-i\mathbf{k}\cdot\mathbf{r}) \int_0^\infty dt \exp(izt) C(\mathbf{r}, t; \mathbf{p}_1\mathbf{p}_2) \quad (9.6.16)$$

The function $\tilde{C}(\mathbf{k}, z; \mathbf{p}_1\mathbf{p}_2)$ can also be written as an inner product in the form of Eqn (9.1.11), i.e.

$$\tilde{C}(\mathbf{k}, z; \mathbf{p}_1\mathbf{p}_2) = \frac{i}{N}(\delta f_{\mathbf{k}}(\mathbf{p}_1), (z+\mathscr{L})^{-1}\delta f_{\mathbf{k}}(\mathbf{p}_2)) \quad (9.6.17)$$

where $\delta f_{\mathbf{k}}(\mathbf{p}) \equiv \delta f_{\mathbf{k}}(\mathbf{p}, t=0)$ and $\delta f_{\mathbf{k}}(\mathbf{p}, t)$ is a Fourier component of $\delta f(\mathbf{r}, \mathbf{p}; t)$:

$$\delta f_{\mathbf{k}}(\mathbf{p}, t) = \sum_{i=1}^{N} \exp[-i\mathbf{k}\cdot\mathbf{r}_i(t)]\delta[\mathbf{p}-\mathbf{p}_i(t)] - N\delta_{\mathbf{k},\mathbf{0}}\phi_0(\mathbf{p}) \quad (9.6.18)$$

Similar relations hold for the self correlation function $\tilde{C}_s(\mathbf{k}, z; \mathbf{p}_1\mathbf{p}_2)$.

The initial values of the spatial Fourier transforms of $C(\mathbf{r}, t; \mathbf{p}_1\mathbf{p}_2)$ and $C_s(\mathbf{r}, t; \mathbf{p}_1\mathbf{p}_2)$ are

$$\begin{aligned} C^{(0)}(\mathbf{k}; \mathbf{p}_1\mathbf{p}_2) &\equiv C(\mathbf{k}, t=0; \mathbf{p}_1\mathbf{p}_2) \\ &= \phi_0(\mathbf{p}_1)[\delta(\mathbf{p}_1-\mathbf{p}_2) + \phi_0(\mathbf{p}_2)\rho\hat{h}(\mathbf{k})] \end{aligned} \quad (9.6.19a)$$

$$\begin{aligned} C_s^{(0)}(\mathbf{k}; \mathbf{p}_1\mathbf{p}_2) &\equiv C_s(\mathbf{k}, t=0; \mathbf{p}_1\mathbf{p}_2) \\ &= \phi_0(\mathbf{p}_1)\delta(\mathbf{p}_1-\mathbf{p}_2) \end{aligned} \quad (9.6.19b)$$

where $\hat{h}(k)$, as usual, is the Fourier transform of the static pair correlation function. The correlation function $C(\mathbf{k}, t; \mathbf{p}_1\mathbf{p}_2)$ and its Laplace transform $\tilde{C}(\mathbf{k}, z; \mathbf{p}_1\mathbf{p}_2)$ may be looked upon as matrices with continuous indices $\mathbf{p}_1$ and $\mathbf{p}_2$. Inverse matrices can then be defined such that

$$\begin{aligned}\int C(\mathbf{k}, t; \mathbf{p}_1\mathbf{p}_3)C^{-1}(\mathbf{k}, t; \mathbf{p}_3\mathbf{p}_2)\,\mathrm{d}\mathbf{p}_3 \\ = \int C^{-1}(\mathbf{k}, t; \mathbf{p}_1\mathbf{p}_3)C(k, t; \mathbf{p}_3\mathbf{p}_2)\,\mathrm{d}\mathbf{p}_3 \\ = \delta(\mathbf{p}_1-\mathbf{p}_2)\end{aligned} \tag{9.6.20}$$

The inverses of the matrices $C^{(0)}(\mathbf{k}; \mathbf{p}_1\mathbf{p}_2)$ and $C_{\mathrm{s}}^{(0)}(\mathbf{k}; \mathbf{p}_1\mathbf{p}_2)$ are obtained from Eqns (9.6.19), (9.6.20) and (5.2.11), with the result:

$$C^{(0)-1}(\mathbf{k}; \mathbf{p}_1\mathbf{p}_2) = \frac{1}{\phi_0(\mathbf{p}_1)}\delta(\mathbf{p}_1-\mathbf{p}_2) - \rho\hat{c}(k) \tag{9.6.21a}$$

$$C_{\mathrm{s}}^{(0)-1}(\mathbf{p}_1\mathbf{p}_2) = \frac{1}{\phi_0(\mathbf{p}_1)}\delta(\mathbf{p}_1-\mathbf{p}_2) \tag{9.6.21b}$$

where $\hat{c}(k)$ is the Fourier transform of the direct correlation function.

We wish, finally, to relate the phase-space correlation functions to the density and current autocorrelation functions of Section 7.4. To do so, we switch from a continuous to a discrete matrix representation of the phase-space correlation functions by introducing a complete set of orthonormal momentum functions $\psi_\nu(\mathbf{p})$; these are generally chosen to be the Hermite polynomials (Mazenko *et al.*, 1972). The first five members of the set are the momentum functions associated with the conserved variables, i.e. density, momentum and kinetic energy (the last of which is conserved only for hard-sphere systems). Thus

$$\begin{aligned}\psi_1(\mathbf{p}) = 1, \qquad \psi_2(\mathbf{p}) = \frac{p_z}{p_0}, \qquad \psi_3(\mathbf{p}) = \frac{1}{\sqrt{6}}\left(\frac{p^2}{p_0^2}-3\right) \\ \psi_4(\mathbf{p}) = \frac{p_x}{p_0}, \qquad \psi_5(\mathbf{p}) = \frac{p_y}{p_0}\end{aligned} \tag{9.6.22}$$

where, by convention, $\mathbf{k}$ lies along the z-axis and $p_0 = (mk_{\mathrm{B}}T)^{1/2}$. Introducing a vector notation, we define an inner product in momentum space with a Maxwell–Boltzmann weight as

$$\langle \nu | \mu \rangle = \int \psi_\nu(\mathbf{p})\psi_\mu(\mathbf{p})\phi_0(\mathbf{p})\,\mathrm{d}\mathbf{p} = \delta_{\nu\mu} \tag{9.6.23}$$

which expresses the orthogonality and normalization of the basis functions. The "matrix elements" or "momentum contractions" of the phase-space correlation functions are defined, for example, as

$$\tilde{C}_{\nu\mu}(\mathbf{k}, z) = \iint \psi_\nu(\mathbf{p}_1)\tilde{C}(\mathbf{k}, z; \mathbf{p}_1\mathbf{p}_2)\psi_\mu(\mathbf{p}_2)\, d\mathbf{p}_1\, d\mathbf{p}_2 \tag{9.6.24}$$

Similar definitions apply in real space and to $\tilde{C}_{s\nu\mu}(\mathbf{k}, z)$. With these definitions, and with the help of elementary properties of the delta-functions contained in Eqn (9.6.18), it is a straightforward matter to verify that the correlation functions $F(\mathbf{k}, t)$, $C_l(\mathbf{k}, t)$ and $C_t(\mathbf{k}, t)$ of Section 7.4 are given in terms of elements of $\boldsymbol{C}(\mathbf{k}, t)$ by

$$F(\mathbf{k}, t) = C_{11}(\mathbf{k}, t) \tag{9.6.25a}$$

$$C_l(\mathbf{k}, t) = \omega_0^2 C_{22}(\mathbf{k}, t) \tag{9.6.25b}$$

$$C_t(\mathbf{k}, t) = \omega_0^2 C_{44}(\mathbf{k}, t) = \omega_0^2 C_{55}(\mathbf{k}, t) \tag{9.6.25c}$$

with analogous identifications holding for the self correlation functions. Although the matrices $\boldsymbol{C}$ and $\boldsymbol{C}_s$ have an infinite number of elements, the most important ones are those corresponding to the 5×5 "hydrodynamic" subspace spanned by the basis functions (9.6.22).

While the two-point functions introduced so far are, in general, the physically relevant ones, since they are measurable by inelastic scattering experiments, correlation functions involving more than two field points play an important part in the formal structure of kinetic theory. These functions form a hierarchy similar to the BBGKY hierarchy of Eqn (2.1.15); their properties are discussed in Appendix E.

9.7 EXACT KINETIC EQUATIONS FOR THE PHASE-SPACE CORRELATION FUNCTIONS

Our main task in this section is to derive exact kinetic equations for the phase-space correlation functions $C(\mathbf{r}, t; \mathbf{p}_1\mathbf{p}_2)$ and $C_s(\mathbf{r}, t; \mathbf{p}_1\mathbf{p}_2)$. First, however, it is useful to show how the distribution-function formulation of kinetic theory is linked to the correlation-function description of equilibrium fluctuations (Ortoleva and Nelkin, 1969). Consider a system that is at equilibrium in the infinite past and to which a weak, time-dependent external field $\phi(\mathbf{r}, \mathbf{p})$ is applied. The field is assumed to couple to the initial phase-space density $f(\mathbf{r}, \mathbf{p})$, giving rise to a perturbation of the form

$$\mathcal{H}' = -\iint f(\mathbf{r}', \mathbf{p}')\phi(\mathbf{r}', \mathbf{p}')\, d\mathbf{r}'\, d\mathbf{p}' \tag{9.7.1}$$

At $t=0$, the perturbing field is switched off. The relaxation of the system towards equilibrium is described by the time evolution of the distribution function $f^{(1)}(\mathbf{r}, \mathbf{p}; t)$, which is a non-equilibrium (n.e.) average of $f(\mathbf{r}, \mathbf{p}; t)$. To first order in $\mathcal{H}'$, the response of the system is given by an obvious generalization of Eqn (7.6.23), i.e.

$$\langle f(\mathbf{r}, \mathbf{p}; t)\rangle_{\text{n.e.}} = \int_{-\infty}^{0} \mathrm{d}s \iint \mathrm{d}\mathbf{r}'\,\mathrm{d}\mathbf{p}'\,\Phi(\mathbf{r}-\mathbf{r}', t-s; \mathbf{pp}')\phi(\mathbf{r}', \mathbf{p}') \quad (9.7.2)$$

where, from Eqn (7.6.17):

$$\Phi(\mathbf{r}-\mathbf{r}', t; \mathbf{pp}') = -\beta\langle \dot{f}(\mathbf{r}, \mathbf{p}; t) f(\mathbf{r}', \mathbf{p}')\rangle \quad (9.7.3)$$

If this result is substituted in Eqn (9.7.2), and use is made of the fact that $\dot{f} = \delta\dot{f}$, then

$$\begin{aligned} f^{(1)}(\mathbf{r}, \mathbf{p}; t) &= \beta \iint \langle \delta f(\mathbf{r}, \mathbf{p}; t) f(\mathbf{r}', \mathbf{p}')\rangle \phi(\mathbf{r}', \mathbf{p}')\,\mathrm{d}\mathbf{r}'\,\mathrm{d}\mathbf{p}' \\ &= \beta \iint C(\mathbf{r}-\mathbf{r}', t; \mathbf{pp}')\phi(\mathbf{r}', \mathbf{p}')\,\mathrm{d}\mathbf{r}'\,\mathrm{d}\mathbf{p}' \end{aligned} \quad (9.7.4)$$

It follows immediately from the example (4.2.8) that the phase-space correlation function in Eqn (9.7.4) is the functional derivative of $f^{(1)}(\mathbf{r}, \mathbf{p}; t)$ with respect to the external field:

$$C(\mathbf{r}-\mathbf{r}', t; \mathbf{pp}') = \frac{\delta f^{(1)}(\mathbf{r}, \mathbf{p}; t)}{\delta\beta\phi(\mathbf{r}', \mathbf{p}')} \quad (9.7.5)$$

This implies that any linearized kinetic equation satisfied by the distribution function $f^{(1)}(\mathbf{r}, \mathbf{p}; t)$ is also an equation that describes the time evolution of the equilibrium phase-space correlation function $C(\mathbf{r}, t; \mathbf{p}_1\mathbf{p}_2)$, and vice versa.

The general kinetic equation obeyed by $C(\mathbf{r}, t; \mathbf{p}_1\mathbf{p}_2)$ can be derived by an extension of the projection-operator methods of Section 9.1 (Akcasu and Duderstadt, 1969). Let $\mathcal{P}_\mathbf{k}$ be an operator that projects an arbitrary phase-space function B onto the variable $\delta f_\mathbf{k}(\mathbf{p})$ of Eqn (9.6.18) according to the rule

$$\mathcal{P}_\mathbf{k} B = \iint \mathrm{d}\mathbf{p}_1\,\mathrm{d}\mathbf{p}_2\,\delta f_\mathbf{k}(\mathbf{p}_1)(\delta f_\mathbf{k}(\mathbf{p}_1), \delta f_\mathbf{k}(\mathbf{p}_2))^{-1}(\delta f_\mathbf{k}(\mathbf{p}_2), B) \quad (9.7.6)$$

The inner product $(\delta f_\mathbf{k}(\mathbf{p}_1), \delta f_\mathbf{k}(\mathbf{p}_2))$ is the static correlation function (9.6.19a). If we now follow the same steps that lead from (9.1.10) to (9.1.15), we arrive at an exact kinetic equation of the form

$$-iz\tilde{C}(\mathbf{k}, z; \mathbf{p}_1\mathbf{p}_2) + \int \mathrm{d}\mathbf{p}_3\,\Sigma(\mathbf{k}, z; \mathbf{p}_1\mathbf{p}_3)\tilde{C}(\mathbf{k}, z; \mathbf{p}_3\mathbf{p}_2) = C^{(0)}(\mathbf{k}; \mathbf{p}_1\mathbf{p}_2) \quad (9.7.7)$$

where the evolution kernel Σ is defined as

$$\Sigma(\mathbf{k}, z; \mathbf{p}_1\mathbf{p}_2) = \int d\mathbf{p}_3 \, (\delta f_{\mathbf{k}}(\mathbf{p}_1), [-i\mathscr{L} + \mathscr{M}(z)]\delta f_{\mathbf{k}}(\mathbf{p}_3))C^{(0)-1}(\mathbf{k}; \mathbf{p}_3\mathbf{p}_2) \quad (9.7.8)$$

with

$$\mathscr{M}(z) = i\mathscr{L}\mathscr{Q}_{\mathbf{k}}(z + \mathscr{Q}_{\mathbf{k}}\mathscr{L}\mathscr{Q}_{\mathbf{k}})^{-1}\mathscr{Q}_{\mathbf{k}}\mathscr{L} \quad (9.7.9)$$

where $\mathscr{Q}_{\mathbf{k}} = 1 - \mathscr{P}_{\mathbf{k}}$.

The kernel Σ of Eqn (9.7.8) separates naturally into three parts. The first two are purely static contributions, i.e. independent of z; these arise from the action of the Liouville operator in (9.7.8). It follows from eqns (9.6.5a) and (9.6.18) that

$$-i\mathscr{L}_{\mathrm{O}}\delta f_{\mathbf{k}}(\mathbf{p}_3) = \frac{i\mathbf{k}\cdot\mathbf{p}_3}{m}\delta f_{\mathbf{k}}(\mathbf{p}_3) \quad (9.7.10)$$

where $\mathscr{L}_{\mathrm{O}}$ is the free-streaming part of the Liouville operator. Substitution of this result in Eqn (9.7.8) and use of the definition (9.6.20) gives the first static contribution to Σ:

$$\Sigma^{\mathrm{O}}(\mathbf{k}; \mathbf{p}_1\mathbf{p}_2) = \frac{i\mathbf{k}\cdot\mathbf{p}_1}{m}\delta(\mathbf{p}_1 - \mathbf{p}_2) \quad (9.7.11)$$

The second static term comes from the interaction part of the Liouville operator, defined by Eqn (9.6.5b). In this case, we find that

$$\begin{aligned}(\delta f_{\mathbf{k}}(\mathbf{p}_1), -i\mathscr{L}_{\mathrm{I}}\delta f_{\mathbf{k}}(\mathbf{p}_3)) &= (i\mathscr{L}_{\mathrm{I}}\delta f_{\mathbf{k}}(\mathbf{p}_1), \delta f_{\mathbf{k}}(\mathbf{p}_3)) \\ &= -\sum_{i=1}^{N}\sum_{j=1}^{N}\left\langle \exp[-i\mathbf{k}\cdot(\mathbf{r}_i - \mathbf{r}_j)]\delta(\mathbf{p}_3 - \mathbf{p}_j)\frac{\partial}{\partial \mathbf{r}_i}V_N\cdot\frac{\partial}{\partial \mathbf{p}_i}\delta(\mathbf{p}_1 - \mathbf{p}_i)\right\rangle\end{aligned} \quad (9.7.12)$$

We now make use of the identity (cf. Eqn (7.2.11)):

$$\left\langle A\frac{\partial}{\partial \mathbf{r}_i}V_N\right\rangle = -k_{\mathrm{B}}T\left\langle\frac{\partial A}{\partial \mathbf{r}_i}\right\rangle \quad (9.7.13a)$$

and the property

$$\frac{\partial}{\partial \mathbf{p}_i}\delta(\mathbf{p}_1 - \mathbf{p}_i) = -\frac{\partial}{\partial \mathbf{p}_1}\delta(\mathbf{p}_1 - \mathbf{p}_i) \quad (9.7.13b)$$

to rewrite Eqn (9.7.12) as

$$\begin{aligned}&(\delta f_{\mathbf{k}}(\mathbf{p}_1), i\mathscr{L}_{\mathrm{I}}\delta f_{\mathbf{k}}(\mathbf{p}_3)) \\ &\quad = ik_{\mathrm{B}}T\mathbf{k}\cdot\frac{\partial}{\partial \mathbf{p}_1}\sum_{i\neq j}^{N}\sum\langle\exp[-i\mathbf{k}\cdot(\mathbf{r}_i - \mathbf{r}_j)]\delta(\mathbf{p}_1 - \mathbf{p}_j)\delta(\mathbf{p}_3 - \mathbf{p}_i)\rangle \\ &\quad = -\frac{i\mathbf{k}\cdot\mathbf{p}_1}{m}\phi_0(\mathbf{p}_1)\phi_0(\mathbf{p}_3)\rho\hat{h}(k)\end{aligned} \quad (9.7.14)$$

When (9.7.14) and (9.6.21a) are substituted in (9.7.8), the integral can be evaluated to give the "mean-field" (or "self-consistent-field") contribution to Σ:

$$\Sigma^{\mathrm{S}}(\mathbf{k};\,\mathbf{p}_1\mathbf{p}_2) = -\frac{i\mathbf{k}\cdot\mathbf{p}_1}{m}\phi_0(\mathbf{p}_1)\rho\hat{c}(k) \tag{9.7.15}$$

Since $\hat{c}(k) \simeq -\beta\hat{v}(k)$ at small k, it is evident that the direct correlation function appears here in the role of an effective pair potential, "renormalized" by short-range correlations (cf. Eqn (7.9.13)).

The frequency-dependent part of Σ arises from the action of the operator $\mathscr{M}(z)$ in Eqn (9.7.8). This term is called the "collision kernel" and is given by

$$\begin{aligned}\Sigma^{\mathrm{C}}&(\mathbf{k}, z;\,\mathbf{p}_1\mathbf{p}_2)\\ &= \int \mathrm{d}\mathbf{p}_3(\delta f_{\mathbf{k}}(\mathbf{p}_1), \mathscr{M}(z)\delta f_{\mathbf{k}}(\mathbf{p}_3))C^{(0)-1}(\mathbf{k};\,\mathbf{p}_3\mathbf{p}_2)\\ &= \frac{i}{\phi_0(\mathbf{p}_2)}(\delta f_{\mathbf{k}}(\mathbf{p}_1), \mathscr{L}\mathscr{Q}_{\mathbf{k}}(z+\mathscr{Q}_{\mathbf{k}}\mathscr{L}\mathscr{Q}_{\mathbf{k}})^{-1}\mathscr{Q}_{\mathbf{k}}\mathscr{L}\delta f_{\mathbf{k}}(\mathbf{p}_2))\\ &\quad -\rho\hat{c}(k)\int (\delta f_{\mathbf{k}}(\mathbf{p}_1), \mathscr{L}\mathscr{Q}_{\mathbf{k}}(z+\mathscr{Q}_{\mathbf{k}}\mathscr{L}\mathscr{Q}_{\mathbf{k}})^{-1}\mathscr{Q}_{\mathbf{k}}\mathscr{L}\delta f_{\mathbf{k}}(\mathbf{p}_3))\,\mathrm{d}\mathbf{p}_3\end{aligned} \tag{9.7.16}$$

Equations (9.6.18) and (9.7.13b) together show that

$$\begin{aligned}\mathscr{L}\delta f_{\mathbf{k}}(\mathbf{p}_3) &= -i\delta\dot{f}_{\mathbf{k}}(\mathbf{p}_3)\\ &= -\frac{\mathbf{k}\cdot\mathbf{p}_3}{m}\delta f_{\mathbf{k}}(\mathbf{p}_3) - i\frac{\partial}{\partial\mathbf{p}_3}\sum_{i=1}^{N}\exp(-i\mathbf{k}\cdot\mathbf{r}_i)\delta(\mathbf{p}_3-\mathbf{p}_i)\frac{\partial V_N}{\partial\mathbf{r}_i}\end{aligned} \tag{9.7.17}$$

Now consider the effect of substituting this result in the integral in Eqn (9.7.16). The first term in (9.7.17) vanishes when operated on by $\mathscr{Q}_{\mathbf{k}}$, while the contribution of the second term also vanishes after integration over $\mathbf{p}_3$. Thus Σ^{C} reduces to

$$\Sigma^{\mathrm{C}}(\mathbf{k}, z;\,\mathbf{p}_1\mathbf{p}_2) = \frac{i}{\phi_0(\mathbf{p}_2)}(\delta f_{\mathbf{k}}(\mathbf{p}_1), \mathscr{L}\mathscr{Q}_{\mathbf{k}}(z+\mathscr{Q}_{\mathbf{k}}\mathscr{L}\mathscr{Q}_{\mathbf{k}})^{-1}\mathscr{Q}_{\mathbf{k}}\mathscr{L}\delta f_{\mathbf{k}}(\mathbf{p}_2)) \tag{9.7.18}$$

with the further simplification that only the second term on the right-hand side of (9.7.17) contributes to the result. In summary, therefore, the evolution kernel can be divided in the form

$$\Sigma(\mathbf{k}, z;\,\mathbf{p}_1\mathbf{p}_2) = \Sigma^{\mathrm{O}}(\mathbf{k};\,\mathbf{p}_1\mathbf{p}_2) + \Sigma^{\mathrm{S}}(\mathbf{k};\,\mathbf{p}_1\mathbf{p}_2) + \Sigma^{\mathrm{C}}(\mathbf{k}, z;\,\mathbf{p}_1\mathbf{p}_2) \tag{9.7.19}$$

and the kinetic equation (9.7.7) re-expressed as

$$\left(-iz+\frac{i\mathbf{k}\cdot\mathbf{p}_1}{m}\right)\tilde{C}(\mathbf{k},z;\mathbf{p}_1\mathbf{p}_2)-\frac{i\mathbf{k}\cdot\mathbf{p}_1}{m}\phi_0(\mathbf{p}_1)\rho\hat{c}(k)\int d\mathbf{p}_3\,\tilde{C}(\mathbf{k},z;\mathbf{p}_3\mathbf{p}_2)$$
$$+\int d\mathbf{p}_3\Sigma^{\mathrm{C}}(\mathbf{k},z;\mathbf{p}_1\mathbf{p}_3)\tilde{C}(\mathbf{k},z;\mathbf{p}_3\mathbf{p}_2)=C^{(0)}(\mathbf{k};\mathbf{p}_1\mathbf{p}_2) \tag{9.7.20}$$

where Σ^{C} is given by Eqn (9.7.18).

As expected, Eqn (9.7.20) has the same basic structure as the generalized Langevin equation (9.1.23), with the static term on the left-hand side of the former playing the part of Ω in the latter, and the collision kernel Σ^{C} playing the part of the memory function $\tilde{M}(z)$. A succession of projection operators can also be introduced, as in Section 9.1, from which a continued-fraction expansion of the phase-space correlation function can be derived (Boley, 1975). The free-streaming motion of non-interacting particles is described by the term in brackets on the left-hand side of (9.7.20). The correlations between interacting particles lead both to a short-time response, contained in the mean-field term, and to non-markovian effects, represented by the collision kernel. If the potentials are continuous, the collision kernel vanishes in the limit $z\to\infty$, just as the memory function does. For hard-core fluids, however, there is a contribution to Σ^{C} that is markovian, i.e. independent of z; this has its origin in the instantaneous nature of the collisions.

The autocorrelation function $C_s(\mathbf{r},t;\mathbf{p}_1\mathbf{p}_2)$ obeys a kinetic equation similar to (9.7.20), but lacking the mean-field term, since a tagged particle does not experience the average restoring force responsible for collective excitations (Akcasu *et al.*, 1970; Desai, 1971). The collision kernel is given by the analogue of (9.7.18) in which $\delta f_{\mathbf{k}}(\mathbf{p})$ is replaced by $\delta f_{s\mathbf{k}}(\mathbf{p})=\exp(-i\mathbf{k}\cdot\mathbf{r}_i)\delta(\mathbf{p}-\mathbf{p}_i)$ and $\mathcal{Q}_{\mathbf{k}}$ is replaced by the projection operator associated with $\delta f_{s\mathbf{k}}$.

The collision kernel Σ^{C} of Eqn (9.7.18) is proportional to the autocorrelation function of the generalized "random force" $\mathcal{Q}_{\mathbf{k}}\mathcal{L}\delta f_{\mathbf{k}}(\mathbf{p})$; this has a projected time evolution determined by the operator $\exp(i\mathcal{Q}_k\mathcal{L}\mathcal{Q}_k t)$, in complete analogy with Eqns (9.1.17), (9.1.18) and (9.1.21). Since the projection operators cannot easily be dealt with in developing microscopic approximations for the collision kernel, it is useful to introduce a different expression for Σ^{C}, involving higher-order phase-space correlation functions. The derivation is given in Appendix E; in the shorthand notation introduced earlier in this section, the real-space form of Σ^{C} (see Eqn (E.27)) is

$$\Sigma^{\mathrm{C}}(12;z)=\frac{-1}{\rho\phi_0(\mathbf{p}_2)}\iint d3\,d4\,L_{\mathrm{I}}(1,3)L_{\mathrm{I}}(2,4)\tilde{G}(13,24;z) \tag{9.7.21}$$

where $\tilde{G}$ is the four-point correlation function defined by Eqn (E.24). Equation (9.7.21) is the starting point for the so-called "fully renormalized" kinetic theory of dense fluids (Mazenko, 1973a, 1974; Mazenko and Yip, 1977).

The collision kernel has a number of properties that follow directly from the formal expression (9.7.18) or, equivalently, from (9.7.21). First, as we have already discussed, $\lim_{z\to\infty} \Sigma^C = 0$, except for systems with hard cores. Secondly, the hermitian property of the operator $\mathcal{Q}_{\mathbf{k}}\mathcal{L}\mathcal{Q}_{\mathbf{k}}$ means that the propagator $(z+\mathcal{Q}_{\mathbf{k}}\mathcal{L}\mathcal{Q}_{\mathbf{k}})^{-1}$ can have poles only on the real axis. Hence Σ^C is an analytic function of z for $\text{Im}\, z \neq 0$, and therefore has a spectral representation of the form (7.1.19), i.e.

$$\Sigma^C(\mathbf{k}, z; \mathbf{p}_1\mathbf{p}_2) = i \int_{-\infty}^{\infty} \frac{\Gamma(\mathbf{k}, \omega; \mathbf{p}_1\mathbf{p}_2)}{z-\omega}\, d\omega \tag{9.7.22}$$

with

$$\begin{aligned}\Gamma(\mathbf{k}, \omega; \mathbf{p}_1\mathbf{p}_2) &= \lim_{\varepsilon\to 0} \frac{1}{\pi} \text{Re}\, \Sigma^C(\mathbf{k}, z=\omega+i\varepsilon; \mathbf{p}_1\mathbf{p}_2) \\ &= \frac{1}{2\pi} \int_{-\infty}^{\infty} \Sigma^C(\mathbf{k}, t; \mathbf{p}_1\mathbf{p}_2) \exp(i\omega t)\, dt \end{aligned} \tag{9.7.23}$$

The spectral function Γ is non-negative; this follows as a special case of Eqn (7.1.24), since Σ^C is the autocorrelation function of the "random force" $\mathcal{Q}_{\mathbf{k}}\delta f_{\mathbf{K}}(\mathbf{p}, t)$. The non-negativity condition can be expressed as

$$\iint d\mathbf{p}_1\, d\mathbf{p}_2\, \psi^*(\mathbf{p}_1)\Gamma(\mathbf{k}, \omega; \mathbf{p}_1\mathbf{p}_2)\psi(\mathbf{p}_2)\phi_0(\mathbf{p}_2) \geqslant 0 \tag{9.7.24}$$

where $\psi(\mathbf{p})$ is any function of momentum. Thirdly, rotational invariance implies that Σ^C depends only on the six scalar combinations of the vectors $\mathbf{k}$, $\mathbf{p}_1$ and $\mathbf{p}_2$. It is also easily shown that

$$\Sigma^C(\mathbf{k}, z; \mathbf{p}_1\mathbf{p}_2) = -\Sigma^C(-\mathbf{k}, -z; \mathbf{p}_1\mathbf{p}_2) = -(\Sigma^C(\mathbf{k}, z^*; \mathbf{p}_1\mathbf{p}_2))^* \tag{9.7.25}$$

and

$$\Sigma^C(\mathbf{k}, z; \mathbf{p}_1\mathbf{p}_2)\phi_0(\mathbf{p}_2) = \Sigma^C(\mathbf{k}, z; \mathbf{p}_2\mathbf{p}_1)\phi_0(\mathbf{p}_1) \tag{9.7.26}$$

Equation (9.7.26) is an expression of the condition of detailed balance. Finally, the short-time expansion of $C(\mathbf{k}, t; \mathbf{p}_1\mathbf{p}_2)$ leads to sum rules for its power spectrum, defined as

$$C(\mathbf{k}, \omega; \mathbf{p}_1\mathbf{p}_2) = \frac{1}{2\pi} \int_{-\infty}^{\infty} C(\mathbf{k}, t; \mathbf{p}_1\mathbf{p}_2) \exp(i\omega t)\, dt \tag{9.7.27}$$

The calculation proceeds along the general lines of Eqns (7.1.25) and (7.1.26). Because $C(\mathbf{k}, \omega; \mathbf{p}_1\mathbf{p}_2)$ can be regarded as a matrix with continuous indices $\mathbf{p}_1$ and $\mathbf{p}_2$, both odd and even frequency moments are non-zero, and their values depend on the indices. The frequency moments lead to sum rules for the spectral function Γ and also determine the coefficients in the high-frequency expansion of Σ^C. The coefficients of z^{-1} and z^{-2} in the expansion of Σ^C have been worked out by Forster (1974); they are expressible solely in terms of the static pair distribution function, whereas later coefficients in the series involve triplet and higher-order functions.

The markovian limit of Σ^C is obtained by taking the long-wavelength, low-frequency limit; it is therefore local both in space and time, which is also true of the collision kernels in the Boltzmann and Fokker–Planck equations. We write the markovian limiting form as $K(\mathbf{p}_1, \mathbf{p}_2)$, where

$$K(\mathbf{p}_1, \mathbf{p}_2) = \lim_{z \to i0+} \lim_{k \to 0} \Sigma^C(\mathbf{k}, z; \mathbf{p}_1\mathbf{p}_2) \tag{9.7.28}$$

It follows from Eqns (9.7.25) and (9.7.26) that $K(\mathbf{p}_1, \mathbf{p}_2)$ is both real and symmetric in its arguments, and from Eqns (9.7.23) and (9.7.24) that its eigenvalues are non-negative. It can be shown, in fact, that the "states" listed in (9.6.22) correspond to the eigenvalue zero, i.e.

$$\int K(\mathbf{p}_1, \mathbf{p}_2)\psi_\nu(\mathbf{p}_1)\, d\mathbf{p}_1 = 0, \qquad \nu = 1 \text{ to } 5 \tag{9.7.29}$$

These five "eigenstates" are the usual five collisional invariants corresponding to the conservation of particle number ($\nu = 1$), momentum ($\nu = 2$, 4 and 5) and kinetic energy ($\nu = 3$) in a binary hard-sphere collision (Résibois and DeLeener, 1977). At non-zero wavenumbers and frequencies, only particle number is conserved; this is expressed formally by the statement that

$$\int \Sigma^C(\mathbf{k}, z; \mathbf{p}_1\mathbf{p}_2)\, d\mathbf{p}_1 = 0 \tag{9.7.30}$$

Equation (9.7.30) is a direct consequence of the continuity equation, which shows that

$$\begin{aligned} \int \mathscr{L}\delta f_{\mathbf{k}}(\mathbf{p}_1)\, d\mathbf{p}_1 &= -i \int \delta \dot{f}_{\mathbf{k}}(\mathbf{p}_1)\, d\mathbf{p}_1 \\ &= -i\dot{\rho}_{\mathbf{k}} \\ &= -\mathbf{k} \cdot \mathbf{j}_{\mathbf{k}} \\ &= -\mathbf{k} \cdot \int \frac{\mathbf{p}_1}{m} \delta f_{\mathbf{k}}(\mathbf{p}_1)\, d\mathbf{p}_1 \end{aligned} \tag{9.7.31}$$

The last line of Eqn (9.7.31) represents a function lying in the single-particle space spanned by the $\delta f_{\mathbf{k}}$, and therefore vanishes when acted upon the operator $\mathscr{Q}_{\mathbf{k}}$ in (9.7.18).

9.8 FROM KINETIC THEORY TO HYDRODYNAMICS

We now explore the relationship between the exact kinetic equations derived in Section 9.7 and the linearized equations of hydrodynamics (Résibois, 1970; Forster and Martin, 1970; Forster, 1974). The connection between a microscopic kinetic equation and the macroscopic equations of hydrodynamics was first established independently by Chapman and Enskog for the case of the Boltzmann equation (2.1.20) (Chapman and Cowling, 1970; Résibois and DeLeener, 1977). These authors were able to derive the Navier–Stokes equation from a systematic expansion of the solution of the Boltzmann equation about the local equilibrium form (2.1.26), and to obtain microscopic expressions for the transport coefficients η, ζ and λ. A similar program can be carried out for the exact kinetic equation (9.7.7) (or (9.7.20)).

We begin by rewriting Eqn (9.7.7) in matrix form as

$$(-iz+\boldsymbol{\Sigma})\tilde{\boldsymbol{C}}=\boldsymbol{C}^{(0)} \tag{9.8.1}$$

which has a solution

$$\tilde{\boldsymbol{C}}=\boldsymbol{\zeta}\boldsymbol{C}^{(0)} \tag{9.8.2}$$

where $\boldsymbol{\zeta}$ is the "resolvent" matrix:

$$\boldsymbol{\zeta}=(-iz+\boldsymbol{\Sigma})^{-1} \tag{9.8.3}$$

We now take matrix elements of both sides of Eqn (9.8.2) with respect to the complete basis of momentum functions $\psi_\nu(\mathbf{p})$ that satisfy the orthogonality and normalization condition (9.6.23). The completeness relation for the basis functions is

$$\sum_\nu |\nu\rangle\langle\nu| = \boldsymbol{I} \tag{9.8.4}$$

This is a shorthand (Dirac) notation for

$$\sum_\nu \psi_\nu^*(\mathbf{p}_1)\psi_\nu(\mathbf{p}_2)\phi_0(\mathbf{p}_1)=\delta(\mathbf{p}_1-\mathbf{p}_2) \tag{9.8.5}$$

Given Eqn (9.8.4), the matrix elements (9.6.24) of $\boldsymbol{C}$ can be re-expressed as

$$\tilde{C}_{\nu\mu}(\mathbf{k},z)=\sum_\lambda \zeta_{\nu\lambda}(\mathbf{k},z)C_{\lambda\mu}^{(0)}(\mathbf{k}) \tag{9.8.6}$$

where the $C_{\lambda\mu}^{(0)}(\mathbf{k})$ are defined as in Eqn (9.6.24) and the matrix elements of $\boldsymbol{\zeta}$ are

$$\begin{aligned}\zeta_{\nu\mu}(\mathbf{k},z)&=\langle\nu|\boldsymbol{\zeta}|\mu\rangle\\&=\int \mathrm{d}\mathbf{p}_1\int \mathrm{d}\mathbf{p}_2\, \psi_\nu^*(\mathbf{p}_1)\zeta(\mathbf{k},z;\mathbf{p}_1\mathbf{p}_2)\psi_\mu(\mathbf{p}_2)\phi_0(\mathbf{p}_2)\end{aligned} \tag{9.8.7}$$

Note the presence of the additional factor $\phi_0(\mathbf{p}_2)$ in the definition (9.8.7) as compared with Eqn (9.6.24); use of the Dirac notation $\langle\nu|\zeta|\mu\rangle$ will be reserved for matrix elements of the type (9.8.7) (or (9.6.23)). The matrix elements $C^{(0)}_{\nu\mu}(\mathbf{k})$ can be calculated from the expression (9.6.19a) with the aid of (9.6.23) and the fact that $\psi_1(\mathbf{p})=1$. The result is

$$C^{(0)}_{\nu\mu}(\mathbf{k})=\delta_{\nu\mu}+\rho\hat{h}(k)\delta_{\lambda 1}\delta_{\mu 1} \tag{9.8.8}$$

Substitution of (9.8.8) in Eqn (9.8.6) gives the final expression for $\tilde{C}_{\nu\mu}(\mathbf{k}, z)$ in the form

$$\tilde{C}_{\nu\mu}(\mathbf{k},z)=\zeta_{\nu\mu}(\mathbf{k}, z)[1+\rho\hat{h}(k)\delta_{\nu 1}\delta_{\mu 1}] \tag{9.8.9}$$

In order to establish the link with hydrodynamics, it is convenient to introduce a projection operator P that acts in the space of momentum functions $\psi_\nu(\mathbf{p})$ and projects any function of momenta onto the "hydrodynamic" subspace spanned by the five conserved variables (9.6.22), i.e.

$$P=\sum_{\nu=1}^{5}|\nu\rangle\langle\nu| \tag{9.8.10}$$

Thus the operator $Q=1-P$ projects on the complementary subspace spanned by all non-hydrodynamic states. The hydrodynamic elements (ν, $\mu=1$ to 5) of the resolvent matrix (9.8.3) can then be written as

$$\begin{aligned}\zeta_{\nu\mu}&=\langle\nu|(-iz+\Sigma P+\Sigma Q)^{-1}|\mu\rangle\\&=\langle\nu|(-iz+\Sigma Q)^{-1}-(-iz+\Sigma Q)^{-1}\Sigma P(-iz+\Sigma)^{-1}|\mu\rangle\end{aligned} \tag{9.8.11}$$

The first term reduces to $(i/z)\langle\nu|\mu\rangle$, since $Q|\mu\rangle=0$ for $\mu=1$ to 5, and the second term can be transformed with the help of Eqn (9.8.10) to give

$$\begin{aligned}\zeta_{\nu\mu}&=\frac{i}{z}\langle\nu|\mu\rangle\\&\quad-\sum_{\lambda=1}^{5}\langle\nu|(-iz+\Sigma Q)^{-1}i\Sigma|\lambda\rangle\langle\lambda|(-iz+\Sigma)^{-1}|\mu\rangle\\&=\frac{i}{z}\delta_{\mu\nu}-\sum_{\lambda=1}^{5}\langle\nu|(-iz+\Sigma Q)^{-1}\Sigma|\lambda\rangle\zeta_{\lambda\nu}\end{aligned} \tag{9.8.12}$$

The hydrodynamic matrix elements are therefore the solutions to a system of linear algebraic equations:

$$-iz\zeta_{\nu\mu}+\sum_{\lambda=1}^{5}W_{\nu\lambda}\zeta_{\lambda\mu}=\delta_{\nu\mu} \tag{9.8.13}$$

where the elements of the "transport" matrix $\mathbf{W}(\mathbf{k}, z)$ are

$$W_{\nu\lambda}=\langle\nu|-iz(-iz+\Sigma Q)^{-1}\Sigma|\lambda\rangle \tag{9.8.14}$$

The operator identity (E.8) can be used to rewrite Eqn (9.8.14) in a more symmetric form as

$$W_{\nu\lambda} = \langle\nu|\mathbf{\Sigma}|\lambda\rangle + \langle\nu|\mathbf{\Sigma}Q(iz - Q\mathbf{\Sigma}Q)^{-1}Q\mathbf{\Sigma}|\lambda\rangle$$
$$= W^{(1)}_{\nu\lambda} + W^{(2)}_{\nu\lambda} \tag{9.8.15}$$

where the first and second terms, respectively, are usually called the "direct" and "indirect" contributions to the transport matrix. By combining Eqns (9.8.13) and (9.8.6), we obtain the corresponding system of five linear equations for the "hydrodynamic" components of the phase-space correlation function:

$$\sum_{\lambda=1}^{5} (-iz\delta_{\nu\lambda} + W_{\nu\lambda})\tilde{C}_{\lambda\mu} = C^{(0)}_{\nu\mu} \tag{9.8.16}$$

The linear system (9.8.16) is reminiscent of the hydrodynamic system of equations (8.3.28), the hydrodynamic matrix in (8.3.28) being replaced here by the transport matrix (9.8.15). Rotational invariance implies that the 5×5 matrix $\boldsymbol{C}$ has the same block-diagonal form as the hydrodynamic matrix; this in turn means that the two equations for the transverse-current correlation functions, C_{44} and C_{55}, are completely decoupled from the equations for the longitudinal modes and from each other. Labelling the 44 and 55 elements of $\boldsymbol{C}$ and $\boldsymbol{W}$ by t, and recalling Eqns (9.8.8) and (9.6.25c), we find that

$$\tilde{C}_t(k, z) = \omega_0^2\tilde{C}_{44}(\mathbf{k}, z) = \omega_0^2\tilde{C}_{55}(\mathbf{k}, z)$$
$$= \frac{\omega_0^2}{-iz + W_t(\mathbf{k}, z)} \tag{9.8.17}$$

which, as expected, is of exactly the same form as Eqn (9.3.10), with the quantity W_t/k^2 appearing in place of the generalized kinematic shear viscosity $\tilde{\nu}(\mathbf{k}, z)$. It follows from the continuity equation for transverse momentum, Eqn (8.4.1), that $W_t(\mathbf{k}, z)$ is proportional to k^2; the ratio $W_t(\mathbf{k}, z)/k^2$ therefore remains finite in the limit $k\to 0$. The shear viscosity is given by

$$\eta = \rho m \lim_{z\to i0+} \lim_{k\to 0} \frac{1}{k^2} W_t(\mathbf{k}, z) \tag{9.8.18}$$

Comparison of Eqns (9.8.15) and (9.8.18) shows that η splits naturally into a "direct" part, $\eta^{(1)}$, and an "indirect" part, $\eta^{(2)}$. For reasons of parity, the free-streaming and mean-field terms $\mathbf{\Sigma}^{\mathrm{O}}$ and $\mathbf{\Sigma}^{\mathrm{S}}$ contribute nothing to $\eta^{(1)}$, which reduces, by virtue of Eqns (9.7.23) and (9.7.24) to

$$\eta^{(1)} = \rho m \lim_{z\to i0+} \lim_{k\to 0} \frac{1}{k^2}\langle t|\Sigma^{\mathrm{C}}|t\rangle$$

$$= \pi\rho m \lim_{\omega\to 0} \lim_{k\to 0} \frac{1}{k^2}\langle t|\boldsymbol{\Gamma}(k,\omega)|t\rangle \tag{9.8.19}$$

The general property (9.7.24) implies that $\eta^{(1)}$ is positive. Action on a transverse state $|t\rangle$ by the mean-field part of the evolution kernel gives zero by symmetry, and the expression for the indirect part of the shear viscosity becomes

$$\eta^{(2)} = \rho m \lim_{z\to i0+} \lim_{k\to 0} \frac{1}{k^2} W_t^{(2)}(\mathbf{k}, z)$$

$$= \rho m \lim_{z\to i0+} \lim_{k\to 0} \frac{1}{k^2}\langle t|(\boldsymbol{\Sigma}^{\mathrm{O}}+\boldsymbol{\Sigma}^{\mathrm{C}})Q(iz-Q\boldsymbol{\Sigma}Q)^{-1}$$

$$\times Q(\boldsymbol{\Sigma}^{\mathrm{O}}+\boldsymbol{\Sigma}^{\mathrm{C}})|t\rangle \tag{9.8.20}$$

This result for $\eta^{(2)}$ can be further reduced and shown to be positive (Forster, 1974; Mazenko, 1974). In practical calculations, the basis of non-hydrodynamic states appearing in the definition of the projection operator Q is restricted to a few Hermite polynomials (Gross and Jackson, 1959; Mazenko *et al.*, 1972). The analysis we have sketched can be extended to the longitudinal modes associated with the basis functions $\psi_\nu (\nu = 1$ to $3)$ (Forster, 1974; Mazenko, 1974). The generalization to multicomponent fluids has been worked out by Linnebur and Duderstadt (1973) and by Castresana *et al.* (1977), while the case of fluids of charged particles, which will be the subject of the next chapter, has been examined in detail by Baus (1977).

The method of hydrodynamic projection can be used to solve a kinetic equation of the form (9.8.1) to any degree of accuracy by transforming it into an $n\times n$ matrix equation. Using the separation of the evolution kernel given in Eqn (E.13), we write the matrix equation (9.8.1) with continuous indices $\mathbf{p}_1$ and $\mathbf{p}_2$ as

$$(-iz+\boldsymbol{\Sigma}^{\mathrm{O}}+\boldsymbol{\Sigma}')\tilde{\boldsymbol{C}} = \boldsymbol{C}^{(0)} \tag{9.8.21}$$

The corresponding equation for the free-particle motion is

$$(-iz+\boldsymbol{\Sigma}^{\mathrm{O}})\tilde{\boldsymbol{C}}_0 = \boldsymbol{C}_0^{(0)} = \boldsymbol{I} \tag{9.8.22}$$

where $\boldsymbol{I}$ is the matrix with elements $\delta(\mathbf{p}_1-\mathbf{p}_2)$ and the subscript 0 refers to free-particle quantities. A combination of Eqns (9.8.21) and (9.8.22) leads to

$$(\tilde{\boldsymbol{C}}_0^{-1}+\boldsymbol{\Sigma}')\tilde{\boldsymbol{C}} = \boldsymbol{C}^{(0)} \tag{9.8.23}$$

We now introduce an incomplete basis of n momentum functions ψ_ν, and approximate the interaction part of the evolution kernel by

$$\mathbf{\Sigma}' = \sum_{\nu=1}^{n} \sum_{\mu=1}^{n} |\nu\rangle\sigma_{\nu\mu}\langle\mu| + \alpha\boldsymbol{I} \tag{9.8.24}$$

where

$$\sigma_{\nu\mu} = \langle\nu|\mathbf{\Sigma}'|\mu\rangle - \alpha\delta_{\nu\mu} \tag{9.8.25}$$

and

$$\alpha(\mathbf{k}, z) = \langle n+1|\mathbf{\Sigma}'|n+1\rangle \tag{9.8.26}$$

The procedure summarized by Eqns (9.8.24) to (9.8.26) is known as "kinetic modelling" (Gross and Jackson, 1959; Mazenko *et al.*, 1972); it amounts to replacing the infinite matrix $\mathbf{\Sigma}'$ by an $n \times n$ matrix, together with a constant in place of all other diagonal elements and zeros elsewhere. Substitution of (9.8.24) in (9.8.23) leads to an $n \times n$ matrix equation with elements

$$\sum_{\lambda} \{[\tilde{\boldsymbol{C}}_0^{-1}(\mathbf{k}, z+i\alpha)]_{\nu\lambda} + \sigma_{\nu\lambda}(\mathbf{k}, z)\}\tilde{C}_{\lambda\mu}(\mathbf{k}, z) = C_{\nu\mu}^{(0)}(\mathbf{k}) \tag{9.8.27}$$

where, from Eqn (9.8.22):

$$[\tilde{\boldsymbol{C}}_0^{-1}(\mathbf{k}, z+i\alpha)]_{\nu\lambda} = [-iz + \alpha(k, z)]\delta_{\nu\lambda} + \Sigma_{\nu\lambda}^{\mathrm{O}}(\mathbf{k}) \tag{9.8.28}$$

Thus the solution of the kinetic equation (9.8.21) has been reduced to the calculation of (n^2+1) matrix elements of the evolution kernel $\mathbf{\Sigma}'$, followed by a matrix inversion. As usual, the static correlation functions appearing in $\boldsymbol{C}^{(0)}(\mathbf{k})$ are assumed to be known quantities that serve as input in the dynamical calculations.

9.9 KINETIC THEORIES OF LIQUIDS

The phase-space description of time-dependent fluctuations developed in Section 9.6 has enabled us to derive the general kinetic equation (9.7.20) and to obtain exact, formal expressions for the collision kernel Σ^{C}. If explicit calculations of time-correlation functions are to be made, however, tractable approximations for Σ^{C} are required. In this context, the phase-space formulation of the problem has several advantages over the simpler, configuration-space approach described in earlier sections of this chapter. First, the exact short-time behaviour is automatically built into the theory by the separation out of the frequency-independent terms Σ^{O} and Σ^{S}. Secondly, the exact properties of Σ^{C} established in Section 9.7 act as constraints in the choice of approximation. In particular, by satisfying the conservation laws (9.7.29) and (9.7.30), it is guaranteed that the correlation functions will have the correct structure in the hydrodynamic limit. Thirdly, the form in which Σ^C is written in Eqn (9.7.21) lends itself naturally to approximations that are similar in character to the mode-coupling theories of Section 9.5. Lastly, a single approximation for Σ^C is sufficient to determine all correlation functions of interest; there is no need to work out a separate treatment of each as is done, for example, in Sections 9.3 and 9.4. In this section, we briefly discuss some of the approximations for Σ^C that have appeared in the literature. We omit the technical details, which can be found in the original papers. A useful review with a similar perspective to ours is that of Yip (1979).

The simplest and most radical approximation is to neglect collisional effects altogether by setting $\Sigma^C = 0$. This corresponds to the generalized Vlasov approximation embodied in Eqns (7.9.12) and (7.9.13); it is valueless as a theory of the dynamics of liquids, despite the inclusion of static correlations through the mean-field term Σ^S (Nelkin and Ranganathan, 1967). The effects of collisions can be incorporated into the theory in a systematic way by expanding Σ^C in powers either of density or of a coupling parameter λ that characterizes the strength of the interactions between particles. The density expansion leads to a generalized Boltzmann equation, as we shall discuss later. In the weak-coupling expansion, the dominant term is proportional to λ^2 (Akcasu and Duderstadt, 1969; Forster and Martin, 1970). To prove this, we first rewrite Eqn (9.7.18) as

$$\Sigma^C(\mathbf{k}, z; \mathbf{p}_1\mathbf{p}_2) = \frac{i}{\phi_0(\mathbf{p}_2)}(R_\mathbf{k}(\mathbf{p}_1), (z + \mathcal{Q}_\mathbf{k}\mathcal{L}\mathcal{Q}_\mathbf{k})^{-1}R_\mathbf{k}(\mathbf{p}_2)) \tag{9.9.1}$$

where the "random force" $R_\mathbf{k}(\mathbf{p})$ is

$$\begin{aligned} R_\mathbf{k}(\mathbf{p}) &= \mathcal{Q}_\mathbf{k} i\mathcal{L}\delta f_\mathbf{k}(\mathbf{p}) \\ &= \mathcal{Q}_\mathbf{k}\frac{\partial}{\partial \mathbf{p}} \cdot \sum_{i=1}^N \exp(-i\mathbf{k}\cdot\mathbf{r}_i)\delta(\mathbf{p}-\mathbf{p}_i)\frac{\partial V_N}{\partial \mathbf{r}_i} \end{aligned} \tag{9.9.2}$$

and is clearly of order λ. The propagator in Eqn (9.9.1) can be expanded in powers of the interaction part of the Liouville operator, and hence of λ, in the form

$$\begin{aligned} (z + \mathcal{Q}_\mathbf{k}\mathcal{L}\mathcal{Q}_\mathbf{k})^{-1} &= (z + \mathcal{Q}_\mathbf{k}\mathcal{L}_\mathrm{O}\mathcal{Q}_\mathbf{k})^{-1} - (z + \mathcal{Q}_\mathbf{k}\mathcal{L}_\mathrm{O}\mathcal{Q}_\mathbf{k})^{-1} \\ &\quad \times \mathcal{L}_\mathrm{I}(z + \mathcal{Q}_\mathbf{k}\mathcal{L}_\mathrm{O}\mathcal{Q}_\mathbf{k})^{-1} + \mathcal{O}(\lambda^2) \end{aligned} \tag{9.9.3}$$

Explicit calculation shows that

$$i\mathcal{L}_\mathrm{O}\mathcal{Q}_\mathbf{k} - \mathcal{Q}_\mathbf{k} i\mathcal{L}_\mathrm{O} \sim \hat{c}(k) \tag{9.9.4}$$

which is also of order λ, and hence that

$$\Sigma^{C}(\mathbf{k}, z; \mathbf{p}_1\mathbf{p}_2) = \frac{i}{\phi_0(\mathbf{p}_2)}(R_{\mathbf{k}}(\mathbf{p}_1), (z+\mathscr{L}_{O})^{-1}R_{\mathbf{k}}(\mathbf{p}_2)) + \mathcal{O}(\lambda^3) \quad (9.9.5)$$

The weak-coupling approximation for Σ^{C} is easily evaluated in the time domain if it is recognized that the effect of the free-streaming evolution operator $\exp(i\mathscr{L}_{O}t)$ is merely to displace particles from positions $\mathbf{r}_i$ to positions $(\mathbf{r}_i + \mathbf{p}_i t/m)$ (Akcasu and Duderstadt, 1969; Forster and Martin, 1970). Like the generalized Vlasov approximation, it is not a good approximation for liquids; because the interparticle forces are strongly repulsive at short distances, the weak-coupling approach breaks down. A useful approximate kernel having a similar structure to the weak-coupling expression can, however, be derived by a more phenomenological approach (Lebowitz *et al.*, 1969; Duderstadt and Akcasu, 1970) in which Σ^{C} is factorized as

$$\Sigma^{C}(\mathbf{k}, t; \mathbf{p}_1\mathbf{p}_2) = \Sigma_0^{C}(\mathbf{k}; \mathbf{p}_1\mathbf{p}_2)M(\mathbf{k}, t) \quad (9.9.6)$$

where $\Sigma_0^{C}(\mathbf{k}; \mathbf{p}_1\mathbf{p}_2) \equiv \Sigma^{C}(\mathbf{k}, t=0; \mathbf{p}_1\mathbf{p}_2)$. The key feature of this approximation is the fact that the time dependence of the collision kernel is contained in a function $M(k, t)$ that is assumed to be independent of the particle momenta. To illustrate how the calculation proceeds, we consider the simpler problem of the self correlation function $C_s(\mathbf{k}, t; \mathbf{p}_1\mathbf{p}_2)$ (Akcasu *et al.*, 1970; Desai, 1971); this obeys a kinetic equation similar to (9.7.20), but without the mean-field term. If we denote the corresponding collision kernel by Σ_s^{C}, we can write the kinetic equation as

$$\left(\frac{\partial}{\partial t} + \frac{i\mathbf{k}\cdot\mathbf{p}_1}{m}\right) C_s(\mathbf{k}, t; \mathbf{p}_1\mathbf{p}_2) + \int_0^t \mathrm{d}s \int \mathrm{d}\mathbf{p}_3\, \Sigma_s^{C}(\mathbf{k}, t-s; \mathbf{p}_1\mathbf{p}_3) C_s(\mathbf{k}, s; \mathbf{p}_3\mathbf{p}_2) = 0 \quad (9.9.7)$$

where, by analogy with (9.9.1) and (9.9.2):

$$\Sigma_s^{C}(\mathbf{k}, t; \mathbf{p}_1\mathbf{p}_2) = \frac{1}{\phi_0(\mathbf{p}_2)}(R_{s\mathbf{k}}(\mathbf{p}_1), \exp(i\mathscr{Q}_{s\mathbf{k}}\mathscr{L}\mathscr{Q}_{s\mathbf{k}}t)R_{s\mathbf{k}}(\mathbf{p}_2)) \quad (9.9.8)$$

with

$$R_{s\mathbf{k}}(\mathbf{p}) = \mathscr{Q}_{s\mathbf{k}} i\mathscr{L}\delta f_{s\mathbf{k}}(\mathbf{p})$$
$$= \exp(-i\mathbf{k}\cdot\mathbf{r}_i)\frac{\partial V_N}{\partial \mathbf{r}_i}\cdot\frac{\partial}{\partial \mathbf{p}}\delta(\mathbf{p}-\mathbf{p}_i) \quad (9.9.9)$$

The projection operator $\mathcal{Q}_{s\mathbf{k}}$ is defined in terms of $\delta f_{s\mathbf{k}}$ rather than $\delta f_{\mathbf{k}}$; $\mathcal{Q}_{s\mathbf{k}}$ is absent in the second line of Eqn (9.9.9) because $\delta \dot{f}_{s\mathbf{k}}$ is orthogonal to $\delta f_{s\mathbf{k}}$. The initial value of Σ_s^C is independent of $\mathbf{k}$ and given by

$$\Sigma_{s0}^C(\mathbf{p}_1, \mathbf{p}_2) \equiv \Sigma_s^C(\mathbf{k}, t=0; \mathbf{p}_1\mathbf{p}_2)$$

$$= \frac{1}{\phi_0(\mathbf{p}_2)}\left\langle \exp(i\mathbf{k}\cdot\mathbf{r}_i)\frac{\partial V_N}{\partial \mathbf{r}_i}\cdot\frac{\partial}{\partial \mathbf{p}_1}\delta(\mathbf{p}_1-\mathbf{p}_i)\right.$$

$$\left.\times \exp(-i\mathbf{k}\cdot\mathbf{r}_i)\frac{\partial V_N}{\partial \mathbf{r}_i}\cdot\frac{\partial}{\partial \mathbf{p}_2}\delta(\mathbf{p}_2-\mathbf{p}_i)\right\rangle$$

$$= \frac{k_B T}{3\phi_0(\mathbf{p}_2)}\langle \nabla_i^2 V_N\rangle \left\langle \frac{\partial}{\partial \mathbf{p}_1}\delta(\mathbf{p}_1-\mathbf{p}_i)\cdot\frac{\partial}{\partial \mathbf{p}_2}\delta(\mathbf{p}_2-\mathbf{p}_i)\right\rangle$$

$$= \Omega_0^2 \frac{\phi_0(\mathbf{p}_1)}{\phi_0(\mathbf{p}_2)}\left(\mathbf{p}_1\cdot\frac{\partial}{\partial \mathbf{p}_1} - mk_B T\frac{\partial}{\partial \mathbf{p}_1}\cdot\frac{\partial}{\partial \mathbf{p}_1}\right)\delta(\mathbf{p}_2-\mathbf{p}_1) \qquad (9.9.10)$$

where Ω_0 is the Einstein frequency (7.2.12). We now make an approximation analogous to (9.9.6), introducing a "memory function" $M_s(\mathbf{k}, t)$. Then equations (9.9.7) and (9.9.10) lead to an approximate kinetic equation of the form

$$\left(-iz + \frac{i\mathbf{k}\cdot\mathbf{p}_1}{m} - m\Omega_0^2 \tilde{M}_s(\mathbf{k}, z)G(\mathbf{p}_1)\right)\tilde{C}_s(\mathbf{k}, z; \mathbf{p}_1\mathbf{p}_2) = C_s^{(0)}(\mathbf{p}_1\mathbf{p}_2) \qquad (9.9.11)$$

where $C_s^{(0)}(\mathbf{p}_1\mathbf{p}_2)$ is given by Eqn (9.6.19b) and

$$G(\mathbf{p}) = \frac{\partial}{\partial \mathbf{p}}\cdot\left(k_B T\frac{\partial}{\partial \mathbf{p}} + \frac{\mathbf{p}}{m}\right) \qquad (9.9.12)$$

is the Fokker–Planck operator (Résibois and DeLeener, 1977). Equation (9.9.11) is a generalized, non-markovian Fokker–Planck equation in which the friction coefficient that appears in the standard Fokker–Planck equation is replaced by the wavenumber and frequency-dependent quantity $m\Omega_0^2\tilde{M}_s(\mathbf{k}, z)$. Exponential and gaussian approximations for $M_s(\mathbf{k}, t)$ have been used to evaluate (9.9.11) and hence to determine the self dynamic structure factor $S_s(k, \omega)$, which is the momentum contraction of $C_s(\mathbf{k}, t; \mathbf{p}_1\mathbf{p}_2)$ (see Eqn (9.6.25a)). Reasonable agreement with molecular dynamics results for liquid argon is obtained if the decay of $M_s(\mathbf{k}, t)$ is described by a wavenumber-dependent relaxation time (Akcasu *et al.*, 1970; Desai, 1971).

Similar methods have been used for the calculation of the collective current correlations (Duderstadt and Akcasu, 1970), but the problem becomes much more complicated. One difficulty is the fact that although the approximation (9.9.6) meets the requirements of conservation of particle number and momentum, it does not conserve energy. Thus the theory does not yield the correct hydrodynamic behaviour in the limit $k, \omega \to 0$. The situation can be remedied by including the energy density, defined as

$$\varepsilon(\mathbf{r}, t) = \int \frac{|\mathbf{p}|^2}{2m} f(\mathbf{r}, \mathbf{p}; t)\, d\mathbf{p} + \tfrac{1}{2} \iiint d\mathbf{r}'\, d\mathbf{p}\, d\mathbf{p}'\, v(\mathbf{r} - \mathbf{r}') f(\mathbf{r}, \mathbf{p}; t) f(\mathbf{r}', \mathbf{p}'; t) \tag{9.9.13}$$

(with $v(\mathbf{0}) = 0$) as an additional dynamical variable in the set of functions $\delta f(\mathbf{r}, \mathbf{p}; t)$ (Jhon and Forster, 1975). The longitudinal-current spectra calculated in this way are in almost perfect agreement with the results of simulations of the Lennard-Jones fluid (Levesque *et al.*, 1973) and with neutron-scattering results for liquid argon (Sköld *et al.*, 1972).

The theory outlined above is similar in spirit to the memory-function analyses of Sections 9.2 to 9.4 insofar as it relies for its implementation on assumptions concerning the time-dependence of the collision kernel. Use of methods that take as their starting point the exact expression for Σ^{C} given by Eqn (9.7.18) (or (9.7.21)) has so far been limited mostly to hard spheres, for which the collisions are strictly binary and the two-particle dynamics are particularly simple. Hard spheres have the further advantage that the energy density (9.9.13) has only a kinetic part and is therefore automatically included in the set of momentum states (9.6.22). Inclusion of a potential-energy term in the energy density poses a severe technical problem, since potential energy is not a linear functional of the single-particle phase-space density.

We have seen in earlier chapters that the Enskog equation provides a satisfactory method of calculating the transport coefficients of hard spheres at intermediate densities ($\rho^* \lesssim 0.3$). The Enskog approximation takes account both of collisional transfer and of short-range, static correlations, but it ignores the sequences of correlated collisions that become important as the density increases. It is therefore tempting to proceed in two stages. The first step is to describe the dynamics of uncorrelated binary collisions rigorously through a generalized Enskog equation, obtained from the exact markovian ($z \to 0$) limit of the collision kernel. In the second stage, non-markovian corrections to Σ^{C} are incorporated in an approximate way that allows for mode-coupling effects of the type discussed in Section 9.5. This approach corresponds to splitting the hard-sphere collision kernel into two parts in the general form

$$\Sigma^{\mathrm{C}}(\mathbf{k}, z; \mathbf{p}_1\mathbf{p}_2) = \Sigma^{\mathrm{C}}_{\mathrm{E}}(\mathbf{k}; \mathbf{p}_1\mathbf{p}_2) + \tilde{M}(\mathbf{k}, z; \mathbf{p}_1\mathbf{p}_2) \tag{9.9.14}$$

where the frequency-independent Enskog term $\Sigma^{\mathrm{C}}_{\mathrm{E}}$ describes the dynamically-uncorrelated collisions and all the more complicated dynamical processes are contained in the "relaxation kernel" $\tilde{M}(\mathbf{k}, z; \mathbf{p}_1\mathbf{p}_2)$. The generalized Enskog kinetic equation corresponds to setting $\tilde{M} = 0$; this approximation preserves the correct short-time behaviour since, by construction, $\Sigma^{\mathrm{C}}_{\mathrm{E}}$ is the exact high-frequency limit of Σ^{C} (Lebowitz *et al.*, 1969). Because the collisions between hard spheres are of infinitesimally short duration, the interaction part of the Liouville operator has a form different from that given in Eqn (9.6.5b). Ernst *et al.* (1969) have shown that a dynamical variable of a hard-sphere system evolves in time under the action of a "pseudo" Liouville operator $\mathscr{L}^{(+,-)} = \mathscr{L}_{\mathrm{O}} + \mathscr{L}_{\mathrm{I}}^{(+,-)}$ that has a different form for positive (+) and negative (−) times. From this starting point it is possible to derive an expression for $\Sigma^{\mathrm{C}}_{\mathrm{E}}$ in terms of the contact value $g(d)$ of the hard-sphere radial distribution function (Lebowitz *et al.*, 1969; Blum and Lebowitz, 1969; Konijnendijk and van Leeuwen 1973; Mazenko and Yip, 1977). The Enskog term acts on the phase-space correlation function $\tilde{C}$ according to the rule

$$\begin{aligned}
&\int \mathrm{d}\mathbf{p}_3\, \Sigma^{\mathrm{C}}_{\mathrm{E}}(\mathbf{k}; \mathbf{p}_1\mathbf{p}_2)\tilde{C}(\mathbf{k}, z; \mathbf{p}_3\mathbf{p}_2) \\
&\qquad = i\rho g(d)\hat{c}_0(k)\phi_0(\mathbf{p}_1)\frac{\mathbf{k}\cdot\mathbf{p}_1}{m}\int \mathrm{d}\mathbf{p}_3\tilde{C}(\mathbf{k}, z; \mathbf{p}_3\mathbf{p}_2) \\
&\qquad\quad -\frac{\rho d^2}{m}g(d)\int \mathrm{d}\mathbf{p}_3\int \mathrm{d}\hat{\mathbf{r}}\,\theta[(\mathbf{p}_3-\mathbf{p}_1)\cdot\hat{\mathbf{r}}](\mathbf{p}_3-\mathbf{p}_1)\cdot\hat{\mathbf{r}} \\
&\quad \times\{\phi_0(\mathbf{p}_3')\tilde{C}(\mathbf{k}, z; \mathbf{p}_1'\mathbf{p}_2) - \phi_0(\mathbf{p}_3)\tilde{C}(\mathbf{k}, z; \mathbf{p}_1\mathbf{p}_2) \\
&\quad +\exp(i\mathbf{k}\cdot\hat{\mathbf{r}}d)\phi_0(\mathbf{p}_1')\tilde{C}(\mathbf{k}, z; p_3'p_2) \\
&\quad -\exp(-i\mathbf{k}\cdot\hat{\mathbf{r}}d)\phi_0(\mathbf{p}_1)\tilde{C}(\mathbf{k}, z; \mathbf{p}_3\mathbf{p}_2)\}
\end{aligned} \tag{9.9.15}$$

where $\theta(\ )$ is the Heaviside step-function, $\hat{\mathbf{r}} \equiv (\mathbf{r}_1 - \mathbf{r}_3)/|\mathbf{r}_1 - \mathbf{r}_3|$ is a unit vector directed along the line joining the centres of the colliding particles 1 and 3, $\mathbf{p}_1'$ and $\mathbf{p}_3'$ are the post-collision momenta, and $\hat{c}_0(k) = 4\pi d \sin(kd)/k^2$ is the Fourier transform of the direct correlation function in the low-density limit. Note that the first term on the right-hand side of (9.9.15) is of the same form as the mean-field term Σ^{S} (see Eqn (9.7.15)). If $g(d)$ is replaced by its low-density limit, $\Sigma^{\mathrm{C}}_{\mathrm{E}}$ reduces to a generalized Boltzmann collision kernel; the latter is the leading term in the expansion of the exact hard-sphere collision kernel in powers of ρ (Mazenko, 1971; Boley, 1972; Mazenko *et al.*, 1972). In the long-wavelength limit, the size of the colliding spheres becomes irrelevant, and the generalized Boltzmann kernel itself reduces to the linearized version of the local collision kernel (2.1.21).

The method of "kinetic modelling", briefly sketched at the end of Section 9.8, has been used to solve both the linearized Boltzmann equation (Yip and Nelkin, 1964; Sugawara *et al.*, 1968) and the generalized Enskog equation (Mazenko *et al.*, 1972; Dufty *et al.*, 1981; Yip *et al.*, 1982), and hence to calculate the density and current autocorrelation functions of the hard-sphere fluid over a wide range of density. The linearized Boltzmann equation is applicable only at low densities ($\rho^* \leqslant 0.01$) and at wavelengths larger than the Enskog mean-free path $l_E = 1/\sqrt{2}\pi\rho d^2 g(d)$. At liquid-like densities, however, the dynamic structure factors calculated from the generalized Enskog equation are in remarkably good agreement with results of molecular dynamics simulations (Alley *et al.*, 1983), even for $\rho^* \simeq 0.9$; a comparison is made in Figure 7.5. At high densities, a three-peak structure of Rayleigh–Brillouin type persists for wavenumbers up to $kd \simeq 0.5$. The discrepancies between theory and simulation are largest in the intermediate wavenumber range, where viscoelastic effects are known to be important, and the theory also fails to predict the appearance of propagating shear waves. To correct these deficiencies, it is necessary to add a relaxation term to the collision kernel, as in Eqn (9.9.14). A mode-coupling calculation of $\tilde{M}(\mathbf{k}, z; \mathbf{p}_1\mathbf{p}_2)$ has been made by Leutheusser (1982a, b) along lines very similar to the configuration-space calculation of Bosse *et al.* (1978a, b, c) described in Section 9.5. Equation (9.7.21) shows that $\tilde{M}$ is related to a four-point correlation function, often called a "two-particle propagator," since it involves the correlated motion of two particles (not necessarily the same at the beginning and the end) from field points 1, 3 to field points 2, 4 in a time t. As is usual in mode-coupling approximations, the two-particle propagator is factorized into products of two one-particle propagators, i.e. two-point correlation functions, in a manner akin to the "disconnected" approximation described in Appendix E. This leads to a non-linear equation for the phase-space correlation function $\tilde{C}(\mathbf{k}, z; \mathbf{p}_1\mathbf{p}_2)$; the equation must be solved self-consistently, taking the generalized Enskog result as input in an iterative procedure. The numerical results for the various time-correlation functions are a significant improvement over those obtained from the generalized Enskog equation. As an illustration, the predictions of the two theories for the wavenumber-dependent shear viscosity $\eta(k)$ introduced in Section 8.7 are compared in Figure 9.6 with the results of molecular dynamics calculations; the superiority of the mode-coupling approximation is evident from the figure. Mode-coupling theory also correctly predicts that a propagating shear wave should appear at intermediate wavenumbers.

Much effort has also been devoted to the simpler problem of self motion and, in particular, to the calculation of the velocity autocorrelation function

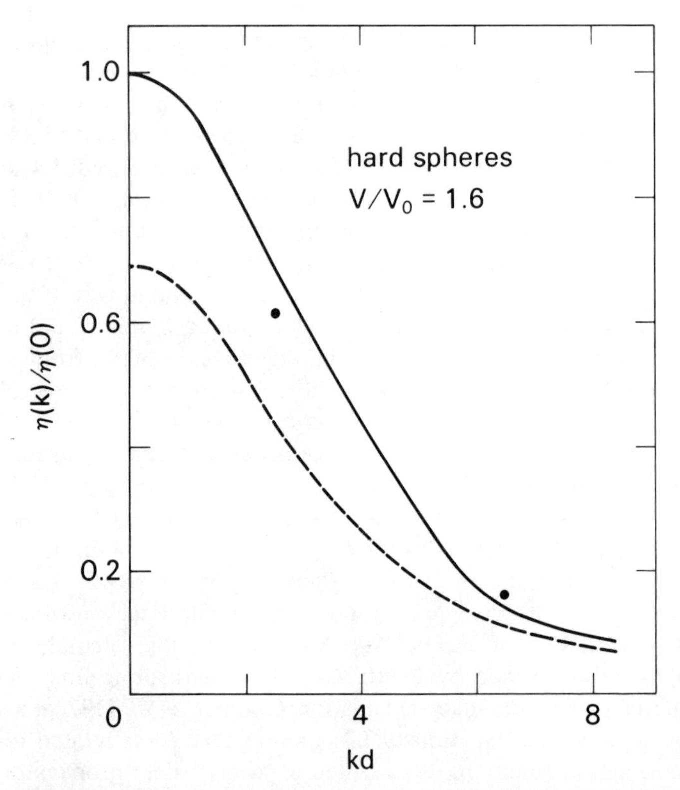

FIG. 9.6. Wavenumber-dependent shear viscosity of the hard-sphere fluid; $V_0 = Nd^3/\sqrt{2}$ is the volume at close packing. The full curve shows the results of molecular dynamics calculations, the dashed curve is calculated from the generalized Enskog theory, and the two points are results of a mode-coupling calculation (Leutheusser, 1982a, b) discussed in the text. After Alley and Alder (1983).

$Z(t)$. In the Enskog approximation, the tagged particles suffer collisions only with bath particles drawn from an equilibrium distribution. No account is therefore taken of the local disturbance induced by the tagged particle in its immediate surroundings, including "backflow" effects. This is a poor approximation at high densities, where recollision events are very important. After an initial collision between the tagged particle and a bath particle, there is a high probability either that the two particles will recollide after a few mean collision times, or that the tagged particle will encounter another bath particle that is dynamically correlated with the initial collision partner. Correlated collisions of this type are the "ring collisions" discussed earlier and pictured in Figure 7.3. The contribution from ring-collision events must be included in the relaxation kernel for self motion if the correct long-time behaviour is to be obtained (Dorfman and Cohen, 1972; Dufty, 1972) (see

Section 8.7). Detailed calculations of the velocity autocorrelation function of the hard-sphere fluid have been carried out in which allowance is made for two correlated binary collisions, i.e. single ring-collision events (Mazenko, 1973a, b; Résibois, 1975; Furtado *et al.*, 1976). The memory function of $Z(t)$, i.e. the generalized friction coefficient $\xi(t)$ of Eqn (7.3.24), is written as the sum of an instantaneous Enskog contribution and a correction due to collisions. Thus

$$\xi(t) = \tfrac{2}{3}\tau_E^{-1}\delta(t) + \Delta\xi(t) \tag{9.9.16}$$

where the Enskog mean collision time τ_E is defined in Eqn (7.2.20). The recollision term $\Delta\xi(t)$ can be expressed as an integral over wavevectors of a momentum contraction of the product of the self and collective phase-space density correlation functions $C_s(\mathbf{k}, t; \mathbf{p}_1\mathbf{p}_2)$ and $C(\mathbf{k}, t; \mathbf{p}_1\mathbf{p}_2)$. If only hydrodynamic states are retained in the calculation of the momentum contractions, the resulting expression for $\Delta\xi(t)$ is very similar to that obtained by mode-coupling theory, Eqn (9.5.16). At the highest densities, the effects of repeated ring collisions can no longer be ignored. This problem has been analysed by Cukier and Mehaffey (1978); their results, shown in Figure 7.2, reproduce satisfactorily the density dependence of the self diffusion coefficient obtained by molecular dynamics. The structure of the theoretical expression for D suggests that the density dependence results from a competition between the coupling of the tagged-particle motion to the collective shear motion of the bath particles (the backflow effect) and the coupling to the density fluctuations (the cage effect). Inclusion of repeated ring collisions is of crucial importance when the size of the tagged particle is much larger than that of the bath particles (brownian motion), since the single-ring theory leads to an unphysical divergence of D beyond a certain ratio of brownian-to-bath particle diameters (Mehaffey and Cukier, 1978). This last calculation also provides a microscopic foundation for Stokes' law (7.3.20). The ring-collision analysis can be generalized to include reorientational variables, and has been applied to the "rough-sphere fluid," a simplified model of a molecular liquid in which colliding spheres exchange both linear and angular momentum (Mehaffey *et al.*, 1977).

The difficulty of extending hard-sphere kinetic theory to systems with continuous potentials lies in the fact that the collisions are no longer instantaneous and the many-particle dynamics cannot be decomposed unambiguously into sequences of collisions involving only two particles. Hence there is no obvious starting point similar to that provided by the generalized Enskog theory in the case of hard spheres. Nonetheless, if we consider first the problem of the velocity autocorrelation function, it seems reasonable to retain a division of the memory function $\xi(t)$ into two parts, one associated with individual binary collisions, and thus descriptive of the short-time behaviour, and a part that corresponds to collective events (Sjögren and Sjölander, 1979). The memory function is therefore written as

$$\xi(t) = \xi_b(t) + \Delta\xi(t) \tag{9.9.17}$$

where the binary-collision part $\xi_b(t)$, though not having the delta-function form of Eqn (9.9.16), still decays on a much faster timescale than the collective contribution $\Delta\xi(t)$. Such a separation is in fact clearly suggested by the results of computer simulations of the Lennard-Jones fluid (Levesque *et al.*, 1973). The binary-collision part can be adequately represented by a gaussian function of t; since it can be shown that $\Delta\xi(t)$ starts as t^4, the two parameters required to specify the gaussian are uniquely determined by the zeroth and second frequency moments of the Fourier transform $\xi(\omega)$ or, equivalently, by the second and fourth moments of the power spectrum of the velocity autocorrelation function. The collective part $\Delta\xi(t)$ is expressed in terms of mode-coupling integrals similar to those used to describe the ring-collision events for hard spheres (Sjögren and Sjölander, 1979). As Figure 9.7 shows, the correlation functions calculated in this way agree well with molecular dynamics results for liquid rubidium and liquid argon, particularly if the width of the gaussian is slightly increased, since the shape of $Z(t)$ at intermediate times is sensitive to the choice of this parameter (Sjögren, 1980a).

The general idea behind the separation contained in Eqn (9.9.17) has been extended to the case of the collective phase-space correlation function $C(\mathbf{k}, t; \mathbf{p}_1\mathbf{p}_2)$ and applied to the study of density fluctuations in liquid rubidium (Sjögren, 1980b, c). The goal of this work is to generalize the relation (7.9.19) (Kerr, 1968) between the self and collective density autocorrelation functions by including backflow effects via a mode-coupling scheme. The theory gives results that agree well with molecular dynamics calculations at intermediate and large wavenumbers (Rahman, 1974a, b). There are discrepancies at small wavenumbers because the theory ignores the coupling to energy fluctuations; this effect plays an important role in the hydrodynamic limit (Jhon and Forster, 1975; Sjödin and Sjölander, 1978).

The general conclusion to be drawn from the last three chapters is that a judicious combination of frequency sum rules and mode-coupling approximations, both of which require static correlation functions as input, leads to a reasonably accurate description of dynamical fluctuations and linear transport in atomic liquids.

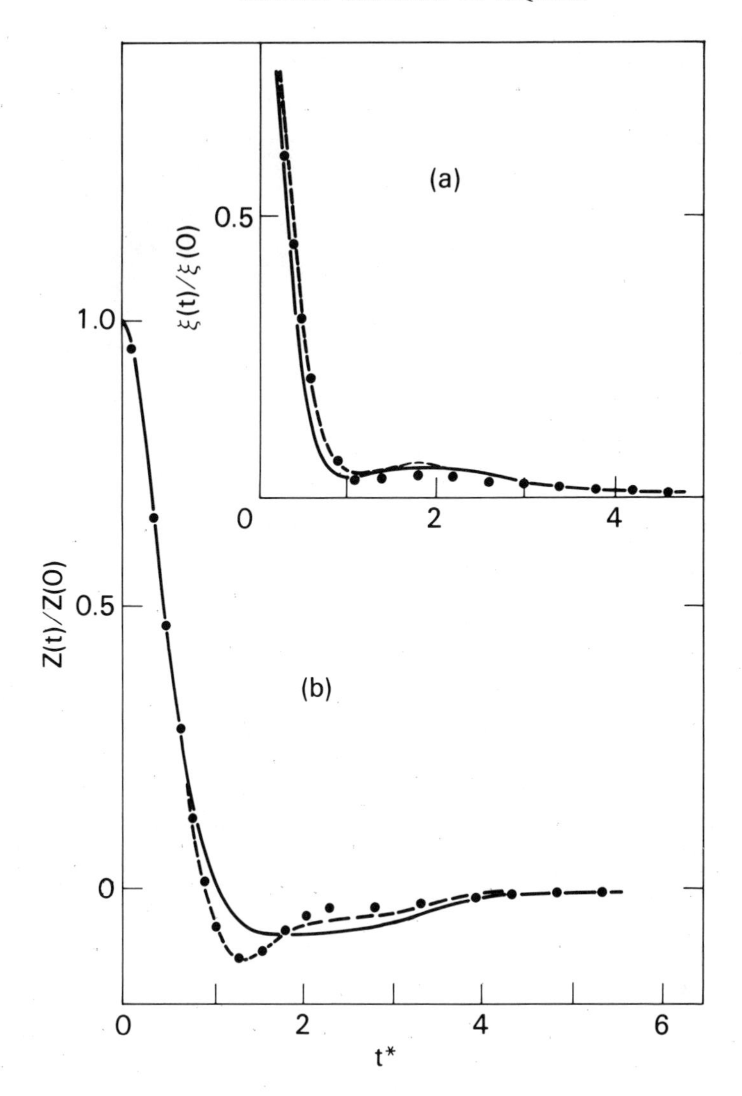

FIG. 9.7. Velocity autocorrelation function and the associated memory function (inset) of the Lennard-Jones fluid near the triple point. The points are molecular dynamics results of Levesque and Verlet (1970), and the curves are calculated from the kinetic theory of Sjögren (1980a) before (full lines) and after (dashed lines) modification of the binary-collision term in the memory function (see text). The unit of time is the quantity τ_0 defined by Eqn (3.3.5). After Sjögren (1980a).

CHAPTER 10

Ionic Liquids

10.1 CLASSES AND MODELS OF IONIC LIQUIDS

We have been concerned so far almost exclusively with fluids in which the range of the interparticle forces is of the order of a few atomic radii. This chapter and the one that follows are devoted to systems in which the particles carry an electric charge. Ionic liquids have certain properties that are absent in fluids composed of neutral particles, and many of their distinguishing characteristics are associated in some way with the slow, r^{-1} decay of the Coulomb potential and the corresponding k^{-2} singularity in its Fourier transform. Our attention will be focussed on three types of system: molten salts, liquid metals and ionic solutions. *Molten salts* are in many respects the simplest class of ionic liquids. We shall consider only the case in which there is a single cation and a single anion species, of which the molten alkali halides are the commonest and best understood examples. Molten salts are characterized by large binding energies and high melting temperatures, and by ionic conductivities of order $1\ \Omega^{-1}\ \mathrm{cm}^{-1}$. There exist also certain crystalline salts that have conductivities comparable with those of the molten phase. These are the so-called "fast-ion" conductors, or "solid electrolytes" in which one of the ionic species becomes liquid-like in its behaviour above a certain temperature (Dieterich *et al.*, 1980). *Liquid metals* are superficially similar in composition to molten salts, the anion of the salt being replaced by electrons from the valence or conduction bands. The analogy is not a good one, however, because the very small mass of the electron leads to a pronounced disymmetry between the two species present in the metal. In particular, whereas the behaviour of the ions can be discussed in the framework of classical statistical mechanics, the electrons form a degenerate Fermi gas for which a quantum-mechanical treatment is required. The presence of "free" electrons is also the origin of the fact that the electrical conductivities of liquid metals are typically three to four orders of magnitude larger than those of molten salts. *Metal–salt solutions* are mixtures of liquid metals and molten salts. The phase diagram of such a system usually has a miscibility gap, and there is a rapid transition from ionic to metallic behaviour as the concentration of metal is increased. Finally, *ionic solutions* are liquids consisting of a solvent formed from neutral, polar molecules, and a solute that dissociates into positive and negative ions. They vary widely in complexity: in the classic *electrolyte solutions*, the cations and anions are of comparable size and absolute charge, whereas macromolecular ionic solutions contain both macroions (charged polymer chains or coils, micelles, charged colloidal particles etc.) and microscopic counterions. Despite their complexity, certain systems of the latter type, including charged colloidal suspensions, can be treated quantitatively by standard methods of liquid-state theory (Dickinson, 1983).

The systems we have listed vary widely in character, but they have two important features in common: first, that of overall, macroscopic charge neutrality and, secondly, the presence of mobile charge carriers. The condition of overall charge neutrality imposes a constraint on the relative concentrations of the ions. If the fluid contains $\rho_\nu = N_\nu / V$ ions per unit volume of species ν and if the charge carried by ions of that species is $q_\nu = z_\nu e$, where e is the elementary charge, overall charge neutrality requires that

$$\sum_{\nu=1}^{n} z_\nu \rho_\nu = 0 \tag{10.1.1}$$

where n is the total number of species. We shall see in Section 10.2 that a tendency towards charge neutrality exists even at the local, microscopic level. This effect gives rise, in turn, to the phenomenon of *screening*. Introduction of an external charge into an ionic fluid causes a rearrangement or polarization of the surrounding charge density of a nature such that the net electrostatic potential due to the external charge and the "polarization cloud" decays much faster than the bare Coulomb potential; in fact, the potential decays exponentially, as we shall show in Section 10.2. Since it is permissible to regard any ion in the fluid as an "external" charge, it follows that the screening mechanism determines the long-range behaviour of the ionic distribution functions; screening also requires that the distribution functions satisfy a number of important sum rules. In liquids of high charge density, such as molten salts, there is a competition between packing effects and screening; this leads to a *charge ordering* of the ions, which manifests itself as an alternation in sign of the charge carried by successive coordination shells around a central ion.

The presence of mobile charge carriers plays an important part in determining the dynamical properties of ionic liquids. It leads, in particular, to new kinds of transport, of which electrical conduction is the most familiar example. In addition, the interplay between Maxwell's equations and the usual equations of hydrodynamics causes the long-wavelength charge fluctuations to relax in a qualitatively different manner from that of concentration fluctuations in mixtures of uncharged particles. Under conditions that are achievable, in particular, in molten salts, fluctuations in charge can give rise to propagating, high-frequency, collective modes; these excitations are similar in character to the optic modes of ionic crystals, and are also closely related to the charge oscillations found in plasmas.

Theories of ionic liquids rely heavily on the use of simple hamiltonian models in which only the essential features of the ionic interactions are retained. One simplifying approximation that is commonly made is to ignore the polarizability of the ions and represent the interactions by a *rigid-ion model.* The total potential energy is then assumed to be pairwise additive, and written as the sum of short-range (superscript S) and coulombic (superscript C) terms in the form

$$V_N(\mathbf{r}^N) = V_N^{\mathrm{S}}(\mathbf{r}^N) + V_N^{\mathrm{C}}(\mathbf{r}^N) = V_N^{\mathrm{S}}(\mathbf{r}^N) + \sum_{i<j}^{N}\sum \frac{z_i z_j e^2}{\varepsilon|\mathbf{r}_i - \mathbf{r}_j|} \tag{10.1.2}$$

where ε is the dielectric constant of the medium in which the ions are embedded. It is often convenient to replace the Coulomb term in (10.1.2) by a sum in reciprocal space. Let $\rho_{\mathbf{k}}^{Z}$ be a Fourier component of the microscopic charge density, defined as

$$\rho_{\mathbf{k}}^{Z} = \sum_{\nu=1}^{n} z_\nu \rho_{\mathbf{k}}^{\nu} = \sum_{\nu=1}^{n} z_\nu \sum_{i=1}^{N_\nu} \exp(-i\mathbf{k}\cdot\mathbf{r}_i^{\nu}) \tag{10.1.3}$$

Then the total Coulomb energy of a periodic system of volume V is

$$V_N^{\mathrm{C}} = \frac{1}{2V}\sum_{\mathbf{k}} \hat{v}(k)\left(\rho_{\mathbf{k}}^{Z}\rho_{-\mathbf{k}}^{Z} - \sum_{i=1}^{N} z_i^2\right) \tag{10.1.4}$$

where the sum on $\mathbf{k}$ runs over wavevectors compatible with the periodic boundary condition and the (negative) second term inside brackets cancels the infinite self-energy of the ions. The function $\hat{v}(k)$ is the Fourier transform of the Coulomb potential between two elementary charges, i.e.

$$\hat{v}(k) = \frac{4\pi e^2}{k^2} \tag{10.1.5}$$

This expression was used earlier in the derivation of the Debye–Hückel result (5.3.23); the k^{-2} singularity in the limit $k \to 0$ is an important characteristic of Coulomb systems. In the thermodynamic limit, the sum over wavevectors in Eqn (10.1.4) becomes an integral over $\mathbf{k}$ divided by $(2\pi)^3$; the equivalence of the expressions for V_N^{C} in (10.1.2) and (10.1.4) is then an immediate consequence of elementary properties of the Fourier transform.

In order to achieve electrical neutrality, an ionic fluid must contain at least two oppositely charged species. The simplest representation of such a system is obtained by replacing one of the species by a uniformly smeared-out, structureless background; the total charge of the background must be equal in magnitude but opposite in sign to that of the discrete ions. In the case when the discrete ions are identical point charges, the resulting model (already discussed in Section 5.3) is called the *one-component plasma* or OCP (Baus and Hansen, 1980). The total potential energy of an OCP in which the ions carry a charge ze is given by the sum on $\mathbf{k}$ in (10.1.4), with $\rho_{\mathbf{k}}^{Z} = z\rho_{\mathbf{k}}$, except that the presence of a neutralizing background means that the term for $\mathbf{k}=0$ must be omitted. The OCP has certain non-physical features. For example, mass and charge fluctuations are proportional to one another, and the system has zero resistivity because conservation of total momentum is equivalent to conservation of the microscopic electric current. Nevertheless, as the prototype ionic fluid, the OCP plays a conceptual role similar to that filled by the hard-sphere model in the theory of simple insulating liquids. In particular, it provides a useful starting point for the study of liquid metals, where the mobile species corresponds to the metal ions and the background represents the conduction electrons.

To illustrate the usefulness of the OCP in qualitative discussion of the properties of ionic liquids, we return briefly to the question of the high-frequency, charge-fluctuation modes mentioned earlier in this section. The characteristic frequency of the longitudinal mode is the plasma frequency ω_{p}. In the case of the OCP, an expression for ω_{p} can be obtained by a simple argument based on a delta-function representation of the dynamic structure factor. We assume that $S(k, \omega)$ consists of a pair of delta-functions located at frequencies $\pm\omega_k$; then the plasma frequency can be identified as $\omega_{\mathrm{p}} = \lim_{k\to 0} \omega_k$. If the spectrum is to satisfy the sum rules (7.4.32) and (7.4.40), ω_k must be such that

$$\omega_k^2 = \frac{\omega_0^2}{S(k)} = \frac{k^2 k_{\mathrm{B}} T}{mS(k)} \tag{10.1.6}$$

The long-wavelength limit of $S(k)$ can be estimated within the random-phase approximation of Section 6.5. If we choose the ideal gas as reference system and make the substitution $\hat{c}(k) = -\beta z^2 \hat{v}(k)$, Eqn (6.5.34) becomes

$$\begin{aligned} S(k) &= \frac{1}{1+\beta\rho z^2 \hat{v}(k)} \\ &= \frac{1}{1+4\pi\beta\rho z^2 e^2/k^2} \underset{k\to 0}{\sim} \frac{k^2}{k_{\mathrm{D}}^2} \end{aligned} \tag{10.1.7}$$

where k_D is the Debye wavenumber defined by Eqn (5.3.20); as we shall see later, Eqn (10.1.7) is exact for the OCP. If we now substitute for $S(k)$ in Eqn (10.1.6), we find that

$$\omega_p = \lim_{k\to 0} \omega_k = \left(\frac{4\pi\rho z^2 e^2}{m}\right)^{1/2} \tag{10.1.8}$$

The frequency of the propagating mode is therefore non-zero even in the limit $k\to 0$; this is a characteristic feature of an optic-type excitation. The fact that ω_p is non-zero is a direct consequence of the k^{-2} singularity in $\hat{v}(k)$, because it is this singularity that determines the small-k behaviour of $S(k)$. Note also that the plasma frequency is independent of temperature.

In a genuine two-component ionic fluid, consisting of discrete cations and anions, a short-range repulsion is essential if the system is to be stable against the collapse of oppositely-charged pairs. This is most easily achieved within a model by imposing a hard-sphere repulsion between ions. The short-range contribution to the pair potential is then

$$\begin{aligned} v^S_{\nu\mu}(r) &= \infty, \qquad r < d_{\nu\mu} = \tfrac{1}{2}(d_{\nu\nu} + d_{\mu\mu}) \\ &= 0, \qquad r > d_{\nu\mu} \end{aligned} \tag{10.1.9}$$

where $d_{\nu\nu}$ is the diameter of ions of species ν. The choice of interaction represented by (10.1.9) corresponds to the *primitive model* of electrolytes and molten salts. The primitive model has been widely used in the study of osmotic properties of ionic solutions (Outhwaite, 1975), the solvent being replaced by a continuum of dielectric constant ε that acts to reduce the Coulomb interaction between ions; the "restricted" version of the model is one in which all ions have the same diameter d and the same absolute valence $|z|$. Improvements to the primitive model of ionic solutions require explicit account to be taken of the molecular nature of the solvent (Patey and Carnie, 1983; Hansen, 1984).

The restricted primitive model with $\varepsilon = 1$ provides the simplest example of a rigid-ion model of a molten salt (Abramo *et al.*, 1978). Alternatively, the short-range interactions in the salt can be modelled by soft-core repulsions characterized by a single length parameter σ. The pair potentials may then be written in the form

$$v_{\nu\mu}(r) = \frac{e^2}{\sigma}\left[\frac{1}{n}\left(\frac{\sigma}{r}\right)^n + z_\nu z_\mu\left(\frac{\sigma}{r}\right)\right] \tag{10.1.10}$$

where $|z_\nu| = |z_\mu| = z$. Equation (10.1.10) defines what we shall call the "simple molten salt". It provides a fair representation of the ionic interactions in the molten alkali halides, particularly of those systems in which the positive and negative ions are of approximately equal size. The values of n appropriate to the alkali halides are in the range $n \simeq 8$ to 10; in the limit $n \to \infty$, the simple molten salt reduces to the restricted primitive model. If the two ionic species have equal masses, the hamiltonian of the simple molten salt is fully symmetric under charge conjugation, i.e. cations and anions play identical roles.

A number of more realistic models of molten salts have also been devised. The best known of these are the potentials derived for salts of the alkali halide family by Tosi and Fumi (1964). In the Tosi–Fumi potentials, the short-range interaction between ions is written as the sum of a Born–Mayer repulsion and attractive van der Waals contributions, i.e.

$$v_{\nu\mu}^{S}(r) = A_{\nu\mu} \exp(-r/\lambda) - \frac{C_{\nu\mu}}{r^6} - \frac{D_{\nu\mu}}{r^8} \qquad (10.1.11)$$

where the parameters λ (assumed to have a common value for a given salt), $A_{\nu\mu}$, $C_{\nu\mu}$ and $D_{\nu\mu}$ are obtained from experimental crystal data. Finally, if the ions are highly polarizable, as is often the case for negative ions, induction forces can no longer be ignored. Ionic polarization has been incorporated into molecular dynamics simulations of molten salts by use of the "shell model" of lattice dynamics (Jacucci *et al.*, 1976; Sangster and Dixon, 1976). In the shell model, the total charge of a polarizable ion is divided between a core and a massless shell, the latter being bound to the core by a harmonic potential. Polarization of the ions corresponds to a bodily shift of the shell (representing the outer electron cloud) relative to the core (representing the nucleus and inner electrons); the shells, being of zero mass, are assumed to adjust themselves instantaneously in such a way as to minimize the total potential energy of the system.

10.2 STATIC STRUCTURE: SCREENING AND CHARGE ORDERING

The microscopic structure of an n-component ionic fluid can be discussed in terms of $\frac{1}{2}n(n+1)$ partial structure factors $S_{\nu\mu}(k)$, with $\nu, \mu = 1$ to n, but it is certain linear combinations of these functions that are of most physical interest. If

$$\rho_{\mathbf{k}}^{N} = \sum_{\nu=1}^{n} \rho_{\mathbf{k}}^{\nu} \qquad (10.2.1)$$

is a Fourier component of the local number density, and if the components of the charge density are defined as in Eqn (10.1.3), fluctuations in the densities are described by three static structure factors of the form

$$S_{NN}(k) = \frac{1}{N}\langle \rho_{\mathbf{k}}^{N} \rho_{-\mathbf{k}}^{N} \rangle = \sum_{\nu}\sum_{\mu} S_{\nu\mu}(k) \qquad (10.2.2a)$$

$$S_{NZ}(k)=\frac{1}{N}\langle\rho_{\mathbf{k}}^{N}\rho_{-\mathbf{k}}^{Z}\rangle=\sum_{\nu}\sum_{\mu}z_{\mu}S_{\nu\mu}(k) \tag{10.2.2b}$$

$$S_{ZZ}(k)=\frac{1}{N}\langle\rho_{\mathbf{k}}^{Z}\rho_{-\mathbf{k}}^{Z}\rangle=\sum_{\nu}\sum_{\mu}z_{\nu}z_{\mu}S_{\nu\mu}(k) \tag{10.2.2c}$$

Of these three functions, the number–number structure factor $S_{NN}(k)$ is closest in significance to the single structure factor of a one-component fluid.

Let $\delta\phi_{\mu}(\mathbf{r})$ be a weak external potential that couples to the number density of species μ. A straightforward generalization of the argument of Section 5.2 shows that the change induced in Fourier components of the single-particle density of species ν is

$$\delta\hat{\rho}_{\nu}^{(1)}(\mathbf{k})=\chi_{\nu\mu}(k)\delta\hat{\phi}_{\mu}(\mathbf{k}) \tag{10.2.3}$$

where $\chi_{\nu\mu}(k)$ is a static density response function. It follows from the fluctuation-dissipation theorem that $\chi_{\nu\mu}(k)$ is related to the corresponding partial structure factor by a generalization of Eqn (5.2.6):

$$\chi_{\nu\mu}(k)=-\beta\rho S_{\nu\mu}(k) \tag{10.2.4}$$

The problem of greatest interest here concerns the response of the fluid to a weak field produced by an external charge density with Fourier components $e\delta\hat{\rho}_{\text{ext}}(\mathbf{k})$; to simplify the discussion, we consider a system of rigid ions *in vacuo* ($\varepsilon=1$). The electric potential due to the external charge density is obtained from the **k**-space version of Poisson's equation, i.e.

$$k^{2}\delta\hat{\phi}_{\text{ext}}(\mathbf{k})=4\pi e\delta\hat{\rho}_{\text{ext}}(\mathbf{k}) \tag{10.2.5}$$

The electric potential couples directly to the microscopic charge density of the fluid, giving rise to a mean induced charge density $\delta\hat{\rho}_{Z}(\mathbf{k})$. The latter is proportional to $e\delta\hat{\phi}_{\text{ext}}(\mathbf{k})$, the constant of proportionality being the charge response function $\chi_{ZZ}(k)$. Thus

$$\delta\hat{\rho}_{Z}(\mathbf{k})=\sum_{\nu}z_{\nu}\delta\hat{\rho}_{\nu}^{(1)}(\mathbf{k})=\chi_{ZZ}(k)e\delta\hat{\phi}_{\text{ext}}(\mathbf{k}) \tag{10.2.6}$$

If we put $\delta\hat{\phi}_{\mu}(\mathbf{k})=z_{\mu}e\delta\hat{\phi}_{\text{ext}}(\mathbf{k})$ in Eqn (10.2.3) and then substitute for $\delta\hat{\rho}_{\nu}^{(1)}(\mathbf{k})$ in (10.2.6), we find that the charge response function can be identified as

$$\chi_{ZZ}(k)=\sum_{\nu}\sum_{\mu}z_{\nu}z_{\mu}\chi_{\nu\mu}(k) \tag{10.2.7}$$

and combination of Eqns (10.2.2c), (10.2.4) and (10.2.7) leads to the charge-response version of the fluctuation-dissipation theorem:

$$\chi_{ZZ}(k)=-\beta\rho S_{ZZ}(k) \tag{10.2.8}$$

The electrostrictive behaviour of the fluid, i.e. the number-density response to an external field that couples to the charge density, is characterized by a cross-response function $\chi_{NZ}(k)$ through an expression analogous to (10.2.6):

$$\delta\hat{\rho}_N(\mathbf{k}) = \sum_\nu \delta\hat{\rho}_\nu^{(1)}(\mathbf{k}) = \chi_{NZ}(k) e \delta\hat{\phi}_{\text{ext}}(\mathbf{k}) \tag{10.2.9}$$

The charge response to the external potential can equally well be described in terms of the static, longitudinal dielectric function $\varepsilon(k)$; the latter is simply a k-dependent generalization of the macroscopic dielectric constant familiar from elementary electrostatics. If **E** is the electric field and **D** is the electric displacement, $\varepsilon(k)$ is given by

$$\frac{1}{\varepsilon(k)} = \frac{\mathbf{k} \cdot \hat{\mathbf{E}}(k)}{\mathbf{k} \cdot \hat{\mathbf{D}}(k)} = 1 + \frac{\delta\hat{\rho}_Z(\mathbf{k})}{\delta\hat{\rho}_{\text{ext}}(\mathbf{k})} \tag{10.2.10}$$

where Maxwell's equations have been used to relate **E** and **D**, respectively, to the total and external charge densities. Equations (10.2.5), (10.2.6) and (10.2.10) can now be combined to yield the fundamental relation between the dielectric and charge-response functions:

$$\frac{1}{\varepsilon(k)} = 1 + \frac{4\pi e^2}{k^2} \chi_{ZZ}(k) \tag{10.2.11}$$

The definition (10.2.2c) shows that $S_{ZZ}(k)$ can never be negative. Equations (10.2.8) and (10.2.11) therefore imply that

$$\frac{1}{\varepsilon(k)} \leqslant 1 \tag{10.2.12}$$

for all k.

It is known experimentally that an external charge distribution is completely screened by a conducting fluid. In other words, the total charge density vanishes in the long-wavelength limit, or

$$\lim_{\mathbf{k}\to 0} [\delta\hat{\rho}_{\text{ext}}(\mathbf{k}) + \delta\hat{\rho}_Z(\mathbf{k})] = 0 \tag{10.2.13}$$

If this result is to be consistent with (10.2.10), it follows that

$$\lim_{k\to 0} \varepsilon(k) = \infty \tag{10.2.14}$$

In combination with equations (10.2.8) and (10.2.11), the assumption of *perfect screening* contained in (10.2.14) implies that the charge structure factor at long wavelength behaves as

$$\lim_{k\to 0} \frac{k_{\text{D}}^2}{k^2} S_{ZZ}(k) = \sum_\nu x_\nu z_\nu^2 \tag{10.2.15}$$

where $x_\nu = \rho_\nu/\rho$ and k_D, the Debye wavenumber, is defined by the multicomponent generalization of Equation (5.3.20):

$$k_D^2 = \sum_\nu k_{D\nu}^2 = \sum_\nu 4\pi\beta\rho_\nu z_\nu^2 e^2 \tag{10.2.16}$$

The quantity $\Lambda_D = 1/k_D$ is the Debye screening length, familiar from ionic-solution theory; in a dilute electrolyte, it is the distance beyond which the electric potential due to an ion is completely screened by the local, induced charge distribution. If Eqn (10.2.15) is compared with the compressibility equation (5.1.13), we see that large-scale (long-wavelength) charge fluctuations are strongly inhibited in comparison with the number-density fluctuations of a neutral fluid. It has been proved rigorously (Martin and Yalcin, 1980) that the fluctuation in the total charge Q_V contained in a volume V, i.e. $(\langle Q_V^2\rangle - \langle Q_V\rangle^2)$, is proportional only to the surface area bounding the volume; by contrast, Eqn (2.4.23) shows that the fluctuation in the number of particles within the volume is proportional to V itself.

Equation (10.2.15) leads directly to two important sum rules for the partial pair distribution functions (Stillinger and Lovett, 1968b). We see from Eqns (5.1.37) and (10.2.2c) that the charge structure factor is related to the partial pair correlation functions $h_{\nu\mu}(r)$ by

$$S_{ZZ}(k) = \sum_n \sum_\mu z_\nu z_\mu \left[x_\nu \delta_{\nu\mu} + 4\pi\rho x_\nu x_\mu \int_0^\infty \frac{\sin kr}{kr} h_{\nu\mu}(r) r^2 \, dr \right] \tag{10.2.17}$$

If the functions $h_{\nu\mu}(r)$ decay sufficiently fast at large r, the Fourier integrals in (10.2.17) may be expanded to order k^2. The sum rules can then be obtained by equating the coefficients of the terms of zeroth and second order in k in (10.2.15) and (10.2.17) and exploiting the condition of overall charge neutrality expressed by Eqn (10.1.1). The two results obtained in this way are

$$\sum_\nu x_\nu z_\nu \rho \int \sum_\mu x_\mu z_\mu g_{\nu\mu}(r) \, d\mathbf{r} = -\sum_\nu x_\nu z_\nu^2 \tag{10.2.18a}$$

and

$$\sum_\nu x_\nu z_\nu \rho \int \sum_\mu x_\mu z_\mu g_{\nu\mu}(r) r^2 \, d\mathbf{r} = -6\Lambda_D^2 \sum_\nu x_\nu z_\nu^2 \tag{10.2.18b}$$

The assumption concerning the large-r behaviour of the correlation functions is equivalent to a "clustering" hypothesis for the particle densities. An n-particle density $\rho^{(n)}(\mathbf{r}^n)$ is said to have a clustering property if, for all $m < n$, it reduces to the product $\rho^{(m)}(\mathbf{r}^m)\rho^{(n-m)}(\mathbf{r}^{(n-m)})$ faster than a prescribed inverse power of the distance between the centres of mass of the clusters $(\mathbf{r}_1, \ldots, \mathbf{r}_m)$ and $(\mathbf{r}_{m+1}, \ldots, \mathbf{r}_n)$ as the clusters become infinitely separated. If the clustering hypothesis is used, the sum rules (10.2.18) can be derived from the Yvon–Born–Green hierarchy of Section 2.7 without making any assumption about the small-k behaviour of $S_{ZZ}(k)$. The derivation is therefore not dependent on the perfect-screening condition (10.2.14); perfect screening appears instead as a consequence of the sum rules (Martin and Gruber, 1983). The sum rules can be extended to the inhomogeneous case including, for example, the electrical double layer near an electrode (Carnie and Chan, 1981).

The sum rule (10.2.18a) is simply a linear combination of local electroneutrality conditions of the form

$$\rho \int \sum_{\mu} x_{\mu} z_{\mu} g_{\nu\mu}(r)\, d\mathbf{r} = -z_{\nu} \tag{10.2.19}$$

The physical meaning of (10.2.19) is that the total charge around a given ion must exactly cancel the charge of the ion. This is the first of a series of sum rules satisfied by the multipole moments of the charge distribution in the vicinity of a given number of fixed ions (Blum *et al.*, 1982). The sum rules can again be derived from the Yvon–Born–Green hierarchy if appropriate clustering assumptions are made. In particular, if correlations are assumed to decay exponentially, it can be shown that the charge distribution around any number of fixed ions has no net multiple moment of any order. The local electroneutrality condition may be re-expressed in terms of the long-wavelength limits of the partial structure factors. In the case of a two-component fluid, Eqn (10.2.19) becomes

$$z_1^2 S_{11}(0) = -z_1 z_2 S_{12}(0) = z_2^2 S_{22}(0) \tag{10.2.20a}$$

or

$$x_2^2 S_{11}(0) = x_1 x_2 S_{12}(0) = x_1^2 S_{22}(0) \tag{10.2.20b}$$

The equivalence of (10.2.20a) and (10.2.20b) is a consequence of overall charge neutrality.

The $k \to 0$ limits of the partial structure factors of a binary ionic fluid are related to the isothermal compressibility via the Kirkwood–Buff formula (5.2.20). The special conditions imposed by charge neutrality mean, however, that direct substitution of (10.2.20) in (5.2.20) leads to an indeterminate result. To avoid this problem, we use the definitions (10.2.2) to rewrite Eqn (5.2.20) in the form

$$\rho k_B T \chi_T = \frac{-z_1 z_2}{x_1 x_2 (z_1 - z_2)^2} \lim_{k \to 0} \frac{S_{NN}(k) S_{ZZ}(k) - S_{NZ}^2(k)}{S_{ZZ}(k)} \tag{10.2.21}$$

The small-k limits of the three structure factors in (10.2.21) can be deduced from the asymptotic behaviour of the partial direct correlation functions

$c_{\nu\mu}(r)$. On the basis of the general arguments of Section 5.2, we may expect the direct correlation functions to behave as

$$c_{\nu\mu}(r) \underset{r\to\infty}{\sim} -\beta v_{\nu\mu}(r) = -\frac{\beta z_\nu z_\mu e^2}{r} \tag{10.2.22}$$

It is therefore natural to separate $c_{\nu\mu}(r)$ into short-range and coulombic parts; in **k**-space, $\hat{c}_{\nu\mu}(k)$ becomes

$$\hat{c}_{\nu\mu}(k) = \hat{c}^{\mathrm{S}}_{\nu\mu}(k) - \frac{4\pi\beta z_\nu z_\mu e^2}{k^2} \tag{10.2.23}$$

where $\hat{c}^{\mathrm{S}}_{\nu\mu}(k)$ is a regular function in the limit $k\to 0$. Substitution of (10.2.23) in the Ornstein–Zernike relation (5.2.18) leads, after some straightforward algebra and use of Eqns (10.1.1) and (10.2.2), to the required results: at small k, $S_{NN}(k)\sim k^0$, $S_{NZ}(k)\sim k^2$ and $S_{ZZ}(k)\sim k^2$; the last result agrees with (10.2.15). The compressibility equation for a two-component ionic fluid is then obtained by combining Eqns (10.2.20) and (10.2.21) in the form

$$\frac{-z_1 z_2}{x_1 x_2 (z_1 - z_2)^2} \lim_{k\to 0} S_{NN}(k) = \lim_{k\to 0} (x_\nu x_\mu)^{-1} S_{\nu\mu}(k) = \chi_{\mathrm{T}}/\chi^0_{\mathrm{T}} \tag{10.2.24}$$

while Eqn (5.2.19) becomes

$$\lim_{k\to 0} \sum_\nu \sum_\mu x_\nu x_\mu \rho \hat{c}^{\mathrm{S}}_{\nu\mu}(k) = 1 - \chi^0_{\mathrm{T}}/\chi_{\mathrm{T}} \tag{10.2.25}$$

The coefficients of the terms of order k^4 in the small-k expansion of $S_{ZZ}(k)$ and those of order k^2 in the expansions of $S_{NZ}(k)$ and $S_{NN}(k)$ can be determined by macroscopic arguments based on linearized hydrodynamics (Giaquinta *et al.*, 1976) or on thermodynamic fluctuation theory (Vieillefosse, 1977). The resulting expressions have been used in the analysis of data obtained from small-angle neutron-scattering experiments on molten salts (Rovere *et al.*, 1979). We give here the equivalent calculation for the OCP, where the problem is simplified by the fact that fluctuations in number are equivalent to fluctuations in charge. In the absence of any flow, the force per unit volume due to the electric field must exactly balance the force due to the pressure gradient. Thus

$$\rho z e \mathbf{E}(\mathbf{r}) = \nabla P(\mathbf{r}) \tag{10.2.26}$$

where ρze is the mean charge density of the mobile point ions; the field $\mathbf{E}(\mathbf{r})$ is related to the sum of external and induced charge densities by Poisson's equation:

$$\nabla \cdot \mathbf{E}(\mathbf{r}) = 4\pi e[\delta\rho_{\mathrm{ext}}(\mathbf{r}) + \delta\rho_Z(\mathbf{r})] \tag{10.2.27}$$

If we assume that the system is in local thermodynamic equilibrium, the pressure change in an isothermal process can be written as

$$\delta P(\mathbf{r}) = P(\mathbf{r}) - P = \left(\frac{\partial P}{\partial \rho}\right)_T \delta\rho(\mathbf{r}) = \frac{1}{z}\left(\frac{\partial P}{\partial \rho}\right)_T \delta\rho_Z(\mathbf{r}) \qquad (10.2.28)$$

Equations (10.2.26) to (10.2.28) may now be combined to give a differential equation for $\delta\rho_Z(\mathbf{r})$ of the form

$$\frac{1}{4\pi z^2 e^2}\left(\frac{\partial P}{\partial \rho}\right)_T \nabla^2 \delta\rho_Z(\mathbf{r}) - \delta\rho_Z(\mathbf{r}) = \delta\rho_{\text{ext}}(\mathbf{r}) \qquad (10.2.29)$$

This equation is easily solved by taking Fourier transforms. The result for $\delta\hat{\rho}_Z(\mathbf{k})$ is

$$\delta\hat{\rho}_Z(\mathbf{k}) = -\frac{\delta\hat{\rho}_{\text{ext}}(\mathbf{k})}{1 + k^2/k_S^2} \qquad (10.2.30)$$

with

$$k_S^2 = 4\pi z^2 e^2 \rho^2 \chi_T \qquad (10.2.31)$$

Comparison of (10.2.30) with (10.2.10) shows that the long-wavelength limit of $\varepsilon(k)$ is

$$\varepsilon(k) = 1 + k_S^2/k^2 \qquad (10.2.32)$$

which clearly satisfies the perfect-screening condition (10.2.14). The corresponding long-wavelength expression for $S_{ZZ}(k)$ $(= z^2 S(k))$, obtained from Eqns (10.2.8) and (10.2.11), is

$$S_{ZZ}(k) = \frac{z^2 k^2/k_D^2}{1 + k^2/k_S^2} \qquad (10.2.33)$$

in agreement with (10.1.7). Equations (10.2.32) and (10.2.33) are also valid for systems of oppositely charged ions with $z_1 = -z_2 = z$, except that k_S is defined as

$$k_S^2 = 4\pi z^2 e^2 \left(\frac{\partial \mu_Z}{\partial \rho_Z}\right)_{P,T}^{-1} \qquad (10.2.34)$$

where $\mu_Z = (m_2\mu_1 - m_1\mu_2)/e(z_1 m_2 - z_2 m_1)$ is the electrochemical potential and m_ν, μ_ν are, respectively, the mass and chemical potential of ions of species ν (Giaquinta *et al.*, 1976; Parrinello and Tosi, 1979).

The Fourier components of the total electrostatic potential $\delta\phi(\mathbf{r})$ are related to the components of the total charge density by the analogue of

Eqn (10.2.5). In the long-wavelength limit, it follows from Eqns (10.2.10) and (10.2.32) that

$$\delta\hat{\phi}(\mathbf{k}) = \frac{4\pi e}{k^2}[\delta\hat{\rho}_{\mathrm{ext}}(\mathbf{k}) + \delta\hat{\rho}_Z(\mathbf{k})]$$

$$= \frac{4\pi e}{k^2\varepsilon(k)}\delta\hat{\rho}_{\mathrm{ext}}(\mathbf{k})$$

$$= \frac{4\pi e}{k^2 + k_{\mathrm{S}}^2}\delta\hat{\rho}_{\mathrm{ext}}(\mathbf{k}) \tag{10.2.35}$$

If an ion of species ν in the fluid is regarded as an "external" charge placed at the origin, the "external" charge density is $e\delta\rho_{\mathrm{ext}}(\mathbf{r}) = z_\nu e\delta(\mathbf{r})$, and Eqn (10.2.35) shows that the effective potential due to the ion decays as

$$\phi_\nu(r) = \frac{z_\nu e}{r}\exp(-k_{\mathrm{S}}r) \tag{10.2.36}$$

The quantity $\phi_\nu(r)$ $(=\delta\phi(r))$ is the potential of mean force for ions of species ν. In the case of the OCP, k_{S} is given by Eqn (10.2.31); this reduces to the Debye wavenumber (5.3.20) in the weak-coupling limit ($\rho\to 0$ or $T\to\infty$), where χ_{T} may be replaced by its ideal-gas value, $\chi_T^0 = \beta\rho$. With these simplifications, Eqn (10.2.36) reduces to the Debye–Hückel result (5.3.22). In the strong-coupling regime, the compressibility of the OCP becomes negative (Hansen, 1973), k_{S} takes on imaginary values, and the potential of mean force develops the oscillations that are characteristic of systems with short-range order (Vieillefosse and Hansen, 1975).

Oscillations of the charge density around a given ion are also a feature of two-component ionic fluids. Such oscillations are a result of competition between hard-core packing and local charge neutrality. In the restricted primitive model, a simple argument based on the sum rules (10.2.18) shows that the radial charge distribution function $[g_{++}(r) - g_{+-}(r)]$ must change sign as a function of r if $k_{\mathrm{D}}d \geqslant \sqrt{6}$ (Stillinger and Lovett, 1968a). Charge ordering of this type is a very strong effect in molten salts, and oscillations in the charge density around a central ion extend over many ionic radii. The effect is related to the behaviour of the "like" (l) and "unlike" (u) pair distribution functions defined, for monovalent salts, as

$$g_{\mathrm{l}}(r) = \tfrac{1}{2}[g_{++}(r) + g_{--}(r)]$$

$$g_{\mathrm{u}}(r) = g_{+-}(r) \tag{10.2.37}$$

Computer simulations show that oscillations in $g_{\mathrm{l}}(r)$ are almost exactly out of phase with those in $g_{\mathrm{u}}(r)$ (Woodcock and Singer, 1971). In the alkali halides, the functions $g_{++}(r)$ and $g_{--}(r)$ are very similar in form. Thus the radial charge distribution functions around either a cation (i.e. $[g_{++}(r) - g_{+-}(r)]$) or an anion (i.e. $[g_{--}(r) - g_{+-}(r)]$) are essentially the same and strongly oscillatory. The similarity between $g_{++}(r)$ and $g_{--}(r)$ gives some support to the use of the "simple molten salt" (Eqn (10.1.10)) as a model for the alkali halides. Some molecular dynamics results for such a model (with $n = 9$) are shown in Figures 10.1 and 10.2. The regular alternation of concentric shells of oppositely charged ions is clearly visible in the curves of $g_{+-}(r)$ and $g_{++}(r)$ $(= g_{--}(r))$ plotted in Figure 10.1. In **k**-space, the effects of charge ordering reflect themselves in the very sharp main peak in the charge structure factor $S_{ZZ}(k)$ (the Fourier transform of $[g_l(r) - g_u(r)]$), as shown in Figure 10.2. By contrast, $S_{NN}(k)$ (the transform of $[g_l(r) + g_u(r)]$) is a relatively structureless function. Because of the symmetry of the model, charge and number fluctuations are completely decoupled, and $S_{NZ}(k) = 0$ at all k; in the general case, the fluctuations are strictly independent only in the long-wavelength limit, since $S_{NZ}(k) \sim k^2$ as $k \to 0$.

The main structural features exhibited by the simple molten salt are also seen in simulations of more realistic rigid-ion models of the alkali halides (Sangster and Dixon, 1976). The effects of ion polarization on the pair

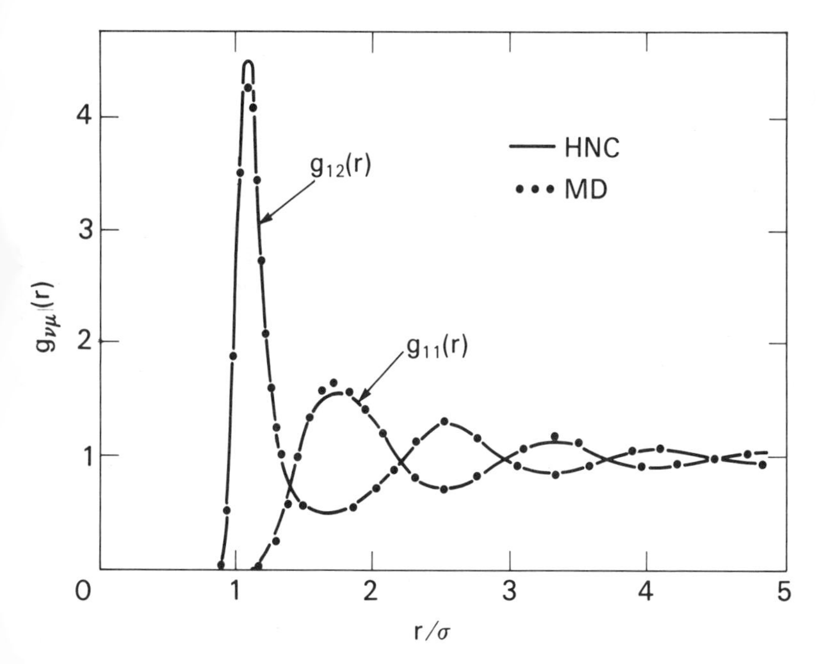

FIG. 10.1. Partial radial distribution functions of the "simple molten salt" defined by Eqn (10.1.10), calculated by molecular dynamics and from the HNC approximation. After Hansen and McDonald (1975).

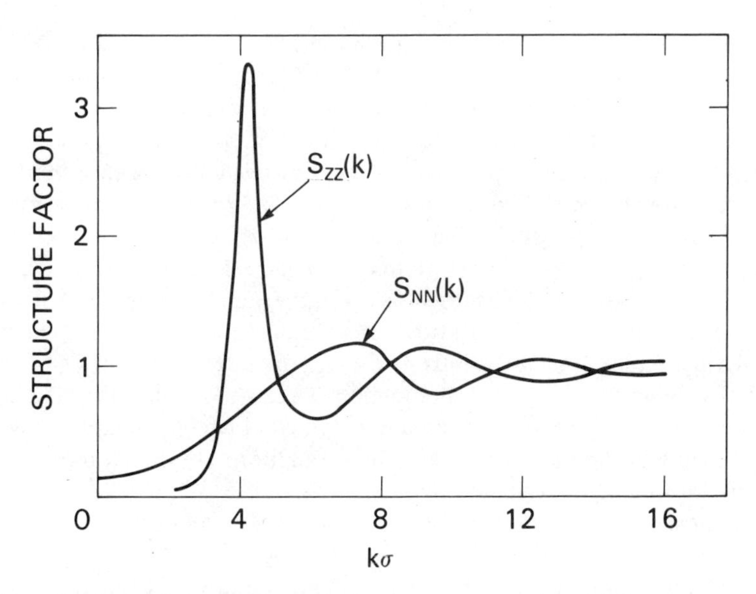

FIG. 10.2. Molecular dynamics results for the number-number and charge-charge structure factors of the "simple molten salt" defined by Eqn (10.1.10) (Hansen and McDonald, 1975).

structure can be significant, however, especially when the size difference between cations and anions is large (Dixon and Gillan, 1981). Inclusion of ion polarization generally leads to a flattening of the pair distribution functions and an enhancement of the difference between $g_{++}(r)$ and $g_{--}(r)$; this behaviour is consistent with results obtained by neutron diffraction. Neutron-scattering experiments rely on the use of isotopic substitution to separate the contributions of the partial structure factors $S_{\nu\mu}(k)$ to the total measured intensity (Enderby and Neilson, 1980); some experimental results for molten sodium chloride are shown in Figure 10.3. Similar experiments have been carried out for the alkaline-earth halides (Biggin and Enderby, 1981) and the results are in good general agreement with those obtained by computer simulations based on rigid-ion models (deLeeuw, 1978). Because the absolute charges of the cations and anions are now different, the marked similarity in $g_{++}(r)$ and $g_{--}(r)$ seen in the alkali halides is lost.

10.3 THEORIES OF IONIC PAIR STRUCTURE

The techniques described in Chapters 5 and 6 provide a number of possible routes to the calculation of thermodynamic and structural properties of simple ionic liquids. Most of the published theoretical work is concerned with the alkali halides, either in the molten phase or in solution, and these

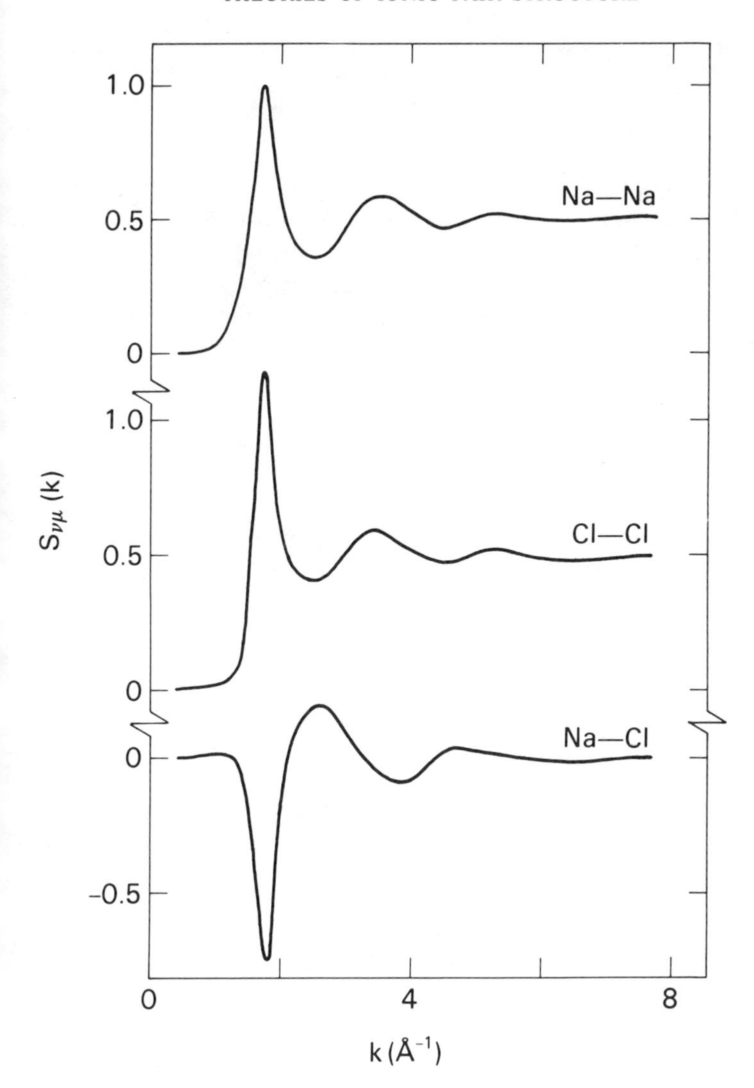

FIG. 10.3. Results from neutron-scattering experiments for the partial structure factors of molten NaCl. After Biggin and Enderby (1982).

are the only systems that we shall discuss here. The physical conditions are, of course, very different in the molten-salt and electrolyte regimes. If we adopt the primitive model of Section 10.1, the thermodynamic state is conveniently characterized by the reduced density $\rho^* = Nd^3/V$, where N is the total number of ions and $d = \frac{1}{2}(d_{11} + d_{22})$ is the mean ionic diameter, and a reduced Coulomb coupling parameter, or inverse temperature, defined as

$$\beta^* = \frac{e^2}{\varepsilon k_B T d} \tag{10.3.1}$$

Near the melting point of an alkali halide, $\rho^* \simeq 0.6$ and $\beta^* \simeq 50$, while for a 1M aqueous solution of the same salt at room temperature, $\rho^* \simeq 0.02$ and $\beta^* \simeq 2$. We must therefore expect the nature of the interionic correlations to be very different in the two cases. The liquid–vapour phase diagram of a molten alkali halide is qualitatively similar to that, say, of a rare gas, but the reduced critical densities of the salts are only about one-third of those of typical insulating liquids.

In order to illustrate the value of different theoretical approaches, we focus attention primarily on systems of charged hard spheres and, in particular, on the restricted primitive model, for which $d_{11} = d_{22} = d$ and $z_1 = -z_2 = 1$. A convenient starting point for the discussion is the mean spherical approximation (MSA) introduced in Section 5.6, since in this case the MSA has a completely analytic solution. The MSA for equisized hard spheres is

$$\begin{aligned} g_{\nu\mu}(r) &= 0, \qquad r < d \\ c_{\nu\mu}(r) &= -\frac{\beta z_\nu z_\mu e^2}{\varepsilon r}, \qquad r > d \end{aligned} \tag{10.3.2}$$

which must be used in conjunction with the Ornstein–Zernike relation for equimolar binary mixtures; this is obtained as a special case of (5.2.18), with $x_1 = x_2 = \frac{1}{2}$. The symmetry of the restricted primitive model allows the Ornstein–Zernike relation to be rewritten as two independent equations for the linear combinations

$$\begin{aligned} h_S(r) &= \tfrac{1}{2}[h_{11}(r) + h_{22}(r)] \\ h_D(r) &= h_{11}(r) - h_{12}(r) \end{aligned} \tag{10.3.3}$$

and the corresponding direct correlation functions $c_S(r)$ and $c_D(r)$; $h_S(r)$ is a number density correlation function and $h_D(r)$ describes the correlation in charge density. Written in terms of the new functions, the MSA becomes

$$h_S(r) = -1, \qquad r < d; \qquad c_S(r) = 0, \qquad r > d \tag{10.3.4}$$

$$h_D(r) = 0, \qquad r < d; \qquad c_D(r) = -\frac{2\beta e^2}{\varepsilon r}, \qquad r > d \tag{10.3.5}$$

The closure relation (10.3.4) is just the PY approximation for hard spheres, for which the solution is known. The solution to Eqn (10.3.5) and the associated Ornstein–Zernike relation between $h_D(r)$ and $c_D(r)$ can also be obtained in closed form by incorporating the sum rules (10.2.18) into generalized versions of the methods used to solve the PY equation for hard spheres (Waisman and Lebowitz, 1972; Blum, 1980a). The result for $c_D(r)$ inside the hard core is

$$c_D(r) = -\frac{2\beta e^2}{\varepsilon d}(2 - Br/d)B \tag{10.3.6}$$

with

$$B = [\xi + 1 - (1+2\xi)^{1/2}]/\xi \tag{10.3.7}$$

where $\xi^2 = k_D^2 d^2 = 4\pi\rho^*\beta^*$ and k_D is the Debye wavenumber defined by Eqn (10.2.16). The excess internal energy has a very simple form:

$$\frac{U^{ex}}{N} = -\frac{e^2}{\varepsilon d} B \tag{10.3.8}$$

and is a function of the single coupling constant ξ and not separately of ρ^* and β^*. Within the MSA, Eqn (10.3.8) is valid for any number of ionic species, with arbitrary charges but equal diameters, provided the global charge-neutrality condition (10.1.1) is satisfied. The virial pressure is given by the sum of a contact term and the contribution of the Coulomb forces as

$$\frac{\beta P^v}{\rho} = 1 + \frac{\beta U^{ex}}{3N} + \frac{2\pi}{3}\rho d^3 g_S(d) \tag{10.3.9}$$

but more reliable results are obtained by integrating (10.3.8) to give the free energy and then differentiating with respect to density; the resulting "energy" equation of state is in good agreement with Monte Carlo calculations in the range of ρ^* and β^* appropriate to concentrated electrolytes (Rasaiah *et al.*, 1972). A comparison of theory and simulation is made in Table 10.1; it shows, not unexpectedly, that the MSA is less satisfactory in the molten-salt regime.

In the high-temperature or low-density (or concentration) limit, i.e. for $\xi \ll 1$, the MSA internal energy reduces to the Debye–Hückel result, namely

$$\frac{U_{DH}^{ex}}{N} = -\frac{e^2}{\varepsilon d}\xi = -\frac{k_B T}{4\pi\rho} k_D^3 \tag{10.3.10}$$

The limiting law (10.3.10) is valid when ion-size effects are negligible; it corresponds to the case when the direct correlation functions $c_{\nu\mu}(r)$ are replaced by their asymptotic form (10.3.2) for all r. In the opposite limit ($\xi \gg 1$), the MSA energy behaves in a manner consistent with an exact lower

TABLE 10.1. *Thermodynamic properties of the restricted primitive model as a function of* $\rho^* = \rho d^3$ *and* $\beta^* = e^2/\epsilon k_B T$; $\xi = k_D d = (4\pi\rho^*\beta^*)^{1/2}$ *is the reduced Debye wavenumber. MC: Monte Carlo results of Card and Valleau* (1970) *and Larsen* (1976)

(*a*) $\beta U^{ex}/N$

ρ^*	β^*	ξ	MC	MSA	HNC	RHNC	TΓ2A
9.591×10^{-3}	1.680	0.45	0.2739 ± 0.0014	0.2673	0.2714		0.2675
3.929×10^{-2}	1.680	0.911	0.4341 ± 0.0017	0.4261	0.4295		0.4282
9.246×10^{-2}	1.680	1.397	0.5516 ± 0.0016	0.5402	0.5447		0.5465
1.819×10^{-1}	1.680	1.960	0.6511 ± 0.0020	0.6358	0.6460		0.6501
1.785×10^{-1}	6.80	3.905	3.522 ± 0.012	3.373	3.396		3.437
0.2861	1.882	2.601	0.839 ± 0.010	0.804	0.832	0.829	0.831
	9.411	5.817	5.465 ± 0.030	5.278	5.431	5.414	5.420
	47.055	13.01	33.62 ± 0.08	31.87			32.52
0.7535	1.363	3.590	0.711 ± 0.026	0.657	0.719	0.713	0.711
	6.817	8.034	4.511 ± 0.035	4.160	4.434	4.392	4.423
	34.086	17.96	25.71 ± 0.02	24.45	25.46	25.26	25.63

(*b*) $\beta P/\rho$

ρ^*	β^*	ξ	MC	MSA(v)	MSA(e)	HNC(v)	RHNC	TΓ2A
9.591×10^{-3}	1.680	0.45	0.9445 ± 0.0012	0.9312	0.9454	0.9453		0.9452
3.929×10^{-2}	1.680	0.911	0.9774 ± 0.0046	0.9446	0.9806	0.9796		0.9786
9.246×10^{-2}	1.680	1.397	1.094 ± 0.005	1.0390	1.0971	1.0926		1.0908
1.819×10^{-1}	1.680	1.960	1.346 ± 0.009	1.2757	1.3581	1.3514		1.3438
1.785×10^{-1}	6.80	3.905		0.352	0.912	0.886		0.848
0.2861	1.882	2.601	1.62 ± 0.15	1.62	1.75	1.76	1.70	1.72
	9.411	5.817	0.79 ± 0.31	1.32	1.13	1.05	0.96	0.99
	47.055	13.01	-0.79 ± 0.37	-8.73	-1.53			-2.18
0.7535	1.363	3.590	6.71 ± 0.75	4.66	6.59	7.88	5.95	6.54
	6.817	8.034	6.55 ± 0.10	4.77	6.17	7.21	5.17	5.90
	34.086	17.96	2.61 ± 0.40	-2.00	4.40	4.51	1.85	3.23

v = from virial equation; e = from energy equation.

bound due to Onsager (1939), i.e.

$$\frac{U^{ex}}{N} \geqslant -\frac{e^2}{\varepsilon d} \tag{10.3.11}$$

According to Gauss' theorem, the total electrostatic energy of a system of charged hard spheres is unaltered if the charge $z_\nu e$ at the centre of each sphere is uniformly distributed over its surface. The electrostatic self-energy $\mathscr{E}^s$ of such a charge distribution in a medium of dielectric constant ε is easily calculated to be $\mathscr{E}^s = e^2/\varepsilon d$. Then, if $\mathbf{E}(\mathbf{r})$ is the total electric field at a point $\mathbf{r}$, it follows from elementary electrostatics that the total potential energy of the system is

$$V_N = \tfrac{1}{2} \sum_{i \neq j}^{N} \sum \frac{z_i z_j e^2}{\varepsilon |\mathbf{r}_i - \mathbf{r}_j|}$$
$$= \tfrac{1}{2} \int |\mathbf{E}(\mathbf{r})|^2 \, d\mathbf{r} - N\mathscr{E}^s \geqslant -\frac{Ne^2}{\varepsilon d} \qquad (10.3.12)$$

Since the inequality holds for any allowed configuration, it also holds for the ensemble average of V_N, i.e. for the excess internal energy U^{ex} in (10.3.11).

We showed in Section 6.5 that the MSA is a special case of the optimized random-phase approximation (ORPA) in which the reference hard-sphere pair distribution function is replaced by its approximate, PY form. The diagrammatic formalism used in the development of the ORPA is easily generalized to a multicomponent system by adopting the convention that each black circle in a diagram implies not only an integration over coordinates but also a summation over all species. In particular, for a two-component system such as the restricted primitive model, the chain bond defined by Eqn (6.5.22) must be calculated from a 2×2 matrix generalization of (6.5.32) (Andersen and Chandler, 1972). Equation (6.5.21) then becomes

$$g_{\nu\mu}(r) = g_0(r) + C_{\nu\mu}(r) \qquad (10.3.13)$$

where the functions $C_{\nu\mu}(r)$ are the elements of the chain-bond matrix and the reference system is assumed to be a one-component fluid of hard spheres. The charge-conjugation symmetry of the restricted primitive model implies that the matrix elements $C_{\nu\mu}(r)$ are all of the form $C_{\nu\mu}(r) = z_\nu z_\mu C(r)$ and hence that

$$g_{11}(r) = g_{22}(r) = g_0(r) + C(r)$$
$$g_{12}(r) = g_0(r) - C(r) \qquad (10.3.14)$$

In the low-density limit, where $S_0(k) = 1$, the chain bond $C(r)$ is obtained from Eqn (6.5.32) as

$$C(r) = -\frac{\beta^* d}{r} \exp(-k_D r) \qquad (10.3.15)$$

If, in addition, the ideal gas is chosen as the reference system, Eqns (10.3.14) reduce to a two-component generalization of the linearized Debye–Hückel result (5.3.24). The Debye–Hückel theory of ionic solutions was originally derived from an entirely different point of view in which the focus of attention is the net electrostatic potential due to the charge distribution around a given ion (Debye and Hückel, 1923); the same approach has been pursued in more recent times for both homogeneous and inhomogeneous (electrical double-layer) systems (Outhwaite, 1975; Blum, 1980a; Torrie and Carnie, 1984).

The thermodynamic results of the OPRA can be improved by inclusion of higher-order terms in the optimized cluster expansion of Section 6.5. The term a_{HTA} in the free-energy expansion (6.5.45) is zero for the restricted primitive model. The ring term is given by the matrix generalization of Eqn (6.5.38); this is minimized with respect to variations in the Coulomb potential inside the hard cores (see Eqn (6.5.40)). The optimized chain-bond matrix can then be used to evaluate the term B_2, given by the multicomponent generalization of (6.5.46), i.e.

$$B_2 = \tfrac{1}{2}\sum_{\nu}\sum_{\mu} \rho_\nu \rho_\mu \left\{ \tfrac{1}{2} \int h_0(r)[C_{\nu\mu}(r)]^2 \,\mathrm{d}\mathbf{r} + \sum_{n=3}^{\infty} \frac{1}{n!} \int g_0(r)[C_{\nu\mu}(r)]^n \,\mathrm{d}\mathbf{r} \right\} \tag{10.3.16}$$

or, when specialized to the restricted primitive model:

$$B_2 = -\tfrac{1}{2}\rho^2 \int \{g_0(r)[1-\cosh C(r)] + \tfrac{1}{2}[C(r)]^2\} \,\mathrm{d}\mathbf{r} \tag{10.3.17}$$

In contrast to the MSA energy (10.3.8), the correction represented by B_2 is dependent separately on ρ^* and β^*. At low densities (or concentrations), $C(r)$ is given by Eqn (10.3.15), $g_0(r)$ may be replaced by the hard-sphere Mayer function, and B_2 then provides the first correction to the Debye-Hückel limiting law (10.3.10) (Friedman, 1962). The low-density limit of B_2 is called the second ionic virial coefficient. Unlike the virial coefficients defined in Section 4.5, the ionic virial coefficients are functions of density through their dependence on the Debye wavenumber k_D. The second, and higher order virial coefficients of Chapter 4 all diverge for Coulomb systems; the density dependence of the ionic coefficients arises from the resummations that are needed in order to obtain convergent results.

An alternative to the optimized cluster expansion approach is to group the corrections to the MSA in ascending powers of the coupling constant β^*; this corresponds to the "Γ-ordering" scheme of Lebowitz *et al.* (1965). The leading correction to the MSA free energy is then given by the first term on the right-hand side of Eqn (10.3.16). In this "Γ2" approximation, the free energy per unit volume $\mathcal{A} = -\beta A^{\mathrm{ex}}/V$ of the restricted primitive model relative to its value $\mathscr{A}_0$ for uncharged hard spheres is given by

$$\Delta\mathscr{A}_{\Gamma 2} = \mathscr{A} - \mathscr{A}_0 = \Delta\mathscr{A}_{\text{MSA}}(\xi) + \tfrac{1}{4}\rho^2 \int h_0(r)[C(r)]^2 \,\mathrm{d}\mathbf{r} \quad (10.3.18)$$

where $\Delta\mathscr{A}_{\text{MSA}}(\xi)$ is obtained by integration of Eqn (10.3.8):

$$\Delta\mathscr{A}_{\text{MSA}}(\xi) = \frac{1}{12\pi d^3}[3\xi^2 + 6\xi + 2 - 2(1+\xi)^{3/2}] \quad (10.3.19)$$

Replacement of $h_0(r)$ in (10.3.18) by its low density limit leads to the "truncated Γ2" approximation ($T\Gamma 2A$), in which the free energy is expressible in closed analytic form (Larsen *et al.*, 1977). As can be seen from Table 10.1, the $T\Gamma 2A$ results for the equation of state are an improvement over the MSA, particularly under conditions appropriate to molten salts (Larsen, 1976).

Although the MSA is an excellent starting point for the calculation of thermodynamic properties of the restricted primitive model, it is much less reliable in predicting the correlation functions, particularly when the coupling is strong. If the density and temperature are such that $\xi \gg 1$, use of the MSA leads to like distribution functions $g_{11}(r)$ (or $g_{22}(r)$) that become negative at separations close to contact. The situation is improved if, at small r, the direct correlation functions $c_S(r)$ and $c_D(r)$ are allowed to deviate from their asymptotic forms. In the "generalized" MSA or GMSA of Høye *et al.* (1974), the deviation from asymptotic behaviour is expressed in terms of Yukawa functions, and the closure relations (10.3.4) and (10.3.5) are replaced by

$$c_S(r) = \frac{A_1}{r}\exp[-t_1(r-d)], \qquad r > d \quad (10.3.20)$$

$$c_D(r) = -\frac{2\beta e^2}{\varepsilon r} - \frac{A_2}{r}\exp[-t_2(r-d)], \qquad r > d \quad (10.3.21)$$

The properties of the Yukawa functions make it possible to formulate the GMSA as a system of non-linear, coupled algebraic equations by the techniques used by Wertheim and Baxter for the solution of the PY equation for hard spheres (Waisman, 1973; Høye *et al.*, 1976; Blum, 1980b). The parameters A_1, t_1, A_2 and t_2 can be related to three thermodynamic properties (the internal energy, the compressibility and the virial pressure) and one non-thermodynamic quantity (the contact value of $g_D(r)$) (Høye and Stell, 1977b). In applications of the GMSA to the restricted primitive model (Larsen *et al.*, 1977), $T\Gamma 2A$ results provide the thermodynamic input and $g_D(d)$ is calculated from the EXP approximation of Eqn (6.5.43). Since all thermodynamic properties are obtained from the same free-energy expression, the theory is thermodynamically self-consistent; the resulting pair distribution functions represent a significant improvement over the MSA.

An analytic solution to the MSA has also been obtained in the general case, i.e. for any number of components and for arbitrary ionic charges and hard-sphere diameters, subject again only to overall charge neutrality (Waisman and Lebowitz, 1972; Blum, 1975; Blum and Høye, 1977). All thermodynamic properties and pair distribution functions are determined by a single parameter, the meaning of which is that of a generalized inverse Debye screening length; it is the physical solution of an algebraic equation of high degree. In the case of the restricted primitive model, the equation is quadratic, and the physical solution, γ say, is related to the parameter ξ of Eqn (10.3.7) by

$$2d\gamma = (1+2\xi)^{1/2} - 1 \tag{10.3.22}$$

Solutions for γ in the asymmetric case must be obtained numerically. The results have been used to calculate the partial structure factors of molten alkali halides; with reasonable choices for the ionic diameters, it is possible to obtain fair agreement with the available neutron and X-ray diffraction data (Abramo *et al.*, 1978). Comparison has also been made between MSA results and Monte Carlo data for binary systems of oppositely charged spheres ($z_1 = -z_2$) with unequal diameters. In this case, the agreement with the simulations is significantly improved if an extension of the GMSA of Eqns (10.3.20) and (10.3.21) is used (Abramo *et al.*, 1983).

The main appeal of theories such as the MSA or GMSA in the calculation of the ionic pair structure is that they can be solved analytically in closed or nearly closed form. Their applicability is limited, however, to systems of charged hard spheres. These "primitive" models display certain structural features that are artefacts of the hard-sphere interaction; in particular, the main peaks in $g_{++}(r)$ and $g_{--}(r)$ show a marked splitting that is absent from the experimental data. The spurious features in the structure disappear when soft-core models of the ionic interactions are used. Many computer simulations have been carried out for rigid-ion potentials of the types represented by Eqns (10.1.10) and (10.1.11), with results that are in generally good agreement with those of neutron scattering experiments on the molten alkali halides. The integral-equation theories of Chapter 5 can also be applied to such potentials. Of the standard theories, the HNC approximation is far better suited to ionic systems than its PY counterpart. As the closure relation (5.4.14) shows, the PY approximation cannot account for the exponential screening of the pair correlations at large separations, since the pair distribution function decays as the pair potential. By contrast, the HNC approximation does describe the long-range correlations correctly; there is also a close connection between HNC theory and the traditional form of the Debye–Hückel approach (Blum, 1980a). The HNC integral equation can be derived by ignoring the bridge diagrams in Eqn (5.3.14); these are expected to be short-range functions even for ionic systems.

When applied to the restricted primitive model, HNC theory, particularly in the "reference" (RHNC) form discusssed in Section 5.8, represents a considerable improvement over the MSA in both electrolyte (Rasaiah *et al.*, 1972) and molten salt (Larsen, 1978) regimes. Some results for thermodynamic properties are shown in Table 10.1. In addition, use of the HNC approximation gives very good results for the pair structure of soft-core systems where the MSA, at least in its conventional form, cannot be applied. In Figure 10.1, the calculated radial distribution functions of the "simple molten salt" defined by Eqn (10.1.10) are compared with results of molecular dynamics simulations (Hansen and McDonald, 1975). The agreement is everywhere excellent, and the deficiencies of the HNC approximation are evident only in the small-k region of $S_{NN}(k)$; the error there means that the compressibility given by the theory is about twice as large as that obtained by simulation. A systematic study of all the alkali chlorides based in pair potentials of the form (10.1.11) has confirmed that the HNC theory is able to reproduce quantitatively all the main features of the pair distribution functions of more realistic rigid-ion models (Abernethy *et al.*, 1981), and still better results can be obtained by including the contributions from the bridge diagrams in a semiempirical way (Ballone *et al.*, 1984).

10.4 FREQUENCY-DEPENDENT ELECTRIC RESPONSE

We have seen in earlier sections of this chapter how the static properties of ionic liquids are affected by the long-range of the Coulomb potential or, equivalently, by the k^{-2} singularity in its Fourier transform. In this section, we consider how the same factors influence the dynamical correlations. We restrict the discussion to two-component systems, and introduce variables that describe fluctuations in mass, density and concentration along the lines of Section 8.6; in the present case, the electroneutrality expressed by Eqn (10.1.1) means that fluctuations in concentration are equivalent to fluctuations in charge density. The charge fluctuations generate a local electric field that acts as a restoring force on the local charge density. The latter responds in a diffusive manner at low frequencies, but at high frequencies there is a reactive behaviour; this gives rise to a propagating mode of the type briefly discussed in Section 10.1.

The linear combinations of microscopic partial densities that occur naturally in a discussion of the collective dynamics are the mass (M) and charge (Z) densities, defined in terms of Fourier components as

$$\rho_{\mathbf{k}}^{M}(t) = m_1\rho_{\mathbf{k}}^{1}(t) + m_2\rho_{\mathbf{k}}^{2}(t) \tag{10.4.1}$$

$$\rho_{\mathbf{k}}^{Z}(t) = z_1\rho_{\mathbf{k}}^{1}(t) + z_2\rho_{\mathbf{k}}^{2}(t) \tag{10.4.2}$$

With each fluctuating density we may associate a current. Thus

$$\mathbf{j}_{\mathbf{k}}^{M}(t) = m_1\mathbf{j}_{\mathbf{k}}^{1}(t) + m_2\mathbf{j}_{\mathbf{k}}^{2}(t) \tag{10.4.3}$$

$$\mathbf{j}_{\mathbf{k}}^{Z}(t) = z_1\mathbf{j}_{\mathbf{k}}^{1}(t) + z_2\mathbf{j}_{\mathbf{k}}^{2}(t) \tag{10.4.4}$$

where the partial currents associated with the two ionic species are defined as in Eqn (7.4.7). The currents can be divided into longitudinal (l) and transverse (t) parts in the general form

$$\mathbf{j}_{\mathbf{k}}^{A} = \mathbf{j}_{\mathbf{k}l}^{A} + \mathbf{j}_{\mathbf{k}t}^{A} = \frac{\mathbf{kk}}{k^2}\cdot\mathbf{j}_{\mathbf{k}}^{A} + \left(\mathbf{I} - \frac{\mathbf{kk}}{k^2}\right)\cdot\mathbf{j}_{\mathbf{k}}^{A} \tag{10.4.5}$$

where $A = M$ or Z; the longitudinal currents satisfy equations of continuity analogous to (7.4.4). The mass current is related to the stress tensor $\boldsymbol{\sigma}_{\mathbf{k}}$ by

$$\dot{\mathbf{j}}_{\mathbf{k}}^{M}(t) + i\mathbf{k}\cdot\boldsymbol{\sigma}_{\mathbf{k}} = 0 \tag{10.4.6}$$

where the components of $\boldsymbol{\sigma}_{\mathbf{k}}$ are given by a two-component generalization of Eqn (8.4.15). The corresponding relation for the charge current (Giaquinta *et al.*, 1978) is

$$\dot{\mathbf{j}}_{\mathbf{k}}^{Z}(t) + i\mathbf{k}\cdot\boldsymbol{\sigma}_{\mathbf{k}}^{Z} = -\tfrac{1}{2}\sum_{i\neq j}^{N}\sum\left(\frac{z_i}{m_i} - \frac{z_j}{m_j}\right)\frac{\mathbf{r}_{ij}}{r_{ij}}\,v'(r_{ij})\exp\left(-i\mathbf{k}\cdot\mathbf{r}_{ij}\right) \tag{10.4.7}$$

with $v'(r) \equiv \mathrm{d}v(r)/\mathrm{d}r$ and

$$\boldsymbol{\sigma}_{\mathbf{k}}^{Z} = \sum_{i=1}^{N}\left[z_i\mathbf{u}_i\mathbf{u}_i + \tfrac{1}{4}\sum_{j\neq i}^{N}\left(\frac{z_i}{m_i} + \frac{z_j}{m_j}\right)\frac{\mathbf{r}_{ij}\mathbf{r}_{ij}}{r_{ij}^2}\,\Phi_{\mathbf{k}}(\mathbf{r}_{ij})\right]\exp\left(-i\mathbf{k}\cdot\mathbf{r}_i\right) \tag{10.4.8}$$

where $\Phi_{\mathbf{k}}(\mathbf{r})$ is defined by Eqn (8.4.14). Equation (10.4.6) shows that the time derivative of $\mathbf{j}_{\mathbf{k}}^{M}$ vanishes in the long-wavelength limit; the mass current is therefore a conserved variable in the sense of Section 8.1. On the other hand, it is clear that in general the term on the right-hand side of Eqn (10.4.7) does not vanish as $k \to 0$; it follows that the charge current is a non-conserved quantity. The difference in behaviour of the two currents reflects the fact that although the total momentum of the ions is conserved, there is a continuous exchange of momentum between the two species; the momentum exchange is the source of the electrical resistivity of the fluid.

The mass and charge densities can be used to construct three independent time-correlation functions $F_{AB}(k, t)$ (with $A, B = M$ or Z), defined in a manner similar to the intermediate scattering function of Eqn (7.4.30). The initial values of the correlation functions are equal to the static structure factors in (10.2.2), but with the number N replaced by the mass M; the corresponding dynamical structure factors are defined as in Eqn (7.4.31). Finally, three longitudinal and three transverse current correlation functions can be defined through straightforward generalizations of Eqn (7.4.34) and (7.4.35):

$$C^{l}_{AB}(k, t) = \frac{k^2}{N} \langle \mathbf{j}^{A}_{\mathbf{k}l}(t) \cdot \mathbf{j}^{B}_{-\mathbf{k}l} \rangle \tag{10.4.9}$$

$$C^{t}_{AB}(k, t) = \frac{k^2}{2N} \langle \mathbf{j}^{A}_{\mathbf{k}t}(t) \cdot \mathbf{j}^{B}_{\mathbf{k}t} \rangle \tag{10.4.10}$$

It is clear from the continuity equations that each $C^{l}_{AB}(k, t)$ is related to the corresponding $F_{AB}(k, t)$ by the analogue of Eqn (7.4.36).

We next consider how the response of the system to an external electric field can be described in terms of the correlation functions introduced above. This requires a generalization to frequency-dependent perturbations of the result in Eqn (10.2.6). As an extension of the linear-response relation (7.7.16), we find that the mean induced charge density is

$$\delta\hat{\rho}_Z(\mathbf{k}, t) = \langle \rho^{Z}_{\mathbf{k}}(t) \rangle = \chi_{ZZ}(k, \omega) e\phi^{\text{ext}}_{\mathbf{k}} \exp(-i\omega t) \tag{10.4.11}$$

The imaginary part of the complex dynamic susceptibility $\chi_{ZZ}(k, \omega)$ is related to the charge-charge dynamic structure factor through a trivial modification of the fluctuation-dissipation theorem (7.7.17). The susceptibility can also be expressed in terms of the complex dielectric function $\varepsilon(k, \omega)$ by a frequency-dependent generalization of Eqn (10.2.11):

$$\frac{1}{\varepsilon(k, \omega)} = 1 + \frac{4\pi e^2}{k^2} \chi_{ZZ}(k, \omega) \tag{10.4.12}$$

The functions $\chi_{ZZ}(k, \omega)$ and $1/\varepsilon(k, \omega)$ measure the linear response of a fluid of charged particles to an external electric field. The external field polarizes the fluid and the local, internal field—or Maxwell field—is the sum of the field due to the external charge distribution and that due to the induced charge density. The local field is, of course, the field experienced by the ions. The response of the system to the local electric potential is described by a screened response function $\chi^{\text{sc}}_{ZZ}(k, \omega)$, defined through the expression

$$\delta\hat{\rho}_Z(\mathbf{k}, t) = \chi^{\text{sc}}_{ZZ}(k, \omega) e[\phi^{\text{ext}}_{\mathbf{k}} \exp(-i\omega t) + \delta\hat{\phi}_{\text{ind}}(\mathbf{k}, \omega)] \tag{10.4.13}$$

where the induced electric potential $\delta\hat{\phi}_{\text{ind}}(k, \omega)$ is related to the induced charge density by Poisson's equation:

$$\delta\hat{\phi}_{\text{ind}}(\mathbf{k}, \omega) = \frac{4\pi e}{k^2}\,\delta\hat{\rho}_Z(\mathbf{k}, \omega) \tag{10.4.14}$$

Comparison of Eqns (10.4.11) and (10.4.13) shows that the relation between the external and screened susceptibilities is

$$\chi_{ZZ}(k, \omega) = \frac{\chi_{ZZ}^{\text{sc}}(k, \omega)}{1 - \dfrac{4\pi e^2}{k^2}\,\chi_{ZZ}^{\text{sc}}(k, \omega)} \tag{10.4.15}$$

and hence, from (10.4.12), that

$$\varepsilon(k, \omega) = 1 - \frac{4\pi e^2}{k^2}\,\chi_{ZZ}^{\text{sc}}(k, \omega) \tag{10.4.16}$$

The response function $\chi_{ZZ}(k, \omega)$ satisfies the Kramers–Kronig relations (7.7.7) and (7.7.8), which are merely consequences of causality. The same is not necessarily true of the screened function $\chi_{ZZ}^{\text{sc}}(k, \omega)$. The latter describes the response of the system to the local field; since the local field depends on the material properties of this system under study, it cannot be controlled at will in an experiment.

The electric response of an ionic fluid can also be discussed in terms of the induced electric current. Let $\mathbf{E}(\mathbf{k}, \omega)$ be a Fourier component of the local electric field. Ohm's Law in its most general form states that the induced electric current $\mathbf{J}^Z$ is linearly related to the field, i.e.

$$\mathbf{J}^Z(\mathbf{k}, \omega) = \boldsymbol{\sigma}(\mathbf{k}, \omega) \cdot \mathbf{E}(\mathbf{k}, \omega) \tag{10.4.17}$$

where $\boldsymbol{\sigma}$ is the conductivity tensor (not to be confused with the stress tensor). The conductivity tensor can be split into longitudinal and transverse parts such that

$$\boldsymbol{\sigma}(\mathbf{k}, \omega) = \frac{\mathbf{kk}}{k^2}\,\sigma_l(\mathbf{k}, \omega) + \left(\boldsymbol{I} - \frac{\mathbf{kk}}{k^2}\right)\sigma_t(\mathbf{k}, \omega) \tag{10.4.18}$$

where $\sigma_l(\mathbf{k}, \omega)$ and $\sigma_t(\mathbf{k}, \omega)$ are scalars. Then the longitudinal and transverse projections of the induced current are related, respectively, to the longitudinal (or irrotational) and transverse (or divergence-free) components of the local electric field. Thus

$$\mathbf{J}_l^Z(\mathbf{k}, \omega) = \sigma_l(\mathbf{k}, \omega)\mathbf{E}_l(\mathbf{k}, \omega) \tag{10.4.19a}$$

$$\mathbf{J}_t^Z(\mathbf{k}, \omega) = \sigma_t(\mathbf{k}, \omega)\mathbf{E}_t(\mathbf{k}, \omega) \tag{10.4.19b}$$

Since $\mathbf{E} = -\nabla \delta\phi$, it follows that the component $\mathbf{E}_l(\mathbf{k}, \omega)$ of the local field is related to the total electric potential by the expression

$$\mathbf{E}_l(\mathbf{k}, \omega) = -i\mathbf{k}\delta\hat{\phi}(\mathbf{k}, \omega) = -i\mathbf{k}[\phi_{\mathbf{k}}^{\text{ext}} \exp(-i\omega t) + \delta\hat{\phi}_{\text{ind}}(\mathbf{k}, \omega)] \quad (10.4.20)$$

Equations (7.4.4), (10.4.13), (10.4.16), (10.4.19a) and (10.4.20) can then be combined to yield the fundamental relation between the dielectric function and the conductivity tensor:

$$\varepsilon(k, \omega) = 1 + \frac{4\pi i}{\omega}\sigma_l(k, \omega) \quad (10.4.21)$$

It should be noted that $\sigma_l(k, \omega)$ is a screened response function, in the same sense as $\chi_{ZZ}^{\text{sc}}(k, \omega)$, since it measures a response to the internal field (see Eqn (10.4.17)); the analytic properties of $\sigma_l(k, \omega)$ are discussed at length by Martin (1967).

In Section 7.8, linear-response theory was used to derive a microscopic expression for the frequency-dependent electrical conductivity; this "external" conductivity measures the response of a fluid to a uniform ($\mathbf{k} = 0$) applied electric field. A uniform field corresponds to a situation in which the boundaries of the system are removed to infinity, thereby avoiding the appearance of a surface polarization. The electric response to an inhomogeneous ($\mathbf{k}$-dependent) applied field is measured by a wavenumber-dependent external conductivity that can be related to the time autocorrelation function of the fluctuating charge current $\mathbf{j}_{\mathbf{k}}^Z(t)$. In the case of the longitudinal component, the required generalization of Eqn (7.8.10) is simply

$$\sigma_l^{\text{ext}}(k, \omega) = \frac{\beta e^2}{V}\int_0^\infty \langle \mathbf{j}_{\mathbf{k}l}^Z(t) \cdot \mathbf{j}_{-\mathbf{k}l}^Z \rangle \exp(i\omega t)\, \mathrm{d}t \quad (10.4.22)$$

An argument similar to the one that leads to Eqn (10.4.15) yields an analogous relation between the external and screened longitudinal conductivities (Kubo, 1966). However, the macroscopic electrical conductivity σ given by Eqn (7.8.11) is not the same as the $k, \omega \to 0$ limit of the external longitudinal conductivity of Eqn (10.4.22). Indeed, it follows from the continuity equation (7.4.4) that the integral in (10.4.22) can be re-expressed as

$$\begin{aligned}\int_0^\infty \langle \mathbf{j}_{\mathbf{k}l}^Z(t) \cdot \mathbf{j}_{-\mathbf{k}l}^Z \rangle \exp(i\omega t)\, \mathrm{d}t \\ &= \frac{1}{k^2}\int_0^\infty \langle \dot{\rho}_{\mathbf{k}}^Z(t)\dot{\rho}_{-\mathbf{k}}^Z \rangle \exp(i\omega t)\, \mathrm{d}t \\ &= \frac{-i\omega N S_{ZZ}(k) + \omega^2 N\tilde{F}_{ZZ}(k, \omega)}{k^2} \quad (10.4.23)\end{aligned}$$

Written in this form, it is easy to see that the integral vanishes as $k, \omega \to 0$, since $S_{ZZ} \sim k^2$ for small k. (Note that $\tilde{F}_{ZZ}(k, \omega)$ is the Laplace transform of $F_{ZZ}(k, t)$, and the latter is bounded above by $S_{ZZ}(k)$; see Eqn (7.1.14).) On the other hand, the rotational invariance of an isotropic fluid implies that the macroscopic longitudinal and transverse conductivities must be the same, i.e.

$$\sigma_l^{\text{ext}}(0, \omega) = \sigma_t^{\text{ext}}(0, \omega) = \sigma(\omega) \tag{10.4.24}$$

Hence σ may be defined as the $k, \omega \to 0$ limit of the spectrum of transverse charge-current fluctuations. The transverse current is not related to the charge density by a continuity equation and is therefore unaffected by the small-k divergence of the longitudinal electric field. Thus

$$\begin{aligned} \sigma &= \lim_{\omega \to 0} \lim_{k \to 0} \frac{\beta e^2}{2V} \int_0^\infty \langle \mathbf{j}_{\mathbf{k}t}^Z(t) \cdot \mathbf{j}_{-\mathbf{k}t}^Z \rangle \exp(i\omega t)\, \mathrm{d}t \\ &= \lim_{\omega \to 0} \lim_{k \to 0} \frac{\beta \rho e^2}{k^2} \tilde{C}_{ZZ}^t(k, \omega) \end{aligned} \tag{10.4.25}$$

The differing behaviour of the longitudinal and transverse charge-current correlation functions is also evident from the frequency sum rules for the corresponding spectra. The short-time expansions of $C_{ZZ}^l(k, t)$ and $C_{ZZ}^t(k, t)$ can be written in a form similar to Eqns (7.4.41) and (7.4.48) as

$$C_{ZZ}^l(k, t) = \omega_0^2 \left(1 - \omega_{1l}^2 \frac{t^2}{2!} + \cdots\right) \tag{10.4.26}$$

$$C_{ZZ}^t(k, t) = \omega_0^2 \left(1 - \omega_{1t}^2 \frac{t^2}{2!} + \cdots\right) \tag{10.4.27}$$

with, in the case when $z_1 = -z_2 = z$:

$$\omega_0^2 = z^2 k^2 \left(\frac{k_B T}{2M}\right) \tag{10.4.28}$$

where $M = m_1 m_2/(m_1 + m_2)$ is the reduced mass of a cation–anion pair. The frequency moments ω_{1l}^2 and ω_{1t}^2 are the charge-current analogues of the quantities defined in Section 7.4. If the z-axis is chosen parallel to $\mathbf{k}$, and if the interionic potentials are separated into their short-range (S) and coulombic (C) parts, the derivation of Eqns (7.4.45) and (7.4.50) can be generalized (Abramo *et al.*, 1973; Adams *et al.*, 1977) to give (for $z_1 = -z_2 = z$) the expressions

$$\omega_{1l}^2 = 3\omega_0^2 \frac{\overline{m^2}}{\bar{m}^2} + \tfrac{2}{3}\omega_p^2$$

$$+\tfrac{1}{2}\rho \sum_\nu \sum_\mu \frac{z_\nu z_\mu M}{m_\nu m_\mu} \left\{ \int g_{\nu\mu}(r)(1-\cos kz) \frac{d^2}{dz^2} v_{\nu\mu}^{S}(r) \, d\mathbf{r} \right.$$

$$\left. + \int [g_{\nu\mu}(r) - 1](1-\cos kz) \frac{d^2}{dz^2} v_{\nu\mu}^{C}(r) \, d\mathbf{r} \right\}$$

$$+ \frac{\rho}{6M} \int g_{12}(r) \nabla^2 v_{12}^{S}(r) \, d\mathbf{r} \tag{10.4.29}$$

and

$$\omega_{1t}^2 = \omega_0^2 \frac{\overline{m^2}}{\bar{m}^2} - \tfrac{1}{3}\omega_p^2$$

$$+\tfrac{1}{2}\rho \sum_\nu \sum_\mu \frac{z_\nu z_\mu M}{m_\nu m_\mu} \left\{ \int g_{\nu\mu}(r)(1-\cos kz) \frac{d^2}{dx^2} v_{\nu\mu}^{S}(r) \, d\mathbf{r} \right.$$

$$\left. + \int [g_{\nu\mu}(r) - 1](1-\cos kz) \frac{d^2}{dx^2} v_{\nu\mu}^{C}(r) \, d\mathbf{r} \right\}$$

$$+ \frac{\rho}{6M} \int g_{12}(r) \nabla^2 v_{12}^{S}(r) \, d\mathbf{r} \tag{10.4.30}$$

where $\overline{m^n} = \frac{1}{2}(m_1^n + m_2^n)$, the summations extend over the two ionic species, and ω_p is the plasma frequency (10.1.8), generalized to the multicomponent case:

$$\omega_p^2 = \sum_\nu \frac{4\pi\rho_\nu z_\nu^2 e^2}{m_\nu} \tag{10.4.31}$$

Equations (10.4.29) and (10.4.30) have been deliberately cast in a form suitable for taking the $k \to 0$ limit; it is easy to show that

$$\lim_{k\to 0} \omega_{1l}^2(k) = \tfrac{2}{3}\omega_p^2 + \frac{\rho}{6M} \int g_{12}(r) \nabla^2 v_{12}^{S}(r) \, d\mathbf{r} \tag{10.4.32}$$

and

$$\lim_{k\to 0} \omega_{1t}^2(k) = -\tfrac{1}{3}\omega_p^2 + \frac{\rho}{6M} \int g_{12}(r) \nabla^2 v_{12}^{S}(r) \, d\mathbf{r} \tag{10.4.33}$$

Thus, in contrast to the results obtained in Section 7.4, the characteristic frequencies of the charge-current fluctuations remain non-zero in the limit $k \to 0$. In addition, the longitudinal and transverse frequencies at $\mathbf{k} = 0$ are split according to the rule

$$\omega_{1l}^2(0) - \omega_{1t}^2(0) = \omega_p^2 \tag{10.4.34}$$

This result is of the same form as the Lyddane–Sachs–Teller relation between the longitudinal and transverse optic frequencies of ionic crystals (Lyddane *et al.*, 1941). The behaviour of $\omega_{1l}(k)$ and $\omega_{1t}(k)$ at finite wavelengths is also similar to that of the corresponding phonon dispersion curves for the crystal; $\omega_{1l}(k)$ falls rapidly with increasing k, but the curve of $\omega_{1t}(k)$ versus k is almost flat.

The nature of the collective modes associated with fluctuations in mass, charge and temperature can be analysed by the methods developed in Sections 9.3 and 9.4. By analogy with the phonon spectra of ionic crystals, the collective modes in ionic liquids are expected to be of acoustic and optic character, corresponding to low-frequency sound waves and high-frequency "plasma" oscillations. The different fluctuations are, in general, strongly coupled, and the memory function associated with, say, the longitudinal charge-current correlation function necessarily has a complicated structure (Giaquinta *et al.*, 1978). A considerable simplification occurs when the anions and cations differ only in the sign of their electrical charge. Under such conditions, charge fluctuations are rigorously decoupled from fluctuations in mass and temperature at all times and for all wavenumbers (Hansen and McDonald, 1975; Giaquinta *et al.*, 1978). In particular, $F_{ZM}(k, t) = C^l_{ZM}(k, t) = 0$. The same result is true for any molten salt in the long-wavelength limit; this makes it possible to calculate the spectrum of charge fluctuations at long wavelengths by the following simple, macroscopic argument (Giaquinta *et al.*, 1976; Vieillefosse, 1977). The Laplace transform of the continuity equation for the induced charge density is

$$-i\omega\delta\tilde{\rho}_Z(\mathbf{k}, \omega) = \delta\hat{\rho}_Z(\mathbf{k}, t=0) + i\mathbf{k}\cdot\mathbf{J}^Z(\mathbf{k}, \omega) \qquad (10.4.35)$$

while Poisson's equation may be written as

$$-i\mathbf{k}\cdot\mathbf{E}(\mathbf{k}, \omega) = 4\pi\delta\tilde{\rho}_Z(\mathbf{k}, \omega) \qquad (10.4.36)$$

These two results can be combined with the longitudinal projection (10.4.19a) of Ohm's Law to give

$$\delta\tilde{\rho}_Z(\mathbf{k}, \omega) = \frac{\delta\hat{\rho}_Z(\mathbf{k}, t=0)}{-i\omega + 4\pi\sigma_l(k, \omega)} \qquad (10.4.37)$$

If we multiply through Eqn (10.4.37) by $\delta\hat{\rho}_Z(-\mathbf{k}, t=0)$ and take the statistical average, we find that

$$\tilde{F}_{ZZ}(k, \omega) = \frac{S_{ZZ}(k)}{-i\omega + 4\pi\sigma_l(k, \omega)} \qquad (10.4.38)$$

Taking the limit $k \to 0$ and using Eqn (10.4.24), we obtain an important result:

$$\lim_{k\to 0} \frac{\tilde{F}_{ZZ}(k, \omega)}{S_{ZZ}(k)} = \frac{1}{-i\omega + 4\pi\sigma(\omega)} \qquad (10.4.39)$$

Comparison with (7.3.25) shows that the frequency-dependent complex conductivity is the memory function for the long-wavelength limit of the charge-density correlation function. The spectrum of charge-density fluctuations can therefore be expressed in terms of the real (σ') and imaginary (σ'') parts of $\sigma(\omega)$ in the form

$$\lim_{k\to 0}\frac{S_{ZZ}(k,\omega)}{S_{ZZ}(k)}=\frac{1}{\pi}\frac{4\pi\sigma'(\omega)}{[\omega-4\pi\sigma''(\omega)]^2+[4\pi\sigma'(\omega)]^2} \tag{10.4.40}$$

In the low-frequency limit, $\sigma'(\omega)\to\sigma$, $\sigma''(\omega)\to 0$, and Eqn (10.4.40) reduces to

$$S_{ZZ}(k,\omega)\underset{k,\omega\to 0}{\sim}\frac{1}{\pi}\frac{4\pi\sigma(k/k_D)^2}{\omega^2+(4\pi\sigma)^2} \tag{10.4.41}$$

Thus the charge fluctuations in the low-frequency, long-wavelength regime are of a non-propagating type. The same is true of concentration fluctuations in mixtures of neutral particles, as described by the hydrodynamic result (8.6.25), but the two cases differ in a significant way. To facilitate the comparison, we ignore the effect of temperature fluctuations in (8.6.25). This is equivalent to setting $D_T=k_T=0$, so that $z_0^+=-iDk^2$ and $z_0^-=0$. The spectrum of concentration fluctuations obtained by Fourier transforming Eqn (8.6.25) is then

$$S_{cc}(k,\omega)\simeq\frac{1}{\pi}\frac{Dk^2}{\omega^2+(Dk^2)^2} \tag{10.4.42}$$

which corresponds to a lorentzian curve centred at $\omega=0$ and having a width that varies as k^2. In contrast, the width of the spectrum (10.4.41) remains non-zero even in the long-wavelength limit. The difference arises from the fact that in the coulombic case the "restoring force" is proportional to the charge-density fluctuation, whereas in the neutral system it is proportional to the laplacian of the concentration fluctuation (see Eqn (8.6.13)).

While the hydrodynamic analysis yields the correct low-frequency behaviour, the possibility that a propagating charge-density oscillation could occur at higher frequencies has to be investigated within the framework of either generalized hydrodynamics (Giaquinta *et al.*, 1978) or of the memory-function formalism (Schofield, 1974; Hansen and McDonald, 1975; Kerr, 1976). In particular, the memory-function representations developed in Sections 9.3 and 9.4 lend themselves easily to a unified treatment of transverse and longitudinal charge fluctuations. We restrict the discussion to long wavelengths and the case $z_1=-z_2=1$, and use the result that

$$\lim_{k\to 0}\frac{\omega_0^2}{S_{ZZ}(k)}=\omega_p^2 \tag{10.4.43}$$

When adapted to the problem in hand, Eqns (9.3.10) (with $\tilde{N}_t = k^2\tilde{\nu}$) and (9.4.6) become

$$\tilde{C}^t_{ZZ}(k, \omega) = \frac{\omega_0^2}{-i\omega + \tilde{N}_t(k, \omega)} \tag{10.4.44}$$

$$\tilde{C}^l_{ZZ}(k, \omega) = \frac{\omega_0^2}{-i\omega + \dfrac{\omega_p^2}{-i\omega} + \tilde{N}_l(k, \omega)} \tag{10.4.45}$$

The fact that the ratio in (10.4.43) approaches a non-zero value in the limit $k \to 0$ means that $\lim_{\omega\to 0} \lim_{k\to 0} \tilde{C}^l_{ZZ}(k, \omega)/C^l_{ZZ}(k, t=0) = 0$, as we have already proved in a different context in the discussion of Eqn (10.4.23). Use of (10.4.23) shows that the expression for the charge-density autocorrelation function corresponding to (10.4.45) is given in terms of Laplace transforms by

$$\tilde{F}_{ZZ}(k, \omega) = \frac{S_{ZZ}(k)}{-i\omega + \dfrac{\omega_p^2}{-i\omega + \tilde{N}_l(k, \omega)}} \tag{10.4.46}$$

From a comparison of Eqn (10.4.46) with (10.4.39) we find that

$$\sigma = \lim_{\omega\to 0} \lim_{k\to 0} \frac{\omega_p^2}{4\pi\tilde{N}_l(k, \omega)} \tag{10.4.47}$$

while combination of Eqns (10.4.25) and (10.4.44) shows that

$$\sigma = \lim_{\omega\to 0} \lim_{k\to 0} \frac{\omega_p^2}{4\pi\tilde{N}_t(k, \omega)} \tag{10.4.48}$$

The crossover from low-frequency to high-frequency behaviour can be studied in an approximate way by assuming that the memory functions are exponentials in time, as in Eqns (9.4.11) and (9.4.20). This leads, for small k, to

$$\tilde{N}_l(k, \omega) = \frac{\omega_{1l}^2 - \omega_p^2}{-i\omega + 1/\tau_l(k)} \tag{10.4.49}$$

$$\tilde{N}_t(k, \omega) = \frac{\omega_{1t}^2}{-i\omega + 1/\tau_t(k)} \tag{10.4.50}$$

Equations (10.4.47) and (10.4.48), together with the long-wavelength sum rule (10.4.34), now imply (Schofield, 1974) that

$$\lim_{k\to 0} \tau_l(k) = \lim_{k\to 0} \tau_t(k) \tag{10.4.51}$$

If, conversely, the correctness of (10.4.51) is accepted on grounds of spatial isotropy, we are led back to the Green–Kubo relation (10.4.25). According to Eqns (7.7.17) and (10.4.23), the charge-response function is directly proportional to $C^{l}_{ZZ}(k, \omega)$; the characteristic frequencies of the charge-fluctuation modes are therefore determined by the poles of the function (10.4.45). If Eqn (10.4.49) is substituted in (10.4.45), we obtain a dispersion relation of the form

$$\omega^2(\omega\tau_l) - i\omega^2 - (\omega\tau_l)\omega_{1l}^2 - i\omega_p^2 = 0 \tag{10.4.52}$$

In the low-frequency limit, i.e. when $\omega\tau_l \ll 1$, the roots of Eqn (10.4.52) are purely imaginary ($\omega = \pm i\omega_p$), corresponding to the charge-relaxation mode embodied in (10.4.41). At high frequencies, on the other hand, i.e. for $\omega\tau_l \gg 1$, the roots are purely real ($\omega = \pm\omega_{1l}$), and correspond to charge fluctuations that propagate at a frequency of the order of the plasma frequency, but renormalized by the short-range interactions between ions; for the alkali halides, ω_{1l} is always larger than ω_p, typically by 20–30%. High-frequency propagating modes have in fact been observed both experimentally and in computer simulations; the results will be discussed in more detail in the next section.

A complete analysis of the longitudinal modes in ionic fluids is contained in the work of Baus (1977), Giaquinta *et al.* (1978) and Bernu (1983). These authors take explicit account of the electrostatic and thermoelectric effects that arise from the coupling between charge fluctuations and fluctuations in mass and temperature; they also give Green–Kubo formulae for the transport coefficients other than electrical conductivity. The thermoelectric coefficients relate the macroscopic charge and heat currents to the applied thermodynamic forces through the following constitutive relations (de Groot and Mazur, 1962):

$$\mathbf{J}^Z = -\Omega_{ZZ}\mathbf{F}^Z - \Omega_{ZQ}\mathbf{F}^Q \tag{10.4.53}$$

$$\mathbf{J}^Q = -\Omega_{QZ}T\mathbf{F}^Z - \Omega_{QQ}\mathbf{F}^Q \tag{10.4.54}$$

The thermodynamic forces $\mathbf{F}^Z$ and $\mathbf{F}^Q$ are defined as

$$\mathbf{F}^Z = \nabla(\mu_Z + \phi) \tag{10.4.55}$$

$$\mathbf{F}^Q = \nabla T \tag{10.4.56}$$

where ϕ is the applied electric potential and μ_Z is the electrochemical potential introduced in Eqn (10.2.34). The Onsager reciprocity relations imply that the cross coefficients are equal, i.e.

$$\Omega_{ZQ} = \Omega_{QZ} = \frac{\alpha}{T} \tag{10.4.57}$$

where α is the thermoelectric coefficient; the coefficient Ω_{ZZ} is simply the electrical conductivity σ. Elimination of $\mathbf{F}^Z$ between Eqns (10.4.53) and (10.4.54) leads to the expression

$$\mathbf{J}^Q = -\left(\Omega_{QQ} - \frac{T\Omega_{QZ}\Omega_{ZQ}}{\Omega_{ZZ}}\right)\mathbf{F}^Q + \frac{T\Omega_{QZ}}{\Omega_{ZZ}}\mathbf{J}^Z \tag{10.4.58}$$

The thermal conductivity λ is defined as the ratio of the heat current to the temperature gradient in the absence of an electric current ($\mathbf{J}^Z = 0$). Thus

$$\lambda = \Omega_{QQ} - \frac{T\Omega_{QZ}\Omega_{ZQ}}{\Omega_{ZZ}} = \Omega_{QQ} - \frac{\alpha^2}{T\sigma} \tag{10.4.59}$$

The thermal conductivity of an ionic liquid is therefore obtained as a combination of coefficients that separately are expressible as time integrals of autocorrelation functions of microscopic currents; the appropriate formula for σ, for example, is given by Eqn (7.8.10). It should be noted, however, that λ itself cannot be expressed in such a way (Bernu and Hansen, 1982; Bernu, 1983); this situation is quite different from that of a neutral, one-component liquid, to which the Green–Kubo relation (8.5.27) applies.

10.5 MICROSCOPIC DYNAMICS IN MOLTEN SALTS

There now exists a large body of computer-simulation results on various aspects of the microscopic dynamics in molten salts. The richness of the observed single-particle and collective behaviour has not only stimulated theoretical analysis, but has also led to experimental efforts to detect high-frequency collective modes by neutron scattering.

Single-particle motion is conveniently discussed in terms of the velocity autocorrelation functions $Z_\nu(t)$ and self-diffusion coefficients D_ν of the two ionic species; D_ν is related to $Z_\nu(t)$ in the manner of Eqn (7.2.8). As in the case of mixtures of neutral particles, it is natural to ask whether the processes of self-diffusion and interdiffusion can be satisfactorily related in a simple way. In an ionic liquid, however, interdiffusion is equivalent to electrical conduction. We have shown in Section 7.8 that the static electrical conductivity σ is proportional to the time integral of the electric-current autocorrelation function $J(t)$, defined as

$$J(t) = \langle \mathbf{j}^Z(t) \cdot \mathbf{j}^Z \rangle$$
$$= \sum_{i=1}^{N} \sum_{j=1}^{N} z_i z_j \langle \mathbf{u}_i(t) \cdot \mathbf{u}_j \rangle \tag{10.5.1}$$

If the self-correlation terms ($i=j$) in (10.5.1) are separated from the cross terms ($i \neq j$), integration over time and use of Eqn (7.8.10) shows that

$$\sigma = \beta e^2(\rho_1 z_1^2 D_1 + \rho_2 z_2^2 D_2)(1-\Delta) \qquad (10.5.2)$$

Equation (10.5.2), with $\Delta = 0$, is called the Nernst–Einstein relation; the value of the deviation factor Δ is a measure of the importance of cross-correlations. If $\Delta = 0$, Eqn (10.5.2) becomes the ionic equivalent of the approximate relation (8.6.36). In practice, at least for the molten alkali halides, Δ is significantly different from zero and is always positive. The importance of cross correlations in systems of this type is illustrated in Figure 10.4, where molecular dynamics results for the velocity and electric-current autocorrelation results of the "simple molten salt" (Eqn (10.1.10))

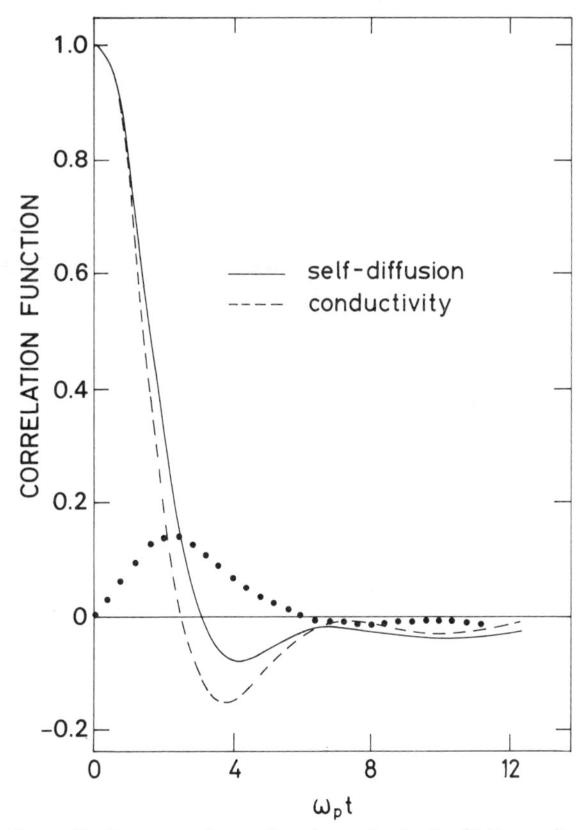

FIG. 10.4. Normalized autocorrelation functions of velocity (full curve) and electric current (dashed curve) from a molecular dynamics simulation of the "simple molten salt" defined by Eqn (10.1.10). The points show the difference between the two curves, and represent the effects of cross correlations (see text). Note the similarity with the results of Figure 8.6. After Hansen and McDonald (1975).

are plotted. The symmetry of the model means that the velocity autocorrelation functions of cations and anions are identical; if cross-correlations of velocities were negligible, the normalized curves of $Z(t)$ and $J(t)$ would also be the same. At short times, however, there are substantial differences between $Z(t)$ and $J(t)$, and the calculated Nernst–Einstein deviation factor for the case shown is $\Delta = +0.19$. The positive value of Δ corresponds to the fact that motion in the same direction by a neighbouring pair of oppositely charged ions contributes to self-diffusion but not to electrical conduction. The numerical result agrees well with experimental data; the mean value of Δ for eight alkali metal salts is $+0.26$. The observed deviations from the Nernst–Einstein relation therefore have a natural explanation in terms of short-lived correlations between the velocities of neighbouring ions. The correlations are of a type that physical intuition would lead one to expect, but it is not necessary to assume the existence of well-defined ion pairs.

Note that the velocity autocorrelation function shown in Figure 10.4 has a negative plateau similar to that seen in argon-like liquids. Both the shape of $Z(t)$ and the value of the diffusion coefficient are reasonably well reproduced by a mode-coupling calculation of the type discussed in Section 9.5 (Sjögren and Yoshida, 1982; Munakata and Bosse, 1983). The mode-coupling results for the electric-current autocorrelation function of the simple molten salt are much less satisfactory, and the theoretical value of the electrical conductivity is about 30% too small. These discrepancies have been attributed to the neglect of temperature fluctuations in the mode-coupling calculations (Sjögren and Yoshida, 1982).

Molecular dynamics results on self-diffusion are also available for more realistic rigid-ion models of the alkali halides in which allowance is made for the differences in mass and size of the two ions (Ciccotti *et al.*, 1976). The computed velocity autocorrelation functions for the cations and anions in molten LiF, RbCl and RbI are plotted in Figure 10.5. Because of the mass differences, the results are significantly different from those obtained in the symmetrical case illustrated in Figure 10.4. The autocorrelation function of the light ion is strongly oscillatory; this effect is the result of a "rattling" motion of the ion in the relatively long-lived cage formed by its heavier neighbours, and is particularly marked in the case of the very light Li^+ ion. The resulting self-diffusion coefficients are systematically smaller (typically by 20%) than the corresponding experimental values, whereas the computed and experimental conductivities agree, in general, very well. It follows that the Nernst–Einstein deviation factor is everywhere too small. This discrepancy may be an inherent defect of the rigid-ion approximation. Simulations of a shell model of molten KI have shown that the effect of polarization is to increase substantially the values of both diffusion

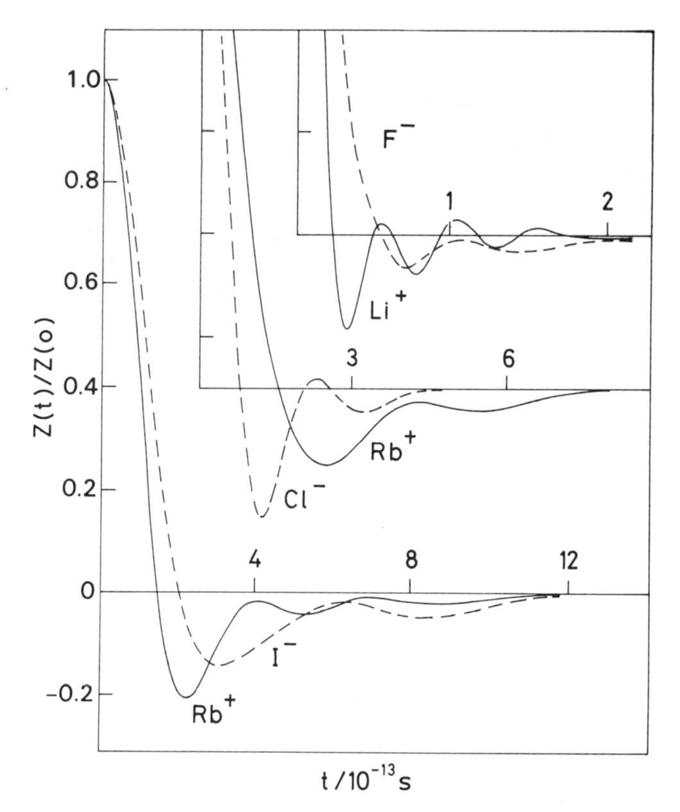

FIG. 10.5. Velocity autocorrelation functions from molecular-dynamics simulations of LiF, RbCl and RbI. Full curves: cations; dashed curves: anions. After Ciccotti *et al.* (1976).

coefficients (particularly of K^+), while leaving the electrical conductivity almost unaltered (Jacucci *et al.*, 1976). The enhancement of ionic self-diffusion by polarization is not unexpected. In a rigid-ion model, local charge neutrality around a diffusing ion can be maintained only by bodily displacement of its neighbours; when polarization is allowed, an additional screening mechanism is present that does not entail movement of the ion cores. The net result is that the cage effect is smaller for polarizable ions, as evidenced by an increased damping of the oscillations in the velocity autocorrelation function, and the diffusion coefficient is therefore larger. Similar conclusions have been drawn from molecular dynamics calculations for a polarizable model of molten NaI (Dixon, 1983a).

The wavenumber-dependent collective motions in molten salts have also been studied by computer simulation. The simple molten salt is particularly well-suited to theoretical investigation of the collective modes, because mass and charge fluctuations are strictly independent at all wavelengths (see Section 10.4). The main interest is in the optic-type modes associated with charge fluctuations, since these are specific to ionic fluids. The simulations show that the charge-density autocorrelation function $F_{ZZ}(k, t)$ is strongly oscillatory at wavelengths up to about twice the mean interionic spacing (Hansen and McDonald, 1975). The oscillations in $F_{ZZ}(k, t)$ give rise to a well-defined "plasmon" peak in the charge-density spectrum $S_{ZZ}(k, \omega)$, as shown in Figure 10.6. The frequency at which the optic peak occurs is comparable with the plasma frequency ω_p, and its dispersion is strongly negative; the peak eventually disappears at a value of k in the neighbourhood of the main peak in the charge–charge structure factor $S_{ZZ}(k)$. The loss of the optic peak is accompanied by a marked narrowing of the spectrum (note the change of frequency scale for $k\sigma = 3.753$). There is also a growth in the weight of the low-frequency modes that manifests itself particularly clearly in a pronounced peak in $S_{ZZ}(k, \omega = 0)$ close to the critical wavenumber; the effect is well illustrated by the results for a model of molten NaI plotted in Figure 10.7. A final noteworthy feature of the

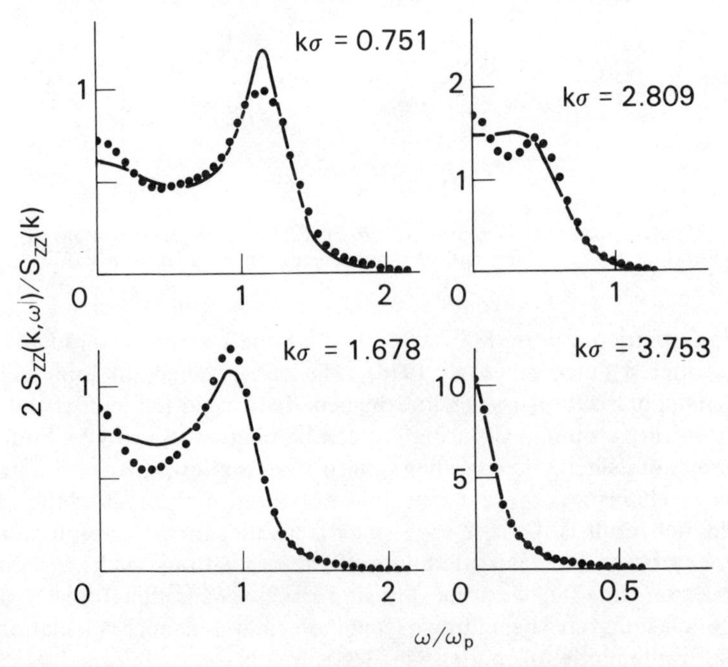

FIG. 10.6. Charge–charge dynamic structure factor for the "simple molten salt" defined by Eqn (10.1.10). The points are molecular-dynamics results and the curves are calculated from the single-relaxation-time approximation (9.4.20) (Hansen and McDonald, 1975). Note the change of scale for $k\sigma = 3.753$. The unit of frequency is the plasma frequency (10.4.31).

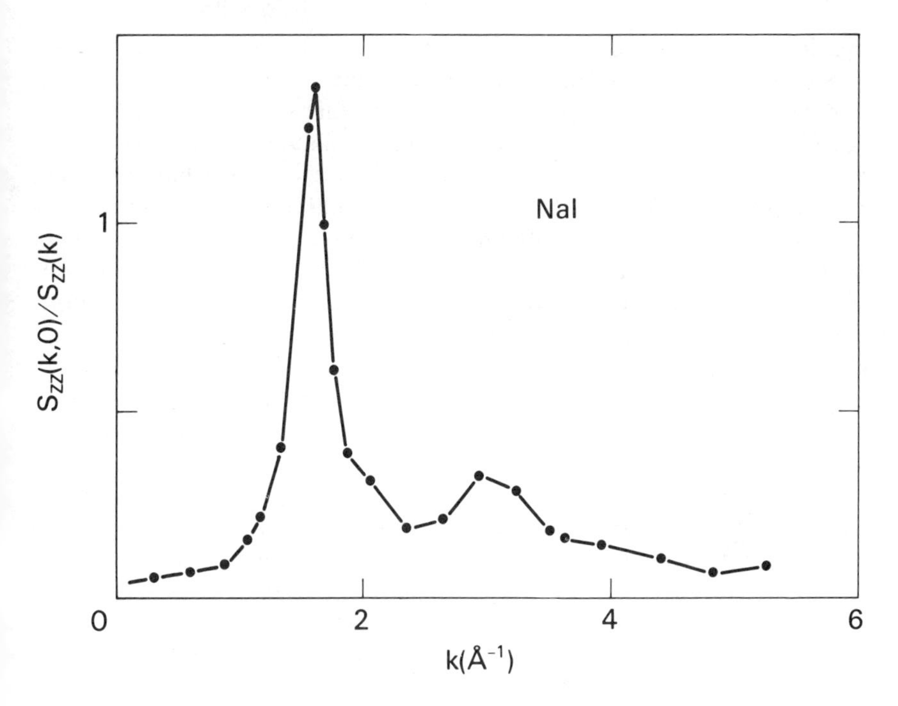

FIG. 10.7. Charge–charge dynamic structure factor at zero frequency, in arbitrary units, relative to the value of the corresponding static structure factor for a model of molten NaI. The points are molecular dynamics results and the line is drawn as an aid to the eye. After Dixon (1983a).

computed charge-fluctuation spectrum is the fact that at small wavenumbers the optic peak sharpens as k increases, i.e. the damping of the plasmon mode becomes weaker. This behaviour is in striking contrast to that of the sound-wave mode; in molten salts, as in systems of neutral particles, the damping of sound waves increases rapidly with k.

The main features of the charge-fluctuation spectrum of the simple molten salt are also seen in simulations of more realistic rigid-ion models (Copley and Rahman, 1976; Adams *et al.*, 1977; Bernu, 1986); the effect of including polarization is to broaden the optic peak and shift it to lower frequencies (Jacucci *et al.*, 1976; Dixon, 1983a). It can be seen from Figure 10.6 that in the case of the simple molten salt, the single-relaxation-time, viscoelastic approximation (9.4.20) adequately describes the dispersion of the optic peak, but it cannot account for the detailed shape of the spectrum, particularly in the region of the central, diffusive peak. Similarly, in their analysis of molecular dynamics results for a model of molten NaCl, Adams *et al.* (1977) have shown that at least two relaxation times, differing by nearly an order of magnitude, are needed in order to fit the computed spectra. The memory function for the longitudinal charge-current correlation function consists of a short-lived initial decay and a long-time, quasi-exponential tail; it is therefore similar in structure to the memory function required to describe the density fluctuations in argon-like liquids (see Section 9.4). A fair description of the spectra of mass and charge fluctuations in the simple molten salt is also obtained by mode-coupling methods along the lines of Section 9.5. Sjögren and Yoshida (1982) reduce the problem of predicting the spectra to that of calculating the memory functions for the velocity and electric-current autocorrelation functions. The memory functions are split, in the manner suggested by the empirical findings of Adams *et al.* (1977), into a binary-collision part and a collective, recollision term; the appearance of two widely different relaxation times can be linked to the existence of fast and slow two-mode decay channels for the charge-density excitations (Chaturvedi *et al.*, 1980). Other mode-coupling calculations have been described by Bosse and Munakata (1982); their analysis predicts that the width of the optic peak should decrease with increasing k in a certain wavenumber range, in qualitative agreement with the unexpected behaviour observed in the simulations (Hansen and McDonald, 1975).

Several attempts have been made to detect a collective, plasmon-like excitation in molten salts by inelastic neutron scattering (Copley and Dolling, 1978; McGreevy *et al.*, 1984). If b_1 and b_2 are the coherent neutron scattering lengths of the two ionic species, a straightforward extension of the derivation given for a one-component fluid in Section 7.5 shows that the coherent inelastic cross-section is

$$\frac{d^2\sigma}{d\Omega\, d\omega} = \frac{1}{4}\left(\frac{k_2}{k_1}\right)[(b_1+b_2)^2 S_{NN}(k, \omega) + 2(b_1^2 - b_2^2)S_{NZ}(k, \omega) + (b_1 - b_2)^2 S_{ZZ}(k, \omega)] \quad (10.5.3)$$

which reduces to (7.5.12) when $b_1 = b_2$. Equation (10.5.3) shows that a single experiment yields only a linear combination of the three dynamic structure factors (number–number, number–charge and charge–charge). Moreover, the relative contribution of the charge-fluctuation component is in general very low at small wavenumbers, because $S_{ZZ}(k)$ (the integral of $S_{ZZ}(k, \omega)$) is proportional to k^2 in the limit $k \to 0$. Only when the scattering lengths are such that $b_1 \simeq -b_2$ does the component $S_{ZZ}(k, \omega)$ dominate, and this situation is not easily achieved with readily available isotopes (Copley and Dolling, 1978). The three contributions in Eqn (10.5.3) can, in principle, be separated by varying the isotopic composition of the melt, thereby changing the scattering lengths of the nuclei (Enderby and Neilson, 1980). Experiments of this type have been carried out on molten RbCl, CsCl and NaI (McGreevy *et al.*, 1984). Broad peaks were detected in the function

$C^l_{ZZ}(k, \omega) = \omega^2 S_{ZZ}(k, \omega)$, but these can hardly be assigned to a plasmon mode because the measurements were restricted to relatively large wavenumbers ($k > 0.4$ Å^{-1}). The only convincing experimental evidence obtained so far for the existence of a plasmon mode in ionic liquids comes from an analysis of infrared reflectivity data for molten LiF (Giaquinta *et al.*, 1978). The resulting spectrum, which corresponds effectively to $S_{ZZ}(k, \omega)$ at zero wavenumber, does indeed display a well-defined plasmon response, at a frequency somewhat above ω_p, which is well separated from a central, diffusive peak.

The autocorrelation functions of the transverse components of the mass and charge currents have been calculated in molecular dynamics simulations of molten NaCl (Adams *et al.*, 1977), NaI (Dixon, 1983b) and KCl (Bernu, 1984). The frequency of the transverse optic mode lies roughly an amount ω_p below that of its longitudinal counterpart, as suggested by the sum rule (10.4.34), and is relatively insensitive to wavenumber. As in the case of the longitudinal modes, an accurate memory-function fit of the transverse-current spectra requires the introduction of two very different relaxation times.

CHAPTER 11

Simple Liquid Metals

Liquid metals differ from the other classes of liquids considered so far primarily through the presence of the conduction electrons. Although the theory of liquid metals has much in common with that of the other ionic liquids discussed in Chapter 10, the problem is complicated by the fact that the electronic component requires a quantum-mechanical treatment. Many elements show metallic behaviour in the liquid state, but their electronic band structures can differ widely. We shall restrict ourselves to the so-called "simple" metals. The class of simple metals comprises those in which the electronic valence states are well separated in energy from the tightly-bound core states; their properties are reasonably well described by the *nearly-free-electron* model. Metals that are classified as simple in this sense include the alkali metals, magnesium, zinc, mercury, gallium and aluminium. Other liquid metals (noble and transition metals, alkaline earths, lanthanides and actinides) have more complicated electronic structures, and the theory of such systems is correspondingly less well advanced. More complete accounts of the properties of liquid metals than we are able to give here can be found in a number of excellent monographs (Faber, 1972; Shimoji, 1977) and review articles (Ashcroft and Stroud, 1978; Evans, 1978), and in the proceedings of regular conferences on the subject (Takeuchi, 1973; Evans and Greenwood, 1977; Cyrot-Lackmann and Desré, 1980).

11.1 ELECTRONS AND IONS

Pure liquid metals are two-component fluids consisting of N_i positive ions and $N_e = zN_i$ conduction electrons; z is the ionic valence, i.e. the positive nuclear charge plus the total negative charge of the core electrons, all in units of the elementary charge. The ionic core radius is usually only a small fraction of the mean interionic spacing, with the result that the ion cores generally occupy less than 10% of the total volume of the metal. In the nearly-free-electron picture, the conduction electrons are assumed to move more or less freely through the liquid, interacting only rarely with the ions; the mean-free path of the electrons is typically ten to a hundred times larger than the separation of neighbouring ions. In the crudest (Sommerfeld) approximation, interactions are neglected altogether and the electrons are treated as an ideal Fermi gas characterized by the energy ε_F of the highest occupied (Fermi) level. The Fermi energy is

$$\varepsilon_F = \frac{|\mathbf{p}_F|^2}{2m_e} = \frac{\hbar^2 k_F^2}{2m_e} = \frac{\hbar^2}{2m_e}(3\pi^2 \rho_e)^{2/3} \tag{11.1.1}$$

where $\rho_e = N_e/V$ is the number density of conduction electrons, m_e is the electron mass and $\mathbf{p}_F = \hbar \mathbf{k}_F$ is the Fermi momentum. Values of the corresponding Fermi temperature $T_F = \varepsilon_F / k_B$ of a number of metals are listed in Table 11.1. As the table shows, T_F is always much larger than the melting temperature T_m. It is therefore a good approximation to assume that the electron gas is completely degenerate; under normal liquid-metal conditions, finite-temperature effects are negligible.

TABLE 11.1. *Characteristic temperatures and molar volumes of some liquid metals at melting (m) and at the critical point (c); r_s is the electron-sphere radius at melting, in atomic units, and T_F is the corresponding Fermi temperature. The critical-point data are taken from Ross and Greenwood* (1969)

Metal	T_m (K)	V_m (cm^3 mol^{-1})	r_s	T_F (K)	T_c (K)	V_c (cm^3 mol^{-1})
Li	452	13.4	3.30	53400	>3000	?
Na	371	24.8	4.05	35500	2600	109
K	337	47.4	5.03	23000	2200	206
Rb	312	58.0	5.38	20100	2100	244
Cs	303	72.2	5.78	17400	2030	302
Mg	924	15.47	2.75	77000	?	?
Hg	234	14.7	2.70	79800	1760	47(?)
Al	932	11.4	2.17	124000	?	?

1 a.u. of length $\simeq$ 0.5292 Å.

The large binding energies typical of liquid metals have their origins in the Coulomb interactions between electrons and ions. The simplest model that takes account of the electron–ion interaction is the "jellium" model of Wigner (Singwi and Tosi, 1981; Ichimaru, 1982). The jellium model is the quantum-mechanical analogue of the one-component plasma (OCP) discussed in Chapter 10; it treats the conduction electrons as an interacting Coulomb gas moving in the uniform background provided by the positive ions. The hamiltonian of the model is

$$\mathcal{H} = K_{N_e} + V_{N_e} \tag{11.1.2}$$

where K_{N_e} is the kinetic-energy operator and the potential energy V_{N_e} is the sum of electron–electron, electron–background and background–background terms. In a **k**-space representation, V_{N_e} is given as a special case of Eqn (10.1.4) by

$$V_{N_e} = \frac{1}{2V} \sum_{\mathbf{k}}{}' \hat{v}_{ee}(k)(\rho_{\mathbf{k}}^{e} \rho_{-\mathbf{k}}^{e} - N_e) \tag{11.1.3}$$

where $\rho_{\mathbf{k}}^{e}$ is a Fourier component of the microscopic electron-gas density, $\hat{v}_{ee}(k) = 4\pi e^2/k^2$ is the Fourier transform of the electron–electron potential, and the prime on the summation means that the term for $\mathbf{k} = 0$ is omitted because of cancellation by the background. The ground-state energy of the jellium model has been calculated by methods of quantum-mechanical many-body theory (Pines and Nozières, 1966). The energy per electron, ε_0, is the sum of three terms: the kinetic energy $\varepsilon_{\text{kin}} = 3\varepsilon_F/5$; the exchange energy ε_{ex}, obtained by a Hartree–Fock calculation based on a Slater determinant of plane waves; and the correlation energy $\varepsilon_{\text{corr}}$, for which an explicit form has been given by Pines and Nozières (1966). The three contributions are conveniently expressed in terms of the dimensionless density parameter $r_s = a_e/a_0$, where $a_e = (3/4\pi\rho_e)^{1/3}$ and $a_0 = \hbar^2/m_e e^2$ is the Bohr radius. The total energy in atomic (Rydberg) units is then

$$\varepsilon_0 = \frac{2.21}{r_s^2} - \frac{0.916}{r_s} + (0.031 \log r_s - 0.115) \tag{11.1.4}$$

where the three terms on the right-hand side represent, successively, ε_{kin}, ε_{ex} and $\varepsilon_{\text{corr}}$. Equation (11.1.4) gives a minimum energy, corresponding to zero pressure, of -0.16 a.u. (-2.2 eV) at $r_s \simeq 4.3$. This result is independent of the chemical nature of the system, but the predicted value of r_s agrees reasonably well with experimental results for the alkali metals, as shown in Table 11.1. We shall see in the next section how the simple jellium model can be improved upon. It is worthwhile noting here, however, that the binding energy calculated from Eqn (11.1.4) is an order of magnitude larger than the internal energies characteristic of simple insulating liquids. Associated with this is the fact that the "liquid range", as measured by the ratio of the critical (T_c) and triple-point (T_t) temperatures, is also much larger for liquid metals than for insulating liquids, as exemplified by the results for sodium and argon given in Table 1.1.

11.2 REDUCTION TO AN EFFECTIVE ONE-COMPONENT PROBLEM

The total hamiltonian of a liquid metal is the sum of a purely electronic term $\mathscr{H}_e$, a purely ionic term $\mathscr{H}_i$, and an electron–ion interaction V_{ei}. Thus

$$\mathscr{H} = \mathscr{H}_e + \mathscr{H}_i + V_{ei} \tag{11.2.1}$$

with

$$\mathscr{H}_e = K_e + V_{ee} = \sum_j \frac{|\mathbf{p}_j|^2}{2m_e} + \tfrac{1}{2}\sum_{j\neq j'}\sum v_{ee}(|\mathbf{r}_j - \mathbf{r}_{j'}|) \tag{11.2.2a}$$

$$\mathscr{H}_i = K_i + V_{ii} = \sum_l \frac{|\mathbf{P}_l|^2}{2m_i} + \tfrac{1}{2}\sum_{l\neq l'}\sum v_{ii}(|\mathbf{R}_l - \mathbf{R}_{l'}|) \tag{11.2.2b}$$

$$V_{ei} = \sum_j \sum_l v_{ei}(|\mathbf{r}_j - \mathbf{R}_l|) \tag{11.2.2c}$$

where the coordinates and momenta of the electrons are denoted by $\{\mathbf{r}_j, \mathbf{p}_j\}$ and those of the ions by $\{\mathbf{R}_l, \mathbf{P}_l\}$. In general, the Coulomb repulsion between ions is sufficiently strong to prevent any short-range repulsion coming into play; van der Waals forces can also be ignored, because the ion cores are only weakly polarizable. It is therefore a good approximation to put $v_{ii}(R) = z^2e^2/R$ in Eqn (11.2.2b). The electron–electron interaction is purely coulombic, i.e. $v_{ee}(r) = e^2/r$ for all r, and the electron–ion potential is also coulombic outside the ion core; we shall see below that inside the core, v_{ei} can be replaced by a weak *pseudopotential*.

Following Ashcroft and Stroud (1978), we now add to and subtract from the hamiltonian the two contributions to the potential energy that would arise if the electrons were replaced by a uniform background of charge density $-e\rho_e$. The terms involved are the ion–background interaction V_{bi} and the background self-energy V_{bb}, given by

$$V_{bi} = -\rho_e \sum_l \int \frac{ze^2}{|\mathbf{r} - \mathbf{R}_l|}\,d\mathbf{r} \tag{11.2.3}$$

and

$$V_{bb} = \tfrac{1}{2}\rho_e^2 \int d\mathbf{r} \int d\mathbf{r}' \frac{e^2}{|\mathbf{r} - \mathbf{r}'|} \tag{11.2.4}$$

The hamiltonian (11.2.1) is therefore rewritten as

$$\mathscr{H} = \mathscr{H}'_e + \mathscr{H}'_i + V'_{ei} \tag{11.2.5}$$

with

$$\mathscr{H}'_e = \mathscr{H}_e - V_{bb} \tag{11.2.6a}$$

$$\mathscr{H}'_i = \mathscr{H}_i + V_{bi} + V_{bb} \tag{11.2.6b}$$

$$V'_{ei} = V_{ei} - V_{bi} \tag{11.2.6c}$$

The **k**-space representation of (11.2.6) is

$$\mathcal{H}'_{\mathrm{e}} = K_{\mathrm{e}} + \frac{1}{2V}\sum_{\mathbf{k}}{}' \frac{4\pi e^2}{k^2}(\rho^{\mathrm{e}}_{\mathbf{k}}\rho^{\mathrm{e}}_{-\mathbf{k}} - N_{\mathrm{e}}) \tag{11.2.7a}$$

$$\mathcal{H}'_{\mathrm{i}} = K_{\mathrm{i}} + \frac{1}{2V}\sum_{\mathbf{k}}{}' \frac{4\pi z e^2}{k^2}(\rho^{\mathrm{i}}_{\mathbf{k}}\rho^{\mathrm{i}}_{-\mathbf{k}} - N_{\mathrm{i}}) \tag{11.2.7b}$$

$$V'_{\mathrm{ei}} = U_0 + \frac{1}{V}\sum_{\mathbf{k}}{}' \hat{v}_{\mathrm{ei}}(k)\rho^{\mathrm{i}}_{\mathbf{k}}\rho^{\mathrm{e}}_{-\mathbf{k}} \tag{11.2.7c}$$

where

$$\begin{aligned} U_0 &= \frac{1}{V}\lim_{k\to 0}\left[\hat{v}_{\mathrm{ei}}(k) + \frac{4\pi z e^2}{k^2}\right]\rho^{\mathrm{i}}_{\mathbf{k}}\rho^{\mathrm{e}}_{-\mathbf{k}} \\ &= N_{\mathrm{i}}\rho_{\mathrm{e}}\int\left[v_{\mathrm{ei}}(r) + \frac{z e^2}{r}\right]\mathrm{d}\mathbf{r} \end{aligned} \tag{11.2.8}$$

The term $\mathcal{H}'_{\mathrm{e}}$ is identical to the jellium hamiltonian (11.1.2), and $\mathcal{H}'_{\mathrm{i}}$ is the hamiltonian of an OCP of ions in a uniform background of charge density $-e\rho_{\mathrm{e}}$. In this formulation of the problem, a liquid metal emerges as a "mixture" of a classical OCP and a quantum-mechanical jellium, the two components being coupled together through the term V'_{ei}.

Inside the ion core, the interaction of the conduction electron with the ion is determined by details of the charge distribution of the core electrons. The true electron–ion interaction is therefore a complicated, non-local function for $r < r_{\mathrm{c}}$, where r_{c} is the ion-core radius. In addition, the potential has a singularity at $r = 0$. Despite these difficulties, it is possible to treat the electron–ion coupling by perturbation theory if the interaction is recast in pseudopotential form (Heine 1970; Cohen and Heine, 1970). The wave function of a conduction electron has a plane-wave character outside the ion core, but inside the core it oscillates so as to orthogonalize itself to the core-electron states. The oscillations are a manifestation of bound states and are irrelevant to the scattering of the valence electron by the ion. The orthogonality requirement can, however, be put to work in the construction of a pseudopotential $v^{\mathrm{ps}}_{\mathrm{ei}}$; the pseudopotential has no bound states, but it retains the same valence-energy spectrum as the true electron–ion potential and the same phase shifts at the core radius. These conditions are not sufficient to define a unique pseudopotential, which moreover may be both non-local and energy dependent. The procedure adopted in practice is to parameterize an assumed (local) functional form for the pseudopotential by fitting to experimental data on quantities that are sensitive to electron–ion collisions. The pseudopotentials obtained in this way are usually sufficiently weak to justify a perturbation calculation. A particularly simple and widely used pseudopotential consists in writing

$$v_{\mathrm{ei}}^{\mathrm{ps}}(r)=0, \qquad r<r_{\mathrm{c}}$$
$$=\frac{-ze^2}{r}, \qquad r>r_{\mathrm{c}} \tag{11.2.9}$$

This is called the "empty-core" pseudopotential (Ashcroft, 1966); values of the parameter r_{c} can be derived from transport and Fermi-surface data and lie close to generally accepted values for the ionic radii of simple metals.

If the pseudopotential is weak, the term V'_{ei} in (11.2.5) can be treated as a perturbation. The reference system is a superposition of a classical OCP (for the ionic component) and a degenerate, interacting electron gas (eg). To lowest order in perturbation theory, the structure of each component in the reference system is unaffected by the presence of the other. In this approximation, assuming the two fluids to be homogeneous:

$$\langle \rho_{\mathbf{k}}^{\mathrm{i}} \rho_{-\mathbf{k}}^{\mathrm{e}} \rangle = \langle \rho_{\mathbf{k}}^{\mathrm{i}} \rangle \langle \rho_{-\mathbf{k}}^{\mathrm{e}} \rangle = 0, \qquad \mathbf{k} \neq 0 \tag{11.2.10}$$

Hence, on averaging the perturbation (11.2.7c), only the structure-independent term survives, and the internal energy of the metal is obtained as

$$U(N_{\mathrm{i}}, N_{\mathrm{e}}, V, T) = U_{\mathrm{OCP}}(N_{\mathrm{i}}, V, T) + U_{\mathrm{eg}}(N_{\mathrm{e}}, V) + U_0(N_{\mathrm{i}}, N_{\mathrm{e}}, V) \tag{11.2.11}$$

The energy of the degenerate electron gas is given by Eqn (11.1.4) and the internal energy of the classical OCP is known from Monte Carlo calculations as a function of the dimensionless coupling constant

$$\Gamma = \frac{z^2 e^2}{a_{\mathrm{i}} k_{\mathrm{B}} T} \tag{11.2.12}$$

where $a_{\mathrm{i}} = (3/4\pi\rho_{\mathrm{i}})^{1/3}$ is the "ion-sphere" radius (Hansen, 1973; Slattery *et al.*, 1980). At densities and temperatures typical of liquid alkali metals near the melting point, $\Gamma \simeq 180$ to 210. In this range, the internal energy of the OCP is accurately given by the simple "ion-sphere" model in which each ion interacts only with the neutralizing background contained in a sphere of radius a_{i} centred on the ion. An elementary electrostatic calculation (Baus and Hansen, 1980; Ichimaru, 1982) shows that

$$U_{\mathrm{OCP}} = 3N_{\mathrm{i}} k_{\mathrm{B}} T \left(1 - \frac{3\Gamma}{10}\right) \tag{11.2.13}$$

Finally, U_0 can be calculated as a function of r_{c} from Eqns (11.2.8) and (11.2.9). On assembling the results, noting that $a_{\mathrm{i}} = z^{1/3} a_{\mathrm{e}}$, and expressing energies in Rydbergs and lengths in Bohr radii, we find that the total internal energy of the liquid metal is

$$\frac{U}{N_i} = -\frac{1.8z^{5/3}}{r_s} + \frac{6z^{5/3}}{\Gamma r_s} + \frac{3zr_c^2}{r_s^3} + z\left(\frac{2.21}{r_s^2} - \frac{0.916}{r_s} + 0.031 \log r_s - 0.115\right) \tag{11.2.14}$$

The corresponding equation of state is

$$\frac{PV}{N_i} = -\frac{0.6z^{5/3}}{r_s} + \frac{3z^{5/3}}{\Gamma r_s} + \frac{3zr_c^2}{r_s^3} + z\left(\frac{1.47}{r_s^2} - \frac{0.305}{r_s} - 0.010\right) \tag{11.2.15}$$

The zero-pressure value of r_s can be calculated from (11.2.15) for the choice of r_c appropriate to a particular metal. The results for the alkali metals are summarized in Table 11.2, together with the corresponding values of the compressibility and excess entropy; the latter quantity is essentially that of the OCP, since the entropy of the electron gas is negligible at the melting temperatures of alkali metals. Given the crudeness of the model, the agreement with experiment is surprisingly good.

A more accurate calculation necessitates taking account of the influence of the ionic component on the structure of the electron gas and vice versa. To do so, we must go to second order in perturbation theory. We also use the adiabatic approximation: the electrons are assumed to adjust themselves instantaneously to the much slower changes in the ionic coordinates. Thus

TABLE 11.2. *Densities and isothermal compressibilities of liquid alkali metals near the melting point as calculated from the first-order perturbation theory of Eqn (11.2.15); r_c is the core radius (Hartmann, 1971) and r_s is the electron-sphere radius, in atomic units, and Γ_m is the Coulomb coupling parameter (11.2.12) at melting. The experimental compressibilities are from Webber and Stephens (1968) and Huijben and van der Lugt (1979)*

		r_s		χ_T (10^{-12} cm^2 dyn^{-1})		
Metal	r_c	Calc.	Expt	Calc.	Expt	Γ_m
Li	1.4	3.44	3.30	8.2	9.3	212
Na	1.69	3.93	4.05	13.4	19.3	210
K	2.23	4.85	5.03	29.2	38.9	186
Rb	2.40	5.14	5.38	36.3	49.2	188
Cs	2.62	5.52	5.78	47.2	63	180

1 a.u. of length ≃ 0.5292 Å.

the problem to be considered is that of an interacting electron gas in the external field produced by a given ionic charge distribution; because the electron–ion pseudopotential is assumed to be weak, the influence of the external field can be treated by linear-response theory. The polarization of the electron gas by the ionic charge distribution leads to a screening of the external field and hence to a new, effective interaction between the ions. To derive the form of the effective interaction, we follow the now standard procedure of reducing the electron–ion problem to that of a one-component fluid of pseudoatoms (Ashcroft and Stroud, 1978; Evans, 1978).

The partition function corresponding to the hamiltonian (11.2.5) is

$$Q_{N_i, N_e}(V, T) = \frac{1}{N_i!\, h^{3N_i}} \int d\mathbf{R}^{N_i}\, d\mathbf{P}^{N_i} \exp(-\beta \mathcal{H}_i') \times Tr_e \exp[-\beta(\mathcal{H}_e' + V_{ei}')] \tag{11.2.16}$$

The trace is taken over a complete set of states of the electron gas in the field due to a fixed ionic configuration; the free energy of this inhomogeneous electron gas is a function of the ionic coordinates $\{\mathbf{R}_l\}$:

$$\beta A_{eg}'(\mathbf{R}_1, \ldots, \mathbf{R}_{N_i}) = -\log\{Tr_e \exp[-\beta(\mathcal{H}_e' + V_{ei}')]\} \tag{11.2.17}$$

If the homogeneous electron gas is taken as a reference system and V_{ei}' is again treated as a perturbation, the free energy (11.2.17) is obtained from the coupling-parameter formula (6.2.5) as

$$A_{eg}' = A_{eg} + \int_0^1 \langle V_{ei}' \rangle_\lambda \, d\lambda \tag{11.2.18}$$

where the subscript λ denotes that the average is to be taken over an ensemble characterized by the hamiltonian $(\mathcal{H}_e' + \lambda V_{ei}')$. From Eqn (11.2.7c), with $\hat{v}_{ei}(k)$ replaced by $\hat{v}_{ei}^{ps}(k)$, we find that for a fixed ionic configuration

$$\langle V_{ei}' \rangle_\lambda = U_0 + \frac{1}{V} \sum_{\mathbf{k}}' \hat{v}_{ei}^{ps}(k) \rho_{\mathbf{k}}^{i} \langle \rho_{-\mathbf{k}}^{e} \rangle_\lambda \tag{11.2.19}$$

The result of first-order perturbation theory corresponds to setting $\lambda = 0$ in (11.2.19); $\langle \rho_{\mathbf{k}}^{e} \rangle_0$ is zero because the reference system is homogeneous. To obtain the second-order result, it is sufficient to calculate the components of the induced electron density to first order in $\lambda V_{ei}'(\{\mathbf{R}_l\})$. If $\chi_e(k)$ is the static electron-density response function, the induced density is

$$\langle \rho_{-\mathbf{k}}^{e} \rangle_\lambda = \chi_e(k) \lambda \hat{v}_{ei}^{ps}(k) \rho_{-\mathbf{k}}^{i} \tag{11.2.20}$$

If we now substitute for $\langle V'_{\mathrm{ei}}\rangle_\lambda$ in Eqn (11.2.18) and integrate over λ, we find that the free energy of the electron gas is given to second order in the electron–ion coupling by

$$A'_{\mathrm{eg}} = A_{\mathrm{eg}} + U_0 + \frac{1}{2V}\sum_{\mathbf{k}}{}' \chi_{\mathrm{e}}(k)[\hat{v}^{\mathrm{ps}}(k)]^2 \rho^{\mathrm{i}}_{\mathbf{k}}\rho^{\mathrm{i}}_{-\mathbf{k}} \tag{11.2.21}$$

By comparing this result with Eqns (11.2.16) and (11.2.17), we see that the problem has been reduced, as required, to that of a one-component fluid for which the total interaction energy is

$$V_{N_{\mathrm{i}}}(\{\mathbf{R}_l\}) = V_0 + \frac{1}{2V}\sum_{k}{}'\{\hat{v}_{\mathrm{ii}}(k) + \chi_{\mathrm{e}}(k)[\hat{v}^{\mathrm{ps}}_{\mathrm{ei}}(k)]^2\}(\rho^{\mathrm{i}}_{\mathbf{k}}\rho^{\mathrm{i}}_{-\mathbf{k}} - N_{\mathrm{i}}) \tag{11.2.22}$$

where

$$V_0 = U_0 + A_{\mathrm{eg}} + \tfrac{1}{2}\rho_{\mathrm{i}}\sum_{\mathbf{k}}{}' \chi_{\mathrm{e}}(k)[\hat{v}^{\mathrm{ps}}_{\mathrm{ei}}(k)]^2 \tag{11.2.23}$$

is a contribution that is independent of the structure of the liquid. If, as is generally the case, $T \ll T_{\mathrm{F}}$, the free energy A_{eg} can be replaced by the ground-state energy of the interacting electron gas, as given by Eqn (11.1.4).

The structure-dependent term in Eqn (11.2.22) can be rewritten as a sum of pair interactions in the form

$$V_{N_{\mathrm{i}}} = V_0 + \tfrac{1}{2}\sum_{l\neq l'}^{N_{\mathrm{i}}}\sum v^{\mathrm{eff}}_{\mathrm{ii}}(|\mathbf{R}_l - \mathbf{R}_{l'}|) \tag{11.2.24}$$

The effective ion–ion potential $v^{\mathrm{eff}}_{\mathrm{ii}}(R)$ is the Fourier transform of the sum of the bare ion–ion interaction and an electron-induced contribution, i.e.

$$\begin{aligned}\hat{v}^{\mathrm{eff}}_{\mathrm{ii}}(k) &= \hat{v}_{\mathrm{ii}}(k) + \hat{v}'_{\mathrm{ii}}(k)\\ &= \frac{4\pi z^2 e^2}{k^2} + [\hat{v}^{\mathrm{ps}}_{\mathrm{ei}}(k)]^2\chi_{\mathrm{e}}(k)\\ &= \frac{4\pi z^2 e^2}{k^2} + \frac{[\hat{v}^{\mathrm{ps}}_{\mathrm{ei}}(k)]^2}{(4\pi e^2/k^2)}\left(\frac{1}{\varepsilon_{\mathrm{e}}(k)} - 1\right)\end{aligned} \tag{11.2.25}$$

where $\varepsilon_{\mathrm{e}}(k)$, the dielectric function of the electron gas, is related to the susceptibility $\chi_{\mathrm{e}}(k)$ by the analogue of Eqn (10.2.11); the properties of the function $\varepsilon_{\mathrm{e}}(k)$ are well documented (Singwi and Tosi, 1981; Ichimaru, 1982). In the long-wavelength limit, $\varepsilon_{\mathrm{e}}(k)$ behaves as

$$\lim_{k\to 0} k^2\varepsilon_{\mathrm{e}}(k) = k^2_{\mathrm{TF}}\frac{\chi_{\mathrm{Te}}}{\chi^0_{\mathrm{Te}}} \tag{11.2.26}$$

where $\chi_{\rm Te}$ and $\chi^0_{\rm Te}$ are the isothermal compressibilities, respectively, of the interacting and non-interacting electron gas, and $k_{\rm TF}=2(k_{\rm F}/\pi a_0)^{1/2}$ is the Thomas–Fermi wavenumber; Eqn (11.2.26) is analogous to the compressibility sum rule (10.2.32) satisfied by the classical OCP. In the same limit, $\hat{v}^{\rm ps}_{\rm ei}(k)\simeq 4\pi ze^2/k^2$. It follows that the effective interaction $\hat{v}^{\rm eff}_{\rm ii}(k)$ defined by Eqn (11.2.25) is a regular function in the limit $k\rightarrow 0$, the k^{-2} singularity in the bare potential $\hat{v}_{\rm ii}(k)$ being cancelled by the same singularity in $\hat{v}'_{\rm ii}(k)$. In other words, the bare ion–ion potential $v_{\rm ii}(R)$ is completely screened by the polarization of the electron gas, and the effective potential $v^{\rm eff}_{\rm ii}(R)$ is a short-range function. A typical effective potential has a soft repulsive core, an attractive well with a depth (in temperature units) of a few hundred degrees kelvin, and a weakly oscillatory tail of the form $R^{-3}\cos 2k_{\rm F}R$; the so-called Friedel oscillations at large R are linked to a logarithmic singularity in $\varepsilon_{\rm e}(k)$ at $k=2k_{\rm F}$ (Pines and Nozières, 1966). An example of a calculated effective potential for liquid potassium is shown in Figure 1.2.

The results of this section can be summarized by saying that we have reduced the liquid-metal problem to one of calculating the classical partition function of a fluid of $N_{\rm i}$ pseudoatoms in which the particles interact through a short-range effective potential $v^{\rm eff}_{\rm ii}(R)$. After integration over momenta, the partition function can be written as

$$Q_{N_{\rm i}}(V,T)=\frac{\exp(-\beta V_0)}{N_{\rm i}!\,\Lambda_{\rm i}^{3N_{\rm i}}}\int\exp\left[-\tfrac{1}{2}\beta\sum_{l\neq l'}^{N_{\rm i}}\sum v^{\rm eff}_{\rm ii}(R_{ll'})\right]{\rm d}\mathbf{R}^{N_{\rm i}}\qquad(11.2.27)$$

where $R_{ll'}\equiv|\mathbf{R}_l-\mathbf{R}_{l'}|$ and $\Lambda_{\rm i}$ is the de Broglie thermal wavelength of the ions. Equation (11.2.27) differs from the usual partition function of a monatomic fluid in two important ways: first, in the appearance of a large structure-independent energy V_0; and, secondly, in the fact that both V_0 and $v^{\rm eff}_{\rm ii}(R)$ are functions of density by virtue of the density dependence of the properties of the electron gas. The reduction of the problem to the form described by Eqn (11.2.27) means that the theoretical methods developed for simple classical liquids can also be applied to liquid metals. Special care is needed only when evaluating volume derivatives of the free energy, because the density dependence of the effective interaction gives rise to extra terms. In particular, the pressure equation (2.5.14) must be replaced by

$$\frac{\beta P}{\rho_{\rm i}}=1+\rho_{\rm i}\frac{{\rm d}V_0(\rho_{\rm i})}{{\rm d}\rho_{\rm i}}$$
$$-\tfrac{1}{2}\beta\rho_{\rm i}\int g_{\rm ii}(R)\left(\frac{R}{3}\frac{\partial}{\partial R}-\rho_{\rm i}\frac{\partial}{\partial\rho_{\rm i}}\right)v^{\rm eff}_{\rm ii}(R)\,{\rm d}\mathbf{R}\qquad(11.2.28)$$

where $\rho_{\rm i}=N_{\rm i}/V$ and $g_{\rm ii}(R)$ is the ion–ion pair distribution function (Ascarelli and Harrison, 1969).

11.3 IONIC STRUCTURE

The microscopic structure of liquid metals can be discussed in terms of the three partial structure factors $S_{ii}(k)$, $S_{ei}(k)$ and $S_{ee}(k)$. The ion–ion structure factor $S_{ii}(k)$ is directly measurable by neutron diffraction, while X-ray diffraction yields a linear combination of the three functions because the photons are scattered by both core and conduction electrons (Egelstaff *et al.*, 1974). The contributions of $S_{ei}(k)$ and $S_{ee}(k)$ to the X-ray structure factor will generally be small, however, because for most liquid metals the number of conduction electrons is much smaller than the number of electrons in the core. It is therefore difficult to obtain reliable experimental estimates of either of these functions, and the discussion in this section will be concentrated on the structure of the ionic component. The partial structure factors, like those of any binary fluid, are defined by Eqn (5.1.33), where $N = N_i + N_e$. The interpretation of a liquid metal as a system of N_i pseudoatoms also makes it convenient to define an ionic structure factor $S(k)$ as

$$S(k) = \frac{1}{N_i}\langle \rho^i_{\mathbf{k}} \rho^i_{-\mathbf{k}} \rangle = \left(\frac{N_i + N_e}{N_i}\right) S_{ii}(k)$$
$$= (1+z) S_{ii}(k) \tag{11.3.1}$$

The long-wavelength limits of the three partial structure factors are related to the compressibility χ_T by the equivalent of Eqn (10.2.24):

$$\lim_{k\to 0} S_{ee}(k) = \lim_{k\to 0} z S_{ei}(k) = \lim_{k\to 0} z^2 S_{ii}(k)$$
$$= \left(\frac{z}{1+z}\right) \rho_i k_B T \chi_T \tag{11.3.2}$$

Theories of the structure and thermodynamics of simple liquid metals have developed along two main lines. In one case, the ions are treated as an OCP (Minoo *et al.*, 1977), and the effects of the electron–ion interaction on the properties of the ionic component are calculated by perturbation theory (Chaturvedi *et al.*, 1981; Tosi, 1983). The alternative approach is based on the methods described in the preceding section. The metal is treated as a one-component system in which the particles interact through an effective, short-range pair potential. The usual integral-equation and perturbation-theory methods can then be applied; if a perturbation approach is adopted, the natural choice of reference system is the hard-sphere fluid. The two points of view are to some extent complementary. Both the OCP and the hard-sphere fluid give a fair overall fit to the ionic structure factors of the alkali metals, as is evident from the results for liquid potassium shown in Fig. 11.1. Each is therefore suitable for use as a reference system. However, whereas the hard-sphere model is particularly well suited to a description of the small-k region, the OCP gives better results at large k, where the details of the oscillations are determined by the essentially coulombic repulsion between ions at short distances.

Use of the OCP as a reference system has the advantage that good results for both structure and thermodynamics are obtained in zeroth order without the introduction of any adjustable parameter: the appropriate value of the coupling constant Γ of Eqn (11.2.12) is determined unambiguously by the density, temperature and ionic valence of the metal of interest. Some results obtained in first order for the thermodynamic properties of liquid alkali metals are compared with experimental data in Table 11.2. To the same order, the ionic structure factor is given by $S(k) \simeq S_{\text{OCP}}(k)$. As we have already noted, this is a good approximation at large k. At small k, however, $S_{\text{OCP}}(k)$ vanishes as k^2, but $S(k)$ approaches a non-zero value, proportional to χ_{T} (see Eqns (10.2.33) and (11.3.2)). This deficiency can be corrected by treating the difference between the bare and effective ion–ion potentials as

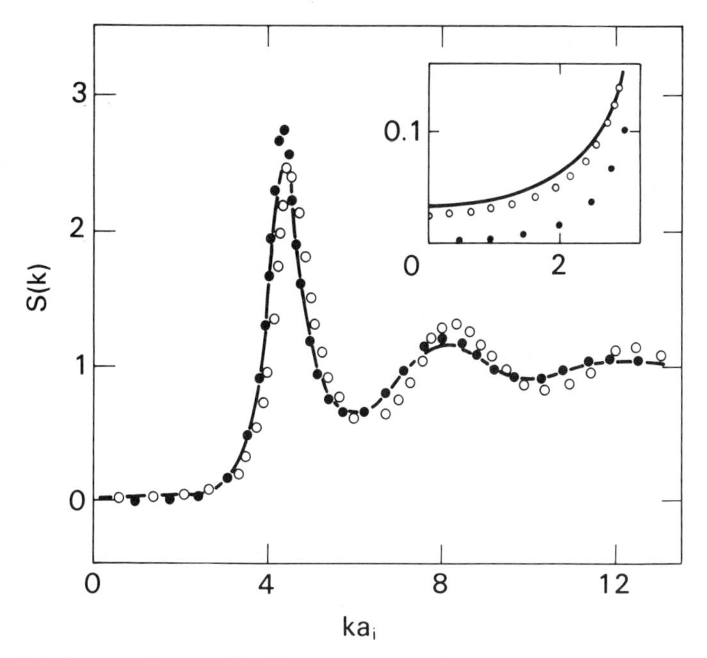

FIG. 11.1. Structure factor of liquid potassium at 135°C. The curve shows the X-ray results of Greenfield *et al.* (1971), the closed circles are Monte Carlo results for the OCP at $\Gamma = 155$, and the open circles are results for the hard-sphere fluid at $\eta = 0.46$. The unit of wavenumber is the inverse ion-sphere radius, defined after Eqn (11.2.12). The inset shows an enlargement of the behaviour at small k. After Minoo *et al.* (1977).

a perturbation. A simple scheme (Chaturvedi *et al.*, 1981) consists of replacing the term $\hat{\Phi}(k)$ in the RPA expression (6.5.33) by $-\beta\hat{v}'_{ii}(k)$, where $\hat{v}'_{ii}(k)$ is defined by Eqn (11.2.25). If the electron–ion pseudopotential is assumed to have the "empty-core" form (11.2.9), the expression obtained for the ionic structure factor in the case of a monovalent metal is

$$S(k) = \frac{S_{OCP}(k)}{1+\left(\frac{k_D}{k}\cos kr_c\right)^2\left(\frac{1}{\varepsilon_e(k)}-1\right)S_{OCP}(k)} \tag{11.3.3}$$

where k_D is the Debye wavenumber, defined by Eqn (5.3.20). The small-k limit of $S(k)$ can now be obtained from the results obtained previously for $S_{OCP}(k)$ (Eqn (10.2.33)) and $\varepsilon_e(k)$ (Eqn (11.2.26)). We rewrite the latter as

$$\lim_{k\to 0}\frac{1}{\varepsilon_e(k)} = \frac{k^2}{k_e^2} \tag{11.3.4}$$

where $k_e^2 = (\chi_{Te}/\chi_{Te}^0)k_{TF}^2$ is the square of the electronic "screening wavenumber". The ionic (OCP) screening wavenumber, previously defined by Eqn (10.2.31), can be written in similar form as $k_s^2 = (\chi_{Ti}/\chi_{Ti}^0)k_D^2$. Then

$$\lim_{k\to 0} S(k) = \left(\frac{k_D^2}{k_s^2} + \frac{k_D^2}{k_e^2} + k_D^2 r_c^2\right)^{-1} \tag{11.3.5}$$

Given that $\chi_{Te}^0 = 3/2\rho_e\varepsilon_F$ (from Eqn (11.1.1)) and $\chi_{Ti}^0 = \beta/\rho_i$, we see from Eqns (11.3.1) and (11.3.5) that the compressibility can be written in a form that emphasizes the two-component nature of the problem:

$$\chi_T^{-1} = \chi_{Ti}^{-1} + \chi_{Te}^{-1} + 4\pi(\rho_i zer_c)^2 \tag{11.3.6}$$

Compressibilities calculated from Eqn (11.3.6) and the known compressibilities of the OCP (Baus and Hansen, 1980) and the electron gas are in good agreement with experimental data on liquid alkali metals (Chaturvedi *et al.*, 1981; Evans and Sluckin, 1981). Note that the alkali metals are significantly less compressible than the rare-gas liquids; close to the melting point, the ratio χ_T/χ_T^0 is typically 0.02 for the metals and 0.05 for the rare gases.

The condition (10.2.12) ($1/\varepsilon_e(k) \leqslant 1$ for all k) implies that the structure factors predicted by Eqn (11.3.3) lie systematically above those of the OCP reference system. In practice, we can expect the main peak in the true structure factor to be less pronounced than in the OCP, since the correlations between ions are weakened by electron screening. This inadequacy in the theory is linked to the fact that the RPA is a satisfactory approximation only at small wavenumbers. Considerable improvement in the wavenumber range around and beyond the main peak can be obtained by going to the "optimized" (ORPA) version of the RPA, as described in Section 6.5. Use of the ORPA gives results in excellent agreement with the measured structure factors of alkali metals (Pastore and Tosi, 1984). The experimental data have a "uniform fluid" property, i.e. the density dependence of $S(k)$ is described (Egelstaff *et al.*, 1980) to a good approximation by the expression

$$\rho_i\left(\frac{\partial S(k)}{\partial \rho_i}\right)_T \simeq -\frac{k}{3}\left(\frac{\partial S(k)}{\partial k}\right)_T \tag{11.3.7}$$

This behaviour is readily understandable in terms of a theory based on the choice of the OCP as a reference system. The particular form of the Coulomb potential means that the OCP structure factors scale with density as

$$S_{OCP}(k;\, T, \rho_i) = S_{OCP}(q = a_i k;\, \Gamma) \tag{11.3.8}$$

If $S(k) \simeq S_{OCP}(k)$, Eqn (11.3.7) is an immediate consequence of (11.3.8), provided the variation of $S_{OCP}(q)$ with Γ over the density range of interest is small enough to be ignored (Hayter *et al.*, 1983).

Although Eqn (11.3.3) and its ORPA modification are useful approximations for monovalent metals, they are less successful in the polyvalent case. The polarization of the electron gas is much stronger in polyvalent metals; the effective repulsion between ions is therefore stiffer, and the natural choice of reference system is the hard-sphere fluid rather than the OCP. Indeed, as we have already discussed in Section 6.1, the structure factors of simple liquid metals near their melting points, including the alkali metals, are reasonably well fitted the PY results for the hard-sphere fluid at a packing fraction $\eta \simeq 0.45$ (Ashcroft and Lekner, 1966). The success of the hard-sphere model is illustrated by the results for liquid potassium plotted in Figure 11.1.

A number of more elaborate calculations have been carried out, based on the Gibbs–Bogoliubov inequality (6.2.21), in which the hard-sphere packing fraction is treated as a variational parameter (Ashcroft and Stroud, 1978; Evans, 1978). In these calculations, the entropy of the liquid metal is equal to that of the hard-sphere fluid at the optimized packing fraction (Silbert *et al.*, 1975). With the exception of liquid gallium, the resulting excess entropies are systematically lower than the experimental values for simple metals; in the case of the alkali metals, better results are obtained from variational calculations in which the OCP is chosen as the reference system and the free energy is minimized with respect to variations in the coupling constant Γ (Galam and Hansen, 1976; Ross *et al.*, 1981; Mon *et al.*, 1981; Iwamatsu *et al.*, 1982). The perturbation theory of Weeks, Chandler and Andersen (1972) discussed in Chapter 6 has also been applied to simple liquid metals (Evans, 1978; Regnaut *et al.*, 1980). The theory gives results for the entropy that are an improvement over those obtained by the variational method, but the calculated structure factors agree well with experiment only for the polyvalent metals. The comparative failure of the WCA approach in the case of the alkali metals is attributable to the poor convergence of the "blip-function" expansion for the soft effective ion–ion potentials that are characteristic of these systems. The results are improved if the blip-function method is used in the modified form represented by Eqn (6.3.21).

The pair distribution functions of a number of liquid metals have been calculated by computer simulation for a variety of effective ion–ion potentials. Excellent agreement with experimental data has been obtained in many cases (see, for example, Day *et al.*, 1979; Jacucci *et al.*, 1981), but this cannot be taken as strong evidence of the adequacy of the models used in the simulations because the calculated structure factors are insensitive to details of the pair potentials (Schiff, 1969). Some metals, including gallium, tin, antimony and bismuth, have an unusual structural feature in the form of a shoulder on the large-k side of the main peak in $S(k)$ (Narten, 1972). This behaviour is reproduced, at least qualitatively, by models in which the particles interact through a "square-mound" potential; this is identical in form to Eqn (1.2.2), except for being *positive* in the interval $d < r < \gamma d$ (Silbert and Young, 1976; Levesque and Weis, 1977). Calculations based on more realistic potentials give similar results (Regnaut *et al.*, 1980).

By an inversion of the usual procedure, experimental structure-factor data can be used as input to a theory of liquid-state structure in order to obtain information on the interparticle potential. Many attempts have been made to use such a scheme to determine the effective ion–ion potential in liquid metals, but the lack of sensitivity of the structure to the pair potential makes the problem a difficult one. Early calculations based on the standard integral-equation theories of Chapter 5 (Born–Green, PY and HNC) were unsuccessful because these approximations are not sufficiently accurate for the purpose of inversion (Johnston *et al.*, 1964; Howells and Enderby, 1972). Later it was recognized that thermodynamic properties provide a more sensitive probe of the potential than does the structure factor; this has led to the use of thermodynamically consistent integral equations (see Section 5.8), by means of which satisfactory results have been obtained for rubidium (Mitra *et al.*, 1976). Other methods have been based on thermodynamic perturbation theory (Jacobs and Andersen, 1975; Mitra, 1978) and on the MSA closure for fluids with continuous potentials, Eqn (5.6.9) (Madden and Rice, 1980). The most accurate procedure devised so far appears to be an iterative predictor–corrector method in which the RHNC approximation is used in combination with repeated simulations (Levesque *et al.*, 1985);

the technique relies for its success on the availability of very accurate structural data, i.e. results for $S(k)$ with experimental errors at the one per cent level over a wide range of k.

11.4 LIQUID ALLOYS

The study of liquid alloys, as opposed to pure metals, requires consideration to be taken of concentration fluctuations. Much of the material in this section is equally applicable to mixtures of non-metallic liquids, but the formalism we describe has mostly been exploited in work on metals. We consider only binary alloys, consisting of N_ν ions of species ν, with $\nu=1$ or 2, plus the conduction electrons. The mole fractions of the two species are therefore $x_1=N_1/N$ and $x_2=1-x_1$, where $N=N_1+N_2$. The Fourier components $\rho_{\mathbf{k}}^N$ of the microscopic number density are defined, as in Eqn (10.2.1), in terms of the partial densities $\rho_{\mathbf{k}}^\nu$, but the charge density used throughout Chapter 10 is here replaced by the concentration density, defined as

$$\rho_{\mathbf{k}}^C=x_2\rho_{\mathbf{k}}^1-x_1\rho_{\mathbf{k}}^2 \tag{11.4.1}$$

Fluctuations in number density and concentration are described by the structure factors $S_{NN}(k)$, $S_{NC}(k)$ and $S_{CC}(k)$ (Bhatia and Thornton, 1970), where $S_{NN}(k)$ is defined by Eqn (10.2.2a) and

$$\begin{aligned} S_{NC}(k)&=\frac{1}{N}\langle\rho_{\mathbf{k}}^N\rho_{-\mathbf{k}}^C\rangle \\ &=x_2S_{11}(k)-x_1S_{22}(k)+(x_2-x_1)S_{12}(k) \end{aligned} \tag{11.4.2a}$$

$$\begin{aligned} S_{CC}(k)&=\frac{1}{N}\langle\rho_{\mathbf{k}}^C\rho_{-\mathbf{k}}^C\rangle \\ &=x_2^2S_{11}(k)+x_1^2S_{22}(k)-2x_1x_2S_{12}(k) \end{aligned} \tag{11.4.2b}$$

A straightforward generalization of the calculation of Section 5.1 shows that the coherent cross-section for neutrons scattered with a momentum transfer $\hbar\mathbf{k}$ is

$$\frac{\mathrm{d}\sigma}{\mathrm{d}\Omega}=N[\bar{b}^2S_{NN}(k)+2\bar{b}(\Delta b)S_{NC}(k)+(\Delta b)^2S_{CC}(k)] \tag{11.4.3}$$

where $\bar{b}=x_1b_1+x_2b_2$, b_ν is the coherent scattering length for nuclei of species ν, and $\Delta b=(b_1-b_2)$. A "zero-alloy" composition is one for which $\bar{b}=0$. This is attainable when one component has a negative scattering length (^{7}Li is an important example); in such a case, the cross-section is proportional to $S_{CC}(k)$.

The $k \to 0$ limits of the three structure factors can be derived either by thermodynamic fluctuation theory, in the manner outlined in Section 2.7, or by a calculation in the grand canonical ensemble that generalizes the one-component results (2.4.23) and (2.5.32) (Kirkwood and Buff, 1951). Both routes lead to the compressibility sum rule (5.2.20); the equivalent results for the partial direct correlation functions defined by the Ornstein–Zernike relation (5.2.18) is

$$\rho \hat{c}_{\nu\mu}(0) = \delta_{\nu\mu} - \frac{(\rho_\nu \rho_\mu)^{1/2}}{k_B T} \left(\frac{\partial \mu_\nu}{\partial \rho_\mu} \right)_{V,T,\rho_{\bar{\mu}}} \tag{11.4.4}$$

where μ_ν is the chemical potential of species ν, $\rho_{\bar{\mu}} = \rho - \rho_\mu$, and $\rho = \rho_1 + \rho_2$ is the total number density. By switching from conditions of constant volume to those of constant pressure, Eqns (5.2.18) and (11.4.2) can be combined with standard thermodynamic relations (Bhatia and Thornton, 1970) to give

$$S_{NN}(0) = \frac{1}{N} \langle (\Delta N)^2 \rangle = \rho k_B T \chi_T + \delta^2 S_{CC}(0) \tag{11.4.5a}$$

$$S_{NC}(0) = \frac{1}{N} \langle (\Delta N)(x_2 \Delta N_1 - x_1 \Delta N_2) \rangle = -\delta S_{CC}(0) \tag{11.4.5b}$$

$$S_{CC}(0) = \frac{1}{N} \langle (x_2 \Delta N_1 - x_1 \Delta N_2)^2 \rangle = \frac{N k_B T}{\left(\dfrac{\partial^2 G}{\partial x_1^2} \right)_{N,P,T}} \tag{11.4.5c}$$

where $\Delta N_\nu = N_\nu - \langle N_\nu \rangle$, $\delta = \rho(v_1 - v_2)$ is the "dilation factor" and $v_\nu = (\partial V / \partial N_\nu)_{N_{\bar{\nu}},P,T}$ is the partial molar volume of species ν. In a strictly substitutional alloy, the dilation factor is zero; number-density and concentration fluctuations are therefore independent in the long-wavelength limit, and the number-fluctuation formula (11.4.5a) reduces to its familiar one-component form (see Eqn (5.1.13)).

The methods of conformal-solution theory developed in Section 6.6 can be used to relate the thermodynamic properties of the alloy to those of a single-component reference fluid. If the excess Gibbs free energy has the simple form given by (6.6.4), Eqn (11.4.5c) becomes

$$S_{CC}(0) = \frac{x_1 x_2}{1 - 2 x_1 x_2 \beta w} \tag{11.4.6}$$

where the "interchange" energy w can be determined from experimental results for the heat of mixing via Eqn (6.6.5). In terms of conformal solution theory, the interchange energy has the meaning contained in Eqns (6.6.23) and (6.6.24). Within the random-phase approximation, on the other hand (see, for example, Eqn (6.5.33)), w has a natural interpretation (Copestake *et al.*, 1983) as the volume integral of the "ordering potential" $w_{\text{ord}}(r)$, defined as

$$w_{\text{ord}}(r) = v_{12}(r) - \tfrac{1}{2}[v_{11}(r) + v_{22}(r)] \tag{11.4.7}$$

Equation (11.4.6) has been used in the analysis of experimental data on moderately symmetric alloys, including KCs (Alblas *et al.*, 1981). In "regular-solution" theory, w is assumed to be temperature independent; if it is positive, it is clear from (11.4.6) that $S_{CC}(0)$ diverges at a critical temperature given by $T_c = w/2k_B$. Below T_c, the alloy separates into two fluid phases; close to phase separation, fluctuations in concentration become very large, in the manner of the density fluctuations close to the liquid–gas critical point. In an ideal solution, $w=0$ and $S_{CC}(0) = x_1x_2$; this provides a fair first estimate of $S_{CC}(0)$ for many alloys.

Conformal-solution theory breaks down for highly disymmetric alloys. Such systems are characterized by large volume ratios of the constituent pseudoatoms and hence by large values of the dilation factor; the alloy NaCs is an example. Size effects can be incorporated into a theory by focusing on the concentrations by volume rather than on mole fractions. This corresponds in practice to the introduction of a composition-dependent interchange energy in Eqn (11.4.6). The symmetry around $x_1 = x_2 = \frac{1}{2}$ is thereby broken (Bhatia and March, 1975), in agreement with the experimental findings.

For $k > 0$, $S_{NN}(k)$ behaves much like the structure factor of a pure metal, whereas the functions $S_{CC}(k)$ and $S_{NC}(k)$ oscillate weakly around x_1x_2 and zero, respectively, at least for alloys that are nearly random in nature; typical examples are plotted in Figure 11.2. The behaviour of $S_{CC}(k)$ therefore contrasts sharply with the strongly oscillatory form of the function $S_{ZZ}(k)$ in molten salts (see Figure 10.2). The oscillations in $S_{ZZ}(k)$ are characteristic of a pronounced heterocoordination (charge ordering) that is absent in most alloys. The main features of $S_{NN}(k)$, $S_{NC}(k)$ and $S_{CC}(k)$ are well reproduced by a model consisting of hard spheres with unequal but additive diameters (Ashcroft and Langreth, 1967). Variational calculations based on the Gibbs-Bogoliubov inequality, with a hard-sphere mixture used as a reference system, give results for excess thermodynamic properties of simple-metal alloys in fair agreement with experiment (Ashcroft and Stroud, 1978). In particular, calculations of this type reproduce semiquantitatively the asymmetric phase-separation curve of the alloy LiNa (Stroud, 1973).

The approach, described in Section 11.3, whereby a liquid metal is treated as an ion–electron plasma is easily extended to the calculation of the ionic structure of liquid alloys (Chaturvedi *et al.*, 1981). An application to the alloy NaK shows, however, that linear electron-screening theory cannot account for the electronic charge transfer from potassium to sodium that

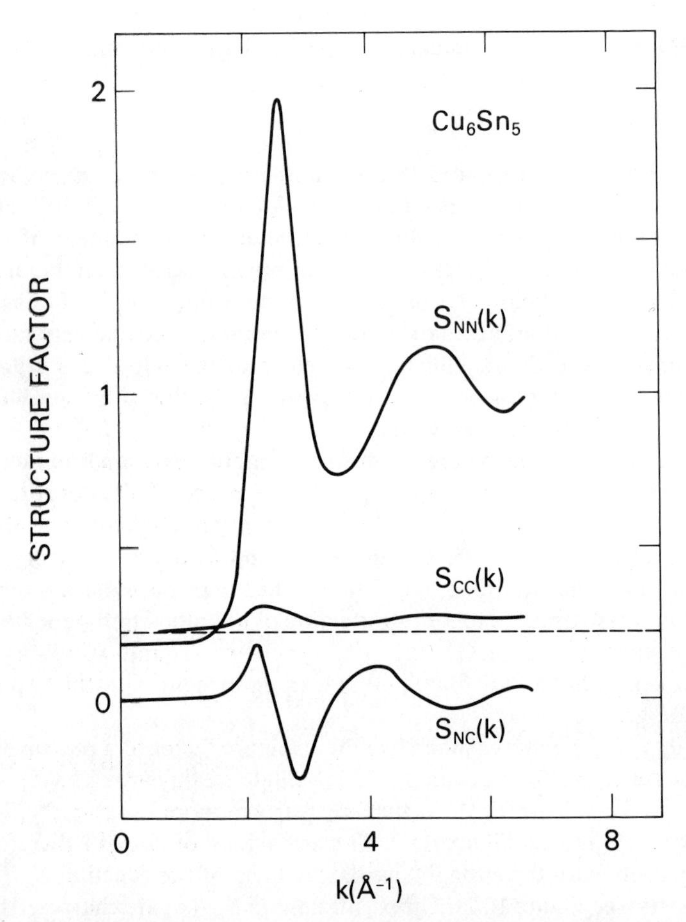

FIG 11.2. Neutron-scattering results for the number–number, concentration–concentration and number–concentration structure factors of the alloy Cu_6Sn_5. Note that $S_{CC}(0)$ (dashed line) differs appreciably from its value in an ideal mixture, i.e. $x_1x_2 \simeq 0.248$ (see text); the latter is also the value of $\lim_{k\to\infty} S_{CC}(k)$. After Bhatia and Thornton (1970).

occurs in the alloying process. This failure of the method shows up most clearly in calculations of long-wavelength concentration fluctuations, whereas number-density fluctuations are reasonably well described by the theory.

11.5. ELECTRON TRANSPORT

The electrical and thermal conductivities of liquid metals are about three orders of magnitude larger than those of molten salts under comparable physical conditions. The reason for this, of course, is the very high mobility of the conduction electrons in metals. On melting, the electrical conductivity of simple metals typically decreases by only a factor of two. This implies that the ionic disorder consequent on melting does not strongly modify the electronic states, although the existence of Bloch states is precluded by the absence of strict lattice periodicity. Photoemission experiments have in fact shown that the electronic density of states in many simple liquid metals is well represented by its free-electron form, namely $\rho(E) \propto E^{1/2}$. On the other hand, some semiconductors, including silicon and germanium, become metallic on melting and their conductivities increase by many orders of magnitude. Resistivities of simple liquid metals lie typically in the range from 10^{-5} to $10^{-4}\ \Omega$ cm.

It has been known experimentally since the last century that the electrical and thermal conductivities of simple metals are related by the Wiedemann–Franz law:

$$\frac{\lambda}{\sigma T} = L \tag{11.5.1}$$

where L, the Lorenz number, is almost a constant. The value of L is in practice close to that predicted by the nearly-free-electron model, i.e. $L_0 = \frac{1}{3}(\pi k_B/e)^2 \simeq 2.44 \times 10^{-8}$ W Ω K^2; this model is applicable in cases where scattering of the electrons by the ions is sufficiently weak to justify the use of the Born approximation, which is true of all simple metals. In addition, the electrons can be treated as if they were scattered independently of each other if the bare electron–ion pseudopotential $\hat{v}_{ei}^{ps}(k)$ is replaced by a suitably screened pseudopotential $\hat{w}(k)$, given by

$$\hat{w}(k) = \hat{v}_{ei}^{ps}(k)/\varepsilon_e'(k) \tag{11.5.2}$$

Note that the function $\varepsilon'(k)$ in (11.5.2) is not the same as the dielectric function in the screened ion–ion potential (11.2.25). The difference between the two functions is made clear by the following linear-response argument. Consider the ionic charge distribution as being an external field with Fourier components $\rho_{\mathbf{k}}^{i}$. The electronic density (or polarization) induced by this field is

$$\langle \rho_{\mathbf{k}}^{e} \rangle = \chi_e(k)\hat{\phi}_{ext}(k) = \chi_e(k)\hat{v}_{ei}^{ps}(k)\rho_{\mathbf{k}}^{i} \tag{11.5.3}$$

where $\chi_e(k)$ is the electron-density response function introduced in Eqn (11.2.20). The total potential $\hat{\phi}(k)$ acting on an electronic test charge is the sum of the external and induced potentials. The latter consists of the Hartree (H) mean field, together with the exchange (x) and correlation (c) contributions that arise because the self-consistent field also acts on other electrons.

Thus

$$\begin{aligned}\hat{\phi}(k) &= \hat{\phi}_{\text{ext}}(k)+\hat{\phi}_{\text{H}}(k)+\hat{\phi}_{\text{xc}}(k)\\ &= \hat{v}_{\text{ei}}^{\text{ps}}(k)\rho_{\mathbf{k}}^{\text{i}}+\hat{v}_{\text{ee}}(k)\langle\rho_{\mathbf{k}}^{\text{e}}\rangle+\hat{v}_{\text{xc}}(k)\langle\rho_{\mathbf{k}}^{\text{e}}\rangle\\ &= \hat{v}_{\text{ei}}^{\text{ps}}(k)\{1+\chi_{\text{e}}(k)\hat{v}_{\text{ee}}(k)[1-G(k)]\}\rho_{\mathbf{k}}^{\text{i}}\\ &= \hat{w}(k)\rho_{\mathbf{k}}^{\text{i}}\end{aligned} \tag{11.5.4}$$

where $G(k)=-\hat{v}_{\text{xc}}(k)/\hat{v}_{\text{ee}}(k)$ is a measure of exchange and correlation effects. Comparison of Eqns (11.5.2) and (11.5.4) shows that

$$\frac{1}{\varepsilon_{\text{e}}'(k)}=1+\frac{4\pi e^2}{k^2}[1-G(k)]\chi_{\text{e}}(k) \tag{11.5.5a}$$

The corresponding expression for $\varepsilon_{\text{e}}(k)$ (see Eqn (11.2.25)) is

$$\frac{1}{\varepsilon(k)}=1+\frac{4\pi e^2}{k^2}\chi_{\text{e}}(k) \tag{11.5.5b}$$

The factor $G(k)$ in (11.5.5a) has been calculated numerically in the range of density ($2\lesssim r_{\text{s}}\lesssim 6$) relevant to liquid metals (Singwi and Tosi, 1981; Ichimaru, 1982).

The introduction of the screened electron–ion potential $\hat{w}(k)$ reduces the many-electron problem to that of a gas of non-interacting electrons in which each electron is scattered independently by fixed ions (a Lorentz gas). The electronic transport coefficients can then be calculated from the linearized Boltzmann equation (Ziman, 1963). Let $f_0(\mathbf{k})$ and $f(\mathbf{k})=f_0(\mathbf{k})+\delta f(\mathbf{k})$ denote, respectively, the equilibrium and non-equilibrium Fermi distributions of the non-interacting electron gas. Each electron is scattered by the total potential $W=\sum_l w(\mathbf{r}-\mathbf{R}_l)$ appropriate to a given ionic charge distribution. The Boltzmann equation describes the balance between the external forces acting on the electrons and the electron–ion collisions. If $\mathbf{E}$ is the external field, the transport equation (see Eqns (2.1.20) and (2.1.21)) in the presence of a temperature gradient is

$$\begin{aligned}&\frac{\hbar\mathbf{k}_0}{m_{\text{e}}}\cdot\nabla_{\mathbf{r}}f(\mathbf{k}_0)-\frac{e\mathbf{E}}{\hbar}\cdot\nabla_{\mathbf{k}_0}f(\mathbf{k}_0)\\ &\quad=\frac{V}{(2\pi)^3}\int P(\mathbf{k}_0,\mathbf{k}_1)\{f(\mathbf{k}_0)[1-f(\mathbf{k}_1)]-f(\mathbf{k}_1)[1-f(\mathbf{k}_0)]\}\;\mathrm{d}\mathbf{k}_1\end{aligned} \tag{11.5.6}$$

where $V/(2\pi)^3$ is the density of states in $\mathbf{k}$-space and $P(\mathbf{k}_0,\mathbf{k}_1)$ is the probability per unit time of a transition between plane-wave states $|\mathbf{k}_0\rangle$ and

$|\mathbf{k}_1\rangle$, with $P(\mathbf{k}_0, \mathbf{k}_1) = P(\mathbf{k}_1, \mathbf{k}_0)$ (microscopic reversibility). The form of the collision term in Eqn (11.5.6) satisfies the Pauli principle, since transitions are allowed only between occupied initial states and empty final states. For the sake of simplicity, we consider only the situation in which there is no temperature gradient; the electron gas is therefore homogeneous. We also ignore ion-recoil effects; this assumption is justified by the large ion-to-electron mass ratio, and implies that the scattering process is essentially elastic, i.e. $|\mathbf{k}_0| \simeq |\mathbf{k}_1|$ and $f_0(\mathbf{k}_0) \simeq f_0(\mathbf{k}_1)$. Since the collision term is linear in δf, we linearize the left-hand side of Eqn (11.5.6), and are left with

$$-\frac{e\mathbf{E}}{\hbar} \cdot \nabla_{\mathbf{k}_0} f_0(\mathbf{k}_0) = \frac{V}{(2\pi)^3} \int P(\mathbf{k}_0, \mathbf{k}_1)[\delta f(\mathbf{k}_0) - \delta f(\mathbf{k}_1)]\, \mathrm{d}\mathbf{k}_1 \quad (11.5.7)$$

The electrons couple to the ionic density $\rho^{\mathrm{i}}_{\mathbf{k}}$ in a manner that is very similar to the scattering of thermal neutrons by density fluctuations. The calculation of the transition probability $P(\mathbf{k}_0, \mathbf{k}_1)$ therefore proceeds along the same lines as the calculation of the neutron scattering cross-section in Section 7.5. (Baym, 1964). According to the "golden rule":

$$P(\mathbf{k}_0, \mathbf{k}_1) = \frac{2\pi}{\hbar} |\langle \mathbf{k}_0 | W | \mathbf{k}_1 \rangle|^2 \delta(\varepsilon_0 - \varepsilon_1)$$

$$= \frac{2\pi}{\hbar} \frac{N_{\mathrm{i}}}{V^2} S(k) |\hat{w}(k)|^2 \delta(\varepsilon_0 - \varepsilon_1) \quad (11.5.8)$$

where $k = |\mathbf{k}_0 - \mathbf{k}_1|$ and is related to the scattering angle θ by Eqn (5.1.19), $\varepsilon = \hbar^2 k^2 / 2m_{\mathrm{e}}$ and $S(k)$ is the ionic structure factor defined by Eqn (11.3.1). The relation between k and θ means that the transition probability (11.5.8) can be expressed as $P(\mathbf{k}_0, \mathbf{k}_1) = P(k_0, \theta)\delta(\varepsilon_0 - \varepsilon_1)$. If, in addition, we make the change of variable

$$\mathrm{d}\mathbf{k}_1 = k_1^2\, \mathrm{d}k_1\, \mathrm{d}\Omega_1 = (m_{\mathrm{e}}/\hbar^2) k_1\, \mathrm{d}\varepsilon_1\, \mathrm{d}\Omega_1 \quad (11.5.9)$$

where $\mathrm{d}\Omega_1$ is an element of solid angle, the transport equation (11.5.7) becomes

$$-\frac{e\mathbf{E}}{\hbar} \cdot \nabla_{\mathbf{k}_0} f_0(\mathbf{k}_0) = -e\mathbf{E} \cdot \frac{\hbar \mathbf{k}_0}{m_{\mathrm{e}}} \frac{\partial f_0(\varepsilon_0)}{\partial \varepsilon_0}$$

$$= \frac{V m_{\mathrm{e}}}{8\pi^3 \hbar^2} \int_0^\infty \mathrm{d}\varepsilon_1 \int \mathrm{d}\Omega_1\, k_1 P(k_0, \theta)[\delta f(\mathbf{k}_1) - \delta f(\mathbf{k}_0)]$$

$$= \frac{V m_{\mathrm{e}} k_0}{8\pi^3 \hbar^2} \int \mathrm{d}\Omega_1\, P(k_0, \theta)[\delta f(\mathbf{k}_1) - \delta f(\mathbf{k}_0)] \quad (11.5.10)$$

Equation (11.5.10) has an exact solution of the form

$$\delta f(\mathbf{k}) = \tau(k)\frac{e\mathbf{E}}{\hbar}\cdot\nabla_{\mathbf{k}} f_0(\varepsilon) = \tau(k)e\mathbf{E}\cdot\frac{\hbar\mathbf{k}}{m_e}\frac{\partial f_0}{\partial\varepsilon} \tag{11.5.11}$$

and substitution of this result in Eqn (11.5.10) shows that the relaxation time $\tau(k)$ is given by

$$\tau^{-1}(k_0) = \frac{Vm_e k_0}{8\pi^3\hbar^2}\int d\Omega(1-\cos\theta)P(k_0,\theta)$$

$$= \frac{m_e k_0 \rho_i}{2\pi\hbar^3}\int_{-1}^{1} d(\cos\theta)S(k)|\hat{w}(k)|^2(1-\cos\theta) \tag{11.5.12}$$

If the electrons are strongly degenerate, i.e. if $T \ll T_F$, then $\partial f_0/\partial\varepsilon \simeq -\delta(\varepsilon-\varepsilon_F)$, where ε_F is the Fermi energy of Eqn (11.1.1). Under these circumstances, it follows from Eqn (11.5.11) that only those electrons near the Fermi surface are scattered by the ions. Thus $k_0 \simeq k_1 \simeq k_F$ and the range of k that contributes to (11.5.12) is limited to $0 \leqslant k \lesssim 2k_F$. These results, together with the relation (5.1.19), allow the expression for the relaxation time to be rewritten as

$$\tau^{-1} \equiv \tau^{-1}(k_F) = \frac{\rho_i m_e}{4\pi\hbar^3 k_F^3}\int_0^{2k_F} k^3 S(k)|\hat{w}(k)|^2\, dk$$

$$= \frac{m_e}{12\pi^3 z\hbar^3}\int_0^{2k_F} k^3 S(k)|\hat{w}(k)|^2\, dk \tag{11.5.13}$$

The solution to the transport equation contained in Eqns (11.5.11) to (11.5.13) can be used in the calculation of the electrical conductivity. The electric current $\langle\mathbf{j}^Z\rangle$ induced by the electric field $\mathbf{E}$ is

$$\langle\mathbf{j}^Z\rangle = -e\frac{2V}{(2\pi)^3}\int d\mathbf{k}\frac{\hbar\mathbf{k}}{m_e}f(\mathbf{k})$$

$$= -\frac{eV}{4\pi^3}\int d\mathbf{k}\frac{\hbar\mathbf{k}}{m_e}\delta f(\mathbf{k})$$

$$= -\frac{e^2 V\hbar^2}{4\pi^3 m_e^2}\int d\mathbf{k}\,\tau(k)\frac{\partial f_0}{\partial\varepsilon}\mathbf{kk}\cdot\mathbf{E}$$

$$= \boldsymbol{\sigma}\cdot\mathbf{E} \tag{11.5.14}$$

where $\boldsymbol{\sigma}$ is the conductivity tensor. Because the unperturbed electron gas is isotropic, the conductivity tensor is of the form $\boldsymbol{\sigma} = \sigma\boldsymbol{I}$, where σ is a scalar and $\boldsymbol{I}$ is the unit tensor. From Eqn (11.5.14) it follows that the scalar electrical condutivity per unit volume is

$$\begin{aligned}\sigma &= -\frac{e^2\hbar^2}{12\pi^3 m_e^2}\int \mathrm{d}\mathbf{k}\ \tau(k)\frac{\partial f_0}{\partial \varepsilon}k^2 \\ &= -\frac{e^2}{3\pi^2 m_e}\int_0^\infty \mathrm{d}\varepsilon\ k^3\tau(k)\frac{\partial f_0}{\partial \varepsilon} \\ &= \frac{e^2 k_F^3 \tau(k_F)}{3\pi^2 m_e} \qquad (T \ll T_F) \\ &= \frac{\rho_e e^2 \tau}{m_e} \end{aligned} \tag{11.5.15}$$

Substitution for τ from Eqn (11.5.13) leads to the Ziman formula for the resistivity $\rho = \sigma^{-1}$ (Ziman, 1961):

$$\rho = \frac{m_e^2}{12\pi^3 z\rho_e e^2\hbar^3}\int_0^{2k_F} k^3 S(k)\,|\hat{w}(k)|^2\,\mathrm{d}k \tag{11.5.16}$$

Equation (11.5.16) shows that the resistivity can be calculated from a knowledge of the ionic structure factor $S(k)$ discussed in Section 11.3 and the screened electron-ion pseudopotential $\hat{w}(k)$.

The calculation of the thermal conductivity λ and thermoelectric coefficient α proceeds along similar lines, except that now the system is subject to a temperature gradient. The effect of the temperature gradient is to add to the left-hand side of the transport Eqn (11.5.7) a term of the form

$$\frac{\hbar\mathbf{k}_0}{m_e}\cdot\nabla_\mathbf{r} f_0(\mathbf{k}_0) = \frac{\partial f_0(\mathbf{k}_0)}{\partial T}\frac{\hbar\mathbf{k}_0}{m_e}\cdot\nabla_\mathbf{r} T \tag{11.5.17}$$

The resulting thermal conductivity (defined for zero electric current) satisfies the Wiedemann–Franz law (11.5.1) and the Lorenz number has its nearly-free-electron value. The thermoelectric power $Q = \alpha/\sigma k_B T$ is

$$Q = \frac{\pi^2 k_B T}{3e}\frac{\partial \log \rho}{\partial \varepsilon_F} \tag{11.5.18}$$

where the derivative corresponds to the rate of change of resistivity with a notional expansion of the Fermi sphere in which the density of the metal is held constant.

The formulae for ρ, λ and α are expected to be valid under conditions such that the mean-free path of the electrons, $l = p_F\tau/m_e$, is large compared with the mean interionic spacing a_i. This "weak-scattering" condition is well satisfied in all simple metals, particularly in the liquid alkali metals where, systematically, $l > 10a_i$ (Faber, 1972). The quantum-mechanical version of linear-response theory can be used to obtain a formal expression for the electrical conductivity in terms of an electric-current autocorrelation function (Chester, 1963); in the weak-coupling limit (electrons colliding with weak, dilute scatterers), the Ziman formula (11.5.16) is recovered. In order to treat the case of strong scatterers, such as the transition metals, the Born approximation underlying the derivation of (11.5.8) must be replaced by a t-matrix formulation that includes multiple scattering (Dunleavy and Jones, 1978).

A large number of calculations based on Eqn (11.5.13) have been reported. There is generally good agreement with experimental data on simple metals, though the results are sensitive to the choice both of structure factor and pseudopotential. This sensitivity is particularly marked in the case of the temperature coefficient of the resistivity at constant volume, defined as

$$\xi = \frac{T_m}{\rho_m}\left(\frac{\partial \rho}{\partial T}\right)_V \tag{11.5.19}$$

where ρ_m is the resistivity at the melting temperature T_m. The temperature dependence of ρ is effectively that of the ionic structure factor, since the properties of the electron gas (including $\varepsilon'(k)$) are almost independent of T for $T \ll T_F$. Excellent agreement between theory and experiment has been achieved in calculations of the temperature coefficients of the alkali metals in which $S(k)$ is approximated by the structure factor of the OCP, corrected at small k to allow for the non-zero compressibility of the liquid metal (Minoo *et al.*, 1977). The resistivity turns out to be an almost linear function of T over a wide range of temperature. The slope is positive (i.e. $\xi > 0$) for all alkali and most other metals; zinc is a notable exception. The thermoelectric power, however, can be either positive or negative, while the volume coefficient of the resistivity, $(\partial \log \rho/\partial \log V)_T$, is always positive, with the possible exception of lithium.

The resistivity of alloys composed of weak scatterers can be calculated from a straightforward generalization of (11.5.16) of the form

$$\rho = \frac{m_e^2}{12\pi^3 \bar{z}\rho_e e^2 \hbar^3} \sum_\nu \sum_\mu (x_\nu x_\mu)^{1/2} \int_0^{2k_F} k^3 S_{\nu\mu}(k) \hat{w}_\nu(k) \hat{w}_\mu^*(k) \mathrm{d}k \tag{11.5.20}$$

where $\bar{z} = \sum_\nu x_\nu z_\nu$ is the mean ionic valence and $\hat{w}_\nu(k)$ is the screened electron–ion pseudopotential for species ν. Calculations based on this equation reproduce correctly both the observed parabolic concentration dependence of the resistivity of alkali-metal alloys and the linear variation seen in alloys of polyvalent simple metals (Shimoji, 1977). For those alloys that undergo phase separation, the strong concentration fluctuations near the critical point of the miscibility gap lead to a significant enhancement of the resistivity.

Certain liquid alloys have unusual electrical properties in a narrow concentration range that are correlated with a number of thermodynamic and magnetic anomalies. A much studied example is the system lithium–lead; this exhibits maximum deviations from ideality near the stoichiometric composition Li_4Pb, where the resistivity peaks and concentration fluctuations are greatly reduced ($S_{CC}(0) \simeq 0.05x_1x_2$). Neutron-diffraction experiments on the "nearly-zero" alloy $^7Li_4Pb$ provide evidence of a strong tendency towards heterocoordination of nearest neighbours, reflected in a pronounced peak in $S_{CC}(k)$ at $k \simeq 1.5$ Å^{-1} (Ruppersberg and Reiter, 1982). Because the observed pair structure has the features commonly associated with charge ordering, it has been suggested that the Li_4Pb alloy may be partly ionic in character. The structural results can in fact be accounted for, at least qualitatively, by a screened-Coulomb potential model. Some electronic charge is transferred from Li to Pb, leaving the former with an effective valence $z_{eff} \simeq 0.5$ while the latter carries a charge $z_{eff} \simeq -2$. The coulombic interactions are assumed to be screened by the conduction electrons, giving rise to a set of effective pairwise potentials of Yukawa form. This model, when used in conjunction with either the MSA or HNC approximations, leads to results in fair agreement with the diffraction data (Copestake *et al.*, 1983).

An extreme case of ionicity is provided by alloys of caesium and gold, the electrical conductivity of which falls by three orders of magnitude in the vicinity of the stoichiometric composition CsAu (Hensel, 1979). A smoother variation of electrical properties with concentration is observed in a class of alloys that undergo a Mott-like, metal/non-metal transition, i.e. solutions of metals in their own salts. The most thoroughly investigated members of this class are the alkali metal-salt solutions of the form $M_x(MX)_{1-x}$, where x is the mole fraction of the metal M and X is a halogen. The electrical conductivity of these mixtures falls rapidly, but continuously, from typically metallic values for $x \simeq 1$ to values typical of pure molten salts as $x \to 0$. In most systems, there is a miscibility gap below a critical temperature T_c; for $T < T_c$, the solution separates into metal-rich and salt-rich liquid phases (Bredig, 1964). The strong concentration fluctuations at near-critical conditions give rise to a characteristic small-k divergence of $S_{CC}(k)$ that has been studied by small-angle neutron-scattering experiments (Jal *et al.*, 1980). Neutron-diffraction methods have also been used to determine the pair structure of the solutions $K_x(KCl)_{1-x}$ and $Rb_x(RbBr)_{1-x}$ as a function of the concentration x (Jal, 1981). Such experiments, which measure the linear combination of structure factors defined in Eqn (11.4.3), have shown that the small-angle scattering remains very intense over a wide range of concentration away from critical conditions, and that the structure of the mixture changes rather abruptly from "metallic" to "ionic" at $x \simeq 0.8$. Both these features are surprisingly well reproduced by a simple extension of the OCP model of pure metals discussed in Section

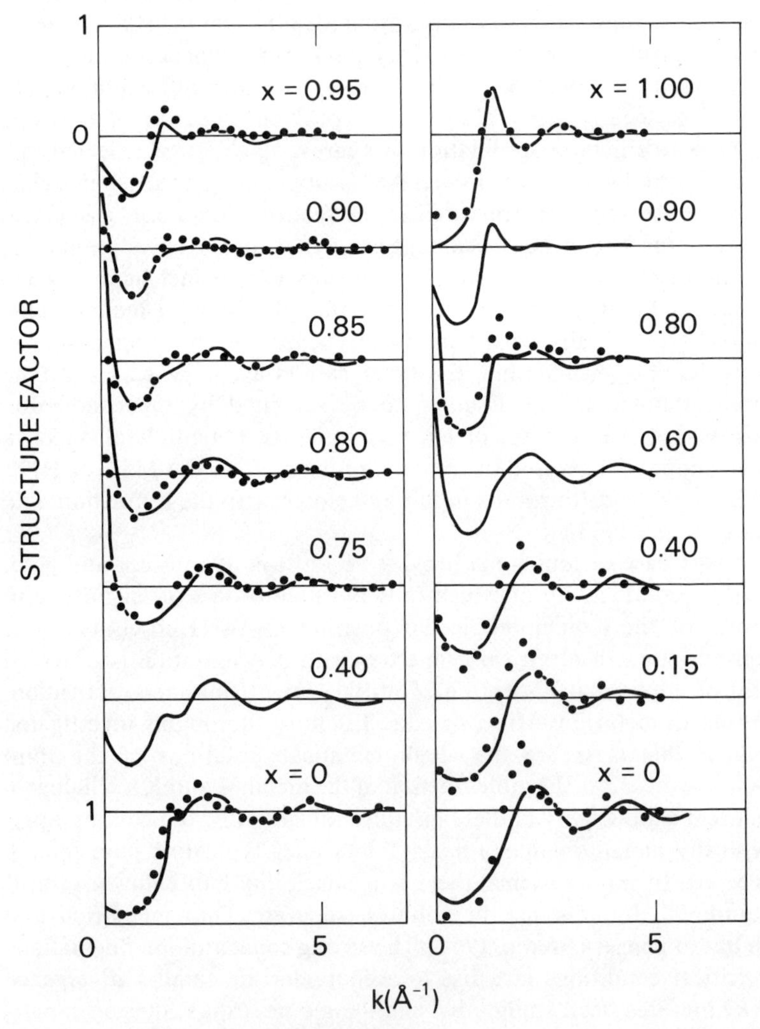

FIG. 11.3. Neutron-weighted structure factors for the solutions $K_x(KCl)_{1-x}$ (left) and $Rb_x(RbBr)_{1-x}$ (right). The points show the results of neutron-scattering experiments (Jal, 1981) at 1073 K (potassium) and 1023 K (rubidium). The curves are calculated from the HNC approximation for the potential model (11.5.21) at temperatures of 1350 K (potassium) and 1300 K (rubidium). After Chabrier and Hansen (1983).

11.3. The cations and anions are assumed to move in a uniform background of conduction electrons and to interact through pair potentials of the form

$$v_{\nu\mu}(r)=\frac{z_\nu z_\mu e^2}{r}+(1-\delta_{\nu\mu})A\exp(-r/\sigma) \tag{11.5.21}$$

where ν, μ label the two ionic species and z_ν, $z_\mu=\pm1$. The coupled HNC equations for this model have been solved numerically to yield the three ionic pair distribution functions. The calculated structure factors are in semiquantitative agreement with the results of neutron-scattering experiments, as illustrated in Figure 11.3 (Chabrier and Hansen, 1983).

In the metal-rich phase, the electrical conductivity of an alkali-metal/alkali-halide solution is strongly affected by the presence of the halogen ions, but it can still be described in the framework of Ziman's weak-scattering theory (Senatore *et al.*, 1982). At low concentration of metal, the valence-electron states become localized and the electrical conductivity is mainly ionic in character. In the very dilute case, the localized electron states are similar to the F-centres of ionic crystals, i.e. to defects in which an electron occupies the site of an anion vacancy (Parrinello and Rahman, 1984).

11.6 IONIC DYNAMICS IN LIQUID METALS AND ALLOYS

The microscopic dynamics of the ions in liquid metals do not differ in any fundamental way from the corresponding motions in simple insulating liquids such as the rare gases. This is not surprising, since the pair potentials for metallic pseudoatoms and rare-gas atoms are qualitatively similar. For the same reason, experimental and theoretical methods that have been successfully used to study the dynamics of argon-like liquids have, for the most part, met with comparable success in their application to simple liquid metals. However, the interactions in, say, potassium and argon do differ considerably in detail, as shown in Figure 1.2, and this gives rise to quantitative differences in dynamic behaviour.

The dynamic structure factor $S(k,\omega)$ has been measured by inelastic neutron-scattering for several simple liquid metals and has also been the subject of study in a number of molecular dynamics calculations. The most complete set of results available at present are those obtained for liquid rubidium by a combination of experimental and simulation techniques (Copley and Rowe, 1974a, b; Rahman, 1974a, b). These results have already been discussed in Sections 8.5 and 9.4; their most striking feature is the persistence of a well-defined sidepeak in $S(k,\omega)$ up to $k\simeq 1$ Å^{-1}, corresponding to approximately two-thirds of the wavenumber at which the main peak occurs in the static structure factor. The persistence of the peak implies that a collective longitudinal mode continues to propagate even when the wavelength is roughly equal to the interionic spacing; in the rare gases, the acoustic waves are already overdamped at wavelengths that are several times large than this. The difference in behaviour has been ascribed to the fact that the repulsive core of the potential is much softer in the alkali metals than it is for rare-gas atoms (Lewis and Lovesey, 1977).

Results similar to those for rubidium have been obtained in molecular dynamics simulations of liquid sodium and potassium (Jacucci and McDonald, 1980a). The same "experiments" show that liquid metals can also sustain propagating shear waves similar to those seen in the Lennard-Jones fluid near its triple point (see Figure 9.1). The comparison with the Lennard-Jones results reveals an interesting difference in the range of wavelength over which longitudinal and transverse modes propagate in the two types of liquid (Jacucci and McDonald, 1980b). If a is the nearest-neighbour spacing in the liquid, defined as the separation at which the radial distribution function has its main peak, the simulations reveal that

$$\lambda_l/a \simeq 6, \qquad \lambda_t/a \simeq 7 \qquad \text{(Lennard-Jones fluid)}$$

$$\lambda_l/a \simeq 1.4, \qquad \lambda_t/a \simeq 14 \qquad \text{(liquid alkali metals)}$$

where λ_l is the *shortest* wavelength for which a propagating longitudinal mode is still observed and λ_t is the *longest* wavelength at which a propagating shear wave occurs. In other words, the longitudinal mode in argon-like liquids vanishes at about the same wavelength at which the shear mode first appears, but in the alkali metals there is a very wide range of wavelength, extending from roughly one or two to about fourteen interatomic spacings, in which the liquid can support propagating collective excitations of both transverse and longitudinal character.

A different description of the dynamics can be obtained through the representation of a liquid metal as an ion–electron plasma, in close analogy with the treatment of static properties in Section 11.2. In this picture of the liquid, the ionic and electronic components are only weakly coupled through the electron–ion pseudopotential, and each component may be regarded as an external perturbation on the other (Postogna and Tosi, 1980). Let $\phi_\nu(\mathbf{k},\omega)$ be an external potential that acts on component ν, where $\nu = 1$ for the ions and $\nu = 2$ for the electrons. Within linear-response theory, the Fourier components of the induced densities are related to the external potentials by a matrix of density response functions:

$$\langle \rho_{\mathbf{k}}^{\nu}(\omega)\rangle = \sum_{\mu} \chi_{\nu\mu}(k,\omega)\phi_\mu(\mathbf{k},\omega) \tag{11.6.1}$$

The response to the internal field is described by a similar matrix of screened response functions, or "proper polarizabilities", $\chi^{\text{sc}}(k,\omega)$. Written in matrix form, the response is

$$\langle \boldsymbol{\rho}_{\mathbf{k}}(\omega) \rangle = \boldsymbol{\chi}^{\mathrm{sc}}(k, \omega)[\boldsymbol{\phi}(\mathbf{k}, \omega) + \hat{\boldsymbol{v}}(k)\langle \boldsymbol{\rho}_{\mathbf{k}}(\omega) \rangle] \tag{11.6.2}$$

where $\hat{\boldsymbol{v}}(k)$ is the matrix of bare potentials $\hat{v}_{\nu\mu}(k)$ and the second term in square brackets is the "polarization potential". Elimination of $\langle \rho_{\mathbf{k}}^{\nu}(\omega) \rangle$ between (11.6.1) and (11.6.2) leads to a matrix generalization of the relation (10.4.15) between the external and screened response functions:

$$\boldsymbol{\chi}(k, \omega) = \boldsymbol{\chi}^{\mathrm{sc}}(k, \omega) + \boldsymbol{\chi}^{\mathrm{sc}}(k, \omega)\hat{\boldsymbol{v}}(k)\boldsymbol{\chi}(k, \omega) \tag{11.6.3}$$

or, in terms of the elements of the inverse matrices:

$$[\boldsymbol{\chi}(k, \omega)]_{\nu\mu}^{-1} = [\boldsymbol{\chi}^{\mathrm{sc}}(k, \omega)]_{\nu\mu}^{-1} - \hat{v}_{\nu\mu}(k) \tag{11.6.4}$$

To lowest order in the ion–electron coupling, the two species respond to the internal field as two independent, one-component plasmas. The off-diagonal element $\chi_{12}^{\mathrm{sc}}(k, \omega)$ is therefore zero, while $\chi_{11}^{\mathrm{sc}}(k, \omega)$ is the screened response function of the classical OCP and $\chi_{22}^{\mathrm{sc}}(k, \omega)$ is the screened response function of the degenerate electron gas in a uniform background ("jellium"). It follows, given Eqns (10.4.15) and (10.4.16), that

$$[\boldsymbol{\chi}^{\mathrm{sc}}(k, \omega)]_{11}^{-1} = \chi_{\mathrm{OCP}}^{-1}(k, \omega) + \hat{v}_{11}(k) \tag{11.6.5a}$$

$$[\boldsymbol{\chi}^{\mathrm{sc}}(k, \omega)]_{22}^{-1} = \frac{\hat{v}_{22}(k)}{1 - \varepsilon_{\mathrm{e}}(k, \omega)} \tag{11.6.5b}$$

and the external susceptibility of the ions is obtained from Eqn (11.6.4) as

$$\chi_{11}(k, \omega) = \frac{\chi_{\mathrm{OCP}}(k, \omega)}{1 - \hat{v}(k, \omega)\chi_{\mathrm{OCP}}(k, \omega)} \tag{11.6.6}$$

where $\hat{v}(k, \omega)$ describes the dynamical screening of the ion–ion interaction by the electrons:

$$\hat{v}(k, \omega) = \frac{k^2 \hat{v}_{12}^2(k)}{4\pi e^2}\left(\frac{1}{\varepsilon_{\mathrm{e}}(k, \omega)} - 1\right) \tag{11.6.7}$$

Equation (11.6.6) is the dynamic generalization of (11.3.3), except that in the latter case the electron–ion interaction is represented by the "empty-core" pseudopotential (11.2.9). Because the frequency scale of the electronic motion is much higher than any frequencies associated with the ions, it is reasonable to use an adiabatic approximation in which $\hat{v}(k, \omega)$ is replaced by $\hat{v}(k, 0)$. The characteristic frequencies of the longitudinal modes of the screened ionic plasma are then given by the roots of the denominators in Eqn (11.6.6) or, in the adiabatic approximation, by the solution to the equation

$$1 - \hat{v}(k, 0)\chi_{\mathrm{OCP}}(k, \omega) = 0 \tag{11.6.8}$$

In the long-wavelength limit, the ratio $\tilde{F}_{\text{OCP}}(k, \omega)/S_{\text{OCP}}(k)$ is related to the frequency-dependent electrical conductivity by Eqn (10.4.39). Thus, from Eqns (7.6.27) and (7.7.18):

$$\begin{aligned}
&\lim_{k\to 0} \chi_{\text{OCP}}(k, \omega) \\
&\quad = -\beta\rho_{\text{i}} \lim_{k\to 0}\lim_{\varepsilon\to 0}[S_{\text{OCP}}(k) + i(\omega + i\varepsilon)\tilde{F}_{\text{OCP}}(k, \omega + i\varepsilon)] \\
&\quad = -\beta\rho_{\text{i}} \lim_{k\to 0} S_{\text{OCP}}(k) \lim_{\varepsilon\to 0} \frac{4\pi\sigma(\omega + i\varepsilon)}{-i(\omega + i\varepsilon) + 4\pi\sigma(\omega + i\varepsilon)}
\end{aligned} \tag{11.6.9}$$

The long-wavelength limit of $S_{\text{OCP}}(k)$ is given by Eqn (10.2.15), and the complex conductivity $\sigma(\omega + i\varepsilon)$ can be expressed, via Eqn (7.8.10), in the form

$$\sigma(\omega + i\varepsilon) = \frac{\beta}{V}\int_0^\infty J(t) \exp[i(\omega + i\varepsilon)t]\, \mathrm{d}t \tag{11.6.10}$$

where $J(t)$ is the electrical-current autocorrelation function. In the OCP, the proportionality of mass and charge currents means that the conservation of total linear momentum is equivalent to the conservation of electric current, i.e. the resistivity is zero. Hence

$$J(t) = J(0) = \frac{N_{\text{i}} z^2 e^2 k_{\text{B}} T}{m_{\text{i}}} \tag{11.6.11}$$

and, from Eqn (11.6.10):

$$\sigma(\omega + i\varepsilon) = \frac{i\omega_{\text{pi}}^2}{4\pi(\omega + i\varepsilon)} \tag{11.6.12}$$

where $\omega_{\text{pi}}^2 = 4\pi\rho_{\text{i}} z^2 e^2/m_{\text{i}}$ is the square of the ionic plasma frequency. Substitution of (11.6.12) in (11.6.9) shows that

$$\lim_{k\to 0} \frac{\chi_{\text{OCP}}(k, \omega)}{S_{\text{OCP}}(k)} = \frac{\beta\rho_{\text{i}}\omega_{\text{pi}}^2}{\omega^2 - \omega_{\text{pi}}^2} \tag{11.6.13}$$

In the small-k limit, $\varepsilon_{\text{e}}(k, 0) \simeq k_{\text{e}}^2/k^2$, from Eqn (11.3.4), and $\hat{v}_{12}(k)$, in the empty-core model, behaves as

$$\begin{aligned}
\hat{v}_{12}(k) &= \frac{4\pi z e^2 \cos kr_{\text{c}}}{k^2} \\
&\underset{k\to 0}{\sim} \frac{4\pi z e^2}{k^2}(1 - \tfrac{1}{2}k^2 r_{\text{c}}^2)
\end{aligned} \tag{11.6.14}$$

so that

$$\lim_{k\to 0} \hat{v}(k, 0) = \frac{4\pi z^2 e^2}{k^2}(1 - k^2 r_{\text{c}}^2)\left(\frac{k^2}{k_{\text{e}}^2} - 1\right) \tag{11.6.15}$$

When the results are gathered together, it is found that to order k^2 the solution to (11.6.8) leads to a typical sound-wave dispersion relation of the form

$$\omega = \omega_{pi}(k_e^{-2} + k_s^{-2} + r_c^2)^{1/2}k = ck \tag{11.6.16}$$

where k_s is the ionic screening number defined by Eqn (10.2.31) and c is the speed of sound. Thus the effect of electron screening is to convert the plasmon mode at frequency ω_{pi} into a sound wave of a frequency that vanishes linearly with k. The expression for c given by Eqn (11.6.16) is consistent with that obtained from the isothermal compressibility in combination with Eqn (11.3.5); c can therefore be identified as the isothermal speed of sound, but this differs little from the adiabatic value, since the ratio of specific heats is close to unity for liquids metals.

The general approach sketched above has also been applied to binary alloys (Postogna and Tosi, 1980). Electron screening again transforms the plasmon mode (Hansen *et al.*, 1979) into a sound mode, but interdiffusion of the two ionic species now provides an additional damping mechanism compared with the one-component case. Interdiffusion also gives rise to a prominent central peak in the density-fluctuation spectrum $S_{NN}(k, \omega)$. This phenomenon is illustrated in Figure 11.4, where results from a molecular dynamics simulation of the alloy NaK are shown, together with the corresponding spectra for the pure metals (Jacucci and McDonald, 1980a). The simulations gave no evidence for the existence of a propagating concentration fluctuation; $S_{CC}(k, \omega)$ was found to have a diffusive structure throughout the wavenumber range that was studied and a width that increases with k, in qualitative agreement with the hydrodynamic results of Section 8.6. In the "nearly-zero" alloy $^7Li_4Pb$, the dominant contribution to the coherent scattering of neutrons comes from concentration fluctuations, making possible a direct experimental measurement of $S_{CC}(k, \omega)$. As discussed in Section 11.5, this system displays a pronounced tendency towards heterocoordination, which can be explained in terms of a partial charge transfer between the two species. The alloy therefore has a partially ionic character, but no trace has been found experimentally (Soltwisch *et al.*, 1983) of a longitudinal "optic" mode, presumably because the Coulomb interaction between the ions is screened by the residual conduction electrons. Nevertheless, the experimental $S_{CC}(k, \omega)$ differs markedly from that observed in the molecular dynamics calculations for NaK. Although the spectrum remains essentially lorentzian at all wavenumbers, its width does not increase monotonically in the manner predicted by the hydrodynamic result (10.4.42), but instead has a structure, reminiscent of de Gennes narrowing, at wavenumbers in the region of the main peak in $S_{CC}(k)$ ($k \simeq 1.5$ Å^{-1}). This behaviour can be accounted for in a phenomenological

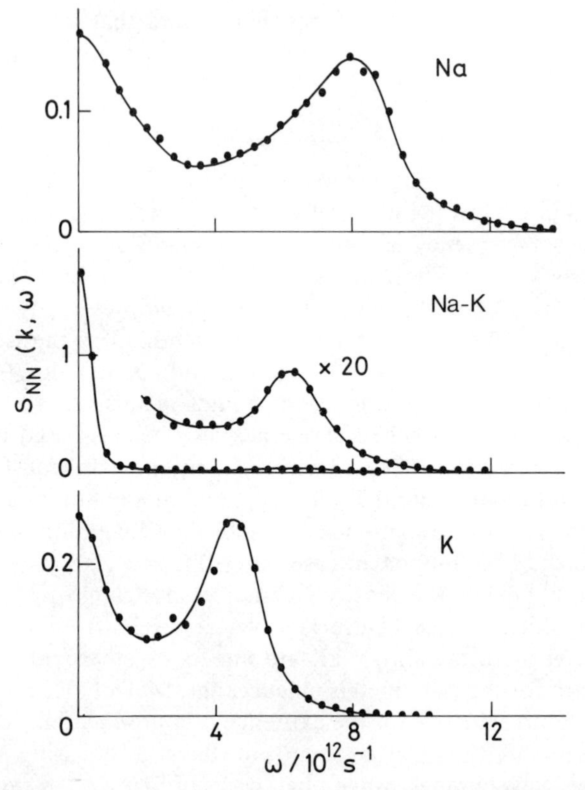

FIG. 11.4. Molecular dynamics results for the number–number dynamic structure factors for models of liquid sodium and potassium and for the equimolar alloy NaK. Note the enlargement of scale for NaK. After Jacucci and McDonald (1980b).

way by introducing a k-dependent interdiffusion constant into Eqn (10.4.42); $D(k)$ is written as a generalization of the approximate relation (8.6.36) in which the thermodynamic factor $(\partial^2(\beta G/N)/\partial x_1^2)_{P,T} = 1/S_{CC}(0)$ (see Eqn (11.4.5c)) is replaced by $1/S_{CC}(k)$. Thus

$$D(k) = \frac{x_1 x_2 (x_1 D_2 + x_2 D_1)}{S_{CC}(k)} \tag{11.6.17}$$

Use of this simple prescription leads to good agreement with the experimental results (Soltwisch *et al.*, 1983).

CHAPTER 12

Molecular Liquids

The earlier part of the book has been concerned largely with atomic systems. In this chapter, we consider some of the new problems that arise when the theory is broadened to include molecular fluids. To keep the material within reasonable bounds, we limit the discussion to molecules that are rigid, and ignore the effects of polarizability.

12.1 THE MOLECULAR PAIR DISTRIBUTION FUNCTION

The description of the structure of a molecular fluid in terms of particle densities and distribution functions can be developed along lines similar to those followed in the atomic case. The main added complication is the fact that the phase-space probability density for particles with rotational degrees of freedom cannot immediately be factorized into kinetic and configurational parts. This problem is very well treated by Gray and Gubbins (1984), and we shall not dwell on it here. The final expressions for the molecular distribution functions resemble closely those obtained for atomic fluids, except that all quantities are now functions of the molecular orientations as well as of the translational coordinates.

In this chapter, we shall be concerned almost exclusively with pair correlations. We therefore take as our starting point a suitably generalized form of the definition (2.5.36) of the pair density. Let $\mathbf{R}_i$ be the translational coordinates of molecule i and let $\mathbf{\Omega}_i$ be the orientation of i in a laboratory-fixed frame of reference. If the molecule is linear, $\mathbf{\Omega}_i \equiv (\theta_i, \phi_i)$, where θ_i, ϕ_i are the usual polar angles; if it is non-linear, $\mathbf{\Omega}_i \equiv (\theta_i, \phi_i, \chi_i)$, where θ_i, ϕ_i, χ_i are the Euler angles. Then the molecular pair density is defined as

$$\rho^{(2)}(\mathbf{R}, \mathbf{R}', \mathbf{\Omega}, \mathbf{\Omega}') = \left\langle \sum_{i \neq j}^{N} \sum \delta(\mathbf{R}_i - \mathbf{R})\delta(\mathbf{R}_j - \mathbf{R}')\delta(\mathbf{\Omega}_i - \mathbf{\Omega})\delta(\mathbf{\Omega}_j - \mathbf{\Omega}') \right\rangle \tag{12.1.1}$$

and the molecular pair distribution function as

$$g(\mathbf{R}_{12}, \mathbf{\Omega}_1, \mathbf{\Omega}_2) = (\Omega/\rho)^2 \rho^{(2)}(\mathbf{R}_{12}, \mathbf{\Omega}_1, \mathbf{\Omega}_2) \tag{12.1.2}$$

where $\Omega \equiv \int d\mathbf{\Omega}_i$. The definition of Ω means that

$$\Omega = \iint d(\cos\theta_i)\, d\phi_i$$

$$= 4\pi, \quad \text{for a linear molecule} \tag{12.1.3a}$$

$$\Omega = \iiint d(\cos\theta_i)\, d\phi_i\, d\chi_i$$

$$= 8\pi^2, \quad \text{for a non-linear molecule} \tag{12.1.3b}$$

The coordinates $\mathbf{R}_i$ are often taken to be those of the molecular centre of mass or of some other point of high symmetry in the molecule, but the choice of molecular "centre" is entirely arbitrary. To simplify the notation, it is convenient to use the symbol $i \equiv (\mathbf{R}_i, \mathbf{\Omega}_i)$ to denote both the coordinates of the molecular centre and the orientation of molecule i. The molecular pair distribution function will therefore frequently be written simply as $g(1,2)$ and the molecular pair correlation function as $h(1,2) = g(1,2) - 1$. The functions $e(1,2) = \exp[-\beta v(1,2)]$, $f(1,2) = e(1,2) - 1$ and $y(1,2) = g(1,2)/e(1,2)$ have the same significance as in the atomic case but now are functions of the orientations $\mathbf{\Omega}_1, \mathbf{\Omega}_2$. Finally, the molecular direct correlation function $c(1,2)$ is related to $h(1,2)$ by a generalization of the Ornstein–Zernike relation (5.2.10):

$$h(1,2) = c(1,2) + (\rho/\Omega) \int c(1,3) h(3,2)\, d3 \tag{12.1.4}$$

Integration of the pair distribution function $g(1,2)$ over the variables $\mathbf{\Omega}_1, \mathbf{\Omega}_2$ yields a function $g_c(R)$ (with $R \equiv |\mathbf{R}_{12}|$) that describes the radial distribution of molecular centres:

$$g_c(R) = (1/\Omega^2) \iint g(\mathbf{R}, \mathbf{\Omega}_1, \mathbf{\Omega}_2)\, d\mathbf{\Omega}_1\, d\mathbf{\Omega}_2$$

$$\equiv \langle g(1,2) \rangle_{\mathbf{\Omega}_1 \mathbf{\Omega}_2} \tag{12.1.5}$$

Here and elsewhere in this chapter we use angular brackets with subscripts $\mathbf{\Omega}_1 \cdots$ to denote an unweighted average over the angles $\mathbf{\Omega}_1 \cdots$. Thus

$$\langle \cdots \rangle_{\mathbf{\Omega}_1} \equiv (1/\Omega) \int \cdots d\mathbf{\Omega}_1 \tag{12.1.6}$$

With this convention, the Ornstein–Zernike relation (12.1.4) may be re-expressed as

$$h(1,2)=c(1,2)+\rho\int\langle c(1,3)h(3,2)\rangle_{\mathbf{\Omega}_3}\,\mathrm{d}\mathbf{R}_3 \tag{12.1.7}$$

If $g(1,2)$ is multiplied by some function of the orientations $\mathbf{\Omega}_1, \mathbf{\Omega}_2$ and then integrated over all coordinates of the pair 1, 2, the result is a quantity that measures the importance of angular correlations of a specific type. Suppose that the molecule has an axis of symmetry and let $\mathbf{u}_i$ be a unit vector along the symmetry axis of molecule i. A set of angular correlation parameters that are of special interest both theoretically and experimentally are those defined as

$$\begin{aligned} G_l &= \rho\int\langle g(\mathbf{R}_{12},\mathbf{\Omega}_1,\mathbf{\Omega}_2)P_l(\mathbf{u}_1\cdot\mathbf{u}_2)\rangle_{\mathbf{\Omega}_1\mathbf{\Omega}_2}\,\mathrm{d}\mathbf{R}_{12} \\ &= \langle (N-1)P_l(\mathbf{u}_1\cdot\mathbf{u}_2)\rangle \end{aligned} \tag{12.1.8}$$

where $P_l(\ )$ denotes a Legendre polynomial. The value of the first-rank order parameter G_1 determines the dielectric constant of a polar fluid, as we shall see in Section 12.5, while G_2 is related to a number of measurable properties, including the integrated intensity of the spectrum observed in depolarized light-scattering experiments (Ladanyi and Keyes, 1977; Battaglia *et al.*, 1979).

When the total potential energy of the fluid is a sum of pair terms, the internal energy and equation of state can both be written as integrals over $g(1,2)$. The excess internal energy, for example, is given by

$$\begin{aligned} \frac{U^{\mathrm{ex}}}{N} &= \frac{1}{N}\langle V_N\rangle \\ &= \tfrac{1}{2}(\rho/\Omega^2)\iiint g(1,2)v(1,2)\,\mathrm{d}\mathbf{R}_{12}\,\mathrm{d}\mathbf{\Omega}_1\,\mathrm{d}\mathbf{\Omega}_2 \\ &= 2\pi\rho\int_0^\infty \langle g(1,2)v(1,2)\rangle_{\mathbf{\Omega}_1\mathbf{\Omega}_2}R_{12}^2\,\mathrm{d}R_{12} \end{aligned} \tag{12.1.9}$$

which is the molecular analogue of Eqn (2.5.12). The corresponding result for the pressure is a generalization of (2.5.14):

$$\frac{\beta P}{\rho}=1-\tfrac{2}{3}\pi\beta\rho\int_0^\infty\langle g(1,2)v'(1,2)\rangle_{\mathbf{\Omega}_1\mathbf{\Omega}_2}R_{12}^3\,\mathrm{d}R_{12} \tag{12.1.10}$$

where the prime denotes differentiation with respect to R_{12} with $\mathbf{\Omega}_1, \mathbf{\Omega}_2$

held constant. Irrespective of whether or not the potential energy is pairwise-additive, the isothermal compressibility is given by

$$\rho k_B T \chi_T = 1 + \rho \int \langle g(1,2) - 1 \rangle_{\mathbf{\Omega}_1 \mathbf{\Omega}_2} \, d\mathbf{R}_{12}$$

$$= 1 + \rho \int [g_c(R) - 1] \, d\mathbf{R} \tag{12.1.11}$$

which is the analogue of (2.5.32). The last result is of particular interest insofar as all reference to angular coordinates has disappeared. The reason why this is so will become clear after the discussion of the molecular structure factor in Section 12.3.

Equations (12.1.9), (12.1.10) and (12.1.11) are identical to their atomic counterparts except for the fact that the pair functions (or products of pair functions) in the integrands are replaced by their unweighted angle averages. Their significance, however, is largely formal. The many-dimensional character of the molecular pair distribution function means that, in general, these results do not represent practical routes to the calculation of thermodynamic properties. The shape of $g(1, 2)$ is difficult even to visualize, and if progress is to be made the basic problem must be cast in simpler form. Two different approaches have been used. In one, which we review in the next section, $g(1, 2)$ (or $h(1, 2)$) is expanded in a series of suitably-chosen, angle-dependent, basis functions; in the other, which we discuss in Section 12.3, the fluid structure is described in terms of *site–site distribution functions*. Use of site–site distribution functions is particularly appropriate in cases where the intermolecular potential is cast in site–site form, as in Eqn (1.2.6).

12.2 EXPANSIONS OF THE PAIR DISTRIBUTION FUNCTION

The pair distribution function for molecules of arbitrary symmetry can be expanded in terms of the Wigner rotation matrices or generalized spherical harmonics (Steele, 1963). The general formalism has not been widely used, however, and the discussion that follows is restricted to linear molecules. In this case, the natural expansion functions are the usual spherical harmonics, which we shall denote as $Y_{lm}(\theta, \phi)$ in the convention of Rose (1957). Let $\mathbf{\Omega}_1$, $\mathbf{\Omega}_2$ be the orientations of molecules 1, 2 in a system of polar coordinates in which the z-axis lies along the vector $\mathbf{R}_{12} = \mathbf{R}_2 - \mathbf{R}_1$ (the "intermolecular" frame). The $g(1, 2)$ may be written as

$$g(1,2) = 4\pi \sum_{l_1} \sum_{l_2} \sum_{m} g_{l_1 l_2 m}(R) Y_{l_1 m}(\mathbf{\Omega}_1) Y_{l_2 \bar{m}}(\mathbf{\Omega}_2) \tag{12.2.1}$$

where $R \equiv |\mathbf{R}_{12}|$, $\bar{m} \equiv -m$, and the sum on m runs from $-l$ to l, where l is the lesser of l_1 and l_2; the indices m of the two spherical harmonics are equal (apart from sign) by virtue of the cylindrical symmetry with respect to the axis $\mathbf{R}_{12}$. A complete and very readable treatment of the definition and properties of the spherical harmonics can be found in Appendix A of the book by Gray and Gubbins (1984). Here we need only to note that the spherical harmonics are normalized and orthogonal, i.e.

$$\int Y^*_{lm}(\mathbf{\Omega})\, Y_{l'm'}(\mathbf{\Omega})\, \mathrm{d}\mathbf{\Omega} = \delta_{ll'}\delta_{mm'} \tag{12.2.2}$$

and that $Y_{l\bar{m}}(\mathbf{\Omega}) = (-1)^m Y^*_{lm}(\mathbf{\Omega})$.

If Eqn (12.2.1) is multiplied through by $Y^*_{l_1\bar{m}}(\mathbf{\Omega}_1)\, Y^*_{l_2m}(\mathbf{\Omega}_2)$ and integrated over $\mathbf{\Omega}_1, \mathbf{\Omega}_2$, it follows from the properties just given that

$$\begin{aligned} g_{l_1l_2m}(R) &= \frac{1}{4\pi}\iint g(1,2)\, Y_{l_1\bar{m}}(\mathbf{\Omega}_1)\, Y_{l_2m}(\mathbf{\Omega}_2)\, \mathrm{d}\mathbf{\Omega}_1\, \mathrm{d}\mathbf{\Omega}_2 \\ &= 4\pi\langle g(1,2)\, Y_{l_1\bar{m}}(\mathbf{\Omega}_1)\, Y_{l_2m}(\mathbf{\Omega}_2)\rangle_{\mathbf{\Omega}_1\mathbf{\Omega}_2} \end{aligned} \tag{12.2.3}$$

The expansion coefficients $g_{l_1l_2m}(R)$ are called the "projections" of $g(1,2)$ onto the corresponding angular functions, and are easily calculated by computer simulation. Certain projections of $g(1,2)$ are closely related to quantities introduced in Section 12.1. First, given that $Y_{00}(\mathbf{\Omega}) = (1/4\pi)^{1/2}$, we see that $g_{000}(R)$ is identical to the centres distribution function $g_c(R)$ defined by Eqn (12.1.5); this is the reason for the inclusion of the factor 4π in (12.2.1). Secondly, the order parameters defined by Eqn (12.1.8) can be re-expressed as

$$G_l = \frac{\rho}{2l+1}\sum_m (-1)^m \int g_{llm}(R)\, \mathrm{d}\mathbf{R} \tag{12.2.4}$$

This result is a consequence of the addition theorem for spherical harmonics, i.e.

$$P_l(\cos\gamma_{12}) = \frac{4\pi}{2l+1}\sum_m Y^*_{lm}(\mathbf{\Omega}_1)\, Y_{lm}(\mathbf{\Omega}_2) \tag{12.2.5}$$

where γ_{12} is the angle between two vectors with orientations $\mathbf{\Omega}_1$ and $\mathbf{\Omega}_2$.

An expansion similar to (12.2.1) can be made of any scalar function of the variables $\mathbf{R}_{12}$, $\mathbf{\Omega}_1$ and $\mathbf{\Omega}_2$, including both the intermolecular potential $v(1,2)$ and its derivative with respect to R_{12}. The corresponding expansion coefficients $v_{l_1l_2m}(R)$ and $v'_{l_1l_2m}(R)$ can be calculated numerically for any pair potential and in many cases are expressible in analytic form (Downs *et al.*, 1979). If we introduce the expansions of $g(1,2)$ and $v(1,2)$ into Eqn (12.1.9) and integrate over angles, the energy equation becomes

$$\frac{U^{\text{ex}}}{N} = 2\pi\rho \sum_{l_1}\sum_{l_2}\sum_{m} \int_0^\infty g_{l_1 l_2 m}(R) v_{l_1 l_2 m}(R) R^2 \, dR \tag{12.2.6}$$

The pressure equation (12.1.10) can similarly be rewritten in terms of the coefficients $g_{l_1 l_2 m}(R)$ and $v'_{l_1 l_2 m}(R)$. The multidimensional integrals that appear on the right-hand sides of Eqns (12.1.9) and (12.1.10) are thereby transformed into an infinite sum of one-dimensional integrals. Whether or not the new expressions are, in a computational sense, improvements on (12.1.9) and (12.1.10) depends on the rate at which the harmonic expansions converge. Streett and Tildesley (1977) have made Monte Carlo calculations for a number of two-site Lennard-Jones fluids and compared the results obtained for energy and pressure with those calculated by summing different numbers of terms in the corresponding spherical-harmonic expansions. On the whole, for the numbers of terms that it is practical to compute, the agreement is poor and becomes rapidly worse as the elongation of the molecule increases. Broadly similar conclusions have been reached by those who have made direct tests of the rate of convergence of the expansion (12.2.1) (Monson and Rigby, 1979; Haile and Gray, 1980).

A different expansion of $g(1,2)$ is obtained if the orientations $\mathbf{\Omega}_1, \mathbf{\Omega}_2$ are referred to a laboratory-fixed frame of reference (the "laboratory" frame). Let $\mathbf{\Omega}_R$ be the orientation of the vector $\mathbf{R}_{12}$ in the laboratory frame. Then $g(1,2)$ may be expanded in the form

$$g(1,2) = \sum_{l_1 l_2 l} g(l_1 l_2 l;\, R) \sum_{m_1 m_2 m} C(l_1 l_2 l;\, m_1 m_2 m) Y_{l_1 m_1}(\mathbf{\Omega}_1) \times Y_{l_2 m_2}(\mathbf{\Omega}_2) Y^*_{lm}(\mathbf{\Omega}_R) \tag{12.2.7}$$

where $C(\;)$ is a Clebsch–Gordan coefficient (Gray and Gubbins, 1984). The coefficients $g(l_1 l_2 l;\, R)$ are linear combinations of the coefficients in the intermolecular-frame expansion (12.2.1), and Eqn (12.2.7) reduces to (12.2.1) (apart from the factor 4π in the latter) when the z-axis of the laboratory frame is parallel to $\mathbf{R}_{12}$. The general relation between the two sets of coefficients is

$$g(l_1 l_2 l;\, R) = \left(\frac{64\pi^3}{2l+1}\right)^{1/2} \sum_m C(l_1 l_2 l;\, m\bar{m}0) g_{l_1 l_2 m}(R) \tag{12.2.8}$$

with, as a special case, $g(000;\, R) = (4\pi)^{3/2} g_{000}(R)$. Equation (12.2.7) is also frequently written in the abbreviated form

$$g(1,2) = \sum_{l_1}\sum_{l_2}\sum_{l} g(l_1 l_2 l;\, R) \Phi^{l_1 l_2 l}(\mathbf{\Omega}_1, \mathbf{\Omega}_2, \mathbf{\Omega}_R) \tag{12.2.9}$$

where $\Phi^{l_1 l_2 l}$ is a *rotational invariant* (Blum and Torruella, 1972).

Use of (12.2.7) in preference to (12.2.1) does not help in resolving the problem of slow convergence, but it does have some advantages, particularly in the manipulation of Fourier transforms. We shall use the notation $\hat{g}(1,2) \equiv \hat{g}(\mathbf{k}, \mathbf{\Omega}_1, \mathbf{\Omega}_2)$ to denote a Fourier transform with respect to $\mathbf{R}_{12}$, i.e.

$$\hat{g}(\mathbf{k}, \mathbf{\Omega}_1, \mathbf{\Omega}_2) = \int g(\mathbf{R}_{12}, \mathbf{\Omega}_1, \mathbf{\Omega}_2) \exp(-i\mathbf{k} \cdot \mathbf{R}_{12}) \, \mathrm{d}\mathbf{R}_{12} \quad (12.2.10)$$

Then $\hat{g}(1,2)$ can be written in terms of laboratory-frame harmonics as

$$\hat{g}(1,2) = \sum_{l_1 l_2 l} g(l_1 l_2 l; k) \sum_{m_1 m_2 m} C(l_1 l_2 l; m_1 m_2 m) Y_{l_1 m_1}(\mathbf{\Omega}_1) Y_{l_2 m_2}(\mathbf{\Omega}_2) Y^*_{lm}(\mathbf{\Omega}_k) \quad (12.2.11)$$

where $\mathbf{\Omega}_k$ is the orientation of $\mathbf{k}$ in the laboratory frame. The reason that this expansion and the corresponding expansions of $\hat{h}(1,2)$ and $\hat{c}(1,2)$ are so useful is the fact that the coefficients $g(l_1 l_2 l; k)$ and $g(l_1 l_2 l; R)$ are related by a generalized Fourier, Fourier–Bessel, or Hankel transform:

$$g(l_1 l_2 l; k) = 4\pi i^l \int_0^\infty j_l(kR) g(l_1 l_2 l; R) R^2 \, \mathrm{d}R \quad (12.2.12)$$

where $j_l(\ )$ is the spherical Bessel function of order l. No equivalent simplification is found in the case of the intermolecular-frame expansion. We shall not give a general proof of (12.2.12), since we are concerned in this book only with the cases $l = 0$ and $l = 2$. The case $l = 0$ corresponds to the usual Fourier transform of a spherically symmetric function, as in Eqn (5.1.9); the case $l = 2$ will be considered in some detail in Section 12.4.

Expansions of $g(1,2)$ and other pair functions along the lines of (12.2.1) and (12.2.7) play a central role in many theories of molecular liquids. The most successful applications have occurred in the theory of polar fluids, as we shall see in Sections 12.6 and 12.7.

12.3 SITE–SITE DISTRIBUTION FUNCTIONS

When an interaction-site model is used to represent the intermolecular potential, the natural way to describe the structure of the fluid is in terms of site-site distribution functions. If the coordinates of site α on molecule i are denoted by $\mathbf{r}_{i\alpha}$, and those of site β on molecule j ($j \neq i$) by $\mathbf{r}_{j\beta}$, then the site–site radial distribution function $g_{\alpha\beta}(r)$ is defined in a manner similar to (2.5.37):

$$\rho^2 g_{\alpha\beta}(\mathbf{r}) = \left\langle \sum_{i \neq j}^{N} \sum \delta(\mathbf{r}_{i\alpha}) \delta(\mathbf{r}_{j\beta} - \mathbf{r}) \right\rangle$$

$$= \frac{1}{V} \langle N(N-1) \delta(\mathbf{r}_{2\beta} - \mathbf{r}_{1\alpha} - \mathbf{r}) \rangle \quad (12.3.1)$$

The corresponding site–site pair correlation function is defined, in the usual way, as $h_{\alpha\beta}(r)=g_{\alpha\beta}(r)-1$. The site–site distribution functions are, of course, of interest in a wider context than that of interaction-site models. In any real molecular fluid, the most important site–site distribution functions are those that describe the distribution of atomic sites.

The definition (12.3.1) can be used to related the site–site distribution functions to the molecular pair distribution function $g(1,2)$. Let $\mathbf{l}_{i\alpha}$ be the vector displacement of site α in molecule i from the molecular centre $\mathbf{R}_i$, i.e.

$$\mathbf{l}_{i\alpha}=\mathbf{r}_{i\alpha}-\mathbf{R}_i \tag{12.3.2}$$

Then $g_{\alpha\beta}(\mathbf{r})$ is given by the integral of $g(1,2)$ over all coordinates, subject to the constraint that the separation of sites α, β is equal to $\mathbf{r}$:

$$\begin{aligned} g_{\alpha\beta}(\mathbf{r}) &= (1/\Omega^2)\iiiint \mathrm{d}\mathbf{R}_1\,\mathrm{d}\mathbf{R}_2\,\mathrm{d}\mathbf{\Omega}_1\,\mathrm{d}\mathbf{\Omega}_2\,g(1,2) \\ &\quad\times\delta[\mathbf{R}_1+\mathbf{l}_{1\alpha}(\mathbf{\Omega}_1)]\delta[\mathbf{R}_2+\mathbf{l}_{2\beta}(\mathbf{\Omega}_2)-\mathbf{r}] \\ &= (1/\Omega^2)\iiint \mathrm{d}\mathbf{R}_{12}\,\mathrm{d}\mathbf{\Omega}_1\,\mathrm{d}\mathbf{\Omega}_2\,g(1,2) \\ &\quad\times\delta[\mathbf{R}_{12}+\mathbf{l}_{2\beta}(\mathbf{\Omega}_2)-\mathbf{l}_{1\alpha}(\mathbf{\Omega}_1)-\mathbf{r}] \end{aligned} \tag{12.3.3}$$

From (12.3.3) it follows that the Fourier transform of $g_{\alpha\beta}(\mathbf{r})$ with respect to $\mathbf{r}$ is

$$\begin{aligned} \hat{g}_{\alpha\beta}(\mathbf{k}) &= (1/\Omega^2)\iiiint \mathrm{d}\mathbf{R}_{12}\,\mathrm{d}\mathbf{\Omega}_1\,\mathrm{d}\mathbf{\Omega}_2\,g(1,2)\delta[\mathbf{R}_{12}+\mathbf{l}_{2\beta}(\mathbf{\Omega}_2)-\mathbf{l}_{1\alpha}(\mathbf{\Omega}_1)-\mathbf{r}] \\ &\quad\times\exp(-i\mathbf{k}\cdot\mathbf{r})\,\mathrm{d}\mathbf{r} \\ &= (1/\Omega^2)\iiint \mathrm{d}\mathbf{R}_{12}\,\mathrm{d}\mathbf{\Omega}_1\,\mathrm{d}\mathbf{\Omega}_2\,g(1,2)\exp(-i\mathbf{k}\cdot\mathbf{R}_{12}) \\ &\quad\times\exp[-i\mathbf{k}\cdot\mathbf{l}_{2\beta}(\mathbf{\Omega}_2)]\exp[i\mathbf{k}\cdot\mathbf{l}_{1\alpha}(\mathbf{\Omega}_1)] \\ &= \langle\hat{g}(1,2)\exp[-i\mathbf{k}\cdot\mathbf{l}_{2\beta}(\mathbf{\Omega}_2)]\exp[i\mathbf{k}\cdot\mathbf{l}_{1\alpha}(\mathbf{\Omega}_1)]\rangle_{\mathbf{\Omega}_1\mathbf{\Omega}_2} \end{aligned} \tag{12.3.4}$$

where $\hat{g}(1,2)$ is defined by Eqn (12.2.10). There is an analogous expression for $\hat{h}_{\alpha\beta}(\mathbf{k})$ in terms of $h(1,2)$.

The site–site distribution functions have a simple physical interpretation. They are also directly related to the structure factors measured in X-ray and neutron-scattering experiments. On the other hand, the integrations in (12.3.3) involve an irretrievable loss of information. One consequence of this is the fact that $g(1,2)$ cannot be reconstructed exactly from any finite set of site–site distribution functions. A number of approximate methods of doing so have been proposed in which $g(1,2)$ is written as some product of the $g_{\alpha\beta}(r)$ (Hsu *et al.*, 1976; Quirke and Tildesley, 1982). These "site-superposition" approximations provide some qualitative insight into the angular structure of $g(1,2)$, but they have found little application in accurate numerical work.

Many of the quantities that are expressible as integrals over $g(1,2)$ can also be written in terms of site–site distribution functions. For example, if the intermolecular potential is of the interaction-site form and the site–site potentials $v_{\alpha\beta}(r)$ are spherically symmetric, the excess internal energy is given by

$$\frac{U^{\mathrm{ex}}}{N} = 2\pi\rho \sum_{\alpha}^{m}\sum_{\beta} \int_0^\infty g_{\alpha\beta}(r) v_{\alpha\beta}(r) r^2 \,\mathrm{d}r \qquad (12.3.5)$$

where m is the number of interaction sites per molecule. Equation (12.3.5) is a straightforward generalization of (2.5.12) and can be derived by the same intuitive approach that is discussed in connection with the earlier result. The generalization of the virial equation (2.5.14) is more complicated, and knowledge of $g_{\alpha\beta}(r)$ for all α, β is not sufficient to calculate the pressure. The equation of state can, however, be determined by integration of the compressibility equation (12.1.11). Because the choice of molecular centre is arbitrary, and need not be the same for each molecule, it is clear that Eqn (12.1.11) may be written as

$$\begin{aligned} \rho k_{\mathrm{B}} T \chi_T &= 1 + \rho \int [g_{\alpha\beta}(r) - 1] \,\mathrm{d}\mathbf{r} \\ &= 1 + \rho \hat{h}_{\alpha\beta}(0) \end{aligned} \qquad (12.3.6)$$

where α, β refer to *any* pair of sites. Finally, the angular correlation parameters G_l defined by Eqn (12.1.8) can be expressed as integrals over combinations of the functions $h_{\alpha\beta}(r)$. A general procedure, valid for all l, has been described by Høye and Stell (1977a). In the case of a heteronuclear but non-polar diatomic molecule with atomic sites α and β, the result for G_1 (Sullivan and Gray, 1981) is

$$G_1 = -\frac{\rho}{2l^2} \int_0^\infty r^2 \Delta h(r) \,\mathrm{d}\mathbf{r} \qquad (12.3.7)$$

where l is the bondlength and

$$\Delta h(r) = h_{\alpha\alpha}(r) + h_{\beta\beta}(r) - 2h_{\alpha\beta}(r) \qquad (12.3.8)$$

If $\hat{h}_{\alpha\beta}(k)$ is expanded in powers of k in the form

$$\begin{aligned} \hat{h}_{\alpha\beta}(k) &= 4\pi \int_0^\infty h_{\alpha\beta}(r) \frac{\sin kr}{kr} r^2 \,\mathrm{d}r \\ &= \hat{h}_{\alpha\beta}(0) + h_{\alpha\beta}^{(2)} k^2 + \cdots \end{aligned} \qquad (12.3.9)$$

we find that G_1 is proportional to the coefficient of k^2 in the small-k expansion of $\Delta\hat{h}(k)$:

$$G_1 = \frac{3\rho}{l^2}\Delta h^{(2)} \tag{12.3.10}$$

Similarly, G_l for $l > 1$ can be written in terms of higher-order coefficients $\Delta h^{(n)}$. The example we have quoted is somewhat artificial, since any real heteronuclear molecule can be expected to have a dipole moment, and in that case Eqn (12.3.7) is no longer correct. Nonetheless, the example serves to illustrate the general form of the results. We shall show in Section 12.5 how the expression for G_1 given by (12.3.7) may be recovered from the result appropriate to polar molecules. If the molecule is homonuclear, all site–site distribution functions are the same and G_1 vanishes, as it must do on grounds of symmetry.

Information on the atom–atom distribution functions of real molecules is gained experimentally from the analysis of radiation-scattering experiments. Consider first the simple case of a homonuclear diatomic molecule. Let $\mathbf{u}_i$ be a unit vector along the internuclear axis of molecule i. Then the coordinates of the atoms α, β relative to the centre of mass $\mathbf{R}_i$ are

$$\mathbf{r}_{i\alpha} = \mathbf{R}_i + \tfrac{1}{2}\mathbf{u}_i l, \qquad \mathbf{r}_{i\beta} = \mathbf{R}_i - \tfrac{1}{2}\mathbf{u}_i l \tag{12.3.11}$$

We define the Fourier components of the atomic density as

$$\rho_{\mathbf{k}} = \sum_{i=1}^{N} [\exp(-i\mathbf{k}\cdot\mathbf{r}_{i\alpha}) + \exp(-i\mathbf{k}\cdot\mathbf{r}_{i\beta})] \tag{12.3.12}$$

and the molecular structure factor as

$$S(k) = \frac{1}{4N}\langle\rho_{\mathbf{k}}\rho_{-\mathbf{k}}\rangle \tag{12.3.13}$$

where N is the number of molecules. The factor $\frac{1}{4}$ is included in (12.3.13) in order to make the definition of $S(k)$ reduce to that of an atomic fluid in the limit $l \to 0$. The statistical average in (12.3.13) may be rewritten in terms of either the atomic or molecular pair distribution functions. In the first case, by exploiting the fact that the atoms α, β in each molecule play equivalent roles, we can write

$$\frac{1}{4N}\langle\rho_{\mathbf{k}}\rho_{-\mathbf{k}}\rangle = \frac{1}{2} + \frac{1}{2N}\sum_{i=1}^{N}\langle\cos(\mathbf{k}\cdot\mathbf{u}_i l)\rangle_{\mathbf{\Omega}_i}$$
$$+ \frac{1}{4N}\left\langle\sum_{i\neq j}^{N}\sum \exp[-i\mathbf{k}\cdot(\mathbf{r}_{i\alpha} - \mathbf{r}_{j\beta})]\right\rangle \tag{12.3.14}$$

The second term on the right-hand side involves only an average over angles, and the third term can be related to any of the four identical distribution functions $g_{\alpha\beta}(r)$ via the definition (12.3.1). Thus

$$S(k) = S_{\text{intra}}(k) + S_{\text{inter}}(k) \tag{12.3.15}$$

where

$$\begin{aligned} S_{\text{intra}}(k) &= \tfrac{1}{2}(1 + \langle \cos \mathbf{k} \cdot \mathbf{u}_i l \rangle_{\mathbf{\Omega}_i}) \\ &= \frac{1}{2}\left(1 + \frac{\sin kl}{kl}\right) \end{aligned} \tag{12.3.16}$$

is the intramolecular contribution. The intermolecular part is given by

$$\begin{aligned} S_{\text{inter}}(k) &= \rho \int h_{\alpha\beta}(r) \exp(-i\mathbf{k} \cdot \mathbf{r})\, \mathrm{d}\mathbf{r} \\ &= S_{\alpha\beta}(k) - 1 \end{aligned} \tag{12.3.17}$$

where $S_{\alpha\beta}(k)$ is the atomic structure factor and a physically unimportant term in $\delta(\mathbf{k})$ has been omitted. The total intensity of scattered radiation at a given wavevector is proportional to the structure factor (12.3.15); the latter can be inverted to yield the atomic pair distribution function provided the intramolecular term is first removed.

In order to relate $S(k)$ to the molecular pair distribution function, we start from the definition (12.3.13) and proceed as follows:

$$\begin{aligned} S(k) &= \frac{1}{4N} \langle \rho_{\mathbf{k}} \rho_{-\mathbf{k}} \rangle \\ &= \frac{1}{N} \left\langle \sum_{i=1}^{N} \sum_{j=1}^{N} \exp(-i\mathbf{k} \cdot \mathbf{R}_{ij}) \cos(\tfrac{1}{2}\mathbf{k} \cdot \mathbf{u}_i l) \cos(\tfrac{1}{2}\mathbf{k} \cdot \mathbf{u}_j l) \right\rangle \\ &= \frac{1}{2}\left(1 + \frac{\sin kl}{kl}\right) \\ &\quad + \frac{1}{N} \left\langle \sum_{i \neq j}^{N}\sum \exp(-i\mathbf{k} \cdot \mathbf{R}_{ij}) \cos(\tfrac{1}{2}\mathbf{k} \cdot \mathbf{u}_i l) \cos(\tfrac{1}{2}\mathbf{k} \cdot \mathbf{u}_j l) \right\rangle \\ &= S_{\text{intra}}(k) + \frac{\rho}{\Omega^2} \iiint \exp(-i\mathbf{k} \cdot \mathbf{R}_{12})[g(1,2) - 1] \\ &\quad \times \cos(\tfrac{1}{2}\mathbf{k} \cdot \mathbf{u}_1 l) \cos(\tfrac{1}{2}\mathbf{k} \cdot \mathbf{u}_2 l)\, \mathrm{d}\mathbf{R}_{12}\, \mathrm{d}\mathbf{\Omega}_1\, \mathrm{d}\mathbf{\Omega}_2 \end{aligned} \tag{12.3.18}$$

Equation (12.3.18) is an exact relation between $S(k)$ and $g(1, 2)$. Comparison with (12.3.4) confirms that the second term on the right-hand side is just $\hat{h}_{\alpha\beta}(k)$, as is also evident from consideration of Eqns (12.3.15) to (12.3.17). To obtain a more tractable expression, we substitute in (12.3.18) the spherical-harmonic expansion of $g(1,2)$ given by Eqn (12.2.1). The structure factor can then be written as

$$S(k) = S_{\text{intra}}(k) + f(k)[S_c(k) - 1] + S_{\text{aniso}}(k) \tag{12.3.19}$$

where

$$\begin{aligned} f(k) &= \langle \cos(\tfrac{1}{2}\mathbf{k}\cdot\mathbf{u}_1 l)\cos(\tfrac{1}{2}\mathbf{k}\cdot\mathbf{u}_2 l)\rangle_{\mathbf{\Omega}_1\mathbf{\Omega}_2} \\ &= \left(\frac{\sin\frac{1}{2}kl}{\frac{1}{2}kl}\right)^2 \end{aligned} \tag{12.3.20}$$

and $S_c(k)$ is the Fourier transform of the centres radial distribution function $g_c(R)$. The term $S_{\text{aniso}}(k)$ in (12.3.19) represents the contribution to $S(k)$ that arises from the angle-dependent terms in $g(1,2)$, i.e. from all spherical harmonic components beyond $(l_1, l_2, m) = (0, 0, 0)$. If the molecules are nearly spherical, $S_{\text{aniso}}(k)$ will be small. In those circumstances, it follows from (12.3.15), (12.3.17) and (12.3.19) that

$$S_{\alpha\beta}(k) \simeq 1 + f(k)[S_c(k) - 1] \tag{12.3.21}$$

Equation (12.3.21) is called the "free-rotation" approximation. In liquid nitrogen, the orientational correlations between molecules are weak, and the free-rotation approximation works well, as shown by the results for $S_{\text{aniso}}(k)$ plotted in Figure 12.1 (Weis and Levesque, 1976). In most cases,

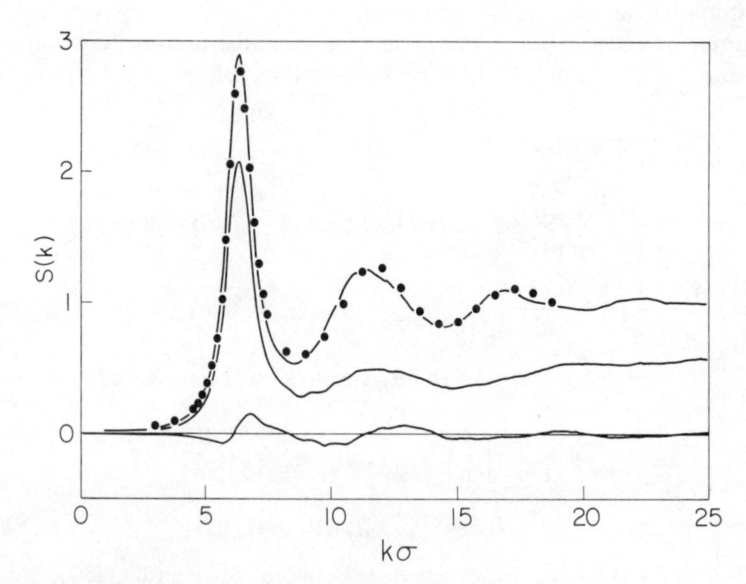

FIG. 12.1. Structure factors calculated by molecular dynamics for a model of liquid nitrogen near the triple point. The curves, reading from top to bottom, show $S_c(k)$, $S(k)$ and $S_{\text{aniso}}(k)$; the points represent the structure factor of a Lennard-Jones fluid with parameters chosen to give the best fit to $S_c(k)$. After Weis and Levesque (1976).

however, even for quasispherical molecules such as carbon tetrachloride, there exists a strong, short-range, angular order that causes the free-rotation approximation to break down (McDonald *et al.*, 1982). At the same time, irrespective of the importance of orientational correlations, the intermolecular contribution to $S(k)$ differs from the structure factor of an atomic fluid because of the modulating role of the function $f(k)$. This effect is clear in the curves drawn in Figure 12.1. In the example shown, the function $S_c(k)$ has a strongly oscillatory character and is very well fitted by the structure factor of an atomic system. By contrast, the oscillations in $S(k)$ are strongly damped beyond the first peak, where $S(k)$ and $S_{\text{inter}}(k)$ are essentially equal apart from a constant. The free-rotation approximation becomes exact in the limit $k \to 0$ because all the cosine terms in Eqn (12.3.18) tend towards unity. Thus

$$\rho k_B T \chi_T = \lim_{k \to 0} S(k) = 1 + \rho \int [g_c(R) - 1] \, d\mathbf{R} \qquad (12.3.22)$$

which is the same result as in Eqn (12.1.11).

For more complicated molecules, there is little value in defining a structure factor through a formula analogous to (12.3.13). It is more useful to focus attention on those combinations of the atomic structure factors that are experimentally accessible (Powles, 1973). In the case of neutron scattering, the measured structure factor $S^N(k)$ can again be written in the form of Eqn (12.3.15), with

$$\left(\sum_\alpha b_\alpha\right)^2 S_{\text{intra}}^N(k) = \sum_\alpha b_\alpha^2 + \sum_{\alpha \neq \beta}\sum b_\alpha b_\beta \frac{\sin kl_{\alpha\beta}}{kl_{\alpha\beta}} \qquad (12.3.23a)$$

$$\left(\sum_\alpha b_\alpha\right)^2 S_{\text{inter}}^N(k) = \rho \sum_\alpha \sum_\beta b_\alpha b_\beta \int \exp(-i\mathbf{k} \cdot \mathbf{r})[g_{\alpha\beta}(r) - 1] \, d\mathbf{r}$$
$$= \sum_\alpha \sum_\beta b_\alpha b_\beta [S_{\alpha\beta}(k) - 1] \qquad (12.3.23b)$$

where the sums run over all nuclei in the molecule, b_α is the coherent neutron scattering length of nucleus α, and $l_{\alpha\beta}$ is the separation of nuclei α, β. These expressions reduce to (12.3.16) and (12.3.17) in the case of a diatomic molecule with $b_\alpha = b_\beta$. After removal of the intramolecular term, Fourier transformation yields a weighted sum of atomic pair distribution functions of the form

$$g^N(r) = \frac{\sum_\alpha \sum_\beta b_\alpha b_\beta g_{\alpha\beta}(r)}{\sum_\alpha b_\alpha^2} \qquad (12.3.24)$$

Isotopic substitution makes it possible to vary the weights with which the different $g_{\alpha\beta}(r)$ contribute to $g^{N}(r)$ and hence, in favourable cases, to determine some or all of the individual atom–atom distribution functions.

A formula similar to (12.3.23) applies also to X-ray scattering, the only difference being that the nuclear scattering lengths are replaced by the atomic form factors (see Section 5.1). Since the latter are functions of k, the weighted distribution function $g^{X}(r)$ obtained by transformation of the measured structure factor $S^{X}(k)$ is not, strictly speaking, a linear combination of the $g_{\alpha\beta}(r)$, but the error involved in ignoring this fact is normally slight. The form factor increases with the number of electrons in the atom, while the nuclear scattering lengths vary irregularly throughout the periodic table. In general, therefore, X-ray and neutron-scattering experiments yield different information about atom–atom correlations. This effect is seen at its most extreme in experiments on hydrogen-containing molecules. In the case of water, for example, $g^{X}(r)$ is to a good approximation equal to the oxygen–oxygen distribution function $g_{OO}(r)$. On the other hand, a neutron-scattering experiment on a deuterated sample yields a weighted sum of O-O, O–D and D–D distribution functions given by

$$g^{N}(r) = 0.09g_{OO}(r) + 0.42g_{OD}(r) + 0.49g_{DD}(r) \qquad (12.3.25)$$

The results of the two types of experiment can, in principle, be combined in order to extract information on individual distribution functions, but in practice it is difficult to do this with demonstrable success.

In Figure 12.2 we show some results obtained by X-ray scattering for the carbon–carbon distribution function $g_{CC}(R)$ in liquid ethylene near its triple point (Narten and Habenschuss, 1981). Although ethylene is a polyatomic molecule, $g_{CC}(r)$ resembles the atomic pair distribution function seen in both simulations and experiments on diatomic systems. The main peak is weaker than in argon-like liquids, and there is a pronounced shoulder on the large-r side. Both these features are a consequence of the interference between inter- and intramolecular correlations. Simple geometry suggests that shoulders might be seen at combinations of distances such that $r_{\alpha\gamma} \simeq |\sigma_{\alpha\beta} \pm l_{\beta\gamma}|$, where σ is an atomic diameter and l is a bondlength, but they are often so smooth as to be undetectable. In the case of fused-hard-sphere models, the shoulders appear as cusps in the site–site distribution functions, i.e. as discontinuities in the derivative of $g_{\alpha\beta}(r)$ with respect to r (Ladanyi and Chandler, 1975). The shoulder seen in Figure 12.2 can be associated with "T-shaped" configurations of the type pictured in Figure 12.3. This particular feature is enhanced for molecules such as bromine that have a large quadrupole moment (Narten *et al.*, 1978), because the quadrupolar interaction strongly favours the "T"-configuration. In nitrogen, by contrast, the shoulder is barely perceptible (Narten *et al.*, 1980).

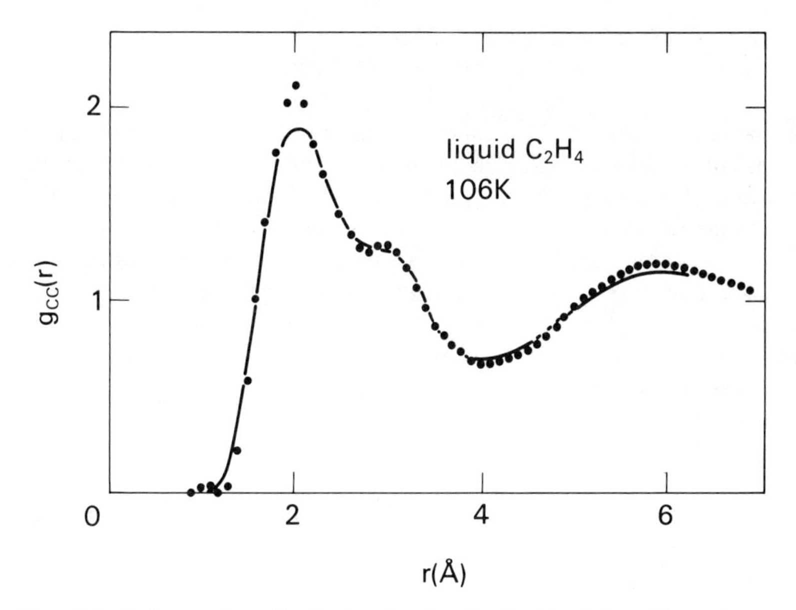

FIG. 12.2. Carbon–carbon distribution function for liquid ethylene. The points are X-ray scattering results (Narten and Habenschuss, 1981) and the curve is obtained from a molecular-dynamics calculation for a six-site model (J. M. Carter, unpublished results).

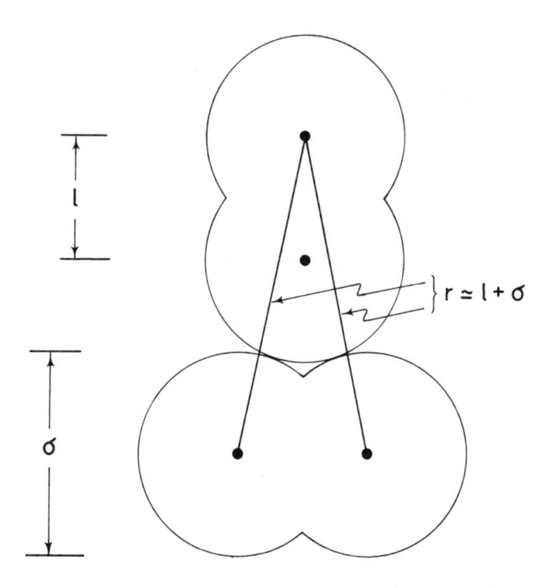

FIG. 12.3. The "T-shaped" configuration of a pair of homonuclear diatomic molecules. After Gray and Gubbins (1984).

12.4 CORRELATION-FUNCTION EXPANSIONS FOR SIMPLE POLAR FLUIDS

In the simplest models of polar fluids, the intermolecular potential can be written as the sum of a small number of spherical-harmonic components. The prospects for success of theories based on harmonic expansions of the pair functions are therefore greater than in situations where the potential contains an infinite number of harmonics and the series expansions are only slowly convergent, as is true, for example, of the Lennard-Jones diatomics studied by Streett and Tildesley (1977) (see Section 12.2). In this section, we discuss some of the general questions that arise in attempts to treat polar fluids in this way.

Consider a polar fluid for which the intermolecular potential is the same as in Eqn (1.2.4), but which we rewrite here as

$$v(1,2) = v_0(R) - \frac{\mu^2}{R^3} D(1,2) \tag{12.4.1}$$

with

$$D(1,2) = 3(\mathbf{u}_1 \cdot \mathbf{s})(\mathbf{u}_2 \cdot \mathbf{s}) - \mathbf{u}_1 \cdot \mathbf{u}_2 \tag{12.4.2}$$

where $R = |\mathbf{R}_{12}|$, $\mathbf{s}$ is a unit vector in the direction of $\mathbf{R}_{12}$, $\mathbf{u}_i$ is a unit vector parallel to the dipole moment of molecule i, $v_0(R)$ is assumed to be spherically symmetric, and the angle-dependent term represents the ideal-dipole interaction. It was first shown by Wertheim (1971), and subsequently elaborated by many others (see, for example, Stell *et al.*, 1981), that an adequate description of the static properties of such a fluid can be obtained with a basis set consisting of only three functions: $S(1,2) = 1$, $\Delta(1,2) = \mathbf{u}_1 \cdot \mathbf{u}_2$, and $D(1,2)$, defined above. The solution for $h(1,2)$ is therefore assumed to be of the form

$$h(1,2) = h_S(R) + h_\Delta(R)\Delta(1,2) + h_D(R)D(1,2) \tag{12.4.3}$$

On multiplying through Eqn (12.4.3) successively by S, Δ and D and integrating over angles, we find that the projections $h_S(R)$, $h_\Delta(R)$ and $h_D(R)$ are given by

$$h_S(R) = \langle h(1,2)\rangle_{\boldsymbol{\Omega}_1\boldsymbol{\Omega}_2} \tag{12.4.4}$$

$$h_\Delta(R) = 3\langle h(1,2)\Delta(1,2)\rangle_{\boldsymbol{\Omega}_1\boldsymbol{\Omega}_2} \tag{12.4.5}$$

$$h_D(R) = \tfrac{3}{2}\langle h(1,2)D(1,2)\rangle_{\boldsymbol{\Omega}_1\boldsymbol{\Omega}_2} \tag{12.4.6}$$

Although it is not immediately apparent, Eqn (12.4.3) is equivalent to an expansion in laboratory-frame harmonics, since the functions Δ and D are equivalent, respectively, to the rotational invariants Φ^{110} and Φ^{112} introduced in Eqn (12.2.9). For the present, however, we retain the more compact notation embodied in (12.4.3).

The direct correlation function can be treated in similar fashion to $h(1,2)$. We therefore write

$$c(1,2)=c_S(R)+c_\Delta(R)\Delta(1,2)+c_D(R)D(1,2) \qquad (12.4.7)$$

and substitute both (12.4.3) and (12.4.7) in the molecular Ornstein–Zernike relation (12.1.4). After taking Fourier transforms, we find that

$$\hat{h}(1,2)=\hat{c}(1,2)+\rho\langle\hat{c}(1,3)\hat{h}(3,2)\rangle_{\Omega_3} \qquad (12.4.8)$$

where, for example:

$$\hat{h}(1,2)=\hat{h}_S(k)+\hat{h}_\Delta(k)\Delta(1,2)+\int h_D(R)D(1,2)\exp(-i\mathbf{k}\cdot\mathbf{R})\,\mathrm{d}\mathbf{R} \qquad (12.4.9)$$

The term in D can be transformed by taking the direction of $\mathbf{k}$ as the z-axis and making the substitution $\mathbf{s}=(\sin\theta\cos\phi, \sin\theta\sin\phi, \cos\theta)$. Two integrations by parts show that

$$\int_{-1}^{1}\int_{0}^{2\pi}(\mathbf{u}_1\cdot\mathbf{s})(\mathbf{u}_2\cdot\mathbf{s})\exp(-i\mathbf{k}\cdot\mathbf{R}\cos\theta)\,\mathrm{d}\phi\,\mathrm{d}(\cos\theta)$$
$$=-4\pi R^2\{3u_{1z}u_{2z}j_2(kR)-\mathbf{u}_1\cdot\mathbf{u}_2[j_0(kR)+j_2(kR)]\} \qquad (12.4.10)$$

where $j_2(x)=(3x^{-3}\sin x-3x^{-2}\cos x-x^{-1}\sin x)$. Thus

$$\int h_D(R)D(1,2)\exp(-i\mathbf{k}\cdot\mathbf{R})\,\mathrm{d}\mathbf{R}=D_k(1,2)\bar{h}_D(k) \qquad (12.4.11)$$

where

$$D_k(1,2)=3u_{1z}u_{2z}-\mathbf{u}_1\cdot\mathbf{u}_2$$
$$=\frac{3(\mathbf{u}_1\cdot\mathbf{k})(\mathbf{u}_2\cdot\mathbf{k})}{k^2}-\mathbf{u}_1\cdot\mathbf{u}_2 \qquad (12.4.12)$$

and the Hankel transform $\bar{h}_D(k)$ is

$$\bar{h}_D(k)=-4\pi\int_0^\infty j_2(kR)h_D(R)R^2\,\mathrm{d}R \qquad (12.4.13)$$

Equation (12.4.11) is a particular case of the general result (12.2.12); the transform of $c_D(R)D(1,2)$ is handled in the same way.

In order to summarize the effect of the angular integrations in (12.4.8), we define the angular convolution of two functions A, B as

$$A*B = B*A = (1/\Omega)\int A(1,3)B(3,2)\,\mathrm{d}\mathbf{\Omega}_3$$

$$\equiv \langle A(1,3)B(3,2)\rangle_{\mathbf{\Omega}_3} \qquad (12.4.14)$$

For the functions of interest here, the following "multiplication table" is easily established (Wertheim, 1971):

	S	Δ	D_k
S	S	0	0
Δ	0	$\frac{1}{3}\Delta$	$\frac{1}{3}\Delta$
D_k	0	$\frac{1}{3}\Delta$	$\frac{1}{3}(D_k+2\Delta)$

Inspection of the table show that the functions S, Δ and D_k form a closed set under the operation (12.4.14) in the sense that convolution of any two functions yields only functions in the same set (or zero). The practical significance of this result is the fact that if $h(1,2)$ is asumed to be of the form (12.4.3), then $c(1,2)$ is necessarily given by (12.4.7), and vice versa. A closure of the Ornstein–Zernike relation is still required, but provided this does not generate any new harmonics, Eqns (12.4.3) and (12.4.7) together form a self-consistent approximation to which an exact solution can be found either analytically (as in the MSA, Section 12.6) or numerically (as in Section 12.7).

At large R, $c(1,2)$ behaves as $-\beta v(1,2)$. It follows that $c_D(R)$ must be long ranged, decaying asymptotically as R^{-3}. It turns out, as we shall see in Section 12.5, that $h_D(R)$ also decays as R^{-3}, the strength of the long-range part being related to the dielectric constant of the fluid, but all other projections of $h(1,2)$ and $c(1,2)$ are short-range functions (Høye and Stell, 1976). The slow decay of $h_D(R)$ and $c_D(R)$ creates difficulties in numerical calculations. It is therefore convenient to introduce two short-range, auxiliary functions $h_D^0(R)$ and $c_D^0(R)$. These are defined in terms, respectively, of $h_D(R)$ and $c_D(R)$ in such a way as to remove the long-range parts. Thus

$$h_D^0(R) = h_D(R) - 3\int_R^\infty \frac{h_D(R')}{R'}\,\mathrm{d}R' \qquad (12.4.15)$$

with an analogous definition of $c_D^0(R)$. We see from (12.4.15) that $h_D^0(R)$ vanishes in any range of R in which $h_D(R)$ has reached its asymptotic form.

The inverse of (12.4.15) is

$$h_D(R) = h_D^0(R) - \frac{3}{R^3}\int_0^R h_D^0(R')R'^2\,\mathrm{d}R' \tag{12.4.16}$$

as can be checked by first differentiating (12.4.16) with respect to R and then integrating from R to $R=\infty$ (where both $h_D(R)$ and $h_D^0(R)$ are zero); this gives back (12.4.15). Equation (12.4.16) shows that $h_D(R)$ behaves asymptotically as

$$\lim_{R\to\infty} h_D(R) = -\frac{3}{4\pi R^3}\lim_{k\to 0}\hat{h}_D^0(k) \tag{12.4.17}$$

The short-range functions $h_D^0(R)$ and $c_D^0(R)$ play an important part in the analytic solution of the MSA for dipolar hard spheres.

We have shown that use of the approximation (12.4.3) has some attractive mathematical features. The solution is of physical interest, however, only because the projections $h_S(R)$, $h_\Delta(R)$ and $h_D(R)$ contain between them all the information needed to calculate both the thermodynamic and static dielectric properties of the fluid. We postpone discussion of the difficult problem of dielectric behaviour until the next section, but expressions for thermodynamic properties are easily derived. If $v_0(R)$ in (12.4.1) is the hard-sphere potential, the excess internal energy is determined solely by the dipole–dipole interaction, and Eqn (12.1.9) becomes

$$\begin{aligned}\frac{U^{\text{ex}}}{N} &= -2\pi\rho\int_0^\infty \frac{\mu^2}{R_{12}}\langle g(1,2)D(1,2)\rangle_{\Omega_1\Omega_2}\,\mathrm{d}R_{12}\\ &= -\frac{4\pi}{3}\mu^2\rho\int_0^\infty \frac{h_D(R)}{R}\,\mathrm{d}R \end{aligned}\tag{12.4.18}$$

where we have used the definition (12.4.6) and the fact that the angle average of $D(1,2)$ is zero. If $v_0(R)$ is the Lennard-Jones potential or some other spherically symmetric but continuous interaction, there will be a further contribution to U^{ex} that can be expressed as an integral over $h_S(R)$. Similarly, Eqn (12.1.10) can be used to relate the equation of state to the projections $h_S(R)$ and $h_D(R)$. Thermodynamic properties are therefore not explicitly dependent on $h_\Delta(R)$.

Examples of $h_\Delta(R)$ and $h_D(R)$ for the dipolar hard-sphere fluid are shown in Figure 12.4. For the state point in question, corresponding to a static dielectric constant ε of approximately 30, the curves retain a pronounced oscillatory structure over a range of three to four molecular diameters. The structure in $h_\Delta(R)$ and $h_D(R)$ disappears as the dipole moment is reduced, but $h_S(R)$ (not shown) is much less sensitive to the value of ε and bears a strong resemblance to the pair correlation function of a fluid of

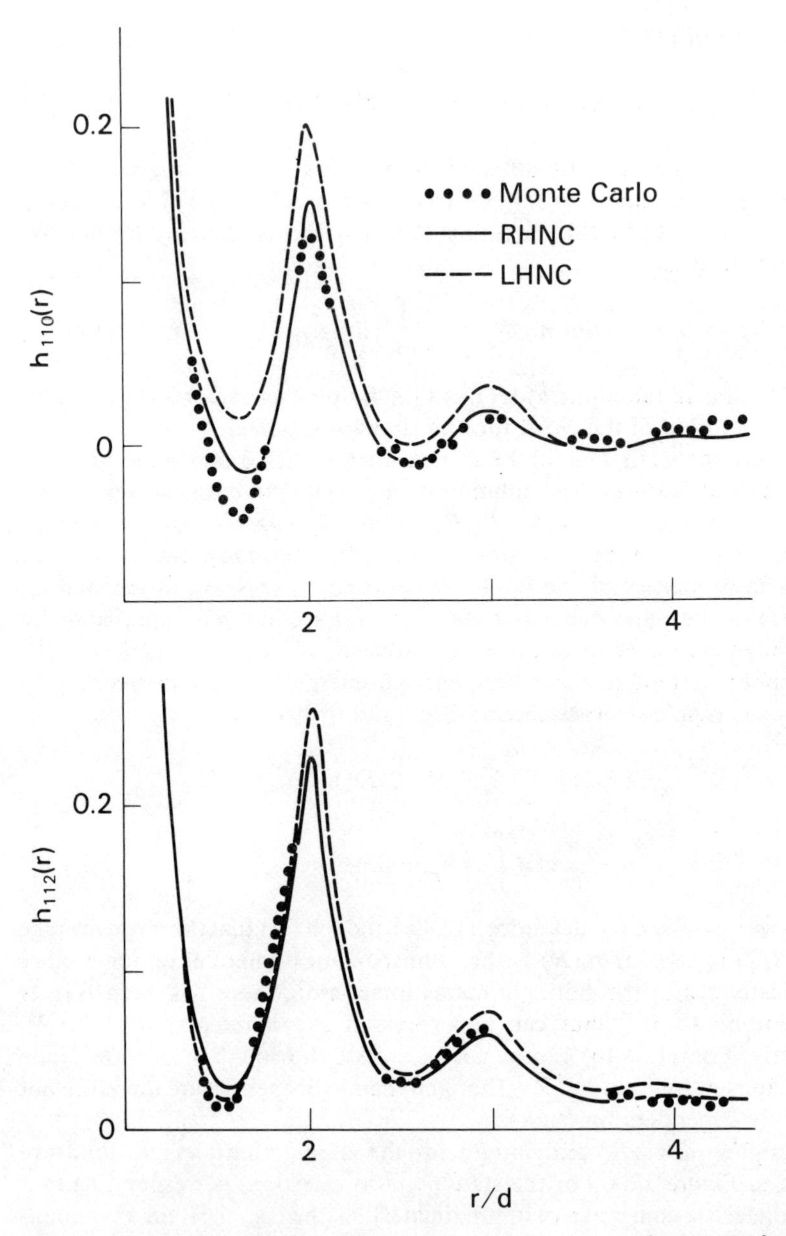

FIG. 12.4. Projections of $h(1, 2)$ for a fluid of dipolar hard spheres at $\rho d^3 = 0.8$ and $\mu^2/k_B T d^3 = 2.0$. The points are Monte Carlo results, and the curves are calculated from the LHNC and RHNC approximations to be discussed in Section 12.7. After Fries and Patey (1985). Note that the 110 and 112 components are equivalent, respectively, to the projections on $\Delta(1, 2)$ and $D(1, 2)$.

non-polar hard shpheres. The structure seen in the Δ and D projections is also depressed by addition of a quadrupole moment, as we shall discuss again in Section 12.7.

12.5 THE STATIC DIELECTRIC CONSTANT

Our goal in this section is to obtain molecular expressions for the static dielectric constant. We shall show, in particular, following in outline the arguments of Madden and Kivelson (1984), that ε is related to the long-wavelength behaviour of each of the functions $\hat{h}_\Delta(k)$ and $\bar{h}_D(k)$ introduced in the previous section. By suitably combining the two results, it is also possible to express ε in terms of site–site distribution functions (Høye and Stell, 1976; Chandler, 1977).

Consider a sample of dielectric material (a polar fluid) placed in an external field. Let $\mathbf{E}(\mathbf{R}, t)$ be the electric field at a point inside the sample (the Maxwell field), let $\mathbf{P}(\mathbf{R}, t)$ be the polarization induced in the sample, and let $\mathbf{E}^0(\mathbf{R}, t)$ be the field that would exist at the same point if the sample were removed (the external field). The polarization is related to the Maxwell field by

$$\mathbf{P}(\mathbf{R}, t) = \int \mathrm{d}\mathbf{R}' \int_{-\infty}^{t} \mathrm{d}t' \, \boldsymbol{\chi}(\mathbf{R}-\mathbf{R}', t-t') \cdot \mathbf{E}(\mathbf{R}', t) \, \mathrm{d}\mathbf{R}' \tag{12.5.1}$$

where the tensor $\boldsymbol{\chi}(\mathbf{R}, t)$ is an after-effect function of the type introduced in Section 7.6. A Fourier–Laplace transform of (12.5.1) (with $z = \omega$) gives

$$\hat{\mathbf{P}}(\mathbf{k}, \omega) = \boldsymbol{\chi}(\mathbf{k}, \omega) \cdot \hat{\mathbf{E}}(\mathbf{k}, \omega) \tag{12.5.2}$$

where the susceptibility $\boldsymbol{\chi}(\mathbf{k}, \omega)$ is related to the dielectric permittivity $\boldsymbol{\varepsilon}(\mathbf{k}, \omega)$ by

$$\boldsymbol{\chi}(\mathbf{k}, \omega) = \frac{1}{4\pi} [\boldsymbol{\varepsilon}(\mathbf{k}, \omega) - \boldsymbol{I}] \tag{12.5.3}$$

The polarization can also be related to the external field via a second susceptibility $\boldsymbol{\chi}^0(\mathbf{k}, \omega)$:

$$\hat{\mathbf{P}}(\mathbf{k}, \omega) = \boldsymbol{\chi}^0(\mathbf{k}, \omega) \cdot \hat{\mathbf{E}}^0(k, \omega) \tag{12.5.4}$$

The external field and Maxwell field will not, in general, be the same, since the polarization of the sample makes a contribution to the Maxwell field. The relation between $\mathbf{E}$ and $\mathbf{E}^0$, and hence also that between $\boldsymbol{\chi}$ and $\boldsymbol{\chi}^0$, is dependent on sample geometry. We shall assume that the system is infinite, thereby avoiding problems linked to surface effects (Fulton, 1975). With this simplification, the relation between $\mathbf{E}$ and $\mathbf{E}^0$ is

$$\mathbf{E}(\mathbf{R}, t) = \mathbf{E}^0(\mathbf{R}, t) + \int \mathrm{d}\mathbf{R}'\, \boldsymbol{T}(\mathbf{R}-\mathbf{R}') \cdot \mathbf{P}(\mathbf{R}', t) \tag{12.5.5}$$

where $\boldsymbol{T}(\mathbf{R})$ is the dipole–dipole interaction tensor defined by Eqn (1.2.5). Integrals involving the dipole–dipole tensor need to be handled with care, since $\boldsymbol{T}(\mathbf{R})$ has a singularity at the origin; the usual procedure is to cut off the integrand inside a sphere of radius σ centred on the origin and take the limit $\sigma \to 0$ after integration (Ramshaw, 1979). The transform of (12.5.5) is then given (Madden and Kivelson, 1984) by

$$\hat{\mathbf{E}}(\mathbf{k}, \omega) = \hat{\mathbf{E}}^0(\mathbf{k}, \omega) - \frac{4\pi}{k^2} \mathbf{kk} \cdot \hat{\mathbf{P}}(\mathbf{k}, \omega) \tag{12.5.6}$$

The relationship between $\boldsymbol{\chi}(\mathbf{k}, \omega)$ and $\boldsymbol{\chi}^0(\mathbf{k}, \omega)$ follows immediately from consideration of Eqns (12.5.2), (12.5.4) and (12.5.6) (cf. Eqn (10.4.15)):

$$\boldsymbol{\chi}^0(\mathbf{k}, \omega) = [\boldsymbol{I} + (4\pi/k^2)\mathbf{kk} \cdot \boldsymbol{\chi}(\mathbf{k}, \omega)]^{-1} \cdot \boldsymbol{\chi}(\mathbf{k}, \omega) \tag{12.5.7}$$

We emphasize again that this result has been derived under the assumption that the system is infinitely large. Different relations would be obtained for other sample geometries.

We shall take it to be an experimental fact that the dielectric permittivity is an intensive property of the fluid, independent of both shape and size of the sample; the same is therefore true of the susceptibility $\boldsymbol{\chi}(\mathbf{k}, \omega)$, since the two quantities are trivially linked by Eqn (12.5.3). It follows, assuming that the system is isotropic, that both $\boldsymbol{\varepsilon}$ and $\boldsymbol{\chi}$ must be independent of the direction of $\mathbf{k}$ in the limit $\mathbf{k} \to 0$. Thus

$$\lim_{\mathbf{k}\to 0} \boldsymbol{\varepsilon}(\mathbf{k}, \omega) = \varepsilon(\omega)\boldsymbol{I} \tag{12.5.8a}$$

$$\lim_{\mathbf{k}\to 0} \boldsymbol{\chi}(\mathbf{k}, \omega) = \chi(\omega)\boldsymbol{I} \tag{12.5.8b}$$

where $\varepsilon(\omega)$ and $\chi(\omega)$ are scalars. On the other hand, the longitudinal (parallel to $\mathbf{k}$) and transverse (perpendicular to $\mathbf{k}$) components of $\boldsymbol{\chi}^0(k, \omega)$ behave differently in the long-wavelength limit; this is inevitable, given that the relation between $\boldsymbol{\chi}$ (an intensive quantity) and $\boldsymbol{\chi}^0$ is shape dependent. Taking the z-axis along the direction of $\mathbf{k}$, we find from (12.5.3), (12.5.7) and (12.5.8) that

$$4\pi \lim_{\mathbf{k}\to 0} \chi^0_{\alpha\alpha}(\mathbf{k}, \omega) = \varepsilon(\omega) - 1, \qquad \alpha = x, y \tag{12.5.9a}$$

$$4\pi \lim_{\mathbf{k}\to 0} \chi^0_{zz}(\mathbf{k}, \omega) = \frac{\varepsilon(\omega) - 1}{\varepsilon(\omega)} \tag{12.5.9b}$$

and

$$4\pi \lim_{\mathbf{k}\to 0} \operatorname{Tr} \boldsymbol{\chi}^0(\mathbf{k}, \omega) = \frac{[\varepsilon(\omega)-1][2\varepsilon(\omega)+1]}{\varepsilon(\omega)} \tag{12.5.10}$$

The statistical-mechanical problem is to obtain expressions for the components of $\boldsymbol{\chi}^0$ in terms of microscopic variables. The microscopic expression for the polarization induced by the external field is

$$\begin{aligned}\mathbf{P}(\mathbf{R}, t) &= \langle \mathbf{M}(\mathbf{R}, t)\rangle_{\mathbf{E}^0} \\ &= \left\langle \mu \sum_{i=1}^{N} \mathbf{u}_i(t)\delta[\mathbf{R}-\mathbf{R}_i(t)]\right\rangle_{\mathbf{E}^0}\end{aligned} \tag{12.5.11}$$

where $\mathbf{M}(\mathbf{R}, t)$ is the dipole-moment density, $\langle\cdots\rangle_{\mathbf{E}^0}$ denotes a statistical average in the presence of the external field, and the other symbols have the same meaning as in earlier sections of this chapter. The susceptibility $\boldsymbol{\chi}^0$ can now be calculated by the methods of linear response theory described in Section 7.6. (Note that $\boldsymbol{\chi}$ cannot be treated in the same way as $\boldsymbol{\chi}^0$, because the Maxwell field is not an "external" field in the required sense.) As an application of Eqn (7.6.27), we find that

$$\boldsymbol{\chi}^0(\mathbf{k}, \omega) = \frac{\beta}{V}\left(\langle \mathbf{M}_{\mathbf{k}}\mathbf{M}_{-\mathbf{k}}\rangle + i\omega \int_0^\infty \mathrm{d}t \langle \mathbf{M}_{\mathbf{k}}(t)\mathbf{M}_{-\mathbf{k}}\rangle \exp(i\omega t)\right) \tag{12.5.12}$$

where the statistical averages are now computed in the absence of the field, $\mathbf{M}_{\mathbf{k}} \equiv \mathbf{M}_{\mathbf{k}}(t=0)$ and

$$\mathbf{M}_{\mathbf{k}}(t) = \mu \sum_{i=1}^{N} \mathbf{u}_i(t) \exp[-i\mathbf{k}\cdot\mathbf{R}_i(t)] \tag{12.5.13}$$

If we take the limit $\omega \to 0$, Eqn (12.5.12) reduces to

$$\chi^0_{\alpha\alpha}(\mathbf{k}, 0) = \frac{\beta}{V}\langle M^\alpha_{\mathbf{k}} M^\alpha_{-\mathbf{k}}\rangle, \qquad \alpha = x, y, z \tag{12.5.14}$$

By combining this result with Eqn (12.5.10), we find that

$$\frac{(\varepsilon-1)(2\varepsilon+1)}{9\varepsilon} = g_{\mathrm{K}} y \tag{12.5.15}$$

where $\varepsilon \equiv \varepsilon(0)$ is the static dielectric constant, y is a molecular parameter defined as

$$y = \frac{4\pi\mu^2\rho}{9k_{\mathrm{B}}T} \tag{12.5.16}$$

and g_{K}, the Kirkwood g-factor, is given by

$$g_{\mathrm{K}} = \frac{\langle |\mathbf{M}|^2\rangle}{N\mu^2} \tag{12.5.17}$$

where $\mathbf{M} \equiv \mathbf{M}_{\mathbf{k}=0}$ is the total dipole moment of the sample. Equation (12.5.17) can be rewritten, with the help of (12.4.5), as

$$\begin{aligned} g_K &= 1 + \langle (N-1)\mathbf{u}_1 \cdot \mathbf{u}_2 \rangle \\ &= 1 + \frac{4\pi\rho}{3} \int_0^\infty h_\Delta(R) R^2 \, \mathrm{d}R \\ &= 1 + \tfrac{1}{3}\rho \hat{h}_\Delta(0) \end{aligned} \tag{12.5.18}$$

where $h_\Delta(R)$ is the function appropriate to an infinite system.

Equation (12.5.15) is the first of the two key results of this section. It was originally derived by Kirkwood (1939) from a calculation of the fluctuation in the total dipole moment of a sphere embedded in a dielectric continuum and is commonly referred to as the Kirkwood formula. By setting $g_K = 1$, we obtain the result known as the Onsager equation; this amounts to ignoring the short-range angular correlations represented by the function $h_\Delta(R)$. Equation (12.5.15) could have been obtained by working throughout in the $\omega = 0$ limit, but the frequency-dependent results are needed for the discussion of dielectric relaxation in Section 12.11.

Our next task is to relate ε to the function $h_D(R)$. To do this, we must consider separately the longitudinal and transverse components of $\boldsymbol{\chi}^0$. For the longitudinal component, we find from Eqns (12.5.13) and (12.5.14) that

$$\begin{aligned} \chi^0_{zz}(\mathbf{k}, 0) &= \frac{\beta}{V} \langle M^z_{\mathbf{k}} M^z_{-\mathbf{k}} \rangle \\ &= \tfrac{1}{3}\mu^2\rho\beta + \mu^2\rho\beta \langle (N-1) u_{1z} u_{2z} \exp(-i\mathbf{k} \cdot \mathbf{R}_{12}) \rangle \\ &= \tfrac{1}{3}\mu^2\rho\beta \\ &\quad + \frac{\mu^2\rho^2\beta}{\Omega^2} \iiint \frac{(\mathbf{k} \cdot \mathbf{u}_1)(\mathbf{k} \cdot \mathbf{u}_2)}{k^2} h(1,2) \\ &\quad \times \exp(-i\mathbf{k} \cdot \mathbf{R}_{12}) \, \mathrm{d}\mathbf{R}_{12} \, \mathrm{d}\boldsymbol{\Omega}_1 \, \mathrm{d}\boldsymbol{\Omega}_2 \\ &= \tfrac{1}{3}\mu^2\rho\beta + \mu^2\rho^2\beta \langle k^{-2} (\mathbf{k} \cdot \mathbf{u}_1)(\mathbf{k} \cdot \mathbf{u}_2) \hat{h}(1,2) \rangle_{\boldsymbol{\Omega}_1 \boldsymbol{\Omega}_2} \end{aligned} \tag{12.5.19}$$

We now substitute for $\hat{h}(1,2)$ from Eqn (12.4.9) and evaluate the angle averages with the help of the following, easily proved results (here $\mathbf{n}$ is a unit vector of fixed orientation):

$$\langle (\mathbf{n} \cdot \mathbf{u}_1)(\mathbf{n} \cdot \mathbf{u}_2)(\mathbf{u}_1 \cdot \mathbf{u}_2) \rangle_{\boldsymbol{\Omega}_1 \boldsymbol{\Omega}_2} = \langle (\mathbf{n} \cdot \mathbf{u}_1)^2 (\mathbf{n} \cdot \mathbf{u}_2)^2 \rangle_{\boldsymbol{\Omega}_1 \boldsymbol{\Omega}_2} = \tfrac{1}{9} \tag{12.5.20}$$

A simple calculation shows that

$$\lim_{\mathbf{k} \to 0} \chi^0_{zz}(\mathbf{k}, 0) = \tfrac{1}{3}\mu^2\rho\beta [1 + \tfrac{1}{3}\rho \hat{h}_\Delta(0) + \tfrac{2}{3}\rho \bar{h}_D(0)] \tag{12.5.21}$$

Although we have used the approximation (12.4.9), Eqn (12.5.21) is an exact result; terms ignored in (12.4.9) make no contribution to the angle average in (12.5.19).

The transverse component can be treated in a similar way. It is possible, however, to take a short-cut, since we are interested only in the $\mathbf{k}\to 0$ limit. Equations (12.5.10), (12.5.15) and (12.5.18) show that the trace of the tensor $\boldsymbol{\chi}^0(k, \omega)$ in the long-wavelength, low-frequency limit is

$$\lim_{\mathbf{k}\to 0} \mathrm{Tr}\, \boldsymbol{\chi}^0(\mathbf{k}, 0) = \mu^2\rho\beta[1+\tfrac{1}{3}\hat{h}_\Delta(0)] \qquad (12.5.22)$$

As the two transverse components are equivalent, we find from (12.5.21) and (12.5.22) that

$$\lim_{\mathbf{k}\to 0} \chi^0_{xx}(\mathbf{k}, 0) = \tfrac{1}{3}\mu^2\rho\beta[1+\tfrac{1}{3}\rho\hat{h}_\Delta(0)-\tfrac{1}{3}\rho\bar{h}_D(0)] \qquad (12.5.23)$$

Use of Eqn (12.5.9) leads to our second main result:

$$\frac{(\varepsilon-1)^2}{\varepsilon} = 4\pi \lim_{\mathbf{k}\to 0} [\chi^0_{xx}(\mathbf{k}, 0) - \chi^0_{zz}(\mathbf{k}, 0)]$$

$$= -3y\rho\bar{h}_D(0) \qquad (12.5.24)$$

It can be shown (Høye *et al.*, 1974) that the Hankel transform in (12.5.24) is also the Fourier transform of the short-range function $h^0_D(R)$ defined by Eqn (12.4.15), i.e. $\bar{h}_D(k) = \hat{h}^0_D(k)$. We may therefore combine Eqns (12.4.17) and (12.5.24) to give

$$\lim_{R\to\infty} h_D(R) = \frac{(\varepsilon-1)^2}{4\pi y\rho\varepsilon}\frac{1}{R^3} \qquad (12.5.25)$$

This calculation establishes both that $h(1,2)$ is long ranged and that the long range of the correlations is responsible for the difference in behaviour of the longitudinal and transverse components of the susceptibility $\boldsymbol{\chi}^0(\mathbf{k}, 0)$.

The expansion of $h(1,2)$ in terms of the functions S, $\Delta(1,2)$ and $D(1,2)$ has been exploited until now primarily in relation to the class of potential models described by Eqn (12.4.1), but its range of applicability is wider than this. It may be used, in particular, to discuss the dielectric properties of linear, interaction-site molecules. Consider a diatomic molecule with charges $\pm q$ located on atoms α, β, and a dipole moment $\mu = ql$, where l is the bondlength. If l_α is the distance of atom α from the molecular centre, Eqn (12.3.4) shows that the Fourier transform of one of the atomic pair correlation functions may be written as

$$\hat{h}_{\alpha\beta}(k) = \langle \hat{h}(1,2) \exp(-i\mathbf{k}\cdot\mathbf{u}_1 l_\alpha) \exp(i\mathbf{k}\cdot\mathbf{u}_2 l_\beta)\rangle_{\mathbf{\Omega}_1\mathbf{\Omega}_2} \qquad (12.5.26)$$

where $l_\alpha + l_\beta = l$, and the plane-wave functions can be replaced by their Rayleigh expansions (Watson, 1966):

$$\exp(-i\mathbf{k}\cdot\mathbf{r}) = \sum_{n=0}^{\infty} (2n+1)i^n j_n(kr) P_n(\mathbf{k}\cdot\mathbf{r}/kr) \qquad (12.5.27)$$

We are interested ultimately only in the behaviour of $\hat{h}_{\alpha\beta}(k)$ to order k^2. This means that we need retain in (12.5.27) only the terms for $n=0$ and $n=1$. If, in addition, we substitute for $\hat{h}(1,2)$ from (12.4.9), Eqn (12.5.26) becomes

$$\begin{aligned}\hat{h}_{\alpha\beta}(k) = \langle\{\hat{h}_S(k) + \hat{h}_\Delta(k)\mathbf{u}_1\cdot\mathbf{u}_2 + \bar{h}_D(k)[3k^{-2}(\mathbf{k}\cdot\mathbf{u}_1)(\mathbf{k}\cdot\mathbf{u}_2) - \mathbf{u}_1\cdot\mathbf{u}_2]\}\\ \times[j_0(-kl_\alpha) + 3ij_1(-kl_\alpha)\mathbf{k}\cdot\mathbf{u}_1/k]\\ \times[j_0(kl_\beta) + 3ij_1(kl_\beta)\mathbf{k}\cdot\mathbf{u}_2/k]\rangle_{\mathbf{\Omega}_1\mathbf{\Omega}_2}\end{aligned} \tag{12.5.28}$$

where $j_0(x) = x^{-1}\sin x$ and $j_1(x) = (x^{-2}\sin x - x^{-1}\cos x)$.

The only terms in (12.5.28) that survive the integration over angles are those of the type shown in (12.5.20). On multiplying out, integrating with the help of (12.5.20), and collecting terms, we find that

$$\begin{aligned}\hat{h}_{\alpha\beta}(k) = \hat{h}_S(k)j_0(-kl_\alpha)j_0(kl_\beta)\\ -[\hat{h}_\Delta(k) + 2\bar{h}_D(k)]j_1(-kl_\alpha)j_1(kl_\beta)\end{aligned} \tag{12.5.29}$$

The functions $\hat{h}_{\alpha\alpha}(k)$, $\hat{h}_{\beta\alpha}(k)$ and $\hat{h}_{\beta\beta}(k)$ may be expressed in a similar way. If we now expand the Bessel functions to order k^2, the result obtained for the Fourier transform of the function $\Delta h(r)$ of Eqn (12.3.8) is

$$\Delta\hat{h}(k) = \frac{k^2l^2}{9}[\hat{h}_\Delta(0) + 2\bar{h}_D(0)] + \mathcal{O}(k^4) \tag{12.5.30}$$

or, from Eqns (12.5.9b) and (12.5.21):

$$\Delta h^{(2)} = \frac{l^2}{9\rho}\left(\frac{\varepsilon-1}{y\varepsilon} - 3\right) \tag{12.5.31}$$

where $\Delta h^{(2)}$ is the coefficient introduced in Eqn (12.3.10). The result in (12.5.31) was obtained by Høye and Stell (1976) and Chandler (1977). It expresses the dielectric constant as a combination of integrals involving only the site–site distribution functions of the molecule and may be rewritten as

$$\sum_\alpha\sum_\beta q_\alpha q_\beta h_{\alpha\beta}^{(2)} = \frac{\mu^2}{9\rho}\left(\frac{\varepsilon-1}{y\varepsilon} - 3\right) \tag{12.5.32}$$

where q_α is the charge associated with site α. In this form, the result is valid for any interaction-site model.

It is clear from (12.5.18) that $\hat{h}_\Delta(0)$ is related to the angular correlation parameter of Eqn (12.3.10) by $G_1 = \frac{1}{3}\rho\hat{h}_\Delta(0)$. This is true whether the molecule has a dipole moment or not, but the analysis that leads to (12.3.10) is valid only in the non-polar case. The difference between polar and non-polar molecules lies in the long-range function $h_D(R)$. The significance of $h_D(R)$

can be seen in the fact that whereas $\hat{h}_\Delta(0)$ contributes equally to the longitudinal and transverse components of the long-wavelength limit of the susceptibility $\boldsymbol{\chi}^0(\mathbf{k}, 0)$, $\bar{h}_D(0)$ does not. The effects of long-range correlations can therefore be suppressed by setting $\bar{h}_D(0)=0$, and Eqn (12.5.30) then reduces to (12.3.10).

In the next two sections, we shall discuss specific examples of theories of polar fluids, the predictions of which can be tested against the results of computer simulations. If the comparisons are to be useful, however, attention must be paid to the role played by the boundary conditions in the simulations. The importance of boundary conditions can be understood by considering the expression obtained for the dipole-moment fluctuations of a spherical sample of dielectric constant ε embedded in a dielectric continuum ε'. The derivation is a generalization of Kirkwood's classic calculation referred to earlier (Kirkwood, 1939); the result (Titulaer and Deutch, 1974; deLeeuw *et al.*, 1980) is

$$\frac{(\varepsilon-1)(2\varepsilon'+1)}{(\varepsilon+2\varepsilon')}=3yg(\varepsilon') \tag{12.5.33}$$

which reduces to the Kirkwood formula (12.5.15) when $\varepsilon=\varepsilon'$. The g-factor is again defined as $\langle|\mathbf{M}|^2\rangle/N\mu^2$, where $\mathbf{M}$ is the dipole moment of the inner region, but is now a function of ε'. On the other hand, $g(\varepsilon')$ is always related to $h_\Delta(R)$ by Eqn (12.5.18), provided the inner region is sufficiently large to justify replacing the upper limit of the integral by infinity. We are therefore forced to the conclusion that the pair correlation function also depends on ε'. An extreme example of this effect occurs in a simulation when the potential is truncated in the manner described in Section 3.1. Spherical truncation corresponds to setting $\varepsilon'=1$ in (12.5.33), giving

$$\frac{\varepsilon-1}{\varepsilon+2}=yg(1) \tag{12.5.34}$$

Thus $yg(1)\rightarrow 1$ as ε becomes large, in striking contrast to the result for the infinite-system limit ($\varepsilon=\varepsilon'$), in which $yg(\varepsilon)$ increases linearly with ε when ε is large. Equation (12.5.34) is therefore useless as a means of calculating the dielectric constant of all except weakly polar fluids, since the estimate made of ε is very sensitive to errors in $g(1)$.

Simulations of polar fluids are convincing, at least insofar as dielectric properties are concerned, only if account is taken of the long range of the dipole–dipole interaction. The question of how best to do this has been the subject of much controversy. In the reaction-field method of Barker and Watts (1973), interactions between close neighbours are calculated with the spherical cut-off convention, while the region outside the truncation sphere is treated as a dielectric continuum. The total potential energy of the system is

$$V_N = \tfrac{1}{2}\sum_{i \neq j}^{N}\sum v(i, j) - \tfrac{1}{2}\sum_{i=1}^{N} \mathbf{E}_\mathrm{R}(i) \cdot \boldsymbol{\mu}_i \tag{12.5.35}$$

where $\mathbf{E}_\mathrm{R}(i)$ is the reaction field acting on molecule i due to polarization of the continuum by the net dipole moment $\mathbf{M}_i$ of the cut-off sphere centred on i. The reaction field (Frölich, 1958) is given by

$$\mathbf{E}_\mathrm{R}(i) = \frac{4\pi}{3V_\mathrm{c}} \frac{2(\varepsilon' - 1)}{2\varepsilon' + 1} \mathbf{M}_i \tag{12.5.36}$$

where V_c is the volume of the truncation sphere. Patey *et al.* (1982) have used some results of Neumann and Steinhauser (1980) to prove that the dielectric constant is then given by the fluctuation formula (12.5.33), provided $\langle|\mathbf{M}|^2\rangle$ is taken as the mean-square moment of the periodic cube and not, as had previously been assumed, that of the cut-off sphere. The natural choice is to take $\varepsilon' = \varepsilon$, in which case the g-factor should be close to that of an infinite system. In general, of course, ε is not known in advance, but the magnitude of the reaction field is only weakly dependent on the choice of ε' when ε' is large.

A second solution to the problem is to calculate the infinite lattice sum given by

$$V_N = \tfrac{1}{2}\sum_{\mathbf{n}}{}' \sum_{i=1}^{N}\sum_{j=1} v(\mathbf{R}_{ij} + L\mathbf{n}, \boldsymbol{\Omega}_i, \boldsymbol{\Omega}_j) \tag{12.5.37}$$

where L is the length of the simulation cube, $\mathbf{n}$ is a vector with unit components, and the prime denotes that the term $\mathbf{n} = 0$ is omitted when $i = j$; it is, of course, only the dipole–dipole contribution to $v(i, j)$ that is relevant here. Use of (12.5.37) means that account is taken of interactions between all images of every particle. Lattice sums of this type can be evaluated by the Ewald method, as already noted in Section 3.1; the form of the Ewald sum appropriate to point dipoles is given by Adams and McDonald (1976). At any stage in the simulation, the system will have a uniform polarization equal to $\mathbf{M}/L^3$, where $\mathbf{M}$ is the instantaneous value of the total dipole moment of any of the periodic array of cells. The uniform polarization makes a contribution to the total potential energy; this corresponds to the $\mathbf{k} = 0$ (macroscopic) term in the reciprocal-space part of the Ewald sum (see Section 3.1). The value of the $\mathbf{k} = 0$ term is shape dependent and is of no physical interest; in conventional applications of the Ewald method it is ignored. So far as fluctuations within the system are concerned, neglect of the $\mathbf{k} = 0$ term is equivalent to supposing that the infinite, periodic lattice of cells is surrounded by a conducting medium (deLeeuw *et al.*,

1980). The correct fluctuation formula to use for the calculation of the dielectric constant can therefore be deduced by setting $\varepsilon' = \infty$ in (12.5.33) to give

$$\varepsilon - 1 = 3yg(\infty) \tag{12.5.38}$$

However, it is unnecessary to invoke the analogy with conducting boundary conditions. That Eqn (12.5.38) is the correct formula for use with the Ewald sum follows rigorously from neglect of the $\mathbf{k} = 0$ term (Felderhof, 1980).

When correctly implemented, the reaction-field and Ewald methods give results that are consistent with each other and probably as close to the infinite-system limit that it is possible to achieve given the underlying periodicity and the small number of particles in the basic cell. The Ewald method yields a substantially larger g-factor ($g(\infty) \simeq 1.5g(\varepsilon)$ when ε is large), but Monte Carlo calculations for dipolar hard spheres have shown that the difference is attributable to the accumulation of small differences in $h_\Delta(R)$ at large R, where $h_\Delta(R)$ itself is small (de Leeuw *et al.*, 1983). At shorter distances, the two methods give essentially the same pair correlation function.

12.6 THE MSA FOR DIPOLAR HARD SPHERES

The expansions of $h(1,2)$ and $c(1,2)$ in terms of S, $\Delta(1,2)$ and $D(1,2)$ was first exploited by Wertheim (1971) in obtaining the analytic solution to the MSA (mean spherical approximation) for dipolar hard spheres. Although the MSA is not a quantitatively satisfactory theory, Wertheim's methods have had a great influence on nearly all later work on simple polar fluids.

The groundwork for the solution has been laid in Section 12.4. The next stage in the calculation consists in substituting for $\hat{h}(1,2)$ and $\hat{c}(1,2)$ in the Ornstein–Zernike relation (12.4.8), integrating over angles with the help of Wertheim's "multiplication table" (p. 456), and equating coefficients of S, Δ and D_k on the two sides of the equation. The terms in S separate from those in Δ and D_k to give

$$\hat{h}_S(k) = \hat{c}_S(k)\hat{h}_S(k) \tag{12.6.1a}$$

$$\hat{h}_\Delta(k) = \hat{c}_\Delta(k) + \tfrac{1}{3}\rho[\hat{c}_\Delta(k)\hat{h}_\Delta(k) + 2\bar{c}_D(k)\bar{h}_D(k)] \tag{12.6.1b}$$

$$\bar{h}_D(k) = \bar{c}_D(k) + \tfrac{1}{3}\rho[\bar{c}_D(k)\bar{h}_D(k) + \bar{c}_D(k)\hat{h}_\Delta(k) + \hat{c}_\Delta(k)\bar{h}_D(k)] \tag{12.6.1c}$$

The Hankel transforms in (12.6.1) are the Fourier transforms of the short-range functions $h_D^0(R)$ and $c_D^0(R)$; this fact has already been used in deriving Eqn (12.5.25). The inverse Fourier transforms of $\hat{h}_S(k)$, $\hat{h}_\Delta(k)$ and

$\bar{h}_D(k)$ can therefore all be written in terms of spatial convolution integrals (denoted by a star):

$$h_S(R) = c_S(R) + \rho c_S * h_S \tag{12.6.2a}$$

$$h_\Delta(R) = c_\Delta(R) + \tfrac{1}{3}\rho(c_\Delta * h_\Delta + 2c_D * h_D) \tag{12.6.2b}$$

$$h_D(R) = c_D(R) + \tfrac{1}{3}\rho(c_D^0 * h_D^0 + c_D^0 * h_\Delta + c_\Delta * h_D^0) \tag{12.6.2c}$$

These equations are to be solved subject to the MSA closure relation (5.6.2). For dipolar hard spheres, Eqn (5.6.2) becomes

$$\begin{aligned} h(1,2) &= -1, \qquad R < d \\ c(1,2) &= \frac{\beta\mu^2 D(1,2)}{R^3}, \qquad R > d \end{aligned} \tag{12.6.3}$$

or, equivalently:

$$\begin{aligned} h_S(R) &= -1, \qquad R < d \\ h_\Delta(R) &= h_D(R) = 0, \qquad R < d \\ c_D(R) &= \frac{\beta\mu^2}{R^3}, \qquad R > d \\ c_S(R) &= c_\Delta(R) = 0, \qquad R > d \end{aligned} \tag{12.6.4}$$

The closure relation involving the projections $h_D(R)$ and $c_D(R)$ may also be written as

$$\begin{aligned} h_D^0(R) &= -3K, \qquad R < d \\ c_D^0(R) &= 0, \qquad R > d \end{aligned} \tag{12.6.5}$$

where K is a dimensionless parameter defined as

$$K = \int_d^\infty \frac{h_D(R)}{R}\,\mathrm{d}R \tag{12.6.6}$$

We now look for a linear combination of functions that will cause Eqns (12.6.2b) and (12.6.2c) to become decoupled. Direct substitution shows that this is achieved by taking

$$\begin{aligned} h_+(R) &= \frac{1}{3K}[h_D^0(R) + \tfrac{1}{2}h_\Delta(R)] \\ h_-(R) &= \frac{1}{3K}[h_D^0(R) - h_\Delta(R)] \end{aligned} \tag{12.6.7}$$

with analogous expressions for $c_+(R)$ and $c_-(R)$. The new variables satisfy the equations

$$
\begin{aligned}
h_+(R) &= c_+(R) + 2K\rho c_+ * h_+ \\
h_-(R) &= c_-(R) - K\rho c_- * h_-
\end{aligned}
\tag{12.6.8}
$$

Equations (12.6.8) are to be solved subject to the closure relations $h_+(R) = h_-(R) = -1$, $R < d$ (this is the reason for the choice of the factor $1/3K$ in (12.6.7)) and $c_+(R) = c_-(R) = 0$, $R > d$.

The original problem has now been greatly simplified. The effect of the decoupling, first in (12.6.2) and then in (12.6.8), is that the molecular Ornstein–Zernike relation has been reduced to three independent equations, namely (12.6.2a) and (12.6.8). These equations, with their corresponding closure relations, are just the PY approximation for hard spheres at densities equal, respectively, to ρ, $2K\rho$ and $-K\rho$. The structure of the hard-sphere fluid is therefore describable in fully analytic form in terms of PY hard-sphere radial distribution functions. The fact that one solution is required at negative density poses no special difficulty.

To complete the analytic solution, it is necessary to relate the quantity K to hard-sphere properties. Given the analogue of Eqn (12.4.16) for $c_D(R)$, the closure relation (12.6.5) requires that

$$
c_D(R) = -\frac{3}{R^3}\int_0^d c_D^0(R')R'^2\,\mathrm{d}R', \qquad R > d \tag{12.6.9}
$$

Because $c_D^0(R)$ is zero for $R > d$, comparison of Eqns (12.6.4) and (12.6.9) shows that

$$
\begin{aligned}
\beta\mu^2 &= -3\int_0^d c_D^0(R)R^2\,\mathrm{d}R \\
&= -\frac{3}{4\pi}\,\bar{c}_D(0)
\end{aligned}
\tag{12.6.10}
$$

The function $c_D^0(R)$ may be expressed as

$$
\begin{aligned}
c_D^0(R) &= K[c_+(R) + c_-(R)] \\
&= K[2c_{\mathrm{PY}}(R; 2K\rho) + c_{\mathrm{PY}}(R; -K\rho)]
\end{aligned}
\tag{12.6.11}
$$

where $c_{\mathrm{PY}}(R;\rho)$ is the PY hard-sphere direct correlation function at a density ρ. Let $Q(\eta) = \beta(\partial P/\partial\rho)_T$ be the PY approximation to the inverse compressibility of the hard-sphere fluid at a packing fraction η. Integrals over $c_{\mathrm{PY}}(R;\rho)$ can be related to $Q(\eta)$ via the general expression (5.2.17) and the approximate result (5.5.8). A short calculation shows that

$$Q(\eta) = 1 - 4\pi\rho \int_0^d c_{\mathrm{PY}}(R;\rho) R^2 \,\mathrm{d}R$$

$$= \frac{(1+2\eta)^2}{(1-\eta)^4} \tag{12.6.12}$$

On combining Eqns (12.6.10) to (12.6.12), we find that

$$\beta\mu^2 = -3K \int_0^d [2c_{\mathrm{PY}}(R;2K\rho) + c_{\mathrm{PY}}(R;-K\rho)] R^2 \,\mathrm{d}R$$

$$= \frac{3}{4\pi\rho}[Q(2K\eta) - Q(-K\eta)] \tag{12.6.13}$$

or

$$3y = Q(2K\eta) - Q(-K\eta) \tag{12.6.14}$$

where y is the parameter defined by (12.5.16). Equations (12.6.12) and (12.6.14) determine K implicitly for given choices of y and η; as y varies from 0 to ∞, $K\eta$ varies from 0 to $\frac{1}{2}$.

As an alternative to (12.6.12), we can write

$$\frac{1}{Q(\eta)} = 1 + 4\pi\rho \int_0^\infty h_{\mathrm{PY}}(R;\rho) R^2 \,\mathrm{d}R \tag{12.6.15}$$

whence, from (12.6.7):

$$\rho\hat{h}_\Delta(0) = 8\pi\rho K \int_0^\infty [h_+(R) - h_-(R)] R^2 \,\mathrm{d}R$$

$$= 8\pi\rho K \int_0^\infty [h_{\mathrm{PY}}(R;2K\rho) - h_{\mathrm{PY}}(R;-K\rho)] R^2 \,\mathrm{d}R$$

$$= \frac{1}{Q(2K\eta)} + \frac{2}{Q(-K\eta)} - 3 \tag{12.6.16}$$

Taken together, Eqns (12.5.15), (12.6.14) and (12.6.16) lead to a remarkably simple expression for the dielectric constant:

$$\varepsilon = \frac{Q(2K\eta)}{Q(-K\eta)} \tag{12.6.17}$$

The same relation is obtained if Eqn (12.5.24) is used instead of (12.5.15).

Although the method of solution is very elegant, comparison with the results of Monte Carlo calculations shows that the MSA does not provide a quantitatively acceptable description of the properties of the dipolar hard-sphere fluid (Stell *et al.*, 1981). As is evident from comparison of Eqns (12.6.14) and (12.6.17), the dielectric constant in the MSA is dependent only on the parameter y and not separately on the two independent parameters $\rho^* = \rho d^3$ and $\mu^{*2} = \beta\mu^2/d^3$ that are required to specify the thermodynamic state of the system. When both these variables are large (for H_2O, $\mu^* \simeq 1.6$), use of the MSA gives values of ε that are much too small, as shown by the results in Figure 12.5.

The MSA, with the necessary modifications, has also been applied to quadrupolar hard spheres, again with results that are quantitatively poor (Patey, 1978). A treatment valid for arbitrary electrostatic multipoles has been worked out by Blum (1972).

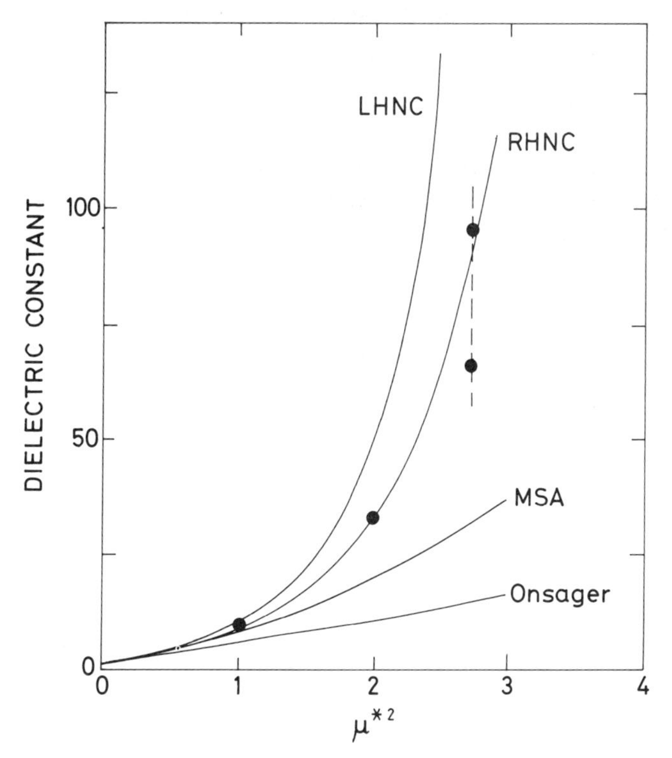

FIG. 12.5. Dielectric constant of the dipolar hard-sphere fluid at $\rho d^3 = 0.8$ as a function of $\mu^{*2} = \mu^2/k_B T d^3$, showing a comparison between Monte Carlo results (points) and the predictions of various theories discussed in the text (curves). The Monte Carlo value at $\mu^{*2} = 2.75$ is subject to large uncertainty. After Stell *et al.* (1981).

12.7 OTHER APPROXIMATIONS FOR POLAR FLUIDS

Of the developments inspired by Wertheim's work on the MSA, the one that has been applied to the widest range of problems is the "linearized HNC" or LHNC approximation of Patey (1977). The LHNC approximation is equivalent to one proposed earlier by Wertheim himself and called by him the "single-superchain" approximation (Wertheim, 1973), but most of the extensions and numerical implementation of the method have been carried out by Patey and his collaborators; much of their work has been reviewed by Stell *et al.* (1981).

In the simplest case, represented by the potential (1.2.4), the LHNC approximation resembles the MSA in basing itself on expansions of $h(1,2)$ and $c(1,2)$ limited to the terms in S, $\Delta(1,2)$ and $D(1,2)$. It improves on the MSA by employing a closure relation that is more realistic and can be applied just as easily, say, to the Stockmayer potential as to dipolar hard spheres. As the name suggests, the LHNC closure corresponds to a linearization of the HNC approximation, which in its general form is

$$c(1,2) = h(1,2) - \log g(1,2) - \beta v(1,2) \tag{12.7.1}$$

The LHNC closure is obtained by substituting for $h(1,2)$ and $c(1,2)$ from Eqns (12.4.3) and (12.4.7) and linearizing with respect to the functions Δ and D. The result is

$$\begin{aligned} c(1,2) = {} & h_S(R) - \log g_S(R) - \beta v_0(R) \\ & + h_\Delta(R)[1 - 1/g_S(R)]\Delta(1,2) \\ & + \{h_D(R)[1 - 1/g_S(R)] + \beta\mu^2/R^3\}D(1,2) \end{aligned} \tag{12.7.2}$$

where $g_S(R) = h_S(R) + 1$. When $v_0(R)$ is the hard-sphere potential, Eqn (12.7.2) becomes

$$h_S(R) = -1, \qquad h_\Delta(R) = h_D(R) = 0; \qquad R < d \tag{12.7.3a}$$

$$\left.\begin{aligned} c_S(R) &= h_S(R) - \log g_S(R) \\ c_\Delta(R) &= h_\Delta(R)[1 - 1/g_S(R)] \\ c_D(R) &= h_D(R)[1 - 1/g_S(R)] + \beta\mu^2/R^3 \end{aligned}\right\} \quad R > d \tag{12.7.3b}$$

Equations (12.7.3) reduce to the MSA closure (12.6.4) if $g_S(R) = 1$ for $R > d$; the MSA may therefore be regarded as the low-density limit of the LHNC approximation.

The linearization involved in (12.7.2) means that the closure relation involves only the harmonics S, Δ and D. This is consistent with the assumed form of $h(1,2)$ and $c(1,2)$ and Eqns (12.6.2) remain valid. In other words, the relation between $h_S(R)$ and $c_S(R)$ remains independent of the other functions, and the results for $h_S(R)$ and $c_S(R)$ correspond to the HNC solution for the potential $v_0(R)$. In contrast to the MSA, however, the projections on Δ and D are influenced by the projections on S through the appearance of $g_S(R)$ in the closure relations for $c_\Delta(R)$ and $c_D(R)$. The method of solution adopted by Patey (1977) for the problem of dipolar hard spheres parallels that used by Wertheim (1971) for the MSA up to the point at which the linear combinations (12.6.7) are introduced. In the LHNC approximation, the functions $h_+(R)$, $c_+(R)$ remain coupled to $h_-(R)$, $c_-(R)$ through the closure relations. The solution must therefore be completed numerically, but the formulation of the problem in terms of the short-range functions $h_D^0(R)$ and $c_D^0(R)$ avoids the difficulties associated with the long range of the dipolar interaction. Indeed, the dipole moment does not appear in the system of equations that has to be solved. What is done instead is to solve for $h_\Delta(R)$, $c_\Delta(R)$, $h_D(R)$ and $c_D(R)$ for a given value of the constant K defined by (12.6.6) and then to determine μ from the requirement that $c_D(R)$ must satisfy Eqn (12.6.10). Strict application of the closure relations requires the use of the HNC results for the hard-sphere radial distribution function. Because the HNC approximation gives poor results for hard spheres, the corresponding distribution function is in practice replaced by the Verlet–Weis parameterization described in Appendix C (Verlet and Weis, 1972b).

The LHNC equations have been solved both for dipolar hard spheres and for the Stockmayer potential over wide ranges of state conditions. Some results for $h_\Delta(R)$ and $h_D(R)$ for the dipolar hard-sphere fluid are compared with those obtained by the Monte Carlo method in Figure 12.4. The Monte Carlo calculations were based on the reaction-field method described in the preceding section and the results should therefore be close to the infinite-system limit. The general agreement between theory and simulation is good and improves significantly as the parameter μ^* is decreased. In Figure 12.5, the dielectric constant of the dipolar hard-sphere fluid is plotted as a function of μ^{*2} for a reduced density $\rho^* = 0.8$. At this high density, the LHNC results lie well above the predictions of both the MSA and the Onsager approximation ($g_K = 1$ in Eqn (12.5.15)). As μ^* is lowered, the LHNC results approach those of the MSA; this is understandable in the light of our earlier identification of the MSA as the low-density limit of the LHNC approximation (Rushbrooke, 1979). For $\mu^{*2} = 2.0$, the LHNC value is also significantly greater than that obtained by simulation, while the MSA result is smaller. At larger values of μ^*, the available Monte Carlo data show a high degree of scatter, but it is clear that the dielectric constant rises rapidly with increasing μ^*, and the inadequacy of the MSA becomes progressively more apparent.

A common feature of the LHNC approximation and the MSA is the fact that $g_S(R)$ is the radial distribution function corresponding to the spherically symmetric potential $v_0(R)$. For the reasons discussed at length in Chapter 6, this is a satisfactory approximation at high densities and moderate values of the dipole moment, but is less successful when the dipole moment is very large or the density is low. Patey (1978) has therefore proposed an extension of the LHNC theory which has the effect of making $h_S(R)$ and $c_S(R)$ dependent on the strength of the dipolar interaction. The same approximate forms of $h(1,2)$ and $c(1,2)$ are retained, but the expansion of the HNC closure relation is now taken to second order in Δ and D. In this "quadratic" or QHNC approximation, the closure relations for $c_\Delta(R)$ and $c_D(R)$ are unaltered, but that for $c_S(R)$ (for the case of dipolar hard spheres) becomes

$$c_S(R) = h_S(R) - \log g_S(R) - B(r)/[g_S(R)]^2 \tag{12.7.4}$$

where

$$B(r) = \tfrac{1}{6}[h_\Delta(R)]^2 + \tfrac{1}{3}[h_D(R)]^2 \tag{12.7.5}$$

The presence of the term in $B(R)$ in (12.7.4) means that $h_S(R)$ and $c_S(R)$ cannot be decoupled from the other projections and therefore cannot be identified with the correlation functions of the pure hard-sphere system. The QHNC equations have been solved numerically (Patey *et al.*, 1979a) and at low densities are a considerable improvement on the LHNC approximation. The two theories give almost identical results for the dielectric constant for $\varepsilon \lesssim 50$, but beyond this their predictions rapidly diverge.

The LHNC and QHNC approximations have also been applied to systems of quadrupolar hard spheres (Patey, 1978) and to fluids of hard spheres carrying both dipoles and quadrupoles (Patey *et al.*, 1979b). In the purely quadrupolar case, the pair potential is

$$v(1,2) = v_0(R) + \frac{3Q^2}{4R^5}\Phi^{224}(1,2) \tag{12.7.6}$$

where $v_0(R)$ is the hard sphere potential, Q is the quadrupole moment, and the rotational invariant $\Phi^{224}(1,2)$ describes the angle dependence of the interaction between two axially-symmetric quadrupoles; explicit forms for Φ^{224} and the other rotational invariants used in this section are given in Appendix B of the review article by Stell *et al.* (1981). The potential (12.7.6) therefore contains the harmonics (rotational invariants) corresponding to $(l_1, l_2, l) = (0, 0, 0)$ ($\equiv S$) and $(2, 2, 4)$; convolution in the Ornstein–Zernike relation yields additionally the harmonics $(2, 2, 0)$ and $(2, 2, 2)$. The assumed solution is therefore of the form

$$h(1,2) = h_S(R) + h^{220}(R)\Phi^{220}(1,2) + h^{222}(R)\Phi^{222}(1,2) + h^{224}(R)\Phi^{224}(1,2) \tag{12.7.7}$$

together with the corresponding expression for $c(1,2)$. The Ornstein–Zernike relation is then solved for a closure relation appropriate to either the LHNC or QHNC versions of the theory. As in the dipolar case, the LHNC results for $h_S(R)$ and $c_S(R)$ depend only on $v_0(R)$. This a a poor approximation even at modest values of the parameter $Q^{*2} = Q^2/k_B T d^5$, and the advantages of the QHNC closure are more apparent than in the purely dipolar problem (Patey, 1978). Good general agreement is found between QHNC and Monte Carlo results for all values of Q^* up to $Q^{*2} = 5/3$ (the largest value studied); note that for N_2, $Q^* \simeq 0.5$, and for CO_2, $Q^* \simeq 0.9$ (Gray and Gubbins, 1984).

In the mixed, dipolar–quadrupolar system, the pair potential contains harmonics corresponding to $(l_1, l_2, l) = (0,0,0), (1,1,2)$ $(\equiv D)$, $(1,2,3)$ and $(2,2,4)$. The extra harmonics generated by the angular convolutions are now $(1,1,0)$ $(\equiv \Delta)$, $(1,2,1)$, $(2,2,0)$ and $(2,2,2)$. Both $h(1,2)$ and $c(1,2)$ are therefore written as the sum of eight terms, and the resulting coupled equations are solved in similar fashion to before. Agreement with the results of computer simulations is less good than for either of the pure systems, The calculations are nonetheless of interest for the light they throw on the way in which the quadrupolar interaction modifies the dipolar correlations in the fluid. The effect on the projection $h_\Delta(R)$ is particularly striking. In the purely dipolar case, at large values of ρ^* and μ^*, $h_\Delta(R)$ is positive nearly everywhere and significantly different from zero out to values of R corresponding to ten or more molecular diameters. Since ε is determined by the integral of $R^2 h_\Delta(R)$ over all R (see Eqn (12.5.18)), these features combine to give very large values for the dielectric constant. The addition of even a small quadrupole moment leads to a marked falling off in both the magnitude and range of $h_\Delta(R)$. The main consequence of the changes in $h_\Delta(R)$ is that ε is a rapidly decreasing function of Q^*, particularly when μ^* is large. Experimentally it is known that the g-factors of many strongly polar liquids are close to unity (water is a notable exception). Any comparison with results for simple models is complicated by the fact that real molecules are polarizable. Nonetheless, the results of Figure 12.5, particularly the very large discrepancies between the Monte Carlo calculations and the predictions of the Onsager approximation ($g_K = 1$), clearly suggest that the dipolar hard-sphere model gives dielectric constants that are unrealistically large; the same comment applies to the Stockmayer fluid (Pollock and Alder, 1980). The role played by quadrupolar forces provides a possible explanation of these observations. It should not be forgotten, however, that real molecules are not spherical, and short-range dipolar correlations will in general also be disrupted by steric effects.

The LHNC/QHNC approach involves two different expansions that are only indirectly linked to each other. In the first stage, the correlation functions $h(1,2)$ and $c(1,2)$ are expanded in a basis set of rotational invariants. For the present, at least, this step seems unavoidable if the Ornstein–Zernike relation is to be reduced to tractable form. Secondly, the HNC closure is further approximated by expanding in powers of the appropriate angular functions and retaining only a small number of terms. The passage from the linear (LHNC) to quadratic (QHNC) forms of the theory is a natural progression, but it does have certain drawbacks. In particular, the self-consistency of the theory is lost. For example, the non-linear nature of the closure relation (12.7.4) means that the S, Δ and D projections of $h(1,2)$ are no longer wholly determined by the corresponding projections of $c(1,2)$. The additional harmonics that are generated by the Ornstein–Zernike relation are, in practice, simply ignored. The theory also becomes computationally more awkward to implement, and the results are not always an improvement on those of the simpler version.

The summary just given suggests that in situations where the LHNC approximation holds it would be better to avoid the second expansion altogether and return to the full HNC closure (12.7.1). The difficulty that arises here is that no analytic expansion in rotational invariants can be made of the term $\log g(1,2)$, but the problem can be bypassed in a surprisingly simple way, as Fries and Patey (1985) have shown. If we differentiate Eqn (12.7.1) with respect to R $(\equiv|\mathbf{R}_{12}|)$, keeping the angular variables constant, we find that

$$\frac{\partial c(1,2)}{\partial R}=\frac{\partial h(1,2)}{\partial R}-\frac{1}{g(1,2)}\frac{\partial g(1,2)}{\partial R}-\beta\frac{\partial v(1,2)}{\partial R}$$

$$=-h(1,2)\frac{\partial x(1,2)}{\partial R}-\beta\frac{\partial v(1,2)}{\partial R} \qquad (12.7.8)$$

where

$$x(1,2)=-\log g(1,2)$$

$$=c(1,2)-h(1,2)+\beta v(1,2) \qquad (12.7.9)$$

Equation (12.7.8) can then be integrated between R and $R=\infty$ (where all the relevant pair functions vanish) to give

$$c(1,2)=-\beta v(1,2)+\int_R^\infty h(1,2)\frac{\partial x(1,2)}{\partial R'}\,\mathrm{d}R' \qquad (12.7.10)$$

This result is equivalent to the original HNC closure relation, but all functions that appear can now be expanded in the required form.

The "reference" HNC or RHNC approximation of Section 5.8 may be treated in a similar fashion. If we put $h = h_0 + \Delta h$, $c = c_0 + \Delta c$, $v = v_0 + \Delta v$ and $d = d_0 + \Delta d \simeq d_0$, Eqn (5.3.14) can be written as

$$\Delta c(1,2) = \Delta h(1,2) - \log [g_0(R) + \Delta h(1,2)] + \log g_0(R) - \beta \Delta v(1,2) \tag{12.7.11}$$

The subscript 0 denotes a property of the reference system, for which the pair potential is the spherically symmetric part of $v(1,2)$. Equation (12.7.11) is exact apart from the characteristic RHNC approximation whereby the true bridge function $d(1,2)$ is replaced by the bridge function $d_0(1,2)$ of the reference system. Differentiation with respect to R followed by integration along the same lines as before now gives

$$\Delta c(1,2) = -\beta \Delta v(1,2) + \int_R^\infty \Delta h(1,2) \frac{\partial \Delta x(1,2)}{\partial R'} \mathrm{d}R' + \int_R^\infty h_0(R) \frac{\partial \Delta x(1,2)}{\partial R'} \mathrm{d}R' - \int_R^\infty \Delta h(1,2) \frac{\partial \log g_0(R')}{\partial R'} \mathrm{d}R' \tag{12.7.12}$$

where $\Delta x(1,2) = x(1,2) - x_0(1,2)$.

Fries and Patey (1985) have solved the RHNC equations for the dipolar hard-sphere fluid over a range of values of density and dipole moment. The closure relation (12.7.10) couples together all harmonic components of $h(1,2)$ and $c(1,2)$. The results obtained are therefore dependent on the basis-set size, i.e. on the number of rotational invariants retained when expanding the pair functions. Fortunately, in practice, this is not a serious problem. Even under the most extreme conditions studied ($\rho^* = 0.8$, $\mu^{*2} = 2.75$), Fries and Patey (1985) found that essentially complete convergence was obtained for the projections $h_S(R)$, $h_\Delta(R)$ and $h_D(R)$ with a basis set consisting of only 16 functions. Results for $h_\Delta(R)$ and $h_D(R)$ are shown in Figure 12.4. While the agreement with Monte Carlo calculations is very good, the most interesting feature of the theoretical results is the fact that they lie consistently below those of the LHNC approximation. The dielectric constant is therefore much reduced with respect to the LHNC value, and now agrees closely with the Monte Carlo work, as shown in Figure 12.5. Excellent results are also obtained for the dielectric constant of the Stockmayer fluid, as illustrated in Figure 12.6 (Lee *et al.*, 1985); the comparison with simulations is more convincing than that for dipolar hard spheres because there are more "experimental" data of high quality available (Pollock and Alder, 1980). Clearly the work of Fries and Patey (1985)

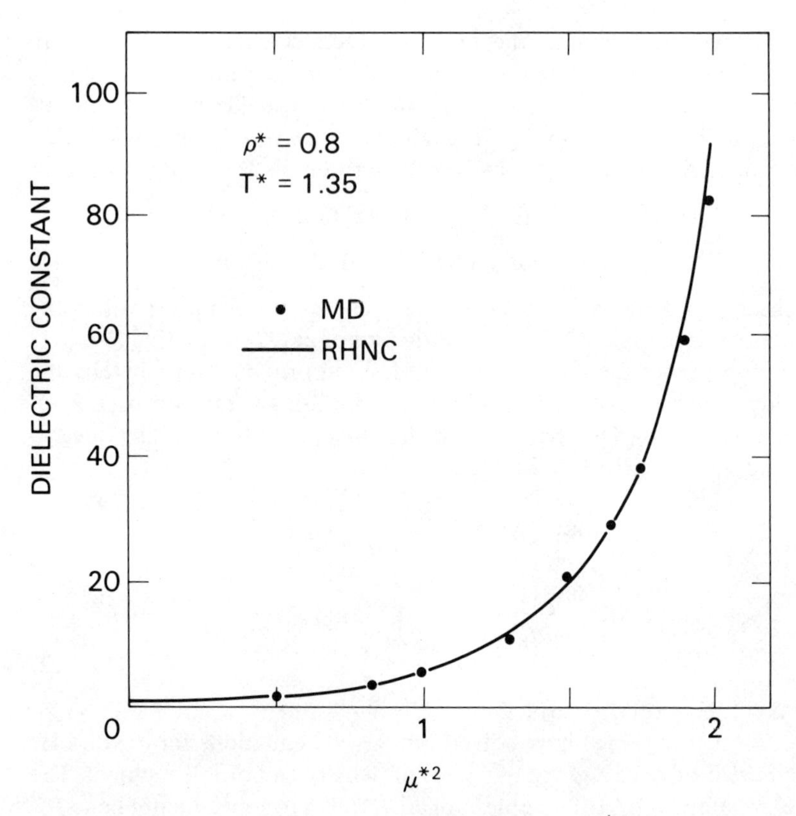

FIG. 12.6. Dielectric constant of the Stockmayer fluid as a function of $\mu^{*2} = \mu^2/\varepsilon\sigma^3$. The points are molecular dynamics results of Pollock and Alder (1980) and the curve is calculated from the RHNC approximation (Lee *et al.*, 1985).

represents a major step towards the construction of a fully quantitative theory of simple polar fluids.

12.8 INTERACTION-SITE DIAGRAMS

The diagrammatic expansions of $h(1,2)$, $c(1,2)$ and $y(1,2)$ given in Chapter 5 are also applicable to molecular fluids if certain minor changes in interpretation are made. First, the circles in a "molecular" diagram are associated with both the translational and orientational coordinates of a molecule and black circles involve integrations over all coordinates. Secondly, black circles carry a weight factor equal to Ω^{-1}, where Ω is defined by Eqn (12.1.3). As an illustration of these rules, the value of one of the diagrams of order ρ in the ρ-circle, f-bond expansion of $h(1,2)$ (see Eqn (5.3.1)) is

$$\underset{1\quad 2}{\wedge} = (\rho/\Omega)\iint f(\mathbf{R}_{13}, \mathbf{\Omega}_1, \mathbf{\Omega}_3)f(\mathbf{R}_{23}, \mathbf{\Omega}_2, \mathbf{\Omega}_3)\,\mathrm{d}\mathbf{R}_3\,\mathrm{d}\mathbf{\Omega}_3 \tag{12.8.1}$$

However, the integral in (12.8.1) is much more complicated to evaluate than in the equivalent atomic problem, and the similarity in formal structure is to some extent deceptive.

The diagrammatic expansion of $h(1,2)$ is not immediately useful in cases where the focus of interest is the set of site–site correlation functions $h_{\alpha\beta}(r)$ rather than $h(1,2)$ itself. Ladanyi and Chandler (1975) have shown how the diagrammatic approach can be adapted to the needs of such a situation, and this section is devoted to a brief review of their main results. We give only a simplified treatment, restricting the detailed discussion to the case of diatomic (or two-site) molecules. The generalization to larger numbers of interaction sites is straightforward, but requires a more complex notation.

The first step is to rewrite the molecular Mayer function $f(1,2)$ as a product of interaction-site Mayer functions $f_{\alpha\beta}(r)$:

$$\begin{aligned} f(1,2) &= \exp[-\beta v(1,2)] - 1 \\ &= \exp\left[-\beta \sum_\alpha \sum_\beta v_{\alpha\beta}(|\mathbf{r}_{1\alpha} - \mathbf{r}_{2\beta}|)\right] - 1 \\ &= -1 + \prod_{\alpha,\beta} [f_{\alpha\beta}(|\mathbf{r}_{1\alpha} - \mathbf{r}_{2\beta}|) + 1] \end{aligned} \tag{12.8.2}$$

The subscripts α, β run over all interaction sites in the molecule; if there are only two sites per molecule, the right-hand side of (12.8.2) consists of 15 separate terms. Equation (12.8.2) can be used to rewrite the integrals occurring in the density expansion of $h(1,2)$. As the simplest possible example, consider the low-density limit of $h(1,2)$, namely

$$\lim_{\rho\to 0} h(1,2) = f(1,2) = \underset{1\qquad 2}{\circ\!\overset{f}{\text{——}}\!\circ} \tag{12.8.3}$$

The corresponding approximation to, say, $h_{\alpha\alpha}(\mathbf{r}, \mathbf{r}')$ is

$$\lim_{\rho\to 0} h_{\alpha\alpha}(\mathbf{r}, \mathbf{r}') = \iint \mathrm{d}1\,\mathrm{d}2\,\delta(\mathbf{r}_{1\alpha} - \mathbf{r})\delta(\mathbf{r}_{2\beta} - \mathbf{r}')f(1,2) \tag{12.8.4}$$

When $f(1,2)$ is replaced by (12.8.2), Eqn (12.8.4) becomes

$$\begin{aligned} \lim_{\rho\to 0} h_{\alpha\alpha}(\mathbf{r}, \mathbf{r}') = f_{\alpha\alpha}(|\mathbf{r} - \mathbf{r}'|) + [1 + f_{\alpha\alpha}(|\mathbf{r} - \mathbf{r}'|)] \\ \times \iint [f_{\alpha\beta}(|\mathbf{r}_{1\alpha} - \mathbf{r}_{2\beta}|) + \text{six other terms}] \\ \times \delta(\mathbf{r}_{1\alpha} - \mathbf{r})\delta(\mathbf{r}_{2\alpha} - \mathbf{r}')\,\mathrm{d}1\,\mathrm{d}2 \end{aligned} \tag{12.8.5}$$

The integrals appearing on the right-hand side of (12.8.5) may be re-expressed in terms of an intramolecular site–site distribution function $s_{\alpha\beta}(\mathbf{x}-\mathbf{y})$, defined as

$$\begin{aligned} s_{\alpha\beta}(\mathbf{x}-\mathbf{y}) &= (1-\delta_{\alpha\beta}) \int \mathrm{d}1\, \delta(\mathbf{R}_1+\mathbf{u}_1 l_\alpha-\mathbf{x})\delta(\mathbf{R}_1-\mathbf{u}_1 l_\beta-\mathbf{y}) \\ &= (1-\delta_{\alpha\beta})\langle \delta(\mathbf{x}-\mathbf{y}-\mathbf{u}_1 l)\rangle_{\mathbf{\Omega}_1} \\ &= \frac{(1-\delta_{\alpha\beta})}{4\pi l^2}\delta(|\mathbf{x}-\mathbf{y}|-l) \end{aligned} \tag{12.8.6}$$

where l_α, l_β and $\mathbf{u}_1$ have the same meaning as in (12.5.26) and $l=l_\alpha+l_\beta$ is the bondlength. The function $s_{\alpha\beta}(\mathbf{r})$ represents the probability of finding site β of a molecule at a position $\mathbf{r}$ given that site α of the same molecule is at the origin. The definition (12.8.6) satisfies the obvious conditions that the interpretation as an intramolecular distribution function requires, i.e. $s_{\alpha\beta}(\mathbf{r})=s_{\beta\alpha}(\mathbf{r})$, $s_{\alpha\alpha}(\mathbf{r})=0$ and

$$\int s_{\alpha\beta}(\mathbf{r})\,\mathrm{d}\mathbf{r}=1, \qquad \alpha\neq\beta \tag{12.8.7}$$

The integral shown explicitly in (12.8.5) can now be transformed as follows:

$$\begin{aligned} &\iint \mathrm{d}1\,\mathrm{d}2\, \delta(\mathbf{r}_{1\alpha}-\mathbf{r})\delta(\mathbf{r}_{2\alpha}-\mathbf{r}')f_{\alpha\beta}(|\mathbf{r}_{1\alpha}-\mathbf{r}_{2\beta}|) \\ &\quad = \int \mathrm{d}\mathbf{x} \int \mathrm{d}1\,\mathrm{d}2\, \delta(\mathbf{r}_{1\alpha}-\mathbf{r})\delta(\mathbf{r}_{2\alpha}-\mathbf{r}')f_{\alpha\beta}(\mathbf{r}_{1\alpha}-\mathbf{r}_{2\beta})\delta(\mathbf{r}_{2\beta}-\mathbf{x}) \\ &\quad = \int \mathrm{d}1\, \delta(\mathbf{r}_{1\alpha}-\mathbf{r}) \int \mathrm{d}\mathbf{x}\, f_{\alpha\beta}(\mathbf{r}-\mathbf{x}) \int \mathrm{d}2\, \delta(\mathbf{r}_{2\alpha}-\mathbf{r}')\delta(\mathbf{r}_{2\beta}-\mathbf{x}) \\ &\quad = \int \mathrm{d}\mathbf{x}\, f_{\alpha\beta}(|\mathbf{r}-\mathbf{x}|)s_{\alpha\beta}(\mathbf{x}-\mathbf{r}') \end{aligned} \tag{12.8.8}$$

All the integrals in (12.8.5) may be treated in the same manner that leads to (12.8.8), and each can then be represented by an *interaction-site diagram.* The circles (white or black) of an interaction-site diagram are associated with the coordinates of interaction sites and the bonds, in the present case, represent functions drawn from the 2×2 matrices $\boldsymbol{f}\equiv[f_{\alpha\beta}]$ and $\boldsymbol{s}\equiv[s_{\alpha\beta}]$. The symmetry number and value of an interaction-site diagram are defined as in the atomic case (see Section 4.3), except that black circles imply a summation over all sites in the molecule in addition to integration over site coordinates. If we denote an f-bond by a solid line and an s-bond by a broken line, Eqn (12.8.5) may be expressed diagrammatically as

$$\lim_{\rho\to 0} h_{\alpha\alpha}(\mathbf{r},\mathbf{r}') = [\text{diagram}] + [\text{diagram}] + [\text{diagram}]$$

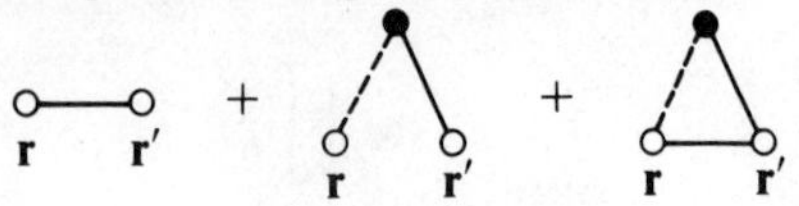

+twelve other diagrams (12.8.9)

The diagrams pictured in (12.8.9) all have a symmetry number of one and all the circles are 1-circles. They are also all of zeroth order in density, since they arise from the molecular diagram (12.8.3); the latter is called the "molecular origin" of the diagrams in (12.8.9). The order in density of any interaction-site diagram in the expansion of a site–site pair correlation function is equal to the number of black circles in its molecular origin, which in turn is equal to the number of black circles in the interaction-site diagram minus the number of *s*-bonds.

The procedure outlined above can be applied to all the integrals appearing in the density expansion of $h(1,2)$. This yields an expansion of any of the functions $h_{\alpha\beta}(r)$ in terms of interaction-site diagrams. As the example (12.8.9) illustrates, each molecular diagram is replaced by a large number of interaction-site diagrams, but the interaction-site diagrams are mathematically simpler objects because all reference to orientational coordinates has disappeared. (Note that the black circles no longer carry the weight factor Ω^{-1} associated with the black circles of a molecular diagram.)

The topology of interaction-site diagrams is restricted in certain ways. What Ladanyi and Chandler (1975) call *allowed* interaction-site diagrams must satisfy the following rules. Diagrams must be simple and connected; white circles must not be connected by an *s*-bond (because different white circles always refer to different molecules); all black circles must be intersected by at least one *f*-bond (otherwise they contribute nothing to the intermolecular correlations); no circle may be intersected by more than one *s*-bond (for reasons to be explained below); and diagrams must be free of articulation circles and articulation *s*-bonds, i.e. *s*-bonds whose removal causes the diagram to separate into two or more components of which at least one contains no white circles. The last restriction is imposed because any such diagram would have as its molecular origin a diagram that contained one or more articulation circles; as we showed in Chapter 5, the expansions of the pair functions of interest here consist entirely of irreducible diagrams.

Given the conditions listed above, the site–site pair correlation functions of a diatomic molecule may be characterized as follows:

$h_{\alpha\beta}(\mathbf{r}_{1\alpha}, \mathbf{r}_{2\beta}) =$ {all allowed interaction-site diagrams consisting of two white 1-circles labelled 1α and 2β, black 1-circles, *f*-bonds and *s*-bonds; each diagram is to be multiplied by ρ^n, where n is the number of black circles minus the number of *s*-bonds} (12.8.10)

The generalization of this result to molecules that contain more than two interaction sites requires the introduction of three-body and higher-order intramolecular distribution functions. It remains true, however, that no circle may be intersected by more than one $\mathfrak{s}$-bond or, indeed, more than one intramolecular bond of any order. Consider the diagram shown in (a) below. For a diatomic molecule, any such diagram is physically meaningless,

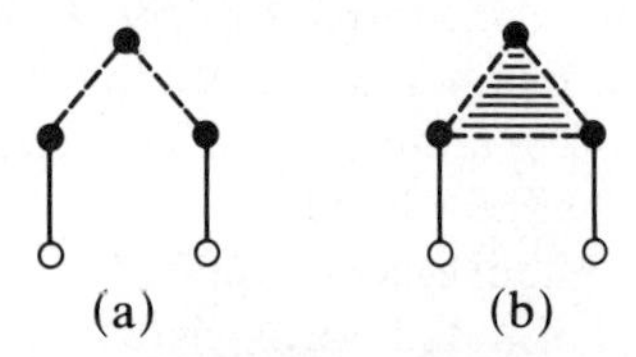

because one site is bonded to two others. But it is also not an allowed diagram even for a triatomic (or larger) molecule, because in that case the three black circles would be linked, as in (b), by a single bond (or "face") representing a three-body intramolecular distribution function.

The interaction-site diagrammatic formalism is not limited to use with rigid molecules. It provides, for example, the basis for a statistical-mechanical theory of conformational equilibrium in liquids (Chandler and Pratt, 1976; Pratt and Chandler, 1977). In such cases, however, a different definition of the intramolecular distribution functions is required.

12.9 INTERACTION-SITE MODELS: THE RISM EQUATIONS

We saw in Section 12.3 that the static structure factors measured in neutron and X-ray scattering experiments on molecular liquids are weighted sums of atomic pair distribution functions. In this section, we describe an approximate integral-equation theory that has been widely used in the interpretation of diffraction experiments and, more generally, in the calculation of site–site distribution functions for interaction-site potential models; this is the "reference interaction-site model" or RISM approximation of Chandler and Andersen (1972). The theory has been used with particular success in calculations on model fluids composed of hard molecules. An accurate treatment of such systems is important for the development of theories of molecular liquids. From experience with atomic systems, we may expect the structure of simple molecular liquids to be dominated by the strongly repulsive part of the pair potential, and an obvious way to represent the short-range repulsions is through an interaction-site model consisting of fused hard spheres of appropriately chosen diameters. There is no doubt that for many molecular liquids this point of view is basically correct. In Figure 12.7, for example, the results of neutron-diffraction measurements of the structure factor of liquid carbon dioxide are compared with theoretical results for a fluid of hard, triatomic molecules (van Tricht *et al.*, 1984). Although the hard-sphere diameters were chosen to give the best fit to the

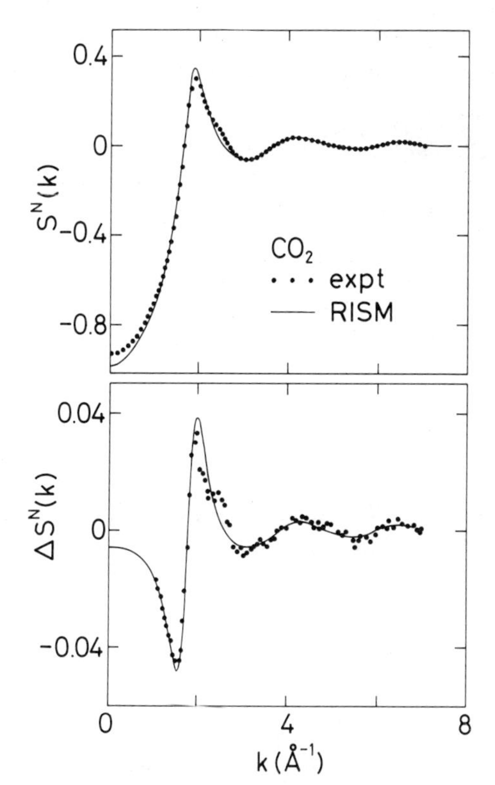

FIG. 12.7. Neutron-weighted structure factor for liquid carbon dioxide. The points are neutron-scattering results and the curves are calculated from the RISM approximation for a three-site model consisting of fused hard spheres. The upper part of the figure shows the intermolecular structure factor at $T = 222$ K, $\rho = 0.0158$ molecules Å^{-3}; the lower part shows the difference between results obtained at $T = 222$ K, $\rho = 0.0158$ molecules Å^{-3} and $T = 239$ K, $\rho = 0.0149$ molecules Å^{-3}. After van Tricht *et al.* (1984).

experimental data, the agreement is nevertheless remarkably good. In particular, the fused-hard-sphere model reproduces almost exactly the subtle changes in structure that accompany a change in density.

The key ingredient of the RISM approximation is an Ornstein–Zernike-like relation between the site–site pair correlation functions $h_{\alpha\beta}(r)$ and a set of direct correlation functions $c_{\alpha\beta}(r)$. In the atomic case, the meaning of the Ornstein–Zernike relation is that the total correlation between particles 1 and 2 is the sum of all possible paths of direct correlations that propagate via the intermediate particles 3, 4, ... etc. The same intuitive idea can be applied to site–site correlations, but allowance must now be made for the fact that correlations also propagate via the intramolecular distribution functions $s_{\alpha\beta}(r)$. Hence, whereas in an atomic fluid $h(1, 2)$ is given diagrammatically by the sum of all simple chains of c-bonds, $h_{\alpha\beta}(r)$ consists of all simple chains formed from $\boldsymbol{c}$-bonds and $\boldsymbol{s}$-bonds. We make this idea precise by writing $h_{\alpha\beta}(r)$ as a sum of interaction-site chain diagrams (Chandler, 1976):

$h_{\alpha\beta}(\mathbf{r}_{1\alpha}, \mathbf{r}_{2\beta})$ = {all allowed interaction-site chain diagrams consisting of two white terminal 1-circles labelled 1α and 2β, black 1-circles, at least one $\boldsymbol{c}$-bond, and $\boldsymbol{s}$-bonds; each diagram is to be multiplied by ρ^{n-1}, where n is the number of $\boldsymbol{c}$-bonds}

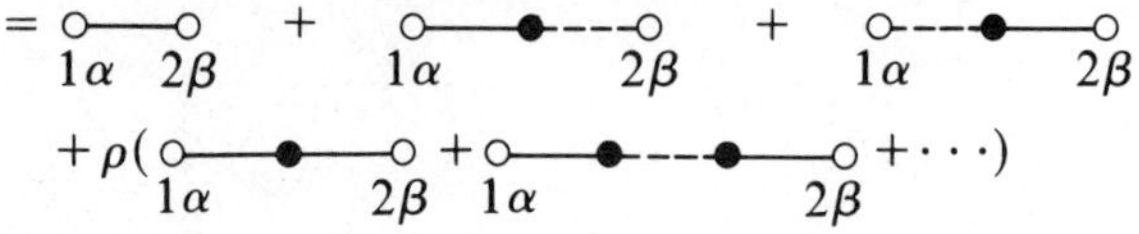

= ○—○ (1α 2β) + ○—●---○ (1α 2β) + ○---●—○ (1α 2β)

+ ρ(○—●—○ (1α 2β) + ○—●---●—○ (1α 2β) + · · ·)

+ · · · (12.9.1)

where a full line denotes a $\boldsymbol{c}$-bond and a broken line denotes an $\boldsymbol{s}$-bond. We recall that in the interaction-site diagrammatic formalism, a black circle implies a summation over all sites in the molecule. For example, the value of the third diagram on the right-hand side of (12.9.1) is

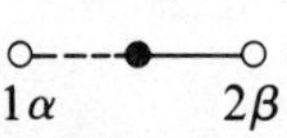

○---●—○ (1α 2β) $= \sum_{\gamma} \int s_{\alpha\gamma}(\mathbf{r}_{1\alpha} - \mathbf{r}_{1\gamma}) f_{\gamma\beta}(|\mathbf{r}_{1\gamma} - \mathbf{r}_{2\beta}|)\, d\mathbf{r}_{1\gamma}$ (12.9.2)

The term for which $\alpha = \gamma$ contributes nothing to the sum because the intramolecular distribution function is zero when the two sites are the same (see Eqn (12.8.6)).

We now have to sum the chain diagrams in (12.9.1). To do so, we use the same techniques as in Section 6.5, because the diagrams in (12.9.1) have the same topology as those in the diagrammatic expansion (6.5.22) of the renormalized potential $C(1, 2)$. We define a matrix of functions $\boldsymbol{\omega} \equiv [\omega_{\alpha\beta}]$ by

$$\omega_{\alpha\beta}(\mathbf{r}) = \delta_{\alpha\beta}\delta(\mathbf{r}) + s_{\alpha\beta}(\mathbf{r}) \tag{12.9.3}$$

and represent $\omega_{\alpha\beta}(\mathbf{r})$ as a hypervertex:

$\omega_{\alpha\beta}(|\mathbf{r}_{1\alpha} - \mathbf{r}_{1\beta}|)$ = 1α ◯ 1β (12.9.4)

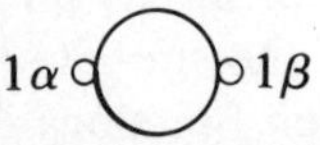

Then the sum of chain diagrams containing n $\boldsymbol{c}$-bonds becomes a single diagram consisting of $(n+1)$ $\boldsymbol{\omega}$-hypervertices and n $\boldsymbol{c}$-bonds. For example:

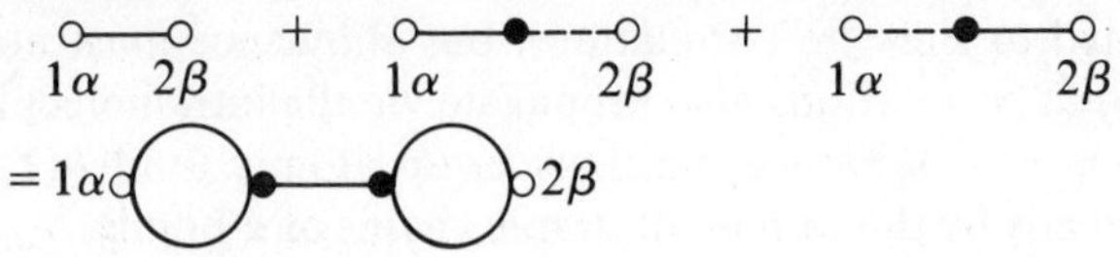

○—○ (1α 2β) + ○—●---○ (1α 2β) + ○---●—○ (1α 2β)

= 1α ◯●—●◯ 2β

$$=\sum_{\gamma}\sum_{\delta}\iint \omega_{\alpha\gamma}(|\mathbf{r}_{1\alpha}-\mathbf{r}_{1\gamma}|)c_{\gamma\delta}(|\mathbf{r}_{1\gamma}-\mathbf{r}_{2\delta}|)\omega_{\delta\beta}(|\mathbf{r}_{2\delta}-\mathbf{r}_{2\beta}|)\,\mathrm{d}\mathbf{r}_{1\gamma}\,\mathrm{d}\mathbf{r}_{2\delta} \tag{12.9.5}$$

Note that each hypervertex represents a single molecule and that the hypervertex incorporates all the intramolecular constraints represented by the s-bonds. The Fourier transform of (12.9.5) is

$$\sum_{\gamma}\sum_{\delta}\hat{\omega}_{\alpha\gamma}(k)\hat{c}_{\gamma\delta}(k)\hat{\omega}_{\delta\beta}(k)=(\hat{\boldsymbol{\omega}}\hat{\boldsymbol{c}}\hat{\boldsymbol{\omega}})_{\alpha\beta} \tag{12.9.6}$$

i.e. it is the $\alpha\beta$-component of the matrix $\hat{\boldsymbol{\omega}}(k)\hat{\boldsymbol{c}}(k)\hat{\boldsymbol{\omega}}(k)$; the components of the matrix $\hat{\boldsymbol{\omega}}(k)$ are

$$\hat{\omega}_{\alpha\beta}(k)=\delta_{\alpha\beta}+(1-\delta_{\alpha\beta})\frac{\sin kl_{\alpha\beta}}{kl_{\alpha\beta}} \tag{12.9.7}$$

where $l_{\alpha\beta}$ is the intramolecular separation of sites α, β. Similarly, the Fourier transform of the sum of all chain diagrams in (12.9.1) containing exactly n c-bonds is $\rho^{n-1}((\hat{\boldsymbol{\omega}}\hat{\boldsymbol{c}})^n\hat{\boldsymbol{\omega}})_{\alpha\beta}$ (cf. Eqn (6.5.31)), and $\hat{h}_{\alpha\beta}(k)$ is the sum of a geometric series (cf. Eqn (6.5.32)). The matrix $\hat{\boldsymbol{h}}(k)$ is therefore given by

$$\hat{\boldsymbol{h}}(k)=\hat{\boldsymbol{\omega}}(k)\hat{\boldsymbol{c}}(k)[\boldsymbol{I}-\rho\hat{\boldsymbol{\omega}}(k)\hat{\boldsymbol{c}}(k)]^{-1}\hat{\boldsymbol{\omega}}(k) \tag{12.9.8a}$$

or

$$\hat{\boldsymbol{h}}(k)=\hat{\boldsymbol{\omega}}(k)\hat{\boldsymbol{c}}(k)\hat{\boldsymbol{\omega}}(k)+\rho\hat{\boldsymbol{\omega}}(k)\hat{\boldsymbol{c}}(k)\hat{\boldsymbol{h}}(k) \tag{12.9.8b}$$

Equation (12.9.8) is the Ornstein–Zernike-like relation. We shall refer to it as the RISM-OZ relation, but we shall see later that its status differs from that of the molecular Ornstein–Zernike relation (12.1.4). If $\hat{\boldsymbol{\omega}}$ is the identity matrix and ρ is appropriately reinterpreted, Eqn (12.9.8) reduces to the Ornstein–Zernike relation for a mixture of atomic fluids, Eqn (5.2.18). Equation (12.9.8) is the general form of the RISM-OZ relation and may be used for any interaction-site model; it can be derived in other ways than the one we have given (Chandler and Andersen, 1972; Chandler, 1973; Topol and Claverie, 1978; Cummings and Stell, 1982), but the diagrammatic method (Chandler, 1976) has a paricularly strong intuitive appeal.

If numerical results are to be obtained, the RISM-OZ relation must be combined with a closure approximation. For systems of fused hard spheres, the obvious choice of closure relation is a generalization of the PY approximation for atomic hard spheres, i.e.

$$\begin{aligned} h_{\alpha\beta}(r)&=-1, \qquad r<d_{\alpha\beta}\\ c_{\alpha\beta}(r)&=0, \qquad r>d_{\alpha\beta}\end{aligned} \tag{12.9.9}$$

where $d_{\alpha\beta}$ is the α–β hard-sphere diameter. When the site–site potentials are continuous, generalizations of either the PY or HNC approximations can be used. A variational method of solution of the RISM integral equation was suggested by Chandler and Andersen (1972) and implemented numerically by Lowden (Lowden and Chandler, 1973, 1974; Hsu *et al.*, 1976); iterative methods suitable for short-range continuous potentials (Johnson and Hazoumé, 1979; Monson, 1982; Enciso, 1985) and refinements needed to handle long-range forces (Hirata and Rossky, 1981; Cummings *et al.*, 1982) have also been described.

Some RISM results based on the PY closure approximation for the atomic pair distribution function of a Lennard-Jones diatomic model of liquid chlorine are shown in Figure 12.8 (Monson, 1982). Although there are some differences in detail, all the main features seen in molecular dynamics calculations for the same potential model (Singer *et al.*, 1977) are well reproduced; note that the shoulder in the ethylene results of Figure 12.2 appears in this case as a well-defined subsidiary peak. The quality of the results shown in Figure 12.8 is representative of that obtained for many other small molecules, although it is also easy to find examples where the agreement with computer simulations is very much better and others where

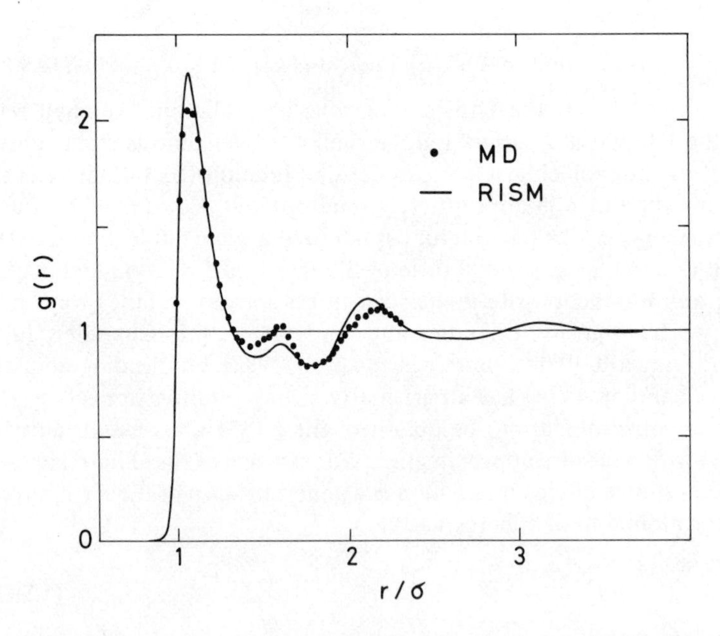

FIG. 12.8. The atom-atom distribution function of a model of liquid chlorine, obtained by molecular dynamics simulation and from calculations based on the RISM approximation. After Monson (1982).

it is significantly worse. The theoretical results in Figure 12.7 were obtained by RISM calculations (with PY closure) for a hard triatomic model with $d_{CC} = 3.3$ Å and $d_{OO} = 2.8$ Å (van Tricht *et al.*, 1984). Similar comparisons with experimental results have been made for a variety of other molecular liquids (see, for example, Montague *et al.*, 1983).

12.10 DEVELOPMENTS BEYOND THE RISM APPROXIMATION

Although successful in many applications, the RISM approximation suffers from a number of major defects. First, it does not lend itself readily to a calculation of the equation of state, and the results that are obtained are thermodynamically inconsistent in the sense of Section 5.4 (Lowden and Chandler, 1973, 1975). Secondly, calculated structural properties show an unphysical dependence on the presence of "auxiliary" sites, i.e. on sites that label points in a molecule but contribute nothing to the intermolecular potential (Hsu *et al.*, 1976; Cummings *et al.*, 1981). Thirdly, use of the RISM approximation leads to trivial and incorrect results for certain quantities descriptive of angular correlations in the fluid. This remarkable fact was first pointed out by Chandler (1978) and later elaborated by Sullivan and Gray (1981). As an example, we prove below by the method of Sullivan and Gray that the order parameter G_1 defined by Eqn (12.1.8) is identically zero for any asymmetric but non-polar diatomic molecule. No assumptions need be made about the form of the site–site potentials except to suppose that they are short ranged.

We note first that all elements of the matrix $\hat{\boldsymbol{\omega}}(k)$ defined by (12.9.7) are unity when $k = 0$. If we define a matrix $\boldsymbol{Q}$ as

$$\boldsymbol{Q} = \boldsymbol{I} - m^{-1}\hat{\boldsymbol{\omega}}(0) = \begin{bmatrix} \frac{1}{2} & -\frac{1}{2} \\ -\frac{1}{2} & \frac{1}{2} \end{bmatrix} \tag{12.10.1}$$

where m is the number of sites (here equal to two), then

$$\boldsymbol{Q}\hat{\boldsymbol{\omega}}(0) = 0 \tag{12.10.2}$$

Next we write the RISM-OZ relation (12.9.8) in the form

$$\hat{\boldsymbol{h}}(k) = \hat{\boldsymbol{\omega}}(k)\boldsymbol{X}(k) \tag{12.10.3}$$

where

$$\boldsymbol{X}(k) = \hat{\boldsymbol{c}}(k)[\hat{\boldsymbol{\omega}}(k) + \rho\hat{\boldsymbol{h}}(k)] \tag{12.10.4}$$

Expanding $\hat{\boldsymbol{\omega}}(k)$ in powers of k, we find to order k^2 that

$$\hat{\boldsymbol{h}}(k) = [\hat{\boldsymbol{\omega}}(0) + k^2\boldsymbol{\omega}^{(2)} + \cdots]\boldsymbol{X}(k) \tag{12.10.5}$$

If multiplied on the left by $\boldsymbol{Q}$, Eqn (12.10.5) reduces, by virtue of the property (12.10.2), to

$$\boldsymbol{Q}\hat{\boldsymbol{h}}(k) = \boldsymbol{Q}[k^2\boldsymbol{\omega}^{(2)} + \cdots]\boldsymbol{X}(k) \tag{12.10.6}$$

We now suppose that $\hat{\boldsymbol{h}}(k)$ and $\hat{\boldsymbol{c}}(k)$ (and hence also $\hat{\boldsymbol{X}}(k)$) have small-k expansions at least up to order k^2. This is plausible, since the site–site potentials are assumed to be short ranged. Then

$$\boldsymbol{Q}[\hat{\boldsymbol{h}}(0) + k^2\boldsymbol{h}^{(2)} + \cdots] = \boldsymbol{Q}[k^2\boldsymbol{\omega}^{(2)} + \cdots][\boldsymbol{X}(0) + k^2\boldsymbol{X}^{(2)} + \cdots] \tag{12.10.7}$$

and by equating coefficients of k^2 we find that

$$\boldsymbol{Q}\boldsymbol{h}^{(2)} = \boldsymbol{Q}\boldsymbol{\omega}^{(2)}\boldsymbol{X}(0) \tag{12.10.8}$$

We have seen in Section 12.3 that all elements of $\hat{\boldsymbol{h}}(0)$ are the same and related to the compressibility by Eqn (12.3.6). Since $\hat{\boldsymbol{\omega}}(0)$ commutes with the other matrices in (12.10.4) (because all its elements are unity), $\boldsymbol{X}(0)$ may be written as

$$\boldsymbol{X}(0) = [1 + \rho\hat{h}_0(0)]\hat{\boldsymbol{c}}(0)\hat{\boldsymbol{\omega}}(0) \tag{12.10.9}$$

where $\hat{h}_0(0)$ (a scalar) is any element of $\hat{\boldsymbol{h}}(0)$. Inserting (12.10.9) in Eqn (12.10.8), multiplying on the right by $\boldsymbol{Q}$, and using again the property (12.10.2), we find that

$$\boldsymbol{Q}\boldsymbol{h}^{(2)}\boldsymbol{Q} = 0 \tag{12.10.10}$$

But every element of the matrix $\boldsymbol{Q}\boldsymbol{h}^{(2)}\boldsymbol{Q}$ is proportional to $\Delta h^{(2)}$, where $\Delta h(r)$ is defined by Eqn (12.3.8). Thus $\Delta h^{(2)} = 0$ and hence, from Eqn (12.3.10), $G_1 = 0$. As we pointed out in Section 12.3, this result is obvious on symmetry grounds for a homonuclear molecule, but in the general case it will be true (except accidentally) only in the ideal-gas limit. Sullivan and Gray (1981) have given a more general proof of (12.10.10); this shows that $G_1 = 0$ for any linear molecule, assuming that all site–site potentials are short ranged. Similarly, by considering terms of order k^4 in Eqn (12.10.7), it is possible to show that $G_2 = 0$ for any symmetric, linear, triatomic molecule.

If the molecule is polar, with the interaction sites carrying point charges, the problem becomes more complicated. When expanding $\hat{\boldsymbol{c}}(k)$, allowance must now be made for a term in k^{-2}, corresponding to an r^{-1} decay of the site–site potential. The k^{-2} term has to be treated separately, but it is then possible to show (Sullivan and Gray, 1981; Morriss and Perram, 1981) that for any interaction-site molecule

$$\rho \sum_{\alpha} \sum_{\beta} q_\alpha q_\beta h_{\alpha\beta}^{(2)} = -\frac{y\mu^2}{1 + 3y} \tag{12.10.11}$$

where q_α is the charge carried by site α. If Eqn (12.10.11) is compared with the exact result (12.5.32), we find that

$$\varepsilon = 1 + 3y \tag{12.10.12}$$

which is a well-known expression for the dielectric constant of an ideal gas of polar molecules (Böttcher, 1973).

The results in Eqns (12.10.10) and (12.10.12) are consequences solely of the RISM-OZ relation (12.9.8). They are independent of the choice of closure relation except insofar as the latter must be consistent with the assumed small-k behaviour of $\hat{\boldsymbol{h}}(k)$ and $\hat{\boldsymbol{c}}(k)$. It follows that the RISM-OZ relation, when combined with a closure approximation of the type used successfully in the theory of atomic fluids, is incapable of describing correctly certain long-wavelength properties of molecular systems, of which G_1, G_2 and ε are important examples.

Attempts to develop a more satisfactory theory within the interaction-site picture have developed along two different lines. The first relies on treating the RISM-OZ relation as providing the definition of the site–site direct correlation functions $c_{\alpha\beta}(r)$. If the theory is to be useful for the calculation of angular correlation parameters, it is clearly necessary to abandon the assumption that $c_{\alpha\beta}(r)$ is a short-range function, even when the corresponding site–site potential is short ranged. For example, a non-zero value of G_1 for an asymmetric diatomic molecule can be obtained if it is assumed that $c_{\alpha\beta}(r) \sim D_{\alpha\beta}/r$ at large r, where $D_{\alpha\beta}$ is some constant. The small-k expansion of $c_{\alpha\beta}(k)$ now begins as in the polar case, i.e.

$$\hat{c}_{\alpha\beta}(k) = \frac{1}{k^2} c_{\alpha\beta}^{(-2)} + \hat{c}_{\alpha\beta}(0) + k^2 c_{\alpha\beta}^{(2)} + \cdots \tag{12.10.13}$$

where $c_{\alpha\beta}^{(-2)} = 4\pi D_{\alpha\beta}$. Thereafter the calculation proceeds as described earlier. By considering terms of order k^{-2}, k^0 and k^2 in the analogue of (12.10.7), it can be shown (Cummings and Stell, 1982) that $D_{11} = D_{22} = -D_{12} = D$, say, and that D is related to G_1 by

$$D = \frac{3}{4\pi\rho l^3} \frac{G_1}{1 + G_1} \tag{12.10.14}$$

If there is no long-range part to $c_{\alpha\beta}(r)$, i.e. if $D = 0$, we recover the previous trivial result, namely that $G_1 = 0$. Similar calculations have been made for other interaction-site molecules with striking and even bizarre results. Cummings and Sullivan (1982), for example, have shown that for a linear, symmetric triatomic molecule, a non-zero value of G_2 is obtained only if $c_{\alpha\beta}(r)$ is assumed to diverge as r at large r. Under such circumstances, the concept of "direct correlation" ceases to have any obvious physical meaning. This is a serious loss, since much of the progress that has been made in liquid-state theory in recent years has its origins precisely in the fact that, at least for atomic fluids, the direct correlation function is mathematically well behaved with a range that is always equal to the range of the interparticle potential.

The methods sketched above establish rigorously the conditions that must be satisfied if solutions to the RISM-OZ relation are to yield physically satisfactory results for certain properties. Their predictive value, however, is unclear. It should also be noted that the failure of the RISM approximation in describing features of the long-wavelength structure is not inconsistent with its ability to reproduce correctly the details of the short-range angular correlations. Rossky *et al.* (1983) have shown numerically for a number of polar diatomic molecules that the addition of the appropriate long-range "tail" to $c_{\alpha\beta}(r)$ has only a very small effect on the behaviour of the site–site distribution functions at all separations for which they have any appreciable structure. Long-range and short-range correlations are therefore only weakly coupled.

In the alternative approach, the view is taken that the RISM-OZ relation, though plausible, does not provide an adequate basis for a complete theory of the structure of molecular fluids. Accordingly, it is there rather than in the closure relation that improvement must be sought (Chandler *et al.*, 1982a). The argument is based on a diagrammatic analysis of the RISM-OZ relation and on the contrast that can be made with the molecular Ornstein–Zernike relation (12.1.4). In the latter case, the function $[h(1,2) - c(1,2)]$ is expressible (see Eqn (5.2.9)) as the sum of all simple molecular chain diagrams containing two or more c-bonds. In any such diagram, every black circle is a nodal circle, whereas $c(1,2)$ consists of the subset of diagrams in the ρ-circle, f-bond expansion of $h(1,2)$ that are free of nodal circles. Any approximate theory that preserves this characterization—PY and HNC, for example—constitutes, in a phrase of Chandler *et al.* (1982a), a "diagrammatically proper" theory. By analogy, it would appear from the formulation of the RISM-OZ relation in Eqn (12.9.1) that the **c**-bonds represent the subset of diagrams in the expansion of $h_{\alpha\beta}(r)$ that are free of nodal circles. This is not the case. For example, the third diagram on the right-hand side of (12.8.9) is a diagram without nodal circles. If this is substituted into the second and third diagrams on the right-hand side of (12.9.1), it yields, respectively, the interaction-site diagrams labelled (a) and (b) below.

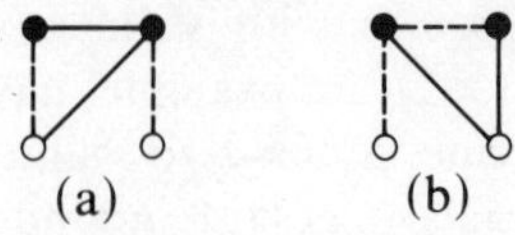

Diagram (a) is a diagram in the exact expansion of $h_{\alpha\beta}(r)$, but (b) is not, because two **s**-bonds intersect the same black circle. It follows that the direct correlation functions in the RISM-OZ relation are not characterized diagrammatically in the same way as in the full molecular case represented by Eqn (12.1.4). While this is not a fundamental objection to the use of the RISM-OZ relation, it leads to the difficulties described earlier, which in turn are resolved only by invoking behaviour in $c_{\alpha\beta}(r)$ at large r that physical intuition cannot easily accommodate.

In order to derive an alternative to the RISM-OZ relation, a more detailed analysis is needed of the diagrams that contribute to the expansion (12.8.10) of $h_{\alpha\beta}(r)$. As illustrated in the example already considered, diagrams allowed in an exact theory are generated only when the diagrams inserted in (12.9.1) contain white circles intersected by an $\boldsymbol{s}$-bond. This suggests that a useful decomposition of $\boldsymbol{h}$ could be obtained by writing

$$\boldsymbol{h} = \boldsymbol{h}_0 + \boldsymbol{h}_{1l} + \boldsymbol{h}_{1r} + \boldsymbol{h}_2 \tag{12.10.15}$$

The subscript numeral denotes the number of white circles that are intersected by an $\boldsymbol{s}$-bond; the subscripts l, r mean that an $\boldsymbol{s}$-bond intersects, respectively, the left- or right-hand circle when the diagram is drawn in conventional fashion. For example, the first diagram on the right-hand side of (12.8.9) contributes to $\boldsymbol{h}_0$ and both the second and third diagrams contribute to $\boldsymbol{h}_{1l}$. A similar decomposition can be made of the matrix $\boldsymbol{c}$ of direct correlation functions defined as the subset of diagrams in $\boldsymbol{h}$ that are free of nodal circles:

$$\boldsymbol{c} = \boldsymbol{c}_0 + \boldsymbol{c}_{1l} + \boldsymbol{c}_{1r} + \boldsymbol{c}_2 \tag{12.10.16}$$

Clearly the functions $c_{\alpha\beta}(r)$ in (12.10.16) are not the same as those appearing in the RISM-OZ relation. If we use an arrowhead to denote intersection by an $\boldsymbol{s}$-bond, Eqn (12.10.16) can be written in diagrammatic form as

$$\boldsymbol{c} = \underset{\circ\!\!-\!\!-\!\!\circ}{c_0} + \underset{\circ\!\!\leftarrow\!\!-\!\!\circ}{c_{1l}} + \underset{\circ\!\!-\!\!\rightarrow\!\!\circ}{c_{1r}} + \underset{\circ\!\!\leftarrow\!\!\rightarrow\!\!\circ}{c_2} \tag{12.10.17}$$

All diagrams in the expansion of $h_{\alpha\beta}(r)$ may now be expressed in terms of allowed chain diagrams containing $\boldsymbol{c}_0$, $\boldsymbol{c}_{1l}$, $\boldsymbol{c}_{1r}$, $\boldsymbol{c}_2$ and $\boldsymbol{s}$-bonds. The class of allowed chain diagrams excludes those shown in (a) to (d) below:

(a) (b) (c) (d)

If, however, Eqn (12.10.17) were substituted in the RISM-OZ relation (12.9.1), diagrams of the types (a) to (d) would be retained. The sum of allowed chain diagrams can be obtained in closed form (Chandler *et al.*, 1982a). We omit the technical details and give only the final result, namely

$$\boldsymbol{\omega} + \rho\boldsymbol{h} = (\boldsymbol{I} - \rho\boldsymbol{\Omega}\boldsymbol{c}_0)^{-1}\boldsymbol{\Omega} \tag{12.10.18}$$

where

$$\boldsymbol{\Omega} = (\boldsymbol{I} - \rho \boldsymbol{c}_{1l})^{-1}(\boldsymbol{\omega} + \rho \boldsymbol{c}_2)(\boldsymbol{I} - \rho \boldsymbol{c}_{1r})^{-1} \tag{12.10.19}$$

An equation different in structure but equivalent to (12.10.18) has been derived by Rossky and Chiles (1984).

In the limit $\rho \to 0$, $\boldsymbol{\Omega}$ and $\boldsymbol{\omega}$ become identical and Eqn (12.10.18) reduces to the RISM-OZ relation if the direct correlation function matrix in (12.9.8) is identified with $\boldsymbol{c}_0$. This fact makes it understandable that the RISM approximation should yield ideal-gas values for certain structural properties. If account is taken of the difference between the matrices $\boldsymbol{\Omega}$ and $\boldsymbol{\omega}$, even in an approximate way, non-trivial results can be obtained for the angular correlation parameters and dielectric constant of interaction-site models without the necessity for assuming any long-range behaviour in the direct correlation functions (Chandler *et al.*, 1982b; Chandler and Richardson, 1983). Calculations of G_2 for a model of liquid carbon disulphide (Richardson and Chandler, 1984) have given results in fair agreement with molecular dynamics simulations (Impey *et al.*, 1981). The important new feature of Eqn (12.10.18) in comparison with the RISM-OZ relation (12.9.8) is the fact that the direct correlation functions in (12.10.18) are well-defined subsets of diagrams in $\boldsymbol{h}$. Integral-equation theories based on (12.10.18) will therefore be "diagrammatically proper" in the sense in which the RISM equations are not. Rossky and Chiles (1984) have formulated the theory in such a way as to yield numerical solutions for both PY and HNC-like closures. The results obtained so far are not an improvement on those of the original RISM approximation, and other forms of closure relation need to be investigated.

12.11 PERTURBATION THEORIES

Given the great success of perturbation theories in treating the properties of atomic liquids, it is not surprising that considerable effort has been devoted to extending the scope of such methods to deal with molecular systems. The basic problem is the same as in the atomic case, but the practical difficulties are much more severe, and the results obtained are, in the main, disappointing. The methods used are conveniently classified according to whether the reference-system potential is spherically symmetric or anisotropic. Theories of the first type are most appropriate when the anisotropy in the full potential is weak and long ranged; those of the second type have a greater physical appeal and a wider range of possible application, but they are more difficult to implement because the calculation of the reference-system properties poses greater problems.

We consider first the case when the reference-system potential $v_0(R)$ is spherically symmetric. The full potential $v(1, 2)$ is then written, in the notation of Chapter 6, as

$$v(1,2) = v_0(R) + w(1,2) \tag{12.11.1}$$

where $R = |\mathbf{R}_{12}|$. The free energy of the fluid with potential $v(1, 2)$ can be calculated by a straightforward generalization of the λ-expansion developed in Section 6.2. The first-order term, Eqn (6.2.13), now involves integrations over $\mathbf{\Omega}_1$ and $\mathbf{\Omega}_2$, but these concern only the perturbation because the reference-system pair distribution function $g_0(R)$ is spherically symmetric. A corresponding simplification occurs at every order in the perturbation. All the results of the λ-expansion for atomic fluids can therefore be taken over to the molecular case by replacing the terms in $w(i, j)$ by their unweighted angle averages. A particularly convenient choice of reference system is one that sets $v_0(R)$ equal to the angle average of the full potential, i.e.

$$v_0(R) = \langle v(1,2) \rangle_{\mathbf{\Omega}_1 \mathbf{\Omega}_2} \tag{12.11.2}$$

Use of (12.11.2) ensures that the first-order correction to the free energy vanishes irrespective of the precise form of the perturbation (Pople, 1952, 1954; Barker, 1953).

Terms of second and higher order in the λ-expansion are most easily evaluated by expanding $w(1, 2)$ in a series of angle-dependent functions, but the problem is greatly simplified if the perturbation is of multipolar type. As an example, consider the case of the potential defined by (12.4.1), and let $w(1, 2)$ be the dipole–dipole term. Because

$$\langle w(1,2) \rangle_{\mathbf{\Omega}_1} = \langle w(1,2) \rangle_{\mathbf{\Omega}_2} = 0 \tag{12.11.3}$$

the second-order correction (6.2.14) reduces to

$$\begin{aligned} \frac{\beta A_2}{N} &= -\frac{\beta^2 \rho}{4} \int g_0(R) \langle [w(1,2)]^2 \rangle_{\mathbf{\Omega}_1 \mathbf{\Omega}_2} \, \mathrm{d}\mathbf{R} \\ &= -\frac{\beta^2 \rho \mu^4}{6} \int g_0(R) R^{-6} \, \mathrm{d}\mathbf{R} \end{aligned} \tag{12.11.4}$$

Similarly, the third-order term, which in general contains integrals over the pair, triplet and four-particle distribution functions of the reference system, is now given (Stell *et al.*, 1972) by

$$\begin{aligned} \frac{\beta A_3}{N} &= \frac{\beta^3 \rho^2}{6} \iint g_0^{(3)}(\mathbf{R}_1, \mathbf{R}_2, \mathbf{R}_3) \langle w(1,2) w(2,3) w(3,1) \rangle_{\mathbf{\Omega}_1 \mathbf{\Omega}_2 \mathbf{\Omega}_3} \, \mathrm{d}\mathbf{R}_{12} \, \mathrm{d}\mathbf{R}_{13} \\ &= \frac{\beta^3 \rho^2 \mu^6}{54} \iint g_0^{(3)}(\mathbf{R}_1, \mathbf{R}_2, \mathbf{R}_3) \left(\frac{1 + 3 \cos \theta_1 \cos \theta_2 \cos \theta_3}{R_{12}^3 R_{23}^3 R_{31}^3} \right) \mathrm{d}\mathbf{R}_{12} \, \mathrm{d}\mathbf{R}_{13} \end{aligned} \tag{12.11.5}$$

where θ_1, θ_2, θ_3 are the interior angles of the triangle formed by molecules 1, 2 and 3; the three-body function whose average gives A_3 is the same as that appearing in the Axilrod–Teller formula for the triple–dipole dispersion energy (see Section 1.2). There is also a considerable simplification of the λ-expansion of $g(1,2)$. From Eqn (D.13) of Appendix D we see that when the condition (12.11.3) is satisfied, $g(1,2)$ is given to first order in the perturbation (Gubbins and Gray, 1971) by

$$g(1,2) = g_0(R)[1-\beta w(1,2)] \tag{12.11.6}$$

This expression is similar in structure to the LIN approximation of Eqn (6.5.49); the two become identical if the renormalized potential in (6.5.49), generalized to include angular variables (Verlet and Weis, 1974; Stell and Weis, 1977), is replaced by its low-density limit.

Equations (12.11.4) and (12.11.5) have been used in calculations of thermodynamic properties of the dipolar hard-sphere and Stockmayer fluids. In both cases, the λ-expansion of A is a series in powers of $\mu^{*2} = \mu^2/k_B T\sigma^3$, where σ is the appropriate length parameter. The numerical results show that the expansions diverge when $\mu^{*2} \gtrsim 1$, which is a range of μ^* that is relevant to many real liquids (Patey and Valleau, 1974; McDonald, 1974; Verlet and Weis, 1974). Divergence is almost inevitable when the anisotropy is strong, because the entire anisotropic part of the potential is contained in the perturbation; it can be avoided by use of a Padé approximant to sum the perturbation series (Stell *et al.*, 1972), but this is a device of doubtful theoretical significance. There are also two other serious deficiencies in the theory. First, the reference-system potential defined by (12.11.2) is suitable only for multipolar-like perturbations; when applied to an interaction-site model, it leads to a division of the potential in which the perturbation becomes very large at small separations. Secondly, the approximation to $g(1,2)$ given by (12.11.6) generates only those harmonics that are contained in the potential. It is therefore useless for a study, say, of the dielectric constant of polar fluids. The situation is improved by going to second order, but application of the theory is then limited to cases in which the perturbation is very weak (Murad, 1980).

An alternative to the use of (12.11.2) is to define a reference-system potential by equating the Boltzmann factor of the potential $v_0(R)$ to the angle-averaged Boltzmann factor of the full potential, i.e.

$$e_0(r) = \langle e(1,2)\rangle_{\Omega_1\Omega_2} \tag{12.11.7}$$

The reference potential is then given explicitly by

$$v_0(R) = -k_B T \log[1+\langle f(1,2)\rangle_{\Omega_1\Omega_2}] \tag{12.11.8}$$

where $f(1,2)$ is the Mayer function for the potential $v(1,2)$. Equation (12.11.8) was originally proposed by Cook and Rowlinson (1953) as an effective central potential for weakly anisotropic fluids, but more recently it has been used as the reference-system potential in what is sometimes called the "reference average Mayer" or RAM perturbation theory (Smith and Nezbeda, 1983). The division of the potential implied by (12.11.8) is of the general form of Eqn (6.2.1) (rather than (6.2.2)), with $w_\lambda(1,2)$ given by

$$w_\lambda(1,2) = -k_B T \log[1 + \lambda f_w(1,2)] \tag{12.11.9}$$

where

$$f_w(1,2) = \frac{e(1,2)}{e_0(R)} - 1 \tag{12.11.10}$$

Use of (12.11.8) has, in common with the theory based on (12.11.2), the convenient feature that the first-order term in the λ-expansion of the free energy is always zero. The second-order term (Smith, 1974) is now

$$\frac{\beta A_2}{N} = -\tfrac{1}{2}\rho^2 \iint g_0^{(3)}(\mathbf{R}_1, \mathbf{R}_2, \mathbf{R}_3)\langle f_w(1,2) f_w(1,3)\rangle_{\Omega_1\Omega_2\Omega_3}\, d\mathbf{R}_{12}\, d\mathbf{R}_{13} \tag{12.11.11}$$

First-order expansion (Perram and White, 1974) of the function $y(1,2) = g(1,2)/e(1,2)$ gives

$$g(1,2) = [1 + f_w(1,2)] \times \left[g_0(R) + \rho \int g_0^{(3)}(\mathbf{R}_1, \mathbf{R}_2, \mathbf{R}_3)\langle f_w(1,3) + f_w(2,3)\rangle_{\Omega_3}\, d\mathbf{R}_3 \right] \tag{12.11.12}$$

The right-hand side of (12.11.12) vanishes wherever the potential $v(1,2)$ is infinite; this is not necessarily true of the corrresponding expansion of $g(1,2)$ itself.

The RAM theory has been applied in a number of slightly different forms to a variety of molecular fluids; the systems studied include several interaction-site models, for which the reference potential defined by (12.11.8) is a better choice than (12.11.2). The papers by Smith and Nezbeda (1983) and Quirke and Tildesley (1983) contain some representative calculations and references to other work. The method has had some successes, but the agreement with simulations is generally poor except when the potential is only weakly anisotropic or when the density is low. The most unsatisfactory feature of the theory is the tendency for results on structural properties to deteriorate in passing from zeroth to first order in the perturbation. This behaviour is strongly suggestive of a poorly convergent series. If a zeroth-order treatment of the structure is adequate, the theory reduces to the use of an effective central potential; such methods are capable of giving useful results with very little computational effort (Johnson *et al.*, 1984; MacGowan *et al.*, 1984), but they have a limited range of applicability.

Perturbation theories based on spherically symmetric reference potentials cannot be expected to work well when (as in most real molecules) the short-range repulsive forces are strongly anisotropic. The natural approach in such cases is to include the strongly varying interactions in the specification of an anisotropic reference potential and to relate the properties of the reference system to those of hard molecules having the same shape (Andersen and Chandler, 1972; Ladanyi and Chandler, 1975). Calculations of this type have been made by Tildesley (1980) for systems of Lennard-Jones diatomic molecules. The method used by Tildesley is a generalization of that of Weeks, Chandler and Andersen (Weeks *et al.*, 1971) described in Section 6.4. The site–site potentials are divided in the form

$$v_{\alpha\beta}(r) = v_{0\alpha\beta}(r) + w_{\alpha\beta}(r) \tag{12.11.13}$$

where $v_{0\alpha\beta}(r)$ represents the purely repulsive part of the potential ($r < 2^{1/6}\sigma_{\alpha\beta}$). The structure of the reference system is then related to that of a fluid of hard dumbbells having the same bondlength and with hard-sphere diameters given by the solution of the equation

$$\int y_{d\alpha\beta}(r)\Delta e_{\alpha\beta}(r)\,d\mathbf{r} = 0 \tag{12.11.14}$$

where a subscript d denotes a property of the hard-dumbbell fluid and the "blip-function" $\Delta e_{\alpha\beta}(r)$ is defined as

$$\Delta e_{\alpha\beta}(r) = e_{0\alpha\beta}(r) - e_{d\alpha\beta}(r) \tag{12.11.15}$$

Equation (12.11.14) is the molecular site–site analogue of (6.3.5). In the atomic case, the relation between the reference system and the underlying hard-sphere fluid is obtained by equating $y_0(r)$ and $y_d(r)$; the structures of the two fluids can be linked in this way because $y_d(r)$ is a continuous function even though the hard-sphere potential is not. The extension to molecular systems is complicated by the presence of cusps in the functions $y_{d\alpha\beta}(r)$. In practice, the site–site distribution functions of the hard-dumbbell fluid are calculated from the RISM equations (12.9.8) and (12.9.9), and in that approximation the sum of all interaction-site diagrams in the expansion of $y_{d\alpha\beta}(r)$ that lead to cusps can be summed in closed form. This diagrammatic result makes it possible to devise a scheme whereby the cusps in $y_{d\alpha\beta}(r)$ are smoothed in a realistic fashion. Finally, the structure of the system of interest is related to that of the reference system by setting $g_{\alpha\beta}(r) = g_{0\alpha\beta}(r)$, and the contributions to thermodynamic properties due to the perturbation $w_{\alpha\beta}(r)$ are calculated to first order in the λ-expansion.

Tildesley (1980) has made calculations along the lines sketched above for two-site models of liquid fluorine and carbon dioxide and compared the results with those obtained by molecular dynamics simulation (Singer *et al.*, 1977). The agreement is comparable with that obtained by the RISM method with the PY closure applied to the full site–site potentials. There is, however, an appreciable improvement in the region of the first peak in $g_{\alpha\beta}(r)$ where, as in the example plotted in Figure 12.8, the RISM results are generally too highly structured. The Barker–Henderson perturbation theory has been generalized in a similar way and applied to models of liquid nitrogen (Lombardero *et al.*, 1981) and chlorine (Abascal *et al.*, 1985), again with good results. The main disadvantage of these methods is the large computational effort that their implementation requires.

12.12 REORIENTATIONAL TIME-CORRELATION FUNCTIONS

The description of the dynamical properties of molecular liquids differs most obviously from that used for atomic systems through the appearance of a class of reorientational time-correlation functions. We end this chapter by briefly considering some of the properties of these functions, limiting ourselves mainly to the case of linear molecules. We begin with the simpler problem of the single-molecule functions, leaving until later the question of collective reorientational properties.

The reorientation of linear molecules can be described in a compact way by introducing a family of time-correlation functions defined as

$$C^{(l)}(t) = \langle P_l[\mathbf{u}_i(t) \cdot \mathbf{u}_i] \rangle \tag{12.12.1}$$

where, as usual, $\mathbf{u}_i$ is a unit vector parallel to the axis of molecule i and $P_l(\ )$ is a Legendre polynomial. The functions $C^{(l)}(t)$ are the time-dependent generalizations of the angular order parameters G_l of Eqn (12.1.8). Apart from their application to linear molecules, they are also the most important functions for the description of the reorientational motion of spherical-top molecules, i.e. those in which all three principal moments of inertia are the same (CCl_4, SF_6, etc.), and of the reorientation of the main symmetry axis of symmetric-top molecules, i.e. those in which two of the principal moments of inertia are equal (NH_3, CH_3I, etc.). The $l=1$ and $l=2$ functions are related to the spectral bandshapes measured in infrared absorption ($l=1$) and Raman scattering ($l=2$) experiments (Bailey, 1974; Rothschild, 1984). Information on the correlation functions can be obtained by Fourier transforming the experimental spectra, but the interpretation of the results is complicated by a variety of factors, including uncertainty about the contribution from vibrational relaxation and from collision-induced effects. Because of these difficulties, quantitative studies of reorientational motion in liquids are heavily dependent on the results of molecular dynamics simulations.

Some typical results taken from simulation and experiment for the $l=1$ and $l=2$ correlation functions are plotted in Figures 12.9 and 12.10. In Figure 12.9, results for $C^{(2)}(t)$ are shown for carbon dioxide in two very different thermodynamic states; the results in Figure 12.10 are for $C^{(1)}(t)$ and $C^{(2)}(t)$ for liquid acetonitrile (a symmetric-top molecule) near its triple point. Under liquid-state conditions, the functions are approximately exponential in form, except at short times, whereas at low densities, as shown in Figure 12.9, oscillations appear. The oscillations seen at low densities can be understood by considering the behaviour of the correlation functions in the ideal-gas limit. Let $\boldsymbol{\omega} = \mathbf{u} \times \dot{\mathbf{u}}$ be the angular velocity of a linear molecule with moment of inertia I. In the absence of any interactions, the angular velocity is a constant of the motion, and in a time t the molecule will rotate through an angle $\omega t = \cos^{-1}[\mathbf{u}(t) \cdot \mathbf{u}(0)]$, where $\omega \equiv |\boldsymbol{\omega}|$. The probability that a molecule will rotate through an angle ωt in time t is therefore determined by the probability that the molecule has an angular

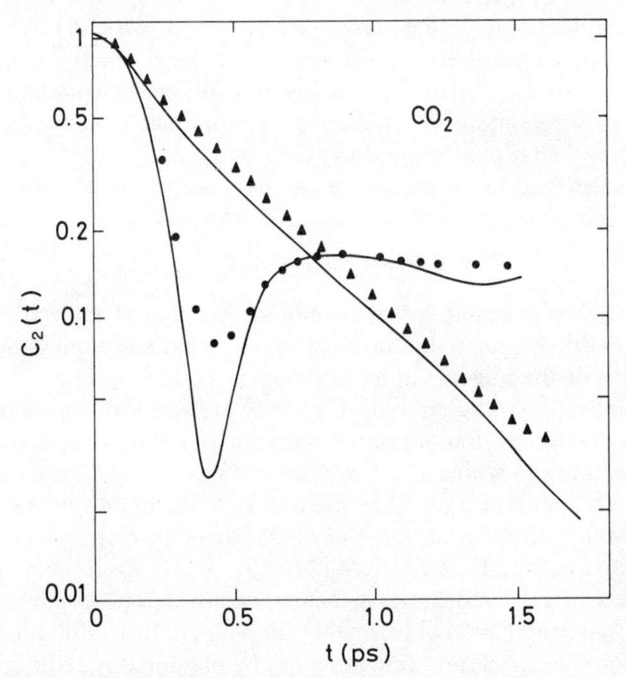

FIG. 12.9. The $l=2$ reorientational correlation function for carbon dioxide. The points are experimental results of Versmold (1981) at $T=313$ K and $\rho=1.1045$ g cm^{-3} (triangles) or $\rho=0.0423$ g cm^{-3} (circles); the curves are obtained from molecular dynamics calculations for a three-site potential model. After Böhm *et al.* (1984b).

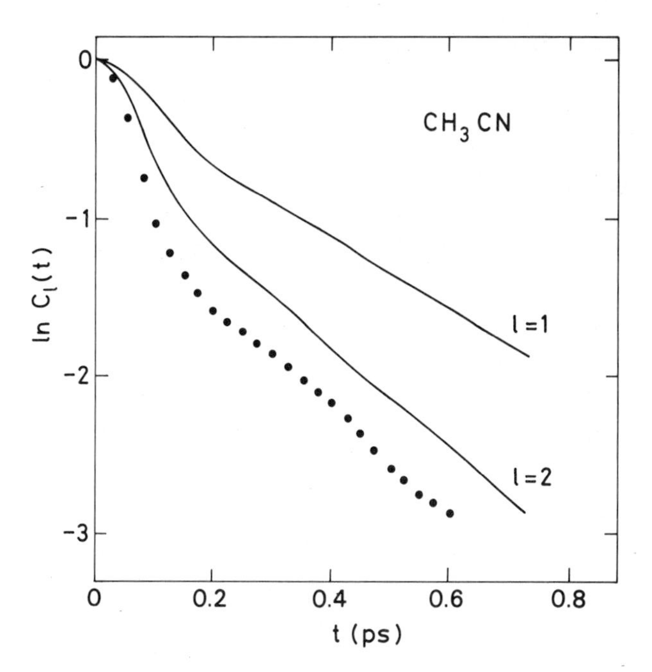

FIG. 12.10. The $l=1$ and $l=2$ reorientational correlation functions for liquid acetonitrile near the triple point. The curves are obtained from molecular dynamics calculations for a six-site potential model; the points are experimental Raman scattering results for the $l=2$ function (Bien *et al.*, 1981). All results relate to the reorientation of the main symmetry axis of the molecule. After Böhm *et al.* (1984a).

velocity of magnitude in the range $\omega \rightarrow \omega + d\omega$. Thus the correlation function $C^{(l)}(t)$ is the value of $P_l(\cos \omega t)$ averaged over a Maxwell–Boltzmann distribution of angular velocities, or

$$C^{(l)}(t) = \beta I \int_0^\infty P_l(\cos \omega t) \exp(-\tfrac{1}{2}\beta I \omega^2) \omega \, d\omega \qquad (12.12.2)$$

Such functions are oscillatory and tend to zero as $t \rightarrow \infty$ only when l is odd. They are commonly called the "free-rotor" correlation functions, and the oscillations seen in gas-phase experimental results are remnants of the free-rotor behaviour. Similar results are obtained for the free-rotor functions of non-linear molecules; the principle of the calculation is the same, but the final expressions have a more complicated form (St. Pierre and Steele, 1981).

The short-time expansion of the Legendre polynomial in (12.12.2) begins as

$$P_l(\cos \omega t) = 1 - \tfrac{1}{4} l(l+1) \omega^2 t^2 + \cdots \qquad (12.12.3)$$

If we expand the correlation functions as

$$C^{(l)}(t) = 1 - M_2^{(l)} \frac{t^2}{2!} + \cdots \tag{12.12.4}$$

a simple integration shows that

$$M_2^{(l)} = l(l+1) \frac{k_B T}{I} \tag{12.12.5}$$

At sufficiently short times, molecules rotate freely. Hence, although Eqn (12.12.5) has been derived only in the free-rotor limit, it is also valid for interacting molecules; there is an analogy here with the short-time behaviour of the mean-square translational displacement. From the general properties of time-correlation functions discussed in Section 7.1, it follows that the coefficient $M_2^{(l)}$ is the second moment of the power spectrum of $C^{(l)}(t)$. The mean-square width of the experimental bandshape is therefore independent of the molecular interactions. The fourth moment, however, contains a contribution proportional to the mean-square torque acting on the molecule.

The quasiexponential behaviour of the correlation functions at high densities can be explained by invoking an approximation similar in spirit to the Langevin equation (7.3.23). We begin by writing a memory-function equation for $C^{(l)}(t)$ and taking the Laplace transform to give

$$\tilde{C}^{(l)}(\omega) = \frac{1}{-i\omega + \tilde{N}^{(l)}(\omega)} \tag{12.12.6}$$

From Eqn (12.12.5) it follows that the memory function $N^{(l)}(t)$ behaves as

$$N^{(l)}(t) = l(l+1)\left(\frac{k_B T}{I}\right) n^{(l)}(t) \tag{12.12.7}$$

with $n^{(l)}(t=0) = 1$. We now suppose that reorientation occurs as the result of a succession of small, uncorrelated steps. This is the Debye approximation or "small-step-diffusion" model. In memory-function language, the Debye approximation is equivalent to the assumption that $\tilde{N}^{(l)}(\omega)$ is independent of frequency. In order to preserve the l-dependence contained in the exact result (12.12.7), we approximate the memory function in the form $\tilde{N}^{(l)}(\omega) \simeq l(l+1)D_R$, where D_R is a "rotational-diffusion constant". Then

$$C^{(l)}(t) = \exp[-l(l+1)D_R t] \tag{12.12.8}$$

In this approximation, the correlation functions are exponential at all times and for all values of l, and the entire family of functions is characterized by the single parameter D_R; for small molecules under triple-point conditions, D_R is typically of order $10^{11}\ s^{-1}$. The characteristic decay times for different values of l are related by the simple rule

$$\frac{\tau_l}{\tau_{l+1}} = \frac{l+2}{l} \tag{12.12.9}$$

The correlation times derived from infrared and Raman measurements should therefore be in the ratio $\tau_1/\tau_2 = 3$. This is approximately true of many real liquids, and also of correlation times obtained by simulation, though the ratio is usually a little less that its ideal Debye value (see, for example, Madden and Tildesley, 1983). The relation (12.12.9) is also valid under the less restrictive assumption that $n^{(l)}(t)$ has the same time dependence for all l, provided that τ_l is interpreted as the integral correlation time, i.e.

$$\tau_l = \int_0^\infty C^{(l)}(t)\, dt \tag{12.12.10}$$

A more demanding test of the Debye approximation is to plot $-[l(l+1)]^{-1} \log C^{(l)}(t)$ as a function of t. If the approximation were exact, the curves for different l would be coincident at all times. In practice, there is a fanning-out of the curves at large t, with the rate of decay becoming slower with increasing l; this is true, for example, of the molecular dynamics results of Figure 12.10 when these are plotted in the appropriate form. Under such circumstances, the right-hand side of Eqn (12.12.9) provides an upper bound on the ratio τ_l/τ_{l+1}.

The other main weakness of the Debye approximation is its neglect of the fact that, on a sufficiently short timescale, molecules rotate freely. It therefore cannot account for the quadratic time dependence of the reorientational correlation functions at small t. A more complete theory must also describe correctly the details of the transition to the long-time, quasiexponential behaviour. In the logarithmic plots of Figure 12.10, the transition region is marked by an "elbow" in the curve of $\log C^{(l)}(t)$ versus t; in other words, the (negative) gradient of the curve approaches its asymptotic value form above. It is possible to discuss the behaviour in short, intermediate and long-time regimes in a unified way by relating the form of the reorientational correlation functions to that of the angular-velocity autocorrelation function. By analogy with Eqns (7.2.7) and (7.2.8), we define the rotational-diffusion constant of a linear molecule as

$$D_R = \frac{k_B T}{I} \lim_{t\to\infty} \int_0^t \left(1 - \frac{s}{t}\right) C_\omega(s)\, ds \tag{12.12.11}$$

where

$$C_\omega(t) = \frac{\langle \boldsymbol{\omega}_i(t) \cdot \boldsymbol{\omega}_i \rangle}{\langle |\boldsymbol{\omega}_i|^2 \rangle} = \frac{I}{2k_B T} \langle \boldsymbol{\omega}_i(t) \cdot \boldsymbol{\omega}_i \rangle \tag{12.12.12}$$

is the normalized angular-velocity autocorrelation function. Then substitution of (12.12.11) in Eqn (12.12.8) gives the required expression for $C^{(l)}(t)$ in terms of $C_\omega(t)$:

$$\log C^{(l)}(t) \simeq -l(l+1)\left(\frac{k_B T}{I}\right) \int_0^t (t-s)\, C_\omega(s)\, ds \tag{12.12.13}$$

Equation (12.12.13) can be derived in a systematic way via a cumulant expansion of the reorientational time propagator (Lynden-Bell, 1984), but the intuitive argument we have used is sufficient for our purposes.

Results obtained from (12.12.13) provide a fair fit to the reorientational correlation functions of a variety of simulated liquids (Lynden-Bell and McDonald, 1981), but the main merit of the approximation is the fact that it contains the correct short-time behaviour, yet also goes over to the Debye model at long times. Let τ_ω be the integral correlation time for the angular velocity, defined as

$$\tau_\omega = \int_0^\infty C_\omega(t)\, dt \tag{12.12.14}$$

At times $t \ll \tau_\omega$, it is legitimate to set $C_\omega(t) \simeq 1$. Then Eqn (12.12.13) becomes

$$\log C^{(l)}(t) \simeq -l(l+1)\left(\frac{k_B T}{I}\right)\frac{t^2}{2} \tag{12.12.15}$$

in agreement with the exact result of Eqn (12.12.5). In the opposite limit, $t \gg \tau_\omega$, Eqn (12.12.13) becomes

$$\log C^{(l)}(t) \simeq -l(l+1)\left(\frac{k_B T}{I}\right)\tau_\omega t \tag{12.12.16}$$

which is equivalent to the Debye approximation (12.12.8) with the identification $D_R = (k_B T/I)\tau_\omega$. Finally, the behaviour at intermediate times can be related to the detailed shape of the function $C_\omega(t)$. Differentiating Eqn (12.12.13) twice with respect to time, we find that

$$\frac{d^2 \log C^{(l)}(t)}{dt^2} = -l(l+1)\left(\frac{k_B T}{I}\right) C_\omega(t) \tag{12.12.17}$$

The results of molecular dynamics calculation show that in liquids in which the intermolecular torques are strong, the function $C_\omega(t)$ decays rapidly at short times and then becomes negative. The change of sign occurs because, on average, the direction of the angular-velocity vector is soon reversed; the behaviour is similar to that seen in the linear-velocity autocorrelation function at high densities and low temperatures (see Figure 7.1). Equation (12.12.17) shows that a change in sign of $C_\omega(t)$ corresponds to a point of inflexion in $\log C^{(l)}(t)$ of the type visible in Figure 12.10. This explains why "elbows" in the reorientational correlation functions are such a common feature of high-torque fluids. At higher temperatures and lower densities, or for molecules that are nearly spherical, $C_\omega(t)$ decays monotonically and roughly exponentially. Under these conditions, the curves of $-\log C^{(l)}(t)$ are expected to be concave downwards until they reach their asymptotic, linear form.

Although Eqn (12.12.13) is useful in rationalizing certain features of the reorientational correlation functions, the description it provides is incomplete to the extent that it involves the unknown function $C_\omega(t)$. For many large molecules, the simple Debye diffusion model provides a fair fit to the experimental results, the rotational-diffusion constant or, for non-linear molecules, the elements of the rotational-diffusion tensor being treated as adjustable parameters. A large number of approximations have been devised to treat those cases in which the Debye model is inadequate. Many of these are expressible in terms of simple approximations for the appropriate memory functions. One possibility that is physically appealing is to write $N^{(l)}(t)$ in a form that interpolates between its free-rotor form and the diffusion limit. There are a number of ways in which this can be achieved. The simplest, which corresponds to the "extended-diffusion" (or "J-diffusion") model of Gordon (1966), is to write $N^{(l)}(t)$ as

$$N^{(l)}(t) = N_0^{(l)}(t) \exp(-\alpha t) \qquad (12.12.18)$$

where, for linear molecules, $N_0^{(l)}(t)$ is the (known) memory function for the free, linear rotor. Equation (12.12.18) can be derived from a physical model in which molecules rotate freely in the intervals between uncorrelated collisions; at each collision the angular velocity is assumed to be randomized in both direction and magnitude (Gordon, 1966; Chandler, 1974). The parameter α is the mean collision frequency, and the assumptions underlying the model imply that the angular-velocity autocorrelation function decays exponentially with a characteristic time equal to α^{-1}; free-rotor behaviour and the Debye approximation are recovered from the model by taking the limits, respectively, of $\alpha \to 0$ and $\alpha \to \infty$. As $C_\omega(t)$ is often far from exponential, some improvement can be anticipated by rewriting Eqn (12.12.18) as

$$N^{(l)}(t) = N_0^{(l)} C_\omega(t) \qquad (12.12.19)$$

where $C_\omega(t)$ is the exact angular-velocity autocorrelation function (Steele,

1981); since a knowledge of $C_\omega(t)$ is assumed, the problem of implementation is the same as for the approximation (12.12.13). In a high-torque fluid, $C_\omega(t)$ decays much faster than $N_0^{(l)}(t)$; it is then permissible to replace the latter by its initial value, given in Eqn (12.12.7). This leads to a new approximation of the form

$$N^{(l)}(t) = l(l+1)\left(\frac{k_B T}{I}\right) C_\omega(t) \tag{12.12.20}$$

which is a result obtained in a different way by Nee and Zwanzig (1970). Equation (12.12.20) implies that the reorientational correlation times for different values of l are related in the manner of the Debye approximation, Eqn (12.12.9), and also that

$$l(l+1)\left(\frac{k_B T}{I}\right) \tau_l \tau_\omega = 1 \tag{12.12.21}$$

Equation (12.12.21) has been tested for the case $l = 2$ against results obtained both experimentally and by simulation for liquid carbon disulphide (Madden and Tildesley, 1983), and is found to be satisfied within 20% throughout the liquid range; experimental estimates of τ_ω are obtained from measurements of NMR spin-rotation times. Finally, a more elaborate memory-function scheme has been proposed (Kivelson and Keyes, 1972) that corresponds to a continued-fraction expansion of $\tilde{N}^{(l)}(\omega)$ truncated at second order. This procedure is equivalent to writing a memory-function equation for a three-component vector built from suitably orthogonalized combinations of the variables $\mathbf{u}_i$, $\dot{\mathbf{u}}_i$ and $\ddot{\mathbf{u}}_i$, and is therefore called the "three-variable" theory.

We have kept the discussion of memory-function models deliberately brief for the good reason that none of the many approximations that have been suggested has proved to be satisfactory either for any large group of molecules or for any particular molecule over a wide range of temperature and density. The failure to develop an adequate theory is striking in view of the apparent simplicity in structure of the reorientational correlation functions.

We have focussed until now on the reorientational motion of single molecules. There are, in addition, a number of collective reorientational correlation functions that are of experimental significance and are also many-particle generalizations of single-molecule functions. There is considerable interest in establishing a relation between the corresponding collective and single-particle functions and, in particular, between the two correlation times, since this allows a connection to be made between the results of very different experiments. As an important example, we consider first the collective motions that determine the frequency-dependent dielectric behaviour of a polar fluid, as described by the complex dielectric permittivity $\varepsilon(\omega)$ introduced in Section 12.5. The quantities of interest in the study of dielectric relaxation are the correlation functions and associated power spectra of the longitudinal (l) and transverse (t) components of the polarization density, i.e.

$$C_l(k, t) = \frac{\langle M_{\mathbf{k}}^z(t) M_{-\mathbf{k}}^z \rangle}{\langle |M_{\mathbf{k}}^z|^2 \rangle} \tag{12.12.22}$$

and

$$C_t(k, t) = \frac{\langle M_{\mathbf{k}}^x(t) M_{-\mathbf{k}}^x \rangle}{\langle |M_{\mathbf{k}}^x|^2 \rangle} \tag{12.12.23}$$

where we have followed the usual convention that $\mathbf{k}$ is parallel to the z-axis. The functions $C_l(k, t)$ and $C_t(k, t)$ are collective analogues, generalized to non-zero k, of the single-molecule function $C^{(1)}(t)$. It follows from Eqns (12.5.9) and (12.5.12) that the long-wavelength limits of the Laplace transforms $\tilde{C}_l(k, \omega)$ and $\tilde{C}_t(k, \omega)$ are related to $\varepsilon(\omega)$ by

$$\frac{4\pi\beta}{V} \lim_{\mathbf{k}\to 0} \langle |M_{\mathbf{k}}^z|^2 \rangle [1 + i\omega \tilde{C}_l(k, \omega)] = \frac{\varepsilon(\omega) - 1}{\varepsilon(\omega)} \tag{12.12.24a}$$

$$\frac{4\pi\beta}{V} \lim_{\mathbf{k}\to 0} \langle |M_{\mathbf{k}}^x|^2 \rangle [1 + i\omega \tilde{C}_t(k, \omega)] = \varepsilon(\omega) - 1 \tag{12.12.24b}$$

Following the methods used by Madden and Kivelson (1984), we begin by writing memory-function equations for $C_l(k, t)$ and $C_t(k, t)$ in the form

$$\tilde{C}_l(k, \omega) = \frac{1}{-i\omega + \tilde{N}_l(k, \omega)} \tag{12.12.25a}$$

$$\tilde{C}_t(k, \omega) = \frac{1}{-i\omega + \tilde{N}_t(k, \omega)} \tag{12.12.25b}$$

The initial values of the memory functions $N_l(k, t)$ and $N_t(k, t)$ in the limit $k \to 0$ can be obtained from the general property (9.1.31) and the limiting behaviour described by (12.12.24), taken for $\omega = 0$:

$$\lim_{\mathbf{k}\to 0} N_l(k, t=0) = \lim_{\mathbf{k}\to 0} \frac{\langle |\dot{M}_{\mathbf{k}}^z|^2 \rangle}{\langle |M_{\mathbf{k}}^z|^2 \rangle} = \frac{4\pi\beta\varepsilon}{3V(\varepsilon - 1)} \langle |\dot{\mathbf{M}}|^2 \rangle \tag{12.12.26a}$$

$$\lim_{\mathbf{k}\to 0} N_t(k, t=0) = \lim_{\mathbf{k}\to 0} \frac{\langle |\dot{M}_{\mathbf{k}}^x|^2 \rangle}{\langle |M_{\mathbf{k}}^x|^2 \rangle} = \frac{4\pi\beta}{3V(\varepsilon - 1)} \langle |\dot{\mathbf{M}}|^2 \rangle \tag{12.12.26b}$$

where $\varepsilon \equiv \varepsilon(0)$. In deriving these results, we have exploited the fact that the different components of $\langle |\dot{\mathbf{M}}_{\mathbf{k}}|^2 \rangle$ (unlike those of $\langle |\mathbf{M}_{\mathbf{k}}|^2 \rangle$) are all equivalent and, in particular, that $\lim_{\mathbf{k}\to 0} \langle |\dot{M}_{\mathbf{k}}^\alpha|^2 \rangle = \frac{1}{3} \langle |\dot{\mathbf{M}}|^2 \rangle$, where $\alpha = x, y$ or z, and $\dot{\mathbf{M}} \equiv \dot{\mathbf{M}}_{\mathbf{k}=0}$.

The form of Eqn (12.12.26) makes it convenient to write the memory functions at long wavelength as

$$\lim_{\mathbf{k}\to 0} \tilde{N}_l(k, \omega) = \frac{\varepsilon \tilde{R}_l(\omega)}{\varepsilon - 1} \tag{12.12.27a}$$

$$\lim_{\mathbf{k}\to 0} \tilde{N}_t(k, \omega) = \frac{\tilde{R}_t(\omega)}{\varepsilon - 1} \tag{12.12.27b}$$

It is clear from comparison of (12.12.26) and (12.12.27) that $R_l(t=0) = R_t(t=0) = (4\pi\beta/3V)\langle|\dot{\mathbf{M}}|^2\rangle$. More generally, if Eqns (12.12.24a) and (12.12.24b) are to be consistent with each other in the sense of giving the same result for $\varepsilon(\omega)$, some straightforward algebra shows that $R_l(t)$ and $R_t(t)$ must be the same for all t. Thus

$$R_l(t) = R_t(t) = R(t), \quad \text{say} \tag{12.12.28}$$

This has the immediate consequence that in the long-wavelength limit the correlation times for the longitudinal and transverse functions differ by a factor ε, i.e. $\lim_{\mathbf{k}\to 0} \tilde{C}_l(k, 0)/\tilde{C}_t(k, 0) = \varepsilon^{-1}$, or

$$\lim_{\mathbf{k}\to 0} \int_0^\infty C_l(k, t)\, \mathrm{d}t = \frac{1}{\varepsilon} \lim_{\mathbf{k}\to 0} \int_0^\infty C_t(k, t)\, \mathrm{d}t \tag{12.12.29}$$

The significance of (12.12.28) is the fact that the memory function $R(t)$, like $\varepsilon(\omega)$ itself, is independent of the shape of the sample.

In the problem of dielectric relaxation, the diffusion approximation analogous to (12.12.8) corresponds to setting

$$R(t) = R(0)\delta(t) = \frac{4\pi\beta}{3V} \langle|\dot{\mathbf{M}}|^2\rangle\delta(t) \tag{12.12.30}$$

so that both $\tilde{N}_l(k, \omega)$ and $\tilde{N}_t(k, \omega)$ are assumed to be independent of frequency in the limit $\mathbf{k} \to 0$. If we define a characteristic time τ_D as

$$\tau_D = \frac{3V}{4\pi\beta} \frac{\varepsilon - 1}{\langle|\dot{\mathbf{M}}|^2\rangle} \tag{12.12.31}$$

it follows from Eqns (12.12.25) and (12.12.27) that

$$\lim_{\mathbf{k}\to 0} C_l(k, t) = \exp(-\varepsilon t/\tau_D) \tag{12.12.32a}$$

$$\lim_{\mathbf{k}\to 0} C_t(k, t) = \exp(-t/\tau_D) \tag{12.12.32b}$$

which represents a special case of the general result in (12.12.29). Computer simulations of strongly polar fluids confirm that the longitudinal and transverse correlation functions at small **k** do decay on very different time scales, and that the ratio of the correlation functions is approximately equal to the value of ε derived from fluctuations in the mean-square moment of the sample (Pollock and Alder, 1981; Edwards *et al.*, 1984). The transverse function is also approximately exponential in form, with a decay time that is only weakly dependent on $\mathbf{k}$, in qualitative agreement with (12.12.32a), but the longitudinal function has oscillations at large t; to describe these oscillations, it is necessary to allow for some frequency dependence of the memory function. The approximation for $\varepsilon(\omega)$ corresponding to (12.12.32) is

$$\frac{\varepsilon(\omega)-1}{\varepsilon(0)-1}=\frac{1}{1-i\omega\tau_{\mathrm{D}}} \qquad (12.12.33)$$

This is an expression much used in the analysis of experimental data on $\varepsilon(\omega)$; in this context, τ_{D} is invariably called the Debye relaxation time. A feature of the approximation (12.12.33) is the fact that the curve, or Cole–Cole plot, of the real versus the imaginary part of $\varepsilon(\omega)$ is a semicircle with a maximum at a frequency such that $\omega\tau_{\mathrm{D}}=1$. Many real liquids have Cole–Cole plots that are approximately semicircular; because of its neglect of short-time, inertial effects, the diffusion approximation is least satisfactory at high frequencies, where the deviations from (12.12.33) are mostly to be found. However, as in the case of the single-molecule problem, it is difficult to develop an alternative theory that has a wide applicability.

One of the main goals of dielectric-relaxation theory is to relate the decay times that characterize the collective functions (12.12.22) and (12.12.23) to the single-molecule correlation time τ_1 defined by Eqn (12.12.10). The necessary link can be established by postulating some relationship between the memory functions $R(t)$ and $N^{(1)}(t)$ in a spirit similar to the approximations for $S(k,\omega)$ discussed in Section 9.2. A simple but useful result is obtained by supposing that the two memory functions have the same time dependence, but also have their correct initial values. It follows from Eqn (9.1.31) that

$$\begin{aligned} N^{(1)}(t=0)=\langle|\dot{\mathbf{u}}_i|^2\rangle&=\frac{1}{N\mu^2}\langle|\dot{\mathbf{M}}|^2\rangle \\ &=\frac{R(0)}{3y} \end{aligned} \qquad (12.12.34)$$

where y is defined by Eqn (12.5.16). If, for simplicity, we adopt the diffusion model, we find immediately from the definition (12.12.31) that

$$\tau_{\mathrm{D}}=\left(\frac{\varepsilon-1}{3y}\right)\tau_1 \qquad (12.12.35)$$

or, after substitution from the Kirkwood formula (12.5.15):

$$\tau_D = \left(\frac{3\varepsilon g_K}{2\varepsilon + 1}\right)\tau_1 \tag{12.12.36}$$

This result relates the dielectric relaxation time to the correlation time measured by infrared spectroscopy in a form determined solely by static dielectric properties of the fluid. If the further approximation is made of setting $g_K = 1$, Eqn (12.12.36) reduces to a result known as the Glarum–Powles relation (Powles, 1953).

The same general approach followed in the dielectric case can be used to discuss the collective effects that are monitored in depolarized light-scattering experiments (Berne and Pecora, 1976). In the absence of any collision-induced contribution, the spectrum of depolarized light scattered by a fluid of linear molecules is determined by the Fourier transform of the correlation function defined as

$$C_2^{\mathrm{coll}}(t) = \frac{1}{N}\left\langle P_2\left[\sum_{i=1}^{N} \mathbf{u}_i(t) \cdot \sum_{j=1}^{N} \mathbf{u}_j(0)\right]\right\rangle \tag{12.12.37}$$

This is the collective analogue of the single-particle function $C^{(2)}(t)$, which is measurable by Raman scattering experiments. Since the total scattered intensity is proportional to the initial value $C_2^{\mathrm{coll}}(t=0)$, the effect of orientational correlations between molecules is to enhance the intensity by a factor $(1+G_2)$, where G_2 is the angular order parameter defined by Eqn (12.1.8). In the diffusion approximation, both $C_2^{\mathrm{coll}}(t)$ and $C^{(2)}(t)$ decay exponentially with characteristic times that we denote, respectively, by τ_2^{coll} and τ_2. If the argument that leads to Eqn (12.12.34) is repeated, the two characteristic times are found to be related by

$$\tau_2^{\mathrm{coll}} = (1+G_2)\tau_2 \tag{12.12.38}$$

Simulations of liquid carbon disulphide (Madden and Tildesley, 1983) reveal an approximate equality between τ_2^{coll} and τ_2 that persists down to the triple point. This would suggest that orientational correlations, as measured by G_2, are weak, but this is not the case. Close to the triple point, $G_2 \simeq 0.2$, and the near equality of τ_2^{coll} and τ_2 arises because the dynamical correlations that are neglected in the derivation of (12.12.38) act in such a way as to offset the effect of the static correlations contained in G_2. It is therefore unlikely that useful information on G_2 can be obtained from a comparison of light-scattering and Raman data based solely on the approximation (12.12.38).

References

Abascal, J. L. F., Martin, C., Lombardero, M., Vazquez, J., Bañon, A. and Santamaria, J. (1985). *J. Chem. Phys.* **82**, 2445.
Abe, R. (1959). *Prog. Theor. Phys.* **22**, 213.
Abernethy, G. M., Dixon, M. and Gillan, M. J. (1981). *Phil. Mag. B* **43**, 1113.
Abraham, F. (1978). *J. Chem. Phys.* **68**, 3713.
Abramo, M. C., Parrinello, M. and Tosi, M. P. (1973). *J. Non-Metals* **2**, 67.
Abramo, M. C., Caccamo, C., Pizzimenti, G., Parrinello, M. and Tosi, M. P. (1978). *J. Chem. Phys.* **68**, 2889.
Abramo, M. C., Caccamo, C. and Pizzimenti, G. (1983). *J. Chem. Phys.* **78**, 357.
Adams, D. J. (1974). *Mol. Phys.* **28**, 1241.
Adams, D. J. (1975). *Mol. Phys.* **29**, 307.
Adams, D. J. (1979). *Chem. Phys. Lett.* **62**, 329.
Adams, D. J. and McDonald, I. R. (1975). *J. Chem. Phys.* **63**, 1900.
Adams, D. J. and McDonald, I. R. (1976). *Mol. Phys.* **32**, 931.
Adams, E. M., McDonald, I. R. and Singer, K. (1977). *Proc. Roy. Soc.* **A357**, 37.
Ailawadi, N. K., Rahman, A. and Zwanzig, R. (1971). *Phys. Rev. A* **4**, 1616.
Akcasu, A. Z. and Daniels, E. (1970). *Phys. Rev. A* **2**, 962.
Akcasu, A. Z. and Duderstadt, J. J. (1969). *Phys. Rev.* **188**, 479.
Akcasu, A. Z., Corngold, N. and Duderstadt, J. J. (1970). *Phys. Fluids* **13**, 2213.
Albano, A. M., Bedeaux, D. and Mazur, P. (1975). *Physica* **80A**, 96.
Alblas, B. P., van der Lugt, W., Mensies, O. and van Dyk, C. (1981). *Physica* **106B**, 22.
Alder, B. J. (1964a). *Phys. Rev. Lett.* **12**, 317.
Alder, B. J. (1964b). *J. Chem. Phys.* **40**, 2724.
Alder, B. J. and Alley, W. E. (1984). *Phys. Today* **37**, 56.
Alder, B. J. and Hecht, C. E. (1969). *J. Chem. Phys.* **50**, 2032.
Alder, B. J. and Wainwright, T. E. (1957). *J. Chem. Phys.* **27**, 1208.
Alder, B. J. and Wainwright, T. E. (1959). *J. Chem. Phys.* **31**, 459.
Alder, B. J. and Wainwright, T. E. (1967). *Phys. Rev. Lett.* **18**, 988.
Alder, B. J. and Wainwright, T. E. (1968). *J. Phys. Soc. Japan Suppl.* **26**, 267.
Alder, B. J. and Wainwright, T. E. (1970). *Phys. Rev. A* **1**, 18.
Alder, B. J., Gass, D. M. and Wainwright, T. E. (1970). *J. Chem. Phys.* **53**, 3813.
Alder, B. J., Young, D. A. and Mark, M. A. (1972). *J. Chem. Phys.* **56**, 2013.
Alder, B. J., Strauss, H. L. and Weis, J. J. (1973). *J. Chem. Phys.* **59**, 1002.
Alley, W. E. and Alder, B. J. (1983). *Phys. Rev. A* **27**, 3158.
Alley, W. E., Alder, B. J. and Yip, S. (1983). *Phys. Rev. A* **27**, 3174.
Andersen, H. C. (1980). *J. Chem. Phys.* **72**, 2384.
Andersen, H. C. and Chandler, D. (1970). *J. Chem. Phys.* **53**, 547.

Andersen, H. C. and Chandler, D. (1972). *J. Chem. Phys.* **57**, 1918.
Andersen, H. C., Weeks, J. D. and Chandler, D. (1971). *Phys. Rev. A* **4**, 1597.
Andersen, H. C., Chandler, D. and Weeks, J. D. (1972). *J. Chem. Phys.* **56**, 3812.
Andersen, H. C., Chandler, D. and Weeks, J. D. (1976). *Adv. Chem. Phys.* **34**, 105.
Ascarelli, P. and Harrison, R. J. (1969). *Phys. Rev. Lett.* **22**, 285.
Ashcroft, N. W. (1966). *J. Phys. C* **1**, 232.
Ashcroft, N. W. and Langreth, D. C. (1967). *Phys. Rev.* **156**, 685.
Ashcroft, N. W. and Lekner, J. (1966). *Phys. Rev.* **145**, 83.
Ashcroft, N. W. and Stroud, D. (1978). *Sol. State Phys.* **33**, 1.
Ashurst, W. T. and Hoover, W. G. (1975). *Phys. Rev. A* **11**, 658.
Axilrod, B. M. and Teller, E. (1943). *J. Chem. Phys.* **11**, 299.
Bailey, R. T. (1974). *In* "Molecular Spectroscopy" (D. A. Long, D. J. Millen and R. F. Barrow, eds), Vol. 2. The Chemical Society, London.
Ballone, P., Pastore, G. and Tosi, M. P. (1984). *J. Chem. Phys.* **81**, 3174.
Bansal, R. and Pathak, K. N. (1977). *Phys. Rev. A* **15**, 2519.
Barker, J. A. (1953). *Proc. Roy. Soc.* **A219**, 367.
Barker, J. A. and Henderson, D. (1967). *J. Chem. Phys.* **47**, 4714.
Barker, J. A. and Henderson, D. (1972). *Ann. Rev. Phys. Chem.* **23**, 439.
Barker, J. A. and Henderson, D. (1976). *Rev. Mod. Phys.* **48**, 587.
Barker, J. A. and Watts, R. O. (1973). *Mol. Phys.* **26**, 789.
Barker, J. A., Fisher, R. A. and Watts, R. O. (1971). *Mol. Phys.* **21**, 657.
Barojas, J., Levesque, D. and Quentrec, B. (1973). *Phys. Rev. A* **7**, 1092.
Bartlett, M. S. (1966). "An Introduction to Stochastic Processes". Cambridge University Press, Cambridge.
Battaglia, M. R., Cox, T. I. and Madden, P. A. (1979). *Mol. Phys.* **37**, 1413.
Baus, M. (1977). *Physica* **88A**, 319, 336 and 591.
Baus, M. and Hansen, J. P. (1980). *Phys. Rep.* **59**, 1.
Baxter, R. J. (1968). *Aust. J. Phys.* **21**, 563.
Baym, G. (1964). *Phys. Rev.* **135**, 1691.
Bell, H., Moeller-Wenghoffer, H., Kollmar, A., Stockmeyer, R., Springer, R. and Stiller, H. (1975). *Phys. Rev. A* **11**, 316.
Bennett, C. H. (1976). *J. Comput. Phys.* **22**, 245.
Berendsen, H. J. C. (1986). *In* "Molecular Dynamics Simulation of Statistical-Mechanical Systems" (G. Ciccotti and W. G. Hoover, eds). North-Holland, Amsterdam.
Bergé, P., Calmettes, P., Dubois, M. and Laj, C. (1970). *Phys. Rev. Lett.* **24**, 89.
Bergeon, R. (1958). *J. Rech. CNRS* **14**, 171.
Berne, B. J. and Pecora, R. (1976). "Dynamic Light Scattering". John Wiley, New York.
Berne, B. J., Boon, J. P. and Rice, S. A. (1966). *J. Chem. Phys.* **45**, 1086.
Bernu, B. (1983). *Physica* **122A**, 129.
Bernu, B. (1984). Thesis, Université Pierre et Marie Curie, Paris.
Bernu, B. and Hansen, J. P. (1982). *Phys. Rev. Lett.* **48**, 1375.
Beysens, D., Garrabos, Y. and Zalczer, G. (1980). *Phys. Rev. Lett.*, **45**, 403.
Bhatia, A. B. and March, N. H. (1975). *J. Phys. F* **5**, 1100.
Bhatia, A. B. and Thornton, D. E. (1970). *Phys. Rev. B* **2**, 3004.
Bhatia, A. B., Hargrove, W. H. and March, N. H. (1973). *J. Phys. C* **6**, 621.
Bien, T., Possiel, M., Döge, G., Yarwood, J. and Arnold, K. E. (1981). *Chem. Phys.* **56**, 203.
Biggin, S. and Enderby, J. E. (1981). *J. Phys. C* **14**, 3577.

Biggin, S. and Enderby, J. E. (1982). *J. Phys. C* **15**, L305.
Blum, L. (1972). *J. Chem. Phys.* **57**, 1862.
Blum, L. (1975). *Mol. Phys.* **30**, 1529.
Blum, L. (1980a). *In* "Theoretical Chemistry: Advances and Perspectives", Vol. 5. Academic Press, New York.
Blum, L. (1980b). *J. Stat. Phys.* **22**, 661.
Blum, L. and Høye, J. S. (1977). *J. Phys. Chem.* **81**, 1311.
Blum, L. and Lebowitz, J. L. (1969). *Phys. Rev.* **185**, 273.
Blum, L. and Torruella, A. J. (1972). *J. Chem. Phys.* **56**, 303.
Blum, L., Gruber, C., Lebowitz, J. L. and Martin, P. A. (1982). *Phys. Rev. Lett.* **48**, 1769.
Bogolyubov, N. N. (1946). *J. Phys. USSR* **10**, 257.
Böhm, H. J., Lynden-Bell, R. M., Madden, P. A. and McDonald, I. R. (1984a). *Mol. Phys.* **51**, 761.
Böhm, H. J., Meissner, C. and Ahlrichs, R. (1984b). *Mol. Phys.* **53**, 651.
Boley, C. D. (1972). *Phys. Rev. A* **2**, 986.
Boley, C. D. (1975). *Phys. Rev. A* **11**, 328.
Boon, J. P. and Yip, S. (1980). "Molecular Hydrodynamics". McGraw-Hill, New York.
Born, M. and Green, M. S. (1949). "A General Kinetic Theory of Liquids". Cambridge University Press, Cambridge.
Bosse, J. and Munakata, T. (1982). *Phys. Rev. A* **25**, 2763.
Bosse, J., Götze, W. and Lücke, M. (1978a). *Phys. Rev. A* **17**, 434.
Bosse, J., Götze, W. and Lücke, M. (1978b). *Phys. Rev. A* **17**, 447.
Bosse, J., Götze, W. and Lücke, M. (1978c). *Phys. Rev. A* **18**, 1176.
Bosse, J., Götze, W. and Zippelius, A. (1978d). *Phys. Rev. A* **18**, 1214.
Bosse, J., Götze, W. and Lücke, M. (1979). *Phys. Rev. A* **20**, 1603.
Böttcher, C. J. F. (1973). "Theory of Electric Polarization". Elsevier, Amsterdam.
Boublik, T. (1983). *In* "Molecular-Based Study of Fluids" (J. M. Haile and G. A. Mansoori, eds), Ch. 8. American Chemical Society, Washington.
Bouiller, A., Boon, J. P. and Deguent, P. (1978). *J. Phys. (Paris)* **39**, 159.
Bredig, M. A. (1964). *In* "Molten Salt Chemistry" (M. Blander, ed.). Wiley Interscience, New York.
Broyles, A. A. (1960). *J. Chem. Phys.* **33**, 456.
Brush, S. G., Sahlin, H. L. and Teller, E. (1966). *J. Chem. Phys.* **45**, 2102.
Buff, F. P., Lovett, R. A. and Stillinger, F. H. (1965). *Phys. Rev. Lett.* **15**, 621.
Buyers, W. J. L., Sears, V. F., Lonngi, P. A. and Lonngi, D. A. (1975). *Phys. Rev. A* **11**, 697.
Cahn, J. W. (1977). *J. Chem. Phys.* **66**, 3667.
Card, D. N. and Valleau, J. P. (1970). *J. Chem. Phys.* **52**, 6232.
Carnahan, N. F. and Starling, K. E. (1969). *J. Chem. Phys.* **51**, 635.
Carneiro, K. (1976). *Phys. Rev. A* **14**, 517.
Carnie, S. L. and Chan, D. Y. C. (1981). *Chem. Phys. Lett.* **77**, 437.
Castresana, J. I., Mazenko, G. F. and Yip, S. (1977). *Ann. Phys.* **103**, 1.
Chabrier, G. and Hansen, J. P. (1983). *Mol. Phys.* **50**, 901.
Chandler, D. (1973). *J. Chem. Phys.* **59**, 2742.
Chandler, D. (1974). *J. Chem. Phys.* **60**, 3508.
Chandler, D. (1976). *Mol. Phys.* **31**, 1213.
Chandler, D. (1977). *J. Chem. Phys.* **67**, 1113.
Chandler, D. (1978). *Faraday Disc. Chem. Soc.* **66**, 74.

Chandler, D. and Andersen, H. C. (1972). *J. Chem. Phys.* **57**, 1930.
Chandler, D. and Pratt, L. R. (1976). *J. Chem. Phys.* **65**, 2925.
Chandler, D. and Richardson, D. M. (1983). *J. Phys. Chem.* **87**, 2060.
Chandler, D., Silbey, R. and Ladanyi, B. M. (1982a). *Mol. Phys.* **46**, 1335.
Chandler, D., Joslin, C. G. and Deutch, J. M. (1982b). *Mol. Phys.* **47**, 871.
Chapela, G. A., Saville, G. S., Thompson, S. M. and Rowlinson, J. S. (1977). *J. Chem. Soc. Faraday Trans. II* **73**, 1133.
Chapman, S. and Cowling, T. G. (1970). "The Mathematical Theory of Non-Uniform Gases", 3rd edn. Cambridge University Press, Cambridge.
Chaturvedi, D. K., Marconi, U. M. B. and Tosi, M. P. (1980). *Nuov. Cim. B* **57**, 319.
Chaturvedi, D. K., Rovere, M., Senatore, G. and Tosi, M. P. (1981). *Physica* **111B**, 11.
Chen, M., Henderson, D. and Barker, J. A. (1969). *Can. J. Phys.* **54**, 703.
Chester, G. V. (1963). *Rep. Prog. Phys.* **26**, 411.
Cheung, P. S. Y. and Powles, J. G. (1975). *Mol. Phys.* **30**, 921.
Chihara, J. (1973). *Prog. Theor. Phys.* **50**, 409.
Chung, C. H. and Yip, S. (1969). *Phys. Rev.* **182**, 323.
Ciccotti, G. and Jacucci, G. (1975). *Phys. Rev. Lett.* **35**, 789.
Ciccotti, G., Jacucci, G. and McDonald, I. R. (1976). *Phys. Rev. A* **13**, 426.
Ciccotti, G., Jacucci, G. and McDonald, I. R. (1979). *J. Stat. Phys.* **21**, 1.
Ciccotti, G., Ferrario, M. and Ryckaert, J. P. (1982). *Mol. Phys.* **47**, 1253.
Cohen, C., Sutherland, J. W. H. and Deutch, J. M. (1971). *Phys. Chem. Liq.* **2**, 213.
Cohen, E. G. D. (1983). *Physica* **118A**, 17.
Cohen, M. L. and Heine, V. (1970). *Sol. State Phys.* **24**, 37.
Cook, D. and Rowlinson, J. S. (1953). *Proc. Roy. Soc.* **A219**, 405.
Copestake, A. P., Evans, R., Ruppersberg, H. and Schirmacher, W. (1983). *J. Phys. F* **13**, 1993.
Copley, J. R. D. and Dolling, G. (1978). *J. Phys. C* **11**, 1259.
Copley, J. R. D. and Lovesey, S. W. (1975). *Rep. Prog. Phys.* **38**, 461.
Copley, J. R. D. and Rahman, A. (1976). *Phys. Rev. A* **13**, 2276.
Copley, J. R. D. and Rowe, J. M. (1974a). *Phys. Rev. Lett.* **32**, 49.
Copley, J. R. D. and Rowe, J. M. (1974b). *Phys. Rev. A* **9**, 1656.
Cukier, R. I. and Mehaffey, J. R. (1978). *Phys. Rev. A* **18**, 1202.
Cummings, P. T. and Stell, G. (1982). *Mol. Phys.* **46**, 383.
Cummings, P. T. and Sullivan, D. E. (1982). *Mol. Phys.* **46**, 665.
Cummings, P. T., Gray, C. G. and Sullivan, D. E. (1981). *J. Phys. A* **14**, 1483.
Cummings, P. T., Morriss, G. P. and Wright, C. C. (1982). *J. Phys. Chem.* **86**, 1696.
Cyrot-Lackmann, F. and Desré, P. (eds) (1980). "Fourth International Conference on Liquid Metals", *J. Phys. (Paris)* **41**, C8.
Dagens, L., Rasolt, M. and Taylor, R. (1975). *Phys. Rev. B* **11**, 2726,
Day, R. S., Sun, F. and Cutler, P. H. (1979). *Phys. Rev. A* **19**, 328.
Debye, P. and Hückel, E. (1923). *Zeit. für Phys.* **24**, 185 and 305.
de Dominicis, C. (1962). *J. Math. Phys.* **3**, 983.
de Dominicis, D. (1963). *J. Math. Phys.* **4**, 255.
de Gennes, P. G. (1959). *Physica* **25**, 825.
deGroot, S. R. and Mazur, P. (1962). "Non-equilibrium Thermodynamics", North-Holland, Amsterdam.
deLeeuw, S. W. (1978). *Mol. Phys.* **36**, 103.
deLeeuw, S. W., Perram, J. W. and Smith, E. R. (1980). *Proc. Roy. Soc.* **A373**, 27.
deLeeuw, S. W., Perram, J. W. and Smith, E. R. (1983). *Proc. Roy. Soc.* **A388**, 177.

Desai, R. C. (1971). *Phys. Rev. A* **3**, 320.
deSchepper, I. M. and Cohen, E. G. D. (1980). *Phys. Rev. A* **22**, 287.
deSchepper, I. M. and Cohen, E. G. D. (1982). *J. Stat. Phys.* **27**, 223.
deSchepper, I. M., Verkerk, P., van Well, A. and deGraaf, L. A. (1983). *Phys. Rev. Lett.* **50**, 974.
Dickinson, E. (1983). Annual Reports C, The Royal Society of Chemistry, London.
Dieterich, W., Fulde, P. and Peschel, I. (1980). *Adv. Phys.* **29**, 527.
Dixon, M. (1983a). *Phil. Mag. B* **47**, 509 and 531.
Dixon, M. (1983b). *Phil. Mag. B* **48**, 13.
Dixon, M. and Gillan, M. J. (1981). *Phil. Mag. B* **43**, 1099.
Doob, J. L. (1942). *Ann. Math.* **43**, 351.
Dorfman, J. R. (1981). *Physica* **106A**, 77.
Dorfman, J. R. and Cohen, E. G. D. (1972). *Phys. Rev. A* **6**, 776.
Dorfman, J. R. and Cohen, E. G. D. (1975). *Phys. Rev. A* **12**, 292.
Dorfman, J. R. and van Beijeren, H. (1977). *In* "Statistical Mechanics", Part B (B. J. Berne, ed.). Plenum Press, New York.
Downs, J., Gubbins, K. E., Murad, S. and Gray, C. G. (1979). *Mol. Phys.* **37**, 129.
Dubey, G. S., Jindal, V. K. and Pathak, K. N. (1980). *Prog. Theor. Phys.* **64**, 1893.
Dubois, M. and Bergé, P. (1971). *Phys. Rev. Lett.* **26**, 121.
Duderstadt, J. J. and Akcasu, A. Z. (1970). *Phys. Rev. A* **1**, 905.
Dufty, J. W. (1972). *Phys. Rev. A* **5**, 2247.
Dufty, J. W., Lindenfield, M. J. and Garland, G. E. (1981). *Phys. Rev. A* **24**, 3212.
Dunleavy, H. N. and Jones, W. (1978). *J. Phys. F* **8**, 1477.
Dymond, J. H. and Smith, E. B. (1980). "The Virial Coefficients of Pure Gases and Mixtures". Clarendon Press, Oxford.
Ebbsjö, I., Schofield, P., Sköld, K. and Waller, I. (1974). *J. Phys. C* **7**, 3891.
Ebner, C. and Saam, W. F. (1977). *Phys. Rev. Lett.* **38**, 1486.
Ebner, C., Saam, W. F. and Stroud, D. (1976). *Phys. Rev. A* **14**, 226.
Edwards, D. M. F., Madden, P. A. and McDonald, I. R. (1984). *Mol. Phys.* **51**, 1141.
Egelstaff, P. and Widom, B. (1970). *J. Chem. Phys.* **53**, 2667.
Egelstaff, P. A., March, N. H. and McGill, N. C. (1974). *Can. J. Phys.* **52**, 1651.
Egelstaff, P. A., Suck, J. B., Glaser, W., McPherson, R. and Teitsma, A. (1980). *J. Phys.* (*Paris*) **41**, C8, 222.
Egelstaff, P. A., Glaser, W., Litchinsky, D., Schneider, E. and Suck, J. B. (1983). *Phys. Rev. A* **27**, 1106.
Enciso, E. (1985). *Mol. Phys.* **56**, 129.
Enderby, J. E. and Neilson, G. W. (1980). *Adv. Phys.* **25**, 323.
Ernst, M. H. and Dorfman, J. R. (1975). *J. Stat. Phys.* **12**, 311.
Ernst, M. H., Dorfman, J. R., Hoegy, W. R. and van Leeuwen, J. M. J. (1969). *Physica* **45**, 127.
Ernst, M. H., Hauge, E. H. and van Leeuwen, J. M. J. (1971). *Phys. Rev. A* **4**, 2055.
Ernst, M. H., Hauge, E. H. and van Leeuwen, J. M. J. (1976a). *J. Stat. Phys.* **15**, 7.
Ernst, M. H., Hauge, E. H. and van Leeuwen, J. M. J. (1976b). *J. Stat. Phys.* **15**, 23.
Erpenbeck, J. J. and Wood, W. W. (1981). *J. Stat. Phys.* **24**, 455.
Erpenbeck, J. J. and Wood, W. W. (1982). *Phys. Rev. A* **26**, 1648.
Evans, D. J. (1980). *J. Stat. Phys.* **22**, 81.
Evans, D. J. (1981). *Phys. Rev. A* **23**, 1988.
Evans, D. J. (1982a). *Phys. Lett.* **91A**, 457.
Evans, D. J. (1982b). *Mol. Phys.* **47**, 1165.
Evans, D. J. and Morriss, G. P. (1984). *Comput. Phys. Rep.* **1**, 297.

Evans, D. J. and Murad, S. (1977). *Mol. Phys.* **34**, 327.
Evans, D. J., Hoover, W. G., Bailor, B. H., Moran, B. and Ladd, A. J. C. (1983). *Phys. Rev. A* **28**, 1016.
Evans, R. (1978). *In* "Microscopic Structure and Dynamics of Liquids" (J. Dupuy and A. J. Dianoux, eds). Plenum Press, New York.
Evans, R. (1979). *Adv. Phys.* **28**, 143.
Evans, R. and Greenwood, D. A. (eds) (1977). "Liquid Metals 1976". Institute of Physics Conference Series No. 30.
Evans, R. and Sluckin, T. J. (1981). *J. Phys. C* **14**, 3137.
Evans, R. and Tarazona, P. (1983). *Phys. Rev. A* **28**, 1864.
Evans, R. and Tarazona, P. (1984). *Phys. Rev. Lett.* **52**, 557.
Evans, R. and Tarazona, P. (1985). *Phys. Rev. A* **32**, 3817.
Evans, R., Tarazona, P. and Marconi, U. M. B. (1983). *Mol. Phys.* **50**, 993.
Faber, T. E. (1972). "An Introduction to the Theory of Liquid Metals". Cambridge University Press, Cambridge.
Felderhof, B. U. (1980). *Physica* **101A**, 275.
Feller, W. (1950). "An Introduction to Probability Theory and its Applications", Vol. I. John Wiley, New York.
Ferrario, M. and Ryckaert, J. P. (1985). *Mol. Phys.* **54**, 587.
Fischer, J. and Methfessell, M. (1980). *Phys. Rev. A* **22**, 2836.
Fisher, M. E. and Fishman, S. (1981). *Phys. Rev. Lett.* **47**, 421.
Fixman, M. (1962). *J. Chem. Phys.* **36**, 310.
Fleury, P. A., Daniels, W. B. and Worlock, J. M. (1971). *Phys. Rev. Lett.* **27**, 1493.
Foiles, S. M., Ashcroft, N. W. and Reatto, L. (1984a). *J. Chem. Phys.* **80**, 4441.
Foiles, S. M., Ashcroft, N. W. and Reatto, L. (1984b). *J. Chem. Phys.* **81**, 6140.
Forster, D. (1974). *Phys. Rev. A* **9**, 943.
Forster, D. and Martin, P. C. (1970). *Phys. Rev. A* **2**, 1575.
Forster, D., Martin, P. C. and Yip, S. (1968). *Phys. Rev.* **170**, 155 and 160.
Fox, R. F. (1983). *Phys. Rev. A* **27**, 3216.
Frenkel, D. (1986). *In* "Molecular Dynamics Simulation of Statistical-Mechanical Systems" (G. Ciccotti and W. G. Hoover, eds). North-Holland, Amsterdam.
Frenkel, J. (1955). "Kinetic Theory of Liquids". Dover, New York.
Friedman, H. L. (1962). "Ionic Solution Theory". Wiley Interscience, New York.
Fries, P. H. and Patey, G. N. (1985). *J. Chem. Phys.* **82**, 429.
Frölich, H. (1958). "Theory of Dielectrics", 2nd edn. Clarendon Press, Oxford.
Fulton, R. L. (1975). *Mol. Phys.* **29**, 405.
Furtado, P. M., Mazenko, G. F. and Yip, S. (1976). *Phys. Rev. A* **14**, 869.
Galam, S. and Hansen, J. P. (1976). *Phys. Rev. A* **14**, 816.
Gantmacher, F. R. (1959). "Applications of the Theory of Matrices". Wiley Interscience, New York.
Gaskell, T. and Miller, S. (1978a). *J. Phys. C* **11**, 3749.
Gaskell, T. and Miller, S. (1978b). *J. Phys. C* **11**, 4839.
Giaquinta, P. V., Parrinello, M. and Tosi, M. P. (1976). *Phys. Chem. Liq.* **5**, 305.
Giaquinta, P. V., Parrinello, M. and Tosi, M. P. (1978). *Physica* **92A**, 185.
Gibbons, R. M. (1969). *Mol. Phys.* **17**, 81.
Gibbons, R. M. (1970). *Mol. Phys.* **18**, 809.
Gibbons, R. M. (1975). *J. Chem. Soc. Faraday II* **71**, 353.
Gillan, M. J. (1979). *Mol. Phys.* **38**, 1781.
Gillan, M. J. and Dixon, M. (1983). *J. Phys. C* **16**, 869.
Gordon, R. G. (1966). *J. Chem. Phys.* **44**, 1830.

Gosling, E. M., McDonald, I. R. and Singer, K. (1973). Mol. Phys. **26**, 1475.
Götze, W. and Lücke, M. (1975). *Phys. Rev. A* **11**, 2173.
Gray, C. G. and Gubbins, K. E. (1984). "Theory of Molecular Liquids". Clarendon, Press, Oxford.
Green, K. A., Luks, K. D. and Kozak, J. J. (1979). *Phys. Rev. Lett.* **42**, 985.
Green, K. A., Luks, K. D., Lee, E. and Kozak, J. J. (1980). *Phys. Rev. A* **21**, 356.
Green, M. S. (1954). *J. Chem. Phys.* **22**, 398.
Greenfield, A. J., Wellendorf, J. and Wiser, N. (1971). *Phys. Rev. A* **4**, 1607.
Greytak, T. J. and Benedek, G. B. (1966). *Phys. Rev. Lett.* **17**, 179.
Griffiths, R. B. (1964). *J. Math. Phys.* **5**, 1215.
Grimson, M. J. and Rickayzen, G. (1981). *Mol. Phys.* **42**, 767.
Gross, E. P. and Jackson, E. A. (1959). *Phys. Fluids* **2**, 432.
Grundke, E. W. and Henderson, D. (1972). *Mol. Phys.* **24**, 269.
Grundke, E. W., Henderson, D., Barker, J. A. and Leonard, P. J. (1973). *Mol. Phys.* **25**, 883.
Gubbins, K. E. and Gray, C. G. (1971). *Mol. Phys.* **23**, 187.
Gubbins, K. E., Shing, K. S. and Streett, W. B. (1983). *J. Phys. Chem.* **87**, 4573.
Guggenheim, E. A. (1960). "Elements of the Kinetic Theory of Gases". Pergamon Press, Oxford.
Haile, J. M. and Gray, C. G. (1980). *Chem. Phys. Lett.* **76**, 583.
Hammersley, J. M. and Handscomb, D. C. (1964). "Monte Carlo Methods". John Wiley, New York.
Hansen, J. P. (1970). *Phys. Rev. A* **2**, 221.
Hansen, J. P. (1973). *Phys. Rev. A* **8**, 3096.
Hansen, J. P. (1984). *J. Phys. (Paris)* **45**, C7, 97.
Hansen, J. P. (1986). *In* "Molecular Dynamics Simulation of Statistical-Mechanical Systems" (G. Ciccotti and W. G. Hoover, eds). North-Holland, Amsterdam.
Hansen, J. P. and McDonald, I. R. (1975). *Phys. Rev. A* **11**, 2111.
Hansen, J. P. and McDonald, I. R. (1981). *Phys. Rev. A* **23**, 2041.
Hansen, J. P. and Schiff, D. (1973). *Mol. Phys.* **25**, 1281.
Hansen, J. P. and Verlet, L. (1969). *Phys. Rev.* **184**, 151.
Hansen, J. P. and Weis, J. J. (1972). *Mol. Phys.* **23**, 853.
Hansen, J. P., McDonald, I. R. and Pollock, E. L. (1975). *Phys. Rev. A* **11**, 1025.
Hansen, J. P., McDonald, I. R. and Vieillefosse, P. (1979). *Phys. Rev. A* **20**, 2590.
Hartmann, W. M. (1971). *Phys. Rev. Lett.* **26**, 1640.
Haymet, A. D. J., Rice, S. A. and Madden, W. G. (1981). *J. Chem. Phys.* **74**, 3033.
Hayter, J. B., Pynn, R. and Suck, J. B. (1983). *J. Phys. F* **13**, 11.
Heine, V. (1970). *Sol. State Phys.* **24**, 1.
Helfand, E. (1960). *Phys. Rev.* **119**, 1.
Hemmer, P. C. (1964). *J. Math. Phys.* **5**, 75.
Hemmer, P. C., Kac, M. and Uhlenbeck, G. E. (1964). *J. Math. Phys.* **5**, 60.
Henderson, D. and Barker, J. A. (1971). *In* "Physical Chemistry: An Advanced Treatise" (H. Eyring, D. Henderson and W. Jost, eds), Vol. VIIIA, Ch. 6. Academic Press, New York and London.
Henderson, D. and Blum, L. (1978). *J. Chem. Phys.* **69**, 5441.
Henderson, D. and Grundke, E. W. (1975). *J. Chem. Phys.* **63**, 601.
Henderson, D. and Leonard, P. J. (1971). *In* "Physical Chemistry: An Advanced Treatise" (H. Eyring, D. Henderson and W. Jost, eds), Vol. VIIIB, Ch. 7. Academic Press, New York, and London.

Henderson, D. and Smith, W. R. (1978). *J. Chem. Phys.* **69**, 319.
Henderson, D., Abraham, F. F. and Barker, J. A. (1976). *Mol. Phys.* **31**, 1291.
Hensel, F. (1979). *Adv. Phys.* **28**, 555.
Heyes, D. M. (1984). *J. Chem. Soc. Faraday Trans. II* **79**, 1741.
Hildebrand, J. H., Prausnitz, J. M. and Scott, R. L. (1970). "Regular and Related Solutions". Prentice-Hall, Englewood Cliffs.
Hill, T. L. (1956). "Statistical Mechanics". McGraw-Hill, New York.
Hirata, F. and Rossky, P. J. (1981). *Chem. Phys. Lett.* **83**, 329.
Hirschfelder, J. O., Curtiss, C. F. and Bird, R. B. (1954). "Molecular Theory of Gases and Liquids". John Wiley, New York.
Hockney, R. W. and Eastwood, J. W. (1981). "Computer Simulation Using Particles". McGraw-Hill, New York.
Hohenberg, P. and Kohn, W. (1964). *Phys. Rev.* **136**, B864.
Holian, B. L. and Evans, D. J. (1983). *J. Chem. Phys.* **78**, 5147.
Hoover, W. G. (1983). *Ann. Rev. Phys. Chem.* **34**, 103.
Hoover, W. G. and Ashurst, W. T. (1975). *Adv. Theor. Chem.* **1**, 1.
Hoover, W. G. and Poirier, J. C. (1962). *J. Chem. Phys.* **37**, 1041.
Hoover, W. G., Evans, D. J., Hickman, R. B., Ladd, A. J. C., Ashurst, W. T. and Moran, B. (1980). *Phys. Rev. A* **22**, 1690.
Hoover, W. G., Ladd, A. J. C. and Moran, B. (1982). *Phys. Rev. Lett.* **48**, 1818.
Howells, W. S. and Enderby, J. H. (1972). *J. Phys. C* **5**, 1277.
Høye, J. S. and Stell, G. (1976). *J. Chem. Phys.* **65**, 18.
Høye, J. S. and Stell, G. (1977a). *J. Chem. Phys.* **66**, 795.
Høye, J. S. and Stell, G. (1977b). *J. Chem. Phys.* **67**, 524.
Høye, J. S., Lebowitz, J. L. and Stell, G. (1974). *J. Chem. Phys.* **61**, 3253.
Høye, J. S., Stell, G. and Waisman, E. (1976). *Mol. Phys.* **32**, 209.
Hsu, C. S., Chandler, D. and Lowden, L. J. (1976). *Chem. Phys.* **14**, 213.
Hubbard, J. and Beeby, J. L. (1969). *J. Phys. C* **2**, 556.
Huijben, M. J. and van der Lugt, W. (1979). *Acta Cryst. A* **35**, 431.
Hutchinson, P. and Conkie, W. R. (1971). *Mol. Phys.* **21**, 881.
Ichimaru, S. (1973). "Basic Principles of Plasma Physics". W. A. Benjamin, Reading.
Ichimaru, S. (1982). *Rev. Mod. Phys.* **54**, 1017.
Impey, R. W., Madden, P. A. and Tildesley, D. J. (1981). *Mol. Phys.* **44**, 1319.
Isihara, A. (1968). *J. Phys. A* **1**, 539.
Iwamatsu, M., Moore, R. A. and Wang, S. (1982). *Phys. Lett.* **92**A, 283.
Jackson, J. L. and Mazur, P. (1964). *Physica* **30**, 2295.
Jacobs, R. E. and Andersen, H. C. (1975). *Chem. Phys.* **10**, 73.
Jacucci, G. and McDonald, I. R. (1975). *Physica* **80**A, 607.
Jacucci, G. and McDonald, I. R. (1980a). *Mol. Phys.* **39**, 515.
Jacucci, G. and McDonald, I. R. (1980b). *J. Phys. F* **10**, L15.
Jacucci, G., McDonald, I. R. and Rahman, A. (1976). *Phys. Rev. A* **13**, 1581.
Jacucci, G., Taylor, R., Tenenbaum, A. and Van Doan, N. (1981). *J. Phys. F* **11**, 793.
Jal, J. F. (1981). Thesis, Université Claude Bernard, Lyon.
Jal, J. F., Chieux, P., Dupuy, J. and Dupin, J. P. (1980). *J. Phys. (Paris)* **41**, 657.
Jhon, M. S. and Forster, D. (1975). *Phys. Rev. A* **12**, 254.
Johnson, E. and Hazoumé, R. P. (1979). *J. Chem. Phys.* **70**, 1599.
Johnson, J. D., Shaw, M. S. and Holian, B. L. (1984). *J. Chem. Phys.* **80**, 1279.
Johnston, M. D., Hutchinson, P. and March, N. H. (1964). *Proc. Roy. Soc.* **A282**, 283.
Jones, G. L., Kozak, J. J., Lee, E., Fishman, S. and Fisher, M. E. (1981). *Phys. Rev. Lett.* **46**, 795.

Kac, M., Uhlenbeck, G. E. and Hemmer, P. C. (1963). *J. Math. Phys.* **4**, 216.
Kadanoff, L. P. and Martin, P. C. (1963). *Ann. Phys.* **24**, 419.
Kadanoff, L. P. and Swift, J. (1968). *Phys. Rev.* **166**, 89.
Kawasaki, K. (1970). *Ann. Phys.* **61**, 1.
Kawasaki, K. and Oppenheim, I. (1965). *Phys. Rev.* **139**, A1763.
Kerr, W. C. (1968). *Phys. Rev.* **174**, 316.
Kerr, W. C. (1976). *J. Chem. Phys.* **64**, 885.
Keyes, T. and Oppenheim, I. (1973). *Physica* **70**, 100.
Kihara, T. (1958). *Adv. Chem. Phys.* **1**, 267.
Kim, K. and Nelkin, M. (1971). *Phys. Rev. A* **4**, 2065.
Kim, S. (1969). *Phys. Fluids* **12**, 2046.
Kim, S. and Henderson, D. (1968). *Phys. Lett.* **27A**, 379.
Kirkpatrick, T. R. (1984). *Phys. Rev. Lett.* **53**, 1735 and 2185.
Kirkpatrick, T. R., Cohen, E. G. D. and Dorfman, J. R. (1982a). *Phys. Rev. A* **26**, 950.
Kirkpatrick, T. R., Cohen, E. G. D. and Dorfman, J. R. (1982b). *Phys. Rev. A* **26**, 972.
Kirkpatrick, T. R., Cohen, E. G. D. and Dorfman, J. R. (1982c). *Phys. Rev. A* **26**, 995.
Kirkwood, J. G. (1935). *J. Chem. Phys.* **3**, 300.
Kirkwood, J. G. (1939). *J. Chem. Phys.* **7**, 911.
Kirkwood, J. G. and Buff, F. P. (1949). *J. Chem. Phys.* **17**, 338.
Kirkwood, J. G. and Buff, F. P. (1951). *J. Chem. Phys.* **19**, 774.
Kivelson, D. and Keyes, T. (1972). *J. Chem. Phys.* **57**, 4599.
Klein, M. (1963). *J. Chem. Phys.* **39**, 388.
Klein, M. (1964). *Phys. Fluids* **7**, 341.
Kollár, J. (1981). *Phys. Rev. A* **23**, 2706.
Konijnendijk, H. H. U. and van Leeuwen, J. M. J. (1973). *Physica* **64**, 342.
Konior, J. and Jedrzejek, C. (1985). *Mol. Phys.* **55**, 187.
Kubo, R. (1957). *J. Phys. Soc. Japan* **12**, 570.
Kubo, R. (1966). *Rep. Prog. Phys.* **29**, 255.
Kugler, A. A. (1973). *J. Stat. Phys.* **8**, 107.
Ladanyi, B. M. and Chandler, D. (1975). *J. Chem. Phys.* **62**, 4308.
Ladanyi, B. M. and Keyes, T. (1977). *Mol. Phys.* **33**, 1063.
Lado, F. (1964). *Phys. Rev.* **135**, A1013.
Lado, F. (1967). *J. Chem. Phys.* **47**, 4828.
Lado, F. (1973). *Phys. Rev. A* **8**, 2548.
Lado, F. (1982). *Phys. Lett.* **89A**, 196.
Lado, F. (1984). *Mol. Phys.* **52**, 871.
Lado, F., Foiles, S. M. and Ashcroft, N. W. (1983). *Phys. Rev. A* **28**, 2374.
Lallemand, P. (1974). *In* "Molecular Motion in Liquids" (J. Lascombe, ed.), p. 517. Reidel, Dordrecht.
Landau, L. D. and Lifshitz, E. M. (1963). "Fluid Mechanics". Pergamon Press, London.
Landau, L. D. and Lifshitz, E. M. (1980). "Statistical Physics", 3rd edn. Pergamon Press, London.
Lane, J. E. and Spurling, T. H. (1980). *Aust. J. Chem.* **33**, 231.
Lane, J. E., Spurling, T. H., Freasier, B. C., Perram, J. W. and Smith, E. R. (1979). *Phys. Rev. A* **20**, 2147.
Lantelme, F., Turq, P. and Schofield, P. (1977). *J. Chem. Phys.* **67**, 3869.
Larsen, B. (1976). *J. Chem. Phys.* **65**, 3531.
Larsen, B. (1978). *J. Chem. Phys.* **68**, 4511.
Larsen, B., Stell, G. and Wu, K. C. (1977). J. Chem. Phys. **67**, 530.

Lebowitz, J. L. (1964). *Phys. Rev.* **133**, A895.
Lebowitz, J. L. and Penrose, O. (1964). *J. Math. Phys.* **5**, 841.
Lebowitz, J. L. and Penrose, O. (1966). *J. Math. Phys.* **7**, 98.
Lebowitz, J. L. and Percus, J. K. (1961). *Phys. Rev.* **122**, 1675.
Lebowitz, J. L. and Percus, J. K. (1966). *Phys. Rev.* **144**, 251.
Lebowitz, J. L. and Rowlinson, J. S. (1964). *J. Chem. Phys.* **41**, 133.
Lebowitz, J. L. and Spohn, H. (1978). *J. Stat. Phys.* **19**, 633.
Lebowitz, J. L., Stell, G. and Baer, S. (1965). *J. Math. Phys.* **6**, 1282.
Lebowitz, J. L., Percus, J. K. and Verlet, L. (1967). *Phys. Rev.* **153**, 250.
Lebowitz, J. L., Percus, J. K. and Sykes, J. (1969). *Phys. Rev.* **188**, 487.
Lee, J. K., Barker, J. A. and Pound, G. M. (1974). *J. Chem. Phys.* **60**, 1976.
Lee, L. L. and Levesque, D. (1973). *Mol. Phys.* **26**, 1351.
Lee, L. Y., Fries, P. H. and Patey, G. N. (1985). *Mol. Phys.* **55**, 751.
Lees, A. W. and Edwards, S. F. (1972). *J. Phys. C* **5**, 1921.
Lekkerkerker, H. N. W. and Laidlaw, W. G. (1973). *Phys. Rev. A* **7**, 1332.
Leland, T. W., Rowlinson, J. S. and Sather, G. A. (1968). *Trans. Faraday Soc.* **64**, 1447.
Leland, T. W., Rowlinson, J. S., Sather, G. A. and Watson, I. D. (1969). *Trans. Faraday Soc.* **66**, 2439.
Leonard, P. J., Henderson, D. and Barker, J. A. (1970). *Trans. Faraday Soc.* **66**, 2439.
Leutheusser, E. (1982a). *J. Phys. C* **15**, 2801.
Leutheusser, E. (1982b). *J. Phys. C* **15**, 2827.
Levesque, D. (1966). *Physica* **32**, 1985.
Levesque, D. and Ashurst, W. T. (1974). *Phys. Rev. Lett.* **33**, 277.
Levesque, D. and Verlet, L. (1969). *Phys. Rev.* **182**, 307.
Levesque, D. and Verlet, L. (1970). *Phys. Rev. A* **2**, 2514.
Levesque, D. and Weis, J. J. (1977). *Phys. Lett.* **60A**, 473.
Levesque, D., Verlet, L. and Kürkijarvi, J. (1973). *Phys. Rev. A* **7**, 1690.
Levesque, D., Weis, J. J. and Reatto, L. (1985). *Phys. Rev. Lett.* **54**, 451.
Lewis, J. W. E. and Lovesey, S. W. (1977). *J. Phys. C* **10**, 3221.
Lighthill, M. J. (1958). "Introduction to Fourier Analysis and Generalized Functions". Cambridge University Press, Cambridge.
Linnebur, E. J. and Duderstadt, J. J. (1973). *Phys. Fluids* **16**, 665.
Lombardero, M., Abascal, J. L. F. and Lago, S. (1981). *Mol. Phys.* **42**, 999.
Longuet-Higgins, H. C. (1951). *Proc. Roy. Soc.* **A205**, 247.
Longuet-Higgins, H. C. and Widom B. (1964). *Mol. Phys.* **8**, 549.
Lovesey, S. W. (1971). *J. Phys. C* **4**, 3057.
Lovesey, S. W. (1973). *J. Phys. C* **6**, 1856.
Lovesey, S. W. (1984a). "Theory of Neutron Scattering from Condensed Matter", Vol. 1. Clarendon Press, Oxford.
Lovesey, S. W. (1984b). *Zeit. Phys. B* **58**, 79.
Lowden, L. J. and Chandler, D. (1973). *J. Chem. Phys.* **59**, 6587.
Lowden, L. J. and Chandler, D. (1974). *J. Chem. Phys.* **61**, 5228.
Lowden, L. J. and Chandler, D. (1975). *J. Chem. Phys.* **62**, 4246.
Lu, B. Q., Evans, R. and Telo da Gama, M. M. (1985). *Mol. Phys.* **55**, 1319.
Luttinger, J. M. (1964). *Phys. Rev.* **135**, A1505.
Lyddane, R. H., Sachs, R. G. and Teller, E. (1941). *Phys. Rev.* **59**, 673.
Lynden-Bell, R. M. (1984). *In* "Molecular Liquids: Dynamics and Interactions" (A. J. Barnes, W. J. Orville-Thomas and J. Yarwood, eds). Reidel, Dordrecht.
Lynden-Bell, R. M. and McDonald, I. R. (1981). *Mol. Phys.* **43**, 1429.

MacGowan, D., Waisman, E. M., Lebowitz, J. L. and Percus, J. K. (1984). *J. Chem. Phys.* **80**, 2719.
MacGowan, D., Lebowitz, J. L. and Waisman, E. M. (1985). *Chem. Phys. Lett.* **114**, 321.
Machida, M. and Murase, C. (1973). *Prog. Theor. Phys.* **50**, 1.
Madden, P. A. and Kivelson, D. (1984). *Adv. Chem. Phys.* **56**, 467.
Madden, P. A. and Tildesley, D. J. (1983). *Mol. Phys.* **48**, 129.
Madden, W. G. (1981). *J. Chem. Phys.* **75**, 1984.
Madden, W. G. and Rice, S. A. (1980). *J. Chem. Phys.* **72**, 4208.
Maitland, G. C., Rigby, M., Smith, E. B. and Wakeham, W. A. (1981). "Intermolecular Forces". Clarendon Press, Oxford.
Mansoori, G. A. (1972). *J. Chem. Phys.* **56**, 5335.
Mansoori, G. A. and Canfield, F. B. (1969). *J. Chem. Phys.* **51**, 4958.
Mansoori, G. A. and Leland, T. W. (1970). *J. Chem. Phys.* **53**, 1931.
Mansoori, G. A., Carnahan, N. F., Starling, K. E. and Leland, T. W. (1971). *J. Chem. Phys.* **54**, 1523.
Martin, P. A. and Gruber, C. (1983). *J. Stat. Phys.* **31**, 691.
Martin, P. A. and Yalcin, T. (1980). *J. Stat. Phys.* **22**, 435.
Martin, P. C. (1967). *Phys. Rev.* **161**, 143.
Martin, P. C. (1968). *In* "Many-Body Physics" (C. DeWitt and R. Balian, eds). Gordon and Breach, New York.
Martin, P. C. and Yip, S. (1968). *Phys. Rev.* **170**, 151.
Mayer, J. E. (1950). *J. Chem. Phys.* **18**, 1426,
Mayer, J. E. and Mayer, M. G. (1940). "Statistical Mechanics". John Wiley, New York.
Mayer, J. E. and Montroll, E. (1941). *J. Chem. Phys.* **9**, 2.
Mazenko, G. F. (1971). *Phys. Rev. A* **3**, 2121.
Mazenko, G. F. (1973a). *Phys. Rev. A* **7**, 209.
Mazenko, G. F. (1973b). *Phys. Rev. A* **7**, 222.
Mazenko, G. F. (1974). *Phys. Rev. A* **9**, 360.
Mazenko, G. F. and Yip, S. (1977). *In* "Statistical Mechanics" (B. J. Berne, ed.), Part B. Plenum Press, New York.
Mazenko, G. F., Wei, T. Y. C. and Yip, S. (1972). *Phys. Rev. A* **6**, 1981.
Mazur, P. (1966). *In* "Cargèse Lectures in Theoretical Physics" (B. Jancovici, ed.). Gordon and Breach, New York.
McDonald, I. R. (1972a). *Mol. Phys.* **23**, 41.
McDonald, I. R. (1972b). *Mol. Phys.* **24**, 391.
McDonald, I. R. (1974). *J. Phys. C* **7**, 1225.
McDonald, I. R. and O'Gorman, S. P. (1978). *Phys. Chem. Liq.* **8**, 57.
McDonald, I. R., Bounds D. G. and Klein, M. L. (1982). *Mol. Phys.* **45**, 221.
McGreevy, R. L., Mitchell, E. W. J. and Margaca, F. M. A. (1984). *J. Phys. C* **17**, 775.
McIntyre, D. and Sengers, J. (1968). *In* "Physics of Simple Liquids" (H. N. V. Temperley, J. S. Rowlinson and G. S. Rushbrooke, eds.), p. 447. North Holland, Amsterdam.
McQuarrie, D. A. and Katz, J. L. (1966). *J. Chem. Phys.* **44**, 2398.
Meeron, E. (1960). *J. Math. Phys.* **1**, 192.
Meeron, E. and Siegert, A. J. F. (1968). *J. Chem. Phys.* **48**, 3139.
Mehaffey, J. R. and Cukier, R. I. (1978). *Phys. Rev. A* **17**, 1181.
Mehaffey, J. R., Desai, R. C. and Kapral, R. (1977). *J. Chem. Phys.* **66**, 1665.

Meister, T. F. and Kroll, D. M. (1985). *Phys. Rev. A* **31**, 4055.
Melnyk, T. W. and Sawford, B. L. (1975). *Mol. Phys.* **29**, 891.
Mermin, N. D. (1964). *Phys. Rev.* **137**, A1441.
Metiu, H., Oxtoby, D. W. and Freed, K. F. (1977). *Phys. Rev. A* **15**, 361.
Metropolis, M., Rosenbluth, A. W., Rosenbluth, M. N., Teller, A. N. and Teller, E. (1953). *J. Chem. Phys.* **21**, 1087.
Michels, A. and Wouters, H. (1941). *Physica* **8**, 923.
Michels, A., Wijker, Hu. and Wijker, H. C. (1949). *Physica* **15**, 627.
Michels, A., deGraaff, W. and ten Sololam, T. A. (1960). *Physica* **26**, 393.
Michels, P. and Trappeniers, N. J. (1978). *Physica* **90A**, 179.
Minoo, H., Deutsch, C. and Hansen, J. P. (1977). *J. de Phys. Lettres* **38**, L191.
Mitra, S. K. (1978). *J. Phys. C* **11**, 3551.
Mitra, S. K., Hutchinson, P. and Schofield, P. (1976). *Phil. Mag.* **34**, 1087.
Miyazaki, J., Barker, J. A. and Pound, G. M. (1976). *J. Chem. Phys.* **64**, 3364.
Mo, K. C., Gubbins, K. E., Jacucci, G. and McDonald, I. R. (1974). *Mol. Phys.* **27**, 1173.
Mon, K. K., Gann, R. and Stroud, D. (1981). *Phys. Rev. A* **24**, 2145.
Monson, P. A. (1982). *Mol. Phys.* **47**, 435.
Monson, P. A. and Rigby, M. (1979). *Mol. Phys.* **38**, 1699.
Montague, D. G., Chowdhury, M. R., Dore, J. C. and Reed, J. (1983). *Mol. Phys.* **50**, 1.
Moore, J. W. and Pearson, R. G. (1981). "Kinetics and Mechanism", 3rd edn. John Wiley, New York.
Mori, H. (1958). *Phys. Rev.* **112**, 1892.
Mori, H. (1965a). *Prog. Theor. Phys.* **33**, 423.
Mori, H. (1965b). *Prog. Theor. Phys.* **34**, 399.
Morita, T. (1960). *Prog. Theor. Phys.* **23**, 829.
Morita, T. and Hiroike, K. (1961). *Prog. Theor. Phys.* **25**, 537.
Morriss, G. P. and Perram, J. W. (1981). *Mol. Phys.* **43**, 669.
Mountain, R. D. (1966). *Rev. Mod. Phys.* **38**, 205.
Mountain, R. D. and Deutch, J. M. (1969). *J. Chem. Phys.* **50**, 1103.
Munakata, T. and Bosse, J. (1983). *Phys. Rev. A* **27**, 455.
Munakata, T. and Igarashi, A. (1977). *Prog. Theor. Phys.* **58**, 1345.
Munakata, T. and Igarashi, A. (1978). *Prog. Theor. Phys.* **60**, 45.
Murad, S. (1980). *Chem. Phys. Lett.* **72**, 194.
Murase, C. (1970). *J. Phys. Soc. Japan* **29**, 549.
Murphy, R. D. (1977). *Phys. Rev. A* **15**, 1188.
Narten, A. H. (1972). *J. Chem. Phys.* **56**, 1185.
Narten, A. H. and Habenschuss, A. (1981). *J. Chem. Phys.* **75**, 3072.
Narten, A. H., Agrawal, R. and Sandler, S. I. (1978). *Mol. Phys.* **35**, 1077.
Narten, A. H., Johnson, E. and Habenschuss, A. (1980). *J. Chem. Phys.* **73**, 1248.
Nee, T. W. and Zwanzig, R. (1970). *J. Chem. Phys.* **52**, 6353.
Nelkin, M. and Gathak, A. (1964). *Phys. Rev.* **135**, A4.
Nelkin, M. and Ranganathan, S. (1967). *Phys. Rev.* **164**, 222.
Neumann, M. and Steinhauser, O. (1980). *Mol. Phys.* **39**, 437.
Nicholson, D. and Parsonage, N. G. (1982). "Computer Simulation and the Statistical Mechanics of Adsorption". Academic Press, London and New York.
Nieminen, R. M. and Ashcroft, N. W. (1981). *Phys. Rev. A* **24**, 560.
Nijboer, B. R. A. and Rahman, A. (1966). *Physica* **32**, 415.
Nijboer, B. R. A. and van Hove, L. (1952). *Phys. Rev.* **85**, 777.

Nixon, J. H. and Silbert, M. (1984). *Mol. Phys.* **52**, 207.
Nordholm, S., Johnson, M. and Freasier, B. C. (1980). *Aust. J. Chem.* **33**, 2139.
Nosé, S. (1984a). *Mol. Phys.* **52**, 255.
Nosé, S. (1984b). *J. Chem. Phys.* **81**, 511.
Nosé, S. and Klein, M. L. (1983a). *Mol. Phys.* **50**, 1055.
Nosé, S. and Klein, M. L. (1983b). *J. Chem. Phys.* **78**, 6928.
Ohbayashi, K., Kohno, T. and Utiyama, H. (1983). *Phys. Rev. A* **27**, 2632.
Onsager, L. (1931a). *Phys. Rev.* **37**, 405.
Onsager, L. (1931b). *Phys. Rev.* **38**, 2265.
Onsager, L. (1939). *J. Phys. Chem.* **43**, 189.
Ornstein, L. S. and Zernike, F. (1914). *Proc. Akad. Sci. (Amsterdam)* **17**, 793.
Ortega, J. M. and Rheinboldt, W. C. (1970). "Iterative Solution of Nonlinear Equations in Several Variables". Academic Press, New York and London.
Ortoleva, P. and Nelkin, M. (1969). *Phys. Rev.* **181**, 429.
Outhwaite, C. W. (1975). *In* "Statistical Mechanics" (K. Singer, ed.), Vol. 2, p. 188. The Chemical Society, London.
Parrinello, M. and Rahman, A. (1980). *Phys. Rev. Lett.* **45**, 1196.
Parrinello, M. and Rahman, A. (1981). *J. Appl. Phys.* **52**, 7182.
Parrinello, M. and Rahman, A. (1984). *J. Chem. Phys.* **80**, 860.
Parrinello, M. and Tosi, M. P. (1979). *Riv. Nuovo Cim.* **2**, No. 6.
Parrinello, M., Tosi, M. P. and March, N. H. (1974). *J. Phys. C* **7**, 2577.
Pastore, G. and Tosi, M. P. (1984). *Physica* **124B & C**, 383.
Patey, G. N. (1977). *Mol. Phys.* **34**, 427.
Patey, G. N. (1978). *Mol. Phys.* **35**, 1413.
Patey, G. N. and Carnie, S. L. (1983). *J. Chem. Phys.* **78**, 5183.
Patey, G. N. and Valleau, J. P. (1974). *J. Chem. Phys.* **61**, 534.
Patey, G. N., Levesque, D. and Weis, J. J. (1979a). *Mol. Phys.* **38**, 219.
Patey, G. N., Levesque, D. and Weis, J. J. (1979b). *Mol. Phys.* **38**, 1635.
Patey, G. N., Levesque, D. and Weis, J. J. (1982). *Mol. Phys.* **45**, 733.
Pathak, K. N. and Singwi, K. S. (1970). *Phys. Rev. A* **2**, 2427.
Paul, G. L. and Pusey, P. N. (1981). *J. Phys. A* **14**, 3301.
Percus, J. K. (1962). *Phys. Rev. Lett.* **8**, 462.
Percus, J. K. (1964). *In* "The Equilibrium Theory of Classical Fluids" (H. L. Frisch and J. L. Lebowitz, eds), p. II-33. W. A. Benjamin, New York.
Percus, J. K. and Yevick, G. J. (1958). *Phys. Rev.* **110**, 1.
Perram, J. W. and White, L. R. (1974). *Mol. Phys.* **28**, 527.
Perram, J. W. and White, L. R. (1975). *Faraday Disc. Chem. Soc.* **59**, 29.
Pines, D. and Nozières, P. (1966). "The Theory of Quantum Liquids". W. A. Benjamin, New York.
Pings, C. J. (1968). *In* "Physics of Simple Liquids" (H. N. V. Temperley, J. S. Rowlinson and G. S. Rushbrooke, eds), Ch. 10. North Holland, Amsterdam.
Pollock, E. L. and Alder, B. J. (1980). *Physica* **102A**, 1.
Pollock, E. L. and Alder, B. J. (1981). *Phys. Rev. Lett.* **46**, 950.
Pomeau, Y. (1972). *Phys. Rev. A*, **6**, 776.
Pomeau, Y. and Résibois, P. (1975). *Phys. Rep.* **19**, 63.
Pople, J. A. (1952). *Proc. Roy. Soc.* **A215**, 67.
Pople, J. A. (1954). *Proc. Roy. Soc.* **A221**, 498.
Postogna, F. and Tosi, M. P. (1980). *Nuovo Cim.* **55B**, 399.
Powles, J. G. (1953). *J. Chem. Phys.* **21**, 633.
Powles, J. G. (1973). *Adv. Phys.* **22**, 1.

Powles, J. G., Evans, W. A. B. and Quirke, N. (1982). *Mol. Phys.* **46**, 1347.
Pratt, L. R. and Chandler, D. (1977). *J. Chem. Phys.* **66**, 147.
Prigogine, I. (1957). "The Molecular Theory of Solutions". North Holland, Amsterdam.
Quirke, N. and Tildesley, D. J. (1982). *Mol. Phys.* **45**, 811.
Quirke, N. and Tildesley, D. J. (1983). *J. Phys. Chem.* **87**, 1972.
Rahman, A. (1964a). *Phys. Rev.* **136**, A405.
Rahman, A. (1964b). *Phys. Rev. Lett.* **12**, 575.
Rahman, A. (1966). *J. Chem. Phys.* **45**, 258.
Rahman, A. (1968). *In* "Neutron Inelastic Scattering", Vol. I, p. 561. IAEA, Vienna.
Rahman, A. (1974a). *Phys. Rev. Lett.* **32**, 52.
Rahman, A. (1974b). *Phys. Rev. A* **9**, 1667.
Rahman, A. and Stillinger, F. H. (1971). *J. Chem. Phys.* **55**, 3336.
Ramshaw, J. D. (1979). *J. Chem. Phys.* **70**, 1577.
Rao, M. (1974). *Phys. Rev. A* **9**, 2220.
Rasaiah, J. C. and Friedman, H. L. (1968). *J. Chem. Phys.* **48**, 2742.
Rasaiah, J. C. and Stell, G. (1970). *Mol. Phys.* **18**, 249.
Rasaiah, J. C., Card, D. N. and Valleau, J. P. (1972). *J. Chem. Phys.* **56**, 248.
Ree, F. H. and Hoover, W. G. (1964). *J. Chem. Phys.* **49**, 639.
Ree, F. H. and Hoover, W. G. (1967). *J. Chem. Phys.* **46**, 4181.
Regnaut, C., Badiali, J. P. and Dupont, M. (1980). *J. Phys. (Paris)* **41**, C8, 603.
Reiss, H., Frisch, H. L. and Lebowitz, J. L. (1959). *J. Chem. Phys.* **31**, 369.
Résibois, P. (1968). "Electrolyte Theory". Harper and Row, New York.
Résibois, P. (1970). *J. Stat. Phys.* **2**, 21.
Résibois, P. (1975). *J. Stat. Phys.* **13**, 393.
Résibois P. and DeLeener, M. (1977). "Classical Kinetic Theory of Fluids". John Wiley, New York.
Richardson, D. M. and Chandler, D. (1984). *J. Chem. Phys.* **80**, 4414.
Rogers, F. J. and Young, D. A. (1984). *Phys. Rev. A* **30**, 999.
Romano, S. and Singer, K. (1979). *Mol. Phys.* **37**, 1765.
Ronis, D., Procaccia, I. and Oppenheim, I. (1979). *Phys. Rev. A* **19**, 1324.
Rose, M. E. (1957). "Elementary Theory of Angular Momentum". John Wiley, New York.
Rosenfeld, Y. and Ashcroft, N. W. (1979). *Phys. Rev. A* **20**, 1208.
Ross, M., DeWitt, H. E. and Hubbard, W. B. (1981). *Phys. Rev. A* **24**, 1016.
Ross, R. G. and Greenwood, D. A. (1969). *Prog. Mater. Sci.* **14**, 173.
Rossky, P. J. and Chiles, R. A. (1984). *Mol. Phys.* **51**, 661.
Rossky, P. J., Pettitt, B. M. and Stell, G. (1983). *Mol. Phys.* **50**, 1263.
Rothschild, W. G. (1984). "Dynamics of Molecular Liquids". John Wiley, New York.
Rovere, M., Parrinello, M., Tosi, M. P. and Giaquinta, P. V. (1979). *Phil. Mag.* **39**, 167.
Rowe, J. M. and Sköld, K. (1972). *In* "Neutron Inelastic Scattering 1972", p. 413. IAEA, Vienna.
Rowley, L. A., Nicholson, D. and Parsonage, N. G. (1976). *Mol. Phys.* **31**, 365.
Rowlinson, J. S. (1964a). *Mol. Phys.* **8**, 107.
Rowlinson, J. S. (1964b). *Contemp. Phys.* **5**, 359.
Rowlinson, J. S. (1965). *Mol. Phys.* **9**, 217.
Rowlinson, J. S. and Swinton, F. (1982). "Liquids and Liquid Mixtures", 3rd edn. Butterworths, London.

Rowlinson, J. S. and Widom, B. (1982). "Molecular Theory of Capillarity". Clarendon Press, Oxford.
Ruppersberg, H. and Reiter, H. (1982). *J. Phys. F* **12**, 1311.
Rushbrooke, G. S. (1960). *Physica* **26**, 259.
Rushbrooke, G. S. (1979). *Mol. Phys.* **37**, 761.
Rushbrooke, G. S. and Scoins, H. I. (1953). *Proc. Roy. Soc.* **A216**, 203.
Ryckaert, J. P. and Bellemans, A. (1975). *Chem. Phys. Lett.* **30**, 123.
Ryckaert, J. P. and Ciccotti, G. (1983). *J. Chem. Phys.* **78**, 7368.
Ryckaert, J. P., Ciccotti, G. and Berendsen, H. J. C. (1977). *J. Comput. Phys.* **23**, 327.
Saam, W. F. and Ebner, C. (1978). *Phys. Rev. A* **17**, 1768.
Sangster, M. J. L. and Dixon, M. (1976). *Adv. Phys.* **25**, 247.
Schamp, H. W., Mason, E. A., Richardson, A. C. B. and Altman, A. (1958). *Phys. Fluids* **1**, 329.
Schiff, D. (1969). *Phys. Rev.* **186**, 151.
Schoen, M. and Hoheisel, C. (1984a). *Mol. Phys.* **52**, 33.
Schoen, M. and Hoheisel, C. (1984b). *Mol. Phys.* **52**, 1029.
Schoen, M. and Hoheisel, C. (1985). *Mol. Phys.* **56**, 653.
Schofield, P. (1966). *Proc. Phys. Soc.* **88**, 149.
Schofield, P. (1968). *In* "Physics of Simple Liquids" (H. N. V. Temperley, J. S. Rowlinson and G. S. Rushbrooke, eds), p. 563. North Holland, Amsterdam.
Schofield, P. (1974). Report of CECAM Workshop on Ionic Liquids, Orsay, France.
Schofield, P. (1975). *In* "Statistical Mechanics" (K. Singer, ed.), Vol. 2. The Chemical Society, London.
Sears, V. F. (1970). *Can. J. Phys.* **48**, 616.
Sears, V. F. (1973). *Phys. Rev. A* **7**, 340.
Senatore, G., Giaquinta, P. V. and Tosi, M. P. (1982). *Physica* **112B**, 360.
Shankar, R. (1980). "Principles of Quantum Mechanics". Plenum Press, New York.
Shimoji, M. (1977). "Liquid Metals". Academic Press, London and New York.
Shing, K. S. and Gubbins, K. E. (1982). *Mol. Phys.* **46**, 1109.
Shinomoto, S. (1983). *J. Stat. Phys.* **32**, 105.
Silbert, M. and Young, W. H. (1976). *Phys. Lett.* **58A**, 469.
Silbert, M., Ulmar, I. H., Watabe, M. and Young,, W. H. (1975). *J. Phys. F* **5**, 1262.
Singer, J. V. L. and Singer, K. (1972). *Mol. Phys.* **24**, 357.
Singer, K., Taylor, A. and Singer, J. V. L. (1977). *Mol. Phys.* **33**, 1757.
Singer, K., Singer, J. V. L. and Fincham, D. (1980). *Mol. Phys.* **40**, 515.
Singwi, K. S. and Tosi, M. P. (1981). *Sol. State Phys.* **36**, 177.
Singwi, K. S., Tosi, M. P., Land, R. H. and Sjölander, A. (1968). *Phys. Rev.* **176**, 589.
Singwi, K. S., Sköld, K. and Tosi, M. P. (1970). *Phys. Rev. A* **1**, 454.
Sjödin, S. and Sjölander, A. (1978). *Phys. Rev. A* **18**, 1723.
Sjögren, L. (1980a). *J. Phys. C* **13**, 705.
Sjögren, L. (1980b). *Phys. Rev. A* **22**, 2866.
Sjögren, L. (1980c). *Phys. Rev. A* **22**, 2883.
Sjögren, L. and Sjölander, A. (1979). *J. Phys. C* **12**, 4369.
Sjögren, L. and Yoshida, F. (1982). *J. Chem. Phys.* **77**, 3703.
Sköld, K., Rowe, J. M., Ostrowski, G. and Randolph, P. D. (1972). *Phys. Rev. A* **6**, 1107.
Slattery, W. L., Doolen, G. D. and DeWitt, H. E. (1980). *Phys. Rev. A* **21**, 2087.

Smith, L. S. and Lee, L. L. (1979). *J. Chem. Phys.* **71**, 4085.
Smith, W. R. (1974). *Can. J. Phys.* **52**, 2022.
Smith, W. R. and Henderson, D. (1970). *Mol. Phys.* **19**, 411.
Smith, W. R. and Nezbeda, I. (1983). *In* "Molecular-Based Study of Fluids" (J. M. Haile and G. A. Mansoori, eds), Ch. 11. American Chemical Society, Washington.
Smith, W. R., Henderson, D. and Tago, Y. (1977). *J. Chem. Phys.* **67**, 5308.
Snook, I. K. and Henderson, D. (1978). *J. Chem. Phys.* **68**, 2134.
Snook, I. K. and van Megen, W. (1982). *Mol. Phys.* **47**, 1417.
Söderström, O., Copley, J. R. D., Suck. J. B. and Dorner, B. (1980). *J. Phys. F* **10**, L151.
Söderström, O., Dahlborg, U. and Davidovič, M. (1983). *Phys. Rev. A* **27**, 470.
Soltwisch, M., Quitmann, D., Ruppersberg, H. and Suck, J. B. (1983). *Phys. Rev. B* **28**, 5583.
Squires, G. L. (1978). "Introduction to the Theory of Thermal Neutron Scattering". Cambridge University Press, Cambridge.
Stanley, H. E. (1971). "Introduction to Phase Transitions and Critical Phenomena". Clarendon Press Oxford.
Steele, W. A. (1963). *J. Chem. Phys.* **39**, 3197.
Steele, W. A. (1974). "The Interaction of Gases with Solid Surfaces". Pergamon Press, Oxford.
Steele, W. A. (1981). *Mol. Phys.* **43**, 141.
Stell, G. (1963). *Physica* **29**, 517.
Stell, G. (1964). *In* "The Equilibrium Theory of Classical Fluids" (H. L. Frisch and J. L. Lebowitz, eds), p. II–171. W. A. Benjamin, New York.
Stell, G. and Weis, J. J. (1977). *Phys. Rev. A* **16**, 757.
Stell, G. and Weis, J. J. (1980). *Phys. Rev. A* **21**, 645.
Stell, G., Rasaiah, J. C. and Narang, H. (1972). *Mol. Phys.* **23**, 393.
Stell, G., Patey, G. N. and Høye, J. S. (1981). *Adv. Chem, Phys.* **48**, 183.
Stillinger, F. H. and Lovett, R. (1968a). *J. Chem. Phys.* **48**, 3858.
Stillinger, F. H. and Lovett, R. (1968b). *J. Chem. Phys.* **49**, 1991.
Stoddard, S. D. and Ford, J. (1973). *Phys. Rev. A* **8**, 1504.
St. Pierre, A. G. and Steele, W. A. (1981). *Mol. Phys.* **43**, 123.
Streett, W. B. and Tildesley, D. J. (1977). *Proc. Roy. Soc.* **A355**, 239.
Stroud, D. (1973). *Phys. Rev. B* **7**, 4405.
Sugawara, A., Yip, S. and Sirovich, L. (1968). *Phys. Fluids* **11**, 925.
Sullivan, D. E. (1979). *Phys. Rev. B* **20**, 3991.
Sullivan, D. E. (1981). *J. Chem. Phys.* **74**, 2604.
Sullivan, D. E. and Gray, C. G. (1981). *Mol. Phys.* **42**, 443.
Sullivan, D. E. and Stell, G. (1978). *J. Chem. Phys.* **69**, 5450.
Sullivan, D. E., Levesque, D. and Weis, J. J. (1980). *J. Chem. Phys.* **72**, 1170.
Takeuchi, S. (ed.) (1973). "The Properties of Liquid Metals". Taylor and Francis, London.
Tarazona, P. (1985). *Phys. Rev.* **31**, 2672.
Tarazona, P. and Evans, R. (1984). *Mol. Phys.* **52**, 847.
Teletzke, G. F., Scriven, L. E. and Davis, H. T. (1982). *J. Colloid Int. Sci.* **87**, 550.
Tenenbaum, A., Ciccotti, G. and Gallico, R. (1982). *Phys. Rev. A* **25**, 2778.
Thiele, E. (1963). *J. Chem. Phys.* **39**, 474.
Tildesley, D. J. (1980). *Mol. Phys.* **41**, 341.

Tildesley, D. J. (1984). *In* "Molecular Liquids: Dynamics and Interactions" (A. J. Barnes, W. J. Orville-Thomas and J. Yarwood, eds). Reidel, Dordrecht.
Titulaer, U. M. and Deutch, J. M. (1974). *J. Chem. Phys.* **60**, 1502.
Tong, E. and Desai, R. C. (1970). *Phys. Rev. A* **2**, 2129.
Tonks, L. (1936), *Phys. Rev.* **50**, 1955.
Topol, R. and Claverie, P. (1978). *Mol. Phys.* **35**, 1753.
Torrie, G. and Carnie, S. L. (1984). *Adv. Chem. Phys.* **56**, 141.
Tosi, M. P. (1983). *In* "Electron Correlations in Solids, Molecules and Atoms" (J. T. Devreese and F. Brosens, eds). Plenum Press, New York.
Tosi, M. P. and Fumi, F. G. (1964). *J. Phys. Chem. Solids* **25**, 45.
Toxvaerd, S. (1976). *J. Chem. Phys.* **64**, 2863.
Tremblay, A., Siggia, E. and Arai, M. (1981). *Phys. Rev. A* **23**, 1451.
Trozzi, Č. and Ciccotti, G. (1984). *Phys. Rev. A* **29**, 916.
Uhlenbeck, G. E. and Ford, G. W. (1962). *In* "Studies in Statistical Mechanics", Vol. I. North-Holland, Amsterdam.
Uhlenbeck, G. E., Hemmer, P. C. and Kac, M. (1963). *J. Math. Phys.* **4**, 229.
Valleau, J. P. and Cohen, L. K. (1980). *J. Chem. Phys.* **72**, 5935.
van Hove, L. (1954). *Phys. Rev.* **95**, 249.
van Kampen, N. G. (1964). *Phys. Rev.* **135**, 362.
van Kampen, N. G. (1971). *Physica Norvegica* **5**, 279.
van Leeuwen, J. M. J., Groeneveld, J. and deBoer, J. (1959). *Physica* **25**, 792.
van Tricht, J. B., Fredrikze, H. and van der Laan, J. (1984). *Mol. Phys.* **52**, 115.
van Well, A. A., Verkerk, P., deGraaf, L. A., Suck, J. B. and Copley, J. R. D. (1985). *Phys. Rev. A* **31**, 3391.
Verlet, L. (1960). *Nuovo Cim.* **18**, 77.
Verlet, L. (1964). *Physica* **30**, 95.
Verlet, L. (1965). *Physica* **31**, 959.
Verlet, L. (1967). *Phys. Rev.* **159**, 98.
Verlet, L. (1968). *Phys. Rev.* **163**, 201.
Verlet, L. and Levesque, D. (1967). *Physica* **36**, 254.
Verlet, L. and Weis, J. J. (1972a). *Mol. Phys.* **24**, 1013.
Verlet, L. and Weis, J. J. (1972b). *Phys. Rev. A* **5**, 939.
Verlet, L. and Weis, J. J. (1974). *Mol. Phys.* **28**, 665.
Versmold, H. (1981). *Mol. Phys.* **43**, 383.
Vieillefosse, P. (1977). *J. de Phys. Lettres* **38**, L43.
Vieillefosse, P. and Hansen, J. P. (1975). *Phys. Rev. A* **12**, 1106.
Vineyard, G. H. (1958). *Phys. Rev.* **110**, 999.
Visscher, W. M. (1974). *Phys. Rev. A* **10**, 2461.
Vlasov, A. A. (1961). "Many Particle Theory and its Application to Plasmas". Gordon and Breach, New York.
Volterra, V. (1959). "Theory of Functionals". Dover, New York.
Waisman, E. (1973). *Mol. Phys.* **25**, 45.
Waisman, E. and Lebowitz, J. L. (1972). *J. Chem. Phys.* **56**, 3086 and 3093.
Waisman, E., Henderson, D. and Lebowitz, J. L. (1976). *Mol. Phys.* **32**, 1373.
Wallenborn, J. and Baus, M. (1978). *Phys. Rev. A* **18**, 173.
Watson, G. N. (1966) "A Treatise on the Theory of Bessel Functions", 2nd edn. Cambridge University Press, Cambridge.
Webber, G. M. B. and Stephens, R. W. B. (1968). *In* "Physical Acoustics" (W. P. Mason, ed.), Vol. IVB. Academic Press, London and New York.

Weeks, J. D., Chandler, D. and Andersen, H. C. (1971). *J. Chem. Phys.* **54**, 5237.
Weis, J. J. and Levesque, D. (1976). *Phys. Rev. A* **13**, 450.
Wertheim, M. S. (1963). *Phys. Rev. Lett.* **10**, 321.
Wertheim, M. S. (1964). *J. Math. Phys.* **5**, 643.
Wertheim, M. S. (1971). *J. Chem. Phys.* **55**, 4291.
Wertheim, M. S. (1973). *Mol. Phys.* **26**, 1425.
Whalley, E. and Schneider, W. G. (1955). *J. Chem. Phys.* **23**, 1644.
Wheeler, J. C. and Chandler, D. (1971). *J. Chem. Phys.* **55**, 1645.
Widom, A. (1971). *Phys. Rev. A* **3**, 1394.
Widom, B. (1963). *J. Chem. Phys.* **39**, 2808.
Widom, B. (1966). *J. Chem. Phys.* **44**, 3888.
Wood, W. W. (1968). *In* "Physics of Simple Liquids" (H. N. V. Temperley, J. S. Rowlinson and G. S. Rushbrooke, eds). North Holland, Amsterdam.
Wood, W. W. and Parker, F. R. (1957). *J. Chem. Phys.* **27**, 720.
Woodcock, L. V. and Singer, K. (1971). *Trans. Faraday Soc.* **67**, 12.
Woods, A. D. B., Martel, P. and Svenson, E. C. (1978). *In* "Neutron Inelastic Scattering", Vol. 2, p. 44. IAEA, Vienna.
Yarnell, J. L., Katz, M. J., Wenzel, R. G. and Koenig, S. H. (1973). *Phys. Rev. A* **7**, 2130.
Yashonath, S. and Rao, C. N. R. (1985). *Mol. Phys.* **54**, 245.
Yip, S. (1979). *Ann. Rev. Phys. Chem.* **30**, 547.
Yip, S. and Nelkin, M. (1964). *Phys. Rev.* **135**, A1241.
Yip, S., Alley, W. E. and Alder, B. J. (1982). *J. Stat. Phys.* **27**, 201.
Yvon, J. (1935). "La Théorie Statistique des Fluides et l'Equation d'Etat", Actualités Scientifiques et Industrielles, Vol. 203. Hermann, Paris.
Yvon, J. (1958). *Suppl. Nuovo Cim.* **9**, 144.
Zerah, G. (1986). *J. Comput. Phys.* (in press).
Zerah, G. and Hansen, J. P. (1986). *J. Chem. Phys.* **84**, 2336.
Ziman, J. M. (1961). *Phil. Mag.* **6**, 1013.
Ziman, J. M. (1963). "Electrons and Phonons". Clarendon Press, Oxford.
Zwanzig, R. (1954). *J. Chem. Phys.* **22**, 1420.
Zwanzig, R. (1961). *In* "Lectures in Theoretical Physics" (W. E. Britton, B. W. Downs and J. Downs eds), Vol. III, p. 135. Wiley Interscience, New York.
Zwanzig, R. (1965). *Ann. Rev. Phys. Chem.* **16**, 667.
Zwanzig, R. (1967), *Phys. Rev.* **156**, 190.
Zwanzig, R. and Ailawadi, N. K. (1969). *Phys. Rev.* **182**, 280.
Zwanzig, R. and Bixon, M. (1970). *Phys. Rev. A* **2**, 2005.
Zwanzig, R. and Mountain, R. D. (1965). *J. Chem. Phys.* **43**, 4464.

Appendix A

LEMMAS ON DIAGRAMS

We give here proofs of Lemmas 1, 2 and 4 of section 4.4; the proofs of Lemmas 3 and 5 are similar to those of 2 and 4, respectively, and are therefore omitted. We follow the arguments of the original papers of Morita and Hirioike (1961) and Stell (1964); more details, with illustrations, can be found in an article by McDonald and O'Gorman (1978).

PROOF OF LEMMA 1

Let $\{g_1, \ldots, g_N\}$ be the set of diagrams in G (N may be infinite). A typical diagram in the set H is the star product of n_1 diagrams g_1, n_2 diagrams g_2, $\ldots$, and n_N diagrams g_N, where some of the numbers n_i may be zero; we express this result symbolically by writing

$$\Gamma = (g_1 ** n_1) * (g_2 ** n_2) * \cdots * (g_N ** n_N). \tag{A.1}$$

To avoid any confusion, the value of any diagram g_i will be denoted by $[g_i]$. By definition, the value of g_i is given by the relation

$$[g_i] = \frac{I_i}{s_i}, \tag{A.2}$$

where s_i is the symmetry number of g_i and I_i is the integral associated with that diagram. The value of the diagram Γ is

$$[\Gamma] = \frac{I}{S} = \frac{1}{S} \prod_{i=1}^{N} I_i^{n_i}, \tag{A.3}$$

where the symmetry number of Γ is given by

$$S = \left(\prod_{i=1}^{N} n_i! \right) \left(\prod_{i=1}^{N} s_i^{n_i} \right). \tag{A.4}$$

The factors $n_i!$ in (A.4) take care of the permutations of the n_i identical diagrams g_i; note that (A.4) is valid only for diagrams that are star-irreducible. Given (A.2) and (A.4), Eqn (A.3) can be rewritten as

$$[\Gamma]=\left(\prod_{i=1}^{N} n_i!\right)^{-1}\left(\prod_{i=1}^{N} I_i^{n_i}S_i^{-n_i}\right)$$

$$=\left(\prod_{i=1}^{N} n_i!\right)^{-1}\left(\prod_{i=1}^{N} [g_i]^{n_i}\right). \tag{A.5}$$

We now sum over all diagrams in H and find

$$\sum[\Gamma]=\sum_{n_1=0}^{\infty}\cdots\sum_{n_N=0}^{\infty}\prod_{i=1}^{N}\frac{[g_i]^{n_i}}{n_i!}-1$$

$$=\prod_{i=1}^{N}\left(\sum_{n_i=0}^{\infty}\frac{[g_i]^{n_i}}{n_i!}\right)-1$$

$$=\prod_{i=1}^{N}\exp\{[g_i]\}-1$$

$$=\exp\left\{\sum_{i=1}^{N}[g_i]\right\}-1. \tag{A.6}$$

We subtract unity in the first line of (A.6) because the case when all $n_i=0$ must be excluded.

PROOF OF LEMMA 2

If S is the symmetry number and m is the number of black circles of Γ, the number of topologically inequivalent diagrams that are generated by attaching the labels $\mathbf{r}_1, \ldots, \mathbf{r}_m$ to the black circles in all possible ways is $\nu=m!/S$. Hence, from the definition (4.3.4):

$$\Gamma=\frac{1}{m!}\{\Gamma_1(\gamma\bullet_1,\ldots,\gamma\bullet_m)+\cdots+\Gamma_\nu(\gamma\bullet_1,\ldots,\gamma\bullet_m)\}. \tag{A.7}$$

We now take the functional derivative of Γ with respect to $\gamma(\mathbf{r})$. It follows from (4.2.11) that

$$\frac{\delta\gamma(\mathbf{r}_i)}{\delta\gamma(\mathbf{r})}=\delta(\mathbf{r}-\mathbf{r}_i), \qquad 1\leqslant i\leqslant m \tag{A.8}$$

The operation therefore amounts to replacing successively each of the m black γ-circles in (A.7) by a white 1-circle. In this way νm diagrams are generated, each containing one white circle and $m-1$ black circles; these we shall denote by $\Gamma_i^{(1)}$. Thus

$$\frac{\delta\Gamma}{\delta\gamma(\mathbf{r})}=\frac{1}{m!}\left\{\sum_{i=1}^{\nu}\sum_{j=1}^{m}\Gamma_i^{(1)}(\gamma\bullet_1,\ldots,\gamma\bullet_{(j-1)}\circ_{\mathbf{r}}\gamma\bullet_{(j+1)},\ldots,\gamma\bullet_m)\right\} \tag{A.9}$$

The coordinates $\mathbf{r}_j$ can be permuted at will because in the original definition of Γ they are all equivalent. We may therefore replace the sum over j in Eqn (A.9) by m times the contribution for $j=1$ and obtain the simpler expression given by

$$\frac{\delta\Gamma}{\delta\gamma(\mathbf{r})}=\frac{1}{(m-1)!}\left\{\sum_{i=1}^{\nu}\Gamma_i^{(1)}(\circ_{\mathbf{r}}\gamma\bullet_2,\ldots,\gamma\bullet_m)\right\}. \tag{A.10}$$

The ν diagrams $\Gamma_i^{(1)}$ can now be subdivided into μ groups, chosen according to the topologically distinct diagrams to which each reduces when the labels of the $m-1$ black circles are removed. If the latter diagrams are denoted by $\Gamma_1',\ldots,\Gamma_\mu'$, definition (4.3.4) implies that

$$\frac{\delta\Gamma}{\delta\gamma(\mathbf{r})}=\Gamma_1'+\cdots+\Gamma_\mu'. \tag{A.11}$$

which is the required result.

PROOF OF LEMMA 4

Let m be the number of black circles in Γ. Any diagram in the set H can be expressed as $h(\Gamma;\{g_i\})$, where $\{g_i\}\equiv\{g_1,\ldots,g_m\}$ is a set of diagrams drawn from G that are attached to the black circles of Γ; some of the g_i may be identical. Two diagrams h obtained from two distinct sets $\{g_i\}$ are not necessarily different. In terms of the diagrams h, Lemma 4 can be expressed in more compact form as

$$\sum_{\{g_i\}}{}' h(\Gamma:\{g_i\})=\{\text{the diagram obtained from }\Gamma\text{ by attaching the function }\mathscr{G}(\mathbf{r})\text{ to each of its black circles}\}. \tag{A.12}$$

The summation in Eqn (A.12) is taken over all sets of diagrams $\{g_i\}$, with the restriction (denoted by the prime) that the diagrams $h(\Gamma;\{g_i\})$ must be topologically distinct.

Let $S(\Gamma)$ be the symmetry number of Γ; obviously $S(\Gamma)$is also the symmetry number of the right-hand side of Eqn (A.12). Similarly, let $S(g_i)$ and

$S(\Gamma; \{g_i\})$ be, respectively, the symmetry numbers of the diagrams in G and H. According to the definition (4.3.5):

$$h(\Gamma; \{g_i\}) = \frac{1}{S(\Gamma; \{g_i\})} h(\Gamma'; \{g_i'\}), \tag{A.13}$$

where $h(\Gamma'; \{g_i'\})$ is a diagram obtained from $h(\Gamma; \{g_i\})$ by attaching labels to its black circles in an arbitrary way. Let $h(\Gamma'; \{g_i\})$ be the diagram obtained from $h(\Gamma'; \{g_i'\})$ by removing the labels of the black circles of the g_i' but retaining the labels of the black circles of Γ', and let $S^*(\Gamma; \{g_i\})$ be the number of permutations of the m labels in $h(\Gamma'; \{g_i\})$ that give rise to topologically equivalent diagrams. For each of these S^* permutations, there are $\prod_{i=1}^{m} S(g_i)$ permutations of the black circles of the g_i that yield diagrams equivalent to $h(\Gamma'; \{g_i'\})$. We can therefore write

$$S(\Gamma; \{g_i\}) = S^*(\Gamma; \{g_i\}) \prod_{i=1}^{m} S(g_i). \tag{A.14}$$

We now require a relation between $S(\Gamma)$ and $S^*(\Gamma; \{g_i\})$. Note that $S(\Gamma) \geqslant S^*(\Gamma; \{g_i\})$, since the process of decorating the black circles of Γ can never lead to an increase in symmetry number. Let $n(\Gamma; \{g_i\})$ be the number of labellings that give rise to diagrams $h(\Gamma'; \{g_i\})$ that are topologically inequivalent, but give diagrams Γ' (i.e. labelled versions of Γ taken on its own) that are equivalent. Consider now the set of $S(\Gamma)$ diagrams that are obtained from $h(\Gamma'; \{g_i\})$ by making the $S(\Gamma)$ permutations that leave Γ' topologically unaltered. This set can be divided into precisely $n(\Gamma; \{g_i\})$ groups, such that the diagrams in different groups are topologically inequivalent to each other. Each of the $n(\Gamma, \{g_i\})$ groups consists of $S^*(\Gamma; \{g_i\})$ topologically inequivalent diagrams. Thus

$$S(\Gamma) = n(\Gamma; \{g_i\}) S^*(\Gamma; \{g_i\}) \tag{A.15}$$

Illustration

$\Gamma =$

$g_1 =$

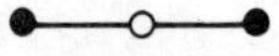

$g_2 = g_3 =$

$h(\Gamma; g_1, g_2, g_3) =$

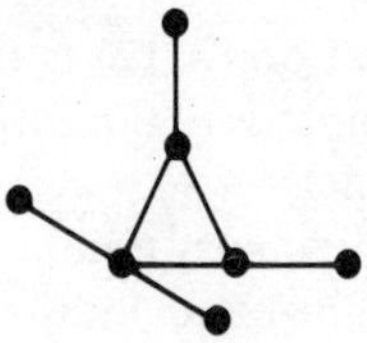

$$h(\Gamma'; g_1, g_2, g_3) =$$

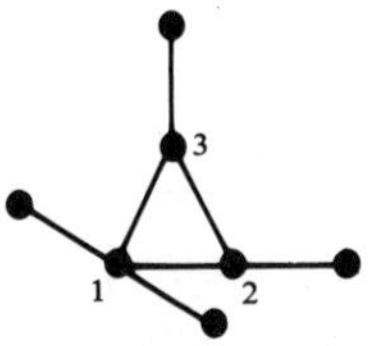

In this example, $S(\Gamma) = 6$; $S^*(\Gamma; g_1, g_2, g_3) = 2$, because labels 2 and 3 can be permuted in $h(\Gamma'; g_1, g_2, g_3)$; and $n(\Gamma; g_1, g_2, g_3) = 3$, because permutation of labels 1 and 2 or 1 and 3 in $h(\Gamma'; g_1, g_2, g_3)$ generates diagrams that are topologically inequivalent.

By combining Eqns (A.14) and (A.15), we find that

$$S(\Gamma; \{g_i\}) = S(\Gamma) \prod_{i=1}^{m} S(g_i)/n(\Gamma; \{g_i\}). \tag{A.16}$$

If use is made of (A.13) and (A.16), the left-hand side of Eqn (A.12) can be rewritten as

$$\sum_{\{g_i\}}{}' \frac{n(\Gamma; \{g_i\})}{S(\Gamma) \prod_{i=1}^{m} S(g_i)} h(\Gamma'; \{g_i'\}), \tag{A.17}$$

or, from the definition (4.3.5):

$$\sum_{\{g_i\}}{}' \frac{n(\Gamma; \{g_i\})}{S(\Gamma)} h(\Gamma'; \{g_i\}). \tag{A.18}$$

Remembering the significance of $n(\Gamma; \{g_i\})$, we see that (A.18) can also be expressed as

$$\sum_{g_1} \cdots \sum_{g_m} \frac{1}{S(\Gamma)} h(\Gamma'; g_1, \ldots, g_m), \tag{A.19}$$

where the m summations over the g_i are now unrestricted, i.e. the summations are taken independently of each other. But (A.19) is just a labelled diagram obtained from Γ' by associating the function $\gamma(\mathbf{r})$ with each black circle and dividing by the symmetry number $S(\Gamma)$. From the definition (4.3.5), it follows that (A.19) is equal to the right-hand side of Eqn (A.12).

Appendix B

SOLUTION OF THE PY EQUATION FOR HARD SPHERES

The PY closure relation for hard spheres is

$$h(r) = -1, \qquad r < d \tag{B.1a}$$

$$c(r) = 0, \qquad r > d \tag{B.1b}$$

When substituted in the Ornstein–Zernike relation (5.2.10), this approximation yields an integral equation that can be solved in closed form. We follow here the method of Baxter (1968), which is based on a transformation of the Ornstein–Zernike relation via a so-called Wiener–Hopf factorization of the function $\hat{A}(k)$ defined as

$$\hat{A}(k) = \frac{1}{S(k)} = 1 - \rho\hat{c}(k) = 1 - \frac{4\pi\rho}{k}\int_0^\infty r \sin(kr) c(r)\, dr \tag{B.2}$$

The three-dimensional Fourier transform of any function f of $r = |\mathbf{r}|$ can be cast in the form

$$\begin{aligned}\hat{f}(k) &= \frac{4\pi}{k}\int_0^\infty r \sin(kr) f(r)\, dr \\ &= 4\pi \int_0^\infty \cos(kr) F(r)\, dr \\ &= 2\pi \int_{-\infty}^\infty \exp(ikr) F(r)\, dr\end{aligned} \tag{B.3}$$

where

$$\begin{aligned}F(r) &= \int_r^\infty s f(s)\, ds \\ &= F(-r)\end{aligned} \tag{B.4}$$

The last line of Eqn (B.4) follows immediately if the convention that $f(r) = f(-r)$ is followed. Substitution of Eqns (B.3), (B.4) and (B.1b) in Eqn (B.2) leads to

$$\hat{A}(k) = 1 - 4\pi\rho \int_0^d \cos(kr) S(r)\, dr$$
$$= \hat{A}(-k) \tag{B.5}$$

where

$$S(r) = \int_r^d t c(t)\, \mathrm{d}t \tag{B.6}$$

Similarly

$$\hat{h}(k) = 2\pi \int_{-\infty}^{\infty} \exp(ikr) J(r)\, \mathrm{d}r \tag{B.7}$$

with

$$J(r) = \int_r^{\infty} s h(s)\, \mathrm{d}s \tag{B.8}$$

Consider now the behaviour of the function $\hat{A}(k)$ in the complex k-plane and set $k = x + iy$. Because $\hat{A}(k)$, as given by Eqn (B.5), is a Fourier transform over a finite interval, it is regular throughout the complex plane. It also has no zeros on the real axis ($y = 0$), since it is the inverse of the static structure factor; the latter is a finite quantity at all wavenumbers. Moreover, according to Eqn (B.5), $\hat{A}(k)$ tends uniformly to unity as $|x| \to \infty$ in any strip $y_1 < y < y_2$. Thus there exists a strip $|y| \leq \varepsilon$ about the real axis within which $\hat{A}(k)$ has no zeros. The function $\log \hat{A}(k)$ is therefore regular within the strip $|y| \leq \varepsilon$ and tends uniformly to zero as $|x| \to \infty$. Integrating around the strip and applying Cauchy's theorem, we find that for any $k = x + iy$ such that $|y| < \varepsilon$:

$$\log \hat{A}(k) = \log \hat{Q}(k) + \log \hat{P}(k) \tag{B.9}$$

where

$$\log \hat{Q}(k) = \frac{1}{2\pi i} \int_{-i\varepsilon - \infty}^{-i\varepsilon + \infty} \mathrm{d}k' \frac{\log \hat{A}(k')}{k' - k} \tag{B.10a}$$

$$\log \hat{P}(k) = -\frac{1}{2\pi i} \int_{i\varepsilon - \infty}^{i\varepsilon + \infty} \mathrm{d}k' \frac{\log \hat{A}(k')}{k' - k} \tag{B.10b}$$

Since $\hat{A}(k)$ is an even function of k, Eqns (B.10) imply that

$$\log \hat{P}(k) = \log \hat{Q}(-k) \tag{B.11}$$

From Eqn (B.10a), we see that $\log \hat{Q}(k)$ is regular in the domain $y > -\varepsilon$. It then follows from Eqns (B.9) and (B.11) that when $|y| < \varepsilon$:

$$\hat{A}(k) = \hat{Q}(k)\hat{Q}(-k) \tag{B.12}$$

where $\hat{Q}(k)$ is regular and has no zeros in the domain $y > -\varepsilon$, since it is the exponential of a function that itself is regular in that domain. Equation (B.12) is the Wiener–Hopf factorization of $\hat{A}(k)$.

When $|x| \to \infty$ within the strip $|y| < \varepsilon$, it follows from Eqn (B.10a) that $\log \hat{Q}(k) \sim x^{-1}$ and hence that $\hat{Q}(k) \sim 1 - \mathcal{O}(x^{-1})$. The function $1 - \hat{Q}(k)$ is therefore Fourier integrable along the real axis, and a function $Q(r)$ can be defined as

$$2\pi\rho Q(r) = \frac{1}{2\pi} \int_{-\infty}^{\infty} \mathrm{d}k \exp(-ikr)[1 - \hat{Q}(k)] \tag{B.13}$$

From Eqn (B.10a) it can be seen that if k is real, the complex conjugate of $\hat{Q}(k)$ is $\hat{Q}(-k)$, from which it follows that $Q(r)$ is a real function. The same equation also shows that when $y \geqslant 0$, $\log \hat{Q}(k)$ tends to zero, and hence $\hat{Q}(k)$ tends to unity, as k tends to infinity. Thus, if $r < 0$, the integration In Eqn (B.13) can be closed around the upper half-plane, where $\hat{Q}(k)$ is regular, to give

$$Q(r) = 0, \qquad r < 0 \tag{B.14}$$

The right-hand side of Eqn (B.10a) is a different analytic function of k according to whether $y > -\varepsilon$ or $y < -\varepsilon$. The analytic continuation of the function $\hat{Q}(k)$ into the lower half-plane is therefore given, not by Eqn (B.10a), but by Eqn (B.12), i.e.

$$\hat{Q}(k) = \hat{A}(k)/\hat{Q}(-k) \tag{B.15}$$

where Eqn (B.10a) can be used to evaluate $\hat{Q}(-k)$ on the right-hand side of Eqn (B.15). Since $\hat{A}(k)$ is regular everywhere, and $\hat{Q}(-k)$ is regular and has no zeros for $y < \varepsilon$, we see from Eqn (B.15) that the function $\hat{Q}(k)$ is also regular for $y < \varepsilon$. Furthermore, since $\hat{Q}(-k) \to 1$ as $y \to -\infty$, it follows from Eqns (B.5) and (B.15) that both $\hat{A}(k)$ and $\hat{Q}(k)$ grow exponentially as $\exp(ikd) = \exp(ixd)\exp(-yd)$ when y becomes large and negative. Thus, when $r > d$, the integration in Eqn (B.13) can be closed around the lower half-plane, giving

$$Q(r) = 0, \qquad r > d \tag{B.16}$$

On inversion of the Fourier transform in Eqn (B.13), Eqns (B.14) and (B.16) together yield

$$\hat{Q}(k) = 1 - 2\pi\rho \int_0^d \exp(ikr)Q(r)\,\mathrm{d}r \tag{B.17}$$

Substitution of the forms (B.5) of $\hat{A}(k)$ and (B.17) of $\hat{Q}(k)$ into Eqn (B.12), followed by multiplication by $\exp(-ikr)$ and integration with respect to k from $-\infty$ to $+\infty$, shows that

$$S(r)=Q(r)-2\pi\rho\int_r^d Q(s)Q(s-r)\,\mathrm{d}s, \qquad 0<r<d \tag{B.18}$$

Equations (B.2), (5.2.11) and (B.12) imply that

$$\hat{Q}(k)[1+\rho\hat{h}(k)]=[\hat{Q}(k)]^{-1} \tag{B.19}$$

where $\hat{h}(k)$ is given by Eqn (B.7). Now multiply both sides of Eqn (B.19) by $\exp(-ikr)$ and integrate with respect to k from $-\infty$ to $+\infty$. The contribution from the right-hand side vanishes when $r>0$, since the integration can then be closed around the lower half-plane, where $\hat{Q}(k)$ is regular, has no zeros, and tends to unity at infinity. On substituting (B.7) and (B.17) into the left-hand side of Eqn (B.19) and carrying out the integration, we obtain a relation between $Q(r)$ and $J(r)$ for $r>0$ of the form

$$-Q(r)+J(r)-2\pi\rho\int_0^d Q(s)J(|r-s|)\,\mathrm{d}s=0 \tag{B.20}$$

It is clear from Eqns (B.6) and (B.18) that $Q(r)\to 0$ as $r\to d$ from below; comparison with Eqn (B.16) then shows that $Q(r)$ is continuous at $r=d$.

Equations (B.18) and (B.20) can be expressed in terms of $c(r)$ and $h(r)$, rather than $S(r)$ and $J(r)$, by differentiating them with respect to r. If we use Eqns (B.6) and (B.8), and the fact that $Q(d)=0$, we find after integration by parts that

$$rc(r)=-Q'(r)+2\pi\rho\int_r^d Q'(s)Q(s-r)\,\mathrm{d}s \tag{B.21}$$

for $0<r<d$, and

$$rh(r)=-Q'(r)+2\pi\rho\int_0^d (r-s)h(|r-s|)Q(s)\,\mathrm{d}s \tag{B.22}$$

for $r>0$, where $Q'(r)\equiv \mathrm{d}Q(r)/\mathrm{d}r$. Equations (B.21) and (B.22) express $h(r)$ and $c(r)$ in terms of the same function $Q(r)$, and constitute a reformulation of the Ornstein–Zernike relation that is applicable whenever $c(r)$ vanishes beyond a range d, which is precisely the case with the PY closure (B.1b). Equation (B.22) is an integral equation for $Q(r)$ that is easy to solve for $0<r<d$, where $h(r)=-1$ and (B.22) reduces to

$$r=Q'(r)+2\pi\rho\int_0^d (r-s)Q(s)\,\mathrm{d}s \tag{B.23}$$

The solution is of the form

$$Q'(r) = ar + b \tag{B.24}$$

with

$$a = 1 - 2\pi\rho \int_0^d Q(s)\, \mathrm{d}s \tag{B.25a}$$

$$b = 2\pi\rho \int_0^d sQ(s)\, \mathrm{d}s \tag{B.25b}$$

Given the boundary condition $Q(d) = 0$, Eqn (B.24) is trivially integrated to yield $Q(r)$. Substitution of the result into the expressions (B.25) for a and b gives two linear equations that can be solved to give

$$a = \frac{1+2\eta}{(1-\eta)^2}, \qquad b = \frac{-3d\eta}{2(1-\eta)^2} \tag{B.26}$$

where $\eta = \pi d^3/6$ is the hard-sphere packing fraction. Thus $Q(r)$ is now a known function of r, and $c(r)$ can therefore be calculated from Eqn (B.21); this leads to the result displayed in Eqns (5.5.5) to (5.5.7). The isothermal compressibility is obtained from Eqns (5.2.17), (B.2) and (B.15) as

$$\chi_T^0/\chi_T = \hat{A}(0) = [\hat{Q}(0)]^2 \tag{B.27}$$

The function $\hat{Q}(0)$ is easily calculated from Eqn (B.17) and the solution for $Q(r)$, leading to the PY compressibility equation of state (5.5.8).

Appendix C

RADIAL DISTRIBUTION FUNCTION OF HARD SPHERES

Verlet and Weis (1972b) have given a simple prescription for the calculation of the hard-sphere radial distribution that has proved very useful in numerical work. Its basis is the solution to the PY equation discussed in Section 5.5 and Appendix B. The PY solution suffers from two main defects (see Figure 5.5): first, the value at contact is too small; secondly, the later oscillations have the wrong phase and are too weakly damped.

To overcome the problem of the errors in phase and amplitude, Verlet and Weis use the PY solution corresponding to a slightly smaller packing fraction η', obtained by decreasing the diameter from d to d', i.e. they set

$$g(r/d;\, \eta) \simeq g'(r/d';\, \eta') \tag{C.1}$$

where d' is such that $\eta' = \eta(d'/d)^3$. The value of d' is calculated by minimizing the absolute difference between $g'(r)$ and the distribution function obtained from computer "experiments", integrated over the interval $[r_1, 3d]$. Verlet and Weis choose $r_1 = 1.6d$, thereby ignoring the region close to contact. In practice, they find that the minimum is achieved for

$$\eta' \simeq \eta - \eta^2/16 \tag{C.2}$$

Use of (C.1) and (C.2) corrects for the defects in the oscillations of the PY solution, but the situation at contact is worsened because the height of the main peak is reduced. A short-range term is therefore added to (C.1), and the final form of the radial distribution function is

$$g(r/d;\, \eta) = g'(r/d';\, \eta') + \delta g_1(r) \tag{C.3}$$

where

$$\delta g_1(r) = \frac{A}{r} \exp\left[-\alpha(r-d)\right] \cos\left[\alpha(r-d)\right] \tag{C.4}$$

The parameter A is chosen to yield the correct value of $g(d)$, i.e. $\rho/4\eta\beta P$, with $\beta P/\rho$ given by the Carnahan–Starling equation of state (4.6.14), and α is determined by forcing the distribution function (C.3) to give the correct isothermal compressibility. This yields for A and α the expressions

$$A/d = g(1;\, \eta) - g'(d/d';\, \eta') \simeq \frac{3}{4} \frac{\eta'^2(1 - 0.7117\eta' - 0.114\eta'^2)}{(1-\eta')^4} \tag{C.5}$$

$$\alpha d \simeq \frac{24A/d}{\eta' g'(1;\, \eta')} \tag{C.6}$$

The resulting $g(r)$ fits the "exact", computer-generated functions to within 1% for all η.

When $r \geqslant d$, the functions $g(r)$ and $y(r) = \exp[\beta v(r)]g(r)$ are identical. The implementation of certain theories also requires an accurate knowledge of $y(r)$ inside the hard core. The scheme proposed by Henderson and Grundke (1975) relies on the fact that the behaviour of $y(r)$ as $r \to 0$ is known exactly for hard spheres (Hoover and Poirier, 1962; Meeron and Siegert, 1968):

$$\log y(0) = \beta\mu^{\text{ex}} \tag{C.7}$$

$$\left.\frac{\mathrm{d} \log y(r)}{\mathrm{d}r}\right|_{r=0} = -6\eta y(d) \tag{C.8}$$

where μ^{ex} is the excess chemical potential. If these results are combined with the Carnahan–Starling formulae (4.6.14) and (4.6.15), we find that

$$\log y(0) = \frac{8\eta - 9\eta^2 + 3\eta^3}{(1-\eta)^3} \tag{C.9}$$

$$\left.\frac{\mathrm{d} \log y(r)}{\mathrm{d}r}\right|_{r=0} = -3\eta \frac{2-\eta}{(1-\eta)^3} \tag{C.10}$$

Henderson and Grundke therefore choose to write $y(r)$ for $r \leqslant d$ as a polynomial of the form

$$\log y(r) = \sum_{n=0}^{3} a_n r^n \tag{C.11}$$

where a_0 and a_1 are given as functions of η by the right-hand sides of (C.9) and (C.10), respectively, while a_2 and a_3 are chosen to match the Verlet–Weis expression for $y(r)$ and $\mathrm{d}y(r)/\mathrm{d}r$ at $r = d$.

The Verlet–Weis results have also been used in the parameterization of $c(r)$ (Henderson and Grundke, 1975) and generalized to hard-sphere mixtures (Grundke and Henderson, 1972).

Appendix D

λ-EXPANSION OF THE PAIR DISTRIBUTION FUNCTION

Let us assume that the particles of the fluid interact through a pair potential of the form $v_\lambda(i,j) = v_0(i,j) + \lambda w(i,j)$ (the notation used is that of Section 6.2). From Eqn (2.5.1) it is a straightforward exercise to show that the derivative of the canonical pair density with respect to λ, at constant number density and temperature, is

$$\frac{\partial \rho_N^{(2)}(1,2;\lambda)}{\partial \lambda} = -\beta\Bigg(w(1,2)\rho_N^{(2)}(1,2;\lambda) + 2\int \rho_N^{(3)}(1,2,3;\lambda) w(2,3)\,\mathrm{d}3 \\ + \tfrac{1}{2}\iint [\rho_N^{(4)}(1,2,3,4;\lambda) \\ - \rho_N^{(2)}(1,2;\lambda)\rho_N^{(2)}(3,4;\lambda)] w(3,4)\,\mathrm{d}3\,\mathrm{d}4\Bigg) \tag{D.1}$$

Since $\rho_N^{(n)}(1,\ldots,n;\lambda) = \rho^n g_N^{(n)}(1,\ldots,n;\lambda)$ for a uniform fluid, a similar relation holds for the pair distribution function. Equation (D.1) is valid for a finite system, but some care is needed in passing to the thermodynamic limit because the asymptotic behaviour of $\rho_N^{(4)}$ when the pairs (1, 2) and (3, 4) are widely separated is given (Lebowitz and Percus, 1961) by

$$\rho_N^{(4)}(1,2,3,4) \sim \rho_N^{(2)}(1,2)\rho_N^{(2)}(3,4) + \frac{1}{N}\Delta(1,2;3,4) + \mathcal{O}(N^{-2}) \tag{D.2}$$

where the quantity $\Delta(1,2;3,4)$ involves density derivatives of $\rho_N^{(2)}(1,2)$ and $\rho_N^{(2)}(3,4)$. When introduced into the last term of (D.1), the $\mathcal{O}(N^{-1})$ term in (D.2) will give, after integration, a finite contribution to $\partial\rho_N^{(2)}/\partial\lambda$. As was pointed out by Henderson and Barker (1971), this contribution is more easily calculated in the grand canonical ensemble, since in that case the asymptotic behaviour of $\rho^{(4)}(1,2,3,4;\lambda) \equiv \rho_\lambda^{(4)}(1,2,3,4)$ for widely separated pairs is simply

$$\rho_\lambda^{(4)}(1,2,3,4) = \rho_\lambda^{(2)}(1,2)\rho_\lambda^{(2)}(3,4) \tag{D.3}$$

In what follows, all derivatives are taken at constant temperature. From the definition (2.5.18), the derivative of $\rho_\lambda^{(2)}$ with respect to λ at constant chemical potential is given by an equation similar to (D.1), except that the canonical particles densities are replaced by their grand canonical counterparts. We are, however, interested in the derivative at constant number density; this is related to the derivative at constant μ by a standard thermodynamic change of variables:

$$\left(\frac{\partial \rho_\lambda^{(2)}(1,2)}{\partial \lambda}\right)_\rho = \left(\frac{\partial \rho_\lambda^{(2)}(1,2)}{\partial \lambda}\right)_\mu + \left(\frac{\partial \rho_\lambda^{(2)}(1,2)}{\partial \mu}\right)_\lambda \left(\frac{\partial \mu}{\partial \lambda}\right)_\rho \tag{D.4}$$

The first term on the right-hand side is given by (D.1); the second term can be transformed, with the help of (2.4.22), to give

$$\begin{aligned}\left(\frac{\partial \rho_\lambda^{(2)}(1,2)}{\partial \mu}\right)_\lambda &= \left(\frac{\partial \rho}{\partial \mu}\right)_\lambda \left(\frac{\partial \rho_\lambda^{(2)}(1,2)}{\partial \rho}\right)_\lambda \\ &= \rho\left(\frac{\partial \rho}{\partial P}\right)_\lambda \left(\frac{\partial \rho_\lambda^{(2)}(1,2)}{\partial \rho}\right)_\lambda\end{aligned} \tag{D.5}$$

The term $(\partial \mu/\partial \lambda)_\rho$ still has to be evaluated. From (2.4.15) we find, for constant $\rho = \langle N\rangle/V$, that

$$\begin{aligned}\mathrm{d}\langle N\rangle &= k_B T\left(\frac{\partial^2 \log \Xi}{\partial \mu^2}\,\mathrm{d}\mu + \frac{\partial^2 \log \Xi}{\partial \lambda\, \partial \mu}\,\mathrm{d}\lambda\right) \\ &= 0\end{aligned} \tag{D.6}$$

and therefore

$$\left(\frac{\partial \mu}{\partial \lambda}\right)_\rho = -\frac{\partial^2 \log \Xi}{\partial \lambda\, \partial \mu}\left(\frac{\partial^2 \log \Xi}{\partial \mu^2}\right)^{-1} \tag{D.7}$$

From the definition (2.4.11) of Ξ, we find successively that

$$\frac{\partial \log \Xi}{\partial \lambda} = -\frac{\beta}{2}\iint \rho_\lambda^{(2)}(1,2) w(1,2)\, \mathrm{d}1\, \mathrm{d}2 \tag{D.8}$$

$$\frac{\partial^2 \log \Xi}{\partial \lambda\, \partial \mu} = \rho\left(\frac{\partial \rho}{\partial P}\right)_\lambda \frac{\partial}{\partial \rho}\left(-\frac{\beta}{2}\iint \rho_\lambda^{(2)}(1,2) w(1,2)\, \mathrm{d}1\, \mathrm{d}2\right) \tag{D.9}$$

$$\frac{\partial^2 \log \Xi}{\partial \mu^2} = \frac{\partial \langle N\rangle}{\partial \mu} = \langle N\rangle \frac{\partial \rho}{\partial P}, \tag{D.10}$$

where use has again been made of (2.4.22). Inserting (D.9) and (D.10) in (D.7), we obtain the required result in the form

$$\left(\frac{\partial \mu}{\partial \lambda}\right)_\rho = \frac{1}{V}\frac{\partial}{\partial \rho}\left(\frac{\beta}{2}\iint \rho_\lambda^{(2)}(1,2) w(1,2)\, \mathrm{d}1\, \mathrm{d}2\right) \tag{D.11}$$

Equations (D.1), (D.5) and (D.11) show that (D.4) can be rewritten as

$$\frac{\partial \rho_\lambda^{(2)}(1,2)}{\partial \lambda} = -\beta\left(w(1,2)\rho_\lambda^{(2)}(1,2) + 2\int \rho_\lambda^{(3)}(1,2,3)w(2,3)\,\mathrm{d}3 \right.$$
$$+\frac{1}{2}\iint [\rho_\lambda^{(4)}(1,2,3,4) - \rho_\lambda^{(2)}(1,2)\rho_\lambda^{(2)}(3,4)]w(3,4)\,\mathrm{d}3\,\mathrm{d}4$$
$$\left. -\rho\left(\frac{\partial \rho}{\partial P}\right)_\lambda \frac{\partial}{\partial \rho}[\rho_\lambda^{(2)}(1,2)]\frac{1}{2V}\frac{\partial}{\partial \rho}\iint \rho_\lambda^{(2)}(1,2)w(1,2)\,\mathrm{d}1\,\mathrm{d}2\right) \tag{D.12}$$

On switching to the pair distribution function, we can write down the following Taylor expansion, limited to first order in λ:

$$g_\lambda(1,2) = g_0(1,2) - \lambda\beta\left\{ w(1,2)g_0(1,2) + \rho\int g_0^{(3)}(1,2,3)w(2,3)\,\mathrm{d}3 \right.$$
$$+\frac{\rho^2}{2}\iint [g_0^{(4)}(1,2,3,4) - g_0(1,2)g_0(3,4)]w(3,4)\,\mathrm{d}3\,\mathrm{d}4$$
$$\left. -\rho\left(\frac{\rho}{\partial P}\right)_{\lambda=0} \frac{\partial}{\partial \rho}[\rho^2 g_0(1,2)]\frac{\partial}{\partial \rho}\left(\frac{\rho^2}{2}\int g_0(1,2)w(1,2)\,\mathrm{d}2\right)\right\}$$
$$+\mathcal{O}(\lambda^2) \tag{D.13}$$

Substitution of (D.13) in Eqn (6.2.7) yields the first and second-order corrections to the free energy, as given by (6.2.9) and (6.2.10a).

Appendix E

THE HIERARCHY OF PHASE-SPACE CORRELATION FUNCTIONS

The definition (9.6.14) of the two-point correlation function

$$C(12;\, t) \equiv C(\mathbf{r}_1 - \mathbf{r}_2,\, t = t_1 - t_2;\, \mathbf{p}_1\mathbf{p}_2) \tag{E.1}$$

can be extended to include higher-order correlation functions in which products of three or more phase-space densities appear. In what follows, we shall need the three-point correlation functions defined by

$$\begin{aligned} C(12, 3;\, t) &= \langle f(1;\, t) f(2;\, t) \delta f(3;\, 0) \rangle \\ &= \langle \delta[f(1;\, t) f(2;\, t)] \delta f(3;\, 0) \rangle \end{aligned} \tag{E.2}$$

$$\begin{aligned} C(1, 23;\, t) &= \langle \delta f(1;\, t) f(2;\, 0) f(3;\, 0) \rangle \\ &= \langle \delta f(1;\, t) \delta[f(2;\, 0) f(3;\, 0)] \rangle \end{aligned} \tag{E.3}$$

where

$$\begin{aligned} \delta[f(1)f(2)] &= f(1)f(2) - \langle f(1)f(2) \rangle \\ &= f(1)f(2) - \phi_0(\mathbf{p}_1)\rho\delta(\mathbf{r}_1 - \mathbf{r}_2)\delta(\mathbf{p}_1 - \mathbf{p}_2) \\ &\quad - \phi_0(\mathbf{p}_1)\phi_0(\mathbf{p}_2)\rho^2 g(\mathbf{r}_1 - \mathbf{r}_2) \end{aligned} \tag{E.4}$$

and the four-point function:

$$C(12, 34;\, t) = \langle \delta[f(1;\, t) f(2;\, t)] \delta[f(3;\, 0) f(4;\, 0)] \rangle \tag{E.5}$$

We shall deal almost exclusively with the Laplace transforms of these functions, which we shall denote, as usual, by a tilde; for the sake of brevity, the argument z will be systematically omitted. By analogy with Eqn (9.6.17),we have, for example, that

$$\tilde{C}(12, 3) = i(\delta f(3), (z + \mathscr{L})^{-1} f(1) f(2)) \tag{E.6}$$

The correlation functions satisfy a hierarchy of equations, closely related to the BBGKY hierarchy (2.1.15). The equation that links the two-point function $C(12)$ to the three-point functions $C(12,3)$ or $C(1,23)$ can be derived immediately from the evolution equation (9.6.6). The corresponding equation for the Laplace transform is readily obtained from the "resolvent" equation

$$\tilde{C}(12) = i(\delta f(2), (z+\mathscr{L})^{-1}\delta f(1)) \tag{E.7}$$

(cf. (9.6.17)) and the operator identity

$$\frac{z}{z+\mathscr{L}} = 1-(z+\mathscr{L})^{-1}\mathscr{L} = 1-\mathscr{L}(z+\mathscr{L})^{-1} \tag{E.8}$$

Use of (E.7) and (E.8) leads to two equivalent equations of the form

$$-iz\tilde{C}(12) = C^{(0)}(12) - (\delta f(2), (z+\mathscr{L})^{-1}\mathscr{L}\delta f(1)) \tag{E.9a}$$

and, from Eqn (9.1.6):

$$-iz\tilde{C}(12) = C^{(0)}(12) - (\mathscr{L}\delta f(2), (z+\mathscr{L})^{-1}\delta f(1)) \tag{E.9b}$$

The action of the Liouville operator on $\delta f(1)$ or $\delta f(2)$ can be written explicitly with the help of Eqn (9.6.6); Eqn (E.9a) is thereby transformed into

$$-iz\tilde{C}(12) = C^{(0)}(12) - \frac{\mathbf{p}_1}{m}\cdot\frac{\partial}{\partial \mathbf{r}_1}\tilde{C}(12) + i\int d3\, L_{\mathrm{I}}(13)\tilde{C}(13,2) \tag{E.10}$$

In deriving Eqn (E.10), we have used (E.6) and the fact that the operators L_{O} and L_{I}, defined by (9.6.8), commute with the resolvent operator, since they act on field, rather than internal variables. Equation (E.10) can be rearranged in the form

$$\left(-iz + \frac{\mathbf{p}_1}{m}\cdot\frac{\partial}{\partial \mathbf{r}_1}\right)\tilde{C}(12) - i\int d3\, L_{\mathrm{I}}(13)\tilde{C}(13,2) = C^{(0)}(12) \tag{E.11a}$$

Similarly, from Eqn (E.9b), we can derive the adjoint equation

$$\left(-iz - \frac{\mathbf{p}_2}{m}\cdot\frac{\partial}{\partial \mathbf{r}_2}\right)\tilde{C}(12) + i\int d3\, L_{\mathrm{I}}(23)\tilde{C}(1,23) = C^{(0)}(12) \tag{E.11b}$$

Written in the same variables, the kinetic equation (9.7.20) becomes

$$\left(-iz + \frac{\mathbf{p}_1}{m}\cdot\frac{\partial}{\partial \mathbf{r}_1}\right)\tilde{C}(12) + \int d3\, \Sigma'(13)\tilde{C}(3,2) = C^{(0)}(12) \tag{E.12}$$

where

$$\begin{aligned}\Sigma' &= \Sigma^{\mathrm{S}} + \Sigma^{\mathrm{C}} \\ &= \Sigma - \Sigma^{\mathrm{O}}\end{aligned} \tag{E.13}$$

is the interaction part of the evolution kernel Σ, i.e. the sum of the mean-field and collisional contributions. Comparison of Eqns (E.11a) and (E.12) shows that

$$\int d3\, \Sigma'(13)\tilde{C}(32) = -i \int d3\, L_I(13)\tilde{C}(13,2) \qquad \text{(E.14)}$$

The three-point correlation function $C(13,2)$ is related to the four-point function (E.5) by an equation that can be derived in a manner entirely analogous to that used to obtain (E.11):

$$\left(-iz - \frac{\mathbf{p}_2}{m}\cdot\frac{\partial}{\partial \mathbf{r}_2}\right)\tilde{C}(13,2) + i\int d4\, L_I(24)\tilde{C}(13,24) = C^{(0)}(13,2) \qquad \text{(E.15)}$$

Equation (E.14) is an integral equation for Σ'. To obtain a solution, we apply the operator

$$-iz - \frac{\mathbf{p}_2}{m}\cdot\frac{\partial}{\partial \mathbf{r}_2}$$

to both sides and use Eqns (E.11b) and (E.15); the result of these manipulations is

$$\begin{aligned}\int d3\, \Sigma'(13)C^{(0)}(32) - i\int d3 \int d4\, \Sigma'(13)L_I(24)\tilde{C}(3,24) \\ = -i\int d3\, L_I(13)C^{(0)}(13,2) \\ - \int d3 \int d4\, L_I(13)L_I(24)\tilde{C}(13,24) \qquad \text{(E.16)}\end{aligned}$$

Equation (E.16) shows that Σ' separates naturally into static (Σ^S) and dynamic (frequency-dependent, Σ^C) parts; these satisfy the integral equations

$$\int d3\, \Sigma^S(13)C^{(0)}(32) = -i\int d3\, L_I(13)C^{(0)}(13,2) \qquad \text{(E.17)}$$

and

$$\begin{aligned}\int d3\, \Sigma^C(13)C^{(0)}(32) \\ = i\int d3\, [\Sigma^S(13) + \Sigma^C(13)] \int d4\, L_I(24)\tilde{C}(3,24) \\ - \int d3 \int d4\, L_I(13)L_I(24)\tilde{C}(13,24) \qquad \text{(E.18)}\end{aligned}$$

Equation (E.17) can be solved for Σ^{S} by multiplying both sides by the inverse of the static correlation function $C^{(0)}(32)$, integrating over the field point 2, and using the reciprocity relation analogous to (9.6.20), i.e.

$$\int C^{(0)}(32)\, C^{(0)-1}(24)\, \mathrm{d}2 = \delta(34)$$
$$\equiv \delta(\mathbf{r}_3 - \mathbf{r}_4)\delta(\mathbf{p}_3 - \mathbf{p}_4) \tag{E.19}$$

This gives

$$\Sigma^{S}(14) = -i \int \mathrm{d}2 \int \mathrm{d}3\, L_{\mathrm{I}}(13) C^{(0)}(13, 2) C^{(0)-1}(24) \tag{E.20}$$

The integral on the right-hand side of Eqn (E.20) can be evaluated with the help of the static analogue of (E.11a) and the reciprocity relation (E.19). The result, as expected, is the **r**-space equivalent of Eqn (9.7.15):

$$\Sigma^{S}(12) = -\frac{\mathbf{p}_1}{m} \cdot \frac{\partial}{\partial \mathbf{r}_1} \phi_0(\mathbf{p}_1) \rho c(\mathbf{r}_1 - \mathbf{r}_2) \tag{E.21}$$

To obtain an explicit expression for Σ^{C}, we begin by multiplying both sides of Eqn (E.14) by $\tilde{C}^{-1}(24)$, integrating over the field variable 2, and making use of the relation (E.19). In this way, we find that

$$\Sigma'(12) = -i \int \mathrm{d}3 \int \mathrm{d}4\, L_{\mathrm{I}}(13) \tilde{C}(13, 4) \tilde{C}^{-1}(24) \tag{E.22}$$

Equation (E.22) can now be substituted in the first integral on the right-hand side of (E.18), leading to a result of the form

$$\int \mathrm{d}3\, \Sigma^{C}(13) C^{(0)}(32)$$
$$= \int \mathrm{d}3 \int \mathrm{d}5 \int \mathrm{d}6\, L_{\mathrm{I}}(15) \tilde{C}(15, 6) \tilde{C}^{-1}(63) \int \mathrm{d}4\, L_{\mathrm{I}}(24) \tilde{C}(3, 24)$$
$$- \int \mathrm{d}3 \int \mathrm{d}4\, L_{\mathrm{I}}(13) L_{\mathrm{I}}(24) \tilde{C}(13, 24)$$
$$= -\int \mathrm{d}3 \int \mathrm{d}4\, L_{\mathrm{I}}(13) L_{\mathrm{I}}(24) \tilde{G}(13, 24) \tag{E.23}$$

where we have introduced a new, four-point correlation function (Mazenko, 1974; Mazenko and Yip, 1977), defined as

$$\tilde{G}(13, 24) = \tilde{C}(13, 24) - \int \mathrm{d}5 \int \mathrm{d}6\, \tilde{C}(13, 5) \tilde{C}^{-1}(56) \tilde{C}(6, 24) \tag{E.24}$$

An expression for Σ^{C} is now obtained from Eqn (E.23) by multiplying through by the inverse $C^{(0)-1}(25)$, integrating over the field variable 2, and applying (E.19):

$$\Sigma^{C}(15) = -\int d2 \int d3 \int d4\, L_{I}(13) L_{I}(24) \tilde{G}(13, 24) C^{(0)-1}(25) \qquad \text{(E.25)}$$

The $\mathbf{r}$-space equivalents of (9.6.19a) and (9.6.21a) are

$$C^{(0)}(12) = \rho\phi_0(\mathbf{p}_1)\delta(12) + \rho^2\phi_0(\mathbf{p}_1)\phi_0(\mathbf{p}_2)h(\mathbf{r}_1 - \mathbf{r}_2) \qquad \text{(E.26a)}$$

$$C^{(0)-1}(12) = \frac{1}{\rho\phi_0(\mathbf{p}_1)}\delta(12) - c(\mathbf{r}_1 - \mathbf{r}_2) \qquad \text{(E. 26b)}$$

The second relation can be substituted in the right-hand side of (E.25). The first term of $C^{(0)-1}(25)$ is trivially integrated, and the second term contributes nothing; the latter result follows from the definition (9.6.8b) of $L_{I}(24)$, together with an integration by parts and the fact that $c(\mathbf{r}_2 - \mathbf{r}_5)$ is independent of the particle momenta. We are then left with

$$\Sigma^{C}(12) = -\frac{1}{\rho\phi_0(\mathbf{p}_2)} \int d3 \int d4\, L_{I}(13) L_{I}(24) \tilde{G}(13, 24) \qquad \text{(E.27)}$$

This is an exact expression for Σ^{C} that is more amenable to approximation than the alternative form (9.7.16), since the latter involves a projected evolution operator, whereas the time evolution of the four-point correlation function (E.24) is governed by the full Liouville operator.

The four-point correlation function in Eqn (E.27) describes the dynamical correlations of two particles (1 and 2) in the presence of the medium represented by particles 3 and 4. If all phase-space densities δf are replaced by their explicit forms (9.6.1) and (9.6.10), we find that no terms proportional to $\delta(13)$ and $\delta(24)$ appear: the field labels correspond to different particles. The correlation function also has important "clustering" properties; it can be split into "connected" and "disconnected" parts such that

$$G(13, 24) = G_{C}(13, 24) + G_{D}(13, 24) \qquad \text{(E.28)}$$

The part G_{C} is completely connected in the sense that it vanishes unless all four field points are mutually correlated. The part G_{D} is disconnected in the sense that it does not vanish when particles 1 and 2 go off to infinity, while 3 and 4 remain fixed; it is singly connected, however, insofar as it vanishes when a single particle is removed to infinity. It can be shown, in fact, that in the time domain G_{D} factorizes into products of two-point correlation functions:

$$G_{D}(13, 24; t) = C(12; t)C(34; t) + C(14; t)C(32; t) \qquad \text{(E.29)}$$

The separation (E.28) leads, via Eqn (E.27), to a corresponding separation of the collision kernel into connected and disconnected parts. According to Eqn (E.29), the disconnected part of Σ^C describes a process in which two particles propagate independently of each other between field points (1, 3) and (2, 4), or (1, 3) and (4, 2), in a time interval t; the connected part describes collision processes of duration t, during which time the particles stay close together. It turns out that retaining only the disconnected part of Σ^C is often a good approximation, particularly for ionic systems (Wallenborn and Baus, 1978). Use of this approximation leads to a non-linear self-consistency problem, since Σ^C, which determines the phase-space correlation function C via the exact kinetic equation (9.7.20), is itself a functional of C.

The expression for Σ^C given by (E.27) involves the pair potential $v(r)$ through the appearance of the operator L_I of Eqn (9.6.8b). The bare potential can be eliminated in favour of a "renormalized" potential, equal to the potential of mean force $\psi(r)$ defined through Eqn (5.3.6), by extracting the initial value $G^{(0)}(13, 24) \equiv G(13, 24; t=0)$ symmetrically from G in the form

$$G(13, 24) = \int d5 \int d6 \int d7 \int d8\ G^{(0)}(13, 56)\mathscr{G}(56, 78)G^{(0)}(78, 24) \tag{E.30}$$

This defining relation for the four-point function $\mathscr{G}$ allows the expression for Σ^C to be cast in the following symmetrical form (Mazenko and Yip, 1977):

$$\Sigma^C(1, 2) = -\frac{1}{\rho\phi_0(\mathbf{p}_2)} \int d5 \int d6 \int d7 \int d8\ \mathscr{V}(1, 56)\mathscr{G}(56, 78)\mathscr{V}(2, 78) \tag{E.31}$$

The "end-point vertices" $\mathscr{V}$ are static quantities, defined as

$$\mathscr{V}(1, 24) = \int d3\ L_I(13)\mathscr{G}(13, 24) \tag{E.32}$$

and expressible in terms of static pair and triplet distribution functions; the pair potential no longer appears explicitly. Manipulation of Eqn (E.31) and its equivalent for the self motion leads to the "fully renormalized" kinetic theory of Mazenko (1973a, b; 1974).

Index